ATOMIC NUMBERS AND ATOMIC WEIGHTS OF THE ELEMENTS[a]

Element	Symbol	Atomic Number	Atomic Weight	Rounded Value
Actinium	Ac	89	(227)	–
Aluminum	Al	13	26.98154	27.0
Americium	Am	95	(243)	–
Antimony	Sb	51	121.75	121.8
Argon	Ar	18	39.948	39.9
Arsenic	As	33	74.9216	74.9
Astatine	At	85	(210)	–
Barium	Ba	56	137.33	137.3
Berkelium	Bk	97	(249)	–
Beryllium	Be	4	9.01218	9.01
Bismuth	Bi	83	208.9804	209.0
Boron	B	5	10.81	10.8
Bromine	Br	35	79.904	79.9
Cadmium	Cd	48	112.41	112.4
Calcium	Ca	20	40.08	40.1
Californium	Cf	98	(251)	–
Carbon	C	6	12.011	12.0
Cerium	Ce	58	140.12	140.1
Cesium	Cs	55	132.9054	132.9
Chlorine	Cl	17	35.453	35.5
Chromium	Cr	24	51.996	52.0
Cobalt	Co	27	58.9332	58.9
Copper	Cu	29	63.546	63.5
Curium	Cm	96	(247)	–
Dysprosium	Dy	66	162.50	162.5
Einsteinium	Es	99	(254)	–
Erbium	Er	68	167.26	167.3
Europium	Eu	63	151.96	152.0
Fermium	Fm	100	(253)	–
Fluorine	F	9	18.998103	19.0
Francium	Fr	87	(223)	–
Gadolinium	Gd	64	157.25	157.3
Gallium	Ga	31	69.72	69.7
Germanium	Ge	32	72.59	72.6
Gold	Au	79	196.9665	197.0
Hafnium	Hf	72	178.49	178.5
Hahnium	Ha	105	(260)	–
Helium	He	2	4.00260	4.00
Holmium	Ho	67	164.9304	164.9
Hydrogen	H	1	1.0079	1.01
Indium	In	49	114.82	114.8
Iodine	I	53	126.9045	126.9
Iridium	Ir	77	192.22	192.2
Iron	Fe	26	55.847	55.8
Krypton	Kr	36	83.80	83.8
Kurchatovium	Ku	104	(247)	–
Lanthanum	La	57	138.9055	138.9
Lawrencium	Lr	103	(257)	–
Lead	Pb	82	207.2	207.2
Lithium	Li	3	6.941	6.94
Lutetium	Lu	71	174.97	175.0
Magnesium	Mg	12	24.305	24.3
Manganese	Mn	25	54.9380	54.9
Mendelevium	Md	101	(256)	–
Mercury	Hg	80	200.59	200.6
Molybdenum	Mo	42	95.94	95.9
Neodymium	Nd	60	144.24	144.2
Neon	Ne	10	20.179	20.2
Neptunium	Np	93	237.0482	237.0
Nickel	Ni	28	58.70	58.7
Niobium	Nb	41	92.9064	92.9
Nitrogen	N	7	14.0067	14.0
Nobelium	No	102	(254)	–
Osmium	Os	76	190.2	190.2
Oxygen	O	8	15.9994	16.0
Palladium	Pd	46	106.4	106.4
Phosphorus	P	15	30.97376	31.0
Platinum	Pt	78	195.09	195.1
Plutonium	Pu	94	(242)	–
Polonium	Po	84	(210)	–
Potassium	K	19	39.0983	39.1
Praseodymium	Pr	59	140.9077	140.9
Promethium	Pm	61	(145)	–
Protactinium	Pa	91	231.0359	231.0
Radium	Ra	88	226.0254	226.0
Radon	Rn	86	(222)	–
Rhenium	Re	75	186.207	186.2
Rhodium	Rh	45	102.9055	102.9
Rubidium	Rb	37	85.4678	85.5
Ruthenium	Ru	44	101.07	101.1
Samarium	Sm	62	150.4	150.4
Scandium	Sc	21	44.9559	45.0
Selenium	Se	34	78.96	79.0
Silicon	Si	14	28.0855	28.1
Silver	Ag	47	107.868	107.9
Sodium	Na	11	22.98977	23.0
Strontium	Sr	38	87.62	87.6
Sulfur	S	16	32.06	32.1
Tantalum	Ta	73	180.9479	180.9
Technetium	Tc	43	98.9062	98.9
Tellurium	Te	52	127.60	127.6
Terbium	Tb	65	158.9254	158.9
Thallium	Tl	81	204.37	204.4
Thorium	Th	90	232.0381	232.0
Thulium	Tm	69	168.9342	168.9
Tin	Sn	50	118.69	118.7
Titanium	Ti	22	47.90	47.9
Tungsten	W	74	183.85	183.9
Uranium	U	92	238.029	238.0
Vanadium	V	23	50.9444	50.9
Xenon	Xe	54	131.30	131.3
Ytterbium	Yb	70	173.04	173.0
Yttrium	Y	39	88.9059	88.9
Zinc	Zn	30	65.38	65.4
Zirconium	Zr	40	91.22	91.2
Name to be determined		106	(263)	–

[a]Based on carbon-12. Numbers in parentheses are the mass numbers of the most stable or best-known isotopes.

Chemistry

Chemistry

AN INTRODUCTION

Second Edition

SYDNEY B. NEWELL

Little, Brown and Company
Boston Toronto

Library of Congress Catalog Card No. 79–84924

ISBN 0-316-604542

9 8 7 6 5 4 3 2

MV

Published simultaneously in Canada
by Little, Brown & Company (Canada) Limited

Printed in the United States of America

Book Design: Susan Creelman-Carr with Wayne Ellis
Art Editor: Tonia Noell-Roberts
Cover Design: Richard Emery
Cover Photo: Karl Faller

To the Instructor

The warm reception accorded the first edition of *Chemistry: An Introduction* has been gratifying. Since its publication many instructors, reviewers, and students have provided helpful comments and suggestions for strengthening the text, and it is largely in response to these that the second edition has been written. Despite the changes that have been made, however, I have kept to the philosophy and purpose of the first edition: to provide beginning students and their instructors with a comprehensive, understandable, and motivating introduction to chemistry.

The most significant alterations are in Chapters 7, 8, and 9—the structure and bonding sequence. Many instructors indicated that they prefer to treat atomic structure before bonding, with the latter condensed to a single chapter. Accordingly, I have placed atomic structure in Chapter 7, followed it with a chapter on the periodic table (Chapter 8), and consolidated ionic and covalent bonding in one chapter (Chapter 9).

Requests for more in-depth treatment of solutions resulted in the addition of a new chapter (14), "Reactions in Solution". In this chapter I thoroughly discuss ionic equations, dilutions, and titrations, and have added material on normality and equivalent weight. I have also simplified the chapters on nomenclature (4) and chemical reactions and equations (5). In the chapter on acids and bases (13), pH is now treated using logarithms, both with a log table and with a calculator. This treatment is supported by a new section on logarithms in Appendix A, which still contains reviews of significant figures, rounding off, use of the calculator, approximate answers, and exponential notation. Appendix B still contains SI units, temperature conversion, and solving problems by the factor-unit method (dimensional analysis). The present edition includes learning objectives at the beginning of each chapter.

Many instructors expressed a strong desire that their students be provided with answers and solutions to all of the exercises. And just as many indicated an equally strong desire for the opposite. To accommodate both groups, I've added a great many more exercises, so that each chapter has two full, equivalent sets. The answers to Set A are at the back of the book; those for Set B are provided in the instructor's manual. The more challenging exercises have been marked with asterisks.

Features of the first edition that were well-received have been retained in the second edition. These include Maxwell's Demon, the scientific cartoon character I use as a lecturer-demonstrator; the Review Questions at the end of each chapter, emphasizing important points section-by-section; the use of italics for defined terms, and their repetition in the margin of the text for emphasis and easy reference; and a comprehensive index. As in the first edition, I include wherever possible (especially in mathematical problems) many examples and applications to everyday life to interest students. A large number of examples have been worked out in the text to give the student ample instruction in problem solving.

As before, a complimentary instructor's manual provides suggestions for lectures, demonstrations, and audiovisual aids; answers and solutions to the Set B exercises; and sample examination questions. In addition, this edition will have a companion laboratory manual (*Experimental Chemistry: An Introduction* by Robert J. Artz) and a student study guide (*Study Guide to Accompany Chemistry: An Introduction,* Second Edition, by Daniel C. Pantaleo and Wayne P. Anderson), both available from Little, Brown and Company.

I am grateful to the many instructors who offered criticism and advice, and to the students who replied to the questionnaire at the back of the book. Their comments were always appreciated and were often very helpful. In addition to the reviewers who participated in the first edition, I want to thank the following reviewers and users for their suggestions and help in the preparation of the second edition: Professor Ronald Backus, American River College; Dr. E. J. Kemnitz, University of Nebraska; Dr. Thomas Knudtson, Northern Illinois University; Dr. Clifton Meloan, Kansas State University; Dr. Gordon Parker, University of Toledo; Dr. George Schenk, Wayne State University; Professor Michael Sevilla, Oakland University; and Dr. Vernon Thielmann, Southwest Missouri State University. As before, Dr. Ronald Rohrer provided helpful suggestions from the technical, nonchemical point of view.

Again, my thanks to the staff of Little, Brown for an efficient handling of the second edition. In particular, Ian Irvine, the Science Editor, provided me with needed reviews and feedback from users, as well as personal consultations that were of great benefit. The production of the second edition went remarkably smoothly, thanks to the skill of Janet Welch, Book Editor, and Rose Sklare, Book Editing Manager. Their early receptiveness to my concerns prevented problems before they arose.

I, and many others, have put our best efforts into enhancing what I feel is a worthwhile introductory text. I offer this second edition to students and their teachers with the hope that they may share my enthusiasm.

To the Student: Introducing Maxwell's Demon

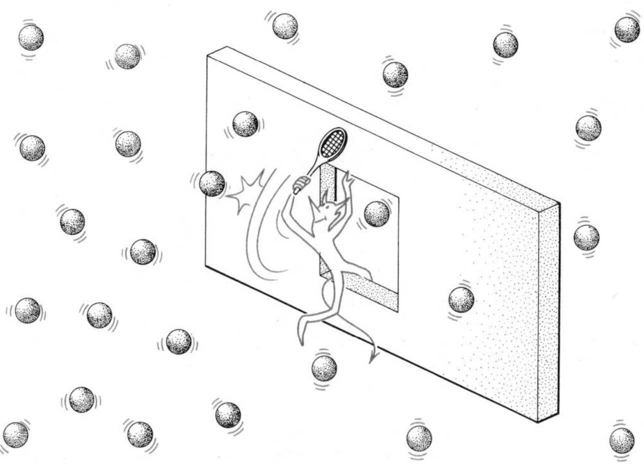

Much of chemistry involves changes—many dramatic, most very fast—from one condition to another. We could see better how chemical processes work if we could slow them down or take them apart and look at them piece by piece. But we mortals can't do that, so in this book I use the services of a cartoon character, Maxwell's Demon.

I didn't invent the demon. He was invented in 1871 by James Clerk Maxwell, a scientist who made contributions in the fields of mathematics, physics, and chemistry. Maxwell conjured up a tiny creature who could reverse natural processes in a way impossible for human beings to do. Many of us would like to undo things that have been done, such as un-breaking a glass, un-burning a forest, or collecting smoke that has spread through the air and stuffing it back into a factory smokestack. We'd also like to make things happen the way we want them to, like having the right horse win the race or making a chemical process happen in a certain way. These are the kinds of things Maxwell's Demon can do.

Maxwell's Demon allows me to be present in this book. Through him, I can talk more conversationally than I feel would be appropriate in the text's running narrative. He'll help us to share the experience of observing chemical principles as he illustrates them in a way that only a demon can.

S.B.N.

Contents

6

CALCULATIONS WITH FORMULAS AND EQUATIONS 109

7

ATOMIC STRUCTURE 137

8

THE PERIODIC TABLE 155

14
REACTIONS IN SOLUTION 309

15
ELECTROCHEMISTRY 337

16
RATES AND EQUILIBRIA OF CHEMICAL REACTIONS 371

19
INTRODUCTION TO BIOCHEMISTRY 455

APPENDIX A
WORKING WITH NUMBERS 489

APPENDIX B
WORKING WITH UNITS 509

Reference List of Selected Tables and Figures

Chemistry

CHAPTER 1

What Is Chemistry?

It is part of human nature to try to understand ourselves and our environment. Wondering what we are, what we're made of, and what else there is occupies a lot of our energy.

There are two kinds of space, outer and inner. Outside our bodies, there's the earth we live on. Beyond the earth, there's the moon. Beyond the moon, our solar system. Beyond our solar system, our galaxy. Beyond our galaxy, the universe. Beyond the universe—who knows? The sky is not the limit. There's no indication that our outer space stops with the universe. This outward approach is used by astronomy and related sciences to explore our outer space.

We can explore our inner space too. Within our bodies, there are tissues. Within the tissues, there are cells. Within the cells, there are molecules. Within the molecules, there are atoms. Within the atoms, there are smaller particles. The inward exploration doesn't stop at the smaller particles, but this book will. Biology, chemistry, and nuclear physics all take this inward approach. Biology usually deals with tissues and cells, and sometimes with molecules. Chemistry deals mostly with molecules and atoms, looking occasionally at the smaller particles. Nuclear physics goes within the atom to the smaller particles and even probes below that level.

This chemistry book is about atoms and molecules—what they are, how they behave, and how chemists study them and manipulate them. We'll see some of our surroundings from the chemical viewpoint, and we'll see how chemistry touches our lives.

1.1 A CHEMICAL VIEW OF THE WORLD

Chemistry is really a language, a special way of describing things that we see and use every day. Just about anything we can think of can be "translated" into chemistry. Figure 1.1 shows some familiar things and their chemical "translations." If someone wrote a paragraph in a language we didn't understand, the words and sentences wouldn't mean any more to us than the chemical "translations" in Figure 1.1. If we studied the language, though, and learned the vocabulary and grammar, then we'd understand the meaning. In this book, we'll learn about these and other chemical "translations." The chemical "translations" in Figure 1.1 are called *formulas*. We'll learn how to read and write formulas and how to work with them. formula

All things are made of matter. *Matter* is anything that takes up space and requires energy to make it move. As far as we know, all matter is made of small particles called atoms. A collection of a single kind of atom is called an *element*. Each of the letters in the formulas of Figure 1.1 stands for a kind of element: C stands for carbon, O for oxygen, N for nitrogen, H for hydrogen, K for potassium, S for sulfur. We already see an advantage to chemical formulas: we can describe these thirteen familiar substances in terms of only six elements.

matter

element

Elements are the building blocks of chemistry, just as they are the

FIGURE 1.1
Familiar things in chemical terms

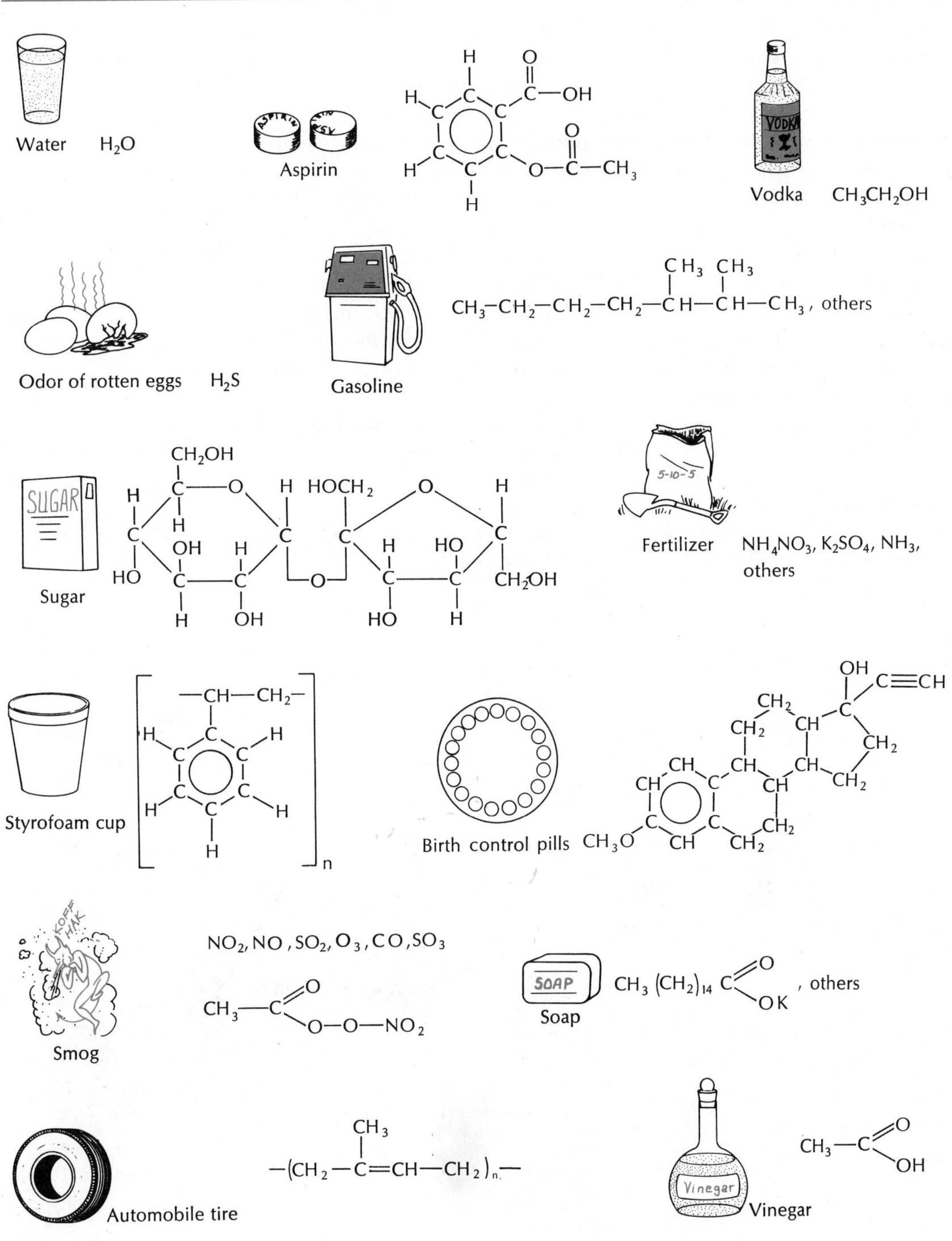

building blocks for the substances shown in Figure 1.1. So far chemists have discovered 106 elements. Our study of chemistry will be simpler than that, though, because only about forty of the 106 elements are abundant enough to deal with at any length in this book. Still, millions of substances can be formed from these forty.

We can see from Figure 1.2A that only ten elements make up over 99 percent of the earth's crust, water, and atmosphere. Oxygen is the most abundant element. It's an important part of the atmosphere and is also found in the earth in combination with other elements. Rocks and sand are mostly made of silicon and oxygen, which is why silicon is the next most abundant element.

When we talk about the earth's crust, water, and atmosphere, we're not including the plants and animals that inhabit it. Of the elements that make up living things or their by-products, oxygen is still the most abundant. Carbon takes second place. The ten most abundant elements in the human body are shown in Figure 1.2B. The human body and other organisms also contain small amounts of other elements as well.

From the chemical viewpoint, chemistry runs our lives. We're made of

FIGURE 1.2
Relative abundance of elements in the earth's crust, waters, and atmosphere (A), and in the human body (B), given in percentages by weight

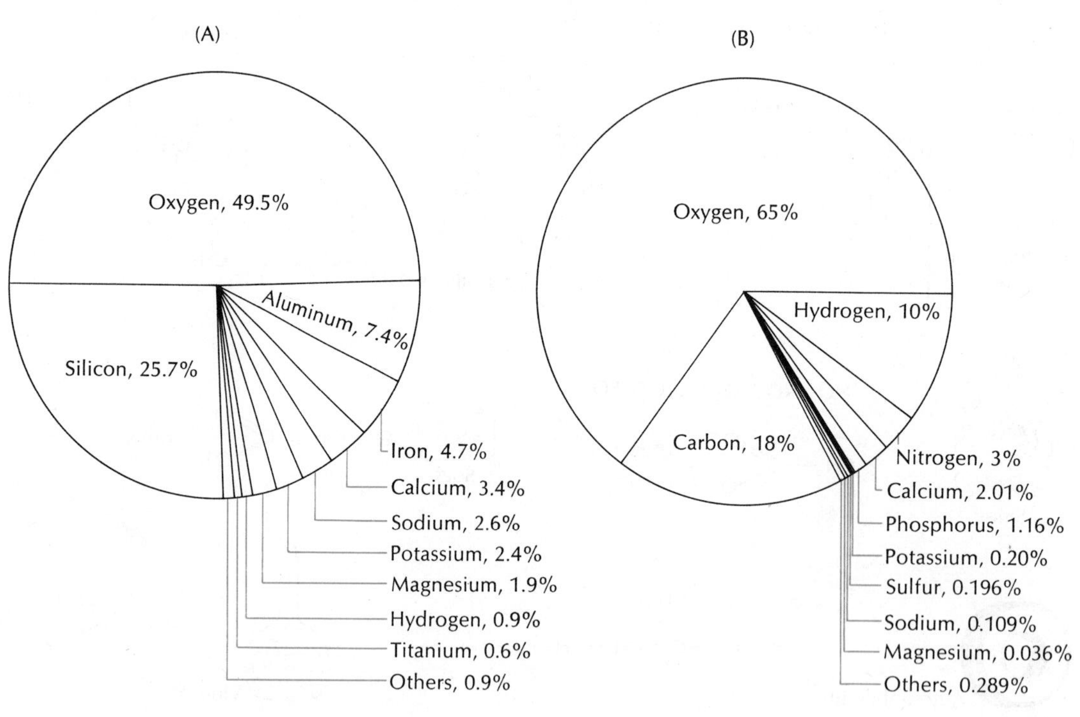

chemicals and so are our surroundings. It's important for us to get to know these substances on which our existence is based, and to know how and why they behave the way they do. That's what chemistry is all about.

1.2 HOW CHEMISTS DISCOVER THINGS

Chemistry is an experimental science, which means that chemical discoveries are made as a result of experiment. This is true of most other sciences as well as chemistry. Many scientific discoveries were made by accident, or as the result of trying to investigate something else. Radioactivity was discovered because someone accidentally left a piece of ore on top of a sealed photographic plate, and then it was found that the film had been exposed through its wrapping. Penicillin was discovered when someone's bacteria became accidentally contaminated with mold. However, when these accidents did lead to discoveries, it was because the discoverers' minds were receptive and alert enough to seize upon these lucky clues and explore them.

Scientists do usually set out to discover something, or to test a hypothesis. A *hypothesis* is a guess about why some physical event happens. The *scientific method* is an approach that involves suggesting hypotheses and then putting them to the test of experiment. To do this, "if . . . then" reasoning is used. "*If* my hypothesis is correct, *then* such and such should happen."

hypothesis

scientific method

It's much easier to disprove a hypothesis than it is to prove it. A positive result of an experiment doesn't prove the hypothesis. It just supports the hypothesis. But a negative result does disprove the hypothesis.

This scientific approach can be illustrated by referring to the early controversy over whether the sun orbits the earth or vice versa. The favored hypothesis was that the sun orbits the earth. We can then make a prediction based on this hypothesis:

> *If* the sun orbits the earth, *then* the sun should rise on one horizon, cross the sky, and set on the other horizon.

This prediction, of course, is true. It is an example of a false hypothesis leading to a true prediction.

For many years, the earth-centered hypothesis was believed in. But finally, mathematical proof showed it to be false, through the same kind of reasoning.

> *If* the sun orbits the earth, *then* the positions of the other planets relative to the earth should agree with the theoretical positions calculated on the basis of the sun orbiting the earth.

This prediction did not come true, and the hypothesis was proven to be false. It was revised to the hypothesis that the earth orbits the sun. This hypothesis has stood the test of many experiments. The more positive the results of experiments that support a hypothesis, the more strongly we believe in the hypothesis. But there's always the chance that someone else will come along and do the disproving experiment we never thought of.

To use the scientific method, the hypothesis has to be testable by experiment. The sun-orbiting-the-earth hypothesis couldn't be disproved until the necessary scientific instruments existed to perform the proper experiments.

Amazingly enough, the ancient Greeks didn't use this method to test their hypotheses. They tried to prove them just by logic and argument. The philosopher Aristotle was especially guilty of impeding scientific progress by upholding his beliefs on religious grounds and refusing to put them to the test of experiment. Many of his hypotheses held up and were believed in for two thousand years before anyone tried to test them. For instance, Aristotle had a hypothesis that heavy objects fall faster than light objects. He never tried to test it, though, and neither did anyone else at the time.

This would seem to be an easy hypothesis to test, but no one did for two thousand years. Then Galileo did test it. He used this logic:

If heavy objects fall faster than light objects, *then* if I drop a heavy object and a light object from a tower at the same time, the heavy object should reach the ground first.

This prediction didn't come true. Both objects arrived at the ground at the same time. Even so, Aristotle's reputation was so great that those who believed in Galileo's experiment (and Galileo himself) were considered crazy. Aristotle's hypothesis came into disfavor only very slowly.

Now scientists try to accept the results of experiments that disprove hypotheses, even though it's sometimes difficult to let go of a pet theory. We'll see some instances later in the book of how the scientific method was used to make discoveries.

CHAPTER 2

Atoms and Elements

LEARNING OBJECTIVES

After studying this chapter, you should be able to:

1. Supply a correct definition, explanation, or example for each of these:

atom	monatomic element
subatomic particle	molecule
proton	diatomic molecule
neutron	chemical reactivity
electron	burning
mass	exploding
atomic mass unit	supporting combustion
static electricity	toxicity
negative charge	physical property
positive charge	physical state
atomic charge unit	solid state
nucleus	liquid state
electron cloud	gas state
atomic number	metallic properties
symbol	metal
mass number	nonmetal
isotopes	metalloid
chemical property	periodic table of the elements
chemical reaction	group (family)
chemical bond	period
compound	alkali metal
triatomic molecule	alkaline earth metal
halogen	transition element
noble gas	allotropic form
rare gas	
inert gas	

2. Write the names when given the symbols, or the symbols when given the names, of the most common elements—the first twenty elements plus Cr, Mn, Fe, Co, Ni, Cu, Zn, Br, Kr, Rb, Sr, Ag, Cd, Sn, I, Cs, Ba, Au, Hg, and Pb.
3. Given any two of mass number, atomic number (or name or symbol), or number of neutrons, write the value of the other.
4. Write or recite from memory those elements that are liquids, solids, gases, and/or diatomic molecules.
5. Name at least two elements that belong to each group, or family, in the periodic table, and name at least one physical or chemical property of each.
6. Use the periodic table of the elements (Figure 2.6) to identify any element as a metal, nonmetal, or metalloid, and to state its group.

In the last chapter, we talked a little bit about atoms and elements without really saying what they are or what they are like. In this chapter, we'll treat them in more detail. We'll see what they're made of and look at some of the different kinds.

Since an element is a collection of many atoms of the same kind, first we'll look at the atoms themselves. Then we'll go on to discuss them in groups, as elements.

2.1 WHAT ATOMS ARE MADE OF

Atoms are small basic particles of matter. At the start of the last chapter, we hinted about particles even smaller than atoms. These are called *subatomic particles*, and there are three main kinds: *protons*, *neutrons*, and *electrons*. Scientists describe these particles with two important quantities, mass and charge, and we have to know what these are before we discuss the particles themselves.

atom

subatomic particle

proton

neutron

electron

MASS. We already know that everything is made of matter, which takes up space and requires energy to move. If we want to know how much matter something has, we must first know how much mass it has. *Mass* is the unvarying amount of matter that any object has. Mass is the measure of matter, just as length is the measure of distance. The basic unit of length is arbitrary (it can be inches, feet, meters, miles, and so forth) and so is that of mass. Now that we're talking about subatomic particles, our unit of mass will be the *atomic mass unit (amu)*, and it's approximately the mass of a proton or a neutron. An electron has very little mass by comparison with a proton or

mass

atomic mass unit (amu)

AN EARLY VIEW OF MATTER

A long time ago, people started wondering what matter was all about. Around 470 B.C., the Greek philosopher Democritus had some ideas about matter. Here's a summary of what he proposed: (1) Everything is made of atoms. (2) Atoms have space between them. (3) Atoms are too small to be seen. (4) Atoms can't be divided. (5) Atoms are the same all the way through. (6) Atoms can't be compressed. (7) Atoms differ from each other only in form, size, and geometry. (8) How matter behaves depends on the arrangement of atoms.

A lot of what Democritus said we still believe to be true. In almost twenty-five centuries, only his ideas that atoms can't be divided and that they are the same throughout have been disproved. Not bad, considering that the ancient Greeks didn't know about the scientific method, didn't do any experiments at all, and didn't have any advanced scientific instruments. They just watched what went on around them, thought about things, and argued with each other about their hypotheses.

neutron. The mass of an electron is about 1/1840 amu, whereas a proton or neutron is about 1 amu. When we add the masses of neutrons, protons, and electrons, we usually neglect the electron's mass altogether.

We compare the masses of two objects by using a balance, which is like a seesaw. We know that a heavy person must sit closer to the middle of a seesaw when trying to balance with a lighter person. Figure 2.1 shows this kind of relationship in comparing the masses of the subatomic particles.

FIGURE 2.1
Relative masses of subatomic particles

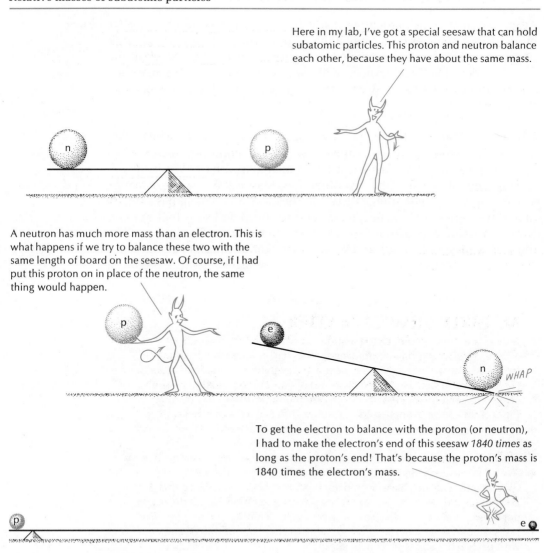

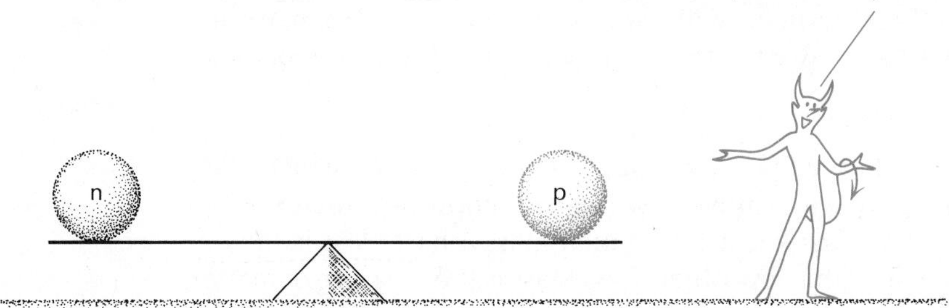

Here in my lab, I've got a special seesaw that can hold subatomic particles. This proton and neutron balance each other, because they have about the same mass.
n
p

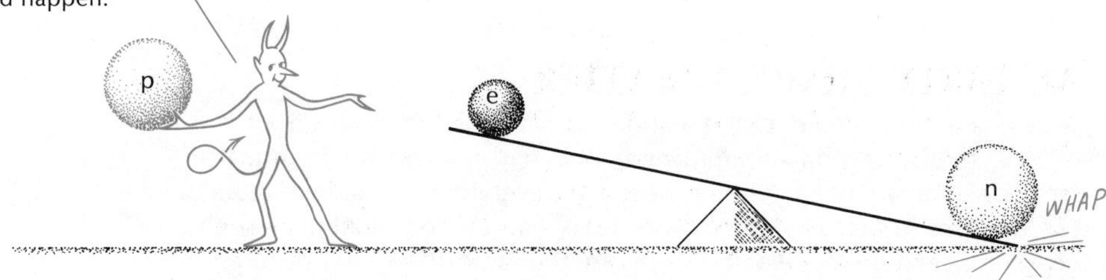

A neutron has much more mass than an electron. This is what happens if we try to balance these two with the same length of board on the seesaw. Of course, if I had put this proton on in place of the neutron, the same thing would happen.
p
e
n
WHAP

To get the electron to balance with the proton (or neutron), I had to make the electron's end of this seesaw *1840 times* as long as the proton's end! That's because the proton's mass is 1840 times the electron's mass.
p
e

To measure how much mass something has, we compare it with another object whose mass we know. In chemistry laboratories, we use a balance that operates on the same principle as a seesaw, although it doesn't look exactly like one.

CHARGE. Electricity is all around us. We see it in lightning. We receive electric shocks when we walk on a nylon rug on a dry day and then touch something (or someone). We can see sparks fly from a cat's fur when we pet it in the dark. We can rub a balloon on a sweater and make the balloon stick to the wall or the ceiling. Our clothes cling together when we take them from the dryer.

These are all examples of *static electricity*. They happen because there is a buildup of one of the two kinds of electrical charge, either positive or negative. Rubbing a rubber rod with a piece of fur gives the rod one kind of charge. Rubbing a glass rod with a piece of silk gives the glass rod the other kind of charge. The two rods will have *unlike* charges and will *attract* each other. Two charged rods with *like* charges will *repel* each other.

static electricity

We define the kind of charge the rubber rod has as a *negative charge*, and the kind of charge the glass rod has as a *positive charge*. Electrons and protons have unlike charges. We define the kind of charge the electron has as a negative charge, and the proton, a positive charge. A neutron has no charge. We say that a neutron is electrically neutral, and that's where its name comes from.

negative charge

positive charge

An object has a negative charge because it has more electrons than it has protons. Electrons are more easily moved than protons. When we rub the rubber rod with the fur, electrons are transferred from the fur to the rod, and they give the rod a negative charge. An object with fewer electrons than protons has a positive charge, like the glass rod. When we rub the glass rod with silk, electrons are removed from the rod by the silk. Figure 2.2, on page 12, summarizes the discussion above.

We use an *atomic charge unit* to say what amount and what kind of charge a subatomic particle has. An atomic charge unit is defined as -1 for an electron. The proton has the same amount of charge as the electron, but is opposite in sign, so its charge is $+1$ atomic charge unit. The neutron has 0 atomic charge units.

atomic charge unit

Table 2.1 summarizes the masses and charges of the subatomic particles.

TABLE 2.1
Summary of charge and mass of subatomic particles

	Atomic Charge Units	*Atomic Mass Units (amu)*
Electron	-1	1/1840 (usually negligible)
Proton	$+1$	1
Neutron	0	1

FIGURE 2.2
Like charges repel, unlike charges attract

Here's a positively charged glass rod that I've hung on a string.

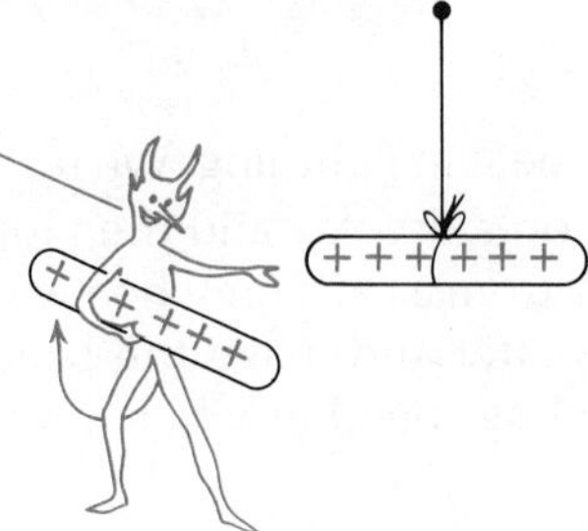

See what happens when I hold another positively charged rod next to it? *Like* charges *repel* each other.

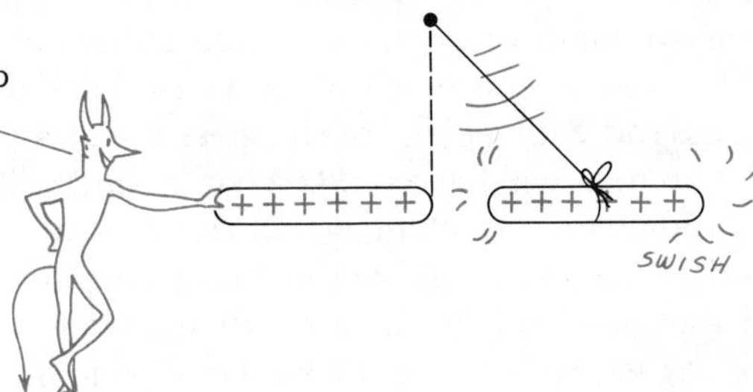

But now if I bring this negatively charged rubber rod close to it, it'll go toward it. *Unlike* charges *attract* each other.

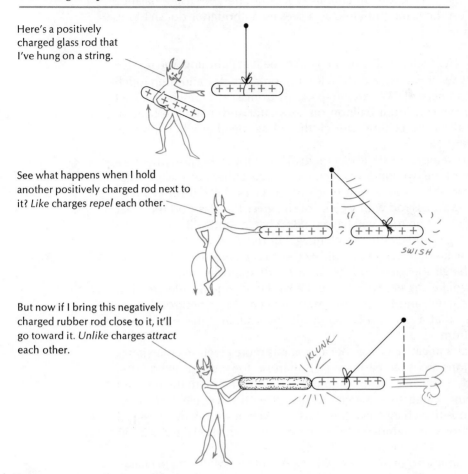

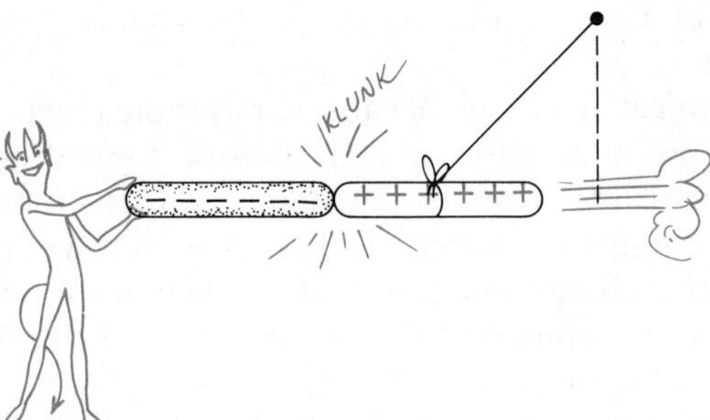

The positively charged rod had more protons on it, and the negatively charged rod had more electrons on it—so it shouldn't surprise us that this single proton is attracted to this single electron! They have unlike charges.

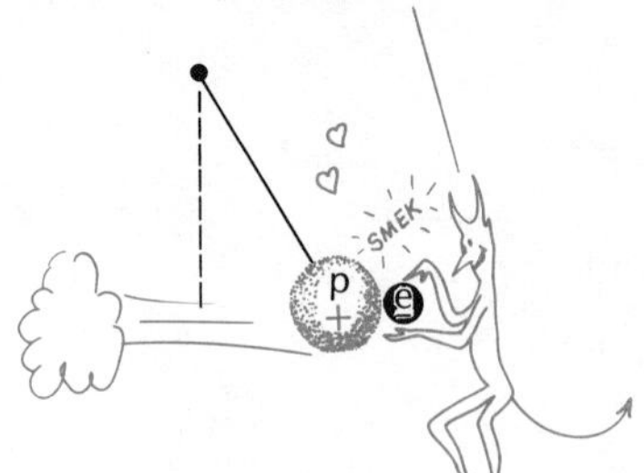

No electrical attraction between a proton and a neutron. A neutron has no charge. It wouldn't feel one way or the other about an electron, either.

2.2
INTRODUCING THE WHOLE ATOM

The smaller particles that atoms are made of—protons, neutrons, and electrons—are only part of the atom's story. The way these particles are arranged in the atom is important too, because the arrangement helps explain why and how matter takes the many forms it does. For now, we'll look only briefly at the most modern ideas about atomic structure. Later, in Chapter 7, we'll examine this topic in more detail.

At the center of every atom is the *nucleus,* which contains the protons and neutrons. These particles in the nucleus give it two characteristics: (1) it contains most of the mass of the atom, since protons and neutrons have much greater masses than electrons; and (2) it is positively charged, since it contains protons with positive charges and neutrons with no charge. The total charge of the nucleus is the sum of the proton charges. nucleus

The nucleus is only a small fraction (about 1/100,000) of the size of an atom. Most of an atom's size is due to its *electron cloud,* which is mostly empty space with electrons rapidly moving about the nucleus. To get some idea of the space occupied by the electron cloud, imagine the outer limits of the cloud to be the fence around a baseball stadium. Compared to that, the atom's nucleus would be the size of a fly on the pitcher's cap! (Of course, atoms are much, much smaller than baseball stadiums. The average atom is not quite 1/250,000,000 of an inch in diameter.) electron cloud

Since electrons are negatively charged, the electron cloud is also negatively charged. The total charge of an electron cloud is the sum of the electron charges. Whole atoms, however, are neither positively charged (like their nuclei) nor negatively charged (like their electron clouds). They are neutral, which means that there must be the same number of electrons as protons. Equal amounts of opposite charges cancel each other.

Figure 2.3 shows an artist's conception of the simple atomic model we've been describing.

To describe an atom, we assign two numbers to it. One is the number of its protons (its atomic number). The other is the total number of protons and neutrons it has (its mass number). We'll look at these two quantities one at a time.

ATOMIC NUMBER. We saw in Chapter 1 that there are 106 kinds of atoms, and that groups of the same kind of atom are called elements. The characteristic of atoms that determines what kind they are is the *atomic number,* symbolized by *Z.* The atomic number is the number of protons an atom has. It's also the number of electrons an atom has, since an atom has the same number of electrons as protons. Later in this book, we'll see that each element behaves the way it does because of the number of electrons it has. The atom's atomic number is the sole factor that causes an element to be the element it is. atomic number (Z)

We also saw in Chapter 1 that elements have names and letters (*symbols*) representing them: H is the symbol for hydrogen; O is the symbol for symbol oxygen, and so forth. Some symbols contain two letters instead of one: Si for silicon, Co for cobalt, Cl for chlorine. These and other symbols come from the English names, but many symbols come from older Latin or

FIGURE 2.3
Pictorial representation of an atom

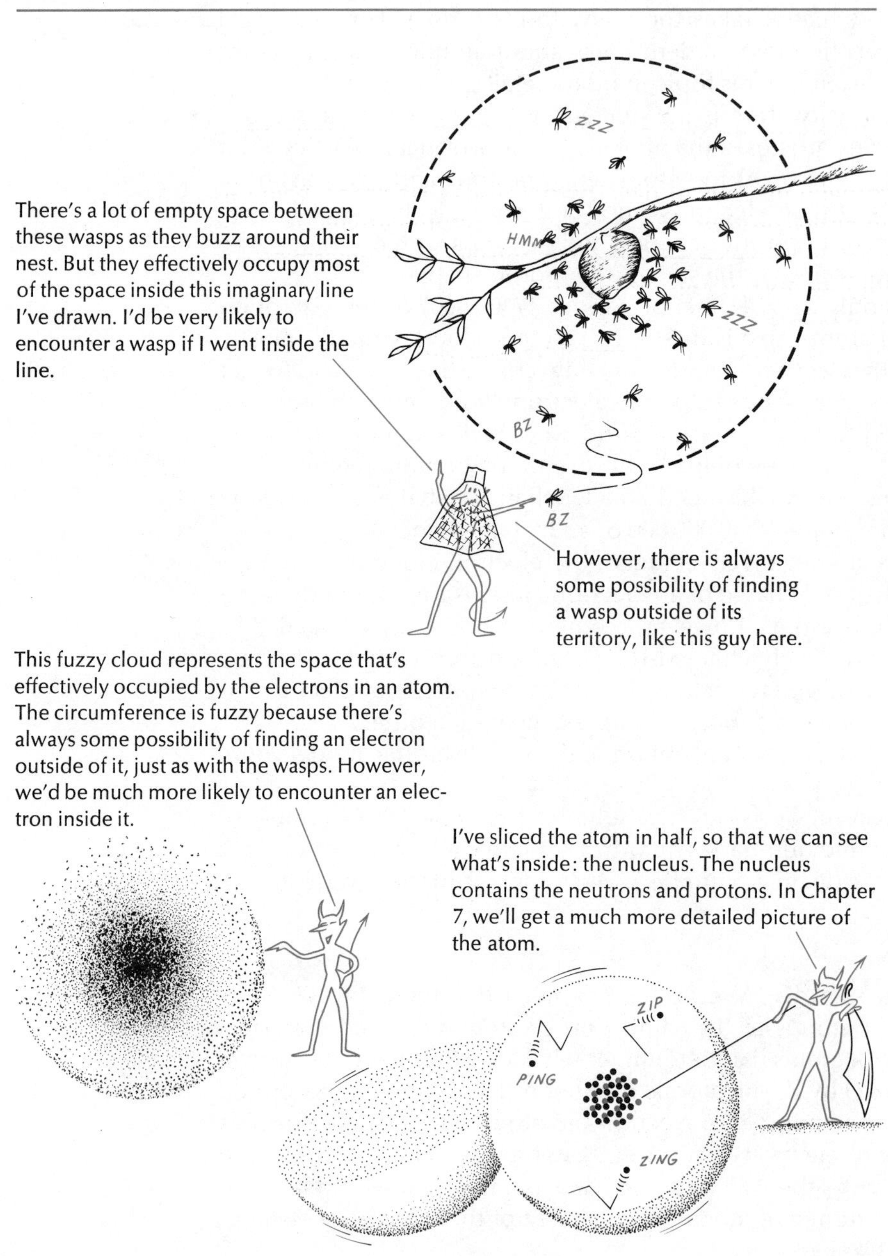

ZZZ
HMM
ZZZ
BZ
BZ
There's a lot of empty space between these wasps as they buzz around their nest. But they effectively occupy most of the space inside this imaginary line I've drawn. I'd be very likely to encounter a wasp if I went inside the line.
However, there is always some possibility of finding a wasp outside of its territory, like this guy here.
This fuzzy cloud represents the space that's effectively occupied by the electrons in an atom. The circumference is fuzzy because there's always some possibility of finding an electron outside of it, just as with the wasps. However, we'd be much more likely to encounter an electron inside it.
I've sliced the atom in half, so that we can see what's inside: the nucleus. The nucleus contains the neutrons and protons. In Chapter 7, we'll get a much more detailed picture of the atom.
ZIP
PING
ZING

German names that were used before the names were changed to English. Each element has its own atomic number, name, and symbol. Table 2.2 gives a list of the elements, with their names, symbols, and atomic numbers.

MASS NUMBER. The sum of the number of neutrons and protons in an atom is the *mass number*, symbolized by *A*. These particles are always together in the atom's nucleus, and it's their combined mass that gives the nucleus its large mass. Notice that protons not only give an element its identity (its atomic number), they also contribute to its mass. Neutrons, however, contribute only to the mass of an atom. The number of neutrons can vary, even among atoms of the same element. (As we'll soon see, it's even possible for a hydrogen nucleus to have *no* neutrons.) Electrons, being so much lighter than either protons or neutrons, contribute virtually nothing to the mass of an atom. As we've seen, though, the electrons determine the atom's atomic number, just as the protons do.

mass number (*A*)

Let's look at hydrogen, which has an atomic number of 1. Its atomic number tells us that it's hydrogen, and that it has one proton and one electron. However, there are three different varieties of hydrogen atoms, each with a different number of neutrons, and thus a different mass number. They're all hydrogen, though, because each has one proton and one electron. Atoms or nuclei having the same atomic number but different mass numbers are called *isotopes*.

isotopes

To tell which isotope of something we're talking about, we call it by its name or symbol and then its mass number. Here are the three isotopes of hydrogen:

hydrogen-1 (H-1): one proton, no neutrons; $Z = 1$, $A = 1$.
hydrogen-2 (H-2): one proton, one neutron; $Z = 1$, $A = 2$ (deuterium).
hydrogen-3 (H-3): one proton, two neutrons; $Z = 1$, $A = 3$ (tritium).

Usually there are, at most, two ways that we can tell the difference between isotopes. The first is obvious: the mass. For instance, we said in Chapter 1 that water is H_2O—that means that water has two hydrogen atoms for every oxygen atom. Water can also be made from hydrogen-2. The kind of water made of hydrogen-2 has more mass and is called "heavy water." Hydrogen-2 itself is called "heavy hydrogen." The second way is the fact that some isotopes are radioactive. We'll learn more about this in Chapter 17.

Hydrogen isn't the only element that has different isotopes. In fact, all the elements can have more than one isotope. Some of them don't occur in nature and have to be made by humans. One example is the uranium isotope U-233, which plays a big role in nuclear power, as we'll see in Chapter 17. One of the most common natural isotopes is carbon-12 (C-12). Since we know from Table 2.2 that carbon's atomic number is 6, we also know that C-12 contains 6 protons and 6 neutrons. Two other carbon isotopes, C-13 and C-14, contain 7 neutrons and 8 neutrons, respectively.

2.3 DESCRIBING THE ELEMENTS

Atoms are very small. They are so small that a 1-carat diamond contains about 10,000,000,000,000,000,000,000 carbon atoms. In chemistry, we work with large quantities of atoms just to be able to see or measure them. Because of this, we usually describe an element instead of an individual atom.

We can describe a person in a lot of ways. If we know the person well enough, we can describe his or her temperament as, say, tranquil, fiery, cold, or irrational. We can mention physical characteristics like body shape, or color of hair, eyes, and skin. We can use numbers to describe height, weight, age, and social security number.

We can describe the elements in ways similar to these. In this chapter,

TABLE 2.2
Names, symbols, and atomic numbers of the elements

Atomic Number, Z	*Name*	*Symbol*	*Atomic Number, Z*	*Name*	*Symbol*
1	Hydrogen	H	27	Cobalt	Co
2	Helium	He	28	Nickel	Ni
3	Lithium	Li	29	Copper (Latin: Cuprum)	Cu
4	Beryllium	Be	30	Zinc	Zn
5	Boron	B	31	Gallium	Ga
6	Carbon	C	32	Germanium	Ge
7	Nitrogen	N	33	Arsenic	As
8	Oxygen	O	34	Selenium	Se
9	Fluorine	F	35	Bromine	Br
10	Neon	Ne	36	Krypton	Kr
11	Sodium (Latin: Natrium)	Na	37	Rubidium	Rb
			38	Strontium	Sr
12	Magnesium	Mg	39	Yttrium	Y
13	Aluminum	Al	40	Zirconium	Zr
14	Silicon	Si	41	Niobium	Nb
15	Phosphorus	P	42	Molybdenum	Mo
16	Sulfur	S	43	Technetium	Tc
17	Chlorine	Cl	44	Ruthenium	Ru
18	Argon	Ar	45	Rhodium	Rh
19	Potassium (Latin: Kalium)	K	46	Palladium	Pd
20	Calcium	Ca	47	Silver (Latin: Argentum)	Ag
21	Scandium	Sc	48	Cadmium	Cd
22	Titanium	Ti	49	Indium	In
23	Vanadium	V	50	Tin (Latin: Stannum)	Sn
24	Chromium	Cr	51	Antimony (Latin: Stibium)	Sb
25	Manganese	Mn			
26	Iron (Latin: Ferrum)	Fe	52	Tellurium	Te

we'll talk about the elements' chemical and physical characteristics, and we'll save the numbers for the next chapter.

CHEMICAL PROPERTIES. The *chemical properties* of elements are like the personalities of people. They describe the way an element behaves in its interaction with other elements or substances, or the way it participates in a chemical reaction. A *chemical reaction* is an interaction involving different atoms, in which chemical bonds are formed, or broken, or both. A *chemical bond* is a strong attractive force that holds two or more atoms together. Both chemical reactions and chemical bonds will be dealt with in greater depth in later chapters, and we'll expand our definitions of them when we have more specific information to build on.

chemical property

chemical reaction

chemical bond

When two elements react with each other, a chemical reaction occurs, and a compound is formed. A *compound* is a substance composed of two or more elements joined by chemical bonds. For example, the element

compound

Atomic Number, Z	*Name*	*Symbol*	*Atomic Number, Z*	*Name*	*Symbol*
53	Iodine	I	80	Mercury (Latin: Hydrargyrum)	Hg
54	Xenon	Xe	81	Thallium	Tl
55	Cesium	Cs	82	Lead (Latin: Plumbum)	Pb
56	Barium	Ba	83	Bismuth	Bi
57	Lanthanum	La	84	Polonium	Po
58	Cerium	Ce	85	Astatine	At
59	Praseodymium	Pr	86	Radon	Rn
60	Neodymium	Nd	87	Francium	Fr
61	Promethium	Pm	88	Radium	Ra
62	Samarium	Sm	89	Actinium	Ac
63	Europium	Eu	90	Thorium	Th
64	Gadolinium	Gd	91	Protactinium	Pa
65	Terbium	Tb	92	Uranium	U
66	Dysprosium	Dy	93	Neptunium	Np
67	Holmium	Ho	94	Plutonium	Pu
68	Erbium	Er	95	Americium	Am
69	Thulium	Tm	96	Curium	Cm
70	Ytterbium	Yb	97	Berkelium	Bk
71	Lutetium	Lu	98	Californium	Cf
72	Hafnium	Hf	99	Einsteinium	Es
73	Tantalum	Ta	100	Fermium	Fm
74	Tungsten (German: Wolfram)	W	101	Mendelevium	Md
75	Rhenium	Re	102	Nobelium	No
76	Osmium	Os	103	Lawrencium	Lr
77	Iridium	Ir	104	Kurchatovium (tentative)	Ku
78	Platinum	Pt	105	Hahnium (tentative)	Ha
79	Gold (Latin: Aurum)	Au	106	(Name to be determined)	

For the ultimate in lists of elements, sing to the tune of "I am the very model of a modern major-general" by Sir Arthur Sullivan, or listen to a recording of this song.

THE ELEMENTS

There's antimony, arsenic, aluminum, selenium
And hydrogen, and oxygen, and nitrogen, and rhenium
And nickel, neodymium, neptunium, germanium
And iron, americium, ruthenium, uranium

Europium, zirconium, lutetium, vanadium
And lanthanum, and osmium, and astatine, and radium
And gold, and protactinium, and indium, and gallium
And iodine, and thorium, and thulium, and thallium

There's yttrium, ytterbium, actinium, rubidium
And boron, gadolinium, niobium, iridium
And strontium, and silicon, and silver, and samarium
And bismuth, bromine, lithium, beryllium and barium

There's holmium and helium and hafnium and erbium
And phosphorus and francium and fluorine and terbium
And manganese and mercury, molybdenum, magnesium
Dysprosium and scandium and cerium and cesium

And lead, praseodymium, and platinum, plutonium
Palladium, promethium, potassium, polonium
And tantalum, technetium, titanium, tellurium
And cadmium and calcium and chromium and curium

There's sulfur, californium, and fermium, berkelium
And also mendelevium, einsteinium, nobelium
And argon, krypton, neon, radon, xenon, zinc, and rhodium
And chlorine, carbon, cobalt, copper, tungsten, tin, and sodium!

These are the only ones[1] *of which the news has come to Ha'vard—*
And there may be many others but they haven't been discavard.

"The Elements," from *An Evening Wasted with Tom Lehrer,* Reprise Records, RS6199.

1. These lyrics list 102 elements. Lawrencium, 103, was discovered in 1961. Element 104 was said to have been discovered in 1964 by Russian scientists and in 1969 by American scientists. Element 105 was discovered in 1970. In 1974 element 106 was discovered.

hydrogen reacts with the element oxygen to form the compound water, which contains hydrogen and oxygen joined by chemical bonds.

Elements exist either as compounds with other elements, or as free elements. Some free elements occur as single atoms, and they are called *monatomic.* (*Mono-,* or sometimes just *mon-,* means "one.") The monatomic elements are helium, neon, argon, xenon, krypton, and radon.

monatomic element

In other elements, atoms of the same element form *molecules,* particles made of two or more atoms joined together by chemical bonds. Some elements form molecules that contain two atoms each, called *diatomic molecules.* (*Di-* means "two.") For diatomic elements, we write the element's symbol with a subscript "2," showing that there are two atoms in a molecule. These elements are fluorine (F_2), chlorine (Cl_2), bromine (Br_2), iodine (I_2), oxygen (O_2), hydrogen (H_2), and nitrogen (N_2). Many other elements have much more complex structures, as we'll see later.

molecule

diatomic molecule

Chemical reactivity is the tendency of an element to participate in chemical reactions. The more reactive an element is, the greater its tendency to combine with other elements. If an element is very reactive, it's more likely to occur in nature combined with other elements—that is, in compounds—than it is to occur as the free element. Elements that are relatively unreactive don't tend to combine with other elements much, and so they are often found in nature as free elements. For example, sodium (Na) is very reactive and never occurs as the free element. On the other hand, gold (Au) is very unreactive and does occur as the free element. When gold is mined, it's in the form of pure, elemental gold.

chemical reactivity

Chemical reactivity is such an important concept that we find it convenient to break it down into a number of subtopics. The following chemical properties all involve chemical reactions and are all just special cases of chemical reactivity.

1. *Reacting with water.* When iron (Fe) reacts with water, it rusts. Some elements—for example, sodium (Na) and potassium (K)—react violently with water. Others, such as silver (Ag), don't react with water at all.
2. *Reacting with air.* Some elements, such as aluminum (Al) or copper (Cu), tarnish just from sitting around in the air. They react slowly with the oxygen (O_2) in the air. Others, such as gold (Au) or mercury (Hg), don't react with air at all.
3. *Burning.* This is reacting with air, too, but at a much higher temperature. When something burns, it combines with the oxygen in the air. For instance, charcoal, which consists largely of the element carbon (C), burns in the presence of oxygen or air. The elements iodine (I_2) and neon (Ne) are examples of elements that don't burn.

   burning
4. *Exploding.* Sometimes an element reacts so violently with air, water, or other substances that an explosion results.

   exploding
5. *Supporting combustion.* This means that something else can burn *in* the substance. Things will burn in air; this is because the oxygen in the air supports combustion. Nitrogen (N_2) supports the combustion of some things, but not many. Helium (He), neon (Ne), argon (Ar), and krypton (Kr) are elements that don't support combustion at all. If we put a lighted match into one of these, the match will go out.

   supporting combustion
6. *Toxicity.* Some elements are poisonous. They react chemically with parts of our body and prevent the body from functioning properly. Chlorine (Cl_2) gas is very poisonous to breathe. Beryllium metal (Be)

   toxicity

is poisonous to touch or to swallow. We'll see, though, that many elements that are toxic as elements are not toxic at all in their compounds. Table salt (NaCl), for example, is a compound of chlorine (Cl_2) and sodium (Na), both toxic as elements.

PHYSICAL PROPERTIES. Anything about an element that we can detect with our senses is a *physical property* of that element. These are some physical properties of the elements:

physical property

1. *Color.* The element copper has a reddish brown color.
2. *Odor.* We can smell the sharp odor of the element chlorine when we open a bottle of chlorine bleach.
3. *Physical state.* The condition of being a solid, a liquid, or a gas is known as the *physical state.* We'll be meeting it several other times in this book, but for now we'll see briefly what it means by using the familiar example of water.

physical state

Ice is solid water; it is water in its *solid state.* The water molecules are fastened firmly together, and they don't move very much. A solid has its own shape and volume, no matter what container you put it in. If we heat ice, it changes to liquid water at its melting point.

solid state

Water in its *liquid state* conforms to the shape, but not the volume, of its container. Its molecules are moving around and aren't firmly fastened together. If we heat liquid water, it changes to steam at its boiling point.

liquid state

Water in its *gas state* is steam. A gas will always occupy all of any container we put it in, so it takes both the shape and the volume of its

gas state

FIGURE 2.4
States of matter

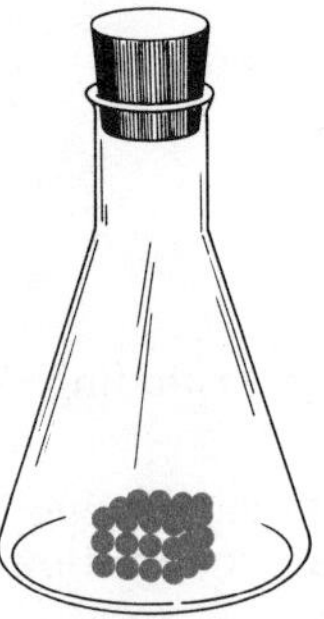

Solid
Atoms or molecules are fastened rigidly together in a fixed order with almost no motion. Solids pay no attention to the shape or volume of the container.

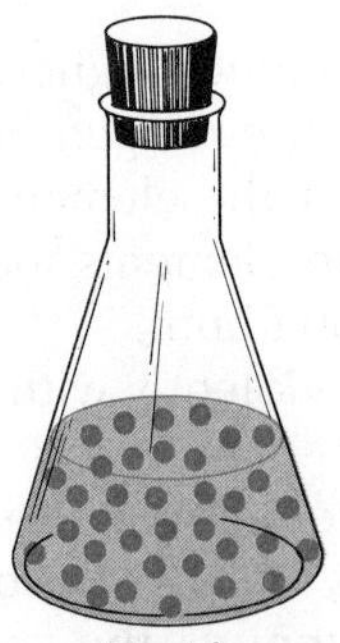

Liquid
Atoms or molecules are farther apart than in the solid state, and move randomly. A liquid takes the shape but not the volume of its container.

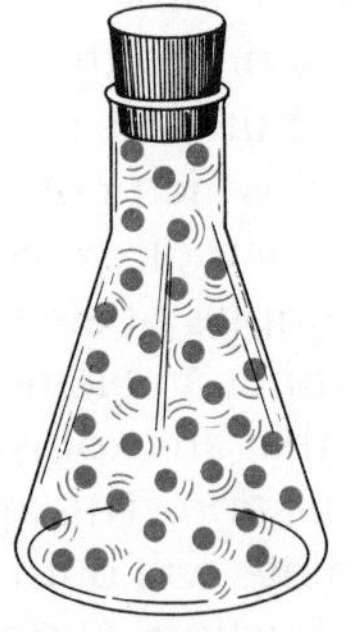

Gas
Atoms or molecules are far apart and move rapidly and randomly. A gas takes up all the room there is.

container. Gas molecules are fast moving and far apart from each other. The three states of matter are illustrated in Figure 2.4.

4. *Metallic properties.* These are thermal (heat) and electrical conductivity; luster (shininess); ductility (can be pulled into a wire without breaking); and malleability (can be pounded out flat without shattering). These properties are illustrated in Figure 2.5. Elements that have all these metallic properties are *metals*. For instance, copper (Cu) is used for electrical wires, so that tells us that copper conducts elec-

metallic properties

metal

FIGURE 2.5
Metallic properties

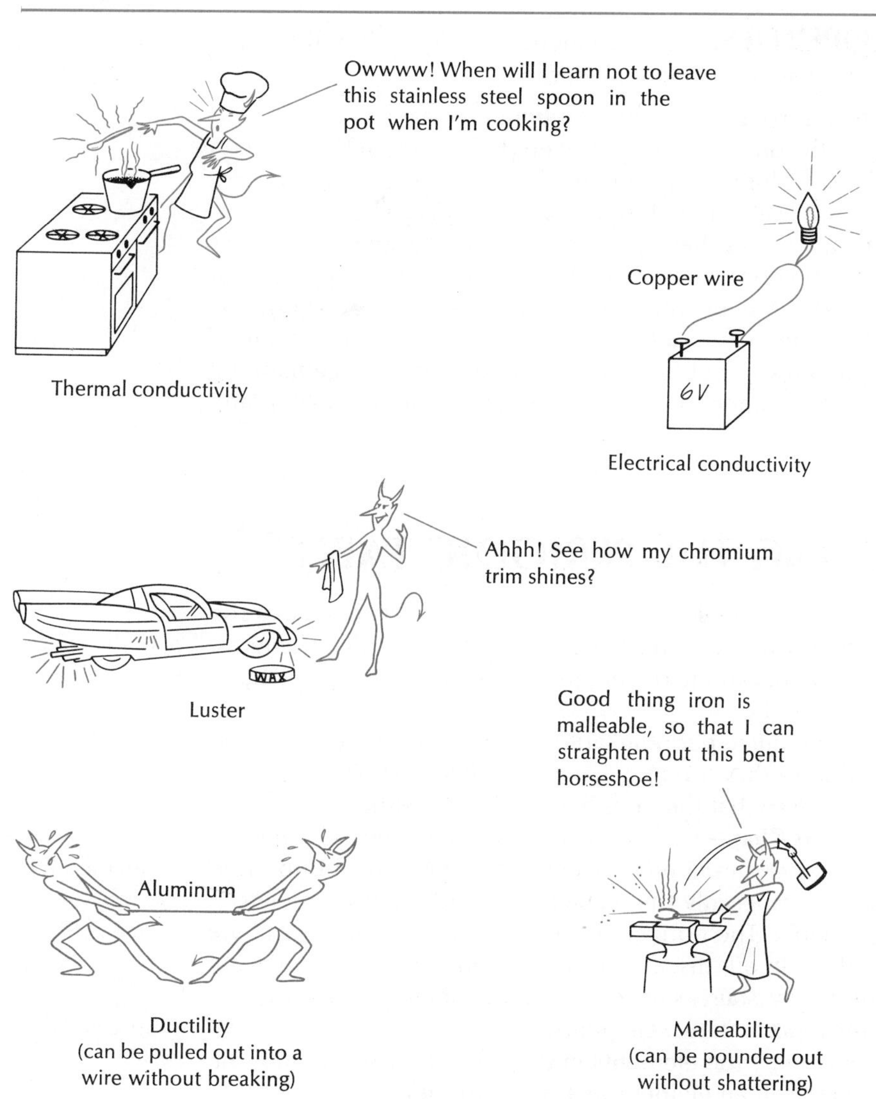

tricity well and is ductile. We know that it's used for the bottom of cooking pots, so it must have good thermal conductivity. Copper is frequently pounded into jewelry, so it must be malleable. When it's clean, it's shiny. It's a metal. Sulfur (S), on the other hand, isn't shiny at all, doesn't conduct electricity or heat, and isn't malleable or ductile. Substances like sulfur that have none of the metallic properties are *nonmetals*. There's a borderline class of elements, called *metalloids*, that have some, but not all, of the metallic properties. Silicon (Si) is shiny like a metal. It's a semiconductor, which means that it does conduct electricity, but not nearly as well as a metal. Silicon is not malleable or ductile. Silicon is a metalloid.

nonmetal

metalloid

GENERAL PROPERTIES. Each element is unique, but there are general statements we can make about all of them.

1. Most elements are solids at room temperature. Mercury (Hg) and bromine (Br_2) are the only liquids. Hydrogen (H_2), nitrogen (N_2), oxygen (O_2), fluorine (F_2), chlorine (Cl_2), helium (He), neon (Ne), argon (Ar), krypton (Kr), xenon (Xe), and radon (Rn) are the only gases.
2. Most elements are metals, having all the metallic properties. Besides the above gases, the nonmetals are carbon (C), phosphorus (P), sulfur (S), selenium (Se), bromine (Br_2), and iodine (I_2), none having any of the metallic properties. Metalloids, having some but not all of the metallic properties, include boron (B), silicon (Si), germanium (Ge), arsenic (As), antimony (Sb), tellurium (Te), polonium (Po), and astatine (At).

2.4 INTRODUCING THE PERIODIC TABLE

Up to now, we've simply listed elements. In chemistry, we keep track of the elements in an orderly way by using the *periodic table of the elements*. The periodic table is the backbone of chemistry, and we'll be using it again and again throughout this book.

periodic table of the elements

Figure 2.6 shows a simplified periodic table (some of the elements, called the "rare earths" because they *are* rare, have been left out). The elements are arranged in the table so that the ones having similar properties are near each other (we'll see in Chapter 8 why this works). The vertical columns are called *groups* or sometimes *families*. Elements in the same group often have similar properties, especially the elements toward the extreme left or extreme right of the table. The horizontal rows are called *periods*, but we won't need to use them in our discussions here. Nonmetallic elements are to the right of the heavy staircase-shaped line. Elements in all the "A" groups are called the *representative elements*.

group (family)

period

representative element

Next, we'll discuss briefly the elements in each group, concentrating on the more usual elements. If an element in any group isn't discussed, that's because the element is rather uncommon.

FIGURE 2.6
Simplified periodic table of the elements

Alkali Metals	Alkaline Earth Metals					Transition Elements						Boron Family	Carbon Family	Nitrogen Family	Oxygen Family	Halogen Family	Noble Gases
IA	IIA	IIIB	IVB	VB	VIB	VIIB	VIIIB	VIIIB	VIIIB	IB	IIB	IIIA	IVA	VA	VIA	VIIA	VIIIA
1 H (g)																	2 He (g)
3 Li (s)	4 Be (s)											5 B (s)	6 C (s)	7 N (g)	8 O (g)	9 F (g)	10 Ne (g)
11 Na (s)	12 Mg (s)											13 Al (s)	14 Si (s)	15 P (s)	16 S (s)	17 Cl (g)	18 Ar (g)
19 K (s)	20 Ca (s)	21 Sc (s)	22 Ti (s)	23 V (s)	24 Cr (s)	25 Mn (s)	26 Fe (s)	27 Co (s)	28 Ni (s)	29 Cu (s)	30 Zn (s)	31 Ga (s)	32 Ge (s)	33 As (s)	34 Se (s)	35 Br (l)	36 Kr (g)
37 Rb (s)	38 Sr (s)	39 Y (s)	40 Zr (s)	41 Nb (s)	42 Mo (s)	43 Tc (s)	44 Ru (s)	45 Rh (s)	46 Pd (s)	47 Ag (s)	48 Cd (s)	49 In (s)	50 Sn (s)	51 Sb (s)	52 Te (s)	53 I (s)	54 Xe (g)
55 Cs (s)	56 Ba (s)	57 La (s)	72 Hf (s)	73 Ta (s)	74 W (s)	75 Re (s)	76 Os (s)	77 Ir (s)	78 Pt (s)	79 Au (s)	80 Hg (l)	81 Tl (s)	82 Pb (s)	83 Bi (s)	84 Po (s)	85 At (s)	86 Rn (g)
87 Fr (s)	88 Ra (s)	89 Ac (s)	104 Ku	105 Ha	106												

Metals Nonmetals Metalloids

(g) Gas
(l) Liquid
(s) Solid

Numbers in boxes are atomic numbers

GROUP IA: THE ALKALI METALS. The elements of group IA are the *alkali metals.* Lithium (Li), sodium (Na), potassium (K), rubidium (Rb), and cesium (Cs) are shiny metals, and they're so soft that they can be cut with a knife. They all react easily with air, so much so that they have to be stored under oil to avoid contact with the air. If a piece of one is cut, the freshly cut surface will be very shiny for an instant but then will tarnish because of the metal's fast reaction with air. These metals also react violently with water. A small piece placed in a container of water will sputter and burst into flame. None of these metals occurs in nature as elements, because they're too reactive. When sodium combines chemically with the element chlorine, the result is table salt.

alkali metal

The first member of this group, hydrogen (H_2), is unique among elements and really defies classification. Its properties are unlike those of the other members of Group IA, but we put it in the group anyway, for reasons we'll see later. Hydrogen is a colorless, odorless, diatomic gas. When hydrogen is burned in air, it combines with the oxygen in the air to form water. If we confine hydrogen in a container and light a match to it, it will explode. Hydrogen was once used in gas balloons, because it's the lightest element and a gas balloon has to be filled with something lighter than air. In 1939, a huge, hydrogen-filled dirigible named the *Hindenburg* exploded, killing many people. Since then, balloonists have switched instead to helium, which is very unreactive. Hydrogen is an important element in biological molecules. Like the other elements in its group, hydrogen doesn't occur on earth as a free element because it is too reactive.

GROUP IIA: THE ALKALINE EARTH METALS. Elements of group IIA are called *alkaline earth metals.* All except magnesium (Mg) are like toned-down versions of the alkali metals. They react with air and water and other elements, but not quite so violently. They're soft, but not quite so soft. Calcium (Ca) we know as an important part of bones and teeth. Magnesium is much harder and less reactive than the other members of its group. It is light and shiny and behaves a lot like aluminum (in Group IIIA). It doesn't have to be stored under oil, but it does burn with a bright flame, and so it comes in handy as the filament in flashbulbs. It's also used in alloys to make light-weight structures, such as bicycles. Radium (Ra), a radioactive element, belongs to this group.

alkaline earth metal

GROUPS IIIB TO IIB: THE TRANSITION ELEMENTS. After Groups IA and IIA, distinctions among groups become less clear. The elements in the groups labeled IIIB to IIB are called *transition elements.* They're all metals, and most of them are good conductors of heat and electricity. Silver (Ag) is the best conductor, followed closely by copper (Cu).

transition element

Many of the structural metals are found among the transition elements. By "structural metal" we mean a metal that we can make a structure out of, such as an automobile or a bridge or a building. Steel, for instance, is mostly iron, with various other metals added, such as chromium (Cr), nickel (Ni), vanadium (V), manganese (Mn), molybdenum (Mo), or zirconium (Zr), depending on what the steel is to be used for.

These groups also include the metals gold (Au), silver (Ag), and platinum (Pt), which are called "precious" because they're scarce, unreactive, and soft. They can be shaped easily (into money or jewelry or fillings), and they aren't very reactive so that there's little danger of their being destroyed by fire or chemicals. People once called these metals "royal" because of their disdain for other elements. A mixture of acids called "aqua regia," or "royal water," will dissolve the precious metals.

One of the most interesting transition elements is mercury (Hg). It's the only metal that's a liquid at room temperature, and it's so heavy that a rock will float on it. Mercury is a very useful element. We see it often in thermometers, where its property of expanding and contracting in a regular way with slight temperature changes makes it ideal. It's also used in switches and thermostats. The main industrial use of mercury in the United States is in processing seawater to make sodium, lye, and chlorine.

Mercury can also be dangerous. Although it's fairly unreactive—it would probably pass right through your body if you happened to swallow some (DON'T!)—some of its compounds can help it accumulate in the body and cause poisoning. In its early stages, mercury poisoning causes impaired vision and weakened muscles. In its later stages, the poisoning can cause blindness, paralysis, insanity, and death. Years ago, a mercury compound

ALCHEMY AND CHEMISTRY

Some of the earliest chemical discoveries were made through greed for gold. During the Middle Ages in Europe, it was the common belief that so-called baser metals like lead could be turned into gold if the right steps were taken. People whose profession it was to attempt this change were called "alchemists." The alchemists were part charlatan, part magician, and part chemist. Kings, hoping to increase their stores of gold, actively bought, sold, and traded their alchemists with other kings—as baseball team owners do today with their players. With so much at stake, it behooved the alchemists to get results—or at least to fake them, which many did. Discovery of their fraud often meant death or, at the very least, lifetime disgrace.

The theory held by most alchemists was that metals grew and matured like plants, until they became "ripe" as gold—considered the "perfect" metal. The other metals were just "unripe" gold, although silver was considered by them to be nearly ripe. One had only to wait until a metal ripened, or to find some way of hastening the process. Alchemists' vain attempts to hasten this ripening ultimately advanced chemical experimentation and knowledge. Some of their laboratory tools are still used today in more sophisticated form: distilling flasks, crucibles, water baths, and balances. They discovered the elements antimony, arsenic, bismuth, phosphorus, and zinc, as well as the substances alum, borax, cream of tartar, ether, plaster of Paris, red lead, and aqua regia. The scientific treatment of disease with chemicals rather than with magic charms was begun by an early alchemist. If we stretch it a little, and equate scientific knowledge with gold, we could say that the alchemists actually succeeded.

was used to process the felt in hats. People who worked in hat factories ("hatters") nearly always got mercury poisoning. That's where we get the expression "mad as a hatter."

We still see the ill effects of the industrial use of mercury. In Japan, it has been used in the manufacture of plastics. No one thought that the mercury compounds left over as waste and dumped into the waters would find their way into the systems of plants and animals, but they did. Since 1953, in the fishing village of Minamata, mercury poisoning has killed fifty-two people, crippled about one hundred, and otherwise harmed almost nine hundred. This tragedy has made the world aware of the dangers of mercury pollution.

GROUP IIIA: THE BORON FAMILY. The elements of Group IIIA—the boron family—don't follow a regular pattern of properties. Boron (B) is a metalloid, whereas all the rest are metals. Boron and the next element, aluminum (Al), aren't a bit alike: boron is a black solid and is not very reactive; aluminum is shiny and quite reactive. Aluminum reacts with air to form a protective coating that prevents it from corroding further. That's why aluminum can be used to make pots and pans, and exterior siding for houses. The Group IIIA elements below aluminum are rather like aluminum, but they're not very common.

GROUP IVA: THE CARBON FAMILY. In Group IVA, carbon (C) is a nonmetal, silicon (Si) and germanium (Ge) are metalloids, and tin (Sn) and lead (Pb) are metals. Diamonds are made of carbon. So are graphite and charcoal (or coke). These are three *allotropic forms* of carbon, which means that the element's atoms are joined in different ways. When we look at a perfect diamond, we are looking at a single molecule. A diamond is made of millions and millions of carbon atoms, each joined to four others by chemical bonds to make a continuous structure. (See Figure 2.7.) A diamond is very hard to cut, because strong chemical bonds have to be broken. Although it's possible to burn a diamond, no one has ever been able to make one melt.

allotropic form

Graphite, on the other hand, is made of a lot of carbon layers stacked one on top of the other, like sheets of paper (see Figure 2.8). Each layer is a molecule. The layers slide over each other, which makes graphite a good lubricant and a good material for pencils. Charcoal is like graphite, except that the layers are broken up into little chunks. Since its particles are much more finely divided than graphite's, charcoal burns much better than graphite. The primary use of charcoal is as a fuel, particularly in making steel.

Carbon is most important to us because it is a basic element in all plant and animal structures. (Recall from Chapter 1 that carbon takes second place in the "top ten" elements important to living organisms.) The last two chapters in this book are given over solely to carbon compounds.

The next member of Group IVA, silicon, is a shiny, gray solid. Silicon is the second most abundant element in the earth's outer layer. Sand and most rocks consist of a compound of silicon and oxygen, called silicon dioxide or silica. Other forms of silicon dioxide are glass and asbestos. People who

FIGURE 2.7
Inside view of diamond structure

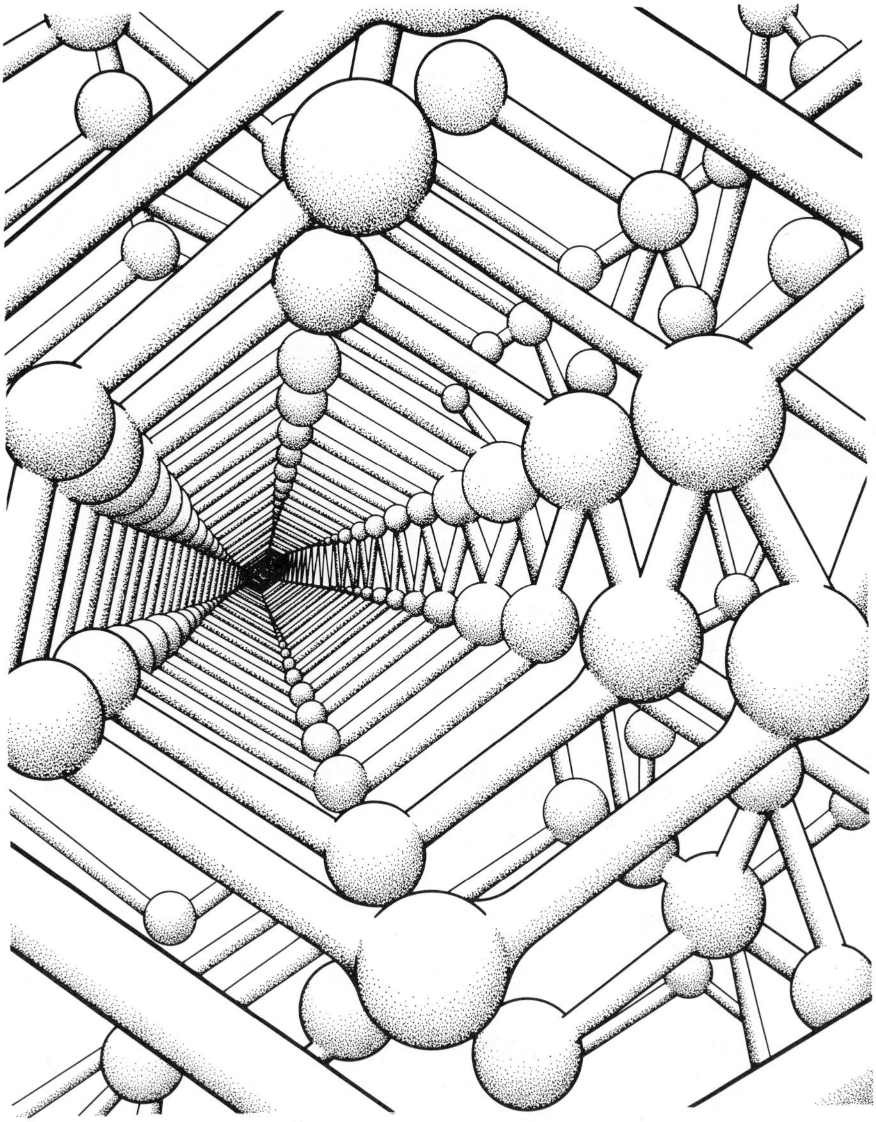

Diamond's carbon atoms are attached by chemical bonds in a structure that goes on and on. If I should fall through a 1-karat diamond at the rate of one carbon atom per minute, it would take me twenty-seven years to get to the other side.

work around and continuously breathe silicon dioxide are subject to a serious disease called "silicosis." Once silicon dioxide gets into the lungs, it stays, and the lungs eventually get so full of sand that they give up. Miners and installers of asbestos insulation or siding usually wear masks these days. Elemental silicon has several allotropic forms, as does carbon. One of these

FIGURE 2.8
Layered structure of graphite

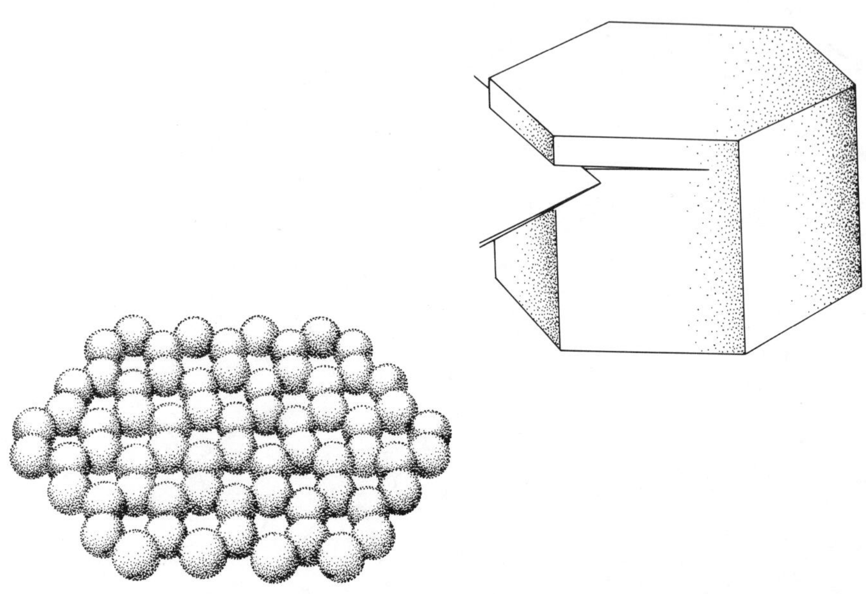

I could put a wedge between the layers and slide the top layer of graphite over the others, because the layers are held together by attractions weaker than chemical bonds.
Within a layer, though, the carbon atoms are attached by chemical bonds. The basic structure is a hexagon, like this one I have in my hand.

is a diamond-type structure, not as hard as diamond, used in making transistors and circuits for televisions, stereo systems, computers, and the like.

The last two Group IVA metals are tin and lead, both of which we're familiar with. Tin is used primarily as a coating for steel, as in tin cans, which are really tin-coated steel cans. The tin coating prevents the steel from reacting with the food in the can. A tin can whose tin layer is broken corrodes rapidly, whereas if the layer is intact, the can will not corrode for many years. Lead's primary use is in storage batteries and in leaded gasoline.

GROUP VA: THE NITROGEN FAMILY. The first two members of Group VA, nitrogen (N_2) and phosphorus (P), are both nonmetals, but they aren't similar in any other ways. Nitrogen occurs naturally as a diatomic, colorless, odorless, unreactive gas. Air is about 79 percent nitrogen, which "dilutes" the oxygen that makes up about 20 percent of the air. All animals and plants contain nitrogen as part of their chemical makeup. Phosphorus—also an important biological element—is a white or red solid. One of its compounds is used in match heads.

Arsenic (As) is an ingredient in mysteries where poisoning is the crime. It behaves somewhat like phosphorus chemically, and it can take the place of phosphorus in important bodily functions, blocking the normal metabolic reactions. Bismuth (Bi) is an ingredient in such medicines as Pepto-Bismol.

GROUP VIA: THE OXYGEN FAMILY. All the Group VIA elements are nonmetals except tellurium (Te) and polonium (Po), which are metalloids.

Oxygen (O_2), the first element, occurs most commonly as a diatomic, odorless, colorless gas. Oxygen supports combustion, and it also supports life. It's the element in air that we must breathe in order to live. We use it to burn up the energy sources that we eat, to make energy to run our bodies. Oxygen is the one element that's very reactive but still occurs in nature as the free element. This is because oxygen is constantly being formed by green plants, which "exhale" oxygen as part of their life processes. As long as we have enough green plants, the oxygen that we use up will be continually replaced, and we won't run out of oxygen.

Besides the diatomic form, oxygen has another allotropic form, called "ozone." Ozone (O_3) is a blue, acrid-smelling gas, consisting of *triatomic molecules* (*tri-* means "three"). Passing an electric spark through oxygen will produce ozone, which is the odor we smell whenever we're around heavy electric machinery. Ozone is also a part of smog, and it can corrode automobile tires and cause serious respiratory damage. It performs a useful function, though, in the stratosphere, which is about twenty miles above the earth. There, it screens out a lot of harmful radiation from the sun.

triatomic molecule

Sulfur (S) is a yellow, solid nonmetal that has several allotropic forms. It burns, is moderately reactive, but doesn't react with water. It occurs in nature both as the free element and in compounds. It's used mostly for manufacturing other chemicals. Selenium (Se) and tellurium (Te) have properties similar to those of sulfur, and are the only elements in this group that are toxic. (Se is the active ingredient in locoweed, a wild plant that grows in southwestern U.S. and causes poisoning in grazing animals.)

GROUP VIIA: THE HALOGENS. At Group VIIA, the elements again have regular properties. These elements are called the *halogens.* They all exist as diatomic molecules, although they're so reactive that none occurs in nature as the free element. Their color becomes more intense going down the group. Fluorine (F_2) is a pale, yellow-green gas. Chlorine (Cl_2) is a greener gas. Bromine (Br_2) is a heavy, reddish brown liquid that causes severe burns if spilled on the skin. Iodine (I_2) is a purplish-black solid. All these are toxic. Fluorine and chlorine support combustion.

halogen

Fluorine has become a familiar element in our lives. One of its compounds, sodium fluoride (NaF), is used in drinking-water "fluoridation" because it helps prevent tooth decay by combining with the calcium in teeth.

We know chlorine from its use in household bleach and in swimming pools, where it disinfects the water. In both cases, chlorine is added to water, sometimes with a stabilizer that keeps the chlorine gas from escaping.

Bromine is used as a disinfectant, and it is also used to make anti-knock gasoline. Iodine is necessary to humans' body chemistry. We need it to make thyroxine, a hormone that regulates our metabolism.

GROUP VIIIA: THE NOBLE GASES. The group VIIIA is a nice, quiet collection of elements with a lot of names. They're called the *noble gases,* the *rare gases,* or the *inert gases.* All these names describe either their scarcity or their lack of reactivity. "Inert" means not reacting with anything at all. Recently some compounds of xenon and krypton have been prepared, so the name "inert gases" isn't used so much any more. This group is easy to classify. All the elements in it are colorless, odorless, monatomic gases that neither burn nor support combustion. Their lack of reactivity kept them from being discovered for a long time. Helium (He) was first discovered on the sun in 1868 (*helios* is Greek for "sun"). However, the element wasn't discovered on earth until 1895. At that time, scientists noticed that all the air's nitrogen, oxygen, water vapor, and carbon dioxide added up to only about 99 percent of air. The remaining gas was completely unreactive, and the scientists named it "argon" (Ar; *argos* is Greek for "lazy"). Later, this remaining 1 percent of air was also proved to contain small amounts of the other noble gases—helium (He), neon (Ne), krypton (Kr), xenon (Xe), and sometimes radon (Rn). Argon makes up 0.93 percent of the air. Neon is next highest, at 0.0018 percent, and xenon occupies a tiny 0.000008 percent.

noble gas

rare gas

inert gas

Helium, the second lightest element, has been used in gas balloons since hydrogen proved too reactive. We know neon best as an ingredient in neon lights. The neon is contained in a glass tube. The tube also contains an electron "gun" that shoots electrons through the gas, knocking other electrons off the neon atoms. The neon's electrons soon return to their atoms, though, and when they do, the atoms give off the light we see. The other noble gases are used in incandescent and fluorescent lights, or wherever an inert atmosphere is needed. None except radon are toxic. Some are used in oxygen mixtures in place of nitrogen (for example, to prevent divers from getting the bends).

REVIEW QUESTIONS

What Atoms Are Made of

1. What is an *element?*
2. What are *subatomic particles?* What are the main kinds?
3. What is *mass?* What do we use it for?
4. What are the masses of the subatomic particles?
5. Give some examples of *static electricity.*
6. Explain the difference between *positive charge* and *negative charge.*
7. What are the *atomic charge units* of the subatomic particles?

Introducing the Whole Atom

8. What is the *nucleus?* What does it contain?
9. Where are the electrons in an atom?
10. Why is an atom electrically neutral?
11. What is an element's *atomic number? Mass number?*

12. What is an element's *symbol?* Why do some symbols seem to have no relation to the element's English name?
13. What are *isotopes?* What are the three isotopes of hydrogen? What other elements can have isotopes?
14. How can we deduce the number of neutrons an atom has, if we know its mass number and its atomic number?

Describing the Elements

15. Why do we describe elements instead of individual atoms?
16. What do we mean by *chemical properties* of elements? What is a *chemical reaction?*
17. What holds the atoms in a compound together?
18. What is a *molecule?* A *diatomic molecule?* What elements exist as diatomic molecules? What elements are *monatomic?*
19. What is *chemical reactivity?* How does it often determine whether elements occur in nature as free elements or in compounds?
20. What are some chemical properties of the elements?
21. If an element is poisonous, does this mean that all of its compounds are poisonous too?
22. What do we mean by *physical properties* of the elements? List them.
23. What are the *physical states?* In what ways do they differ?
24. How can we change a liquid to a gas? A solid to a liquid? A liquid to a solid? A solid to a gas?
25. What are the *metallic properties?* What does each one mean?
26. What is a *metalloid?* What elements are metalloids?
27. What two elements are liquids? Which elements are gases?

Introducing the Periodic Table

28. What are the vertical columns of the periodic table called? The horizontal rows?
29. How can we tell from the periodic table whether an element is a metal, a nonmetal, or a metalloid? What is a *representative element?*
30. What are the Group IA metals called? List some of their properties.
31. Name some properties of hydrogen.
32. What are the Group IIA metals called? List some of their properties.
33. What are the *transition elements?* Give some important transition elements, with their properties.
34. What familiar metal is found in Group IIIA?
35. What are the similarities and differences between diamond and graphite? What do we call such different forms?
36. What are sand, rocks, and glass made of?
37. What important metals are found in Group IVA? What are some uses of these metals?
38. What part of the air do we need to breathe? Why doesn't it get used up?
39. What are the *halogens?* List some of their properties.
40. What are the *noble gases?* What are some other names for them? List some of their properties.

EXERCISES

Set A (Answers at back of book.)

1. It has been suggested that a neutron is a combination of a proton and an electron. Does this idea make sense, when you take into account the particles' masses and charges? Explain.
2. Suppose we were building atoms according to this incomplete table. Fill in the blanks.

No. Protons	*No. Electrons*	*No. Neutrons*	*Atomic Mass Units (amu)*
2	___	2	___
___	3	___	6
3	___	4	___
4	___	___	7

3. With the help of Table 2.2, give the number of protons, neutrons, and electrons in each of these atoms.
   a. Al-27 b. Ar-40 c. Cu-64 d. Au-197 e. Ni-59 f. Co-60
4. Give names and symbols for these atoms.
   a. mass no. 107, 60 neutrons
   b. mass no. 204, 123 neutrons
   c. mass no. 232, 142 neutrons
   d. mass no. 238, 146 neutrons
   e. mass no. 91, 51 neutrons
5. Match up the isotopes of the same elements. Give the name of each element.
   a. $Z = 1, A = 3$
   b. $Z = 3, A = 6$
   c. $Z = 4, A = 6$
   d. $Z = 2, A = 5$
   e. $Z = 3, A = 7$
   f. $Z = 4, A = 7$
   g. $Z = 3, A = 5$
   h. $Z = 4, A = 8$

6. Three isotopes of helium have mass numbers of 3, 4, and 5. How many neutrons, protons, and electrons does each contain?
7. With the help of the periodic table (Figure 2.6), classify each of these as a metal, nonmetal, or metalloid: arsenic, oxygen, barium, zirconium, astatine, sulfur, antimony, aluminum.
8. Find the element that doesn't belong in each of these sets, and state why it doesn't belong.
   a. Na, K, Li, Mg; b. K, Ca, Cr, Br; c. Ne, N, O, F
9. If we had samples of each, how could we tell the difference between each of these pairs?
   a. Na and Mg d. Mg and Ca g. N and P
   b. H and Li e. Al and Si
   c. Hg and Br f. Br and Cl
10. Sodium is a very soft, malleable metal. Why, then, isn't it used to make jewelry?
11. Which would make the best electrical wire: Ca, Fe, or Se? Why?
12. Which can you see: N, Ar, Cl?
13. The study of chemistry is made easier if you can recognize the names and symbols of the most common elements on sight. Try to supply either the name or the symbol, whichever is missing, consulting the list of elements only when necessary.
   a. sodium d. Li g. Sn j. lead
   b. manganese e. K h. Cu k. iron
   c. mercury f. P i. Ag l. Br

Set B (Answers not given.)

1. Suppose we were building atoms according to this incomplete table. Fill in the blanks.

No. Protons	*No. Electrons*	*No. Neutrons*	*Atomic Mass Units (amu)*
5	___	5	___
___	4	___	10
___	___	6	12
___	7	8	___

2. With the help of Table 2.2, give the number of protons, neutrons, and electrons in each of these atoms.
   a. Sr-90 c. C-14 e. O-18
   b. U-235 d. Na-23 f. Sn-116
3. Give names and symbols for these atoms.
   a. mass no. 19, 10 neutrons
   b. mass no. 200, 120 neutrons
   c. mass no. 53, 27 neutrons
   d. mass no. 10, 5 neutrons
   e. mass no. 32, 17 neutrons
4. Match up the isotopes of the same elements. Give the name of each element.
   a. $Z = 5, A = 10$ e. $Z = 19, A = 39$
   b. $Z = 10, A = 19$ f. $Z = 9, A = 20$
   c. $Z = 5, A = 9$ g. $Z = 10, A = 20$
   d. $Z = 9, A = 19$ h. $Z = 18, A = 39$
5. Three isotopes of carbon have mass numbers of 12, 13, and 14. How many neutrons, protons, and electrons does each contain?
6. With the help of the periodic table (Figure 2.6), classify each of these as a metal, nonmetal, or metalloid: hydrogen, antimony, bromine, tin, silicon, gold, iodine, lead.
7. Find the element that doesn't belong in each of these sets, and state why it doesn't belong.
   a. Ar, He, O, I c. Ag, Au, Cu, Ba
   b. N, P, Br, Hg
8. If we had samples of each, how could we tell the difference between the elements of these pairs?
   a. S and Ar d. Br and I g. Fe and Mg
   b. O and H e. Ca and Pb
   c. Cu and I f. Au and Hg
9. Would calcium metal be good for making cooking pots? Explain.
10. Which would be better for building a bridge: C, Fe, Ar?
11. Which would make the best electrical wire: Hg, Si, or Cu? Why?
12. Supply the name and symbol for each of these:
   a. the second lightest element
   b. an allotropic form of oxygen (O_2)
   c. air is mostly made of this
   d. diamond is one form of this element
   e. ingredient of flashbulbs
   f. a semiconductor
   g. the best electrical conductor
   h. forms a protective coating

CHAPTER 3

Measuring Atoms

LEARNING OBJECTIVES

After studying this chapter, you should be able to:

1. Supply a correct definition, example, or explanation for each of these:

mole	calorie
weight	kilocalorie
Avogadro's number	joule
gram-atomic weight	melting point
atomic mass unit	freezing point
atomic weight	heat of fusion
energy	boiling point
heat	condensation point
temperature	specific heat
Celsius scale	density
Kelvin scale	specific gravity
Fahrenheit scale	hydrometer
degree	

2. Explain how weighing can be a means of counting, and the relationship between mass and weight.
3. Explain why gram-atomic weights are not whole numbers.
4. Convert the mass of a given element to moles or the number of moles to mass, given the atomic weight.
5. Calculate the energy in calories released or absorbed when a given mass of water or other substance having a given specific heat undergoes a given temperature change.
6. Given the melting and boiling points of a substance, identify it as a gas, liquid, or solid at room temperature.
7. Given the specific heat of a substance, calculate the temperature change caused by supplying a given amount of heat to a given mass of the substance.
8. Given the mass and volume of a substance, calculate its density, and use density as a conversion factor between mass and volume.
9. Given a table of density data, predict what substances would float or sink in what other substances.
10. Describe how to measure specific gravity, and convert specific gravity to density or density to specific gravity.

In Chapter 2, we described some physical and chemical properties of elements. Now, we'll look at some more physical properties, but these we usually measure with numbers. (For a review of working with numbers, see Appendix A.) Every element has numbers representing atomic number, atomic weight, boiling point, melting point, specific heat, and density. We already know what atomic number is. We'll explore the others in the pages that follow.

3.1 WEIGHING AS A MEANS OF COUNTING

Suppose you're engaged in a building project: laying a floor. You're going to put down 25 boards with 10 nails apiece: that's at least 250 nails you'll need. When you go to the hardware store and ask for 250 nails, the clerk doesn't count out the nails for you. Nails are sold by the pound. You need eight-penny nails, and there are about 130 of them in a pound. The clerk sells you 2 pounds, so you'll be sure to have enough. When we need a large number of a small item, it doesn't make sense to count them out; weighing is more efficient. We probably won't get exactly the number we want, but if we're dealing in large numbers it won't make much difference if there are a few more or less.

In chemistry, we're stuck with using just this method to count atoms and molecules, because it's the only way we *can* do it. Atoms are too small to count out one-by-one, and yet we do need to know how many we have in a bunch so we can study their behavior, which is what chemistry is all about.

Weighing atoms to count atoms is no more mistake-proof than weighing nails to count nails. But there are so many atoms in any sample of an element that even a million atoms more or less will hardly make any difference. (For a discussion of the significance of very small numbers compared with very large numbers, see Appendix A.2, p. 490.)

mole

To count atoms, scientists use a number called the *mole,* which is 602,000,000,000,000,000,000,000 things. Just as with a number called the "dozen," which is twelve things, we can use the mole to count any kind of thing: atoms, molecules, doughnuts, eggs, billiard balls. We usually write the number for the mole in exponential notation, as 6.02×10^{23}. Using exponential notation, we can easily express and work with very large numbers and very small numbers, without having to write out all the zeros involved (Appendix A.4, pp. 494–498, contains a review of writing exponents and using them in problems.)

Even though a mole is like a dozen, the actual number is much larger because atoms are much smaller than the eggs or oranges we usually measure by the dozen. Thus, to weigh—and count—the much smaller atoms and molecules, we have to know the mass of a mole of them. We could also weigh oranges in dozens—if we knew that all oranges weighed the same. Suppose we knew that a dozen oranges weighed 4 pounds and we had a paper bag of oranges that weighed 2 pounds. We'd know, without counting, that we had half a dozen, or 0.5 dozen, or 6 oranges in the bag.

Scientists count atoms the same way, only they don't use the pound. The pound is a unit of weight in the English system of weights and measurements, which is still being used in the United States. Scientists the world over (including American scientists) use the International System of Units (SI), which is a metric system. A few basic SI units are shown in Table 3.1. (A more complete discussion of SI and other units is in Appendixes B.1 and B.2, pp. 509–513.)

Scientists find it convenient to use the grams and kilograms of the metric system to measure mass, instead of using the pounds of the English system to measure weight. Mass is a measure of the amount of matter an object has; the *weight* of an object is a measure of the amount of gravitational attraction on that object. An object's weight can vary, because the gravitational attraction on it can vary from one location to another. Mass doesn't vary, because it doesn't depend on gravitational attraction. However, we do determine the mass of an object by weighing, because weight is proportional to mass. When we weigh an object on a balance, we compare the weight of its mass with the weight of a known mass. Since the same gravitational attraction is acting on both masses, this attraction cancels. As yet, there's no verb "to mass" in the scientific vocabulary, although perhaps there should be. Unfortunately, weight and mass are often used interchangeably. We might hear "such and such weighs 3.2 kilograms," but this is incorrect. Kilograms, grams, and metric tons are all units of mass and not of weight.

weight

TABLE 3.1
Some commonly used SI units and their English equivalents

Quantity	*SI Unit*	*English Equivalent*
Mass	**Basic Unit: Kilogram (kg)**	2.20 pounds (lb)
	Gram (g) = 10^{-3} kg	1/454 pounds (lb)
Length	**Basic Unit: Meter (m)**	3.28 feet (ft)
	Centimeter (cm)[a] = 10^{-2}m	0.394 inches (in)
	Kilometer (km) = 10^{3}m	0.621 miles (mi)
Volume	**Basic Unit: Cubic Meter (m^3)**[b]	35.3 cubic feet (ft^3)
	Liter (ℓ) = 10^{-3} cubic meters (m^3)	1.06 quarts (qt)
	Milliliter (mℓ) = 10^{-3} liters (ℓ)	0.0340 liquid ounces (oz)
	1 mℓ = 1 cubic centimeter (cm^3)	
Amount of Substance	**Basic Unit: Mole (mol)** 1 mole = the amount of any substance that contains the same number of units of that substance as there are atoms in 12.000 . . . grams of C-12.	None

[a]The centimeter, not an officially approved SI unit, is used by chemists because of its convenient size.
[b]Although the cubic meter is the basic SI unit, it is too large to be practical for chemists, who use the smaller units of volume instead.

We know the mass of a mole of carbon-12 atoms is 12.0 grams. If we had 6.00 grams of carbon, we'd know that we had half a mole, or 0.500 moles. And just as we'd know we had 6 oranges if we had half a dozen oranges, so we'd also know we had 3.01×10^{23} carbon-12 atoms, because that's half a mole. (See Figure 3.1.) But how do we know that the mass of a mole of carbon-12 atoms is 12.0 grams?

FIGURE 3.1
Counting oranges or atoms by weighing

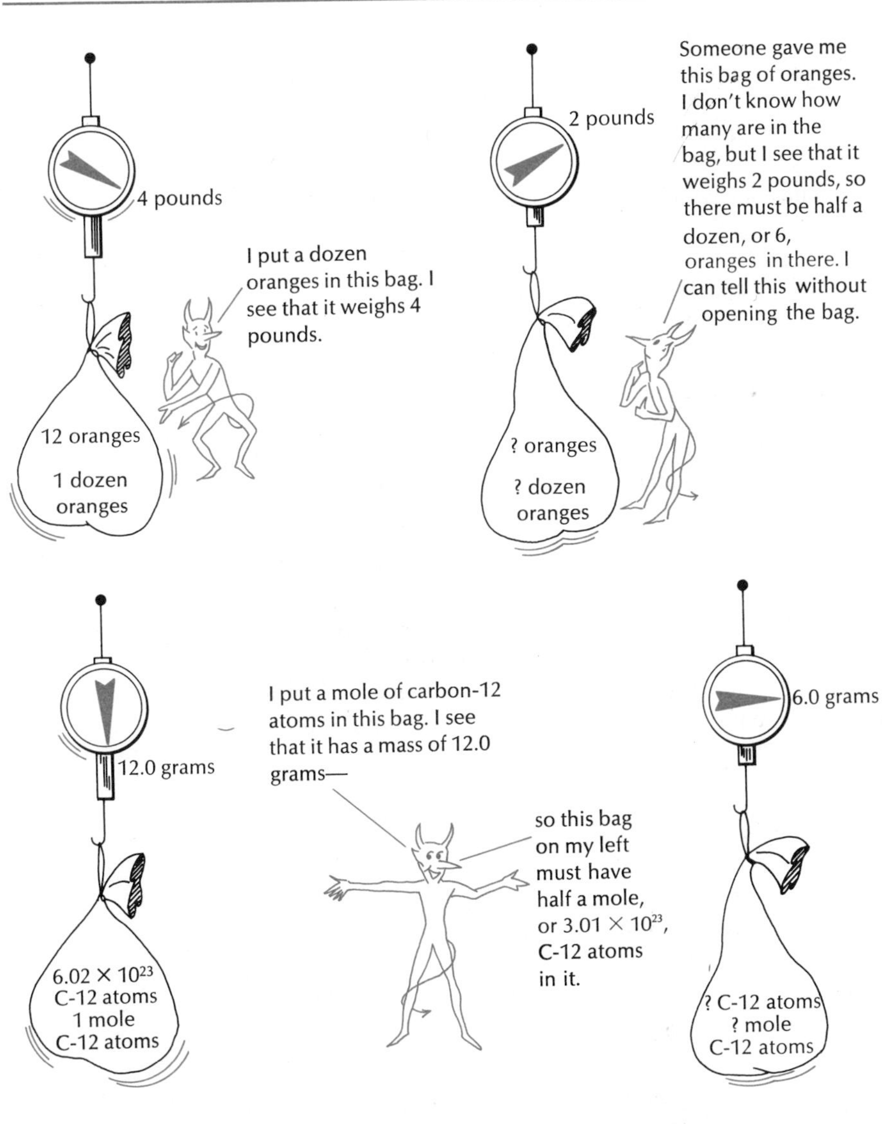

3.2
GRAM-ATOMIC WEIGHT

We've seen that atoms of different elements have different mass numbers, and therefore different masses. Then a mole of each kind of atom will have its own mass, which will be different from the mass of a mole of any other kind of atom. So, too, will a dozen oranges have a different weight from a dozen grapefruit or a dozen kumquats.

(We're about to begin working problems with numbers. In the examples that follow, we'll observe rules of significant figures as described in Appendix A.2, pp. 490–493.)

EXAMPLE 3.1: Given that the mass of an average egg white is 30 grams, an average egg yolk, 20 grams, and an average eggshell, 4 grams, calculate the mass of an egg. (Since eggs vary in mass, values with only one significant figure are given.)

Solution: An egg contains one white, one yolk, and one shell. We add up the parts:

$$\begin{aligned} \text{1 egg white, each 30 grams} &= 30 \text{ grams} \\ \text{1 egg yolk, each 20 grams} &= 20 \text{ grams} \\ \text{1 eggshell, each 4 grams} &= \underline{\ \ 4 \text{ grams}} \\ & \quad\ 54 \text{ grams} \end{aligned}$$

Here, we must round off the 54 grams to 50 grams, since we're not allowed to have any significant figures beyond the first column that contains an uncertain digit, which in this case is the left column. The result is that the 4 grams is negligible (that is, we can neglect its mass) compared to the 50 grams. 50 + 4 is still 50.

Answer: 50 grams. ■

The next example is the first of many we'll be doing that involve units and conversion factors. (Units and conversion factors are fully discussed in Appendix B, pp. 509–522. Working problems with conversion factors involves a few logical steps, which we'll always use. These steps are detailed in Section B.3, p. 513.)

EXAMPLE 3.2: Using the information in Example 3.1, calculate the mass, in grams, of one dozen eggs.

Solution:

Step 1: We want to convert one dozen eggs to grams.

$$1 \text{ doz eggs} = \underline{\quad\quad} \text{ g}$$

Step 2: We need a conversion factor to convert dozen eggs to eggs (12 eggs/doz) and another to convert eggs to grams (50 g/egg).

Step 3:

$$1 \cancel{\text{doz eggs}} \times 12 \frac{\cancel{\text{eggs}}}{\cancel{\text{doz eggs}}} \times 50 \frac{\text{g}}{\cancel{\text{egg}}} = \underline{\quad\quad} \text{ g}$$

Notice that we may cancel "egg" with "eggs." It doesn't matter whether a given unit is singular or plural. The units check.

Answer: 600 grams. ■

EXAMPLE 3.3: If the mass of one dozen eggs is 600 grams, how many dozen eggs are in 1700 grams of eggs?

Solution:
Step 1: We want to convert 1700 grams of eggs to dozen.

$$1700 \text{ g eggs} = \underline{\quad\quad} \text{doz}$$

Step 2: One dozen eggs has a mass of 600 grams. That means that our conversion factor is 600 g eggs/doz. We'll want to invert this factor to 1 doz/600 g eggs, because we want g eggs to cancel. (For information on inverting conversion factors, see Appendix B.3, p. 514.)

Step 3:

$$1700 \cancel{\text{g eggs}} \times \frac{1 \text{ doz}}{600 \cancel{\text{g eggs}}} = \underline{\quad\quad} \text{doz}$$

Our units check, so we solve for the answer. The answer is 2.83, but we must round it to one significant figure.

Answer: 3 dozen. ■

Now, we're ready to do exactly the same kind of calculations with a carbon-12 atom. Then we'll see how we know that the mass of a mole of this kind of atom is 12.0 grams.

EXAMPLE 3.4: The mass of a proton is 1.67×10^{-24} grams (1 amu), a neutron, 1.67×10^{-24} grams (1 amu), and an electron, 9.11×10^{-28} grams (5.49×10^{-4} amu). What, then, is the mass in grams of one carbon-12 atom?

Solution: This is like the egg problem in Example 3.1, except that we have more significant figures and a carbon-12 atom has more than one of each component in it. If we look back at Table 2.2, we see that carbon has an atomic number of 6. That means it has 6 protons. Since the problem gives us the mass number of 12, this atom must also have 6 neutrons. The number of electrons in an atom is the same as the number of protons (also 6). Now, we multiply each part by 6 and add up the masses:

$$\begin{aligned}
\text{6 protons, each } 1.67 \times 10^{-24} \text{ g} &= 10.0 \times 10^{-24} \text{ g} \\
\text{6 neutrons, each } 1.67 \times 10^{-24} \text{ g} &= 10.0 \times 10^{-24} \text{ g} \\
\text{6 electrons, each } 9.11 \times 10^{-28} \text{ g} &= \underline{0.00547 \times 10^{-24} \text{ g}} \\
& 20.00547 \times 10^{-24} \text{ g}
\end{aligned}$$

Again, we see that one of the components—this time, the electrons—has such a small mass compared with the rest that we neglect it. The number above has to be rounded off to one decimal place. This illustrates what we said in Chapter 2 about most of the mass of an atom consisting of the neutrons and protons.

Answer: 20.0×10^{-24} grams. ■

EXAMPLE 3.5: Given that the mass of a carbon-12 atom is 20.0×10^{-24} grams, calculate the mass in grams of 1.00 mole of carbon-12 atoms.

Solution: This is like the dozen problem of Example 3.2

Step 1: We want to convert 1.00 mole of carbon-12 atoms to grams.

$$1.00 \text{ mole C-12} = \_\_\_\_ \text{ g}$$

Step 2: We need a conversion factor to convert moles of carbon-12 atoms to numbers of carbon-12 atoms (6.02×10^{23} C-12 atoms/mole C-12) and another to convert C-12 atoms to grams (20.0×10^{-24} g/C-12 atom).

Step 3:

$$1.00 \, \cancel{\text{mole C-12}} \times 6.02 \times 10^{23} \frac{\cancel{\text{C-12 atoms}}}{\cancel{\text{mole C-12}}} \times 20.0 \times 10^{-24} \frac{\text{g}}{\cancel{\text{C-12 atom}}} = \_\_\_\_ \text{ g}$$

The units check, so we solve for our answer.

Answer: 12.0 grams. ■

The number of grams in a mole of carbon-12 atoms (12.0 g) turns out to be the same numerically as the mass number (12.0 amu). This is not an accident. The mole was chosen so that the mass number of any atom is the same numerically as the number of grams in a mole of the same kind of atom. The number 6.02×10^{23} is called *Avogadro's number,* in honor of an early chemist, Amadeo Avogadro, whom we'll say more about in Chapter 11.

Avogadro's number

Gram-atomic weight is the mass, in grams, of a mole of atoms of any element as it occurs in nature. By definition, the gram-atomic weight of carbon-12 is taken to be exactly 12.0000 grams, and an *atomic mass unit (amu)* is exactly 1/12 the mass of a carbon-12 atom. Then the gram-atomic weights of the elements are given in reference to this standard. These have been determined experimentally and are shown in Table 3.2.

gram-atomic weight

atomic mass unit (amu)

One thing we notice right away is that the gram-atomic weight of the element carbon is 12.011, and not 12.0000. This is because any mole-sized sample of naturally occurring carbon atoms contains about 1 percent carbon-13 and a trace of carbon-14, in addition to carbon-12. When we calculated the mass of a mole of carbon-12 atoms, we knew that each carbon atom in our sample had the same mass. Atoms as they occur in nature don't have the same mass, though, any more than all eggs have exactly the same mass. We took care of this problem in the case of the eggs by using only one significant figure. With atoms, though, we'd like to be more exact. Masses of eggs vary randomly, but there are only a few isotopes of each element, and we know their masses. For any element, we always have a random assortment of isotopes with exactly the same proportion of each isotope. The element chlorine is always 75.5 percent chlorine-35 and 24.5 percent chlorine-37, no matter where we get the sample from or how large it is.

EXAMPLE 3.6: What is the gram-atomic weight of naturally occurring chlorine?

Solution: We know what the percentages are, and we know what the mass of a mole of each isotope is (the same as its mass number). Then:

$$(0.755 \times 35.0 \text{ g}) + (0.245 \times 37.0 \text{ g})$$
$$= \quad 26.4 \text{ g} \quad + \quad 9.07 \text{ g}$$

Answer: 35.5 grams. ■

The gram-atomic weight given in Table 3.2 for chlorine is 35.453, whereas we got 35.5. This is because we carried our calculations to only three significant figures. When we're using gram-atomic weights, we'll usually find it convenient to round them to one decimal place, except for elements having atomic weights less than 10. We'll round the latter to two decimal places. Table 3.2 gives rounded values as well as full measured values. Gram-atomic weights can also be found in the tables inside the covers of this book.

The mass of a single "average" atom is expressed in atomic mass units, and is called the *atomic weight*. We can obtain the atomic weights of the elements just by taking the values in Table 3.2 to be atomic mass units. Chemists often speak of "atomic weight" when they mean "gram-atomic weight." This usually doesn't cause confusion, because their values are the same. Of course, the term "weight" is incorrect in both cases, since both refer to mass. However, "atomic weight" and "gram-atomic weight" have become so common in the working vocabulary of chemistry that any attempt to change would be futile. In these circumstances, we understand that "weight" means "mass."

atomic weight

Now that we know the mass of a mole of each element, we can do problems like the egg problem of Example 3.3.

EXAMPLE 3.7: How many moles are in 156 grams of carbon?

Solution:

Step 1: We want to convert 156 grams of carbon to moles.

$$156 \text{ g C} = \_\_\_ \text{ moles}$$

Step 2: Our conversion factor is the atomic weight of carbon, which we get from Table 3.2. This is 12.0 g C/mole. We invert it to 1.00 mole/12.0 g C.

Step 3:

$$156 \cancel{\text{g C}} \times \frac{1.00 \text{ mole}}{12.0 \cancel{\text{g C}}} = \_\_\_ \text{ moles}$$

The units are correct.

Answer: 13.0 moles. ■

EXAMPLE 3.8: How many grams are in 0.652 moles of chlorine atoms?

Solution:

Step 1: We want to convert 0.652 moles of Cl to grams.

$$0.652 \text{ moles Cl} = \_\_\_ \text{ g}$$

Step 2: Our conversion factor, from Table 3.2, is 35.5 g/mole Cl.

Step 3:

$$0.652 \cancel{\text{moles Cl}} \times 35.5 \frac{\text{g}}{\cancel{\text{mole Cl}}} = \_\_\_ \text{ g}$$

The units check.

Answer: 23.1 grams. ■

TABLE 3.2
Atomic numbers and atomic weights of the elements[a]

Element	Symbol	Atomic Number	Atomic Weight	Rounded Value
Actinium	Ac	89	(227)	–
Aluminum	Al	13	26.98154	27.0
Americium	Am	95	(243)	–
Antimony	Sb	51	121.75	121.8
Argon	Ar	18	39.948	39.9
Arsenic	As	33	74.9216	74.9
Astatine	At	85	(210)	–
Barium	Ba	56	137.33	137.3
Berkelium	Bk	97	(249)	–
Beryllium	Be	4	9.01218	9.01
Bismuth	Bi	83	208.9804	209.0
Boron	B	5	10.81	10.8
Bromine	Br	35	79.904	79.9
Cadmium	Cd	48	112.41	112.4
Calcium	Ca	20	40.08	40.1
Californium	Cf	98	(251)	–
Carbon	C	6	12.011	12.0
Cerium	Ce	58	140.12	140.1
Cesium	Cs	55	132.9054	132.9
Chlorine	Cl	17	35.453	35.5
Chromium	Cr	24	51.996	52.0
Cobalt	Co	27	58.9332	58.9
Copper	Cu	29	63.546	63.5
Curium	Cm	96	(247)	–
Dysprosium	Dy	66	162.50	162.5
Einsteinium	Es	99	(254)	–
Erbium	Er	68	167.26	167.3

Element	Symbol	Atomic Number	Atomic Weight	Rounded Value
Mendelevium	Md	101	(256)	–
Mercury	Hg	80	200.59	200.6
Molybdenum	Mo	42	95.94	95.9
Neodymium	Nd	60	144.24	144.2
Neon	Ne	10	20.179	20.2
Neptunium	Np	93	237.0482	237.0
Nickel	Ni	28	58.70	58.7
Niobium	Nb	41	92.9064	92.9
Nitrogen	N	7	14.0067	14.0
Nobelium	No	102	(254)	–
Osmium	Os	76	190.2	190.2
Oxygen	O	8	15.9994	16.0
Palladium	Pd	46	106.4	106.4
Phosphorus	P	15	30.97376	31.0
Platinum	Pt	78	195.09	195.1
Plutonium	Pu	94	(242)	–
Polonium	Po	84	(210)	–
Potassium	K	19	39.0983	39.1
Praseodymium	Pr	59	140.9077	140.9
Promethium	Pm	61	(145)	–
Protactinium	Pa	91	231.0359	231.0
Radium	Ra	88	226.0254	226.0
Radon	Rn	86	(222)	–
Rhenium	Re	75	186.207	186.2
Rhodium	Rh	45	102.9055	102.9
Rubidium	Rb	37	85.4678	85.5
Ruthenium	Ru	44	101.07	101.1

Europium	Eu	63	151.96	152.0
Fermium	Fm	100	(253)	–
Fluorine	F	9	18.998103	19.0
Francium	Fr	87	(223)	–
Gadolinium	Gd	64	157.25	157.3
Gallium	Ga	31	69.72	69.7
Germanium	Ge	32	72.59	72.6
Gold	Au	79	196.9665	197.0
Hafnium	Hf	72	178.49	178.5
Hahnium	Ha	105	(260)	–
Helium	He	2	4.00260	4.00
Holmium	Ho	67	164.9304	164.9
Hydrogen	H	1	1.0079	1.01
Indium	In	49	114.82	114.8
Iodine	I	53	126.9045	126.9
Iridium	Ir	77	192.22	192.2
Iron	Fe	26	55.847	55.8
Krypton	Kr	36	83.80	83.8
Kurchatovium	Ku	104	(247)	–
Lanthanum	La	57	138.9055	138.9
Lawrencium	Lr	103	(257)	–
Lead	Pb	82	207.2	207.2
Lithium	Li	3	6.941	6.94
Lutetium	Lu	71	174.97	175.0
Magnesium	Mg	12	24.305	24.3
Manganese	Mn	25	54.9380	54.9
Samarium	Sm	62	150.4	150.4
Scandium	Sc	21	44.9559	45.0
Selenium	Se	34	78.96	79.0
Silicon	Si	14	28.0855	28.1
Silver	Ag	47	107.868	107.9
Sodium	Na	11	22.98977	23.0
Strontium	Sr	38	87.62	87.6
Sulfur	S	16	32.06	32.1
Tantalum	Ta	73	180.9479	180.9
Technetium	Tc	43	98.9062	98.9
Tellurium	Te	52	127.60	127.6
Terbium	Tb	65	158.9254	158.9
Thallium	Tl	81	204.37	204.4
Thorium	Th	90	232.0381	232.0
Thulium	Tm	69	168.9342	168.9
Tin	Sn	50	118.69	118.7
Titanium	Ti	22	47.90	47.9
Tungsten	W	74	183.85	183.9
Uranium	U	92	238.029	238.0
Vanadium	V	23	50.9444	50.9
Xenon	Xe	54	131.30	131.3
Ytterbium	Yb	70	173.04	173.0
Yttrium	Y	39	88.9059	88.9
Zinc	Zn	30	65.38	65.4
Zirconium	Zr	40	91.22	91.2
Name to be determined		106	(263)	–

[a]Based on carbon-12. Numbers in parentheses are the mass numbers of the most stable or best-known isotopes.

3.3
TEMPERATURE AND HEAT

The elements' numerical properties of melting point, boiling point, and specific heat all depend on the key concepts of temperature and heat. The two go hand in hand: we can't change the temperature of something without also changing the amount of heat it has. But even though temperature and heat depend on each other, there is a difference between how hot something is (temperature) and how much heat it has (heat). The filament in an electric light bulb is at a much higher temperature than a red-hot electric stove burner, but we'd cook something on the burner and not on the light bulb filament. This is because the burner has more heat than the filament, although the filament is at a higher temperature than the stove burner. We'll see a little later that this is because the burner has more mass than the filament.

Heat is easier to define than temperature. Heat is a kind of energy. We haven't really defined energy yet, but earlier we defined matter as anything that takes up space and requires energy to be moved. Thus energy is what we need to move matter around. Sometimes *energy* is defined as the ability to do work. When we move matter around, we're doing work. There's a direct connection between heat and moving matter around. We define *heat* as the energy associated with the random motions of atoms and molecules. Things are hot because their atoms or molecules are moving fast.

energy

heat

We usually define temperature in terms of heat. The relationship between temperature and heat is that heat will flow from a body at a higher temperature to a body at a lower temperature. If we fill a hot frying pan with cold water, the frying pan cools off and the water warms up. If we left them alone, pretty soon they would both be at the same temperature. Heat would flow from the frying pan to the water because the frying pan was at a higher temperature than the water. We can define *temperature* as that quality of matter which causes heat to flow to or from it.

temperature

MEASURING TEMPERATURE WITH DEGREES. Scientists measure temperature with two scales: the *Celsius* (or centigrade) *scale* (C), and the *Kelvin* (or thermodynamic) *scale* (K). On the Celsius scale, the temperature at which water freezes is the 0° point, and the temperature at which water boils is the 100° point. (The symbol ° stands for "degree.") Between these two points are 100 divisions that measure degrees Celsius. A Celsius degree is thus 1/100 of the interval between the freezing point and the boiling point of water, and that's why it's also called the centigrade scale (*centi-* means "one hundredth"). The Kelvin scale is the official SI temperature scale. It's based on absolute zero, which is the lowest possible temperature and the temperature at which all motion stops (−273.16°C). The size of the kelvin has been chosen to be the same size as the Celsius degree. As we'll see later, this makes it easy to convert between Kelvin and Celsius temperatures. Since the Kelvin scale has the lowest possible temperature as

Celsius scale (C)

Kelvin scale (K)

its zero point, we see that there's no such thing as a negative, or below zero, Kelvin temperature.

The temperature scale still commonly used in the United States is the *Fahrenheit scale* (F). On it, the temperature at which water freezes is 32°F, and the temperature at which water boils is 212°F.

Fahrenheit scale (F)

Right now we'll use the Celsius scale to describe melting and boiling points. The Kelvin scale will be used in Chapter 11 with gases. (A comparison of the three temperature scales, how to convert among them, and some practice problems are given in Appendix B.4, pp. 516–521.)

There's a difference between *temperature* and *degrees*. Degrees are used to measure temperature. If (or when) the United States changes to the Celsius scale, so that freezing weather starts at 0°C instead of 32°F, of course our weather won't get colder. Water freezes at the same point no matter what scale we use to measure that point. We do, however, express a certain temperature with different numbers of degrees according to the kind of thermometer we're using. The reason a temperature scale is necessary at all is so we can say something is at 10°C, or whatever, and people will understand us.

degree

MEASURING HEAT WITH CALORIES. In chemistry, we find it convenient to measure heat in calories. We use water as our standard of reference. One *calorie* (*cal* for short) is the amount of heat it takes to raise the temperature of one gram of water by one degree Celsius. In other words, if we had a sample of water weighing exactly 1.00 gram at a temperature of 14.5°C, we'd have to supply it with exactly 1.00 calorie to raise its temperature to 15.5°C. If we had 10.0 calories of heat available, we could either raise 1.00 gram of water 10.0°C or raise 10.0 grams of water 1.00°C. A larger unit of heat is the *kilocalorie* (*kcal* for short). A kilocalorie is 10^3 calories, or the amount of heat it takes to raise one kilogram of water by one degree Celsius. Neither of these units is an approved SI unit, but they are still used by chemists for convenience. The proper SI unit is the *joule* (J), which is 0.239 calories (or, 1 calorie = 4.184 joules).

calorie (cal)

kilocalorie (kcal)

joule (J)

We can use a formula to determine the number of calories needed to change the temperature of a given number of grams of water.

$$\text{heat in calories} = \text{grams of water} \times \text{temperature change in °C}$$

EXAMPLE 3.9: You want to make a cup of tea. Your tap water is 21°C and you need to heat it to about 99°C. Your cup holds 230 grams of water. How many calories will it take to make the tea?

Solution: Use the formula.

$$\begin{aligned}\text{Heat in calories} &= \text{grams of water} \times \text{temperature change in °C}\\ \text{Heat in calories} &= 230\text{ g water} \times (99 - 21)\text{°C}\\ &= 230\text{ g water} \times 78\text{°C}\end{aligned}$$

Answer: 1.8×10^4 calories, or 18 kilocalories. ■

The Calories that we count when we're trying to lose weight are really kilocalories. (Calories with a capital C means kilocalories.) Caloric values of foods are a measure of their available energy, and these values are measured by burning a sample of known weight. The amount of Calories given off when a sample of known weight is burned is the sample's caloric value. At first it might seem strange to find that something we think of as a unit of fat is really a unit of heat. This makes sense, though, since we "burn" the food in our bodies to make energy. If we consume more Calories' worth of food than we need to burn for energy, the extra energy is stored as future Calories (fat).

3.4 NUMERICAL PROPERTIES RELATED TO TEMPERATURE AND HEAT

Now that we know something about temperature and heat, we're ready to talk about melting point, boiling point, and specific heat—three numerical properties of the elements and of all substances.

MELTING POINT. The particles (atoms or molecules) of a solid are held together by attractive forces, which we'll be talking about in Chapter 10. Heating up a solid, such as a piece of ice, gives its molecules more energy and makes them move. Pretty soon they are moving fast enough to overcome the attractive forces that were holding them rigidly together in the solid. The temperature at which this happens is the *melting point* of the solid. When a liquid, such as water, is cooled, the reverse process happens. We take energy away from the molecules, and pretty soon the molecules are moving slowly enough for their attractive forces to hold them rigidly together again and form a solid. The temperature at which this happens is the *freezing point* of the liquid. Melting point and freezing point are really the same thing, approached from opposite directions. To melt a substance, we supply heat; to freeze it, we remove heat. While a solid is melting, its temperature stays constant at its melting point. Even though we keep heating a solid as it melts, we won't increase its temperature until all of the solid has changed to liquid. When a solid starts to melt, all of the heat that is put into it from then on goes into breaking up the attractive forces that hold the atoms or molecules together in the solid. When the solid is all melted, then the heat that is put in can once more go into increasing the temperature of the substance. The amount of heat that it takes to melt one gram of any substance at its melting point is called the *heat of fusion*. If we let the substance freeze, then it will give off heat in the amount of the heat of fusion. Freezing is a process that releases energy.

melting point

freezing point

heat of fusion

Every substance has a melting (or freezing) point except diamond, which no one has been able to melt yet. The stronger the attractive forces that hold atoms or molecules together in the solid, the higher its melting (or freezing) point will be. The forces holding a diamond together in the solid state are so strong that they can't be overcome by heating. Most elements are solids at "room temperature," a vague term meaning a range of about 20°C to 30°C.

A substance that's a solid at room temperature has a melting point higher than room temperature. Some substances are borderline, and they can be either liquids or solids depending on the weather: we've all seen tar melt on a hot day. Olive oil will solidify (freeze) on a cold day. Table 3.3 on pages 48–49 shows the melting points of some elements.

BOILING POINT. In a liquid, the atoms or molecules are moving around, but there is still quite a bit of attraction among them. Heating a liquid, such as water, gives its molecules still more energy and makes them move even faster. When they are moving fast enough to overcome the attractive forces that keep them as a liquid, then they escape into the gas state. A liquid boils when its molecules escape into the gas state at the bottom of its container and form bubbles, which then rise to the top. This happens at the liquid's *boiling point*. While a liquid is boiling, its temperature stays at the boiling point no matter how much additional heat we supply. (See Figure 3.2.) As with melting, the heat that's put in goes into overcoming the

boiling point

FIGURE 3.2
Melting and boiling take place at constant temperatures

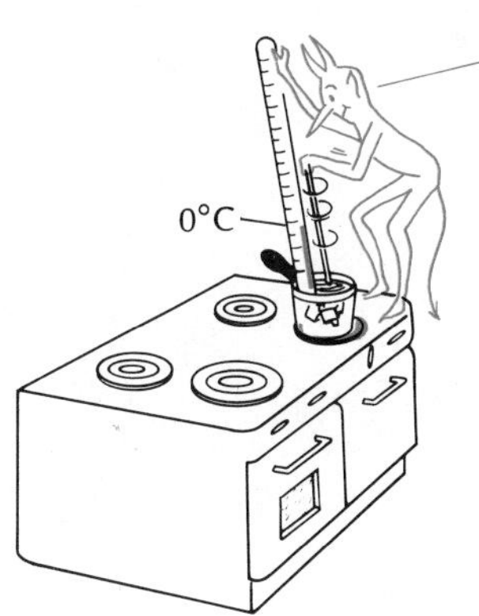

Here's an experiment you can try at home, just like I'm doing. I'm melting some ice in this saucepan and taking its temperature. Right now I have a mixture of ice and water, and I'm stirring really well to make sure that it's always mixed. I'm really blasting it with heat—this burner is on full. But, as long as there is any ice left, the temperature stays at the melting point: 0°C.

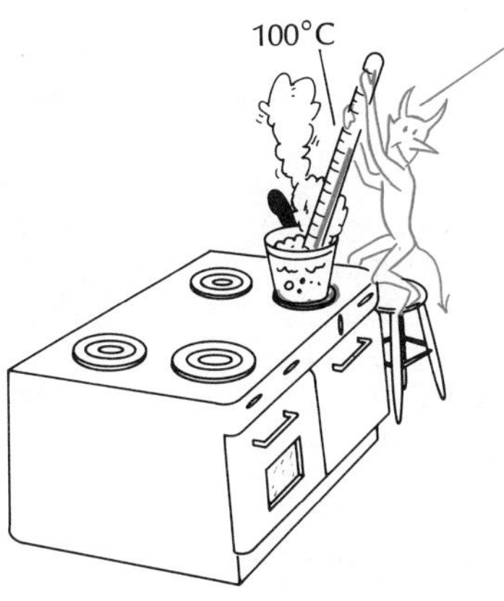

This time I'm boiling some water. I still have the burner on full blast, and I see that I can't get the temperature above 100°C, which is the boiling point of water. I know that if I want to get water hotter than that, I have to use a pressure cooker. We'll see in Chapter 10 how a pressure cooker lets water boil at a temperature higher than 100°C.

molecules' attraction for each other. The amount of heat it takes to change one gram of a liquid to a gas at its boiling point is the substance's *heat of vaporization.* The fact that energy is required to vaporize something lets you cool off on a hot day. You can get wet and let the water evaporate from your body. When the water changes to a gas (steam), it takes heat from your body, and your body cools off.

heat of vaporization

When a gas is cooled, it condenses (changes to a liquid) at its *condensation point,* which is the same temperature as the boiling point. When a gas condenses, it gives back the energy that it took to become a gas: its heat of vaporization. Thus both condensation and freezing are processes that give off heat. A steam burn is worse than a boiling water burn because steam gives off its heat of vaporization in addition to the heat it has because of being at 100°C.

condensation point

If a substance's melting point is lower than room temperature, then it will be a liquid at room temperature. Table 3.3 shows that only two elements—

TABLE 3.3
Numerical properties of selected elements

Element	*Melting Point, °C*	*Boiling Point, °C*	*Specific Heat, cal/(g × °C)*	*Density, g/cm³ (or g/ℓ for gases)*
Aluminum	660	2447	0.215	2.70
Argon	−189	−186	0.124	1.78[a]
Arsenic	817	613	0.0704	5.72
Barium	710	1537	0.068	3.59
Beryllium	1285	2970	0.436	1.86
Bismuth	271	1560	0.0303	9.80
Boron	2074	3675	0.252	2.48
Bromine	−7	59	0.107	3.12
Cadmium	321	767	0.0549	8.64
Calcium	851	1487	0.149	1.55
Carbon (graphite)	3550	3850	0.160	2.27
Cesium	29	670	0.048	1.88
Chlorine	−101	−34	0.115	2.98[a]
Chromium	1900	2640	0.110	7.20
Cobalt	1495	3550	0.109	8.90
Copper	1083	582	0.092	8.92
Fluorine	−220	−188	0.197	1.58[a]
Gallium	30	1980	0.080	5.91
Germanium	937	2830	0.074	5.32
Gold	1063	2707	0.031	19.3
Helium	−272	−269	1.24	0.18[a]
Hydrogen	−259	−253	3.41	0.0899[a]
Iodine	114	184	0.054	4.66
Iron	1530	3000	0.113	7.86
Krypton	−157	−153	0.059	3.74[a]

[a]Gas.

bromine (melting point, −7°C; boiling point, 59°C) and mercury (melting point, −39°C; boiling point, 356°C)—are liquids at room temperature.

Boiling points are usually measured relative to sea level. This is because they depend on atmospheric pressure, which varies at different altitudes (We'll learn more about this in Chapter 10.) It takes longer to hard-boil an egg in the mountains than at the seashore, because water boils at a lower temperature at high altitudes. This is true of all other liquids as well. Figure 3.3 illustrates the changes from solid to liquid to gas.

SPECIFIC HEAT. If you're at the beach on a hot day, the sand might be so hot that you'd have to run to the water to keep from burning the bottoms of your feet. But the water might still be very cold. The water and the sand both received heat from the same sun, but the result is that the water is cold and the sand is hot. Water and sand have different abilities to store heat.

Element	*Melting Point, °C*	*Boiling Point, °C*	*Specific Heat, cal/(g × °C)*	*Density, g/cm³ (or g/ℓ for gases)*
Lead	327	1751	0.031	11.3
Lithium	179	1336	0.850	0.55
Magnesium	650	1117	0.243	1.74
Manganese	1244	2120	0.114	7.30
Mercury	−39	357	0.0331	13.6
Neon	−249	−246	0.246	0.900[a]
Nickel	1455	2840	0.106	8.90
Nitrogen	−210	−196	0.249	1.17[a]
Osmium	2727	4100	0.0313	22.6
Oxygen	−219	−183	0.219	1.33[a]
Phosphorus	44	280	0.190	1.82
Platinum	1774	3800	0.032	21.5
Potassium	64	758	0.188	0.87
Rubidium	39	700	0.086	1.53
Silicon	1415	2680	0.168	2.33
Silver	960	2177	0.0566	10.5
Sodium	98	883	0.293	0.97
Strontium	774	1366	0.0719	2.60
Sulfur	112	445	0.175	1.96
Tin	232	2687	0.0510	7.28
Titanium	1672	3260	0.142	4.51
Tungsten	3415	5000	0.034	19.4
Uranium	1132	3818	0.0276	19.1
Xenon	−112	−108	0.0378	5.90[a]
Zinc	419	907	0.0928	7.14

[a]Gas.

FIGURE 3.3
Conversion of solid to liquid to gas requires energy

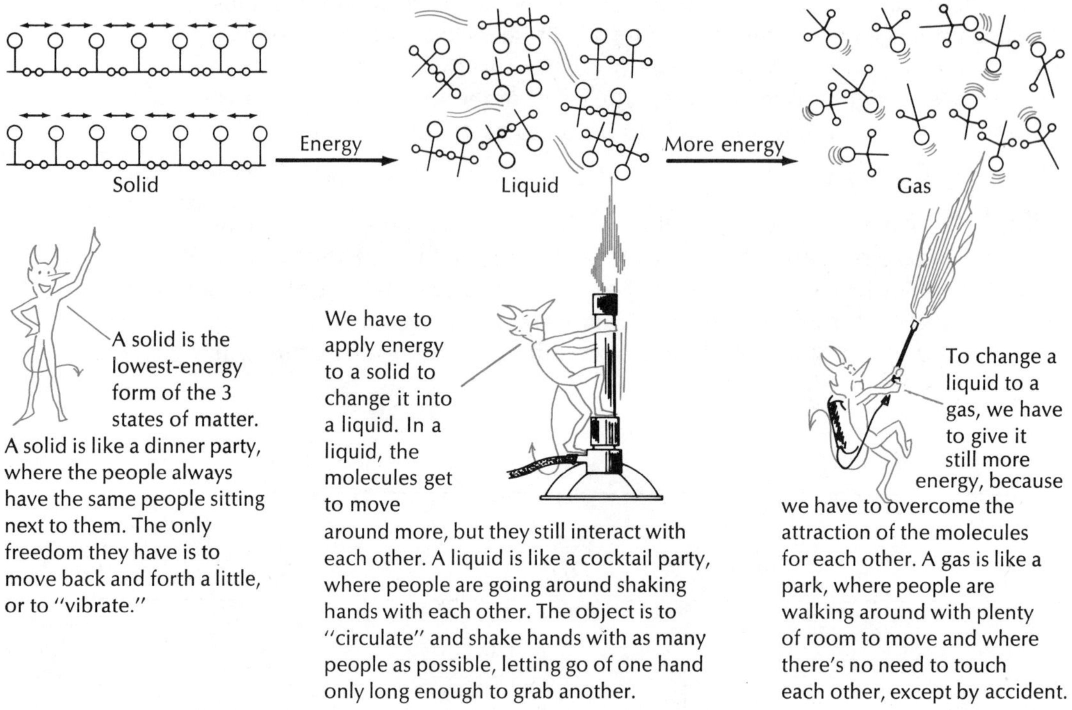

Energy
More energy
Solid
Liquid
Gas
A solid is the lowest-energy form of the 3 states of matter. A solid is like a dinner party, where the people always have the same people sitting next to them. The only freedom they have is to move back and forth a little, or to "vibrate."
We have to apply energy to a solid to change it into a liquid. In a liquid, the molecules get to move around more, but they still interact with each other. A liquid is like a cocktail party, where people are going around shaking hands with each other. The object is to "circulate" and shake hands with as many people as possible, letting go of one hand only long enough to grab another.
To change a liquid to a gas, we have to give it still more energy, because we have to overcome the attraction of the molecules for each other. A gas is like a park, where people are walking around with plenty of room to move and where there's no need to touch each other, except by accident.

Water can store more heat than sand without raising its temperature. Specific heat describes a substance's ability to store heat.

Specific heat is the amount of heat that must be supplied to raise the temperature of a specific mass of a substance by a specified amount. In the metric system, the specific heat in calories is the number of calories needed to raise one gram of a substance one degree Celsius. The formula for finding specific heat is: specific heat

$$\text{specific heat} = \frac{\text{heat in calories}}{(\text{mass in grams}) \times (\text{temperature difference in °C})}$$

The units of specific heat are more interesting than some other units we've had so far. If we write the above equation in a more condensed form:

$$\text{specific heat} = \frac{\text{cal}}{\text{g} \times \text{°C}}$$

we see that the units are calories over grams times degrees Celsius, which we read, "calories per gram-degree."

To find the specific heat of water, we substitute 1.00 cal, 1.00 g, and 1.00°C into the formula and get 1.00 cal/(1.00 g × 1.00°C). This is because we used water to define the calorie in the first place. But other substances can have different specific heats, as shown in Table 3.3.

We can also express specific heat in terms of kilocalories and kilograms, as "kilocalories per kilogram-degree." The values in Table 3.3 can also be expressed as kilocalories per kilogram-degree if this is more useful in solving a problem.

We can rearrange the specific heat equation like this:

$$\text{heat in calories} = (\text{specific heat}) \times (\text{mass in grams}) \times (\text{temperature change in °C})$$

This equation says that the amount of heat a substance can store depends on its specific heat, its mass, and the temperature difference. Now we can understand why the red-hot stove burner has more heat that the light bulb filament, even when the light bulb filament is at a higher temperature.

EXAMPLE 3.10: A light bulb filament made of tungsten has a mass of 1.00 gram and has a temperature of 1500°C. Calculate the amount of heat in calories released when the filament is cooled to 20°C.

Solution: In Table 3.3, the specific heat of tungsten is 0.034 cal/(g × °C). Substituting into the equation gives us:

$$\text{Heat in calories} = 0.034 \frac{\text{cal}}{\cancel{\text{g}} \times \cancel{°\text{C}}} \times 1.00\,\cancel{\text{g}} \times (1500 - 20)\cancel{°\text{C}}$$

The units below the line cancel with those above the line (g, °C).

Answer: 50 calories. ■

EXAMPLE 3.11: An electric stove burner made of iron has a mass of 750 grams and has a temperature of 550°C. Calculate the amount of heat in calories released when the burner is cooled to 20°C.

Solution: In Table 3.3, the specific heat of iron is 0.113 cal/(g × °C). Substituting into the equation gives us:

$$\text{Heat in cal} = 0.113 \frac{\text{cal}}{\cancel{\text{g}} \times \cancel{°\text{C}}} \times 750\,\cancel{\text{g}} \times (550 - 20)\cancel{°\text{C}}$$

Answer: 4.5×10^4 calories, or 45 kilocalories. ■

The stove burner releases roughly a thousand times more calories than the tungsten filament. In this case, the larger mass of the burner is what gives it more heat. If we were comparing two substances of the same mass, then the one with the higher specific heat would have more heat.

Why is water used as a cooling agent—in automobile radiators, electronics industries, and chemical plants? Is it just because it's plentiful, cheap, and nontoxic? These are all good reasons, but another really good reason is water's high specific heat. Of the elements, only hydrogen and helium have higher specific heats than water. Water is a better cooling agent than antifreeze (ethylene glycol). We can illustrate this with still another rearrangement of the specific heat equation:

$$\text{temperature change in °C} = \frac{\text{heat in cal (or kcal)}}{\text{specific heat} \times \text{mass in g (or kg)}}$$

EXAMPLE 3.12: Calculate the temperature change in 5 kg of water used to fill a radiator when 100 kcal of heat is supplied to it.

Solution: We substitute into the equation:

$$\text{temp. change in °C} = \frac{100\,\cancel{\text{kcal}}}{[1.00\,\cancel{\text{kcal}}/(\cancel{\text{kg}} \times \text{°C})] \times 5\,\cancel{\text{kg}}} \quad \left[\text{Note that } \frac{1}{(1/\text{°C})} = \text{°C}\right]$$

Answer: 20°C. (See Appendix B, p. 514, for a discussion of inverting reciprocals.) ■

EXAMPLE 3.13: Calculate the temperature change in 5 kg of ethylene glycol [specific heat 0.544 kcal/(kg × °C)] used to fill a radiator when 100 kcal of heat is supplied to it.

Solution: The specific heat of ethylene glycol is 0.544 kcal/(kg × °C). We substitute into the equation:

$$\text{temp. change in °C} = \frac{100\,\cancel{\text{kcal}}}{[0.544\,\cancel{\text{kcal}}/(\cancel{\text{kg}} \times \text{°C})] \times 5\,\cancel{\text{kg}}}$$

Answer: 40°C. ■

We see that water is about twice as good a cooling agent as straight antifreeze.

Places that are near oceans or large lakes have more moderate climates than places far away from bodies of water. Why? Again, the high specific heat of water. Water's high specific heat causes it to resist changes in temperature. Globally, this serves the purpose of keeping our planet at relatively constant temperatures in spite of the alternation of sunlight and darkness. Later, we'll see what molecular characteristics of water cause it to have this conveniently high specific heat.

3.5 DENSITY

Fifty people jammed into an elevator will be more crowded than ten people in the same elevator. The elevator will also weigh more with fifty people in it than with ten people. The density of people in the first elevator is greater than the density of people in the second elevator. Also, if we compared two elevators—one with fifty fat people in it, and one with fifty skinny people in it—the one with the fifty fat people in it would weigh more.

The density of an element or other substance describes how closely its mass is packed into a given volume. This depends on the masses of the individual atoms and on how closely packed they are. When we talk about something being "light" or "heavy," we're talking about its density (see Figure 3.4). We define *density* as mass per unit volume. Or for liquids and solids, according to the metric system: density

$$\text{density} = \frac{\text{mass in grams}}{\text{volume in cm}^3}$$

FIGURE 3.4
How density works

I've poured 50 cubic centimeters of water and of mercury into each of two beakers. Of course, they don't balance, because the density of mercury is 13.6 times the density of water. Equal volumes don't have the same mass.

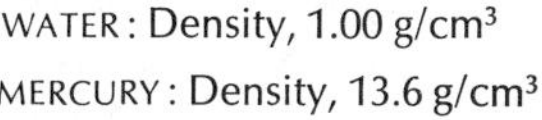

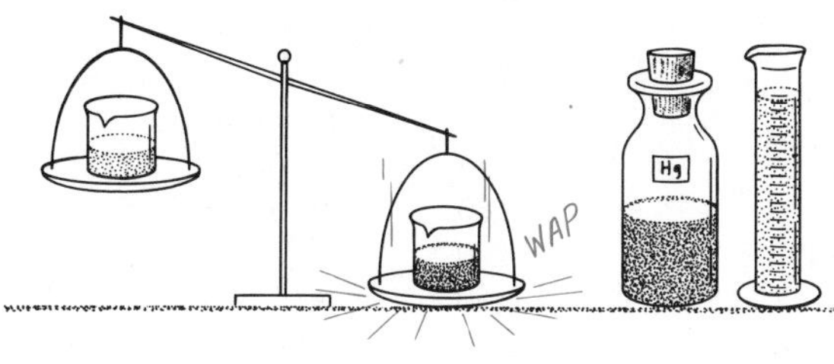

Mercury, though, is still a "lightweight" compared with osmium here. With its density of 22.6, osmium is the most dense element. Its atoms are very tightly packed together. More of them can get into a space.

OSMIUM
Atomic mass: 190.2 amu
Density: 22.6 g/cm³

Even though mercury atoms have more mass than osmium atoms, mercury is less dense than osmium because its atoms are less closely packed.

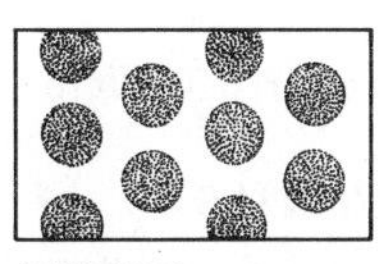

MERCURY
Atomic mass: 200.6 amu
Density: 13.6 g/cm³

For gases, in metric system measurement, we can express density in the following way:

$$\text{density} = \frac{\text{mass in grams}}{\text{volume in liters}}$$

The densities of liquids and solids are expressed in grams per cubic centimeter (g/cm³) or grams per milliliter (g/mℓ). The two are the same,

TABLE 3.4
Densities of some common substances[a]

Solids	*Density, g/cm³*	*Liquids*	*Density, g/cm³*	*Gases*	*Density, g/ℓ*
Cork	0.21	Gasoline	0.70	Hydrogen	0.090
Lithium	0.54	Ethyl alcohol	0.79	Helium	0.18
Paper	0.70	Water (4°C)	1.00	Methane	0.72
Butter	0.86	Glycerine	1.26	Ammonia	0.77
Ice	0.92	Carbon		Neon	0.90
Sugar	1.59	tetrachloride	1.60	Carbon monoxide	1.25
Salt	2.16	Sulfuric acid	1.84	Nitrogen	1.17
Granite	2.60	Mercury	13.6	Air	1.29
Cement	2.70			Oxygen	1.33
Glass	2.50			Hydrogen chloride	1.63
Aluminum	2.70			Argon	1.78
Iron	7.86			Carbon dioxide	1.96
Copper	8.92			Chlorine	2.98
Lead	11.3				
Gold	19.3				
Osmium	22.6				

[a]Densities of liquids and solids are at 20°C, except for water and ice. Gases are at 0°C and atmospheric pressure.

because a cubic centimeter equals a milliliter. The densities of gases, which are much lighter than liquids or solids, are given in grams per liter (g/ℓ).

EXAMPLE 3.14: What is the density of copper if 11.8 cm³ of it has a mass of 105.2 g?

Solution: Use the formula, density = mass/volume. Then:

$$\text{Density} = \frac{105.2\ \text{g}}{11.8\ \text{cm}^3}$$

Answer: 8.92 g/cm³. ■

EXAMPLE 3.15: What is the density of chlorine gas if 2.60 liters of it have a mass of 7.75 grams?

Solution: Use the density formula:

$$\text{Density} = \frac{7.75\ \text{g}}{2.60\ \ell}$$

Answer: 2.98 g/ℓ. ■

Densities of some of the elements appear in Table 3.3, and Table 3.4 compares the densities of various elements with densities of other substances. Since a gram is defined as the mass of 1 cm³ of water (at 4°C), the density of water, by definition, is exactly 1 g/cm³. The densities of other substances tell how much more dense or less dense the substance is than water. Anything less dense than water ("lighter" than water) will float on water; anything more dense ("heavier") will sink. It's the same with air, whose density

FIGURE 3.5
Less dense substances float on more dense substances

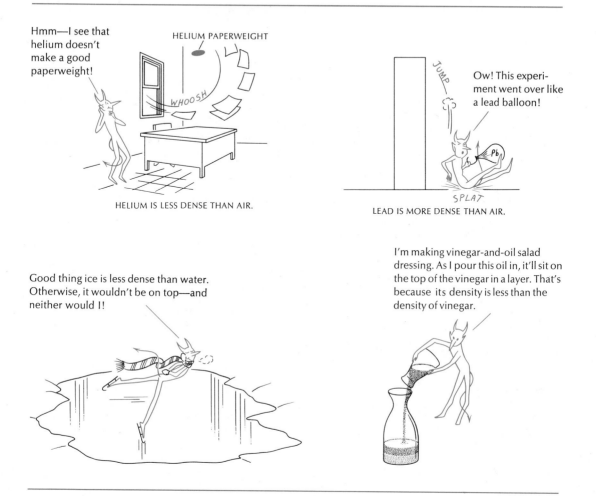

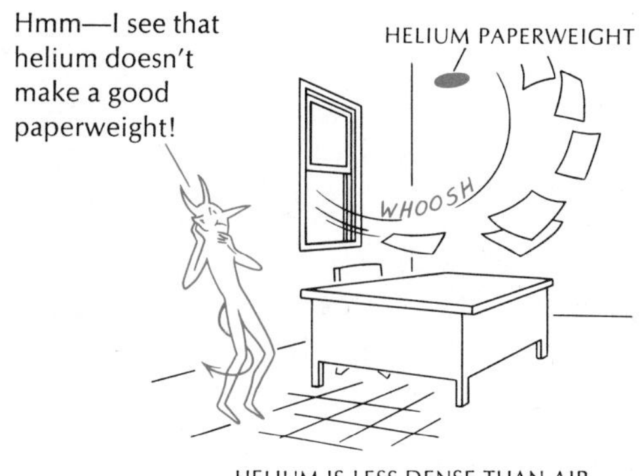

HELIUM IS LESS DENSE THAN AIR.

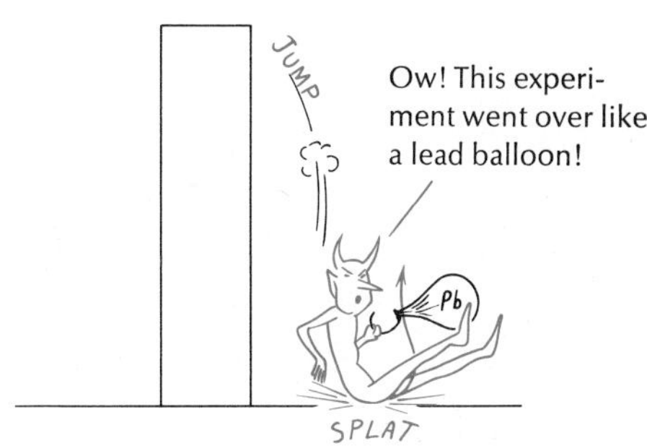

LEAD IS MORE DENSE THAN AIR.

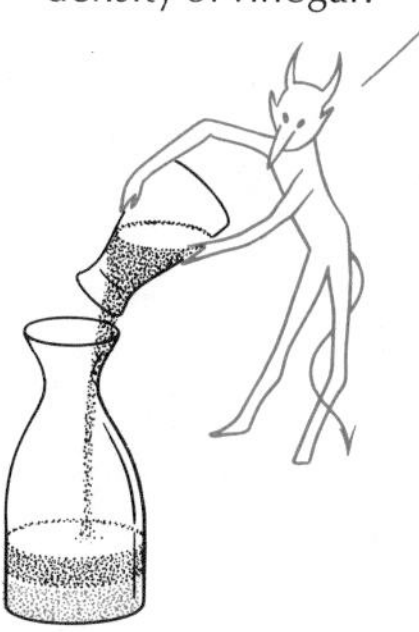

is 1.29 g/ℓ. Anything less dense than air ("lighter" than air) will rise in air; anything more dense ("heavier") will sink. (See Figure 3.5.)

Density is really another conversion factor. It lets us convert mass to volume and vice versa if we know what the substance is. We can invert it, too, as we can any conversion factor. For example, a density of 2.25 g/cm^3 can be inverted as 1.00 cm^3/2.25 g.

EXAMPLE 3.16: A backpacker wants to take a half liter (500 milliliters) of ethyl alcohol on his trip to see him through those cold mountain evenings. How much mass, in grams, will this add to his pack, excluding the container?

Solution:

Step 1: We want to convert 500 milliliters to grams.

$$500 \text{ m}\ell = \_\_\_\_ \text{ g}$$

Step 2: Our conversion factor, from Table 3.4, is 0.79 g/cm³, the density of ethyl alcohol. Also, we need to recognize that a cubic centimeter is the same as a milliliter, so we use the conversion factor 1.00 cm³/mℓ.

Step 3:

$$500\,\cancel{\text{m}\ell} \times 1.00\,\frac{\cancel{\text{cm}^3}}{\cancel{\text{m}\ell}} \times 0.79\,\frac{\text{g}}{\cancel{\text{cm}^3}} = \underline{\quad}\ \text{g}$$

The units are correct.

Answer: 400 g. ■

A useful feature of the metric system is that a density that's expressed in grams per cubic centimeter (g/cm³) can also be expressed in kilograms per liter (kg/ℓ), with the same number. This is because we're multiplying the top and bottom of the factor by 10^3.

EXAMPLE 3.17: Express a density of 3.73 g/cm³ in kilograms/liter (kg/ℓ).

Solution:

Step 1: We want to convert 3.73 g/cm³ to kilograms/liter (kg/ℓ).

$$3.73\,\frac{\text{g}}{\text{cm}^3} = \underline{\quad}\,\frac{\text{kg}}{\ell}$$

Step 2: Our conversion factors (from Appendix Table B.4, p. 512) are 1.00 cm³/mℓ, 10^3 mℓ/ℓ, and 1 kg/10^3 g.

Step 3:

$$3.73\,\frac{\cancel{\text{g}}}{\cancel{\text{cm}^3}} \times 1.00\,\frac{\cancel{\text{cm}^3}}{\cancel{\text{m}\ell}} \times 10^3\,\frac{\cancel{\text{m}\ell}}{\ell} \times \frac{1.00\ \text{kg}}{10^3\,\cancel{\text{g}}} = \underline{\quad}\,\frac{\text{kg}}{\ell}$$

The units are correct.

Answer: 3.73 kg/ℓ. ■

EXAMPLE 3.18: In early times, it was customary for underlords to pay homage to their king once a year by giving him his weight in gold. Kings often stuffed themselves with food just to be ready for this event. How many liters would be occupied by enough gold to buy the good graces of a 105-kilogram king?

Solution:

Step 1: We want to convert 105 kilograms of gold to liters of gold.

$$105\ \text{kg} = \underline{\quad}\,\ell$$

Step 2: Our conversion factor is the density of gold, 19.3 g/cm³. As we showed in Example 3.17, this can also be written as 19.3 kg/ℓ. We invert it to 1.00 ℓ/19.3 kg to make the proper units cancel.

Step 3:

$$105\,\cancel{\text{kg}} \times \frac{1.00\,\ell}{19.3\,\cancel{\text{kg}}} = \underline{\quad}\,\ell$$

The units are correct.

Answer: 5.44 ℓ. ■

Related to density is *specific gravity* (sp gr), which is the density of a substance divided by the density of water. specific gravity

EXAMPLE 3.19: Calculate the specific gravity of ethyl alcohol.

Solution: Looking at Table 3.4, we find that the density of ethyl alcohol is 0.79 g/cm³, and the density of water is 1.00 g/cm³.

$$\text{sp gr (ethyl alcohol)} = \frac{\text{density ethyl alcohol}}{\text{density of water}} = \frac{0.79\ \cancel{\text{g/cm}^3}}{1.00\ \cancel{\text{g/cm}^3}}$$

Answer: 0.79. ■

We can notice two things about specific gravity from Example 3.19. First, it has no units because they all cancel. Second, density in the metric system is numerically the same as specific gravity because water has a density of 1.00 g/cm³.

A *hydrometer* is a device used to measure specific gravity (Figure 3.6). hydrometer

FIGURE 3.6
A hydrometer measures specific gravity

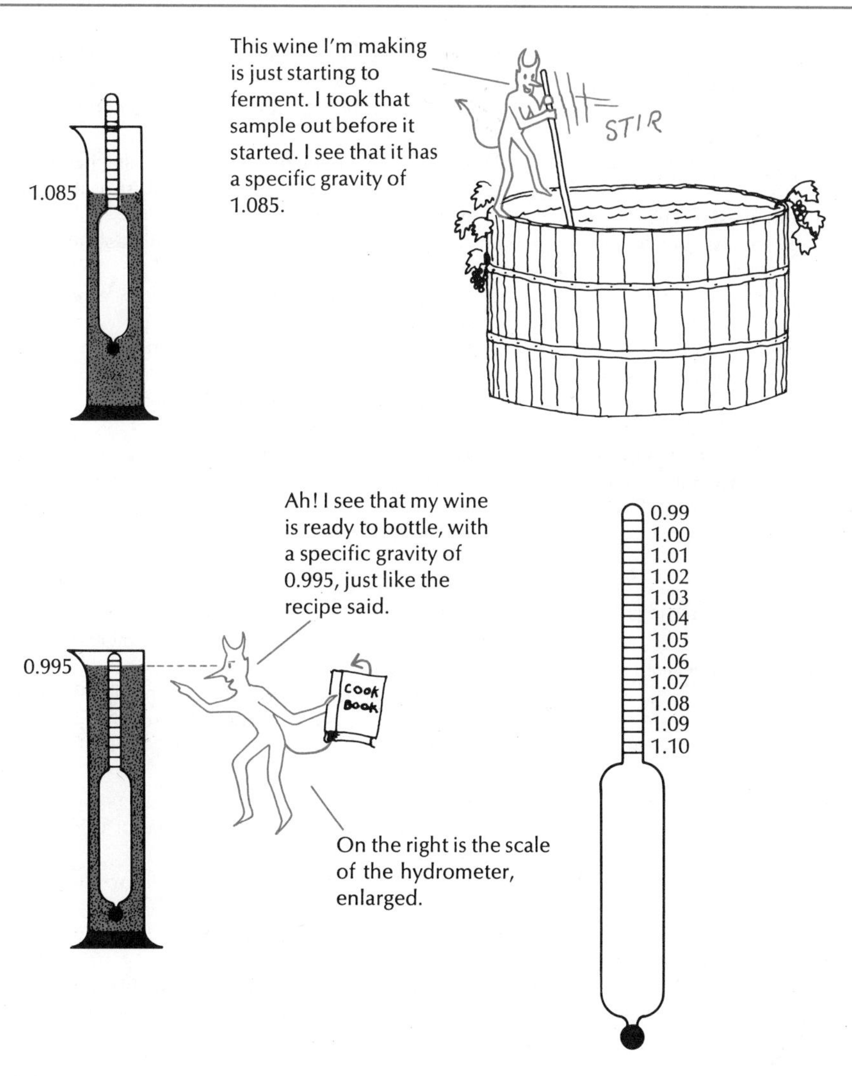

1.085
This wine I'm making is just starting to ferment. I took that sample out before it started. I see that it has a specific gravity of 1.085.
STIR
Ah! I see that my wine is ready to bottle, with a specific gravity of 0.995, just like the recipe said.
0.995
COOK BOOK
On the right is the scale of the hydrometer, enlarged.
0.99
1.00
1.01
1.02
1.03
1.04
1.05
1.06
1.07
1.08
1.09
1.10

When it is placed in a liquid, the specific gravity can be read on the scale at the surface of the liquid. Specific gravity is a useful measure in wine and beer making, because it gives an indication of how far along the fermentation is. An unfermented wine solution has a specific gravity of about 1.085—higher than that of water, because the sugar dissolved in the solution is more dense than water. When the wine has finished fermenting, its specific gravity should be about 0.995—lower than that of water, because most of the sugar has changed into ethyl alcohol, which is less dense than water.

REVIEW QUESTIONS

Weighing as a Means of Counting

1. Why must we weigh atoms in order to count them?
2. What is a *mole?* Why do we need it in chemistry?
3. Express the mole both with and without exponential notation.
4. What are some metric units used in chemistry?
5. Explain how we can know the number of something without counting.

Gram-Atomic Weight

6. What is *gram-atomic weight?*
7. What is a unit? What is a conversion factor? How are they used?
8. What is *Avogadro's number?*
9. What is the chemist's *atomic weight?* Why are these atomic weights often not whole numbers?
10. Explain how atomic weight may be used as a conversion factor.

Temperature and Heat

11. How are *temperature* and *heat* related? How are they different?
12. What is *energy?* In what sense is heat a kind of energy?
13. Describe the three types of temperature scales.
14. What is the difference between temperature and *degrees?*
15. Which temperature scale is used to describe melting and boiling points?
16. What is a *calorie?* How are calories determined?
17. What are the Calories we count when we're dieting?

Numerical Properties Related to Temperature and Heat

18. What is *melting point?* What happens when something melts?
19. What is *freezing point?* What happens when something freezes?
20. Explain why heat is given off when something freezes.
21. What is room temperature? If something has a melting point above room temperature, is it a solid or a liquid at room temperature?
22. Are most elements liquids, solids, or gases at room temperature?
23. What is *boiling point?* What happens when something boils?
24. Why does the temperature remain constant at the melting and boiling points?
25. Explain why heat is given off when a gas condenses to a liquid.
26. How can you tell from a substance's melting and boiling points whether or not it is a liquid at room temperature?
27. Explain the difference between a solid, a liquid, and a gas.
28. What is *specific heat?* How is it related to the definition of the calorie?
29. What three factors determine how much heat an object has?
30. Why is water a good cooling agent? How does it help to keep the earth's climate relatively constant?

Density

31. What is *density?* Give the formula for calculating density.
32. In the metric system, what units do we use to express the density of solids, liquids, and gases?
33. What two factors determine the density of an element?
34. By definition, what is the density of water?
35. Will an object less dense than water float or sink in water?

36. Will an object more dense than air rise or sink in air?
37. Explain why we can express density in kilograms/liter with the same number we use for grams/cubic centimeter.
38. What is *specific gravity*? How is it related to density?
39. How do we use a *hydrometer* to measure specific gravity?

EXERCISES

Set A (Answers at back of book. Asterisks mark the more difficult exercises.)

1. Calculate the mass, in grams, of a dozen doughnuts, if each doughnut has a mass of 225 grams.
2. Calculate the mass, in grams, of a mole of He-4 atoms, if each atom has a mass of 6.64×10^{-24} grams.
3. The weight of a dozen apples is 1.4 kg. How many kg will 2.5 dozen apples weigh? How many apples are in 2.5 dozen?
4. A mole of hydrogen atoms has a mass of 1.01 grams. What will be the mass of 2.50 moles of hydrogen atoms? How many hydrogen atoms are in 2.50 moles?

*5. Since hydrogen exists as diatomic molecules, what will be the mass of a mole of H_2 molecules?

6. Using the table of atomic weights (Table 3.2) and/or Avogadro's number, make these conversions.
   a. 10.0 moles of mercury to grams
   b. 4.00 grams of calcium to moles
   c. 3.82×10^{10} helium atoms to moles
   d. 2.50 moles of iron to number of atoms

*7. Using the value of the atomic weight, calculate the average mass of a gold atom.

8. Naturally occurring boron is 19.6% B-10 and 80.4% B-11. Calculate its atomic weight.
9. A swimming pool contains 8×10^4 kg of water at 10°C. How many kilocalories are needed to heat the water to 25°C?

*10. A sample of butter having a mass of 0.566 g liberated 4.05 kcal when it was burned. Calculate its Caloric content in Calories/g. How many Calories are in a pound of butter?

Consult Table 3.3 to answer Exercises 11–20.

11. What element in Table 3.3 has the largest liquid range, that is, the largest temperature span between its solid and gas states? What element has the smallest liquid range?
12. Why do you think tungsten, and no other metal, is used for electric light bulb filaments?
13. Compare the melting points of the Group IA and IIA metals.
14. If an iron frying pan weighing 2.5 kg is cooled from 210°C to 23°C, how many kilocalories will be released from it?

*15. Find the answer to Exercise 14 in kilojoules (kJ). (1 cal = 4.184 J.)

16. What element in Table 3.3 has the lowest specific heat? The highest?
17. A 5.2-gram piece of lead solder at room temperature (25°C) is subjected to 30. calories with a soldering iron. Calculate the final temperature of the lead. Is it hot enough to melt?
18. A 5.49-gram sample of a metal occupies 3.16 cm^3. What is its density? What is the metal?
19. How many grams would 10.5 cm^3 of bromine weigh?
20. How many liters would 4.00 grams of helium occupy?
21. If 952 calories are supplied to 155 grams of water at 20°C, what will its final temperature be?

Consult both Tables 3.3 and 3.4 to answer Exercises 22–25.

22. Choose some gases that could be used to float a balloon in air.
23. Explain why you can float in salt water better than you can in pure water.
24. Oil spills on oceans can be cleaned up (with difficulty). How do the relative densities of oil and seawater explain this?
25. An airplane is in trouble and the crew has to dump 2 metric tons (each 1000 kg) of its cargo, which is ethyl alcohol. About how many liters of ethyl alcohol will they have to dump?

*26. A method of storing heat from nuclear power plants during low-demand periods and using it in high-demand periods has been proposed that takes advantage of a substance's heat of fusion. The process involves the melting and freezing of sodium nitrate. Explain how it might work.

Set B (Answers not given. Asterisks mark the more difficult exercises.)

1. Calculate the mass, in grams, of a dozen apples, if each apple has a mass of 150. grams.
2. Calculate the mass, in grams, of a mole of Li-7 atoms, if each atom has a mass of 1.16×10^{-23} grams.
3. The weight of a dozen eggs is 1.1 kg. How many kg will 3.5 dozen eggs weigh? How many eggs are in 3.5 dozen?
4. A mole of argon atoms has a mass of 39.9 grams. What will be the mass of 3.5 moles of argon atoms? How many atoms are in 3.5 moles?

*5. A mole of diatomic molecules of an unknown gaseous element has a mass of 32.0 g. What is the atomic weight of the element? What is the element?

6. Using the table of atomic weights (Table 3.2) and/or Avogadro's number, make these conversions.
   a. 0.500 moles of potassium to grams
   b. 354 grams of chlorine atoms to moles
   c. 2.1×10^{28} barium atoms to moles
   d. .450 moles of neon to number of atoms

*7. The mass of an "average" tin atom is 1.97×10^{-22} g. Calculate the atomic weight of tin.

8. Naturally occurring lithium is 94.1% Li-7 and 5.9% Li-6. Calculate its atomic weight.
9. A 200-gallon water bed is filled with water at 18°C. How many kcal will it take to heat it to 70°C?

*10. The Caloric content of marshmallows is 35 Calories per gram. How many kcal of heat would be given off by a marshmallow weighing 7.1 g if it were accidentally dropped into a campfire?

Consult Table 3.3 to answer Exercises 11–21.

11. What element in Table 3.3 has the highest melting point? The lowest?
12. Thermometers containing mercury can't be used below −39°C. (Alcohol is used for lower temperatures.) Explain.
13. Five hundred calories of heat are supplied to 10-gram samples of both copper and aluminum. How many degrees Celsius does the temperature of each element change? Which element is better able to maintain a constant temperature (for example, in providing even heat in a cooking pot)?
14. How many calories would be needed to heat 155 grams of zinc from 35.2°C to 58.4°C?

*15. Find the answer to Exercise 14 in kilojoules (kJ).

16. What element in Table 3.3 has the highest boiling point? The lowest?
17. A silver spoon weighing 25.0 grams was left in a cooking pot and heated to 100°C. How many calories are given off when it cools to body temperature (37°C)?
18. A 6.41-gram sample of a certain gas occupies 1.71 liters. What is its density? What is the gas?
19. How many cubic centimeters would be occupied by 95.2 grams of osmium?
20. Calculate the weight of 22.4 liters of argon.
21. If 15.1 grams of mercury at 25°C is supplied with 245 calories, what will its final temperature be?

Consult both Tables 3.3 and 3.4 to answer Exercises 22–27.

22. Name a substance that will not float in mercury.
23. If a fire extinguisher, containing carbon dioxide, were fired straight ahead, would the vapors rise to the ceiling or sink to the floor? Explain.
24. When water and carbon tetrachloride are poured together, they form two layers. Which is which? How do you know?
25. To make her cooking easier, Grandma said, "A pint's a pound, the world around." In the metric system, we could translate this saying into "A liter's a kilo, from Oshkosh to Hilo." Considering the densities of butter, water, salt, and sugar, prove or disprove Grandma's statement.

*26. If carbon monoxide escapes into a house through a faulty heating system, will its inhabitants be better off sleeping on the first or second floor? Explain.

*27. An old apartment house can support no more than one ton of weight in each room. Can we put a 200-gallon water bed into this apartment? Explain.

CHAPTER 4

Formulas and Names of Compounds

LEARNING OBJECTIVES

After studying this chapter, you should be able to:

1. Supply a correct definition, explanation, or example for each of these:

common name	fixed-charge ion
formula	variable-charge ion
systematic name	binary salt
ionic compound	ternary compound
ion	ternary ionic compound
negative ion	ternary covalent compound
positive ion	polyatomic ion
electrical neutrality	ternary salt
covalent compound	metal hydroxide
binary compound	oxyacid
binary covalent compound	hydrate
oxide	water of hydration
acid	
binary acid	
binary ionic compound	

2. Identify a compound as ionic or covalent by referring to a periodic table that shows the division between metals and nonmetals.
3. Write names and formulas for the two most important oxides of hydrogen, carbon, nitrogen, and sulfur.
4. Know the Greek prefixes and how to use them in naming binary covalent compounds.
5. Classify a metal as forming fixed-charge or variable-charge ions.
6. Referring to a periodic table, give the charge of any fixed-charge ion, positive or negative.
7. List the charges of any variable-charge ion given in this chapter.
8. Classify a substance as belonging in one or more of these categories: binary covalent compound, binary acid, binary ionic compound, metal oxide, nonmetal oxide, binary salt, ternary covalent compound, ternary ionic compound, polyatomic ion, oxyacid, ternary salt, metal hydroxide, mixed compound, or hydrate.
9. Given the formula, supply the name of any substance in the categories of item 8; or, given the name, supply the formula.

We saw in Chapter 2 that a compound is a substance made of more than one element, in which the atoms are joined by chemical bonds. There are over three million chemical compounds known today, and new ones are constantly being prepared and catalogued. Each compound has a name and a formula that describe it and let us know what particular compound we're talking about.

When we're talking about one person to another person, we'll have better luck getting across who we're talking about if we give the person's name, "Harry Robinson," instead of saying, "You know, the guy with blond hair who was talking to Sadie Smith at Louie's party." In the same way, if we wanted to buy a certain chemical compound, or discuss it in any way, we'd be better off giving its name, "hydrogen chloride," instead of saying, "You know, the colorless gas that burns your lungs and dissolves so well in water."

No one who is studying chemistry looks forward to the exercise of learning names and formulas of compounds. But, it could be worse. If names of chemicals were assigned as arbitrarily as names of people, learning them all would be a really discouraging and time-consuming activity. In the early days of chemistry, names were assigned according to how chemicals looked or what they did, or for various other reasons. As the number of known compounds grew, though, chemists had to figure out a more systematic way of naming them. This systematic way is what we use today. A lot of chemicals are still called by their nonsystematic or *common names*. We already know a lot of these common names. Table 4.1 shows some of them with their systematic names and formulas.

common name

The *formula* of a compound tells what elements are in it and the proportion, atom-for-atom or mole-for-mole, among those elements. For instance, the formula H_2SO_4 means that any sample of this particular compound (called sulfuric acid) contains the elements hydrogen, sulfur, and oxygen in the proportion two atoms of hydrogen to one atom of sulfur to four atoms of oxygen. It also means that one mole of the compound contains two moles of hydrogen, one mole of sulfur, and four moles of oxygen. If there is no subscript, "1" is understood.

formula

TABLE 4.1
Common names, formulas, and systematic names of some familiar substances

Common Name	*Formula*	*Systematic Name*
Baking soda (bicarbonate of soda)	$NaHCO_3$	Sodium hydrogen carbonate
Washing soda	$Na_2CO_3 \cdot 10\,H_2O$	Sodium carbonate decahydrate
Milk of magnesia	$Mg(OH)_2$	Magnesium hydroxide
Lye	$NaOH$	Sodium hydroxide
Potash	K_2CO_3	Potassium carbonate
Borax	$Na_2B_4O_7 \cdot 10\,H_2O$	Sodium tetraborate decahydrate
Cream of tartar	$KHC_4H_4O_6$	Potassium hydrogen tartrate
Lime	CaO	Calcium oxide
Slaked lime	$Ca(OH)_2$	Calcium hydroxide
Table salt	$NaCl$	Sodium chloride

The *systematic name* of a compound tells what its formula is by using a certain set of rules, which we'll learn in this chapter. A different set of rules is used according to whether the compound we want to name is ionic or covalent.

systematic name

4.1 IONIC AND COVALENT COMPOUNDS

We'll be learning a lot more about ionic and covalent compounds in Chapters 8 and 9. Right now we want to know the basic differences between them and how to tell which is which, so we can name them properly.

Ionic compounds are made of *ions*. An ion is an electrically charged particle originating from one or more atoms. The charge comes about when an atom gains or loses electrons, so that it has more or fewer electrons than protons. *Negative ions* have more electrons than protons; *positive ions* have fewer electrons than protons. An ion may have a charge of more than one, and we write the charge in the upper right-hand corner after its symbol. A lone plus or minus sign, with no number, means a charge of one. Some examples are Na^+, Mg^{2+}, Al^{3+}, Cl^-, S^{2-}, N^{3-}. Usually metals form positive ions and nonmetals form negative ions.

ionic compound

ion

negative ion

positive ion

An ionic compound is made of both positive ions (usually metals) and negative ions (usually nonmetals). The attractive force between unlike-charged particles is what holds an ionic compound together. The ions combine in such a way that the positive charges and negative charges exactly cancel (or to be precise, neutralize) each other. We say that such a compound is *electrically neutral*. For instance, the compound made of the ions Na^+ and Cl^- is NaCl. One plus charge neutralizes one minus charge. If we were given the formula NaCl (sodium chloride, or table salt), we'd know that it was ionic because we'd locate sodium on the periodic table and find it among the metals, whereas chlorine is found among the nonmetals.

electrical neutrality

Covalent compounds consist of molecules. Molecules are electrically neutral, and they are made of two or more atoms joined together. To make a compound, the atoms have to be different. For instance, water molecules, H_2O, each contain two hydrogen atoms and one oxygen atom. Covalent compounds usually form between two nonmetals. We know that H_2O is covalent because we find both hydrogen and oxygen among the nonmetals.

covalent compound

All of the rules for naming compounds depend on being able to tell whether a compound is ionic or covalent.

4.2 NAMES AND FORMULAS OF BINARY COMPOUNDS

A *binary compound* is one that's made of only two different elements (*bi-* means "two"). The examples NaCl and H_2O are both binary compounds.

binary compound

H_2O contains two atoms of H and one atom of O (a subscript in a formula means the number of atoms of that kind in a formula). Although H_2O has three atoms per molecule, it has only two kinds of atoms, H and O. We can have binary ionic compounds or binary covalent compounds. There are differences in how these two kinds are named, but they have one thing in common: All binary compounds, whether ionic or covalent, end in the suffix *-ide*.

BINARY COVALENT COMPOUNDS (OXIDES, ACIDS, AND OTHERS).

A *binary covalent compound* is composed of two nonmetals, so elements that participate in these compounds are only those to the right of the staircase-shaped line on the periodic table. There are several systems for naming covalent compounds. For now, we'll introduce just one. This system uses Greek prefixes to tell how many of each kind of atom there are. Table 4.2 shows the prefixes with some examples.

binary covalent compound

Using this Greek system, the name of the first element in the formula comes first, preceded by an appropriate prefix (where there is no chance for ambiguity, the prefix *mono-* is usually left off). Then comes the prefix for the second element, followed by its root name, and finished off by the suffix *-ide*. Table 4.3 gives some nonmetallic elements with their roots shown in italics, and also their roots plus *-ide*.

EXAMPLE 4.1: The compound P_4S_3 is used commercially to make match heads. Write its name.

Solution: We see that both phosphorus and sulfur are nonmetals. Looking in Tables 4.2 and 4.3 for appropriate prefixes and roots, and remembering that the name must end in *-ide*, we can write the name.

Answer: Tetraphosphorus trisulfide. ■

TABLE 4.2
Greek prefixes and their use

Number of Atoms	*Prefix*	*Sample*	*Name*
1	*mono-*	CO	Carbon monoxide
2	*di-*	CO_2	Carbon dioxide
3	*tri-*	BF_3	Boron trifluoride
4	*tetra-*	N_2O_4	Dinitrogen tetroxide[a]
5	*penta-*	PCl_5	Phosphorus pentachloride
6	*hexa-*	SF_6	Sulfur hexafluoride
7	*hepta-*	Cl_2O_7	Dichlorine heptoxide[a]
Examples beyond seven are rare.			

[a]Where an element root begins with a vowel, the a on the end of the Greek prefix is dropped for ease of pronunciation.

TABLE 4.3
Some nonmetal names and roots

Element; Root in Italics	*Root*	*Root + -ide*
*Hydr*ogen	Hydr-	Hydride
*Ox*ygen	Ox-	Oxide
*Phosph*orus	Phosph-	Phosphide
*Sulf*ur	Sulf-	Sulfide
*Nitr*ogen	Nitr-	Nitride
*Chlor*ine	Chlor-	Chloride
*Fluor*ine	Fluor-	Fluoride
*Brom*ine	Brom-	Bromide
*Iod*ine	Iod-	Iodide
*Carb*on	Carb-	Carbide

EXAMPLE 4.2: The compound carbon tetrachloride was used as a cleaning fluid until breathing the fumes was shown to cause liver damage. Write its formula.

Solution: We write first the symbol for carbon, C, and then the symbol for chlorine, Cl. The prefix *tetra-* means that there are four chlorines.

Answer: CCl_4. ■

One important class of binary covalent compounds is the set of the oxides of various nonmetals. An *oxide* is a compound of any element with oxygen. As we go through a few of them, we'll see that some are still known to us by their common names. Chemists would rather that everyone used systematic names, but many nonchemists (especially advertisers) don't pay any attention. We'll also see that the use of the prefix *mono-* usually depends on how many oxides of a particular element there are and what their numbers are. oxide

EXAMPLE 4.3: There are two oxides of hydrogen: water, H_2O; and hydrogen peroxide, H_2O_2. Both of these names are common names. What are the systematic names?

Solution: Looking at the chart of Greek prefixes, we select the prefixes *mono-* and *di-* to distinguish these two oxides.

Answer: H_2O, dihydrogen monoxide; H_2O_2, dihydrogen dioxide. ■

Table 4.4 shows some examples, with their systematic and common names.

Except for water, hydrogen peroxide, and ammonia, many binary compounds of hydrogen with a nonmetal can be named in two ways. One way is according to the system we've just learned. Another way is to name them as *acids*. We'll be talking much more about acids later, but for now we'll just say that an acid is a compound (usually dissolved in water) of hydro- acid

TABLE 4.4
Common and systematic names of some binary covalent compounds

Formula	*Common Name*	*Systematic Name*	*Comments*
H_2O	Water	Dihydrogen monoxide	Vital earth fluid
H_2O_2	Hydrogen peroxide	Dihydrogen dioxide	Disinfectant, bleach
CO	–	Carbon monoxide	Poisonous gas in car exhaust
CO_2	–	Carbon dioxide	Gas produced by burning most fuels
N_2O	Nitrous oxide	Dinitrogen monoxide	"Laughing gas," anesthetic
NO	Nitric oxide	Nitrogen monoxide	Colorless gas; air pollutant
NO_2	–	Nitrogen dioxide	Brownish gas; air pollutant
N_2O_3	–	Dinitrogen trioxide	–
N_2O_4	–	Dinitrogen tetroxide	Always found with NO_2
N_2O_5	–	Dinitrogen pentoxide	–
SO_2	–	Sulfur dioxide	Choking gas; air pollutant
SO_3	–	Sulfur trioxide	Air pollutant; used for making sulfuric acid
NH_3	Ammonia	Nitrogen trihydride	Refrigerant; used to make fertilizers and household ammonia

gen with one or more nonmetals. (The reverse isn't true, however. Not *all* hydrogen-nonmetal compounds are acids.) Here, we're talking about *binary acids,* so it's just hydrogen plus one other nonmetal. Binary acids are always written with the hydrogen first. The name of a binary acid always begins with the prefix *hydro-,* followed by the root of the nonhydrogen element, then the suffix *-ic* and *acid.* For instance, the gas HCl would be called hydrogen chloride, according to the Greek system, and its water solution would be called hydrochloric acid. Table 4.5 gives some binary acids, with their names in both systems. Notice that HCN, hydrocyanic acid, follows these rules even though it's not a binary acid. Every now and then we'll come across an exception like this.

binary acid

EXAMPLE 4.4: Dihydrogen sulfide is a foul-smelling, colorless gas that is characteristic of rotten eggs. The odors of sewage gas and intestinal gas are also caused largely by dihydrogen sulfide. Write its formula and its acid name.

Solution: The prefix *di-* means that there are two hydrogen atoms. No prefix on "sulfide" means only one sulfur. We can determine the acid name by following the rules or by referring to Table 4.5.

Answer: H_2S, hydrosulfuric acid. ■

BINARY IONIC COMPOUNDS CONTAINING FIXED-CHARGE IONS.

A *binary ionic compound* is a compound formed from a metal and a nonmetal. Therefore the elements that will participate are found one on each side of the staircase-shaped line in the periodic table. The metal forms the positive ion and the nonmetal forms the negative ion. A *fixed-charge ion* has only one possible charge. Some positive monatomic ions and all negative monatomic ions have fixed charges.

binary ionic compound

fixed-charge ion

Naming binary ionic compounds from fixed-charge ions is easy. The name of the metal comes first, then the root of the nonmetal plus *-ide,* as in the naming of binary covalent compounds. But we don't use any Greek prefixes. The compound NaCl is named simply "sodium chloride." And $CaBr_2$ is named "calcium bromide" without any *di-* even though there are two bromide ions in the formula.

Table 4.6 identifies the fixed-charge ions and gives their charges, along with the charges and rules for negative ions. Table 4.7 gives some examples of naming this type of compound. Notice that there are three ions whose compounds are named with the rules given here, even though their compounds aren't binary: NH_4^+, OH^-, and CN^-.

When we're naming this type of compound, the charges of the individual ions don't enter in. But we do need to use the charges to do the reverse: to write a formula from a name. The chief rule here is that ionic compounds must be electrically neutral; that is, the number of positive

TABLE 4.5
Formulas and names of some binary acids

Formula	*Name, as an Acid*	*Name, Greek System*
HF	Hydrofluoric acid	Hydrogen fluoride
HCl	Hydrochloric acid	Hydrogen chloride
HBr	Hydrobromic acid	Hydrogen bromide
HI	Hydriodic acid[a]	Hydrogen iodide
HCN[b]	Hydrocyanic acid	Hydrogen cyanide
H_2S	Hydrosulfuric acid	Dihydrogen sulfide (hydrogen sulfide)

[a]Here, we drop the *o* before the vowel to aid pronunciation.
[b]HCN is not a binary acid, but it is named according to the rules for binary acids.

HCN appears in this list because it was named when chemists thought it was a binary acid. The name stuck.

TABLE 4.6
Fixed-charge ions

Element	*Charge*	*Element*	*Charge*
Group IA	1+	Group VIIA	1−
Group IIA	2+	Group VIA	2−
Al	3+	H	1−
Zn, Cd	2+	N	3−
Ag	1+	C	4−

Compounds of the following ions with a monatomic ion also take the ending *-ide:* NH_4^+ Ammonium ion; OH^- Hydroxide ion; CN^- Cyanide ion.

charges must exactly balance the number of negative charges. When we write the formula, the metal (or positive ion) always comes first.

EXAMPLE 4.5: Sodium fluoride is a compound used in water fluoridation. Write its formula.

Solution: Looking at Table 4.6, we see that sodium, being a Group IA metal, has a charge of 1+. Fluorine, in Group VIIA, forms an ion with a 1− charge. Therefore, one of each makes the formula electrically neutral. Sodium comes first in the formula as well as in the name.

Answer: NaF. ■

EXAMPLE 4.6: Magnesium hydroxide is the ingredient in "milk of magnesia." Write its formula.

Solution: Things get a bit stickier here. The single negative charge of OH^- isn't enough to balance the two positive charges of Mg^{2+}. We still need

TABLE 4.7
Naming binary ionic compounds from fixed-charge ions

Formula	*Name*	*Comment*
NH_4Cl	Ammonium chloride[a]	"Sal ammoniac."
CaF_2	Calcium fluoride	
ZnO	Zinc oxide	Used in zinc oxide ointment (for sunburn).
CdS	Cadmium sulfide	Used on some TV screens.
NaOH	Sodium hydroxide[a]	This is lye.
KCN	Potassium cyanide[a]	A common poison.
Al_2O_3	Aluminum oxide	A common aluminum ore.
Ag_2S	Silver sulfide	Black tarnish on silver.

[a]These are named like binary compounds, even if they are not binary.

another negative charge. The solution is to combine two hydroxide ions with a single magnesium ion. We enclose the hydroxide ion in parentheses and use a subscript 2.

Answer: $Mg(OH)_2$.

An easy way of writing formulas is just to switch the charge numbers between the positive and the negative ion, and use them for subscripts. (Remember that the subscript "1" is understood and not written.) Above, we used the 2 from the charge of magnesium ion for the subscript of OH, and the 1 from the charge of the hydroxide ion for the subscript of Mg. When the subscript is not 1, we enclose the hydroxide, ammonium, and cyanide ions in parentheses because they contain more than one element. We wouldn't need parentheses for the other ions on the chart.

To write the formula for potassium oxide, we switch the numbers (K^+, O^{2-}) and the formula is K_2O. Aluminum chloride is $AlCl_3$. Calcium nitride is Ca_3N_2 (two of the threes and three of the twos). Figure 4.1 shows why this works.

To write the formula of a binary ionic compound from its name, first decide what the charges are on the positive and negative ions. Then, switch

FIGURE 4.1
Formulas of ionic compounds must show that the compound is electrically neutral

Ca^{2+} F^- = CaF_2 2+ / 2(1−) / 0

K^+ O^{2-} = K_2O 2(1+) / 2− / 0

Al^{3+} Cl^- = $AlCl_3$ 3+ / 3(1−) / 0

Ca^{2+} N^{3-} = Ca_3N_2 3(2+) / 2(3−) / 0

Switching the charge numbers between 2 ions and making them subscripts is a simple way to write correct formulas, and it's easy to see why it works. I've kept score for each one, and we can see that the amount of pluses always equals the amount of minuses. It has to.

I've taken the most complicated case—calcium nitride—to work this out. What we're doing is finding the lowest common multiple of the 2 numbers, just as we find the lowest common denominator in fractions. Here, the lowest common multiple of 3 and 2 is 6. To get 6 pluses, I need 3 calcium ions—and to get 6 minuses, I need 2 nitride ions.

Ca^{2+} Ca^{2+} Ca^{2+} } 6 +

6 − { N^{3-} N^{3-}

Now our compound is electrically neutral, and that's where the formula Ca_3N_2 comes from.

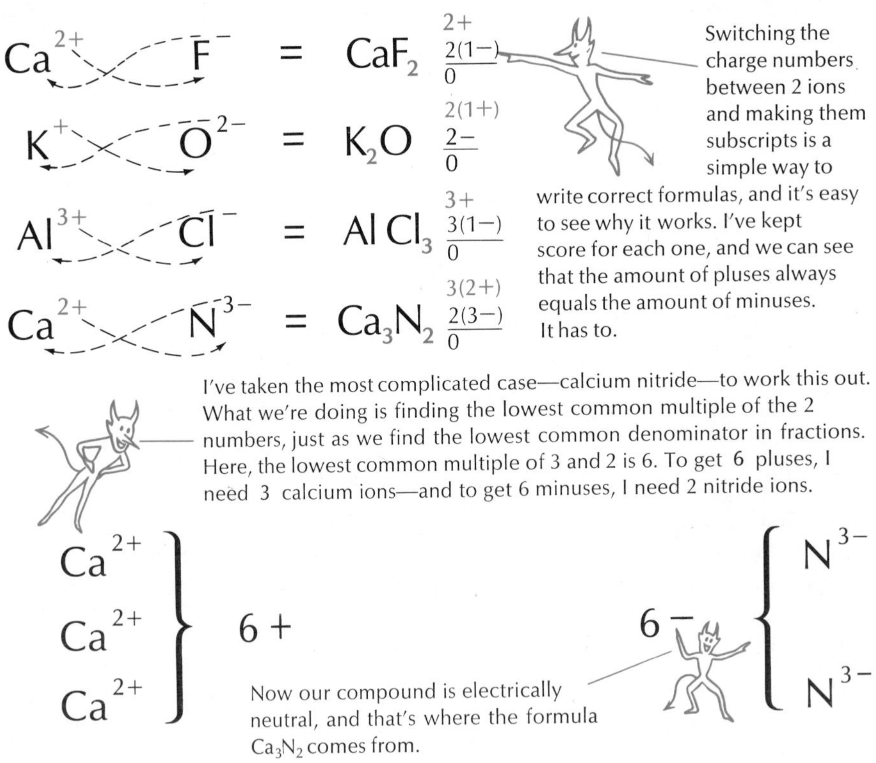

the numbers and write them as subscripts. Sometimes it's necessary to reduce subscripts to the lowest common multiple: CaO and not Ca_2O_2. Exceptions are H_2O_2 and Hg_2Cl_2, which are correct formulas.

EXAMPLE 4.7: Write the formula for aluminum sulfide.

Solution: Aluminum is 3+, sulfide is 2−. We switch the charge numbers around and make them subscripts.

Answer: Al_2S_3. ■

BINARY IONIC COMPOUNDS FROM VARIABLE-CHARGE IONS.

Variable-charge ions are ions that may have several possible charges. Usually the transition metals form variable-charge ions. The most important ones are listed below, along with their charges.

variable-charge ion

Iron (Fe)	2+, 3+	Copper (Cu)	1+, 2+
Chromium (Cr)	2+, 3+, 6+	Tin (Sn)	2+, 4+
Manganese (Mn)	2+, 3+, 4+, 7+	Lead (Pb)	2+, 4+
Cobalt (Co)	2+, 3+	Mercury (Hg)	1+, 2+
Nickel (Ni)	2+, 3+		

There are two systems for naming variable-charge ions: the old system and the Stock system. In the old system, the root of the metal's name is used, sometimes the root of its Latin name if it has one. The suffix *-ic* is added to the root if the ion has the higher charge of two possibilities; the suffix *-ous* is added for the lower charge. Here's part of the above list of metals in the old system:

Fe^{2+}	Ferrous ion	Cr^{3+}	Chromic ion
Fe^{3+}	Ferric ion	Cr^{6+}	What do we do with Cr^{6+}?
Cr^{2+}	Chromous ion		

The truth is that the old system breaks down with metals that can have more than two charges. For these, a metal's ions after the first two have to be ignored. The system was invented before all the ions were discovered, and it's called "old" because chemists no longer use it.

The system chemists use now, the Stock system, is actually easier than the old system, as well as being more practical. All it requires is the name of the metal, followed by the number of the charge in Roman numerals in parentheses, and finished off with the word "ion." Table 4.8 shows the Stock system names for some of the most important variable-charge ions. To read these names, we say, for example, "iron-two ion" for iron(II) ion. Since the Stock system is relatively simple and a lot more useful than the old system, it should come as no surprise that it's the one we'll use in this book. Figure 4.2 shows fixed- and variable-charge positive ions, as well as negative ions.

Table 4.9 shows some examples of binary ionic compounds from variable-charge ions, and their names. We specify the variable-charge ion by the

TABLE 4.8
Names and formulas of some variable-charge ions

Formula	Variable-Charge Ion	Formula	Variable-Charge Ion
Fe^{2+}	Iron(II) ion	Fe^{3+}	Iron(III) ion
Cr^{2+}	Chromium(II) ion	Cr^{3+}	Chromium(III) ion
Cr^{6+}	Chromium(VI) ion		
Mn^{2+}	Manganese(II) ion	Mn^{3+}	Manganese(III) ion
Mn^{4+}	Manganese(IV) ion	Mn^{7+}	Manganese(VII) ion
Cu^{+}	Copper(I) ion	Cu^{2+}	Copper(II) ion
Sn^{2+}	Tin(II) ion	Sn^{4+}	Tin(IV) ion
Pb^{2+}	Lead(II) ion	Pb^{4+}	Lead(IV) ion
Hg_2^{2+}	Mercury(I) ion[a]	Hg^{2+}	Mercury(II) ion

[a]The reason why Hg_2^{2+} is called mercury(I) ion is given in Section 4.3, page 75.

Roman numeral in parentheses. Binary ionic compounds in which the negative ion comes from a nonmetal other than oxygen are called *binary salts*. (Sodium chloride, called "salt," is one specific compound in this category.) If the nonmetal ion is oxygen, the compound is called an *oxide*.

binary salt

TABLE 4.9
Names and formulas of some binary ionic compounds

Formula	Name	Comment
Oxides		
FeO	Iron(II) oxide	
Fe_2O_3	Iron(III) oxide	This is rust.
CrO	Chromium(II) oxide	
Cr_2O_3	Chromium(III) oxide	Green dye used in manufacture of paper money.
PbO	Lead(II) oxide	This is "white lead."
PbO_2	Lead(IV) oxide	This is "red lead," used in lead-based industrial paint pigments.
Binary Salts		
Cu_2S	Copper(I) sulfide	Chalcocite ore.
CuS	Copper(II) sulfide	
SnF_2	Tin(II) fluoride	Ingredient in fluoride toothpaste.
SnF_4	Tin(IV) fluoride	
$HgCl_2$	Mercury(II) chloride	

Here's a compound you've heard of by its old system name, stannous fluoride. It's the ingredient in some fluoride toothpastes, and in advertising it's never called "tin(II) fluoride." Sometimes it's called "Fluoristan." Chemists are bad enough when it comes to naming, but there's no telling what name an advertiser will invent!

FIGURE 4.2
Periodic table showing charges of the ions of important elements

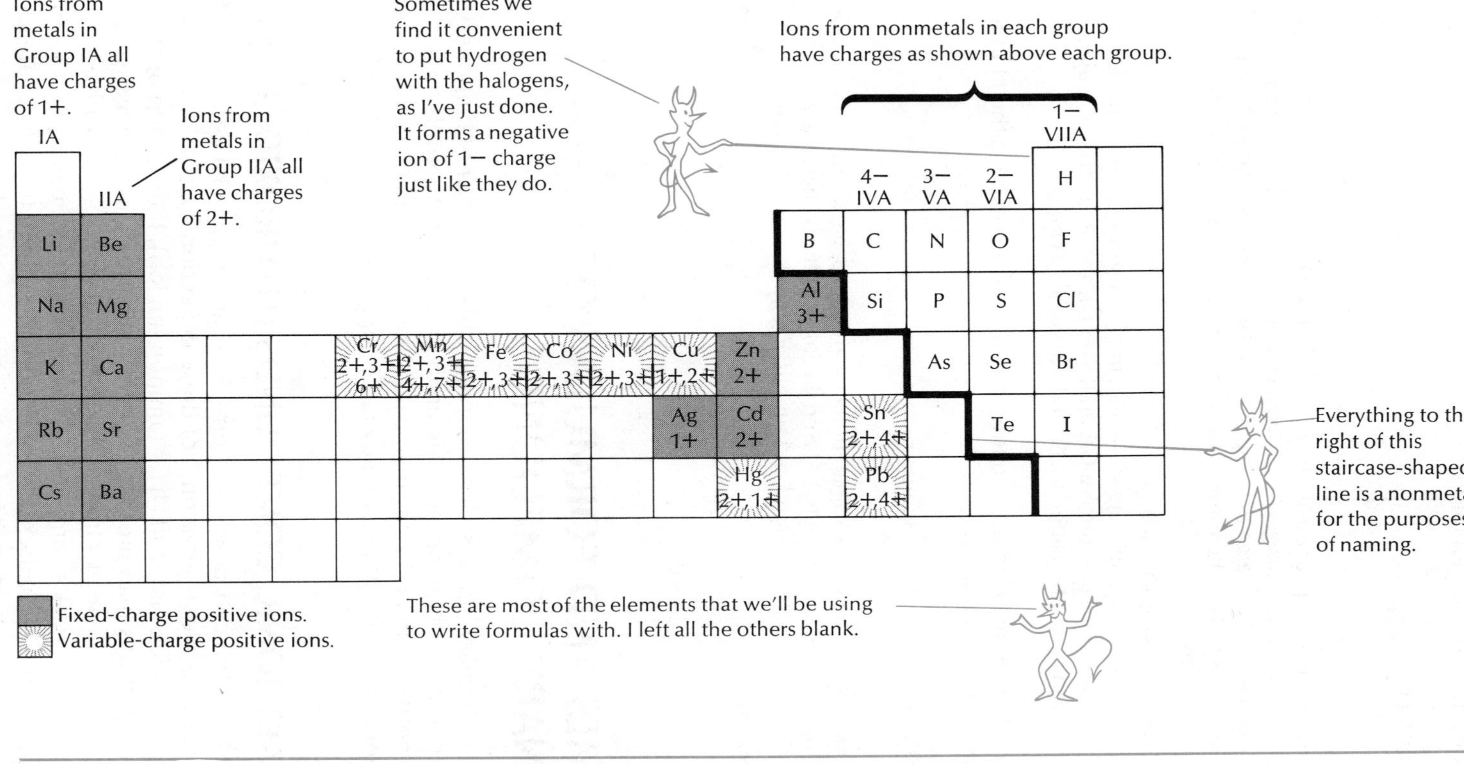

Ions from metals in Group IA all have charges of 1+.
IA
IIA
Ions from metals in Group IIA all have charges of 2+.
Sometimes we find it convenient to put hydrogen with the halogens, as I've just done. It forms a negative ion of 1− charge just like they do.
Ions from nonmetals in each group have charges as shown above each group.
4− IVA
3− VA
2− VIA
1− VIIA
H
Li
Be
B
C
N
O
F
Na
Mg
Al 3+
Si
P
S
Cl
K
Ca
Cr 2+,3+ 6+
Mn 2+,3+ 4+,7+
Fe 2+,3+
Co 2+,3+
Ni 2+,3+
Cu 1+,2+
Zn 2+
As
Se
Br
Rb
Sr
Ag 1+
Cd 2+
Sn 2+,4+
Te
I
Cs
Ba
Hg 2+,1+
Pb 2+,4+
Everything to the right of this staircase-shaped line is a nonmetal, for the purposes of naming.
Fixed-charge positive ions.
Variable-charge positive ions.
These are most of the elements that we'll be using to write formulas with. I left all the others blank.

To write the name of one of these compounds from its formula, we have to decide which of the variable-charge ions is being used.

EXAMPLE 4.8: Manganese is an ingredient in some types of steel. One source of manganese is the ore pyrolusite, which contains the compound MnO_2. Write the name for this compound.

Solution: We find manganese in the list of variable-charge ions: 2+, 3+, 4+, or 7+. We know oxygen has a fixed charge of 2−, so we use it to figure out what the charge of manganese is. Two oxide ions, at 2− each, give us a total of 4−. The formula has only one manganese ion, so its charge must be 4+.

Answer: Manganese(IV) oxide.

Note: Because their names are so similar, manganese and magnesium are easily confused. Magnesium is a Group IIA metal, and so its charge is always 2+. Manganese is a transition metal and has variable charge. ■

Writing the formula from the name is a bit easier, because the charge is given by the Roman numeral.

EXAMPLE 4.9: Galena, a lead ore, contains the compound lead(II) sulfide. Write its formula.

Solution: Because there is a Roman numeral after lead, we know without looking it up that its charge is 2+. And we know sulfur's charge is 2−. Switching the numbers around and simplifying gives us the formula.

Answer: PbS. ■

4.3 NAMES AND FORMULAS OF TERNARY COMPOUNDS

The word "ternary" means "having three parts." A *ternary compound* is one that contains three or more elements. Ternary compounds have two parts: a positive part and a negative part. When the positive part is a metal or ammonium ion, we have a *ternary ionic compound*.

ternary compound

ternary ionic compound

TERNARY IONIC COMPOUNDS (HYDROXIDES AND SALTS). A bonded group of atoms with an overall charge is called a *polyatomic ion*. We've seen some of these in Section 4.2: OH^- (hydroxide ion); CN^- (cyanide ion); and NH_4^+ (ammonium ion). If the group of atoms has more electrons than protons, then the polyatomic ion is negative. If it has more protons than electrons, then the polyatomic ion is positive.

polyatomic ion

A *metal hydroxide* is a compound of a metal with hydroxide ion. Like the oxides in Section 4.2, hydroxides are important enough to be in a category of their own, apart from other ternary ionic compounds. Hydroxides are

metal hydroxide

easy to name. We've already seen in the section on binary ionic compounds that the metal is named first, followed by "hydroxide." Thus $Ca(OH)_2$ is calcium hydroxide, and $Fe(OH)_3$ is iron(III) hydroxide.

ternary salt

A *ternary salt* is a compound of a metal (or ammonium ion) and any negative polyatomic ion except hydroxide ion. These salts are named with the metal (or ammonium ion) first and the negative polyatomic ion second. Table 4.10 shows some of the more important polyatomic ions with the names and formulas of some of their salts.

The table shows only two positive polyatomic ions: mercury(I) and ammonium. The mercury(I) ion is two Hg^+ ions joined together to form a single ion with an overall charge of 2+. The ammonium ion behaves very much like the Group IA metal ions.

The table also shows some *-ite-ate* combinations: nitrite and nitrate, sulfite and sulfate, and phosphite and phosphate. In all these, the *-ate* ion has one more oxygen than the *-ite* ion. The series of ions containing Cl has more than two possibilities, so we add the prefixes *hypo-* (one less oxygen than the *-ite* ion) and *per-* (one more oxygen than the *-ate* ion). (Though they're not as important, Br and I, but not F, form polyatomic ions similar to those of Cl.) The three ions MnO_4^-, CrO_4^{2-}, and $Cr_2O_7^{2-}$ are the only polyatomic ions on the table that contain metals.

As before, we put parentheses around a polyatomic ion when it's used with a subscript other than 1. We read the formula $Cu(NO_3)_2$ as "C-u" (pause) "N-O-three-taken-twice" and $Al_2(SO_4)_3$ as "A-l-two"(pause) "S-O-four-taken-three-times."

TABLE 4.10
Some polyatomic ions with examples of their salts

Ion	*Name*	*Salt*	*Name*
Hg_2^{2+}	Mercury(I) ion	Hg_2Cl_2	Mercury(I) chloride
NH_4^+	Ammonium ion	$(NH_4)_2S$	Ammonium sulfide
$C_2H_3O_2^-$	Acetate ion	$NaC_2H_3O_2$	Sodium acetate
CO_3^{2-}	Carbonate ion	$NiCO_3$	Nickel(II) carbonate
NO_2^-	Nitrite ion	$LiNO_2$	Lithium nitrite
NO_3^-	Nitrate ion	$Cu(NO_3)_2$	Copper(II) nitrate
SO_3^{2-}	Sulfite ion	Ag_2SO_3	Silver sulfite
SO_4^{2-}	Sulfate ion	$Cr_2(SO_4)_3$	Chromium(III) sulfate
PO_3^{3-}	Phosphite ion	K_3PO_3	Potassium phosphite
PO_4^{3-}	Phosphate ion	$FePO_4$	Iron(III) phosphate
ClO^-	Hypochlorite ion	$Ca(ClO)_2$	Calcium hypochlorite
ClO_2^-	Chlorite ion	$Zn(ClO_2)_2$	Zinc chlorite
ClO_3^-	Chlorate ion	$KClO_3$	Potassium chlorate
ClO_4^-	Perchlorate ion	NH_4ClO_4	Ammonium perchlorate
MnO_4^-	Permanganate ion	$KMnO_4$	Potassium permanganate
CrO_4^{2-}	Chromate ion	$PbCrO_4$	Lead(II) chromate
$Cr_2O_7^{2-}$	Dichromate ion	$Na_2Cr_2O_7$	Sodium dichromate
BO_3^{3-}	Borate ion	Na_3BO_3	Sodium borate
O_2^{2-}	Peroxide ion	Na_2O_2	Sodium peroxide
CN^-	Cyanide ion	$Mg(CN)_2$	Magnesium cyanide

TERNARY COVALENT COMPOUNDS (OXYACIDS).

A *ternary covalent compound* contains three or more nonmetals. When the positive part of a ternary compound is hydrogen and the negative part is an oxygen-containing polyatomic ion, we have an *oxyacid*. There are ternary covalent compounds that aren't oxyacids, but we won't discuss them here.

ternary covalent compound

oxyacid

We arrive at the formula of an oxyacid by taking a corresponding polyatomic ion, putting as many hydrogens in front of it as the ion has negative charges, and dropping the charge. For example, the acid corresponding to the sulfate ion, SO_4^{2-}, is H_2SO_4, and the acid corresponding to the phosphate ion, PO_4^{3-}, is H_3PO_4.

The name of an oxyacid is related to the name of the corresponding polyatomic ion. If the ion's name ends in *-ate,* its corresponding oxyacid is named by dropping the *-ate* and adding *-ic acid.* Thus CO_3^{2-} is carbon*ate,* so H_2CO_3 is carbon*ic acid.* If the ion's name ends in *-ite,* the oxyacid ends in *-ous acid;* ClO^- is hypochlor*ite,* HClO is hypochlor*ous acid.* (If the ion has a prefix, the acid keeps it.) Table 4.11 shows some oxyacids and their salts.

Oxyacids, like binary acids, can be named as hydrogen compounds as well as acids. We could call H_2SO_4 hydrogen sulfate, and this would mean the pure compound rather than its water solution. In practice, this system of naming isn't used much.

The list in Table 4.11 isn't as long as the list of polyatomic ions. In theory, we can make an acid out of any polyatomic ion. In practice, the ones we left out just aren't important as oxyacids. To write the name or formula of an oxyacid, just refer to the list, or memorize them.

We can see one feature of the hydroxide ion that sets it apart from the rest of the polyatomic ions. If we replace its negative charge with hydrogen, we get HOH, or a rearranged formula for water, H_2O.

TABLE 4.11
Some common oxyacids and typical salts

Oxyacid	*Name*	*Salt*	*Name*
$HC_2H_3O_2$	Acetic acid	$NaC_2H_3O_2$	Sodium acetate
H_2CO_3	Carbonic acid	$CaCO_3$	Calcium carbonate
HNO_2	Nitrous acid	$NaNO_2$	Sodium nitrite
HNO_3	Nitric acid	$Cd(NO_3)_2$	Cadmium nitrate
H_2SO_3	Sulfurous acid	$CrSO_3$	Chromium(II) sulfite
H_2SO_4	Sulfuric acid	$HgSO_4$	Mercury(II) sulfate
H_3PO_3	Phosphorous acid	$(NH_4)_3PO_3$	Ammonium phosphite
H_3PO_4	Phosphoric acid	Na_3PO_4	Sodium phosphate
HClO	Hypochlorous acid	NaClO	Sodium hypochlorite
$HClO_2$	Chlorous acid	$KClO_2$	Potassium chlorite
$HClO_3$	Chloric acid	$LiClO_3$	Lithium chlorate
$HClO_4$	Perchloric acid	$Zn(ClO_4)_2$	Zinc perchlorate
H_3BO_3	Boric acid	$AlBO_3$	Aluminum borate

4.4
NAMING COMPOUNDS WITH MORE THAN THREE ELEMENTS

We've seen that we can make the oxyacid H_2SO_4 by combining the sulfate ion (SO_4^{2-}) with two hydrogens. Also, we can make the ternary salt Na_2SO_4 by combining SO_4^{2-} with two sodium ions (Na^+). We can also combine SO_4^{2-} with one hydrogen and one sodium ion to get $NaHSO_4$. This compound is partly an acid, because it still has a hydrogen atom, and partly an ionic compound, because it contains a metal ion and a negative ion. Mixed compounds like this are easy to name. We name the metal ion, then hydrogen, then the negative ion: sodium hydrogen sulfate. We see that this compound is made of a sodium ion, Na^+, and an HSO_4^-, a negative ion derived from the sulfate ion, and called a "hydrogen sulfate ion." Any polyatomic ion (like SO_4^{2-}) with more than one negative charge can take fewer hydrogen atoms than it needs to form an oxyacid, and form instead an intermediate ion (HSO_4^-).

Another possibility is to combine both Na^+ and K^+ with SO_4^{2-}, to get $NaKSO_4$. This is called simply sodium potassium sulfate. In short, we can have any combination of metal ions or hydrogen that adds up to the charge on the negative ion. Some mixed compounds are shown in Table 4.12.

TABLE 4.12
Some mixed compounds

Acid	*Ion*	*Name*	*Mixed Compound*	*Name*
H_2CO_3	HCO_3^-	Hydrogen carbonate ion	$NaHCO_3$	Sodium hydrogen carbonate
			$NaKCO_3$	Sodium potassium carbonate
H_2SO_4	HSO_4^-	Hydrogen sulfate ion	$KHSO_4$	Potassium hydrogen sulfate
			$LiNH_4SO_4$	Lithium ammonium sulfate
H_2SO_3	HSO_3^-	Hydrogen sulfite ion	$Ca(HSO_3)_2$	Calcium hydrogen sulfite
H_3PO_4	HPO_4^{2-}	Monohydrogen phosphate ion	K_2HPO_4	Potassium monohydrogen phosphate
			$CaHPO_4$	Calcium monohydrogen phosphate
	$H_2PO_4^-$	Dihydrogen phosphate ion	LiH_2PO_4	Lithium dihydrogen phosphate
			$Ca(H_2PO_4)_2$	Calcium dihydrogen phosphate
			$MgNH_4PO_4$	Magnesium ammonium phosphate
			$LiNaKPO_4$	Lithium sodium potassium phosphate
H_2S[a]	HS^-	Hydrogen sulfide ion	NaHS	Sodium hydrogen sulfide

[a]H_2S is a binary acid, but it's included here because it forms the same kind of mixed compounds as the ternary acids.

4.5 NAMING HYDRATES

Hydrates are ionic compounds that have water molecules in their crystal structures. Since there is a fixed amount of this water, called *water of hydration,* we can write formulas for hydrates. A dot or an "x" is used to show water of hydration. Copper(II) sulfate forms a hydrate that has this formula:

hydrate

water of hydration

$$CuSO_4 \cdot 5\ H_2O \quad \text{or} \quad CuSO_4 \times 5\ H_2O$$

This is named copper(II) sulfate pentahydrate, or copper(II) sulfate 5-hydrate. First comes the name of the salt itself, and then either Greek prefixes or numbers can be used to say how many water molecules there are.

EXAMPLE 4.10: Plaster of paris hardens to form gypsum, which has the formula $CaSO_4 \cdot 2\ H_2O$. Name this substance.

Solution: First, we must name the salt. Then we add the two water molecules.

Answer: Calcium sulfate dihydrate, or calcium sulfate 2-hydrate. ■

4.6 NAMES AND FORMULAS: GENERAL EXAMPLES

We've set down a lot of rules for naming each kind of substance. When we're faced with a compound to name, or a formula to write from a name, the compound won't be categorized for us. Our first problem will be to decide which class of compounds it belongs in. Having done that, then we can follow the rules for that particular class.

EXAMPLE 4.11: KCN is an important ingredient in murder mysteries. Write its name.

Solution: First, we classify the compound. We see that it's a metal plus two nonmetals, so it's a ternary salt. We find in Table 4.10 that the name for CN^- is cyanide ion.

Answer: Potassium cyanide. ■

EXAMPLE 4.12: The compound responsible for the sour taste in vinegar is $HC_2H_3O_2$. What is its name?

Solution: Since the formula begins with hydrogen, we classify it as an acid and find its name in Table 4.11.

Answer: Acetic acid, or hydrogen acetate. ■

EXAMPLE 4.13: Glass is SiO_2. Give the systematic name of this substance.

Solution: This compound is made of two nonmetallic elements, so it's a binary covalent compound. We name it according to those rules.

Answer: Silicon dioxide. ■

EXAMPLE 4.14: Sodium hypochlorite is an ingredient in chlorine bleach. Write its formula.

Solution: This compound contains a metal ion (sodium), so it must be a salt. The prefix *hypo* and suffix *-ite* in "hypochlorite" tell us that it's a polyatomic ion. We find the hypochlorite ion in Table 4.10 with the formula ClO^-.

Answer: NaClO. ■

EXAMPLE 4.15: Baking soda is sodium hydrogen carbonate. Write its formula.

Solution: We see that this compound has more than three elements, so we follow the rules for these. Table 4.12 gives the hydrogen carbonate ion as HCO_3^-.

Answer: $NaHCO_3$. ■

EXAMPLE 4.16: Carbon disulfide is a smelly, inflammable liquid. Write its formula.

Solution: The name ends in *-ide,* and both carbon and sulfur are nonmetals, so we must be dealing with a binary covalent compound.

Answer: CS_2. ■

REVIEW QUESTIONS

1. What does the *formula* of a compound tell us?
2. Why do we need *systematic names* for compounds?
3. What is the difference between a *common name* and a systematic name?

Ionic and Covalent Compounds

4. What is an *ionic compound?*
5. What is an *ion?*
6. How do we show the charge of an ion? Give an example.
7. Do metals usually form *positive ions* or *negative ions?* What about nonmetals?
8. How can we tell whether or not a compound is ionic?
9. What do we mean by *electrical neutrality?*
10. What are *covalent compounds?*
11. How can we tell whether we have a covalent compound or not?

Names and Formulas of Binary Compounds

12. What is a *binary compound?* What ending goes on the name of all binary compounds?
13. What elements participate in a *binary covalent compound?*
14. How do we use Greek prefixes and roots to name binary covalent compounds?
15. What are *oxides?* Can an element have more than one oxide?
16. How many oxides of carbon are there? Give their names and formulas.
17. Give the formulas and the common and systematic names for the two oxides of hydrogen.
18. Give the names and formulas of the two oxides of nitrogen that are air pollutants.
19. What is the formula for ammonia?
20. What is a *binary acid?* State the rules for naming binary acids.
21. Why does HCN appear among the binary acids?
22. What is a *binary ionic compound?* What elements participate in these?
23. What is a *fixed-charge ion?* How do we name binary ionic compounds formed from fixed-charge ions?

24. How are the charges of negative ions and some positive fixed-charge ions related to their positions in the periodic table?
25. Name three ions that don't form binary compounds but whose compounds nevertheless end in *-ide*.
26. Explain how to write the formula of a binary ionic compound from its name.
27. What are *variable-charge ions?* List them and their charges.
28. Explain the Stock system of naming variable-charge ions. Why do we use it rather than the old system?
29. What is a *binary salt?* How does it differ from an oxide?
30. What is the difference between magnesium and manganese?
31. Explain how to name and write formulas of binary ionic compounds formed from variable-charge ions.

Names and Formulas of Ternary Compounds

32. What is a *ternary compound?* What is a *ternary ionic compound?*
33. What is a *polyatomic ion?* A *metal hydroxide?* A *ternary salt?*
34. How are metal hydroxides named?
35. What are the names and formulas of the positive polyatomic ions given in this chapter?
36. What are some endings for names of ternary salts?
37. Why isn't the mercury(I) ion written Hg^+?
38. What polyatomic ion is often grouped with the Group IA metal ions?
39. What do the prefixes *hypo-* and *per-* mean?
40. How do we name and write the formula of a ternary ionic compound? Explain the use of parentheses.
41. What are *oxyacids?* How are they related to polyatomic ions? How are they named? Give examples.

Naming Compounds with More than Three Elements

42. What are the rules for naming compounds with more than three elements? Give an example.

Naming Hydrates

43. What is a *hydrate?* Give an example.
44. How do we designate the number of water molecules in a hydrate?

EXERCISES

Set A (Answers at back of book.)

1. Identify the following compounds as ionic or covalent.
   a. carborundum, SiC
   b. washing soda, $Na_2CO_3 \cdot 10\ H_2O$
   c. table salt, NaCl
   d. phosphine, PH_3
   e. ammonia, NH_3
   f. potash, K_2CO_3
2. Write formulas for these binary covalent compounds.
   a. nitrogen trichloride
   b. carbon tetrafluoride
   c. sulfur dichloride
   d. diboron hexahydride
3. Give the names of these binary covalent compounds.
   a. OF_2
   b. ClI
   c. N_2H_4
   d. SF_6
4. Give names for each binary acid.
   a. H_2S
   b. HI
5. Write formulas for these binary acids.
   a. hydrogen bromide
   b. hydrogen selenide
6. Name these binary ionic compounds.
   a. $ZnCl_2$
   b. LiBr
   c. BaO
   d. Al_2S_3
7. Write these formulas.
   a. cadmium fluoride
   b. ammonium sulfide
   c. potassium oxide
   d. strontium chloride
8. Classify these binary compounds as ionic or covalent and give their names or formulas.
   a. Cl_2O
   b. ZnS
   c. magnesium nitride
   d. BBr_3
   e. tetraphosphorus hexoxide
   f. SO_2
   g. Ca_2C
   h. ammonium chloride
   i. dihydrogen telluride
   j. BaF_2
9. Name these binary ionic compounds formed from variable-charge ions.
   a. HgI_2
   b. $PbCl_2$
   c. Fe_2O_3
   d. SnO_2
10. Write these formulas.
    a. copper(I) chloride
    b. mercury(II) sulfide
    c. cobalt(III) oxide
    d. manganese(IV) fluoride
11. Name these binary ionic compounds.
    a. LiH
    b. BeO
    c. FeS
    d. $NiCl_2$
    e. K_3N
    f. $HgBr_2$
12. Write formulas for these binary ionic compounds.
    a. tin(IV) oxide
    b. barium sulfide
    c. potassium bromide
    d. chromium(II) chloride
    e. cadmium fluoride
    f. lead(II) iodide

Try to answer Exercises 13 through 24 from memory, consulting appropriate tables only when necessary.

13. Give names for these polyatomic ions.
    a. SO_4^{2-}
    b. NO_2^-
    c. ClO_4^-
    d. MnO_4^-
14. Write formulas for these polyatomic ions.
    a. nitrate ion
    b. chromate ion
    c. hypochlorite ion
    d. phosphite ion
15. Name these ternary ionic compounds.
    a. $KClO_4$
    b. $FeSO_4$
    c. $(NH_4)_2Cr_2O_7$
    d. $AgC_2H_3O_2$
    e. Na_3PO_4
    f. $CuCO_3$
16. Write formulas for these ternary ionic compounds.
    a. calcium hypochlorite
    b. ammonium nitrate
    c. barium chromate
    d. magnesium hydroxide
    e. sodium cyanide
    f. ammonium carbonate
17. Name these oxyacids.
    a. HNO_3
    b. H_3PO_4
    c. $HBrO_2$
    d. H_2CO_3
18. Write formulas for these oxyacids.
    a. sulfuric acid
    b. nitrous acid
    c. iodic acid
    d. phosphorous acid
19. Name these mixed compounds.
    a. $LiAl(SO_4)_2$
    b. $KHCO_3$
    c. $Mg(HSO_3)_2$
    d. Li_2HPO_4
20. Write formulas for these mixed compounds.
    a. sodium dihydrogen phosphate
    b. ammonium hydrogen sulfate
    c. sodium potassium sulfite
    d. aluminum hydrogen carbonate
21. Classify and write formulas for these compounds.
    a. sodium peroxide
    b. aluminum sulfide
    c. phosphoric acid
    d. nitrogen monoxide
    e. ammonia
    f. cadmium fluoride
    g. carbon dioxide
    h. copper(II) sulfate
22. Classify and name these compounds.
    a. $CrSO_4 \cdot 4\,H_2O$
    b. $HClO_4$
    c. $NaKSO_3$
    d. PF_5
    e. SiO_2
    f. NH_4CN
    g. $BaCrO_4$
    h. SO_3

23. Write formulas for compounds formed between these positive and negative ions (use the appropriate lines to write the formulas).

	OH^-	CN^-	SO_4^{2-}	PO_4^{3-}	HCO_3^-	NO_3^-	ClO_2^-	O_2^{2-}
NH_4^+	____	____	____	____	____	____	____	____
Co^{2+}	____	____	____	____	____	____	____	____
K^+	____	____	____	____	____	____	____	____
Hg_2^{2+}	____	____	____	____	____	____	____	____
Mn^{2+}	____	____	____	____	____	____	____	____
Pb^{4+}	____	____	____	____	____	____	____	____
Sn^{2+}	____	____	____	____	____	____	____	____
Ca^{2+}	____	____	____	____	____	____	____	____
Li^+	____	____	____	____	____	____	____	____

24. Write formulas for binary ionic compounds formed between these metals and nonmetals (use the appropriate lines to write the formulas).

	O	F	N	S	Cl	Br	I
Zn	____	____	____	____	____	____	____
Li	____	____	____	____	____	____	____
Na	____	____	____	____	____	____	____
Al	____	____	____	____	____	____	____

(continued)

24. *(continued)*

	O	F	N	S	Cl	Br	I
Ca	___	___	___	___	___	___	___
Fe(III)	___	___	___	___	___	___	___
Sr	___	___	___	___	___	___	___
Mg	___	___	___	___	___	___	___
K	___	___	___	___	___	___	___
Cr(II)	___	___	___	___	___	___	___
Hg(II)	___	___	___	___	___	___	___

Set B (Answers not given.)

1. Identify these compounds as ionic or covalent.
   a. acetylene, C_2H_2
   b. alum, $(Al)_2(SO_4)_3$
   c. hydrogen peroxide, H_2O_2
   d. baking soda, $NaHCO_3$
   e. fool's gold, FeS_2
   f. grain alcohol, C_2H_5OH
2. Write formulas for these binary covalent compounds.
   a. carbon disulfide
   b. dinitrogen tetroxide
   c. silicon tetrafluoride
   d. sulfur trioxide
3. Name these binary covalent compounds.
   a. CO_2
   b. SiC
   c. PH_3
   d. Cl_2O
4. Give two names for each binary acid.
   a. HCN
   b. HCl
5. Write formulas for these binary acids.
   a. hydrosulfuric acid
   b. hydrogen fluoride
6. Name these binary ionic compounds.
   a. Na_2O
   b. AlF_3
   c. MgS
   d. KCl
7. Write formulas.
   a. ammonium fluoride
   b. zinc sulfide
   c. barium hydroxide
   d. sodium hydride
8. Classify these binary compounds as ionic or covalent, and give their names or formulas.
   a. K_2O
   b. calcium chloride
   c. PCl_5
   d. NO_2
   e. H_2S
   f. carbon monoxide
   g. magnesium oxide
   h. silver bromide
9. Name these binary ionic compounds formed from variable-charge ions.
   a. CuO
   b. Cr_2S_3
   c. FeF_2
   d. $SnCl_4$
10. Write these formulas.
    a. manganese(II) oxide
    b. lead(IV) sulfide
    c. mercury(I) bromide
    d. iron(III) chloride
11. Name these binary ionic compounds.
    a. $MgCl_2$
    b. Hg_2O
    c. CaS
    d. $MnCl_2$
    e. CrO_3
    f. AlN
12. Write formulas for these binary ionic compounds.
    a. barium hydride
    b. manganese(IV) oxide
    c. potassium sulfide
    d. copper(I) chloride
    e. tin(II) fluoride
    f. zinc bromide

Try to answer Exercises 13–24 from memory, consulting appropriate tables only when necessary.

13. Give names for these polyatomic ions.
    a. NO_3^-
    b. SO_3^{2-}
    c. CrO_4^{2-}
    d. $C_2H_3O_2^-$
14. Write formulas for these polyatomic ions.
    a. carbonate ion
    b. nitrite ion
    c. phosphate ion
    d. permanganate ion
15. Name these ternary ionic compounds.
    a. $Al(OH)_3$
    b. K_2O_2
    c. NaCN
    d. $(NH_4)_3PO_4$
    e. $Fe(NO_3)_3$
    f. $KClO_3$
16. Write formulas for these ternary ionic compounds.
    a. mercury(I) nitrate
    b. sodium dichromate
    c. cobalt(II) sulfate
    d. potassium hypochlorite
    e. silver cyanide
    f. copper(II) hydroxide
17. Name these oxyacids.
    a. H_2SO_4
    b. H_3BO_3
    c. $HClO_4$
    d. HNO_2
18. Write these formulas.
    a. acetic acid
    b. phosphoric acid
    c. hypochlorous acid
    d. nitric acid
19. Name these mixed compounds.
    a. $NaHSO_3$
    b. $Ca(HCO_3)_2$
    c. $NaKCO_3$
    d. NaH_2PO_4

20. Write formulas for these mixed compounds.
    a. lithium ammonium hydrogen phosphate
    b. potassium hydrogen phosphate
    c. barium hydrogen sulfate
    d. sodium hydrogen carbonate
21. Classify and write formulas for these compounds.
    a. barium sulfate
    b. nitric acid
    c. calcium dihydrogen phosphate
    d. dinitrogen monoxide
    e. iron(III) sulfide
    f. nitrogen triiodide
    g. magnesium sulfate heptahydrate
    h. aluminum oxide
22. Classify and name these compounds.
    a. H_3PO_3
    b. NO
    c. K_2SO_4
    d. $NiCO_3$
    e. $SnCl_2 \cdot 2H_2O$
    f. LiH
    g. $Pb(NO_3)_2$
    h. HI

23. Write formulas for compounds formed between these positive and negative ions (use the appropriate lines to write the formulas).

	CO_3^{2-}	$C_2H_3O_2^-$	NO_3^-	ClO^-	MnO_4^-	SO_4^{2-}	OH^-	PO_4^{3-}
Na^+	____	____	____	____	____	____	____	____
Mg^{2+}	____	____	____	____	____	____	____	____
Cu^+	____	____	____	____	____	____	____	____
Ag^+	____	____	____	____	____	____	____	____
Zn^{2+}	____	____	____	____	____	____	____	____
Al^{3+}	____	____	____	____	____	____	____	____
Co^{2+}	____	____	____	____	____	____	____	____
Hg^{2+}	____	____	____	____	____	____	____	____
Fe^{3+}	____	____	____	____	____	____	____	____

24. Write formulas for binary ionic compounds formed between these metals and nonmetals (use the appropriate lines to write the formulas).

	S	Cl	Br	O	I	N	F
NH_4^+	____	____	____	____	____	____	____
Ba	____	____	____	____	____	____	____
Cr(III)	____	____	____	____	____	____	____
Ni(II)	____	____	____	____	____	____	____
Na	____	____	____	____	____	____	____
Sn(II)	____	____	____	____	____	____	____
Pb(IV)	____	____	____	____	____	____	____
Cd	____	____	____	____	____	____	____
Al	____	____	____	____	____	____	____
K	____	____	____	____	____	____	____
Mn(II)	____	____	____	____	____	____	____

CHAPTER 5

Chemical Reactions and Equations

LEARNING OBJECTIVES

After studying this chapter, you should be able to:

1. Supply a correct definition, explanation, or example for each of these:

chemical reaction	precipitate
chemical equation	oxidation
reactant	burning in air
product	basic oxide
yields	basic anhydride
Law of Conservation of Matter	acidic oxide
balancing	acid anhydride
aqueous solution	anhydrous salt
thermal decomposition	activity series
	neutralization
	monoprotic acid

2. Perform, in writing or orally, English-chemistry and chemistry-English translations.
3. Balance an equation comparable in complexity to those in this chapter.
4. Describe and give examples for combination, decomposition, single replacement, and double replacement reactions.
5. Given the reactants, complete and balance equations in each type of reaction in item 4.
6. Given the reactants, classify a chemical reaction as to type, and predict the products and balance the equation.

In this chapter, we're going to look at some of the ways elements and compounds react with each other to free other elements and form other compounds. Before we start, though, we need to know something about the word "react."

We had a crude definition of chemical reaction in Chapter 2. Now we'll modify it to make it a little more specific. We'll have occasion to modify it again later when we have more knowledge to build on. At this point, we'll define a *chemical reaction* as a process by which one or more chemical substances are converted into different chemical substances by breaking and/or forming chemical bonds. When chemicals react, they participate in a chemical reaction. chemical reaction

Chemical reactions are the basis of chemistry and of life as we know it. Chemical reactions are going on around us all the time. The burning of fuel, the rusting of iron, our bodily processes—all these are chemical reactions. In later chapters, we'll look at the nature of chemical bonds and see why they break and form the way they do. For now, we're interested in the *what* before the *why*.

We saw in the last chapter that formulas are a shorthand way of telling what a compound's made of. We also have a shorthand way of describing chemical reactions: with chemical equations.

For instance, in the steel industry, coke, which is largely carbon, is used as a fuel. We could describe the burning of coke in this way: "Carbon reacts with oxygen (burns) to form carbon dioxide." We've already seen, though, that chemists don't like to write things out if they don't have to. Instead of describing a reaction in a sentence, they'd rather translate it into symbols and formulas (their "words"), and equations (their "sentences").

In this chapter, we'll learn how to translate chemistry to English and vice versa. We'll learn how to write complete, correct equations and predict the products of some types of reactions.

5.1 READING AND WRITING EQUATIONS

A *chemical equation* is the chemist's shorthand for describing what happens in a chemical reaction. The reaction we talked about earlier is an easy one that we can translate into chemistry from English. chemical equation

English: "Carbon reacts with oxygen to yield carbon dioxide."
Chemistry: $C + O_2 \longrightarrow CO_2$

The "before" substances on the left, C and O_2, are the *reactants*. The "after" substance on the right, CO_2, is the *product* of the reaction. The arrow between them means *yields*. reactant product yields

$$\text{reactants} \underset{\text{yield}}{\longrightarrow} \text{products}$$

Now we'll see how to do the English-chemistry translation. It involves several steps. Here's another simple example.

English: "Iron reacts with sulfur to yield iron(II) sulfide."

We'll translate this sentence into chemistry, using the steps that follow.

Step 1. *Find out what all the formulas are.*
The formula for iron is Fe; for sulfur, S; and for iron(II) sulfide, FeS. If we're not sure of a formula, we should always go back to Chapter 4 and check. The equation won't be right if the formulas aren't right, any more than a sentence would be right if the words weren't right.

Step 2. *Decide what the reactants are and what the products are.*
The English sentence says, "Iron reacts with sulfur . . . ," so iron and sulfur must be the reactants. The English goes on: ". . . to yield iron(II) sulfide." Iron(II) sulfide must therefore be the product.

Step 3. *Write the equation.*
The reactants go on the left, and the products go on the right. The arrow goes in between them in place of the word "yields."

$$Fe + S \longrightarrow FeS$$

EXAMPLE 5.1: Translate this English into chemistry: "Sodium hydroxide reacts with hydrochloric acid to yield sodium chloride and water."

Solution:

Step 1: Sodium hydroxide is NaOH; hydrochloric acid is HCl; sodium chloride is NaCl; water is H_2O.

Step 2: Sodium hydroxide and hydrochloric acid are reactants; sodium chloride and water are products.

Step 3: NaOH and HCl go on the left; NaCl and H_2O go on the right.

Answer: $NaOH + HCl \longrightarrow NaCl + H_2O$ ■

5.2 BALANCING EQUATIONS

The equations we've looked at so far are simpler than most equations used in chemistry. Here's an English-chemistry translation that presents a new problem.

English: "Nitrogen reacts with hydrogen to yield ammonia."
Chemistry: $N_2 + H_2 \longrightarrow NH_3$

The equations we've had before were correct at this point. This one, though, isn't quite right. As it stands now, it's violating the *Law of Conservation of Matter,* which states that matter can neither be destroyed nor created in a chemical reaction. In a chemical reaction, atoms begin by being joined in a certain way. During the reaction, the atoms are taken apart, and at the end of the reaction, they end up being joined in a different way. But the Law of Conservation of Matter says that none of them can be lost in the process, nor can new ones be introduced. We have to have the same number of each kind of atom at the end as we had at the beginning of the reaction. However, at the end, the atoms will probably belong to different elements or com-

Law of Conservation of Matter

pounds than they belonged to at the beginning. Also, one kind of atom can't change to another kind.

Let's see how this applies to the equation we've written.

$$N_2 + H_2 \longrightarrow NH_3$$

This equation says that one molecule of nitrogen, containing two atoms of nitrogen, reacts with one molecule of hydrogen, containing two atoms of hydrogen, to yield one molecule of ammonia, containing one atom of nitrogen and three atoms of hydrogen. Somehow we've lost an atom of nitrogen, which we're not allowed to do, and we've gained an atom of hydrogen, which we're not allowed to do either. We can't fix things by changing a nitrogen to a hydrogen. That's against the rules too.

We can fix things by making another ammonia molecule out of the extra nitrogen atom on the left. We were already short one hydrogen atom, so we'll have to use more molecules of hydrogen to do the job. Altogether, to make two molecules of ammonia at three hydrogen atoms apiece, we need six hydrogen atoms, or three hydrogen molecules each containing two hydrogen atoms. This is illustrated in Figure 5.1.

FIGURE 5.1
A balanced equation obeys the law of conservation of matter

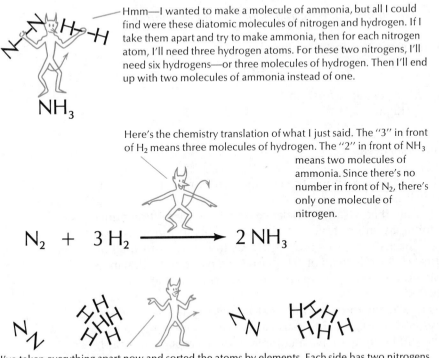

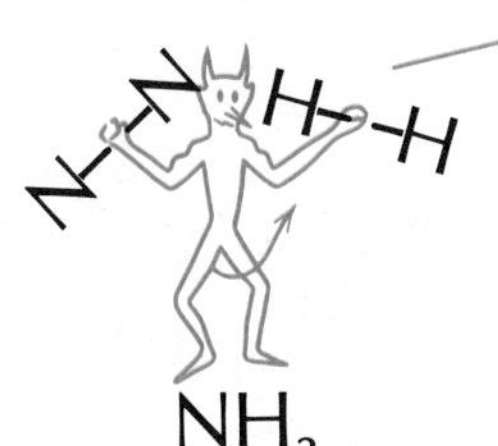

Hmm—I wanted to make a molecule of ammonia, but all I could find were these diatomic molecules of nitrogen and hydrogen. If I take them apart and try to make ammonia, then for each nitrogen atom, I'll need three hydrogen atoms. For these two nitrogens, I'll need six hydrogens—or three molecules of hydrogen. Then I'll end up with two molecules of ammonia instead of one.

NH_3

Here's the chemistry translation of what I just said. The "3" in front of H_2 means three molecules of hydrogen. The "2" in front of NH_3 means two molecules of ammonia. Since there's no number in front of N_2, there's only one molecule of nitrogen.

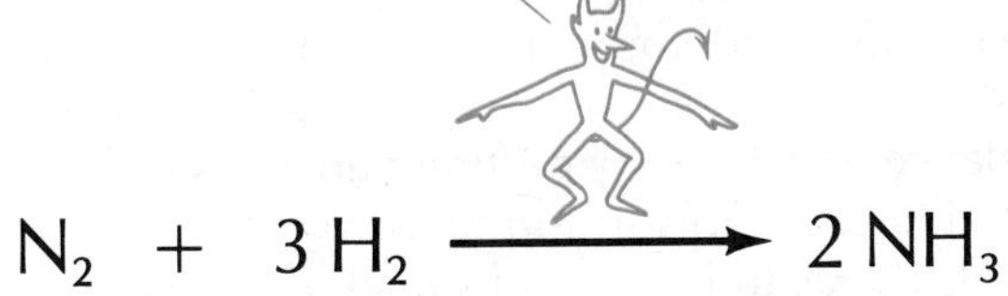

$$N_2 + 3\,H_2 \longrightarrow 2\,NH_3$$

I've taken everything apart now and sorted the atoms by elements. Each side has two nitrogens and six hydrogens. Matter has been neither created nor destroyed, only rearranged.

We can reflect all this very simply in the chemistry translation.

Chemistry: $N_2 + 3\,H_2 \longrightarrow 2\,NH_3$

English: "One molecule of nitrogen reacts with three molecules of hydrogen to yield two molecules of ammonia."

balancing

The original equation needed *balancing,* because it violated the Law of Conservation of Matter. We balanced it by putting numbers in front of the formulas so that the equation would have the same number of each kind of atom on both sides. A number in front of a formula means that everything in that formula is multiplied by the number. If there is no number, "1" is understood.

There are four main steps involved in balancing an equation. To illustrate them, we'll use this unbalanced equation as an example:

$$H_2 + Cl_2 \nrightarrow HCl$$

From now on, we'll use this kind of an arrow to show an unbalanced equation:

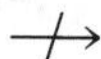

Step 1. *Count the atoms on each side of the equation.*

Left:	2 H	Right:	1 H
	2 Cl		1 Cl

Step 2. *Decide which atom or atoms are unbalanced.*

Both H and Cl are.

Step 3. *Balance the equation by putting appropriate numbers in front of the formulas.*

Put a "2" in front of HCl on the right:

$$H_2 + Cl_2 \longrightarrow 2\,HCl$$

Step 4. *Count atoms again, as a final check.*

Left:	2 H	Right:	2 H
	2 Cl		2 Cl

The equation is balanced.

Of course, if the final check shows the equation *still* to be unbalanced, then more steps will be involved.

It's important to note that we always balance equations by putting numbers in front of formulas and not by changing subscripts in formulas. We might be tempted, for instance, to write "H_2Cl_2" as the product instead of "2 HCl," or to write "H + Cl" instead of "$H_2 + Cl_2$," in an attempt to balance the equation. This would be contrary to another rule, which is that *formulas must not be violated.*

A formula tells how many of each kind of atom belong in a set, and we can't break up the set. A deck of cards is also a set—of fifty-two. If you lose the jack of hearts, you no longer have a complete set. To replace the card, you have to buy a whole new deck. You'll have more than you need of every card except the jack of hearts, but that's the way it goes. A deck of cards is a set, and that's that. In nature, hydrogen and chlorine also come in sets—

two each, or "two-packs"—and that's how they enter into chemical reactions. Since we have to use two of each, we end up with two molecules of hydrogen chloride instead of just one. We can't tamper with the formula for hydrogen chloride either. HCl is HCl: one molecule contains one atom of hydrogen and one atom of chlorine. H_2Cl_2 would mean that one molecule contained two atoms of hydrogen and two atoms of chlorine, and that's wrong.

The only way to balance an equation is to change the total quantity of one substance or another, by placing a number *before* it.

Now, let's follow through on some more examples of balancing equations.

EXAMPLE 5.2: Calcium, a rather active metal, soon tarnishes if it is left out in the air. Here's the equation for that reaction:

$$Ca + O_2 \not\rightarrow CaO$$

Balance this equation.

Solution:

Step 1: Left: 1 Ca Right: 1 Ca
2 O 1 O

Step 2: O is unbalanced.

Step 3: Put a "2" in front of CaO.

$$Ca + O_2 \not\rightarrow 2\ CaO$$

Left: 1 Ca Right: 2 Ca
2 O 2 O

Now Ca is unbalanced, so more work is needed. Put a "2" in front of Ca on the left.

$$2\ Ca + O_2 \longrightarrow 2\ CaO$$

Step 4: Final atom check.

Left: 2 Ca Right: 2 Ca
2 O 2 O

Answer: $2\ Ca + O_2 \longrightarrow 2\ CaO$ ■

EXAMPLE 5.3: Potassium chlorate gives off oxygen when it is heated. This makes it particularly useful in the fireworks industry and in the space industry where a solid source of oxygen is desired. Here's what happens when potassium chlorate is heated:

$$KClO_3 \not\rightarrow KCl + O_2$$

Balance the equation.

Solution:

Step 1: Left: 1 K Right: 1 K
1 Cl 1 Cl
3 O 2 O

Step 2: O is unbalanced.

Step 3: To balance a 3 with a 2, use the lowest common multiple, which is 6 (two of the threes and three of the twos). Put a "2" in front of $KClO_3$ and a "3" in front of O_2.

$$2\ KClO_3 \not\rightarrow KCl + 3\ O_2$$

Left: 2 K Right: 1 K
2 Cl 1 Cl
6 O 6 O

K and Cl are unbalanced now. Put a "2" in front of KCl.

$2\ KClO_3 \longrightarrow 2\ KCl + 3\ O_2$

Step 4: Final atom check.

Left:		Right:	
	2 K		2 K
	2 Cl		2 Cl
	6 O		6 O

Answer: $2\ KClO_3 \longrightarrow 2\ KCl + 3\ O_2$ ■

We should remember that a number in front of a formula means that everything in the formula is multiplied by that number to get the atom count. We multiply the number times a subscript if there is one, as in $KClO_3$ above. The number of oxygens in 2 $KClO_3$ is $2 \times 3 = 6$ oxygens.

EXAMPLE 5.4: Aluminum occurs naturally as Al_2O_3 in bauxite ore. A new process for getting the aluminum out of the ore involves treating the ore with sulfuric acid as a first step. Here's the equation for what happens:

$$Al_2O_3 + H_2SO_4 \not\longrightarrow Al_2(SO_4)_3 + H_2O$$

Balance the equation.

Solution:

Step 1:

Left:		Right:	
	2 Al		2 Al
	3 O		1 O
	2 H		2 H
	1 SO_4		3 SO_4

Notice that when a polyatomic ion like sulfate (SO_4^{2-}) remains intact from one side of the equation to the other, we count the ion as a whole instead of counting its atoms (1 S, 4 O).

Step 2: O and SO_4 are unbalanced.

Step 3: Where more than one thing is unbalanced, it's a good idea to start with the most complicated difference in subscripts, which in this case is SO_4 with a difference of 3 and 1. Start by putting a "3" in front of H_2SO_4 to balance the sulfates.

$$Al_2O_3 + 3\ H_2SO_4 \not\longrightarrow Al_2(SO_4)_3 + H_2O$$

Left:		Right:	
	2 Al		2 Al
	3 O		1 O
	6 H		2 H
	3 SO_4		3 SO_4

Now H and O are unbalanced. Again, we choose the most complicated difference, which is H with 6 and 2. Put a "3" in front of H_2O to balance H.

$$Al_2O_3 + 3\ H_2SO_4 \longrightarrow Al_2(SO_4)_3 + 3\ H_2O$$

Step 4:

Left:		Right:	
	2 Al		2 Al
	3 O		3 O
	6 H		6 H
	3 SO_4		3 SO_4

The equation is balanced. Note that when we balanced H, we also automatically balanced O. This often happens.

Answer: $Al_2O_3 + 3\ H_2SO_4 \longrightarrow Al_2(SO_4)_3 + 3\ H_2O$ ■

EXAMPLE 5.5: Balance this equation:

$$Fe_2(SO_4)_3 + Ca(OH)_2 \not\longrightarrow Fe(OH)_3 + CaSO_4$$

Solution:

Step 1: Left: 2 Fe Right: 1 Fe
3 SO_4 1 SO_4
1 Ca 1 Ca
2 OH 3 OH

Step 2: Everything is unbalanced except Ca.

Step 3: We start with the most complicated difference in subscripts, which is OH, with 2 on the left and 3 on the right. Find the lowest common multiple (6). Put a "3" in front of $Ca(OH)_2$ and a "2" in front of $Fe(OH)_3$.

$$Fe_2(SO_4)_3 + 3\ Ca(OH)_2 \not\longrightarrow 2\ Fe(OH)_3 + CaSO_4$$

Left: 2 Fe Right: 2 Fe
3 SO_4 1 SO_4
3 Ca 1 Ca
6 OH 6 OH

Ca and SO_4 are now unbalanced. Each is 3 to 1. A careful look will show us that we can fix both at once by putting a "3" in front of $CaSO_4$.

$$Fe_2(SO_4)_3 + 3\ Ca(OH)_2 \longrightarrow 2\ Fe(OH)_3 + 3\ CaSO_4$$

Step 4: Left: 2 Fe Right: 2 Fe
3 SO_4 3 SO_4
3 Ca 3 Ca
6 OH 6 OH

Answer: $Fe_2(SO_4)_3 + 3\ Ca(OH)_2 \longrightarrow 2\ Fe(OH)_3 + 3\ CaSO_4$ ■

We can see from these last examples that some unbalanced equations need several tries before they'll balance out. If we go at it systematically enough, we'll always triumph in the end.

Let's summarize the guidelines for balancing a chemical equation.

1. Adjust the numbers in front of each formula so that the same number of each kind of atom appears on each side of the equation.
2. In general, first balance the most complicated difference in subscripts, and thereafter, the most complicated difference in atoms.
3. Consider polyatomic ions as single entities as long as they stay that way on both sides of the equation.
4. Count atoms on each side of the equation as a final check.
5. Remember that matter must be conserved, formulas must not be violated, and a number in front of a formula applies to the whole formula.

5.3 OTHER SYMBOLS USED IN EQUATIONS

Often, we want to convey more in a chemical equation than just the chemical species involved. We want to specify whether each is a gas, a liquid, or a solid, or is dissolved in water. (A substance dissolved in water is in *aqueous solution* and is often called "aqueous.") We might also want

aqueous solution

TABLE 5.1
Some symbols used in writing equations

Meaning	*Symbol*[a]	*Other Symbol*
Substance is a solid	(s)	↓
Substance is a liquid	(l)	
Substance is a gas	(g)	↑
Aqueous solution (substance dissolved in water)	(aq)	
Heat is supplied	Δ	

[a]These symbols are the ones used in this book. The others may be used elsewhere and are listed for information only.

to say that heat is being supplied in the reaction. These symbols are shown in Table 5.1.

We'll now go back through all the equations we've had so far, in order, and get some practice both in understanding these new symbols and in doing chemistry-English translations.

EXAMPLE 5.6: Translate into English:

$$C(s) + O_2(g) \longrightarrow CO_2(g)$$

Solution: We say "solid carbon." We can say either "gaseous oxygen" or "oxygen gas" (same for carbon dioxide).

Answer: "Solid carbon reacts with oxygen gas to yield (or to form) carbon dioxide gas." ■

EXAMPLE 5.7: Translate into English:

$$Fe(s) + S(s) \longrightarrow FeS(s)$$

Solution: For a metal, like iron, we can also say "iron metal" instead of "solid iron." Both are correct.

Answer: "Solid iron (or iron metal) reacts with solid sulfur to yield solid iron(II) sulfide." ■

EXAMPLE 5.8: Translate into English:

$$NaOH(aq) + HCl(aq) \longrightarrow NaCl(aq) + H_2O(l)$$

Solution: Everything besides water is in aqueous solution, and liquid water itself is produced. That means that the whole thing takes place dissolved in water. We can say that just once for the whole reaction. Also, we'll call HCl "hydrochloric acid" and not "hydrogen chloride." (Remember that naming it as an acid means that it's dissolved in water, as we saw in Chapter 4, p. 66.)

Answer: "Sodium hydroxide and hydrochloric acid react in aqueous solution to yield sodium chloride and water." ■

EXAMPLE 5.9: Translate into English:

$$N_2(g) + 3\,H_2(g) \longrightarrow 2\,NH_3(g)$$

Answer: "Nitrogen gas (gaseous nitrogen) reacts with hydrogen gas (gaseous hydrogen) to yield ammonia gas (gaseous ammonia)." ■

EXAMPLE 5.10: Translate into English:

$$H_2(g) + Cl_2(g) \longrightarrow 2\,HCl(g)$$

Solution: We have HCl again, but this time we're told that it's a gas, not in aqueous solution as it was in Example 5.8. When it's a gas, we call it hydrogen chloride and not hydrochloric acid.

Answer: "Hydrogen gas reacts with chlorine gas to yield hydrogen chloride gas." ■

EXAMPLE 5.11: Translate into English:

$$2\,Ca(s) + O_2(g) \longrightarrow 2\,CaO(s)$$

Answer: "Calcium metal reacts with oxygen gas to produce solid calcium oxide." ■

EXAMPLE 5.12: Translate into English:

$$2\,KClO_3(s) \xrightarrow{\Delta} 2\,KCl(s) + 3\,O_2(g)$$

Solution: $KClO_3$ is decomposing (see Section 5.4, page 98) into simpler substances, with the aid of heat. We call this *thermal decomposition*.

thermal decomposition

Answer: "Solid potassium chlorate undergoes thermal decomposition (or, decomposes thermally) to produce solid potassium chloride and gaseous oxygen." ■

EXAMPLE 5.13: Translate into English:

$$Al_2O_3(s) + 3\,H_2SO_4(aq) \longrightarrow Al_2(SO_4)_3(aq) + 3\,H_2O(l)$$

Solution: Here we have another reaction in aqueous solution. Al_2O_3 is said to be a solid, though, so we have to say so. Some things don't dissolve in water, and Al_2O_3 is one of them. We're not supposed to know instinctively which things dissolve in water and which don't; that's what the symbol (*s*) is there to tell us.

Answer: "Solid aluminum oxide reacts with sulfuric acid (not hydrogen sulfate) to yield aluminum sulfate and water in aqueous solution." ■

EXAMPLE 5.14: Translate into English:

$$Fe_2(SO_4)_3(aq) + 3\,Ca(OH)_2(aq) \longrightarrow 2\,Fe(OH)_3(s) + 3\,CaSO_4(aq)$$

Solution: Here one of the products is a solid, so we have to say so. A solid that is formed from a reaction in solution is called a *precipitate*. We could say, "a precipitate of $Fe(OH)_3$ is formed," or, "$Fe(OH)_3$ precipitates out." This latter is the most common usage, but it wouldn't be wrong to say simply "solid $Fe(OH)_3$."

precipitate

Answer: "Iron(III) sulfate reacts with calcium hydroxide in aqueous solution to form calcium sulfate and a precipitate of iron(III) hydroxide." ■

Note from these examples that we can use "forms," "is formed," "produces," or "is produced" interchangeably with "yields" or "to yield."

5.4 TYPES OF CHEMICAL REACTIONS AND THEIR EQUATIONS

It's usually easier for us to learn things if we know something about the classes they belong to. We'll divide chemical reactions into four categories: combination, decomposition, single replacement, and double replacement.

I like to call them "dances"!

COMBINATION REACTIONS. In combination reactions, two elements, two compounds, or an element and a compound react to form a single compound. That is, two substances combine to yield a single different substance.

$$A + B \longrightarrow AB$$

Here are some reactions that fit into this category:

1. Metal + oxygen ⟶ metal oxide

$$2\,Mg(s) + O_2(g) \longrightarrow 2\,MgO(s)$$
$$2\,Fe(s) + O_2(g) \longrightarrow 2\,FeO(s)$$

Here, we figure out the formula of the product by applying the methods learned in Chapter 4. For a variable-charge ion (like Fe above), we assume the lowest charge unless we're told otherwise.

Sometimes we describe reactions like those above by saying that a metal has been oxidized. *Oxidation* means combining with oxygen to form an oxide. (In Chapter 15, we'll discover a broader meaning for the word "oxidation.") Usually the source of oxygen for such reactions is the oxygen of the air. When a substance *burns in air,* it reacts with the oxygen of the air to form an oxide. Some substances will react with oxygen even without burning, among them Group IA and IIA metals.

oxidation

burning in air

2. Nonmetal + oxygen ⟶ nonmetal oxide

$$2\,C(s) + O_2(g)\ \text{(insufficient)} \longrightarrow 2\,CO(g)$$
$$C(s) + O_2(g)\ \text{(excess)} \longrightarrow CO_2(g)$$

To know what the product is here, first we must know what oxides of the nonmetal there are (Chapter 4). If insufficient oxygen is available, we assume that the oxide having the least oxygen (a "low" oxide) is formed; with excess oxygen, a "higher" oxide is formed. It follows that treating a low oxide with more oxygen (for instance, by burning it) will get us a "higher" oxide.

Swing your partner!

CRUNCH

$$2\,CO(g) + O_2(g) \longrightarrow 2\,CO_2(g)$$

3. Metal + nonmetal ⟶ binary salt

$$2\,Na(s) + Cl_2(s) \longrightarrow 2\,NaCl(s)$$
$$Ca(s) + S(s) \longrightarrow CaS(s)$$

Again, we use the methods of Chapter 4 to figure out the formula of the product. If a metal that forms variable-charge ions is involved, then the problem has to tell us which one is formed.

4. Water + metal oxide $\longrightarrow$ metal hydroxide (base)

$$H_2O(l) + Na_2O(s) \longrightarrow 2\ NaOH(aq)$$
$$H_2O(l) + MgO(s) \longrightarrow Mg(OH)_2(ag)$$

We figure out the formulas of the hydroxides as in Chapter 4. Since they form bases in water, metal oxides are sometimes called *basic oxides* or *basic anhydrides* ("anhydride" means "without water"). basic oxide basic anhydride

5. Water + nonmetal oxide $\longrightarrow$ oxyacid

$$H_2O(l) + SO_3(g) \longrightarrow H_2SO_4(aq)$$
$$H_2O(l) + CO_2(g) \longrightarrow H_2CO_3(aq)$$

We figure out the formula of the oxyacid by adding two H's and one O to the formula of the oxide. Sometimes it's more complicated than that; then the problem has to tell us the formula of the oxyacid that's formed. Because they form acids in water, nonmetal oxides are sometimes called *acidic oxides* or *acid anhydrides.* acidic oxide acid anhydride

6. Metal oxide + nonmetal oxide $\longrightarrow$ ternary salt

$$Na_2O(s) + CO_2(g) \longrightarrow Na_2CO_3(s)$$
$$CaO(s) + SO_2(g) \longrightarrow CaSO_3(s)$$

To get the formula of the ternary salt, we add the atoms of the metal oxide and the nonmetal oxide together. Above, the O from Na_2O plus the two O's from CO_2 make 3 O's, which we combine with C and write as Na_2CO_3. Similarly, CaO plus SO_2 gives us $CaSO_3$.

7. Acid + ammonia $\longrightarrow$ ammonium salt

$$HCl(g) + NH_3(g) \longrightarrow NH_4Cl(s)$$
$$H_2SO_4(aq) + NH_3(g) \longrightarrow (NH_4)_2SO_4(aq)$$

This reaction can happen with a gas (such as HCl) or in solution. We figure out the formula for the ammonium salt in the usual way.

8. Hygroscopic substance + water $\longrightarrow$ hydrate

$$CuSO_4(s) + 5\ H_2O(g) \longrightarrow CuSO_4 \cdot 5\ H_2O$$

A *hygroscopic* substance is one that absorbs water vapor from the air to form a hydrate. The dry substance, without any water of hydration, is called *anhydrous* (without water). Anhydrous copper(II) sulfate is white. If we put some on a dish and leave it out in the air, after a time it will turn to blue $CuSO_4 \cdot 5\ H_2O$. Not all substances are hygroscopic, but those that are can be useful as drying agents; anhydrous calcium chloride is frequently used as a drying agent in chemistry laboratories to remove water from other substances. There is no way to tell beforehand how many waters a hydrate has. The problem has to tell us that. hygroscopic anhydrous

DECOMPOSITION REACTIONS. In a decomposition reaction, a compound decomposes either into the elements that make it up, or into simpler compounds, or into some of each.

$$AB(C \cdots) \longrightarrow A + B\,(+\,C + \cdots)$$

Decomposition usually (though not always) occurs with the help of heat. We've seen in Section 5.3 that the Greek letter delta (Δ) is used to show that heat is supplied.

Here are some examples of decomposition reactions:

That number's over. Time to sit down again.

$$\text{1. Metal carbonate} \xrightarrow{\Delta} \text{metal oxide} + CO_2$$
$$CaCO_3(s) \xrightarrow{\Delta} CaO(s) + CO_2(g)$$
$$NiCO_3(s) \xrightarrow{\Delta} NiO(s) + CO_2(g)$$

The first reaction is that used to make lime (CaO) by roasting limestone ($CaCO_3$) in a lime kiln. In the second reaction, we assume that the variable-charge metal Ni has the same charge (2+) after the reaction as it did before.

$$\text{2. Metal hydrogen carbonate} \xrightarrow{\Delta} \text{metal carbonate} + CO_2 + H_2O$$
$$2\,NaHCO_3(s) \xrightarrow{\Delta} Na_2CO_3(s) + CO_2(g) + H_2O(g)$$

This reaction happens when we use baking soda to put out a kitchen fire. The CO_2 and water vapor that are formed help to smother the fire.

$$\text{3. Hydrate} \xrightarrow{\Delta} \text{anhydrous salt} + H_2O \text{ vapor}$$
$$CuSO_4 \cdot 5\,H_2O(s) \xrightarrow{\Delta} CuSO_4(s) + 5\,H_2O(g)$$

We see a change taking place as we heat blue $CuSO_4 \cdot 5\,H_2O$ and watch it change to white anhydrous $CuSO_4$. Some hydrates lose their water even without heating. Those that do are said to be *efflorescent*.

efflorescence

4. Miscellaneous.

$$2\,HgO(s) \xrightarrow{\Delta} 2\,Hg(l) + O_2(g)$$
$$(NH_4)_2Cr_2O_7(s) \xrightarrow{\Delta} N_2(g) + Cr_2O_3(s) + 4\,H_2O(g)$$

There aren't any rules here. The problem has to tell us what's formed in the reaction.

SINGLE REPLACEMENT REACTIONS. In a single replacement reaction, a compound reacts with an element to yield another compound and another element. Either the positive part or the negative part of the compound can be replaced.

$$AB + C \longrightarrow AC + B$$

Here are some examples of reactions where the positive part of the compound is replaced. To show what replaces what, we'll use the *activity series* on the right. In Chapter 2 we learned that the alkali and alkaline earth metals are the most reactive. The activity series is a list of some metals, with hydrogen, in order of decreasing reactivity.

activity series

Li
K
Ba
Ca
Na

Mg
Al
Zn
Fe
Cd
Ni
Sn
Pb

H

Cu
Hg
Ag
Au

$$\text{1. Metal + water} \longrightarrow \text{metal hydroxide + hydrogen gas}$$
$$2Na(s) + 2\,H_2O(l) \longrightarrow 2\,NaOH(aq) + H_2(g)$$
$$Ca(s) + 2\,H_2O(l) \longrightarrow Ca(OH)_2(aq) + H_2(g)$$

Only the most active metals—those above the first dashed line—will react with cold water. Some others will react with steam, but the problem will have to tell us that.

$$\text{2. Metal + acid} \longrightarrow \text{salt + hydrogen gas}$$
$$Zn(s) + 2\,HCl(aq) \longrightarrow ZnCl_2(aq) + H_2(g)$$
$$Ni(s) + H_2SO_4(aq) \longrightarrow NiSO_4(aq) + H_2(g)$$

Here, only metals above H in the activity series will react with acids. Nothing would happen (no reaction) if we added HCl to Cu:

$$Cu(s) + 2HCl \longrightarrow \text{N.R. (no reaction)}$$

$$\text{3. Metal A + salt B} \longrightarrow \text{metal B + salt A}$$
$$Zn(s) + Cu(NO_3)_2(aq) \longrightarrow Cu(s) + Zn(NO_3)_2(aq)$$
$$Fe(s) + CdSO_4(aq) \longrightarrow Cd(s) + FeSO_4(aq)$$

Here, a metal will replace any metal that's below it in the activity series: zinc will replace copper, but copper won't replace zinc and we would write $Cu(s) + ZnSO_4(aq) \longrightarrow$ N.R. Copper will, however, replace silver. For metals with variable charge, we assume that the lower-charged ion is formed, as with $FeSO_4$ above.

The last type of reaction is one in which the negative part of the compound is replaced.

$$\text{4. Nonmetal + binary salt} \longrightarrow \text{nonmetal + binary salt}$$
$$Cl_2(g) + 2\,NaI(aq) \longrightarrow I_2(aq) + 2\,NaCl(aq)$$
$$F_2(g) + CaCl_2(aq) \longrightarrow Cl_2(aq) + CaF_2(aq)$$

Here, a halogen will replace any halogen that's below it in the periodic table but not one that's above it. So we would write $I_2(s) + 2\,NaF(aq) \longrightarrow$ N.R.

DOUBLE REPLACEMENT REACTIONS. In a double replacement reaction, two compounds exchange their positive or negative parts.

Everybody change partners!

$$AB + CD \longrightarrow AD + BC$$

Here are some examples:

1. Neutralization reactions.

$$\text{Acid} \quad + \quad \text{base} \quad \longrightarrow \quad \text{a salt} \quad + \quad \text{water}$$
$$2\,HCl(aq) + Mg(OH)_2 \longrightarrow MgCl_2(aq) + 2\,H_2O(l)$$

This reaction takes place when we take milk of magnesia [$Mg(OH)_2$] to neutralize excess stomach acid (HCl). The reaction of an acid with a base is called *neutralization*.

neutralization

Acids having only one hydrogen, such as HCl, are said to be *monoprotic* or *monobasic*. Some acids have more than one hydrogen. An acid such as H_2SO_4 that has two hydrogens is called a *diprotic*, or *dibasic*, acid. And *triprotic*, or *tribasic*, acids are those, like H_3PO_4, with three hydrogens.

monoprotic (monobasic) acid

diprotic (dibasic) acid

triprotic (tribasic) acid

The hydrogens of di- or tri-protic acids can react one at a time, or all at once. For instance, sulfuric acid can react with one mole of NaOH:

$$H_2SO_4(aq) + NaOH(aq) \longrightarrow NaHSO_4(aq) + H_2O(l)$$

Or, it can react with two moles of NaOH:

$$H_2SO_4(aq) + 2NaOH(aq) \longrightarrow Na_2SO_4(aq) + 2H_2O(l)$$

If we have no information, we assume that di- and tri-protic acids react using all of their hydrogens.

When we say "base," we include the basic (nonmetal) oxides, or basic anhydrides:

$$FeO(s) + H_2SO_4(aq) \longrightarrow FeSO_4(aq) + H_2O(l)$$

Here, the Fe and the H_2 have just changed places. This reaction is used in the steel industry to remove FeO from newly forged steel.

Similarly, when we say "acid," we include the acidic (nonmetal) oxides, or acid anhydrides:

$$SO_3(g) + Ca(OH)_2(aq) \longrightarrow CaSO_4(aq) + H_2O(l)$$

Sulfur trioxide is the anhydride of H_2SO_4, so the salt that's formed is a sulfate. Water is left. The above reaction, used to remove harmful SO_3 from the smokestacks of smelters and refineries, is not strictly a double replacement reaction. It is shown here for comparison with the basic oxide reaction.

2. Precipitate formation.

$$\begin{array}{ccccccc} \text{solution} & + & \text{solution} & \longrightarrow & \text{precipitate} & + & \text{solution} \\ 2\,Na_3PO_4(aq) & + & 3\,Ca(OH)_2(aq) & \longrightarrow & Ca_3(PO_4)_2(s) & + & 6\,NaOH(aq) \end{array}$$

This reaction was used at Lake Tahoe, California, to remove phosphate pollution from sewage; the precipitate of calcium phosphate can be removed once it's formed. To write this kind of equation, we have to know whether the products will be insoluble (form precipitates). We switch the positive and negative ions around, then look at the solubility chart in Appendix D (p. 547) to see if either new substance is insoluble. A look at the chart for the above equation reveals that $Ca_3(PO_4)_2$ is insoluble, so the precipitation reaction will take place. The other product, NaOH, is soluble and remains in solution (*aq*). If no insoluble substance is formed, there will be no precipitation reaction. Here are a few more examples:

$$HCl(aq) + AgNO_3(aq) \longrightarrow AgCl(s) + HNO_3(aq)$$

$$BaCl_2(aq) + Na_2SO_4(aq) \longrightarrow BaSO_4(s) + 2\ NaCl(aq)$$
$$ZnSO_4(aq) + KCl(aq) \longrightarrow \text{N.R. (No Reaction)}$$

You should consult the solubility chart and verify these equations.

A precipitation reaction can take place between an acid and a salt, a base and a salt, or two salts.

3. Reactions forming a gas.

$$\begin{matrix} \text{carbonate or bicarbonate} & + & \text{acid} & \longrightarrow & \text{a salt} & + & \text{water} & + & \text{carbon dioxide} \\ NaHCO_3(s) & + & HC_2H_3O_2(aq) & \longrightarrow & NaC_2H_3O_2 & + & H_2O(l) & + & CO_2(g) \end{matrix}$$

Here, the salt is formed with the negative ion from the acid. This reaction happens when vinegar (acetic acid) is added to sodium bicarbonate, which causes fizzing.

This is the most important type of the reactions that form a gas in a double replacement reaction. There are many others, but we'll deal with them as they come up.

5.5 PREDICTING PRODUCTS AND WRITING EQUATIONS

Sometimes we have to write a complete equation when we know only the reactants. This means that we'll have to figure out what the products are. If we have a set of rules to follow, we can do this for the types of simple reactions we've listed above. Chemical reactions have many more types than these, however, and the rules aren't meant to cover everything. Nor are they meant to be memorized. But by referring to them when we need to, we'll be able to write a lot of equations.

When writing chemical equations, always put in the physical state (*g*, *l*, *s*, *aq*) if you know it. Knowing this for the elements is no problem; Chapter 2 outlined which elements were gases, liquids, and solids at room temperature. Assume that all ionic compounds are solids; if they're aqueous, the problem will say so. Covalent compounds can be solids, liquids, or gases, and there's no good rule right now for telling what a given covalent compound will be. We know a few of them, though, to be gases: hydrogen chloride, ammonia, and the important oxides of carbon, nitrogen, and sulfur. We know that water and hydrogen peroxide are liquids. Compounds named as acids are aqueous. If a compound is encountered that isn't easily classified by any of the above comments, then leave its physical state out.

In the examples that follow, we'll always put all the physical states in.

To complete an equation when you have only the reactants, first try to classify it into one of the reaction types listed in Section 5.4. Then look at the examples in that section to complete the equation.

EXAMPLE 5.15: Write an equation that shows sulfur burning in insufficient oxygen.

Solution: The two reactants are elements: S and O_2. We find this type of reaction under "combination." We choose SO_2 as the product, and not SO_3, because we're told that there is insufficient oxygen.

Answer: $S(s) + O_2(g) \longrightarrow SO_2(g)$ ■

EXAMPLE 5.16: Will sulfuric acid dissolve an aluminum pan? Write the balanced equation for the reaction, if any.

Solution: The reactants are H_2SO_4 and Al: a compound reacting with an element, found under "single replacement." Furthermore, it's a reaction between a metal and an acid, and we follow that format. In the activity series, Al is above H, so the reaction will take place. (If not, we write "N.R.")

Answer: Yes: $2\ Al(s) + 3\ H_2SO_4(aq) \longrightarrow Al_2(SO_4)_3(aq) + 3\ H_2(g)$ ■

EXAMPLE 5.17: Write a balanced equation for the reaction between aqueous solutions of silver nitrate and sodium sulfide.

Solution: The reactants are $AgNO_3$ and Na_2S: two compounds, double replacement. And they're both salts, so it must be a precipitation reaction. First, we trade the positive and negative parts and balance the equation:

$$2\ AgNO_3(aq) + Na_2S(aq) \longrightarrow Ag_2S(?) + 2\ NaNO_3(?)$$

Next, we look at the solubility chart in Appendix D to see if either, neither, or both of these products are insoluble. We find that $NaNO_3$ is soluble, but Ag_2S is insoluble, so the reaction will take place. If both were soluble, no reaction would take place and we'd write "N.R." We add (*s*) and (*aq*) to the above, and balance the equation.

Answer: $2\ AgNO_3(aq) + Na_2S(aq) \longrightarrow Ag_2S(s) + 2\ NaNO_3(aq)$ ■

EXAMPLE 5.18: Write a balanced equation for the reaction that occurs when solutions of KOH and H_3PO_4 are mixed.

Solution: There are two compounds, so it must be double replacement. And one is a base (KOH) and one an acid (H_3PO_4): neutralization.

Answer: $3\ KOH(aq) + H_3PO_4(aq) \longrightarrow K_3PO_4(aq) + 3\ H_2O(l)$ ■

EXAMPLE 5.19: Sodium metal burns in air to form sodium peroxide. Write the equation.

Solution: Here we'd expect sodium oxide to be the product, unless we had information to the contrary, which we do. The fact is that when sodium is burned, sodium peroxide and not sodium oxide is the product.

Answer: $2\ Na(s) + O_2(g) \longrightarrow Na_2O_2(s)$ ■

REVIEW QUESTIONS

1. What is a *chemical reaction?*
2. What is the chemist's shorthand way of describing chemical reactions?

Reading and Writing Equations

3. What are *reactants* and *products?* On which side of a chemical equation is each found?

4. What is the symbol for *yields* in a chemical equation?
5. What are the steps involved in translating English into chemistry?

Balancing Equations

6. What is the *Law of Conservation of Matter?*
7. What do we mean by a *balanced* equation?
8. Why must we balance equations?
9. What does a number in front of a formula mean in an equation?
10. What are the steps involved in balancing equations?
11. Why may we not balance equations by putting numbers in as subscripts?
12. What should always be the final step in balancing an equation?
13. How do we deal with polyatomic ions when balancing equations?
14. When several atoms are unbalanced, how do we choose which to balance first?
15. What are the guidelines for balancing a chemical equation?

Other Symbols Used in Equations

16. What are the symbols for gas, solid, liquid, aqueous solution, and heat?
17. What does "gaseous oxygen" mean?
18. How do we name HCl to specify that it's in aqueous solution?
19. What do we mean by *thermal decomposition?*
20. What is a *precipitate?*
21. What are some synonyms for "yields"?

Types of Chemical Reactions and Their Equations

22. What are the four basic types of chemical reactions?
23. Give an example for each of the four types of reactions.
24. How do we show that heat is required in a reaction?
25. In your own words, tell what the difference is between the following: (a) a combination reaction and a decomposition reaction; (b) a single replacement reaction and a double replacement reaction.
26. What products are usually formed by these combination reactions?
    a. a metal reacting with oxygen
    b. a nonmetal reacting with oxygen
    c. the lower oxide of a nonmetal reacting with oxygen
    d. a metal reacting with a nonmetal
    e. water reacting with a metal oxide
    f. water reacting with a nonmetal oxide
    g. a metal oxide reacting with a nonmetal oxide
    h. acid reacting with ammonia
27. In the reactions above, how do we figure out the formula of the product?
28. What products are usually formed by these decomposition reactions?
    a. a metal carbonate decomposing
    b. a metal hydrogen carbonate decomposing
    c. a hydrate decomposing
29. What products are usually formed by these single replacement reactions?
    a. a metal reacting with water
    b. a metal reacting with an acid
    c. a metal reacting with a salt
    d. a nonmetal reacting with a binary salt
30. How do we use the activity series to decide whether a reaction will take place in question 29?
31. What products are usually formed by these double replacement reactions?
    a. an acid reacting with a metal hydroxide
    b. an acidic oxide reacting with a metal hydroxide
    c. an acid reacting with a metal oxide
32. In a reaction involving precipitate formation, how do we decide which, if any, product is a precipitate?
33. What happens when an acid is added to a carbonate or bicarbonate?

Predicting Products and Writing Equations

34. How do we know the physical state of a substance?
35. What class of compounds are mostly solids?
36. How do we know whether a substance is in aqueous solution?
37. What are the steps in completing an equation given only the reactants?

EXERCISES

Set A (Answers at back of book. Asterisks mark the more difficult exercises.)

1. State which of the following English-chemistry translations are incorrect and correct them:
    a. *English:* "Calcium reacts with chlorine to form calcium chloride."
    Chemistry: $Ca + Cl_2 \longrightarrow CaCl_2$.
    b. *English:* "Sodium reacts with bromine to yield sodium bromide."
    Chemistry: $Na + Br_2 \longrightarrow NaBr_2$.

c. *English:* "Silicon reacts with oxygen to yield silicon dioxide."
Chemistry: $S + O_2 \longrightarrow SO_2$.
d. *English:* "Mercury(II) oxide reacts to form mercury and oxygen."
Chemistry: $HgO \longrightarrow Hg + O$.
e. *English:* "Sodium chloride reacts with silver nitrate to yield silver chloride and sodium nitrate."
Chemistry: $NaCl + AgNO_3 \longrightarrow AgCl + NaNO_3$.

2. Translate the following English sentences into chemistry.
   a. "Lithium reacts with nitrogen to yield lithium nitride."
   b. "Ammonia reacts with oxygen to yield nitrogen monoxide and water."
   c. "Zinc reacts with copper(II) nitrate to yield copper and zinc nitrate."
   d. "Phosphorus reacts with oxygen to yield tetraphosphorus hexoxide."
   e. "Ammonium sulfide reacts with mercury(II) bromide to yield ammonium bromide and mercury(II) sulfide."
3. Which of the following equations are balanced?
   a. $2\,K + Br_2 \longrightarrow 2\,KBr$
   b. $N_2 + O_2 \longrightarrow NO$
   c. $CaCO_3 + 2\,HCl \longrightarrow CaCl_2 + CO_2 + H_2O$
   d. $3\,Fe + 2\,O_2 \longrightarrow Fe_2O_3$
   e. $H_2O_2 \longrightarrow H_2O + O_2$
4. Balance each of the unbalanced equations in Exercise 3.
5. What is wrong with each of these "balanced" equations?
   a. $2\,H + O \longrightarrow H_2O$
   b. $FeS + HBr \longrightarrow FeBr + HS$
   c. $H_2 + F_2 \longrightarrow H_2F_2$
   d. $Na_2CO_3 + CaCl_2 \longrightarrow CaCO_3 + Na_2Cl_2$
   e. $BCl_3 \longrightarrow B + Cl_3$
   f. $Li_2O + H_2O \longrightarrow Li_2(OH)_2$
   g. $Al + 2\,HCl \longrightarrow AlCl_2 + H_2$
   h. $PbNO_3 + KCl \longrightarrow PbCl + KNO_3$
   i. $KClO_4 \longrightarrow KCl + O_4$
6. Write correct, balanced equations for each of the incorrect equations in Exercise 5.
7. Balance these equations.
   a. $P + Cl_2 \not\rightarrow PCl_3$
   b. $Mg_3N_2 + H_2O \not\rightarrow Mg(OH)_2 + NH_3$
   c. $Fe + O_2 \not\rightarrow Fe_2O_3$
   d. $Ni(OH)_2 + H_2SO_4 \not\rightarrow NiSO_4 + H_2O$
   e. $Fe + AgNO_3 \not\rightarrow Fe(NO_3)_2 + Ag$
   f. $BaCl_2 + (NH_4)_2CO_3 \not\rightarrow BaCO_3 + NH_4Cl$
   g. $NaNO_2 + H_2SO_4 \not\rightarrow HNO_2 + Na_2SO_4$
   h. $NaOH + CO_2 \not\rightarrow Na_2CO_3 + H_2O$
   i. $B + F_2 \not\rightarrow BF_3$
   j. $HNO_2 \not\rightarrow N_2O_3 + H_2O$
8. Translate the following from English to chemistry, and balance the equations.
   a. "Ammonia reacts with sulfuric acid to yield ammonium sulfate."
   b. "Methane, CH_4, burns in air to yield carbon dioxide and water."
   c. "Dinitrogen trioxide reacts with water to yield nitrous acid."
   d. "Barium carbonate decomposes with heat to yield barium oxide and carbon dioxide."
   e. "Silicon reacts with chlorine to form silicon tetrachloride."
9. Translate from chemistry into English.
   a. $Ca(s) + 2\,HBr(aq) \longrightarrow CaBr_2(aq) + H_2(g)$
   b. $2\,HI(g) \longrightarrow H_2(g) + I_2(s)$
   c. $2\,Be(s) + O_2(g) \longrightarrow 2\,BeO(s)$
   d. $Mn(NO_3)_2(aq) + Na_2S(aq) \longrightarrow MnS(s) + 2\,NaNO_3(aq)$
   e. $Cl_2O_7(l) + H_2O(l) \longrightarrow 2\,HClO_4(aq)$
10. Translate from English to chemistry (put in *s, l, g, aq,* Δ), and balance the equations.
   a. "Solid boron reacts with oxygen gas to yield solid diboron trioxide."
   b. "Lead metal reacts with solid sulfur to yield solid lead(II) sulfide."
   c. "Nitrogen gas reacts with chlorine gas to yield nitrogen trichloride gas."
   d. "Hydrogen gas reacts with liquid bromine to yield gaseous hydrogen bromide."
   e. "Aqueous potassium hydroxide reacts with nitric acid to yield water and aqueous potassium nitrate."
   f. "Solid zinc sulfide reacts with sulfuric acid to yield dihydrogen sulfide gas and aqueous zinc sulfate."
   g. "Solid ammonium nitrite decomposes thermally to yield nitrogen gas and steam."
   h. "Aqueous chlorine reacts with aqueous potassium bromide to yield aqueous bromine and potassium chloride."
   i. "Solid sodium hydrogen carbonate reacts with hydrochloric acid to yield aqueous sodium chloride, water, and carbon dioxide gas."
   j. "Aqueous barium chloride reacts with aqueous ammonium sulfate to yield aqueous ammonium chloride and a precipitate of barium sulfate."
11. Classify the following reactions as combination, decomposition, single replacement, or double replacement.
   a. $4\,Al(s) + 3\,O_2(g) \longrightarrow 2\,Al_2O_3(s)$
   b. $CaI_2(aq) + Hg(NO_3)_2(aq) \longrightarrow Ca(NO_3)_2(aq) + HgI_2(s)$
   c. $Mg(s) + 2\,HC_2H_3O_2(aq) \longrightarrow Mg(C_2H_3O_2)_2(aq) + H_2(g)$
   d. $CoSO_3(s) \xrightarrow{\Delta} CoO(s) + SO_2(g)$
   e. $FeO(s) + C(s) \longrightarrow Fe(s) + CO(g)$
   f. $2\,Cu(s) + S(s) \longrightarrow Cu_2S(s)$
   g. $2\,AgNO_3(aq) + K_2CrO_4(aq) \longrightarrow Ag_2CrO_4(s) + 2\,KNO_3(aq)$

h. $Sr(OH)_2(aq) + H_2SO_4(aq) \longrightarrow SrSO_4(s) + 2\,H_2O(l)$

i. $Pb(NO_3)_2(s) \xrightarrow{\Delta} PbO(s) + 2\,NO_2(g)$

j. $2\,NH_3(g) + H_2SO_4(aq) \longrightarrow (NH_4)_2SO_4(aq)$

12. Translate each of the equations in Exercise 11 into English.

13. Complete and balance each of these equations for combination reactions.
    a. $Li(s) + O_2(g) \longrightarrow$
    b. $Be(s) + O_2(g) \longrightarrow$
    c. $N_2(g) + O_2(g, \text{excess}) \longrightarrow$
    d. $H_2O(l) + CaO(s) \longrightarrow$
    e. $H_2O(l) + SO_2(g) \longrightarrow$
    f. $BaO(s) + SO_3(g) \longrightarrow$
    g. $NH_3(g) + HC_2H_3O_2(aq) \longrightarrow$
    h. $Mg(s) + N_2(g) \longrightarrow$

14. Complete and balance each of these equations for decomposition reactions.
    a. $FeCO_3 \xrightarrow{\Delta}$
    b. $KHCO_3 \xrightarrow{\Delta}$
    c. $MgSO_4 \cdot 7\,H_2O \xrightarrow{\Delta}$

15. Complete and balance these equations for single replacement reactions.
    a. $Li(s) + H_2O(l) \longrightarrow$
    b. $Fe(s) + H_2SO_4(aq) \longrightarrow$
    c. $SnCl_2(aq) + Fe(s) \longrightarrow$
    d. $Br_2(l) + KI(aq) \longrightarrow$
    e. $Ba(s) + H_2O(l) \longrightarrow$
    f. $Cl_2(g) + CaBr_2(aq) \longrightarrow$
    g. $Mg(s) + HCl(aq) \longrightarrow$
    h. $AgNO_3(aq) + Cu(s) \longrightarrow$

16. Complete and balance these equations for double replacement reactions.
    a. $H_2SO_4(aq) + Al(OH)_3(s) \longrightarrow$
    b. $H_2CO_3(aq) + NaOH(aq) \longrightarrow$
    c. $CaO(s) + HCl(aq) \longrightarrow$
    d. $BaCl_2(aq) + H_2SO_4(aq) \longrightarrow$
    e. $Fe(NO_3)_3(aq) + KOH(aq) \longrightarrow$
    f. $Pb(NO_3)_2(aq) + Na_2CrO_4(aq) \longrightarrow$
    g. $CaCO_3(s) + HCl(aq) \longrightarrow$

17. Classify by reaction type, then complete and balance. (If no reaction, write "N.R.")
    a. $BaCO_3(s) + HNO_3(aq) \longrightarrow$
    b. $Fe_2O_3(s) + H_3PO_4(aq) \longrightarrow$
    c. $Cu(s) + MgCl_2(aq) \longrightarrow$
    d. $PbCl_2(aq) + H_2S(aq) \longrightarrow$
    e. $Ni(s) + O_2(g) \longrightarrow$
    f. $Li_2CO_3(s) \xrightarrow{\Delta}$
    g. $Ag(s) + H_2SO_4(aq) \longrightarrow$
    h. $Al(s) + Cl_2(g) \longrightarrow$
    i. $Mg(s) + HCl(aq) \longrightarrow$

*18. Complete and balance. (Assume that a reaction does take place.)
    a. $Cu(OH)_2 \xrightarrow{\Delta} CuO(s) + \_\_\_\_$
    b. $Ag_2SO_3(s) \xrightarrow{\Delta} Ag_2O(s) + \_\_\_\_$
    c. $K_2S(aq) + I_2(s) \longrightarrow$
    d. $(NH_4)_2S(aq) + HCl(aq) \longrightarrow$
    e. $PCl_5(l) \longrightarrow PCl_3(l) + \_\_\_\_$
    f. $N_2O_3(l) + H_2O(l) \longrightarrow$
    g. $C_2H_2(g) + O_2(g, \text{excess}) \xrightarrow{\Delta}$

*19. Many metals occur in ores as sulfides and are recovered by roasting the ore in air (burning). When zinc sulfide is roasted, zinc oxide and sulfur dioxide are obtained. When mercury(II) sulfide is roasted, free mercury and sulfur dioxide are the products. Write balanced equations for these two reactions and point out the difference in reaction type.

*20. The Solvay process is an important commercial process of manufacturing sodium bicarbonate (sodium hydrogen carbonate). (a) Limestone ($CaCO_3$) is heated to give lime (CaO) and carbon dioxide. (b) The lime is added to water to form slaked lime [$Ca(OH)_2$]. (c) Ammonia is brought in and allowed to react with water and carbon dioxide to form aqueous ammonium hydrogen carbonate. (d) Aqueous sodium chloride is brought in and allowed to react with the ammonium hydrogen carbonate solution to give aqueous sodium hydrogen carbonate (the product) and aqueous ammonium chloride. (e) The ammonium chloride is changed back into ammonia gas, which can be reused, by this reaction: Ammonium chloride reacts with the slaked lime solution to produce ammonia, calcium chloride, and water. Write balanced equations for all reactions and state the reaction type for each.

*21. In Exercise 19, we saw that zinc sulfide was converted to zinc oxide by roasting. The free zinc metal is obtained from zinc oxide by treating it with carbon monoxide, which is converted to carbon dioxide. Write a complete, balanced equation.

*22. A good way to clean silver is to put the silver into an aluminum pan full of salt water. The aluminum reacts with the silver tarnish, Ag_2S, and replaces the silver. Write a complete, balanced equation for this reaction. (The salt only facilitates the reaction; it doesn't enter into the equation.)

Set B (Answers not given. Asterisks mark the more difficult exercises.)

1. State which of these English-chemistry translations are incorrect and correct them.
    a. *English:* "Barium reacts with oxygen to form barium oxide."
       Chemistry: $Ba + O_2 \longrightarrow BaO_2$
    b. *English:* "Lithium reacts with sulfur to yield lithium sulfide."
       Chemistry: $Li + S \longrightarrow LiS$

c. *English:* "Nitrogen reacts with oxygen to form nitrogen dioxide."
Chemistry: $Ni + O_2 \longrightarrow NiO$

d. *English:* "Iron(III) hydroxide decomposes thermally to yield iron(III) oxide and water."
Chemistry: $Fe(OH)_3 \longrightarrow Fe_2O_3 + H_2O$

e. *English:* "Magnesium oxide reacts with water to form magnesium hydroxide."
Chemistry: $MnO + H_2O \longrightarrow Mn(OH)_2$

2. Translate these following English sentences into chemistry.
   a. "Cadmium reacts with sulfur to yield cadmium sulfide."
   b. "Nitrogen triiodide decomposes to yield nitrogen and iodine."
   c. "Aluminum hydroxide reacts with sulfuric acid to form aluminum sulfate and water."
   d. "Hydrogen sulfide burns in oxygen to form water and sulfur dioxide."
   e. "Iron reacts with silver nitrate to form silver and iron(II) nitrate."
3. Which of these equations are balanced?
   a. $C_3H_8 + O_2 \xrightarrow{\Delta} CO_2 + H_2O$
   b. $HNO_3 + NaOH \longrightarrow NaNO_3 + H_2O$
   c. $MnO_2 + Al \longrightarrow Mn + Al_2O_3$
   d. $CaCO_3 + H_2SO_4 \longrightarrow CaSO_4 + CO_2 + H_2O$
   e. $Sr + H_2O \longrightarrow Sr(OH)_2 + H_2$
4. Balance each of the unbalanced equations in Exercise 3.
5. What is wrong with each of these "balanced" equations?
   a. $Ca + N_2 \longrightarrow CaN_2$
   b. $AgNO_3 + Zn \longrightarrow ZnNO_3 + Ag$
   c. $H_2O \longrightarrow 2\,H + O$
   d. $N + H_3 \longrightarrow NH_3$
   e. $CrCl_2 + AgNO_3 \longrightarrow CrNO_3 + AgCl_2$
   f. $Cl_2 + NaBr \longrightarrow Br + NaCl_2$
   g. $Zn + H_2Cl_2 \longrightarrow H_2 + ZnCl_2$
   h. $NaOH + H_2CO_3 \longrightarrow NaCO_3 + H_2OH$
   i. $Na + H_2O \longrightarrow NaOH + H$
6. Write correct, balanced equations for each of the incorrect equations in Exercise 5.
7. Balance these equations.
   a. $LiHCO_3 \not\longrightarrow Li_2CO_3 + CO_2 + H_2O$
   b. $CaC_2 + H_2O \not\longrightarrow Ca(OH)_2 + C_2H_2$
   c. $Al + H_2SO_4 \not\longrightarrow Al_2(SO_4)_3 + H_2$
   d. $NH_4NO_3 \overset{\Delta}{\not\longrightarrow} N_2O + H_2O$
   e. $N_2O_5 + H_2O \not\longrightarrow HNO_3$
   f. $Ca + HCl \not\longrightarrow CaCl_2 + H_2$
   g. $BaCl_2 + (NH_4)_2CO_3 \not\longrightarrow BaCO_3 + NH_4Cl$
   h. $HC_2H_3O_2 + Al(OH)_3 \not\longrightarrow Al(C_2H_3O_2)_3 + H_2O$
   i. $P + Cl_2 \not\longrightarrow PCl_5$
8. Translate these from English to chemistry, and balance the equations.
   a. "Sulfur tetrafluoride reacts with water to form sulfur dioxide and hydrogen fluoride."
   b. "Iron(III) oxide reacts with hydrochloric acid to form iron(III) chloride and water."
   c. "Lead(II) carbonate decomposes with heat to form lead(II) oxide and carbon dioxide."
   d. "Sodium bromide reacts with phosphoric acid to form hydrogen bromide and sodium phosphate."
   e. "Phosphorus burns in air to form tetraphosphorus hexoxide."
9. Translate from chemistry into English.
   a. $2\,H_2S(g) + 3\,O_2(g) \longrightarrow 2\,SO_2(g) + 2\,H_2O(g)$
   b. $2\,SO_2(g) + O_2(g) \longrightarrow 2\,SO_3(g)$
   c. $3\,NH_3(g) + H_3PO_4(aq) \longrightarrow (NH_4)_3PO_4(aq)$
   d. $Zn(OH)_2(s) + 2\,HClO_4(aq) \longrightarrow Zn(ClO_4)_2(aq) + 2\,H_2O(l)$
   e. $Pb(NO_3)_2(aq) + Na_2SO_4(aq) \longrightarrow PbSO_4(s) + 2\,NaNO_3(aq)$
10. Translate from English into chemistry (put in *s, l, g, aq*, Δ), and balance the equations.
    a. "Solid cadmium oxide reacts with hydrofluoric acid to form water and aqueous cadmium fluoride."
    b. "Sodium sulfite reacts with hydrochloric acid in aqueous solution to form sodium chloride, sulfur dioxide gas, and water."
    c. "Hydrogen peroxide decomposes to form oxygen gas and water."
    d. "Propane gas, C_3H_8, burns in limited air to form carbon monoxide and water."
    e. "Copper metal reacts with sulfur to form copper(I) sulfide."
    f. "Chlorine gas reacts with water to form hydrochloric acid and hypochlorous acid."
    g. "Nitrogen monoxide reacts with oxygen to form nitrogen dioxide."
    h. "Mercury(I) nitrate reacts with calcium chloride in aqueous solution to form calcium nitrate and a precipitate of mercury(I) chloride."
    i. "Solid boron reacts with fluorine gas to form gaseous boron trifluoride."
    j. "Acetic acid reacts with solid sodium hydrogen carbonate to form carbon dioxide, water, and sodium acetate."
11. Classify these reactions as combination, decomposition, single replacement, or double replacement.
    a. $2\,(NH_4)_2S(aq) + SnCl_4(aq) \not\longrightarrow 4\,NH_4Cl(aq) + SnS_2(s)$
    b. $F_2(g) + CaI_2(aq) \not\longrightarrow CaF_2(s) + I_2(s)$
    c. $6\,Na(s) + N_2(g) \not\longrightarrow 2\,Na_3N(s)$
    d. $4\,Al(s) + 3\,O_2(g) \longrightarrow 2\,Al_2O_3(s)$
    e. $Fe(s) + CuSO_4(aq) \longrightarrow Cu(s) + FeSO_4(aq)$
    f. $BaO(s) + 2\,HNO_3(aq) \longrightarrow Ba(NO_3)_2(aq) + H_2O(l)$
    g. $2\,K(s) + 2\,H_2O(l) \longrightarrow 2\,KOH(aq) + H_2(g)$
    h. $2\,Na(s) + H_2(g) \longrightarrow 2\,NaH(s)$
    i. $2\,KClO_3(s) \xrightarrow{\Delta} 2\,KCl(s) + 3\,O_2(g)$
    j. $LiOH(aq) + HCl(aq) \longrightarrow LiCl(aq) + H_2O(l)$

12. Translate each of the equations in Exercise 11 into English.
13. Complete and balance each of these equations for combination reactions.
    a. $Li(s) + Cl_2(g) \longrightarrow$
    b. $Cr(s) + O_2(g) \longrightarrow$
    c. $NO(g) + O_2(g) \longrightarrow$
    d. $CO_2(g) + H_2O(l) \longrightarrow$
    e. $K_2O(s) + H_2O(l) \longrightarrow$
    f. $BeO(s) + CO_2(g) \longrightarrow$
    g. $ZnO(s) + SO_2(g) \longrightarrow$
    h. $Rb_2O(s) + SO_3(g) \longrightarrow$
14. Complete and balance each of these equations for decomposition reactions.
    a. $CaCl_2 \cdot 2\,H_2O(s) \xrightarrow{\Delta}$
    b. $Ag_2CO_3(s) \xrightarrow{\Delta}$
    c. $Mg(HCO_3)_2(s) \xrightarrow{\Delta}$
15. Complete and balance these equations for single replacement reactions.
    a. $Ca(s) + H_2O(l) \longrightarrow$
    b. $Rb(s) + H_2O(l) \longrightarrow$
    c. $Cr(s) + HCl(aq) \longrightarrow$
    d. $Cd(s) + H_3PO_4(aq) \longrightarrow$
    e. $Ni(s) + Hg(NO_3)_2(aq) \longrightarrow$
    f. $Zn(s) + CuCl_2(aq) \longrightarrow$
    g. $KBr(aq) + F_2(aq) \longrightarrow$
    h. $BaI_2(aq) + Br_2(aq) \longrightarrow$
16. Complete and balance these equations for double replacement reactions.
    a. $HBr(aq) + KOH(aq) \longrightarrow$
    b. $Zn(OH)_2(s) + HClO_4(aq) \longrightarrow$
    c. $CdO(s) + HF(aq) \longrightarrow$
    d. $MgCO_3(s) + H_3PO_4(aq) \longrightarrow$
    e. $NaHSO_3(aq) + HCl(aq) \longrightarrow$
    f. $AgNO_3(aq) + Na_2SO_4(aq) \longrightarrow$
    g. $SnCl_2(aq) + (NH_4)_2S(aq) \longrightarrow$
17. Classify by reaction type, then complete and balance. (If no reaction, write "N.R.")
    a. $K(s) + S(s) \longrightarrow$
    b. $H_2SO_4(aq) + Sn(s) \longrightarrow$
    c. $N_2(g) + H_2(g) \longrightarrow$
    d. $Hg(l) + AgNO_3(aq) \longrightarrow$
    e. $NaCl(aq) + Br_2(l) \longrightarrow$
    f. $Na_2CO_3(aq) + CuSO_4(aq) \longrightarrow$
    g. $KCl(aq) + Mg(NO_3)_2(aq) \longrightarrow$
    h. $MgO(s) + H_2O(l) \longrightarrow$
    i. $Hg_2CO_3(s) \xrightarrow{\Delta}$

*18. Complete and balance. (Assume that a reaction does take place.)
    a. $CuS(s) + O_2(g) \longrightarrow CuO(s) +$ ____
    b. $Mg(CN)_2(aq) + H_2SO_4(l) \longrightarrow$
    c. $P_4O_{10}(s) + H_2O(l) \longrightarrow$
    d. $Na_2O_2(s) \xrightarrow{\Delta} O_2(g) +$ ____
    e. $NaCl(aq) + H_2SO_4(aq) \longrightarrow$
    f. $Sn(OH)_4(s) \xrightarrow{\Delta} SnO_2(s) +$ ____
    g. $FeO(s) + C(s) \longrightarrow Fe(s) +$ ____

*19. "Nitrogen fixation" is important for the survival of plant life. Nitrogen in the atmosphere is in the form N_2, elemental nitrogen. Plants can't use it in this form, yet nitrogen is an essential part of their nutrition. To absorb nitrogen, the plants must receive it in some combined form, as a chemical compound. One way that nitrogen fixation occurs in nature begins with electrical storms. (a) Lightning causes nitrogen to react with oxygen and form nitrogen monoxide. (b) The nitrogen monoxide reacts further with oxygen in the air to form nitrogen dioxide. (c) Water vapor combines with nitrogen dioxide to form nitric acid and nitrogen monoxide. The nitric acid washes down with rain and becomes part of the soil. Write balanced equations for these reactions and state the reaction type for each.

*20. Hydrazine, N_2H_4, is a rocket fuel. Write a balanced equation for its burning in oxygen. Nitrogen monoxide is one product.

*21. One form of iron ore is Fe_2O_3. Iron can be obtained as the free metal by heating with coke (C). Write a complete, balanced equation for this reaction.

CHAPTER 6

Calculations with Formulas and Equations

LEARNING OBJECTIVES

After studying this chapter, you should be able to:

1. Supply a correct definition, explanation, or example for each of these:

gram-molecular weight	percentage-by-weight
molecular weight	percent composition
gram-formula weight	empirical formula
formula weight	simplest formula
mole ratio	molecular formula
exothermic reaction	percent purity
endothermic reaction	limiting reactant
heat of reaction	theoretical yield
percent	actual yield
	percent yield

2. Calculate the molecular weight or formula weight of a substance, using a table of atomic weights, and use molecular weight or formula weight as a conversion factor to find moles, grams, or molecules of a sample of a substance.
3. Given a chemical equation and a table of atomic weights, perform correct mole-to-mole, mass-to-mole, mole-to-mass, mass-to-mass, mass-to-heat, or heat-to-mass conversions.
4. Using a table of atomic weights, supply the percent composition, given a formula, or a formula, given a percent composition. Determine the percent water in a hydrate, given its formula.
5. Given a compound's percent composition and its molecular weight, supply its empirical formula and its molecular formula.
6. Given a substance's percent purity, a chemical equation, and a table of atomic weights, perform mass-to-mass calculations involving impure substances.
7. In problems with one reactant present in excess, determine the limiting reactant, the amount of product formed, and the amount of excess reactant left over.
8. Calculate the percent yield or actual yield from appropriate data.

We encounter problems every day that involve knowing or being able to tell how much of something it takes to make how much of something else. How much dry cement does it take to mix up enough concrete to mortar a 6-foot-high brick wall 20 feet wide? How many eggs should we buy to make enough cookies for fifty girl scouts? Problems in chemistry are just like these. A chemist in an industrial plant might have to figure out, say, how much calcium carbonate will absorb enough harmful sulfur dioxide from the plant's emissions to comply with the Environmental Protection Agency's standards for emission.

To solve these problems, the bricklayer, the baker, and the chemist need recipes. The chemist's recipe is a chemical equation, which tells what amounts of calcium carbonate and sulfur dioxide react to give what amounts of calcium sulfate and carbon dioxide.

$$2\ CaCO_3 + 2\ SO_2 + O_2 \longrightarrow 2\ CaSO_4 + 2\ CO_2$$

Once the equation is known, the chemist can change the quantities of reactants and products to get (or in this case, to get rid of) the desired amounts of each.

In the last chapter, we interpreted chemical equations in terms of atoms and molecules. Now we want to interpret them in terms of moles and finally in terms of grams, so that we can work with them in a practical way. The chemist in an industrial plant can't count out atoms one by one. We have to make use of a concept we've introduced earlier: weighing as a means of counting. (See Section 3.1, p. 35.) We'll see that both formulas and equations can be expressed in terms of weight with numbers, and that's a chemistry-arithmetic translation. We'll learn how to do chemistry-arithmetic translations, and how to solve problems with conversion factors that we get from balanced chemical equations.

6.1 THE QUANTITATIVE MEANING OF FORMULAS

We saw that a formula means how many of each kind of atom are in a given chemical compound. For instance, the formula H_2SO_4 means that one molecule of the compound contains two atoms of hydrogen, one atom of sulfur, and four atoms of oxygen. That's one chemistry-English translation. We need to look at another chemistry-English translation, and learn how to get to a chemistry-arithmetic translation.

THE MOLE REVISITED. A mole is 6.02×10^{23} things: in particular, we're interested in counting atoms, molecules, and ions with this number. The phrase "one mole of neon atoms" means 6.02×10^{23} neon atoms. It also means 20.2 grams, because the gram-atomic weight of neon is 20.2 grams per mole. So, one arithmetic translation of the English phrase "one mole of neon," is "20.2 grams of neon."

In the same way, "two moles of neon" translates into "40.4 grams of neon." "One-half mole of neon" translates to "10.1 grams of neon," and so on. Since these mean, respectively, 12.0×10^{23} neon atoms and 3.01×10^{23} neon atoms, we can use either moles or grams to count atoms. When the chips are down and we actually have to make a measurement, we use grams instead of moles, because we don't have any kind of scale or balance that lets us measure moles directly. The gram-atomic weight of an element is one arithmetic translation of its symbol.

MOLECULAR WEIGHT. What if we have a diatomic gas, such as oxygen (O_2)? A mole of O_2 means a mole of diatomic O_2 molecules: 6.02×10^{23} molecules. Since each molecule contains two atoms, then a mole of O_2 molecules will contain 12.0×10^{23} O atoms. The mass of a mole of diatomic O_2 molecules is twice the mass of a mole of O atoms. Since each mole of O atoms has a mass of 16.0 grams, a mole of O_2 molecules has a mass of 32.0 grams ($2 \times 16.0 = 32.0$). The arithmetic translation of "one mole of O_2" is "32.0 grams."

We can reinterpret the formula H_2SO_4. Another translation is this: "One mole of H_2SO_4 contains two moles of H, one mole of S, and four moles of O." To understand how this works, let's look at a simple analogy.

A red wagon has these parts: four wheels, one handle, and one body. Two red wagons have eight wheels, two handles, and two bodies. One dozen red wagons have four dozen wheels, one dozen handles, and one dozen bodies. *And,* one mole of red wagons has four moles of wheels, one mole of handles, and one mole of bodies. If we knew what each part weighed, we could calculate the weight of one mole of red wagons by adding up the weight of four moles of wheels, one mole of handles, and one mole of bodies.

In the same way, we can get the mass of a mole of H_2SO_4 molecules by adding up the masses of two moles of H, one mole of S, and four moles of O. The mass, in grams, of a mole of molecules is called the *gram-molecular weight,* or simply *molecular weight.* We get the molecular weight of a substance by adding up the gram-atomic weights of all the atoms that are in it. Molecular weight of a substance is the arithmetic translation of its molecular formula.

gram-molecular weight

molecular weight

EXAMPLE 6.1: What is the molecular weight of H_2SO_4?

Solution: Add up the gram-atomic weights.

$$\begin{aligned} \text{2 H atoms, each 1.01 g/mole} &= 2.02 \text{ g/mole} \\ \text{1 S atom, each 32.1 g/mole} &= 32.1 \text{ g/mole} \\ \text{4 O atoms, each 16.0 g/mole} &= \underline{64.0 \text{ g/mole}} \\ & 98.12 \text{ g/mole} \end{aligned}$$

The answer must be rounded off to one decimal place.

Answer: 98.1 g/mole. ■

Just as with gram-atomic weight, molecular weight is a conversion factor that lets us convert between grams and moles. Numbers of atoms in a formula are pure numbers. The formula H_2SO_4 means that *exactly* one mole contains *exactly* two moles of H, one mole of S, and four moles of O—no more and no less. Our molecular weight will be limited in significant figures only by the individual atomic weights. We'll usually agree to carry out molecular weights to one decimal place.

EXAMPLE 6.2: How many grams are in 2.50 moles of H_2SO_4?

Solution: We want to convert 2.50 moles of H_2SO_4 to grams. We saw above that the molecular weight of H_2SO_4 is 98.1 g/mole. This is our conversion factor. Here's the setup:

$$2.50 \cancel{\text{moles } H_2SO_4} \times 98.1 \frac{\text{g } H_2SO_4}{\cancel{\text{mole } H_2SO_4}} = \_\_\_\_ \text{ g } H_2SO_4$$

Answer: 245 g H_2SO_4. ■

EXAMPLE 6.3: Given a sample of 545 g H_2SO_4, calculate: (a) the number of moles; and (b) the number of molecules.

Solution: (a) We want to convert 545 grams of H_2SO_4 to moles. Our conversion factor is still 98.1 g/mole, which we invert to 1.00 mole/98.1 g to do this conversion.

$$545 \text{ g } \cancel{H_2SO_4} \times \frac{1.00 \text{ mole } H_2SO_4}{98.1 \text{ g } \cancel{H_2SO_4}} = 5.56 \text{ moles } H_2SO_4$$

(b) We want to convert 5.56 moles H_2SO_4 to molecules. We use Avogadro's number, 6.02×10^{23} molecules/mole, as a conversion factor.

$$5.56 \cancel{\text{moles}} \times 6.02 \times 10^{23} \frac{\text{molecules}}{\cancel{\text{mole}}} = \_\_\_\_ \text{ molecules}$$

Answer: (a) 5.56 moles; (b) 3.35×10^{24} molecules. ■

FORMULA WEIGHT.

Since ionic compounds are not molecules, we can't very well talk about their molecular weight. Instead, we talk about *gram-formula weight.* Gram-formula weight, or simply *formula weight,* means the mass, in grams, of one mole of any substance as determined by its formula. The formula weight of a compound is one arithmetic translation of its formula.

gram-formula weight

formula weight

For instance, "one mole of the ionic compound Na_2SO_4" can be translated as "two moles of sodium ions and one mole of sulfate ions" as well as "two moles of Na, one mole of S, and four moles of O." We get the formula weight in exactly the same way as we get the molecular weight: add up the atomic weights of all the atoms in the compound. If the compound is a hydrate, the water is added in too.

EXAMPLE 6.4: What is the formula weight of $CuSO_4 \cdot 5\ H_2O$?

Solution: Here, the "5 H_2O" means five water molecules, or $5 \times 2 = 10$ H atoms and $5 \times 1 = 5$ O atoms. We add these into the formula weight.

1 Cu, each 63.5 g/mole =	63.5 g/mole
1 S, each 32.1 g/mole =	32.1 g/mole
4 O, each 16.0 g/mole =	64.0 g/mole
10 H, each 1.01 g/mole =	10.1 g/mole
5 O, each 16.0 g/mole =	80.0 g/mole
Answer:	249.7 g/mole

■

Formula weight and molecular weight are really the same, since they're derived in the same way. Formula weight is the more common usage, and it can also include molecular weight. We'll usually find it more convenient to use formula weight. Again, as with gram-atomic weight, we find "weight" used incorrectly to mean "mass."

6.2 THE QUANTITATIVE MEANING OF EQUATIONS

We said that chemical equations were the chemist's recipes. For instance, the following equation is the chemist's recipe for making water:

$$2\ H_2 + O_2 \longrightarrow 2\ H_2O$$

(In this chapter, we won't use the symbols showing the physical state [*g, l, s, aq*]; they might confuse or obscure the arithmetic translations we're trying to emphasize.) We can translate the above equation like this: "Two moles of hydrogen react with one mole of oxygen to yield two moles of water." The numbers in front of the formulas in a chemical equation are pure numbers, too, like the subscripts in the formulas themselves.

To do the chemistry-arithmetic translation, we use the arithmetic translations of the individual formulas, namely, the formula weights. By using the formula weight of each substance, we can translate the equation into arithmetic as follows: "4.04 grams of hydrogen react with 32.0 grams of oxygen to yield 36.0 grams of water."

How do we get this translation? The formula weight of H_2 is 2×1.01 g/mole $=$ 2.02 g/mole; O_2 is 2×16.0 g/mole $=$ 32.0 g/mole; H_2O is 2.02 g/mole $+$ 16.0 g/mole $=$ 18.0 g/mole. In the balanced equation, the quantities of both hydrogen and water are doubled (two moles of each). The masses of hydrogen and water are also doubled, so that the total mass on one side of the equation (36.0 grams) equals the total mass on the other side (36.0 grams). Figure 6.1 shows these different translations of the equation for water. Figure 6.2 shows a comparison between chemical equations and nonchemical recipes.

6.3 CALCULATIONS WITH CHEMICAL EQUATIONS

We've seen how to get the arithmetic translations of chemical equations, and we've also seen that formula weights are conversion factors. Now we'll see how to put them together to do calculations.

FIGURE 6.1
Translations of a chemical equation

These are all different ways of writing the recipe for water. In *A*, I've written the equation in terms of molecules. Then, in *B*, I multiplied the whole thing by 6.02 x 10^{23}, and I ended up with *C*. *D* is just *C* divided by 6.02 x 10^{23} molecules/mole, which gives us moles.

$$2\,H_2 + O_2 \longrightarrow 2\,H_2O$$

A	2 molecules H_2	+	1 molecule O_2	----→	2 molecules H_2O
B	6.02×10^{23}(2 molecules H_2	+	1 molecule O_2	----→	2 molecules H_2O)
C	(12.04×10^{23} molecules H_2)	+	(6.02×10^{23} molecules O_2)	----→	(12.04×10^{23} molecules H_2O)
D	2 moles H_2	+	1 mole O_2	----→	2 moles H_2O
E	2(1.01 g + 1.01 g)	+	(16.0 g + 16.0 g)	----→	2(1.01 g + 1.01 g + 16.0 g)
F	4.04 g H_2	+	32.0 g O_2	----→	36.0 g H_2O
G	40.4 g H_2	+	320 g O_2	----→	360 g H_2O
H	20 moles H_2	+	10 moles O_2	----→	20 moles H_2O

In *E* and *F*, I used formula weights to get number of grams from number of moles. In *E*, I got the formula weight by adding the gram-atomic weights and multiplying by the number of moles. That gives us *F*, where we can see that the total mass of matter on both sides is 36.0 grams.

In *G* and *H*, I fooled with the recipe to show how it can be increased. In *G*, I needed 360 grams of water, not 36.0 grams, so I increased all the ingredients 10 times. In H, I did the same thing using moles.

EQUATIONS AS SOURCES OF CONVERSION FACTORS.

We're going to use the information contained in chemical equations for conversion factors. Let's look at an example.

Chlorine gas is made commercially by the electrolytic decomposition of salt from seawater, according to this unbalanced equation:

$$NaCl \not\longrightarrow Na + Cl_2$$

We want to know how many moles of sodium chloride must decompose to get 5.00 moles of chlorine. There are five steps involved in solving problems like this one.

FIGURE 6.2
Comparison of nonchemical recipes and chemical equations

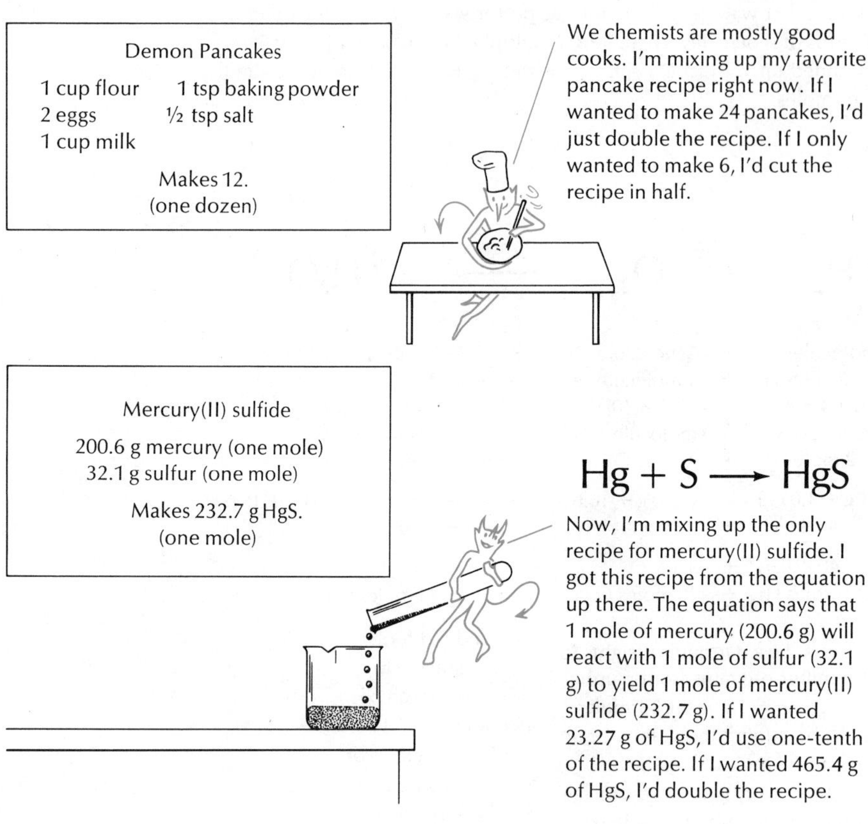

Demon Pancakes
1 cup flour
2 eggs
1 cup milk
1 tsp baking powder
½ tsp salt
Makes 12.
(one dozen)
We chemists are mostly good cooks. I'm mixing up my favorite pancake recipe right now. If I wanted to make 24 pancakes, I'd just double the recipe. If I only wanted to make 6, I'd cut the recipe in half.
Mercury(II) sulfide
200.6 g mercury (one mole)
32.1 g sulfur (one mole)
Makes 232.7 g HgS.
(one mole)
Hg + S ⟶ HgS
Now, I'm mixing up the only recipe for mercury(II) sulfide. I got this recipe from the equation up there. The equation says that 1 mole of mercury (200.6 g) will react with 1 mole of sulfur (32.1 g) to yield 1 mole of mercury(II) sulfide (232.7 g). If I wanted 23.27 g of HgS, I'd use one-tenth of the recipe. If I wanted 465.4 g of HgS, I'd double the recipe.

Step 1. *Check to be sure that the equation is balanced.* (You can't get a correct answer unless the equation is balanced.) The balanced equation:

$$2\ NaCl \longrightarrow 2\ Na + Cl_2$$

Step 2. *Figure out what units are to be converted to what other units.* We want to convert 5.00 moles of chlorine to moles of sodium chloride.

Step 3. *Write the number, with its units, on the left side of the calculation space. Write the units of the desired answer on the right.*

$$5.00 \text{ moles } Cl_2 = \_\_\_\_ \text{ moles NaCl}$$

Step 4. *Pick a conversion factor from the balanced equation.* We want a conversion factor that contains Cl_2 and NaCl, so we only look at these two quantities. We see that two moles of NaCl yield one mole of Cl_2. We want to trade "moles NaCl" for "moles Cl_2," so we arrange our conversion factor so that "moles Cl_2" is on the bottom and will cancel out.

Our conversion factor is:

$$\frac{2 \text{ moles NaCl}}{\text{mole } Cl_2}$$

This is the *mole ratio*. The numbers in front of formulas in equations are pure numbers. We write them as digits and understand that they won't limit the number of significant figures that our answer can have. mole ratio

Step 5. *Set up the problem and check units. Find the answer.*

$$5.00 \text{ moles } Cl_2 \times \frac{2 \text{ moles NaCl}}{\text{mole } Cl_2} = \_\_\_\_ \text{ moles NaCl}$$

The units check. The answer is 10.0 moles NaCl.

MOLE-TO-MOLE CONVERSIONS. The preceding example was a mole-to-mole conversion. We converted moles of one substance to moles of another, using the mole ratio as a conversion factor. We'll do a few more examples of this kind of conversion.

EXAMPLE 6.5: If 6.87 moles of chlorine are produced by the above equation, how many moles of sodium are also produced?

Solution:

Step 1: The equation is already balanced.

Step 2: We want to convert 6.87 moles Cl_2 to moles Na.

Step 3: 6.87 moles Cl_2 = ____ moles Na.

Step 4: Our conversion factor is 2 moles Na/mole Cl_2.

Step 5:

$$6.87 \text{ moles } Cl_2 \times \frac{2 \text{ moles Na}}{\text{mole } Cl_2} = \_\_\_\_ \text{ moles Na}$$

The units check.

Answer: 13.7 moles Na. ■

EXAMPLE 6.6: Nitric acid is prepared commercially by this equation:

$$NO_2 + H_2O \not\rightarrow HNO_3 + NO$$

How many moles of nitric acid can be made from 0.831 moles of NO_2?

Solution:

Step 1: The equation is not balanced, so we balance it.

$$3\,NO_2 + H_2O \longrightarrow 2\,HNO_3 + NO$$

Step 2: We want to convert 0.831 moles of NO_2 to moles of HNO_3.

Step 3: 0.831 moles NO_2 = ____ moles HNO_3.

Step 4: From the equation, our conversion factor is 2 moles HNO_3/3 moles NO_2.

Step 5:

$$0.831 \text{ moles } NO_2 \times \frac{2 \text{ moles } HNO_3}{3 \text{ moles } NO_2} = \_\_\_\_ \text{ moles } HNO_3$$

The units check.

Answer: 0.554 moles HNO_3. ■

MOLE-TO-MASS AND MASS-TO-MOLE CONVERSIONS. In these slightly more complex conversions, we begin to use formula weights as conversion factors, in addition to the mole ratio.

EXAMPLE 6.7: In the preparation of nitric acid (Example 6.6), how many grams of nitric acid can be made from 0.831 moles of NO_2?

Solution:

Step 1: The equation is already balanced.

Step 2: We want to convert 0.831 moles NO_2 to grams of nitric acid.

Step 3: 0.831 moles NO_2 = ____ grams HNO_3.

Step 4: We need two conversion factors. The mole ratio, 2 moles HNO_3/3 moles NO_2, gets us from moles of NO_2 to moles of HNO_3, as in Example 6.6. We need another conversion factor to get from moles of HNO_3 to grams of HNO_3: the formula weight of HNO_3.

H:	1(1.01 g/mole) =	1.01 g/mole
N:	1(14.0 g/mole) =	14.0 g/mole
O_3:	3(16.0 g/mole) =	48.0 g/mole
HNO_3:		63.0 g/mole

This conversion factor is 63.0 g HNO_3/mole HNO_3.

Step 5: Our setup:

$$0.831\ \cancel{\text{moles NO}_2} \times \frac{2\ \cancel{\text{moles HNO}_3}}{3\ \cancel{\text{moles NO}_2}} \times \frac{63.0\text{ g HNO}_3}{\cancel{\text{mole HNO}_3}} = \underline{\quad}\text{ g HNO}_3$$

The units check.

Answer: 34.9 g HNO_3. ■

EXAMPLE 6.8: The rusting of iron is represented by the following unbalanced equation:

$$Fe + O_2 \not\longrightarrow Fe_2O_3$$

11.5 grams of rust are scraped off a rusty tool. How many moles of iron are lost from the tool?

Solution:

Step 1: The equation is not balanced, so we balance it:

$$4\,Fe + 3\,O_2 \longrightarrow 2\,Fe_2O_3$$

Step 2: We want to convert 11.5 grams of Fe_2O_3 to moles of Fe.

Step 3: 11.5 g Fe_2O_3 = ____ moles Fe.

Step 4: Again, we need two conversion factors. This time, we first need the formula weight of Fe_2O_3 to change grams of Fe_2O_3 to moles of Fe_2O_3; then we need the mole ratio to change moles of Fe_2O_3 to moles of Fe. Our formula weight:

Fe_2:	2(55.8 g/mole) =	111.6 g/mole
O_3:	3(16.0 g/mole) =	48.0 g/mole
Fe_2O_3:		159.6 g/mole

The formula weight is 159.6 g Fe_2O_3. The mole ratio, from the balanced equation, is 4 moles Fe/2 moles Fe_2O_3, which simplifies to 2 moles Fe/mole Fe_2O_3.

Step 5: Our setup:

$$11.5\ \text{g Fe}_2\text{O}_3 \times \frac{1\ \text{mole Fe}_2\text{O}_3}{159.6\ \text{g Fe}_2\text{O}_3} \times \frac{2\ \text{moles Fe}}{\text{mole Fe}_2\text{O}_3} = \_\_\_\ \text{moles Fe}$$

The units check.

Answer: 0.144 moles Fe. ■

MASS-TO-MASS CONVERSIONS.

In these calculations, we'll use two formula weight conversion factors plus a mole ratio.

EXAMPLE 6.9: In the rusting of iron (Example 6.8), how many grams of rust are formed from 2.36 grams of iron lost from the tool?

Solution:

Step 1: The equation is already balanced.

Step 2: We want to convert 2.36 g Fe to grams of Fe_2O_3.

Step 3: 2.36 g Fe = ___ g Fe_2O_3.

Step 4: We need a conversion factor to get from grams of iron to moles of iron (the atomic weight of iron); one to get from moles of iron to moles of Fe_2O_3 (the mole ratio); and one to get from moles of Fe_2O_3 to grams of Fe_2O_3 (the formula weight). Our conversion factors are 55.8 g Fe/mole Fe; 2 moles Fe/mole Fe_2O_3; and 159.6 g Fe_2O_3/mole Fe_2O_3.

Step 5: Our setup:

$$2.36\ \text{g Fe} \times \frac{1\ \text{mole Fe}}{55.8\ \text{g Fe}} \times \frac{1\ \text{mole Fe}_2\text{O}_3}{2\ \text{moles Fe}} \times \frac{159.6\ \text{g Fe}_2\text{O}_3}{\text{mole Fe}_2\text{O}_3} = \_\_\_\ \text{g Fe}_2\text{O}_3$$

The units check.

Answer: 3.38 g Fe_2O_3. ■

EXAMPLE 6.10: A major source of sulfur dioxide pollution comes from the smelting of copper sulfide ore. One equation for this is:

$$CuS + O_2 \longrightarrow Cu + SO_2$$

How many metric tons of sulfur dioxide will be released into the atmosphere as a result of smelting 555 metric tons of CuS?

Solution:

Step 1: The equation is balanced.

Step 2: We want to convert 555 metric tons of CuS to metric tons of SO_2.

Step 3: 555 metric tons CuS = ___ metric tons SO_2.

Step 4: We're given a new unit here: a metric ton (t) is 1000 kg; it is 10 percent heavier than the 2000-pound English ton. However, since the answer is asked for in the same units, we don't have to know anything about metric tons to solve the problem. We can express formula weights in metric tons instead of grams if we want to, because formula weights can have any units. We usually understand them to be gram-formula weights, but here we'll express them in ton-formula weights, or ton moles (t-mole). When we add up the formula weights, we can use the same numbers we would use for grams. We can get away with this if we stick to the same units all the way through the problem. (See Figure 6.3.)

CuS:	Cu:	63.5 t/t-mole	SO_2:	S:		32.1 t/t-mole
	S:	32.1 t/t-mole		O_2:	$(2 \times 16.0) =$	32.0 t/t-mole
	CuS:	95.6 t/t-mole		SO_2:		64.1 t/t-mole

Our conversion factors are 95.6 t CuS/t-mole CuS; 1 t-mole CuS/t-mole SO_2; and 64.1 t SO_2/t-mole SO_2.

Step 5: Our setup:

$$555 \text{ t CuS} \times \frac{1 \text{ t-mole CuS}}{95.6 \text{ t CuS}} \times \frac{1 \text{ t-mole } SO_2}{\text{t-mole CuS}} \times \frac{64.1 \text{ t } SO_2}{\text{t-mole } SO_2} = \_\_\_ \text{ t } SO_2$$

The units check.

Answer: 372 t SO_2. ■

FIGURE 6.3
Formula weights may have any units

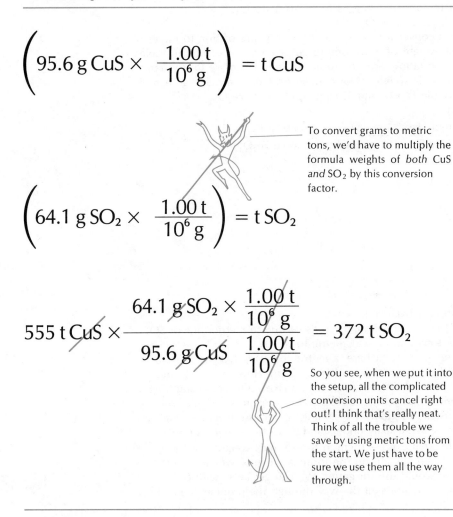

$$\left(95.6 \text{ g CuS} \times \frac{1.00 \text{ t}}{10^6 \text{ g}}\right) = \text{t CuS}$$

To convert grams to metric tons, we'd have to multiply the formula weights of *both* CuS *and* SO_2 by this conversion factor.

$$\left(64.1 \text{ g } SO_2 \times \frac{1.00 \text{ t}}{10^6 \text{ g}}\right) = \text{t } SO_2$$

$$555 \text{ t CuS} \times \frac{64.1 \text{ g } SO_2 \times \dfrac{1.00 \text{ t}}{10^6 \text{ g}}}{95.6 \text{ g CuS} \; \dfrac{1.00 \text{ t}}{10^6 \text{ g}}} = 372 \text{ t } SO_2$$

So you see, when we put it into the setup, all the complicated conversion units cancel right out! I think that's really neat. Think of all the trouble we save by using metric tons from the start. We just have to be sure we use them all the way through.

EXAMPLE 6.11: In the beginning of this chapter, we discussed the removal of SO_2 pollution from smokestack emissions as follows:

$$2\,CaCO_3 + 2\,SO_2 + O_2 \longrightarrow 2\,CaSO_4 + 2\,CO_2$$

How many metric tons of $CaCO_3$ will it take to absorb the 372 metric tons of SO_2 produced in Example 6.10?

Solution:

Step 1: The equation is balanced.

Step 2: We want to convert metric tons of SO_2 to metric tons of $CaCO_3$.

Step 3: 372 metric tons SO_2 = ____ metric tons $CaCO_3$.

Step 4: We add up the formula weights.

SO_2:	S:	32.1 t/t-mole
	O_2: 2(16.0) =	32.0 t/t-mole
	SO_2:	64.1 t/t-mole
$CaCO_3$:	Ca:	40.1 t/t-mole
	C:	12.0 t/t-mole
	O_3: 3(16.0) =	48.0 t/t-mole
	$CaCO_3$:	100.1 t/t-mole

Our conversion factors are 64.1 t SO_2/t-mole SO_2; 2 t-moles $CaCO_3$/2 t-moles SO_2 (simplifies to 1 t-mole $CaCO_3$/t-mole SO_2); and 100.1 t $CaCO_3$/t-mole $CaCO_3$.

Step 5: Our setup:

$$372\ \text{t}\ \cancel{SO_2} \times \frac{1\ \text{t-mole}\ \cancel{SO_2}}{64.1\ \text{t}\ \cancel{SO_2}} \times \frac{1\ \text{t-mole}\ \cancel{CaCO_3}}{\text{t-mole}\ \cancel{SO_2}} \times \frac{100.1\ \text{t}\ CaCO_3}{\text{t-mole}\ \cancel{CaCO_3}} = \underline{\quad\quad}\ \text{t}\ CaCO_3$$

The units check.

Answer: 581 t $CaCO_3$. ■

6.4 HEAT AS PART OF CHEMICAL REACTIONS

In Chapter 5, we saw that some chemical reactions need heat to make them go. We used the Greek letter delta (Δ) to show this in a qualitative way. Heat can also be given off in a chemical reaction, although we don't have any simple way of showing that. In fact, most chemical reactions involve heat, either by needing heat to make them go or by giving off heat. We can measure how much heat is involved in any chemical reaction, and we can use these numbers just like formula weights when we're doing chemical calculations.

A reaction that gives off heat is called an *exothermic reaction,* and the amount of heat is written on the right side of the equation. In this case, heat is a product of the reaction. A reaction that requires heat is called an *endothermic reaction,* and the amount of heat is written on the left side of the

exothermic reaction

endothermic reaction

equation. In that case, heat is a reactant. Here are examples of these reactions:

$$2\,H_2 + O_2 \longrightarrow 2\,H_2O + 136\text{ kcal} \quad \text{(exothermic)}$$
$$N_2 + O_2 + 43.2\text{ kcal} \longrightarrow 2\,NO \quad \text{(endothermic)}$$

The amount of heat involved is called the *heat of reaction* and is expressed in kilocalories (kcal). The heat of reaction is fixed for a given reaction, and we always get this heat, whether we want it or not. Some chemical reactions are used for the sole purpose of generating heat. The burning of natural gas, coal, oil, and other fuels are examples of this.

heat of reaction

EXAMPLE 6.12: When natural gas (methane, CH_4) is burned, 213 kcal/mole of heat is produced according to this equation:

$$CH_4 + 2\,O_2 \longrightarrow CO_2 + 2\,H_2O + 213\text{ kcal}$$

How much heat will be produced by the burning of 454 grams of natural gas?

Solution:
Step 1: The equation is balanced.
Step 2: We want to convert 454 grams of methane to kilocalories.
Step 3: 454 g CH_4 = ____ kcal.
Step 4: Our conversion factors are the formula weight of CH_4 and the ratio 1 mole CH_4/213 kcal from the equation. We add up the formula weight of CH_4:

C:		12.0 g/mole
H_4:	$(4 \times 1.01) =$	4.04 g/mole
CH_4:		16.0 g/mole

Step 5: Our setup:

$$454\text{ g }CH_4 \times \frac{1\text{ mole }CH_4}{16.0\text{ g }CH_4} \times \frac{213\text{ kcal}}{\text{mole }CH_4} = \_\_\_\_\text{ kcal}$$

The units check.

Answer: 6.04×10^3 kcal. ■

Sometimes we don't want *all* the heat of reaction. In a gasoline engine, heat energy is converted into mechanical energy that pushes the pistons and makes the car run. But not all the heat is converted, and that extra heat of reaction is a nuisance. We need radiators or other cooling systems to get rid of this excess heat. The heat of reaction is just as much a part of the reaction as the chemical substances are.

EXAMPLE 6.13: One ingredient in gasoline is octane, C_8H_{18}. It burns according to this equation:

$$2\,C_8H_{18} + 25\,O_2 \longrightarrow 16\,CO_2 + 18\,H_2O + 2614\text{ kcal}$$

How many grams of octane would have to be burned to produce enough heat to boil the water in a 12-liter radiator (about 6480 kcal)?

Solution:
Step 1: The equation is balanced.
Step 2: We want to convert 6480 kcal to grams of octane.
Step 3: 6480 kcal = ____ g octane.

Step 4: We add up the formula weight of octane:

$$\begin{array}{lrl} C_8: & 8(12.0) = & 96.0\ g/mole \\ H_{18}: & 18(1.01) = & 18.2\ g/mole \\ \hline C_8H_{18}: & & 114.2\ g/mole \end{array}$$

Our conversion factors are 2 moles C_8H_{18}/2614 kcal and 114.2 g C_8H_{18}/mole C_8H_{18}.

Step 5: Our setup:

$$6480\ \cancel{kcal} \times \frac{2\ \text{moles}\ \cancel{C_8H_{18}}}{2614\ \cancel{kcal}} \times \frac{114.2\ g\ C_8H_{18}}{\cancel{\text{mole}\ C_8H_{18}}} = \_\_\_\ g\ C_8H_{18}$$

The units check.

Answer: 566 g C_8H_{18}. ■

In chemical plants, many chemical reactions are going on at the same time, some giving off heat, and some requiring heat. A challenging problem that chemical engineers face is to design the plant so that the heat from the exothermic reactions is used to make the endothermic reactions go. This is done through heat exchangers. Usually the numbers don't exactly match up, though, and so some heat has to be supplied or gotten rid of.

6.5 PROBLEMS INVOLVING PERCENTAGES

Percent means parts per hundred parts. We have a number of instances where we need to use percent in chemical calculations. We can express a compound in terms of percentages by weight of its elements. We can talk about the percent purity of a compound. In a chemical reaction, we might not actually obtain all of the product that our calculations led us to believe we would. In that case, we would calculate a percent yield. We'll learn to work with all of these things in this section. percent

PERCENTAGES AND FORMULAS. A chemical formula can be used to calculate the percentages of all the elements in it. For instance, the compound iron(II) sulfate is 36.77% iron, 21.10% sulfur, and 42.13% oxygen, by weight. We'll see a little later how we calculate this composition, but first let's look at weight percentage with a simple example.

Suppose we own the following creatures:

2 elephants, at 4055 pounds apiece
2 aardvarks, at 50.5 pounds apiece

We sell these animals to two different parties in the same city. A zoo buys the elephants, and a private collector buys the aardvarks. They agree to pay the shipping charge, by railroad. How should we divide the shipping bill between the two purchasers?

One way to look at it would be to divide the bill in half. We could say that the load is 50 percent elephants and 50 percent aardvarks. But if our shipping charge is computed by weight (which it is), we have to look at the situation differently, because elephants weigh so much more than aardvarks. To determine the fair amount of the bill to be paid by each buyer, we should calculate the *percentage-by-weight* of each pair of animals. To do this, we compute the total weight, divide the sum into each of the individual weights, and multiply by 100. Like this:

percentage-by-weight

2 elephants, at 4055 pounds each =	8110 pounds
2 aardvarks, at 50.5 pounds each =	101 pounds
Total	8211 pounds

Percentage-by-weight of elephants:

$$\frac{8110 \text{ lb}}{8211 \text{ lb}} \times 100. = 98.8\%$$

Percentage-by-weight of aardvarks:

$$\frac{101 \text{ lb}}{8211 \text{ lb}} \times 100. = 1.2\%$$

The answers total 100 percent. We always add up the percentages when the problem is finished, just to check the answers. If they're right, they'll always add up to 100 percent.

So, the answer is that the recipient of the aardvarks should pay 1.2 percent of the shipping bill; the new owner of the elephants gets stuck with 98.8 percent.

Now we're ready to do the same thing in chemistry. Let's go back to iron(II) sulfate and calculate the percentage-by-weight of iron, sulfur, and oxygen in it. A list of the percentages-by-weight of all the elements in a compound is the compound's *percent composition*.

percent composition

EXAMPLE 6.14: Iron pills, for iron-deficiency anemia, often contain iron in the form of iron(II) sulfate. Calculate its percent composition to two decimal places.

Solution: The formula of iron(II) sulfate is $FeSO_4$. The formula weight is:

Fe:	55.85
S:	32.06
4 O:	64.00
$FeSO_4$:	151.91

Compute the percent composition of each element.

$$Fe = \frac{55.85}{151.91} \times 10^2 = 36.77\%$$

$$S = \frac{32.06}{151.91} \times 10^2 = 21.10\%$$

$$O = \frac{64.00}{151.91} \times 10^2 = 42.13\%$$

The percentages add up to 100.00%.

Answer: 36.77% Fe; 21.10% S; 42.13% O. ■

Sometimes we're concerned with the percent of a portion of a compound, and not just the individual elements. For instance, we can calculate the percent water in a hydrate if we know its formula.

EXAMPLE 6.15: In Example 6.4, we calculated the formula weight of $CuSO_4 \cdot 5\,H_2O$. Calculate the percent water in this hydrate, to one decimal place.

Solution: The percent water is calculated like this:

$$\%\,H_2O = \frac{\text{weight } H_2O \text{ in hydrate}}{\text{total weight of hydrate}} \times 10^2$$

Step 1: Calculate the formula weight of the hydrate. We did this in Example 6.4: 249.7 g/mole.

Step 2: Calculate the weight of H_2O in one mole of the hydrate. Each mole contains 5 moles of H_2O, so we multiply the formula weight of H_2O by 5:

$$5\ \cancel{\text{moles}} \times 18.0\,\frac{\text{grams}}{\cancel{\text{mole}}} = 90.0 \text{ grams}$$

Step 3: Calculate the percent water.

$$\frac{90.0 \text{ g } H_2O}{249.7 \text{ g hydrate}} \times 10^2 = \underline{\qquad}\ \%\,H_2O$$

Answer: 36.0% H_2O. ■

Sometimes we are given percent composition and asked to determine the formula of a compound. This means we must figure out the whole-number ratio of elements in the compound. This is easy to do, as we'll see in the following example.

EXAMPLE 6.16: A compound has the following percent composition: 56.68% K, 8.68% C, and 34.73% O. What is its formula?

Solution: Remember that percent means parts per hundred, and in this case it means grams per 100 grams. If we keep in mind that we're talking about 100 grams of the compound, then we may express the percentages as grams and divide each by the atomic weight of the element.

$$\text{K:}\quad \frac{56.68 \text{ g}}{39.10 \text{ g/mole}} = 1.450 \text{ moles K}$$

$$\text{C:}\quad \frac{8.68 \text{ g}}{12.0 \text{ g/mole}} = 0.723 \text{ moles C}$$

$$\text{O:}\quad \frac{34.73 \text{ g}}{16.00 \text{ g/mole}} = 2.171 \text{ moles O}$$

This gives us the relative numbers of moles of each element in the formula. We *could* write $K_{1.450}C_{0.723}O_{2.171}$, but it wouldn't be very useful. Instead, we get a whole-number answer by dividing each number of moles by the smallest one. In this case, the smallest number is 0.723 moles C. When we do this, we're saying, "Okay, what if we have *one* carbon atom? Then how many of the others do we have?"

$$\frac{1.450 \text{ moles K}}{0.723 \text{ moles C}} = 2.00 \text{ moles K/mole C}$$

$$\frac{0.723 \text{ moles C}}{0.723 \text{ moles C}} = 1.00 \text{ moles C/mole C}$$

$$\frac{2.171 \text{ moles O}}{0.723 \text{ moles C}} = 3.00 \text{ moles O/mole C}$$

For every one mole of carbon, we have two moles of potassium and three of oxygen. We use this ratio to write the formula.

Answer: K_2CO_3. ■

For covalent compounds, we have to make a distinction between empirical formula and molecular formula. The *empirical,* or *simplest, formula* is a formula that shows the simplest ratio of the elements in a compound. Formulas that we calculate from percent composition as in Example 6.16 are empirical: they show the simplest ratio of elements.

empirical formula

simplest formula

The compound acetylene is 92.24% C and 7.76% H by weight. A calculation like that in Example 6.16 gives us a ratio of one carbon atom to one hydrogen atom, and the empirical formula is CH. Measurement of the molecular weight of acetylene, by another experiment, gives the value 26.0 g/mole. This is twice what we'd expect if there were just one C and one H in the molecule (13.0 g/mole). So, the molecular formula must be twice the empirical formula: C_2H_2. The *molecular formula* tells how many of each kind of atom are in a molecule. A molecule of acetylene contains two carbon atoms and two hydrogen atoms, but the simplest ratio is one carbon atom to one hydrogen atom.

molecular formula

For ionic compounds, there is no such thing as a molecule and therefore no such thing as a molecular formula. For some covalent compounds, such as H_2O, CO_2, and NH_3, the empirical and molecular formulas are the same. For others, such as C_2H_2, H_2O_2 and N_2O_4, the empirical formulas are different: CH, HO, and NO_2. To determine a substance's molecular formula, we need to know the molecular weight in addition to the percent composition.

EXAMPLE 6.17: A compound has this percent composition: 92.24% C, 7.76% H. Its molecular weight is 78.06. What is its molecular formula?

Solution: First, find the empirical formula.

$$\text{C:} \quad \frac{92.24 \text{ g}}{12.01 \text{ g/mole}} = 7.680 \text{ moles C}$$

$$\text{H:} \quad \frac{7.76 \text{ g}}{1.01 \text{ g/mole}} = 7.68 \text{ moles H}$$

Since these two numbers are the same, we have a 1-to-1 ratio of carbon to hydrogen. This gives us the empirical formula of CH, which we need to change to the molecular formula. If the empirical formula, CH, were the molecular formula, it would have a molecular weight of 12.01 g/mole + 1.01 g/mole = 13.02 g/mole. It doesn't, though. The molecular weight is 78.06 g/mole. The molecular formula has to be some multiple of the empirical formula. To find out what multiple this is, we divide the weight of the empirical formula into the molecular weight:

$$\frac{78.06 \text{ g/mole}}{13.02 \text{ g/mole}} = 6$$

Our molecular formula is thus 6 times the empirical formula.

Answer: C_6H_6. ■

PERCENT PURITY. Chemicals usually occur in nature mixed with other chemicals. Separating the chemicals wanted from the chemicals not wanted provides many real problems for chemists to solve. To know what we're up against, we use *percent purity* to tell us how many parts of the stuff we want is contained in 100 parts of the total mixture, including the stuff we don't want. Sometimes we don't bother to separate it, but just use it the way it is, in its impure state. If we do that, then we have to use more of the impure substance than we'd have to use if it were pure.

percent purity

EXAMPLE 6.18: A silver ore is 21.0 percent silver. How many grams of ore are needed to produce 10.0 grams of pure silver?

Solution:

Step 1: We want to convert 10.0 grams of silver to grams of ore.

Step 2: 10.0 g Ag = ____ g ore.

Step 3: We need a conversion factor to go from grams of silver to grams of ore. The percentage is our conversion factor. It means that a 100-gram sample of ore contains 21.0 grams of silver, or 21.0 g Ag/100. g ore. This is the same as 0.210 grams of silver per 1 gram of ore, and we'll use this figure. Since we want the "g Ag" to cancel out, we'll put it on the bottom. Then our conversion factor is:

$$\frac{1.00 \text{ g ore}}{0.210 \text{ g Ag}}$$

Step 4:

$$10.0 \text{ g Ag} \times \frac{1.00 \text{ g ore}}{0.210 \text{ g Ag}} = \_\_\_ \text{ g ore}$$

Answer: 47.6 g ore. ■

EXAMPLE 6.19: Coal is mostly carbon. If a given coal sample is 75.0 percent carbon, how many kilograms of carbon dioxide will be formed when 1.00 kg of coal is burned?

Solution:

Step 1: The balanced equation is:

$$C + O_2 \longrightarrow CO_2$$

Step 2: We want to convert kilograms of coal to kilograms of CO_2.

Step 3: 1.00 kg coal = ____ kg CO_2.

Step 4: We need *two* conversion factors: one to convert kilograms of coal to kilograms of carbon, and another to convert kilograms of carbon to kilograms of CO_2. The percent purity is our first conversion factor. 75.0 percent carbon means that 100 parts of coal contain 75.0 parts of carbon, so our conversion factor is 75.0 kg C/100 kg coal, or 0.750 kg C/kg coal. Here's what we have so far:

$$1.00 \text{ kg coal} \times \frac{0.750 \text{ kg C}}{\text{kg coal}} = \_\_\_ \text{ kg } CO_2$$

We get the other conversion factors from the chemical equation and from the atomic and formula weights, as we did earlier in this chapter. Remember that formula weights may be expressed in any units we choose so long as we use the same units throughout the problem. We get the formula weights in kilograms (kg-moles):

C: 12.0 kg/kg-mole

CO_2:		
C:		12.0 kg/kg-mole
O_2:	2(16.0) =	32.0 kg/kg-mole
CO_2:		44.0 kg/kg-mole

Our conversion factors are 12.0 kg C/kg-mole C, 1 kg-mole C/kg-mole CO_2, and 44.0 kg CO_2/kg-mole CO_2.

Step 5: Our finished setup:

$$1.00\ \cancel{\text{kg coal}} \times \frac{0.750\ \cancel{\text{kg C}}}{\cancel{\text{kg coal}}} \times \frac{1\ \cancel{\text{kg-mole C}}}{12.0\ \cancel{\text{kg C}}} \times \frac{1\ \cancel{\text{kg-mole CO}_2}}{\cancel{\text{kg-mole C}}} \times \frac{44.0\ \text{kg CO}_2}{\cancel{\text{kg-mole CO}_2}} = \underline{\quad}\ \text{kg CO}_2$$

The units check.

Answer: 2.75 kg CO_2. ■

LIMITING REACTANT. In the practice of chemistry, it often isn't necessary or even desirable to weigh out exact quantities of all substances that participate in a reaction. What is commonly done is to weigh out one substance and then to use an excess of another. (An *excess* means that there is more than enough of this reactant.) Then, at the end of the reaction, some of this excess reactant is left over. The reactant that isn't in excess is called the *limiting reactant,* because it limits the number of moles of product that can be obtained.

excess reactant

limiting reactant

There can be more than one excess reactant, but only one limiting reactant. Suppose we have one lime, one bottle of tequila, and one bottle of Triple Sec. How many margaritas could we make, if our recipe called for the juice of 1/2 lime, 1/2 ounce of Triple Sec, and 1 ounce of tequila per margarita? We could make only two margaritas, because we're limited by having only one lime: the limiting reactant. We have excess tequila and Triple Sec, and those will just be left over.

EXAMPLE 6.20: In an experiment, a student heats together 5.52 grams of powdered copper and 10.1 grams of powdered sulfur to carry out this reaction:

$$2\,\text{Cu} + \text{S} \longrightarrow \text{Cu}_2\text{S}$$

(a) Which is the limiting reactant? (b) How much Cu_2S will be obtained? (c) How much of the excess reactant will be left over?

Solution:

Step 1: Calculate the number of moles of product that would be obtained from each reactant. The reactant that produces the smallest number of moles of product is the limiting reactant.

$$5.52\ \cancel{g\ Cu} \times \frac{1\ \cancel{mole\ Cu}}{63.5\ \cancel{g\ Cu}} \times \frac{1\ mole\ Cu_2S}{2\ \cancel{moles\ Cu}} = 0.0435\ moles\ Cu_2S$$

$$10.1\ \cancel{g\ S} \times \frac{1\ \cancel{mole\ S}}{32.1\ \cancel{g\ S}} \times \frac{1\ mole\ Cu_2S}{\cancel{mole\ S}} = 0.315\ moles\ Cu_2S$$

Cu is the limiting reactant.

Step 2: From the smaller number of moles, calculate the weight of product in the usual way.

$$0.0435\ \cancel{moles\ Cu_2S} \times 159.1\ \frac{g\ Cu_2S}{\cancel{mole\ Cu_2S}} = 6.92\ g\ Cu_2S$$

Step 3: Calculate the grams of excess reactant needed to produce the number of moles of product.

$$0.0435\ \cancel{moles\ Cu_2S} \times \frac{1\ \cancel{mole\ S}}{\cancel{mole\ Cu_2S}} \times \frac{32.1\ g\ S}{\cancel{mole\ S}} = 1.40\ g\ S$$

This is the amount of S that has reacted.

Step 4: Subtract the value obtained in Step 3 from the original amount of excess reactant. This is the amount left over.

$$10.1\ g - 1.4\ g = 8.7\ g\ S\ \text{left over.}$$

Answer: (a) Cu; (b) 6.92 g Cu_2S; (c) 8.7 g S. ■

PERCENT YIELD. We've done calculations that tell us how much of something will be produced by a certain amount of something else in a chemical reaction. This number is called the *theoretical yield*. This is the maximum amount that we should expect. However, in the working applications of chemistry, we hardly ever get as much product as the equation says we should. There are lots of reasons for this. We might lose some of the material in the process of separating it from other products. The reaction may not have gone to completion (we'll talk about this more in Chapter 16). Maybe the starting materials weren't as pure as we thought. This doesn't mean that the Law of Conservation of Matter is being violated. That extra material is *somewhere*—we just don't have it. The amount that we do get is the *actual yield*. We get the *percent yield* by dividing the actual yield by the theoretical yield.

theoretical yield

actual yield

percent yield

EXAMPLE 6.21: Acetylene (C_2H_2) is made commercially by adding calcium carbide, CaC_2, to water according to this equation:

$$CaC_2 + 2\ H_2O \longrightarrow C_2H_2 + Ca(OH)_2$$

2550 kg of calcium carbide is treated with an excess of water, and 867 kg of acetylene is obtained. Calculate the percent yield.

Solution: First, "an excess of water" means that we can have as much water as we want. The theoretical yield of acetylene thus depends on the amount of calcium carbide we have; calcium carbide is the limiting reactant.

To calculate percent yield, first we must calculate the theoretical yield in the usual way.

Step 1: The equation is balanced.

Step 2: We want to convert kilograms of CaC_2 to kilograms of C_2H_2.

Step 3: 2550 kg CaC_2 = ____ kg C_2H_2.

Step 4:

$$\begin{array}{lr} Ca: & 40.1\ kg/kg\text{-}mole \\ C_2:\ 2(12.0) = & 24.0\ kg/kg\text{-}mole \\ \hline CaC_2: & 64.1\ kg/kg\text{-}mole \end{array} \qquad \begin{array}{lr} C_2:\ 2(12.0) = & 24.0\ kg/kg\text{-}mole \\ H_2:\ 2(1.01) = & 2.02\ kg/kg\text{-}mole \\ \hline C_2H_2: & 26.0\ kg/kg\text{-}mole \end{array}$$

Our conversion factors are 64.1 kg CaC_2/kg-mole CaC_2; 1 kg-mole CaC_2/kg-mole C_2H_2; and 26.0 kg C_2H_2/kg-mole C_2H_2.

Step 5: Our setup:

$$2550\ \text{kg } \cancel{CaC_2} \times \frac{1\ \cancel{\text{kg-mole } CaC_2}}{64.1\ \cancel{\text{kg } CaC_2}} \times \frac{1\ \cancel{\text{kg-mole } C_2H_2}}{\cancel{\text{kg-mole } CaC_2}} \times \frac{26.0\ \text{kg } C_2H_2}{\cancel{\text{kg-mole } C_2H_2}} = \_\_\_\ \text{kg } C_2H_2$$

The units check. The answer here is 1030 kg C_2H_2. But the problem told us we got 867 kg. Thus we have to calculate the percent yield.

$$\frac{867\ \text{kg}}{1030\ \text{kg}} \times 10^2 = \_\_\_\ \text{percent yield}$$

Answer: 84.2 percent yield. ■

EXAMPLE 6.22: The calcium carbide in the previous reaction is made by heating lime (calcium oxide) in an excess of coke (carbon):

$$CaO + 3\,C \xrightarrow{\Delta} CaC_2 + CO$$

This reaction gives 67.9 percent yield. What would be the actual yield of calcium carbide if 9550 kg of lime were used?

Solution: Again, we must first calculate the theoretical yield.

Step 1: The equation is balanced.

Step 2: We want to convert kilograms of calcium oxide to kilograms of calcium carbide.

Step 3: 9550 kg CaO = ___ kg CaC_2.

Step 4: We know that the formula weight of CaC_2 is 64.1 kg.

$$\begin{array}{lr} Ca: & 40.1\ kg/kg\text{-}mole \\ O: & 16.0\ kg/kg\text{-}mole \\ \hline CaO: & 56.1\ kg/kg\text{-}mole \end{array}$$

Our conversion factors are 64.1 kg CaC_2/kg-mole CaC_2 (from Example 6.20); 56.1 kg CaO/kg-mole CaO; and 1 kg/mole CaC_2/kg-mole CaO.

Step 5: Our setup:

$$9550\ \text{kg } \cancel{CaO} \times \frac{1\ \cancel{\text{kg-mole } CaO}}{56.1\ \cancel{\text{kg } CaO}} \times \frac{1\ \cancel{\text{kg-mole } CaC_2}}{\cancel{\text{kg-mole } CaO}} \times \frac{64.1\ \text{kg } CaC_2}{\cancel{\text{kg-mole } CaC_2}} = \_\_\_\ \text{kg } CaC_2$$

The units check. The answer, 1.09×10^4 kg CaC_2, is the theoretical yield. Since:

$$\text{percent yield} = \frac{\text{actual yield}}{\text{theoretical yield}} \times 10^2$$

then

$$\text{actual yield} = \frac{\text{percent yield} \times \text{theoretical yield}}{10^2}$$

$$\text{actual yield} = \frac{67.9 \times 1.09 \times 10^4 \text{ kg } CaC_2}{10^2}$$

$$= \underline{\quad} \text{ kg } CaC_2$$

Answer: 7.40×10^3 kg CaC_2. ■

REVIEW QUESTIONS

1. What is the chemist's recipe?

The Quantitative Meaning of Formulas

2. Why must we use grams to count atoms?
3. What is the arithmetic translation of an element's symbol?
4. What do we mean by *gram-molecular weight?* How do we calculate it?
5. How do we convert grams to moles? To molecules?
6. What is *gram-formula weight?* How does it compare with gram-molecular weight?
7. How do we calculate formula weight?

The Quantitative Meaning of Equations

8. Explain how a chemical equation can be translated into arithmetic.
9. How is a chemical equation like a nonchemical recipe? How is it different?
10. Why must the total mass on one side of an equation equal the total mass on the other side?

Calculations with Chemical Equations

11. How is a chemical equation a source of conversion factors?
12. What are the steps involved in doing calculations with chemical equations?
13. What conversion factors are used in mole-to-mole conversions?
14. What conversion factors are used in mole-to-mass conversions? How do we get these conversion factors?
15. What conversion factors are used in mass-to-mass conversions?
16. Can we use other mass units besides grams in mass-to-mass conversions? How?
17. How can we simplify a conversion factor when the two formulas have the same mole numbers in the equation?

Heat as Part of Chemical Reactions

18. Is heat always needed to make a chemical reaction go?
19. What are *exothermic* reactions and *endothermic* reactions?
20. What is *heat of reaction?* What units do we normally measure it in?
21. How do we treat heat of reaction in chemical calculations?
22. Can we change the heat of reaction for a given equation?
23. How is heat of reaction useful? How is it not useful?

Problems Involving Percentages

24. What does *percent* mean?
25. What is *percentage-by-weight?* What units can it have?
26. How is a chemical formula used to calculate percentage-by-weight of the elements in it?
27. How can we figure out the formula for a substance from the *percent composition?*
28. What is the difference between *empirical formula* and *molecular formula?* Are these always different?
29. What information do we need to find a compound's molecular formula? How do we use this information?
30. What is *percent purity?*
31. How is percentage a conversion factor?
32. When an impure substance is used as a reactant, how does this affect the amount of it that must be used to get a specific amount of product?
33. What are the steps involved in doing a chemical calculation where the starting material is impure?
34. What do we mean by an "excess" of a reactant? What is a *limiting reactant?*
35. What steps are involved in determining which is the limiting reactant, how much product will be obtained, and the amount of excess reactant left over?
36. What is *theoretical yield?* How do we get it?
37. Do we always obtain the theoretical yield? Why?
38. What is *percent yield?*
39. What are the steps involved in calculating the percent yield for a reaction where the *actual yield* is given?

EXERCISES

Set A (Answers at back of book. Asterisks mark the more difficult exercises.)

1. How many atoms are contained in each of the following?
   a. 1/3 mole of helium c. 6 moles of argon
   b. 3.21 grams of sulfur
2. How many molecules are in each of the following? (Assume that fractional and whole numbers of moles are pure numbers and do not limit the number of significant figures in the answer.)
   a. 1/2 mole of NO c. 1/4 mole O_3
   b. 2 moles SO_2
3. How many oxygen atoms are contained in each part of Exercise 2?
4. Calculate the following.
   a. the number of moles of table legs in one mole of tables
   b. the number of moles of legs, wings, and antennae in one-half mole of butterflies
   c. the number of moles of phosphorus atoms in one mole of P_4 molecules
   d. the number of moles of each kind of atom in one mole of H_3PO_4
   e. the number of total moles of ions in one-half mole of $Al_2(SO_4)_3$
5. Calculate the molecular weight or formula weight of each of the following.
   a. N_2 c. Na_2SO_4
   b. HNO_3
6. Calculate the number of moles in the following.
   a. 22.4 grams of neon b. 107 grams of NaBr
7. Calculate the number of grams in the following.
   a. 2.00 moles of I_2 c. 0.12 moles of Fe
   b. 1.57 moles of $HClO_3$
8. State the number of moles of each kind of atom for each of the following.
   a. 2 $Fe(OH)_3$ b. 3 Ca_3N_2
9. Give the six mole-to-mole conversion factors that can be found from each equation.
   a. $2\ C_2H_2 + 5\ O_2 \longrightarrow 4\ CO_2 + 2\ H_2O$
   b. $Fe_2O_3 + 3\ CO \longrightarrow 2\ Fe + 3\ CO_2$
10. How many moles of the first product in each of the following equations would be obtained by using 0.175 moles of the first reactant? (Equations are not necessarily balanced.)
   a. $CuO + HCl \longrightarrow CuCl_2 + H_2O$
   b. $Al + Cl_2 \longrightarrow AlCl_3$
11. In 1970, ammonia was second in the U.S. to sulfuric acid in amount produced, which was 1.18×10^{13} grams. Ammonia is produced by the Haber process:

$$N_2 + 3\ H_2 \longrightarrow 2\ NH_3$$

   a. How many moles of ammonia were produced in 1970?
   b. How many moles of hydrogen and of nitrogen would have to be used to produce this many moles of ammonia?
12. How many moles of the first reactant in each of the following equations would be needed to produce 785 grams of the first product? (Check that equation is balanced.)
   a. $Zn + Pb(NO_3)_2 \longrightarrow Pb + Zn(NO_3)_2$
   b. $AgNO_3 + H_2S \longrightarrow Ag_2S + HNO_3$
13. How many grams of the first reactant in each of the following equations would be needed to react with 2.34 grams of the second reactant?
   a. $S + Cl_2 \not\longrightarrow SCl_2$
   b. $Ba(NO_3)_2 + (NH_4)_2CO_3 \not\longrightarrow BaCO_3 + NH_4NO_3$
14. The decomposition of ammonium dichromate occurs as follows:

$$(NH_4)_2Cr_2O_7 \longrightarrow Cr_2O_3 + N_2 + 4\ H_2O$$

   The green chromium(III) oxide is the pigment used in paper money. How many grams of ammonium dichromate would a counterfeiter have to use to produce 50.0 grams of chromium(III) oxide?
15. Sugar ferments to produce alcohol according to this equation:

$$C_6H_{12}O_6 \longrightarrow 2\ C_2H_5OH + 2\ CO_2$$

   How many grams of alcohol and of carbon dioxide are produced from the fermentation of 5.00 kg of sugar?
16. Dihydrogen sulfide is a common problem in industrial waste. One source of dihydrogen sulfide is bacterial conversion of sulfates, which frequently appear in industrial wastes, according to this equation:

$$H_2SO_4 \xrightarrow{\text{bacteria}} H_2S + 2\ O_2$$

   However, the dihydrogen sulfide can be eliminated by treating it with hydrogen peroxide:

$$H_2O_2 + H_2S \longrightarrow 2\ H_2O + S$$

   a. How many metric tons of dihydrogen sulfide would be produced by 5.00 metric tons of sulfuric acid?
   b. How many metric tons of hydrogen peroxide would it take to get rid of the resulting dihydrogen sulfide?
17. Calculate the amount of heat, in kilocalories, that would be produced in each of the following exothermic reactions if 75.0 grams of the first reactant are used.

a. $Zn + S \longrightarrow ZnS + 148.5$ kcal
b. $2\ C_2H_6 + 7\ O_2 \longrightarrow 4\ CO_2 + 6\ H_2O + 736$ kcal

18. In each of the following endothermic reactions, calculate the number of grams of the first product that would be produced if 855 kcal were supplied.
a. $2\ KClO_3 + 21.4\ \text{kcal} \longrightarrow 2\ KCl + 3\ O_2$
b. $CaCO_3 + 42\ \text{kcal} \longrightarrow CaO + CO_2$

19. Natural gas (CH_4), propane (C_3H_8), and butane (C_4H_{10}) are all used as fuels in the home. They burn according to these equations:

$$CH_4 + 2\ O_2 \longrightarrow CO_2 + 2\ H_2O + 192\ \text{kcal}$$
$$C_3H_8 + 5\ O_2 \longrightarrow 3\ CO_2 + 4\ H_2O + 489\ \text{kcal}$$
$$2\ C_4H_{10} + 13\ O_2 \longrightarrow 8\ CO_2 + 10\ H_2O + 1270\ \text{kcal}$$

Calculate the number of kilocalories per gram of each fuel and decide, on that basis, which is the best fuel.

20. Determine the percent composition of HgI_2 and $Ca_3(PO_4)_2$ to two decimal places.

21. Which has the greater percentage of water, $SnCl_2 \cdot 2\ H_2O$ or $NaC_2H_3O_2 \cdot 3\ H_2O$? What is the value for each?

22. Determine the formulas from these percent compositions.
a. 41.82% Na, 58.18% O
b. 1.48% H, 51.78% Cl, 46.74% O

23. Compute the percent phosphate (PO_4^{3-}) in ammonium phosphate and in calcium phosphate. On that basis, which is the more efficient source of phosphate?

24. A gaseous fuel is 85.59% C and 14.41% H. Its molecular weight is 56.08. Give its empirical formula and its molecular formula.

25. "Fool's gold" is pyrite ore, and it contains beautiful gold crystals of FeS_2. This is also a source of iron metal, and so it is somewhat valuable even though it isn't gold. If a certain pyrite ore is 32.8% FeS_2, how much of this ore must be used to obtain 255 kg of pure iron?

*26. A sample of cinnabar, an ore containing mercury in the form of HgO, was heated to decompose the HgO according to this equation:

$$2\ HgO \longrightarrow 2\ Hg + O_2$$

A 5.120-gram sample of the ore weighed 4.970 grams after heating. Assuming the weight loss to be all from the oxygen driven off, calculate: (a) the number of grams of HgO in the sample; and (b) the percent HgO in the ore.

*27. Limestone and marble are mostly made of calcium carbonate. When the atmospheric pollutant sulfur trioxide combines with rainwater, the resulting sulfuric acid can damage buildings and statues made out of limestone or marble. This is because sulfuric acid dissolves calcium carbonate. The equation for the reaction of sulfur trioxide and rainwater to produce sulfuric acid is:

$$SO_3 + H_2O \longrightarrow H_2SO_4$$

The equation for sulfuric acid's dissolving of calcium carbonate is:

$$CaCO_3 + H_2SO_4 \longrightarrow CaSO_4 + H_2O + CO_2$$

If, over a period of time, rain containing 1.00 kg of sulfur trioxide falls on a statue weighing 545 kg, what percentage of the statue will be dissolved (assuming the statue is 100% $CaCO_3$)?

28. Determine the limiting reactant in these reactions if 25.0 grams of each reactant are used. Calculate the amount of underlined product that would be formed, and the amount of excess reactant that would be left over.
a. $H_2SO_4 + BaCl_2 \longrightarrow \underline{BaSO_4} + 2\ HCl$
b. $4\ NH_3 + 3\ O_2 \longrightarrow 2\ \underline{N_2} + 6\ H_2O$

*29. One way to get a free metal from its oxide is to heat it with coke (carbon). How much zinc can be obtained by heating 21.3 grams of carbon with 123 grams of ZnO? How much of which reactant will be left over?

$$ZnO + C \longrightarrow Zn + CO$$

30. Magnesium is extracted from seawater by precipitating it as $Mg(OH)_2$ according to this equation:

$$MgCl_2 + Ca(OH)_2 \longrightarrow Mg(OH)_2 + CaCl_2$$

500. kg of $Ca(OH)_2$ was added to an excess of seawater, and 245 kg of $Mg(OH)_2$ was obtained. Calculate the percent yield.

31. "Superphosphate" fertilizer is made by treating phosphate rock, $Ca_3(PO_4)_2$, with sulfuric acid according to this equation:

$$Ca_3(PO_4)_2 + 2\ H_2SO_4 \longrightarrow Ca(H_2PO_4)_2 + 2\ CaSO_4$$

If this reaction has a 52.3% yield, how much $Ca(H_2PO_4)_2$ could be obtained from 5.2 metric tons of phosphate rock?

Set B (Answers not given. Asterisks mark more difficult exercises.)

1. How many atoms are contained in each of the following?
a. 20.0 gramsof argon
b. 2 moles of O_2
c. 0.100 moles of sodium

2. How many molecules are in each of these? (Assume that fractional and whole numbers of moles are pure numbers and do not limit the number of significant figures in the answer.)
a. 1 mole N_2O_5
b. 1/10 mole H_2O
c. 1/3 mole of P_4O_{10}

3. How many oxygen atoms are in each part of Exercise 2?

4. Calculate these.
   a. the number of moles of wings in one mole of chickens
   b. the number of moles of doors, wheels, and headlights in 1/4 mole of two-door sedans
   c. the number of moles of O atoms in 2.00 moles of ozone (O_3) molecules
   d. the number of moles of each kind of atom in 0.500 moles of $HC_2H_3O_2$
   e. the number of moles of ions in 2.00 moles of Na_3PO_4
5. Calculate the molecular weight or formula weight of each of the following.
   a. $C_6H_{12}O_6$ c. $(NH_4)_3PO_4$
   b. $CaSO_4 \cdot 2\,H_2O$
6. Calculate the number of moles in these.
   a. 14.9 grams of Cl_2
   b. 46.2 grams of Li_2SO_4
7. Calculate the number of grams in these.
   a. 10.0 moles of NO_2 c. 0.500 moles of C_3H_8
   b. 5.93 moles of Cr_2O_3
8. State the number of moles of each kind of atom for each of these.
   a. 4 N_2O_5 b. 3 NH_4KCO_3
9. Give the six mole-to-mole conversion factors that can be found from each equation.
   a. $3\,Cl_2 + CH_4 \longrightarrow CHCl_3 + 3\,HCl$
   b. $2\,Na + 2\,H_2O \longrightarrow 2\,NaOH + H_2$
10. How many moles of the first product in each of the following equations would be obtained by using 0.175 moles of the first reactant? (Equations are not necessarily balanced.)
   a. $NH_3 + H_2SO_4 \longrightarrow (NH_4)_2SO_4$
   b. $NaNO_3 \longrightarrow NaNO_2 + O_2$
11. Butane, C_4H_{10}, is one kind of bottled gas. How many moles of oxygen are used when 1050 grams of butane are burned to yield carbon dioxide and water?
12. How many moles of the first reactant in each of the following equations would be needed to produce 785 grams of the first product? (Check that equation is balanced.)
   a. $P + O_2 \longrightarrow P_4O_{10}$ b. $PbO_2 \longrightarrow Pb + O_2$
13. How many grams of the first reactant in each of the following equations would be needed to react with 2.34 grams of the second reactant?
   a. $Cr(OH)_3 + HCl \not\longrightarrow CrCl_3 + H_2O$
   b. $Zn + AgNO_3 \not\longrightarrow Zn(NO_3)_2 + Ag$
14. An art student runs out of yellow pigment on a Sunday. A chemistry major friend offers to make some yellow pigment in the form of lead(II) chromate ("chrome yellow"):

$$K_2CrO_4 + Pb(NO_3)_2 \longrightarrow PbCrO_4 + 2\,KNO_3$$

   How many grams of potassium chromate and of lead(II) nitrate does the chemistry major need to produce 3.50 grams of lead(II) chromate?
15. Part of a water purification technique uses this reaction (equation is not balanced):

$$Al_2(SO_4)_3(aq) + Ca(OH)_2(aq) \not\longrightarrow Al(OH)_3(s) + CaSO_4(aq)$$

   The aluminum hydroxide is a fluffy, white precipitate that can trap tiny particles suspended in water. When the aluminum hydroxide is removed from the water, the particles are removed too.
   a. How many kilograms of aluminum sulfate are needed to produce 255 kg of aluminum hydroxide?
   b. How many kilograms of the by-product, calcium sulfate, will also be produced?
16. Carbon dioxide can be removed from air by circulating the air over lithium hydroxide, which reacts with the carbon dioxide according to this equation:

$$2\,LiOH + CO_2 \longrightarrow Li_2CO_3$$

   a. How much LiOH should be packed aboard a spacecraft to take care of two astronauts on a two-day mission? Assume that each astronaut exhales about 1 kg of CO_2 per day.
   b. The compound NaOH reacts with CO_2 in the same way as LiOH does. Why might LiOH be a better choice than NaOH for a space voyage?
17. Calculate the number of grams of the first reactant that would have to be used to release 100. kcal of heat in these exothermic reactions.
   a. $2\,H_2 + O_2 \longrightarrow 2\,H_2O + 136.3$ kcal
   b. $CaO + H_2O \longrightarrow Ca(OH)_2 + 15.6$ kcal
18. In these endothermic reactions, calculate the number of kcal of heat that would have to be supplied to get 50.0 grams of the first product.
   a. $2\,HgO + 43.4$ kcal $\longrightarrow 2\,Hg + O_2$
   b. $2\,MnO_2 + 65.0$ kcal $\longrightarrow 2\,MnO + O_2$

*19. Carbon monoxide, a by-product of some industrial processes, can be used as a fuel according to this reaction:

$$2\,CO + O_2 \longrightarrow 2\,CO_2 + 135 \text{ kcal}$$

   An industry wants to use its carbon monoxide to heat a lime kiln, where the reaction is:

$$CaCO_3 + 42.0 \text{ kcal} \longrightarrow CaO + CO_2$$

   How many tons of carbon monoxide will be needed for every ton of lime (CaO) produced?

20. Determine the percent composition of $FeCl_3$ and $(NH_4)_2SO_4$ to two decimal places.
21. Two bottles containing different hydrates have accidentally lost their labels, which read "$CaCl_2 \cdot 2\,H_2O$" and "$MgSO_4 \cdot 7\,H_2O$." The contents of one of the bottles was found to be 24.49% water. Which hydrate is it? Prove your answer.
22. Determine the formulas from these percent compositions.
   a. 60.05% K, 18.43% C, 21.52% N
   b. 46.31% S, 53.69% Fe

23. Ammonium nitrate is used as a fertilizer; ammonia is, too. Compute the percent nitrogen in each. If both cost the same amount per kilogram, which is the better buy in nitrogen content?
24. A poisonous gas called cyanogen is 46.14% C and 53.86% N. Its molecular weight is 52.02. What are its empirical formula and its molecular formula?
25. Magnetite ore is 27.0% Fe_3O_4, and it is a source of iron metal.
    a. How much pure Fe_3O_4 is contained in 1.00 metric ton of magnetite?
    b. How many metric tons of iron are contained in 1.00 metric ton of magnetite?
26. An older method of manufacturing chlorine gas was to treat pyrolusite ore with hydrochloric acid. The active ingredient in pyrolusite ore is MnO_2, and this is the reaction:

$$MnO_2 + 4\,HCl \longrightarrow Cl_2 + MnCl_2 + 2\,H_2O$$

If the pyrolusite ore is 55.7% MnO_2, how much chlorine can be obtained from 5.54 metric tons of pyrolusite ore?

*27. In a laboratory experiment, a student heated a sample of hydrated $CoCl_2$ to drive off all the water of hydration. Before heating, the sample weighed 10.21 grams; after heating, it weighed 4.759 grams. Calculate the number of waters of hydration (that is, complete this formula: $CoCl_2 \cdot (?)\ H_2O$).

28. Determine the limiting reactant in each of these if 100. grams of each reactant are used. Calculate the amount of underlined product that would be formed, and the amount of excess reactant that would be left over.
    a. $3\,CaCO_3 + 2\,H_3PO_4 \longrightarrow \underline{Ca_3(PO_4)_2} + 3\,CO_2 + 3\,H_2O$
    b. $Zn + 2\,HCl \longrightarrow ZnCl_2 + \underline{H_2}$

*29. Plaster of Paris is a hydrate, $(CaSO_4)_2 \cdot H_2O$. When water is added to it, it takes on more waters of hydration to form gypsum (a "plaster cast") according to this equation:

$$(CaSO_4)_2 \cdot H_2O + 3\,H_2O \longrightarrow 2\,CaSO_4 \cdot 2\,H_2O$$

If one liter of water is mixed with 1 kg of plaster of Paris, will there be too much water, too much plaster of Paris, or just the right amount of each? Prove your answer.

30. Chlorine gas is made commercially by electrolyzing salt from seawater according to this equation:

$$2\,NaCl \longrightarrow 2\,Na + Cl_2$$

If 500. kg of chlorine is obtained from 910. kg of NaCl, calculate the percent yield.

31. The Ostwald process for making nitric acid begins with converting ammonia to nitrogen monoxide:

$$4\,NH_3 + 5\,O_2 \xrightarrow{\Delta} 4\,NO + 6\,H_2O$$

This reaction occurs with an 80.3 percent yield. What is the actual yield of nitrogen monoxide from 10.7 metric tons of ammonia and an excess of oxygen?

CHAPTER 7

Atomic Structure

LEARNING OBJECTIVES

After studying this chapter, you should be able to:

1. Supply a correct definition, explanation, or example for each of these:

quantum mechanical atom	shell
energy	orbital
potential energy	sublevel
quantized	spin
principal energy level	electronic configuration
	noble gas core
	Aufbau Diagram

2. Explain how charged particles have potential energy with respect to each other.
3. Describe each orbital type and tell how many orbitals are in each sublevel and each principal energy level.
4. State the maximum number of electrons in a given orbital, sublevel, or principal energy level
5. Write full electronic configurations of any element using Table 7.2.
6. Write electronic configurations of an element given its atomic number, using the Aufbau Diagram.

So far, we've seen that atoms contain neutrons and protons in a nucleus, and that the nucleus is surrounded by electrons. In this chapter, we'll look more closely at the modern model of the atom, and especially at how an atom's electrons are arranged outside the nucleus. The way the electrons are arranged is important in our study of chemistry, because it's this arrangement that determines an element's behavior.

First, we'll see briefly how the modern model of the atom evolved from early discoveries.

7.1 DEVELOPMENT OF TODAY'S ATOMIC THEORY

Around the turn of the century, scientists knew that atoms are made of negatively charged particles (electrons) and positively charged particles (protons). (We know now that neutrons are also part of atoms, but they don't contribute any charge and therefore were hard to discover.) Over the years, models of the atom were developed as scientists tried to explain how these particles are put together. Each model was subjected to experiment, according to the scientific method; some parts of each were discarded, others kept and improved. (See Box on pp. 140–141.)

The result is the *quantum mechanical atom*. This model hasn't been disproved by experiment yet, but that doesn't mean that it might not be some day. Until it is disproved, the quantum mechanical atom works very well to explain the results of experiments on atoms, as well as everything we'll discuss in this book.

quantum mechanical atom

7.2 THE QUANTUM MECHANICAL ATOM

This present-day model of the atom describes the arrangement of an atom's electrons in terms of their energies. First, then, we need to know a bit more about energy.

ENERGY. In Chapter 3, we said that energy is what's needed to move matter. Moving matter requires work, so we can define *energy* as the ability to do work. This doesn't mean that energy *has* to do work; only that it *can*. We usually talk about energy according to whether it is available now or later.

energy

Stored energy, energy that's being saved for later, is *potential energy*. Water above a dam has potential energy. It has the potential to do work at a later time. We know that if we open the spillway and let the water go through, we can get the water to do work for us. When we do open the spillway, the water will flow to the bottom spontaneously (all by itself).

potential energy

A BRIEF HISTORY OF THE ATOM

Time	*Persons Involved*	*Events*
500 B.C.	**Leucippus** **Democritus**	Proposed that matter is made of tiny particles (atoms) that are fundamental to all substances and that cannot be divided.
400 B.C.	**Aristotle**	Refuted the idea of atoms. Proposed instead that matter is made of four "elements": earth, air, fire, and water. His influence and reputation caused the idea of atoms to be submerged for many centuries.
400 B.C.– 1661 A.D.		The "dark ages" of atomic theory. Alchemy and superstition prevailed. The alchemists recognized three "elements": sulfur, mercury, and salt. Although many discoveries were made that eventually contributed to chemical knowledge, the idea of atoms as fundamental particles remained in obscurity.
1661	**Robert Boyle**	Expressed skepticism about both Greek and alchemical "elements," and suggested that there might be much more fundamental particles.
1804	**John Dalton**	Formalized the atomic theory of Democritus, and improved upon it by experimentation. Reawakened the concept of atoms as fundamental, small, indivisible particles that compose all matter.
1874	**William Crookes**	Discovered negatively-charged particles, but didn't know they were electrons.
1897	**J. J. Thomson**	Characterized the electron. Realized that electrons are part of atoms, and that atoms also have a positive part that is left when the electrons are stripped away. Showed, therefore, that atoms are not indivisible. Proposed his "plum pudding" model of the atom.
1898	**William Wien**	Discovered the proton and established it as part of an atom.
1900	**Max Planck**	Introduced his quantum theory, which states that some kinds of energy are given off not continuously but in discrete lumps (quanta).
1908	**Albert Einstein**	Extended Planck's quantum theory to describe electrons in atoms. Proposed that electrons in atoms occupy quantized "energy levels."

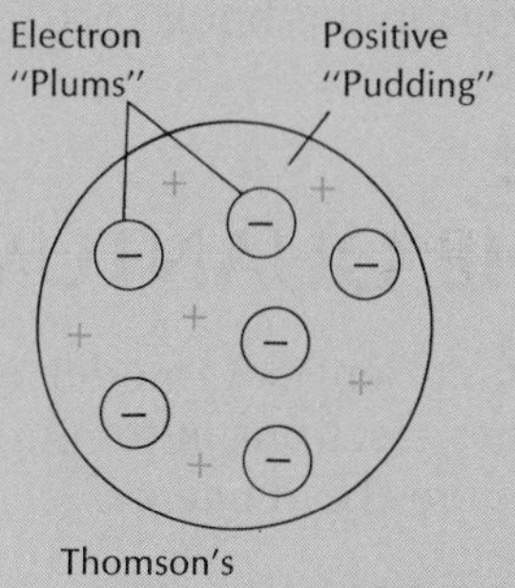

Thomson's "Plum Pudding" model

1911	**Ernest Rutherford**	Disproved Thomson's "plum pudding" model. Proposed, instead, that most of an atom's mass and all of its positive charge are concentrated in a very small center: the nucleus. The electrons were seen to occupy a large volume of mostly empty space, but it wasn't known how the electrons were arranged in that space. 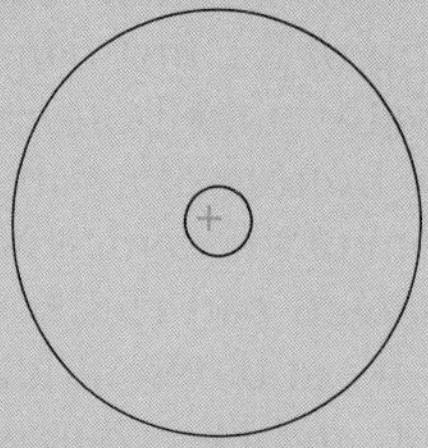 Rutherford's nuclear atom
1913	**Niels Bohr**	Proposed a planetary model of the atom. The nucleus was in the center, and the electrons occupied orbits. The orbits had certain energies and were specific distances from the nucleus. The electrons were supposed to be held in place by electrostatic attraction, which in turn was balanced by centrifugal force. The orbits were calculated on the basis of the electrons' momentum as they orbited the nucleus. Unfortunately, the Bohr atom did not adequately explain atoms more complex than hydrogen. 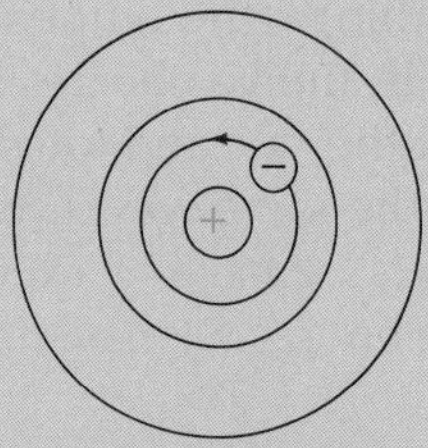 Bohr's model of the hydrogen atom
1924	**Louis de Broglie**	Discovered that moving electrons behave like waves. Therefore, he proposed that electrons in motion around nuclei must be treated as waves.
1925	**Werner Heisenberg**	Developed his uncertainty principle, which says in essence that it is not possible to know exactly where rapidly moving particles like electrons are at any time. One can only know their probable positions.
1926	**Erwin Schrödinger**	Combined de Broglie's and Heisenberg's findings into the development of the Schrödinger equation. This equation treated electrons in atoms like waves. The solutions to his equation give regions around atoms that are likely to have electrons in them. These regions—called orbitals—are quantized. 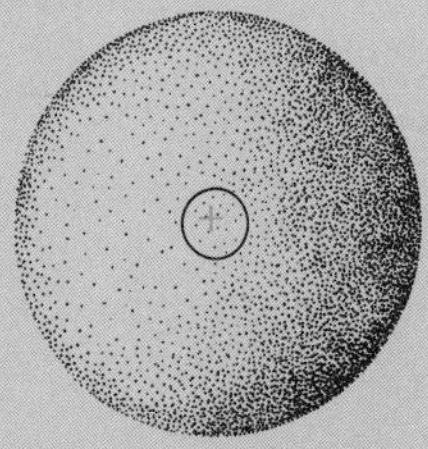 Model of the hydrogen atom resulting from the Schrödinger equation
1932	**James Chadwick**	Established existence of the neutron. Although neutrons had been assumed to exist, the proof was not supplied until this late date in atomic history.

Figure 7.1 illustrates relative potential energy states. The water at the bottom of the first dam is in a state of lower potential energy than the water at the top of the first dam, but it is in a state of higher potential energy than the water at the bottom of the second dam. When water goes from the top to the bottom of a dam, it releases its potential energy. When it gets to the bottom of the dam, it has lower energy than it had at the top.

Charged particles have potential energy in relation to each other, just as water at different levels does. We saw in Chapter 2 that unlike charges attract, and like charges repel. Now we can say it in another way: unlike charges that are apart have higher potential energy than unlike charges that are together. (And like charges that are together have higher potential energy than like charges that are apart.) How much energy is involved depends on two things: the amount of the charges, and the distance between them. That is, unlike charges have higher potential energy the farther from each other they are and lower potential energy the closer together they are.

In the same way, if we want to separate two unlike charges that are together, we have to expend energy. And the farther we separate them, the more energy we have to expend. Figure 7.2 shows how two unlike charges

FIGURE 7.1
Relative potential energy states

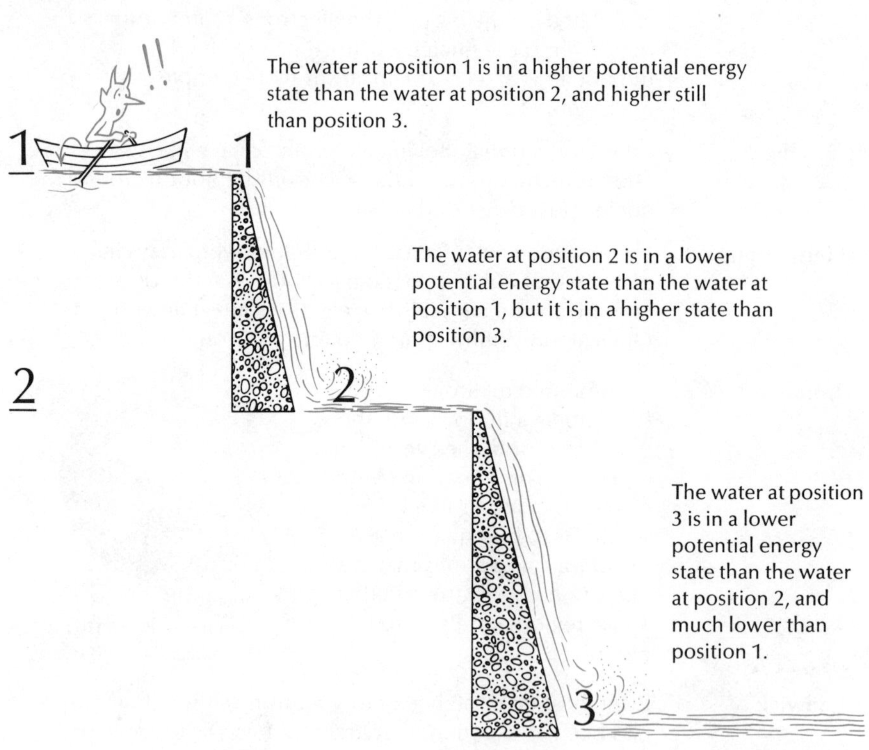
!!
1
1
The water at position 1 is in a higher potential energy state than the water at position 2, and higher still than position 3.
The water at position 2 is in a lower potential energy state than the water at position 1, but it is in a higher state than position 3.
2
2
The water at position 3 is in a lower potential energy state than the water at position 2, and much lower than position 1.
3

can have more or less energy with respect to each other because of the distance between them.

The quantum mechanical atom explains, mathematically, how electrons are arranged about a positive nucleus so that their repulsions for each other are just balanced by their attraction for the nucleus. We won't discuss the mathematical part, but we can look at the results. According to this model, electrons are arranged in quantized energy levels.

FIGURE 7.2
Like-charged particles have higher energy and unlike charged particles have lower energy the closer they are together

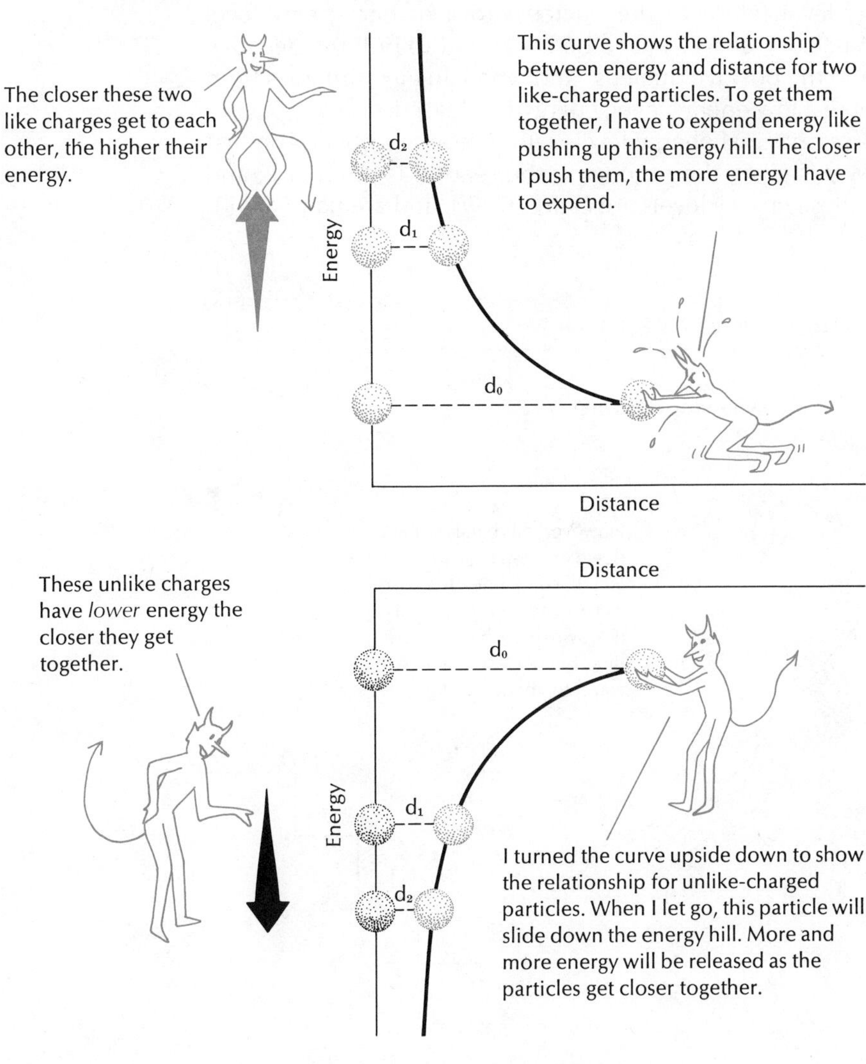

The closer these two like charges get to each other, the higher their energy.
This curve shows the relationship between energy and distance for two like-charged particles. To get them together, I have to expend energy like pushing up this energy hill. The closer I push them, the more energy I have to expend.
Energy
Distance
d_2
d_1
d_0
These unlike charges have *lower* energy the closer they get together.
I turned the curve upside down to show the relationship for unlike-charged particles. When I let go, this particle will slide down the energy hill. More and more energy will be released as the particles get closer together.
Distance
Energy
d_0
d_1
d_2

QUANTIZED ENERGY LEVELS. Something that exists as finite parts is said to be *quantized*. The opposite of quantized is continuous. A ramp is continuous, whereas a staircase is quantized, because it has steps that are a specific distance apart (see Figure 7.3).

quantized

Electrons are arranged in *principal energy levels*. These energy levels correspond to different distances from the nucleus. The closer two unlike-charged particles can get to each other, the lower the energy they have, as we saw in Figure 7.2. Therefore, an electron in an energy level that's closest to the nucleus will have the lowest energy. And electrons in levels farther and farther out have more and more energy.

principal energy level

Quantum mechanics explains that electrons can go from one energy level to another, but they can't stop anywhere between. When an electron goes from a low energy level (close to the nucleus) to a higher energy level (farther from the nucleus), energy has to be expended to pull the negative electron farther from the positive nucleus. And when an electron goes from a high energy level to a low energy level, energy is released.

We describe these principal energy levels with whole numbers. The first (lowest) energy level is given the number 1; the second is number 2; and so on. Sometimes these energy levels are called *shells*, and are named with

shell

FIGURE 7.3
Continuous and quantized energy

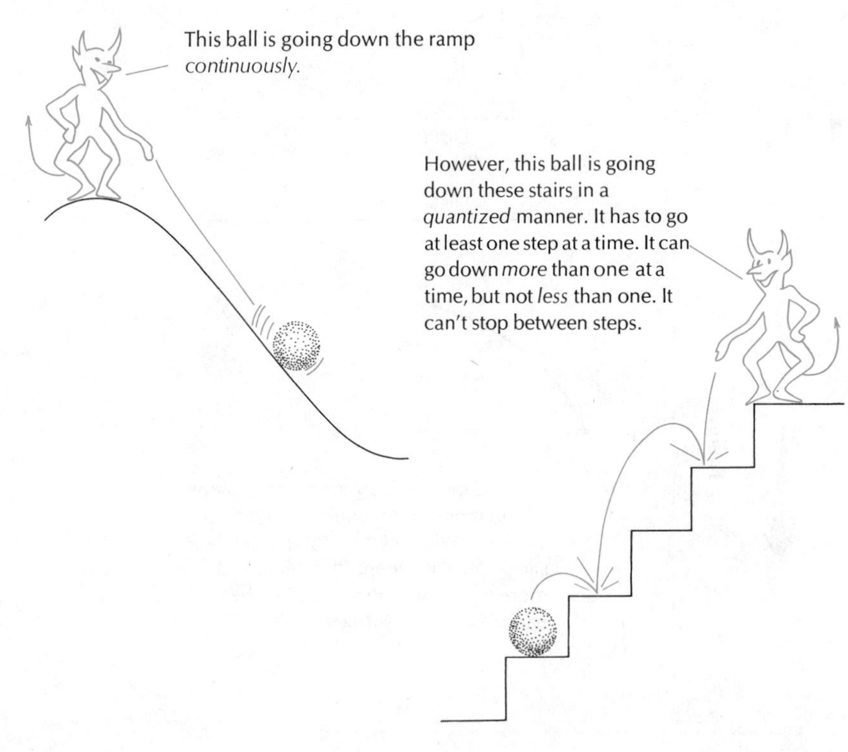

This ball is going down the ramp *continuously*.
However, this ball is going down these stairs in a *quantized* manner. It has to go at least one step at a time. It can go down *more* than one at a time, but not *less* than one. It can't stop between steps.

capital letters starting with K ($K = 1, L = 2, M = 3$, etc.). Like the layers of flavors inside a jawbreaker, energy levels are nested one inside another, as shown in Figure 7.4.

ORBITALS. The uncertainty principle (see page 141) says that we can never know a particle's position *exactly*. We can say that an electron is in the first energy level of an atom, but if we try to pin it down more specifically within that level, we have to begin talking about where we would be *most likely* to find it. Let's look at a simple analogy.

In Figure 7.5 we see that we can read the label on an automobile tire when the tire is standing still. When the tire is moving, though, the writing becomes a fuzzy band. If we want to describe exactly where a letter—say, a "G"—is at any instant on the moving tire, we have a hard time doing it. But we can say that it is more likely to be within the fuzzy band than anywhere else.

Electrons in atoms never stand still. We have just as hard a time saying exactly where any electron is at any instant as we have trying to pinpoint the "G" on the tire. Like the white band on the tire, though, an *orbital* is a region in which we are most likely to find an electron. Orbitals are fuzzy, orbital

FIGURE 7.4
Electrons in atoms are found in quantized energy levels

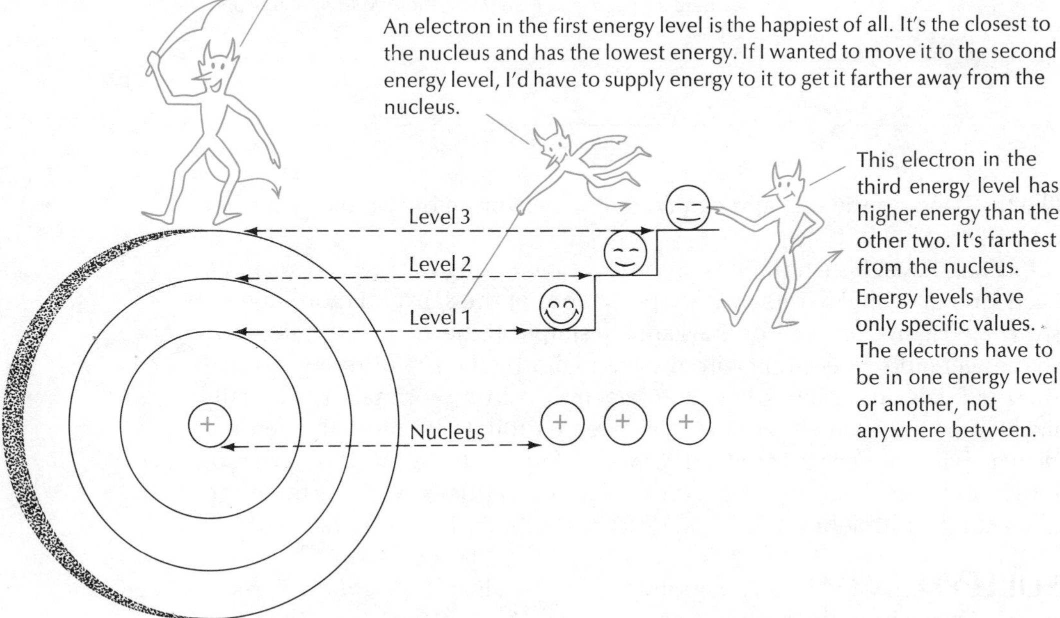

I've sliced an atom in half again, so that we can see the shells, or energy levels. There are really a lot more than this, as we'll see later. But this will give us an idea of how they work. Over on the other side, I've lined them up like stairs.
An electron in the first energy level is the happiest of all. It's the closest to the nucleus and has the lowest energy. If I wanted to move it to the second energy level, I'd have to supply energy to it to get it farther away from the nucleus.
This electron in the third energy level has higher energy than the other two. It's farthest from the nucleus.
Energy levels have only specific values. The electrons have to be in one energy level or another, not anywhere between.
Level 3
Level 2
Level 1
Nucleus

FIGURE 7.5
An orbital is a piece of space most likely to contain an electron

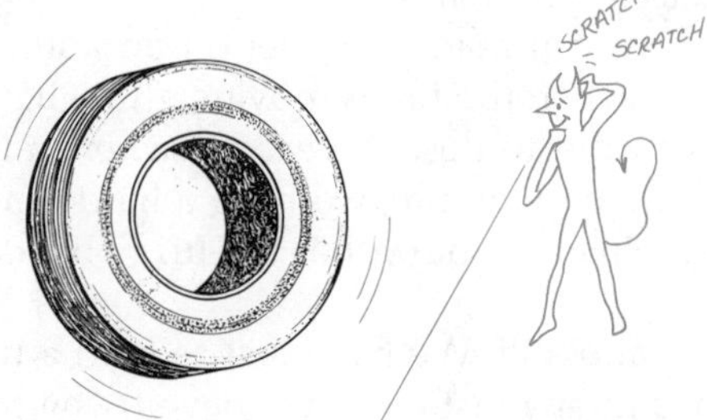

like the band on the tire, and can contain a maximum of two electrons per orbital.

Orbitals have these notations, in order of increasing energy: *s, p, d,* and *f.* Each kind of orbital has its own shape. Shapes of the *s, p,* and *d* orbitals are shown in Figure 7.6. The *s* orbitals are the simplest: they're just spheres. The shapes get more interesting with the *p* and *d* orbitals. The different parts of these orbitals are called *lobes.* If an electron is in a *p* orbital, it is equally likely to be found in either lobe. The three *p* orbitals are mutually perpendicular, which means that they're 90° apart. Most of the *d* orbitals have four lobes, and one electron in a *d* orbital would be equally likely to be found in any of the four lobes. We won't describe *f* orbitals in this book. lobes

SUBLEVELS. Orbitals are grouped into sublevels. A *sublevel* is an energy level within the principal energy level. The sublevels have the same sublevel

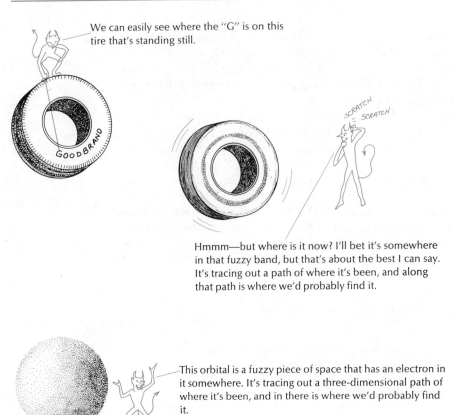

FIGURE 7.6
Shapes of s, p, and d orbitals

Orbitals are much more interesting than the band on the tire. This *s* orbital in my right hand is like a ball, and the *p* orbital in my left is like a dumbbell.

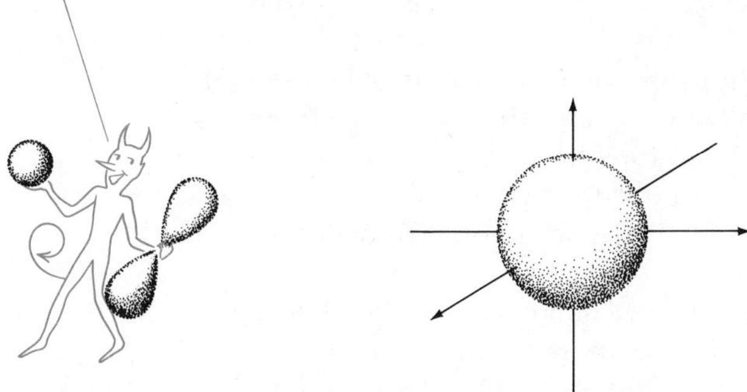

Each *p* orbital has two parts. We call these *lobes*. The three *p* orbitals are arranged on different axes, so they're all perpendicular to each other.

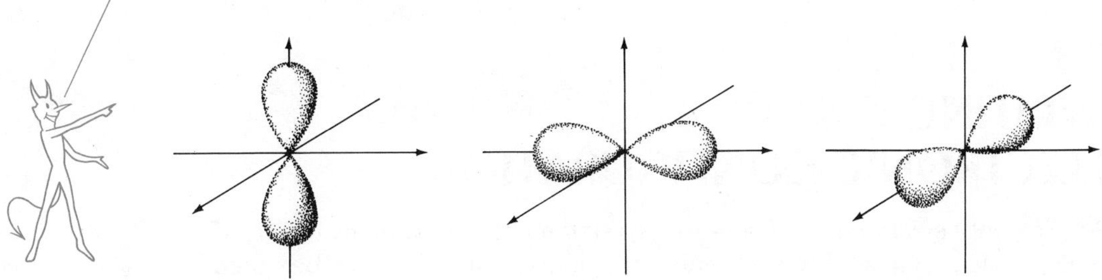

These *d* orbitals are even more interesting! Four of them have four lobes each and they look like cloverleaves. The last one looks like a *p* orbital with a collar.

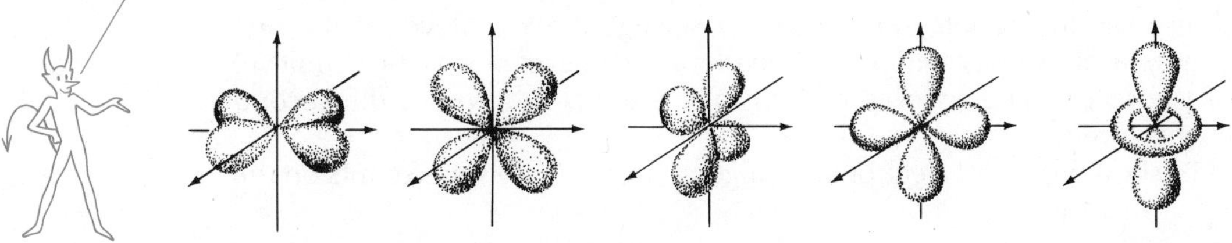

letter-names as the orbitals they contain. Each sublevel contains these numbers of orbitals:

s sublevel: one *s* orbital
p sublevel: three *p* orbitals
d sublevel: five *d* orbitals
f sublevel: seven *f* orbitals

Each principal energy level has the same number of sublevels as its own number. Thus the first energy level (*K* shell) has one sublevel: the *s*. The second (*L* shell) has two sublevels: *s* and *p*. The third (*M* shell) has three sublevels: *s, p,* and *d*. And the fourth and subsequent sublevels have all four: *s, p, d,* and *f*.

There can be as many as two electrons in an orbital. Electrons spin, and this *spin* lets them occupy orbitals in pairs. Two electrons that have opposite spins have a magnetic attraction for each other that overcomes the repulsion they would otherwise have for each other. Two electrons may thus occupy the same orbital provided they have opposite spins, and are said to be "paired." Given the choice of occupying an orbital singly or with another electron, though, an electron will usually choose to be alone in an orbital. spin

We know that two electrons can be in a single orbital. We know how many orbitals are in each sublevel, and we know how many and what kind of sublevels are at each energy level. Therefore, we can calculate how many electrons can be in each principal energy level. Table 7.1 shows these figures.

7.3 WRITING ELECTRONIC CONFIGURATIONS

An element's *electronic configuration* is a statement of how many electrons it has in each sublevel. The electronic configuration of carbon can be stated like this: "Carbon has two electrons in the *s* sublevel of the first energy level, two electrons in the *s* sublevel of the second energy level, and two electrons in the *p* sublevel of the second energy level." This is an obvious place to use shorthand notation. Instead of saying "the *s* sublevel of the first energy level," we say simply 1*s*. Similarly, a 2*s* sublevel means the *s* sublevel of the second energy level, and 2*p* means the *p* sublevel of the second energy level. electronic configuration

To show an atom's electronic configuration, we write superscripts on the

TABLE 7.1
Distribution of electrons in sublevels

Principal Energy Level, n	*Maximum Possible No. of Electrons*	*Maximum Possible No. of Electrons in Sublevel*			
		s	*p*	*d*	*f*
1 (*K* shell)	2	2			
2 (*L* shell)	8	2	6		
3 (*M* shell)	18	2	6	10	
4 (*N* shell)	32	2	6	10	14

number-letter designations, giving the number of electrons in each sublevel. Carbon's configuration is written like this in shorthand notation:

$1s^22s^22p^2$ (Read "one-s-two, two-s-two, two-p-two")

Each individual designation has this meaning:

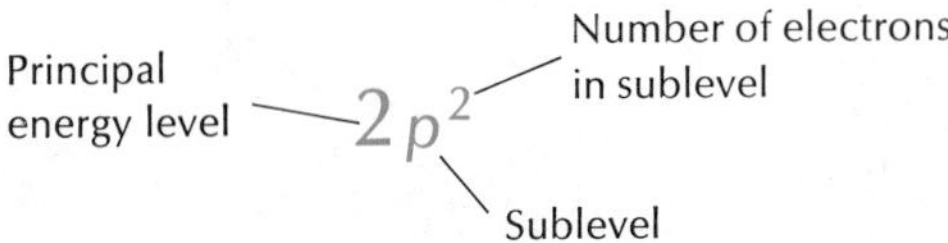

Table 7.2 shows electronic configurations for all the elements. We notice a few things about electronic configurations. First, the sum of the superscripts is the element's atomic number. (Carbon's subscripts add up to six, its atomic number.) This is no surprise, because an element's atomic number is the total number of its electrons. Second, as we proceed from H through Ne, each element has the same configuration as the one just before it, with the addition of the last electron. We're building up configurations for the elements, one electron at a time, in order of increasing atomic number. Third, writing configurations becomes redundant. The configuration of fluorine is the same as that of oxygen, except that fluorine has one more p electron than oxygen ($2p^5$ instead of $2p^4$). Starting after neon, to avoid writing all the electrons each time, we use *noble gas cores*. The core configuration of sodium is (Ne-10)$3s^1$. Sodium then has the neon configuration ($1s^22s^22p^6$) plus an additional electron at the 3s level. Potassium has the argon configuration ($1s^22s^22p^63s^23p^6$) plus one 4s electron. And so on. noble gas core

TABLE 7.2
Electronic configurations of the elements

Element	*Atomic Number*	*Electronic Configuration*	*Element*	*Atomic Number*	*Electronic Configuration*
H	1	$1s^1$	K	19	(Ar-18)$4s^1$
He	2	$1s^2$	Ca	20	(Ar-18)$4s^2$
Li	3	$1s^22s^1$	Sc	21	(Ar-18)$4s^23d^1$
Be	4	$1s^22s^2$	Ti	22	(Ar-18)$4s^23d^2$
B	5	$1s^22s^22p^1$	V	23	(Ar-18)$4s^23d^3$
C	6	$1s^22s^22p^2$	Cr	24	(Ar-18)$4s^13d^5$
N	7	$1s^22s^22p^3$	Mn	25	(Ar-18)$4s^23d^5$
O	8	$1s^22s^22p^4$	Fe	26	(Ar-18)$4s^23d^6$
F	9	$1s^22s^22p^5$	Co	27	(Ar-18)$4s^23d^7$
Ne	10	$1s^22s^22p^6$	Ni	28	(Ar-18)$4s^23d^8$
Na	11	(Ne-10)$3s^1$	Cu	29	(Ar-18)$4s^13d^{10}$
Mg	12	(Ne-10)$3s^2$	Zn	30	(Ar-18)$4s^23d^{10}$
Al	13	(Ne-10)$3s^23p^1$	Ga	31	(Ar-18)$4s^23d^{10}4p^1$
Si	14	(Ne-10)$3s^23p^2$	Ge	32	(Ar-18)$4s^23d^{10}4p^2$
P	15	(Ne-10)$3s^23p^3$	As	33	(Ar-18)$4s^23d^{10}4p^3$
S	16	(Ne-10)$3s^23p^4$	Se	34	(Ar-18)$4s^23d^{10}4p^4$
Cl	17	(Ne-10)$3s^23p^5$	Br	35	(Ar-18)$4s^23d^{10}4p^5$
Ar	18	(Ne-10)$3s^23p^6$	Kr	36	(Ar-18)$4s^23d^{10}4p^6$

(*continued*)

Table 7.2 (continued)

Element	*Atomic Number*	*Electronic Configuration*	*Element*	*Atomic Number*	*Electronic Configuration*
Rb	37	(Kr-36)$5s^1$	Hf	72	(Xe-56)$6s^25d^24f^{14}$
Sr	38	(Kr-36)$5s^2$	Ta	73	(Xe-56)$6s^25d^34f^{14}$
Y	39	(Kr-36)$5s^24d^1$	W	74	(Xe-56)$6s^25d^44f^{14}$
Zr	40	(Kr-36)$5s^24d^2$	Re	75	(Xe-56)$6s^25d^54f^{14}$
Nb	41	(Kr-36)$5s^14d^4$	Os	76	(Xe-56)$6s^25d^64f^{14}$
Mo	42	(Kr-36)$5s^14d^5$	Ir	77	(Xe-56)$6s^25d^74f^{14}$
Tc	43	(Kr-36)$5s^14d^6$	Pt	78	(Xe-56)$6s^15d^94f^{14}$
Ru	44	(Kr-36)$5s^14d^7$	Au	79	(Xe-56)$6s^15d^{10}4f^{14}$
Rh	45	(Kr-36)$5s^14d^8$	Hg	80	(Xe-56)$6s^25d^{10}4f^{14}$
Pd	46	(Kr-36)$4d^{10}$	Tl	81	(Xe-56)$6s^25d^{10}4f^{14}6p^1$
Ag	47	(Kr-36)$5s^14d^{10}$	Pb	82	(Xe-56)$6s^25d^{10}4f^{14}6p^2$
Cd	48	(Kr-36)$5s^24d^{10}$	Bi	83	(Xe-56)$6s^25d^{10}4f^{14}6p^3$
In	49	(Kr-36)$5s^24d^{10}5p^1$	Po	84	(Xe-56)$6s^25d^{10}4f^{14}6p^4$
Sn	50	(Kr-36)$5s^24d^{10}5p^2$	At	85	(Xe-56)$6s^25d^{10}4f^{14}6p^5$
Sb	51	(Kr-36)$5s^24d^{10}5p^3$	Rn	86	(Xe-56)$6s^25d^{10}4f^{14}6p^6$
Te	52	(Kr-36)$5s^24d^{10}5p^4$	Fr	87	(Rn-86)$7s^1$
I	53	(Kr-36)$5s^24d^{10}5p^5$	Ra	88	(Rn-86)$7s^2$
Xe	54	(Kr-36)$5s^24d^{10}5p^6$	Ac	89	(Rn-86)$7s^26d^1$
Cs	55	(Xe-55)$6s^1$	Th	90	(Rn-86)$7s^26d^2$
Ba	56	(Xe-56)$6s^2$	Pa	91	(Rn-86)$7s^26d^15f^2$
La	57	(Xe-56)$6s^25d^1$	U	92	(Rn-86)$7s^26d^15f^3$
Ce	58	(Xe-56)$6s^24f^2$	Np	93	(Rn-86)$7s^26d^15f^4$
Pr	59	(Xe-56)$6s^24f^3$	Pu	94	(Rn-86)$7s^26d^15f^5$
Nd	60	(Xe-56)$6s^24f^4$	Am	95	(Rn-86)$7s^25f^7$
Pm	61	(Xe-56)$6s^24f^5$	Cm	96	(Rn-86)$7s^26d^15f^7$
Sm	62	(Xe-56)$6s^24f^6$	Bk	97	(Rn-86)$7s^26d^15f^8$
Eu	63	(Xe-56)$6s^24f^7$	Cf	98	(Rn-86)$7s^26d^15f^9$
Gd	64	(Xe-56)$6s^25d^14f^7$	Es	99	(Rn-86)$7s^26d^15f^{10}$
Tb	65	(Xe-56)$6s^24f^9$	Fm	100	(Rn-86)$7s^26d^15f^{11}$
Dy	66	(Xe-56)$6s^24f^{10}$	Md	101	(Rn-86)$7s^26d^15f^{12}$
Ho	67	(Xe-56)$6s^24f^{11}$	No	102	(Rn-86)$7s^25f^{14}$
Er	68	(Xe-56)$6s^24f^{12}$	Lr	103	(Rn-86)$7s^26d^15f^{14}$
Tm	69	(Xe-56)$6s^24f^{13}$	Ku	104	(Rn-86)$7s^26d^25f^{14}$
Yb	70	(Xe-56)$6s^24f^{14}$	Ha	105	(Rn-86)$7s^26d^35f^{14}$
Lu	71	(Xe-56)$6s^25d^14f^{14}$	—	106	(Rn-86)____

EXAMPLE 7.1: Write the full electronic configuration of rubidium, Rb.

Solution: Looking for rubidium in Table 7.2, we see that it has a krypton core plus $5s^1$. Krypton has an argon core plus $4s^23d^{10}4p^6$. Argon has a neon core plus $3s^23p^6$. And neon's configuration is $1s^22s^22p^6$. Putting them all together gives us rubidium's configuration.

Answer: $1s^22s^22p^63s^23p^64s^23d^{10}4p^65s^1$. ■

If we want to know an element's electronic configuration and don't happen to have Table 7.2 handy, we can still write most of the configurations.

We need to know the order in which sublevels are filled. A device called the *Aufbau Diagram,* shown in Figure 7.7, shows the sublevels' order of filling. Starting at the bottom of each arrow, going to the end, then going to the bottom of the next arrow, and repeating until we run out of electrons, gives us the order of filling. We see that the order doesn't always follow the order of principal energy level numbers. For instance, 4*s* comes before 3*d,* 5*s* comes before 4*d,* and so on.

Aufbau Diagram

If we don't have Table 7.2 handy, we probably won't have Figure 7.7 handy either. But the Aufbau Diagram is easy to write from memory. Each row of the diagram represents a principal energy level with all its sublevels in increasing order from left to right. Jotting these down and then drawing diagonal arrows through the sublevels gives us the Aufbau Diagram quickly.

EXAMPLE 7.2: Write the electronic configuration of rubidium (atomic number 37) without looking at any figures or tables.

Solution: First, jot down the Aufbau Diagram from memory. Then start at the lowest arrow and fill sublevels until thirty-seven electrons have been used. Remember that the *s* sublevel can contain two electrons, the *p* six, and the *d* ten. Put two electrons into the 1*s* orbital, then go to the bottom of the next arrow. The 2*s* orbital gets two. Then go to the bottom of the next arrow, where the 2*p* gets six and the 3*s* gets two. Next, $3p^6$, $4s^2$. Then $3d^{10}$, $4p^6$—and if you've been counting, you know that thirty-six electrons have been used: one to go. Put it into the next sublevel: $5s^1$.

Answer: $1s^2 2s^2 2p^6 3s^2 3p^6 4s^2 3d^{10} 4p^6 5s^1$ ■

The Aufbau Diagram works well for the representative elements. It doesn't work well, however, for all the transition elements. (Try writing the configuration of copper and comparing it with copper's configuration as shown in Table 7.2.) We won't consider exceptions to the Aufbau Diagram or the reasons for them.

FIGURE 7.7
The Aufbau Diagram

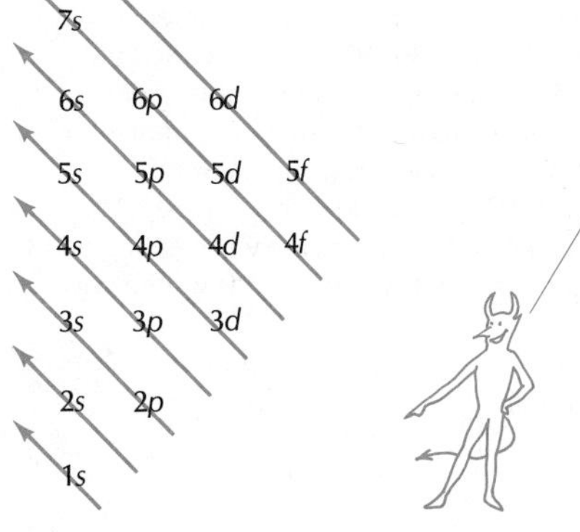
7s
6s 6p 6d
5s 5p 5d 5f
4s 4p 4d 4f
3s 3p 3d
2s 2p
1s
This Aufbau Diagram shows the order in which electrons fill sublevels. To use it, start at the bottom of the bottom arrow and read up. Then start at the bottom of the next arrow. And so on. Reading the first four arrows shows this order:
1s 2s 2p 3s 3p 4s

REVIEW QUESTIONS

Development of Today's Atomic Theory

1. What did scientists know about atoms around the turn of the century? How was our present model of the atom developed?

The Quantum Mechanical Atom

2. What is *energy?* What is *potential energy?* Give an example of potential energy.
3. Explain how charged particles can have potential energy relative to each other. How does distance affect this potential energy?
4. What do we mean by *quantized* energy levels?
5. What are *principal energy levels?* What are *shells?* What notations do we label them with?
6. Can we know exactly where an electron is at a given instant? Explain.
7. What is an *orbital?* How many electrons can an orbital contain? List the letter-names of the orbitals.
8. Sketch the shapes of the *s, p,* and *d* orbitals. How many of each are there?
9. What is a *sublevel?* List the sublevels; how many of which kind of orbital does each contain?
10. List the sublevels for principal energy levels 1 through 4.
11. What is *spin?* How does it let electrons occupy orbitals in pairs?

Writing Electronic Configurations

12. What is an element's *electronic configuration?*
13. Why must all the electrons used in an electronic configuration add up to the element's atomic number?
14. What are *noble gas cores?* Give an example, and state why they are used.
15. What is the order in which sublevels are filled? How can we know the order without memorizing it?
16. Are sublevels always filled in order of their principal energy-level numbers?
17. Does the Aufbau Diagram always work? Explain.

EXERCISES

Set A (Answers at back of book.)

1. Which in the following pairs has the lowest energy?
   a. a bowling ball on the fourth floor, or one on the eighth floor of a building
   b. an electron and a proton that are close together, or an electron and a proton that are far apart
   c. two electrons that are close together, or two electrons that are far apart
   d. an electron and a proton that are far apart, or two electrons and two protons that are far apart
   e. an electron in the first energy level (*K* shell), or an electron in the third energy level (*N* shell)
   f. an electron in the *s* sublevel, or an electron in the *d* sublevel (same principal energy level)
   g. two electrons close together that have the same spin, or two electrons close together that have opposite spin
2. How many electrons can be in each of these?
   a. *s* orbital　　d. *p* sublevel
   b. *p* orbital　　e. second energy level
   c. *d* sublevel　　f. 3*d* orbital
3. Explain what each number and letter means.
   a. $3d^3$　　c. $5p^4$
   b. $2s^2$　　d. $6f^{10}$
4. Using Table 7.2, write full electronic configurations for these elements.
   a. Hg　　b. I　　c. Fe
5. Which sublevel would fill with electrons first? (Consult the Aufbau Diagram, Figure 7.7.)
   a. the 2*p* or the 3*p* sublevel
   b. the 4*s* or the 3*d* sublevel
   c. the 5*s* or the 4*f* sublevel
6. Consulting only the Aufbau Diagram, write full electronic configurations for these elements.
   a. Zn (30)　　b. Sn (50)　　c. Xe (54)
7. Without consulting any figures or tables, write full electronic configurations for these elements (jot down the Aufbau Diagram from memory).
   a. Al (13)　　b. Ba (56)　　c. As (33)
8. Some of these configurations are wrong. Single out the wrong ones and describe what is wrong with each.
   a. $1s^22s^22p^7$　　d. (Ar-18)$4s^3$
   b. (Ne-10)$3s^23p^5$　　e. (Ne-10)$2s^1$
   c. $1s^21p^6$

Set B (Answers not given.)

1. Which in the following pairs has the highest energy?
   a. an empty carbon dioxide cartridge, or a full one
   b. two electrons that are 1 nm apart, or two electrons that are 10 nm apart
   c. an electron and a proton that are 1 nm apart, or an electron and a proton that are 10 nm apart
   d. two electrons that are close together, or four electrons that are close together
   e. an electron in the second energy level (*L* shell), or an electron in the first energy level (*K* shell)
   f. an electron in the *p* sublevel, or an electron in the *d* sublevel (same principal energy level)
   g. two electrons together in a single orbital, or two electrons in each of two separate orbitals
2. How many electrons can be in each of these?
   a. first energy level　　d. third energy level
   b. 2*s* sublevel　　e. 4*p* sublevel
   c. *d* orbital　　f. *N* shell
3. Explain what each number and letter means.
   a. $4p^1$　　c. $3d^{10}$
   b. $1s^2$　　d. $7f^5$
4. Using Table 7.2, write full electronic configurations for these elements.
   a. Cr　　b. Cd　　c. Cs
5. Which sublevel would fill with electrons first? (Consult the Aufbau Diagram, Figure 7.7.)
   a. the 4*p* or the 3*d* sublevel
   b. the 4*s* or the 3*p* sublevel
   c. the 4*s* or the 4*f* sublevel
6. Consulting only the Aufbau Diagram, write full electronic configurations for these elements.
   a. Cl (17)　　b. Bi (83)　　c. W (74)
7. Without consulting any figures or tables, write full electronic configurations for these elements (jot down the Aufbau Diagram from memory).
   a. Ca (20)　　b. Te (52)　　c. Mn (25)
8. Some of these configurations are wrong. Single out the wrong ones and describe what is wrong with each.
   a. (Kr-36)$5s^24d^2$　　d. $1s^22s^22p^23s^2$
   b. (Ne-10)$3s^23p^63d^6$　　e. $1s^22s^22p^62d^1$
   c. $1s^22s^22p^2$

CHAPTER 8

The Periodic Table

LEARNING OBJECTIVES

After studying this chapter, you should be able to:

1. Supply a correct definition, explanation, or example for each of these:

periodic law	inner transition element
s, *d*, *p*, and *f* blocks	Lewis structure
representative element	octet
transition element	atomic radius

2. Write an element's outer configuration given its position in the periodic table, or its position given its configuration.
3. Write the Lewis structure of any representative element given a periodic table or the element's group number.
4. Given the periodic table, state trends in atomic radii, reactivity, and metallic properties going across a period or down a group.
5. Explain the observed trends in atomic radius.
6. Given an unknown element's atomic number, locate it on the periodic table and state its group number, outer configuration, charge of ion, and formula of common compound(s), and predict whether its physical properties (melting and boiling points, density) are higher or lower than those of other members of its group.

In Chapter 2, we saw that elements in the same group (family) often have similar physical and chemical properties, especially in the groups at either end of the periodic table. And in Chapter 4, we used elements' group numbers to write correct formulas for some compounds. Elements of a group don't show this similar behavior by chance. They do so because their atomic structures are similar. In this chapter, we'll see how the periodic table was formulated and how atomic structure is related to an element's behavior.

8.1 DEVELOPMENT OF THE MODERN PERIODIC TABLE

In the last half of the nineteenth century, scientists noticed that some elements' properties are similar to those of others, and they began arranging the elements according to their similar properties. (See Box, page 158.)

MENDELEEV'S PERIODIC TABLE. Using the scientific method, Dmitri Mendeleev concluded that elements' properties are periodic functions of their atomic weights. The word "periodic," in this sense, means occurring at regular intervals.

Mendeleev listed the elements by increasing atomic weight and found that their physical and chemical properties begin to repeat themselves after regular periods. Eight elements after lithium on his list was the element sodium, with properties similar to those of lithium. And eight elements after sodium was another similar element: potassium. The alkaline earth metals, the halogens, and other families fell into this same pattern. Mendeleev constructed a table by placing similar elements in vertical rows and arranging the atomic weights in increasing order from left to right and from top to bottom of the table.

Blank spaces appeared in Mendeleev's table, because at that time not all the naturally occurring elements had been discovered. Using his periodic law, Mendeleev predicted properties of missing elements and even suggested how and where to look for them. Three times he was proven correct when experiments done by others uncovered elements he had predicted.

THE MODERN PERIODIC LAW. Studying the scientific method in Chapter 1, we found that a million correct predictions don't *prove* a hypothesis—and that it takes only one incorrect prediction to *disprove* a hypothesis. More than one of Mendeleev's predictions was faulty. Iodine, for instance, was predicted to be in the halogen family because of its properties, but its atomic weight dictated that it should be where tellurium is. Clearly, Mendeleev's table wasn't quite correct. Then H. G. J. Moseley demonstrated that elements should be arranged by atomic number and not atomic weight. The periodic table that resulted is the one we use today. Together, Mendeleev and Moseley gave us our modern *Periodic Law:* The properties of the elements are periodic functions of their atomic numbers.

Periodic Law

A BRIEF HISTORY OF THE PERIODIC TABLE

Time	*Persons Involved*	*Event*
1864	**John Newlands**	Developed the idea that when the elements are arranged in order of increasing atomic weights, every eighth element has similar chemical and physical properties. Newland's proposal was highly ridiculed by other scientists.
1869	**Dmitri Mendeleev**	Formulated a law stating that the properties of the elements are periodic functions of their atomic weights. On the basis of holes in his periodic table, predicted undiscovered elements—"eka-aluminum," "eka-silicon," and "eka-boron" ("eka" is Sanskrit for "next to")—and stated what properties they should have.
1870	**Lothar Meyer**	Independently formulated the same periodic law as Mendeleev's. However, Mendeleev is usually given credit for the discovery, probably because his was more highly publicized and because he was more dramatic in predicting the existence of undiscovered elements.
1875	**Lecoq de Boisbaudran**	Discovered Mendeleev's "eka-aluminum" and named it gallium. It had all of the properties predicted by Mendeleev.
1876	**Clemens Winkler**	Discovered Mendeleev's "eka-silicon" and named it germanium. This discovery strengthened Mendeleev's periodic law.
1877	**Lars Fredrik Nilson**	Discovered Mendeleev's "eka-boron" and named it scandium. With this discovery, the scientific world was convinced that Mendeleev's periodic law was correct.
1914	**H. G. J. Moseley**	Discovered that each element has a different nuclear charge, which he named the atomic number. Proposed that the elements be ordered by atomic number, not atomic weight. Revised Mendeleev's periodic law to state that the physical and chemical properties of the elements are periodic functions of their atomic numbers. This cleared up a few inconsistencies in the Mendeleev periodic table. Moseley's Periodic Law and table stay with us today.

8.2 ELECTRONS AND THE PERIODIC TABLE

Figure 8.1 shows the periodic table with the representative elements' outer *s* and *p* electronic configurations, and the transition elements' outer *s* and *d* electrons. Elements in the same group have similar configurations. All the elements in Group IA have the configuration s^1; all the Group IIA elements have the configuration s^2; and so forth.

FIGURE 8.1
Periodic table with electronic configurations

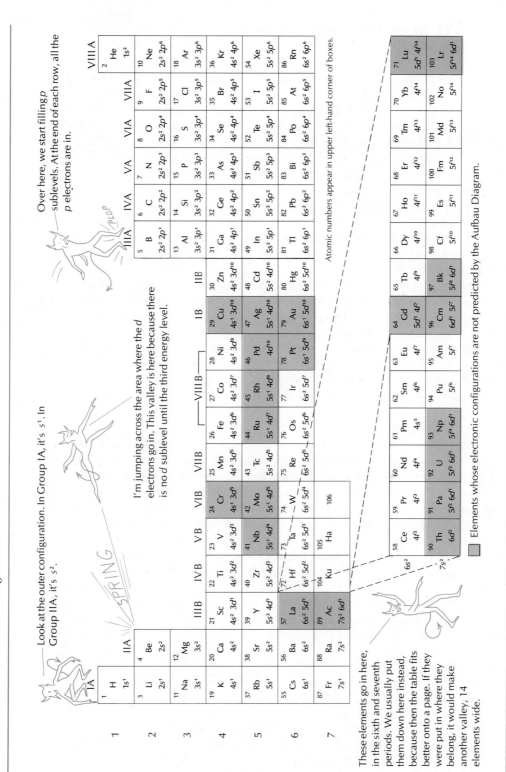

Look at the outer configuration. In Group IA, it's s^1. In Group IIA, it's s^2.

SPRING

I'm jumping across the area where the *d* electrons go in. This valley is here because there is no *d* sublevel until the third energy level.

Over here, we start filling *p* sublevels. At the end of each row, all the *p* electrons are in.

PLOP

	IA	IIA	IIIB	IVB	VB	VIB	VIIB	VIIIB	VIIIB	VIIIB	IB	IIB	IIIA	IVA	VA	VIA	VIIA	VIIIA
1	1 H $1s^1$																	2 He $1s^2$
2	3 Li $2s^1$	4 Be $2s^2$											5 B $2s^2\ 2p^1$	6 C $2s^2\ 2p^2$	7 N $2s^2\ 2p^3$	8 O $2s^2\ 2p^4$	9 F $2s^2\ 2p^5$	10 Ne $2s^2\ 2p^6$
3	11 Na $3s^1$	12 Mg $3s^2$											13 Al $3s^2\ 3p^1$	14 Si $3s^2\ 3p^2$	15 P $3s^2\ 3p^3$	16 S $3s^2\ 3p^4$	17 Cl $3s^2\ 3p^5$	18 Ar $3s^2\ 3p^6$
4	19 K $4s^1$	20 Ca $4s^2$	21 Sc $4s^2\ 3d^1$	22 Ti $4s^2\ 3d^2$	23 V $4s^2\ 3d^3$	24 Cr $4s^1\ 3d^5$	25 Mn $4s^2\ 3d^5$	26 Fe $4s^2\ 3d^6$	27 Co $4s^2\ 3d^7$	28 Ni $4s^2\ 3d^8$	29 Cu $4s^1\ 3d^{10}$	30 Zn $4s^2\ 3d^{10}$	31 Ga $4s^2\ 4p^1$	32 Ge $4s^2\ 4p^2$	33 As $4s^2\ 4p^3$	34 Se $4s^2\ 4p^4$	35 Br $4s^2\ 4p^5$	36 Kr $4s^2\ 4p^6$
5	37 Rb $5s^1$	38 Sr $5s^2$	39 Y $5s^2\ 4d^1$	40 Zr $5s^2\ 4d^2$	41 Nb $5s^1\ 4d^4$	42 Mo $5s^1\ 4d^5$	43 Tc $5s^2\ 4d^5$	44 Ru $5s^1\ 4d^7$	45 Rh $5s^1\ 4d^8$	46 Pd $4d^{10}$	47 Ag $5s^1\ 4d^{10}$	48 Cd $5s^2\ 4d^{10}$	49 In $5s^2\ 5p^1$	50 Sn $5s^2\ 5p^2$	51 Sb $5s^2\ 5p^3$	52 Te $5s^2\ 5p^4$	53 I $5s^2\ 5p^5$	54 Xe $5s^2\ 5p^6$
6	55 Cs $6s^1$	56 Ba $6s^2$	57 La $6s^2\ 5d^1$	72 Hf $6s^2\ 5d^2$	73 Ta $6s^2\ 5d^3$	74 W $6s^2\ 5d^4$	75 Re $6s^2\ 5d^5$	76 Os $6s^2\ 5d^6$	77 Ir $6s^2\ 5d^7$	78 Pt $6s^1\ 5d^9$	79 Au $6s^1\ 5d^{10}$	80 Hg $6s^2\ 5d^{10}$	81 Tl $6s^2\ 6p^1$	82 Pb $6s^2\ 6p^2$	83 Bi $6s^2\ 6p^3$	84 Po $6s^2\ 6p^4$	85 At $6s^2\ 6p^5$	86 Rn $6s^2\ 6p^6$
7	87 Fr $7s^1$	88 Ra $7s^2$	89 Ac $7s^2\ 6d^1$	104 Ku	105 Ha	106												

Atomic numbers appear in upper left-hand corner of boxes.

$6s^2$	58 Ce $4f^2$	59 Pr $4f^3$	60 Nd $4f^4$	61 Pm $4s^5$	62 Sm $4f^6$	63 Eu $4f^7$	64 Gd $5d^1\ 4f^7$	65 Tb $4f^9$	66 Dy $4f^{10}$	67 Ho $4f^{11}$	68 Er $4f^{12}$	69 Tm $4f^{13}$	70 Yb $4f^{14}$	71 Lu $5d^1\ 4f^{14}$
$7s^2$	90 Th $6d^2$	91 Pa $5f^2\ 6d^1$	92 U $5f^2\ 6d^1$	93 Np $5f^4\ 6d^1$	94 Pu $5f^6$	95 Am $5f^7$	96 Cm $6d^1\ 5f^7$	97 Bk $5f^8\ 6d^1$	98 Cf $5f^{10}$	99 Es $5f^{11}$	100 Fm $5f^{12}$	101 Md $5f^{13}$	102 No $5f^{14}$	103 Lr $5f^{14}\ 6d^1$

These elements go in here, in the sixth and seventh periods. We usually put them down here instead, because then the table fits better onto a page. If they were put in where they belong, it would make another valley, 14 elements wide.

■ Elements whose electronic configurations are not predicted by the Aufbau Diagram.

SUBLEVEL BLOCKS. The part of the periodic table that has elements with outer configurations of s^1 and s^2 is called the *s block*. It's two elements wide because there can be two electrons in the *s* sublevel. The next set of elements, the transition elements, make up the *d block* because they are filling their *d* sublevels. This block is ten elements wide because there can be ten electrons in the *d* sublevel. On the far right is the *p block,* where the *p* sublevel is being filled and the outer configurations go from s^2p^1 to s^2p^6. This block is six elements wide, because the *p* sublevel can contain a total of six electrons. Down at the bottom is the *f block*. The *f* block is fourteen elements wide, because there can be fourteen *f* electrons.

s block

d block

p block

f block

Figure 8.2 shows these sublevel blocks. Now we can understand the shape of the periodic table from the numbers of electrons in each sublevel. We redefine a *representative element* as one whose *s* or *p* sublevel is being filled. A *transition element* (or *transition metal*) is one whose *d* sublevel is being filled. And an *inner transition element* is one whose *f* sublevel is being filled.

representative element

transition element

inner transition element

The horizontal rows, or periods, of the periodic table are numbered according to the highest principal energy level being filled by the elements in that period. Thus, the first period is for the first energy level. Since the first energy level contains only one sublevel with one orbital, the period contains only two elements, hydrogen and helium. We usually put helium in the *p* sublevel block, even though it's an *s* element, because helium completes the filling of the first energy level and belongs with the rest of the noble gases. Sometimes we find it convenient to put hydrogen over next to helium in the *p* block. In these and other ways, the first period is a little different from all the others.

The second period, lithium through neon (elements 3 through 10) is for the second energy level. The third, sodium through argon, is for the third energy level. In the fourth period, we find we're filling 3*d* after 4*s*, and then we fill 4*p*. This is the order given by the Aufbau Diagram. Notice, though, that the period number still corresponds to the highest energy level containing electrons. The same is true of the fifth period. In the sixth period, with the inner transition elements, we begin to fill the 4*f* sublevel between 6*s* and 5*d*. The *f* sublevel will always be two energy levels below the period number, and the *d* sublevel will always be one energy level below the period number.

Figure 8.3 shows the periodic table with periods and sublevels.

Now, we can write an element's configuration just from the periodic table, as illustrated by the examples below.

EXAMPLE 8.1: Write the core configuration for yttrium.

Solution: First, locate the element on the periodic table. Second, find the core by locating the noble gas in the period above it: this is krypton. Third, decide which block the element is in: yttrium is in the *d* block. Fourth, count over from the left of the block to find how many electrons are in that sublevel: yttrium is the first element in the *d* block, so it has one electron in the 4*d* sublevel. Fifth, decide whether there are any filled sublevels: yes, the 5*s* sublevel is filled because it comes right before the 4*d*.

FIGURE 8.2
Periodic table showing sublevel blocks

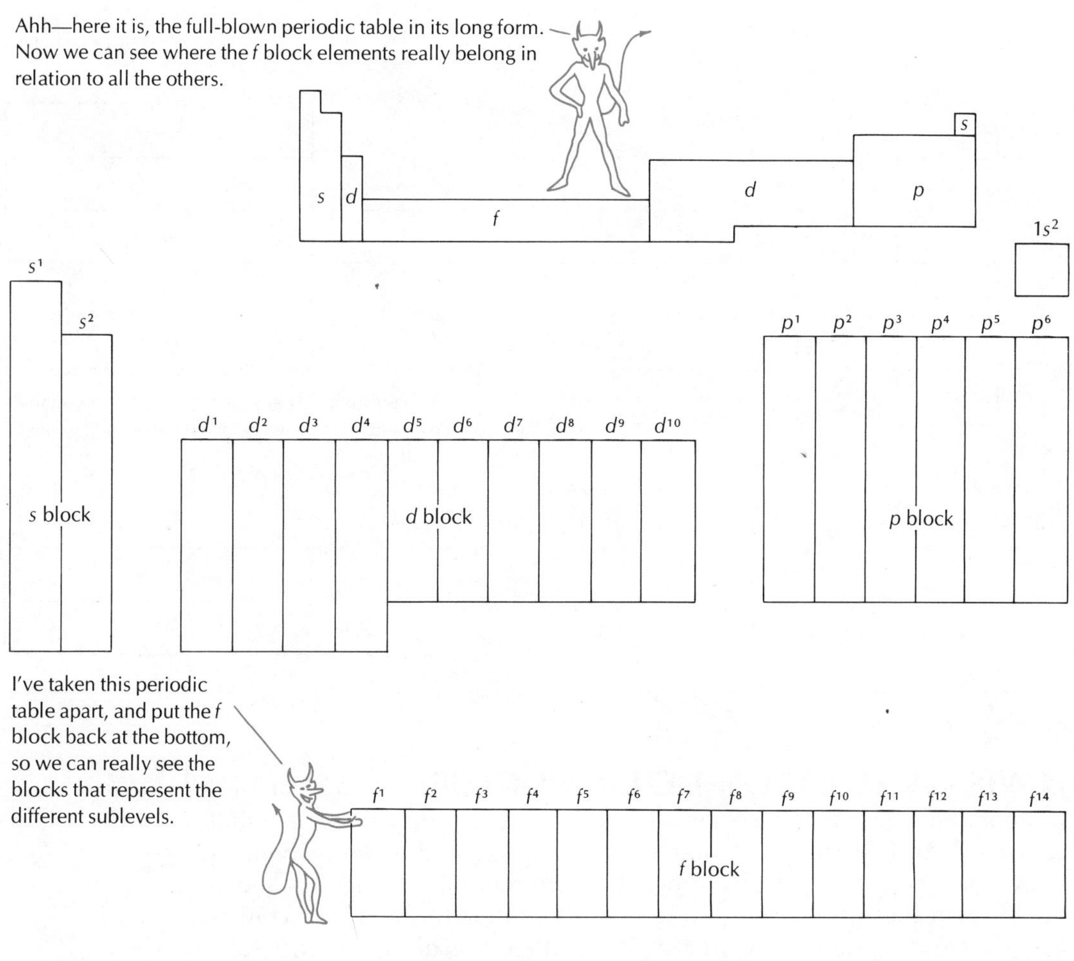
Ahh—here it is, the full-blown periodic table in its long form. Now we can see where the f block elements really belong in relation to all the others.
s
d
f
d
p
s
1s2
s1
s2
s block
d1 d2 d3 d4 d5 d6 d7 d8 d9 d10
d block
p1 p2 p3 p4 p5 p6
p block
I've taken this periodic table apart, and put the f block back at the bottom, so we can really see the blocks that represent the different sublevels.
f1 f2 f3 f4 f5 f6 f7 f8 f9 f10 f11 f12 f13 f14
f block

Answer: (Kr-36)$5s^2 4d^1$. ■

EXAMPLE 8.2: Write the core configuration for phosphorus.

Solution: First, we locate phosphorus. Second, neon is the nearest lower noble gas. Third, phosphorus is in the *p* block. Fourth, three electrons in the *p* block. Fifth, the 3*s* sublevel is filled.

Answer: (Ne-10)$3s^2 3p^3$. ■

This method, like that of the Aufbau Diagram, doesn't work for some of the transition elements (such as copper). Again, we won't attempt to explain these exceptions.

FIGURE 8.3
Periodic table with periods and sublevels

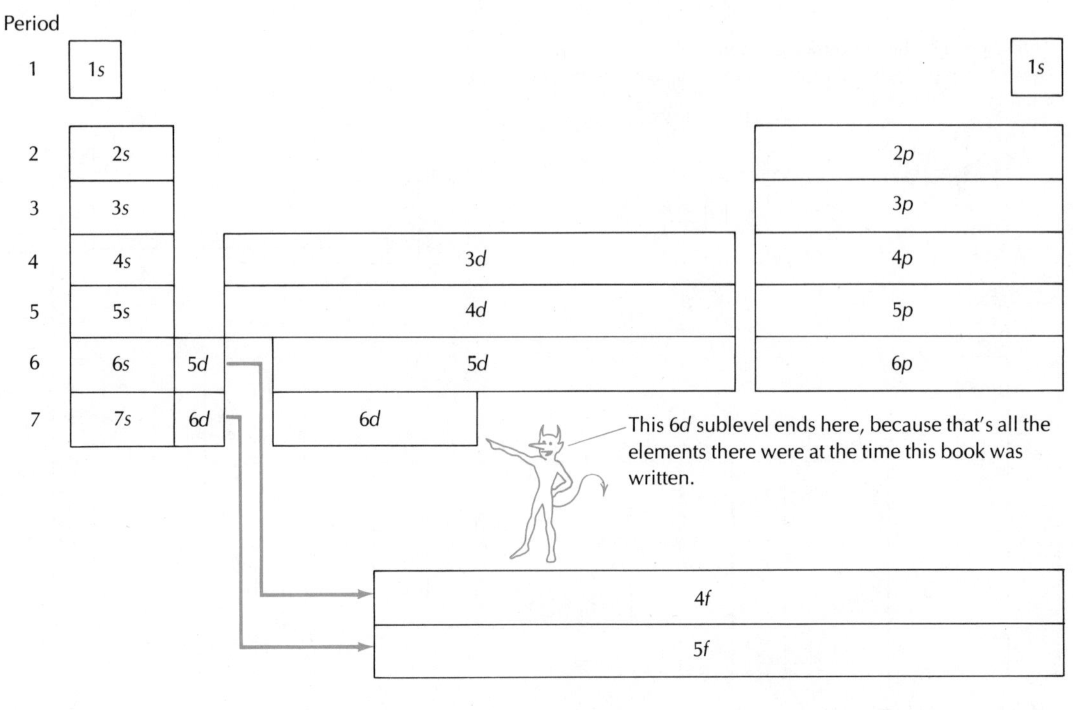

LEWIS, OR ELECTRON-DOT, STRUCTURES. In Chapter 2 we found that the noble gases are all very unreactive. And we see now that all, except for helium, have outer *s* and *p* configurations s^2p^6. Chemists know that this noble gas electronic structure—completely filled *s* and *p* sublevels in the highest energy level—is a particularly stable one. Other elements react chemically in an attempt to attain this noble gas structure. For the representative elements, it's these outer *s* and *p* electrons that determine reactivity. Now we'll introduce a more convenient way of showing these electrons.

In 1916, the American chemist G. N. Lewis invented what he called "electron-dot" structures, and we still find them useful today. A *Lewis structure* consists of the element's symbol plus dots to represent the number of outer *s* and *p* electrons (those outside the element's noble gas core). Lewis structure

According to Lewis, each element can be thought of as having a possible *octet* (*oct-* means "eight") of electrons around it. Imagine that the element's symbol has a square around it, with places for two electrons on each side, like this: octet

Each element's *s* and *p* electrons are placed around its symbol. Each side receives one electron until the four sides are used up; then the electrons are paired. The helium electronic structure is an exception to the rule of eight electrons to a period. Helium has only two electrons and is represented by an electron-dot structure like this:

He:

Lewis structures of the representative elements are shown in Figure 8.4. We've put the noble gases in twice: once on the left side and once on the right. Some chemists prefer to call the noble gases Group 0; others prefer Group VIIIA. Right now we prefer both. We'll see why by looking at a simple analogy.

The week is a period of time seven days long. When one week is finished, a new week starts, and the period repeats. Some people think Sunday is the

FIGURE 8.4
Lewis structures of the representative elements

I've put the noble gases both at the beginning and the end of the periods. At the beginning, we think of them as starting all over again, with no outer electrons—

and at the end, they've got a complete set of eight (or, in the case of helium, two) electrons.

	0	IA	IIA	IIIA	IVA	VA	VIA	VIIA	VIIIA
Period 1		H ·							He :
Period 2	He	Li ·	Be ·	· B ·	· C ·	· N ·	: O ·	: F ·	: Ne :
Period 3	Ne	Na ·	Mg ·	· Al ·	· Si ·	· P ·	: S ·	: Cl ·	: Ar :
Period 4	Ar	K ·	Ca ·	· Ga ·	· Ge ·	· As ·	: Se ·	: Br ·	: Kr :
Period 5	Kr	Rb ·	Sr ·	· In ·	· Sn ·	· Sb ·	: Te ·	: I ·	: Xe :
Period 6	Xe	Cs ·	Ba ·	· Tl ·	· Pb ·	· Bi ·	: Po ·	: At ·	: Rn :
Period 7	Rn	Fr ·	Ra ·						

Metals Nonmetals Metalloids

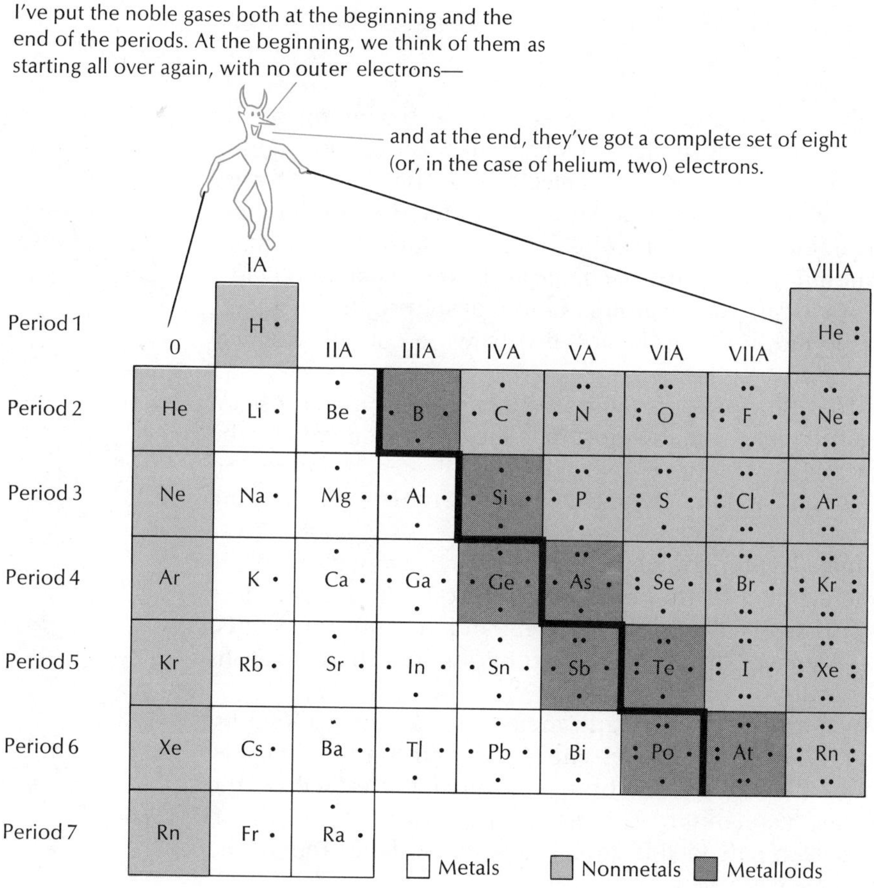

end of the week; others think it is the beginning. It doesn't matter, because either way there are always seven days from one Sunday to the next.

Because the periodic table lists the elements by increasing atomic number, we can think of the table as a building up of electrons, just as a week is a building up of days. That is, any element has one more electron than the element just before it. In Figure 8.4, a horizontal row is a period eight electrons long. When a period is finished, a new one starts, and the period repeats. Some people think the noble gases are the end of the period; others think they are the beginning. Either way, there are eight electrons from one noble gas to the next. We can say that a noble gas has a stable electronic arrangement of either eight electrons or none, depending on whether we're considering it to be at the end or at the beginning of a period.

8.3 CHEMICAL PROPERTIES AND THE PERIODIC TABLE

We'll use Figure 8.4, as well as electronic configurations, to explain some chemical properties.

METALS. For the metals, the noble gases are the beginning of the periods. By definition, metals lose electrons to form positive ions. Representative metals usually lose as many outer electrons as they have, because in this way each metal atom can get down to the outer electron arrangement (none) of the noble gas that comes before it. Of course, the ion that's formed has the same amount of positive charge as the metal had outer electrons, and that's why we can use the group number of some of the elements (mainly the metals found in Group IA and IIA) to write correct chemical formulas.

Elements on the left of the periodic table are more metallic than those on the right, and the alkali metals are the most metallic of all. Elements on the left of the table have fewer electrons to lose for a noble gas structure than the elements on the right, and the fewer electrons, the more easily they are lost.

Within a group, metallic properties increase from each element to the one below it. From one element to the next within a group, more and more electron shells come between the positive nucleus and the outer electron(s) that are to be lost. A greater distance between two unlike charges means they're easier to separate.

Some groups—such as the alkaline earth metals—are all metals. Others, such as Groups IIIA, IVA, and VA, have some metals and some nonmetals. Group IVA starts at the top with carbon, a nonmetal. Next come the metalloids silicon and germanium. The group ends with the metals tin and lead. The lower elements in this group have more shells and so the outer electrons are most easily lost.

Another property of metals is high electrical conductivity. It is caused by a metal's outer electrons moving through the metal. The same factors that cause a metal to lose electrons also cause it to conduct electricity. The more loosely an element's outer electrons are held, the more freely they can move through the metal.

Though we haven't discussed the structure of metals, we can say in a general way that heat conductivity, malleability, and ductility are all related to the looseness of a metal's hold on its outer electrons.

We saw in Chapter 4 that the representative metals tin and lead have variable charges of 2+ and 4+. They have these charges because their outer *s* and *p* configuration is s^2p^2, and therefore they have two kinds of outer electrons: *s* and *p*. It's easier to remove *p* electrons than to remove *s* electrons from the same energy level. The *p* electrons are farther from the nucleus and less tightly held. (Recall the orbital shapes in Figure 7.6, p. 147.) The two *p* electrons are lost first to let tin and lead form the 2+ ions. Losing the two *s* electrons in addition to the two *p* electrons gives rise to the 4+ ions for tin and lead.

TRANSITION ELEMENTS. The transition elements all have either one or two *s* electrons in their outer shells. The ones that have two *s* electrons lose both to form ions with 2+ charges (for example, Mn^{2+}, Fe^{2+}, Co^{2+}, and Ni^{2+}). Many transition elements have variable charges because they can also lose some of the *d* electrons in the lower principal energy level. For example, Mn^{3+}, Fe^{3+}, Co^{3+}, and V^{3+} would all result from the loss of two *s* electrons and one *d* electron. Copper, silver, and gold lose their one *s* electron to form the 1+ ions. Copper also loses one of its *d* electrons to form the 2+ as well as the 1+ ion.

At the end of the transition elements block, zinc, cadmium, and mercury have completely filled *d* sublevels. Each leaves the *d* sublevel intact by losing only its two *s* electrons for the 2+ state. Mercury, in addition, has the 1+ state, in which two atoms each lose an *s* electron and then bond to each other to form diatomic Hg_2^{2+}.

NONMETALS AND METALLOIDS. For the nonmetals, the noble gases are the end of the periods. By definition, nonmetals gain electrons to form negative ions. In this way, a nonmetal can have the electronic arrangement (eight, or for helium, two) of the noble gas that comes after it. The negative ion that's formed has the same amount of negative charge as the number of electrons the atom gained. We see now that the Group VIIA elements form ions with 1− charge because those elements have one electron to gain for the noble gas structure. And Group VIA elements form 2− charged ions because they have two electrons to gain.

The metalloids, along the staircase-shaped line that separates metals and nonmetals, are in an awkward position when it comes to losing or gaining electrons. They have too many to lose and too many to gain. It is difficult to classify them. Sometimes they do lose or gain electrons, and sometimes they share electrons to form covalent bonds, which we'll discuss in Chapter 9.

REACTIVITY. The elements at either end of the periodic table (Group IA and Group VIIA) are the most reactive. This is because of the amount of energy involved in losing or gaining only one electron. Groups IIA and VIA are reactive, but less so. Losing or gaining one or two electrons is fairly easy to do, but losing or gaining more than that is more difficult. Elements toward the center of the periodic table are less reactive than those toward either end.

8.4 OTHER TRENDS IN THE PERIODIC TABLE

Other properties vary in a regular way because of electronic configurations. We'll discuss just a few of these.

ATOMIC RADIUS. If you have ever bought a helium-filled balloon at a circus and when you went to bed left it floating against the ceiling with its string hanging down, you probably found that it sank to the floor overnight. A helium-filled balloon loses some of its helium gradually because helium atoms are very small—small enough to escape through the pores of the rubber in the balloon. A balloon filled with air wouldn't lose its contents so easily, because the nitrogen and oxygen molecules in an air-filled balloon are larger.

We describe an atom's size by its *atomic radius,* the distance from the center of its nucleus to its outer electrons. We measure atomic radii in nanometers (nm; 1 nm = 10^{-9} meters). atomic radius

Within a period, we find that each atom has a smaller atomic radius than the atom to its left. These are values of atomic radii in nanometers for the Period 2 elements:

Li	Be	B	C	N	O	F
0.123	0.089	0.08	0.077	0.074	0.074	0.072

Each element in a period has one more proton and one more electron than the element to its left. The number of electrons in the principal energy level (shell) is increasing. So too is the number of protons in the nucleus. The more of both kinds of charge there are, the greater the attraction between nucleus and electron cloud. The greater the attraction between nucleus and electron cloud, the more the electron cloud will be pulled in, and the smaller the atom will be.

Atomic radius increases as we go down a group. In the halogens (Group VIIA), the atomic radii are (in nanometers):

F	Cl	Br	I
0.072	0.099	0.114	0.133

FIGURE 8.5
Trends in atomic radii

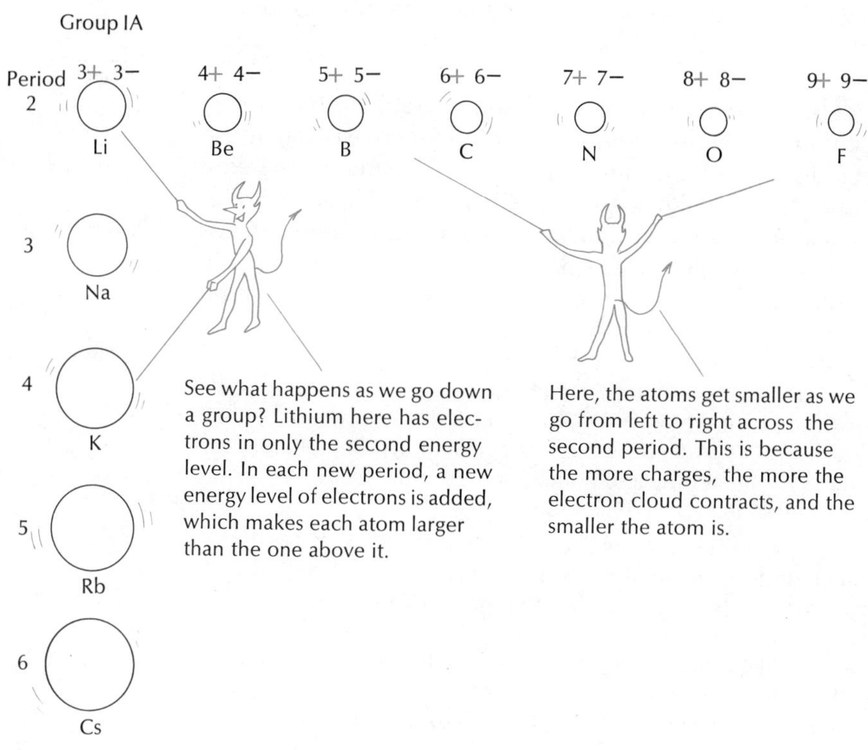

Group IA
Period
2
3
4
5
6
3+ 3−
4+ 4−
5+ 5−
6+ 6−
7+ 7−
8+ 8−
9+ 9−
Li
Be
B
C
N
O
F
Na
K
Rb
Cs
See what happens as we go down a group? Lithium here has electrons in only the second energy level. In each new period, a new energy level of electrons is added, which makes each atom larger than the one above it.
Here, the atoms get smaller as we go from left to right across the second period. This is because the more charges, the more the electron cloud contracts, and the smaller the atom is.

As we go down a group, each element farther down has one more electron shell than the one above it, and therefore one more layer of electrons farther out. Even though we're also increasing the numbers of protons and electrons, a jump from one energy level to another has a much greater effect: it makes an atom larger. Figure 8.5 illustrates the trends in atomic radius going across a period and down a group.

PHYSICAL PROPERTIES. Elements in a group usually show regular changes in physical properties, such as melting point, boiling point, and density. If we look back at Table 3.3 (pp. 48–49), we find this information for the Group VIIIA elements:

	He	*Ne*	*Ar*	*Kr*
Melting point, °C	−272	−249	−189	−157
Boiling point, °C	−269	−246	−186	−153
Density, g/ℓ	0.18	0.90	1.78	3.74

In this series, the melting and boiling points and the density all increase with increasing atomic number. Without knowing these properties for radon, the next lower rare gas in the group, we could predict that radon's melting and boiling points and density would be higher than those of any other member of the group.

NEW ELEMENTS. When Mendeleev devised his table, it had holes; some naturally occurring elements had not been discovered. Now all these elements have been discovered, but new "man-made" elements are being formed by nuclear reactions we'll talk about in Chapter 17. These new elements all have higher atomic numbers than any of the natural elements. We can think of the modern periodic table as having a "hole" in it: at the end.

We can predict some properties of a new element by using trends in the periodic table.

EXAMPLE 8.3: Predict the outer electronic configuration, group number, and ionic charge of the undiscovered element number 119. Write the formula for its oxide, using "X" to represent its symbol. State whether its atomic radius, density, and melting and boiling points would be higher or lower than those of other members of its group.

Solution: Looking at the periodic table, we count over from element 106, the last known element, until we locate the position of element 119. We find that it would fall directly below francium, Fr. That puts element 119 in Group IA, Period 8, so it has an outer configuration of $8s^1$, an ionic charge of 1+, and an oxide X_2O, like other members of its group. Its atomic radius would be larger than that of any other member of the group.

For the physical properties, we go to Table 3.3 for information about other members of the group:

	Li	*Na*	*K*	*Rb*	*Cs*
Melting point, °C	179	98	64	39	29
Boiling point, °C	1336	883	758	700	670
Density, g/cm³	0.55	0.97	0.87	1.53	1.88

Here, we see that the melting and boiling points decrease, and so we'd predict that element 119's melting and boiling points would be lower than those of any other member of the group. In fact, it probably would be a liquid at room temperature (even cesium would be a liquid on a warm day). We'd predict the density of element 119 to be greater than that of any other member of the group.

Answer: $8s^1$, Group IA, X^+, X_2O. Atomic radius and density greater, melting and boiling points lower than those of any other member of the group. ■

REVIEW QUESTIONS

Development of the Modern Periodic Table

1. State Mendeleev's Periodic Law in your own words. What predictions did he make using his law?
2. Was Mendeleev's Periodic Law entirely correct? Explain.
3. What is our modern Periodic Law? How did it come about?

Electrons and the Periodic Table

4. How are the elements' outer electronic configurations related to their positions in the periodic table?
5. Describe the sublevel blocks of the periodic table. What is happening in each block? How wide is each, and why?
6. Describe or sketch the long form of the periodic table, and explain how it is related to the form we usually see.
7. How are the periods related to the principal energy levels?
8. Why does the first period contain only two elements?
9. Explain the steps in writing an element's electronic configuration based on its position in the periodic table.
10. How do we now define *representative element, transition element,* and *inner transition element?*
11. What is a *Lewis structure?* What do the dots in a Lewis structure represent?
12. Write Lewis structures for one sample element in each of the "A" groups of the periodic table.
13. Write the Lewis structure for helium. Why is it different from the Lewis structures of the other noble gases?
14. Why are the noble gases sometimes called Group 0 and sometimes Group VIIIA?

Chemical Properties and the Periodic Table

15. How can metals achieve a noble gas arrangement of electrons?
16. What are the most metallic elements? How do metallic properties vary within the periodic table, and why?
17. What group(s) in the periodic table contain all three: metal, metalloid, and nonmetal?
18. Explain why tin and lead have variable charges of 2+ and 4+.
19. Explain why transition elements have variable charges.
20. How can nonmetals achieve a noble gas arrangement of electrons?
21. Why is it difficult to classify the metalloids?
22. Why are the elements at either end of the periodic table the most reactive?

Other Trends in the Periodic Table

23. What is an atom's *atomic radius?* Explain how and why atomic radii vary across a period from left to right and down a group from top to bottom.
24. Describe trends in melting point, boiling point, and density within a group of the periodic table.
25. Explain how we could predict the electronic structure and some chemical and physical properties of an undiscovered element.

EXERCISES

Set A (Answers at back of book. Asterisks mark the more difficult exercises.)

1. In which group and sublevel block are these electronic configurations found?
   a. s^2 c. s^2p^5
   b. s^2p^2 d. s^2d^3

Consult Figures 8.1, 8.2, and 8.3 for Exercises 2–4.

2. What are the symbols for all elements that have the following outer configurations?
   a. s^2p^3 c. s^2d^1 e. $s^2d^{10}p^6$
   b. s^2f^7 d. s^1d^{10} f. s^1
3. In what sublevel block are each of the following elements found?
   a. Mg d. Sn g. Ar
   b K e. Hg h. W
   c. Ge f. P i. U
4. Write symbols for the element(s) that fit these descriptions.
   a. Period 1, *s* block d. Group VIIA, Period 2
   b. Period 2, *p* block e. Group VB, Period 4
   c. Period 3, *s* block f. Group IB, Period 5

Do Exercises 5–10 without looking at any figures or periodic tables.

5. Element 55 is in Group IA, in the sixth period. Write its outer configuration. Which energy level is being filled?
6. Element 18 is a noble gas. What group is element 17 in?
7. Element 11 has the outer configuration $3s^1$. What is the atomic number of an element having an outer configuration of $3s^23p^2$?
8. Element 15 has an outer configuration of $3s^23p^3$. What would be the outer configuration of element 20?
9. Element 49 is in Group IIIA, in the fifth period. Write the outer configuration of element 50 and state its group number.
10. Element 35 has the outer configuration of $4s^24p^5$. What group and what period is it in?

11. Consult a periodic table, but not Figure 8.1, and write electronic configurations, with noble gas cores, for these.
    a. Sr d. Se g. Tc
    b. Cs e. Ga h. Fe
    c. Zr f. I
12. Write Lewis structures, using the symbol "X," for elements having these electronic configurations.
    a. s^2p^1 b. s^1 c. s^2p^3
13. Without consulting Figure 8.4, and using the letter "E" for element, write Lewis structures that represent elements in each A group of the periodic table. Example: Group IA, Ė.
14. Which in these pairs would be the more metallic?
    a. Rb or Sr c. C or Sn
    b. Cl or I
*15. How many *s* and how many *d* electrons would have to be lost to form each of these transition metal ions?
    a. Ag^+ c. Mn^{4+}
    b. Sc^{3+} d. Cr^{3+}
16. Which in these pairs would be the more reactive?
    a. Na or Si c. F_2 or B
    b. Cs or Pt
17. Which in these pairs would have the larger atomic radius?
    a. Be or Ca c. N or Bi
    b. K or Br
18. Consider the undiscovered element 114, which we'll call "X." Assume that it obeys all rules and trends.
    a. Which group would it be in?
    b. What other elements would it resemble?
    c. What would its outer configuration be?
    d. Write its Lewis structure.
    e. Would it be a metal or a nonmetal? Would it be more or less metallic than other members of its group?
    f. What would be the likely formula(s) of its oxide(s) and chloride(s)?
    *g. Predict its approximate atomic weight.
    h. Would it be a solid, liquid, or gas at room temperature?
    i. Would it be more or less dense than other members of its group?

Set B (Answers not given. Asterisks mark the more difficult exercises.)

1. In which group and sublevel block are these electronic configurations found?
    a. s^2p^1 c. s^2p^3
    b. s^1 d. s^2f^{14}

Consult Figures 8.1, 8.2, and 8.3 for Exercises 2–4.

2. What are the symbols for all elements that have these outer configurations?
    a. s^2d^{10} d. s^2p^4
    b. s^2 e. s^2f^{12}
    c. s^2d^5 f. s^2p^1
3. In which sublevel block are each of these elements found?
    a. Cs d. N g. Cu
    b. Ne e. Pb h. H
    c. Th f. Zn i. O
4. Write symbols for the element(s) that fit these descriptions.
    a. Group IIA, Period 2 d. Group VIIB, Period 4
    b. Period 5, *p* block e. Group VIIA
    c. Group IIIA, Period 4 f. Period 2

Do Exercises 5–10 without looking at any figures or periodic tables.

5. Element 32 is in Group IVA in the fourth period. Write its outer configuration. Which energy level is being filled?
6. Element 12 has the outer configuration $3s^2$. Which group and which period is it in?
7. Element 55 is an alkali metal. What group is element 54 in?
8. Element 28 has an outer configuration of $4s^23d^8$. What is the outer configuration of element 31?
9. Element 14 has an outer configuration of $3s^23p^2$. What is the atomic number of an element having an outer configuration of $3s^23p^5$?
10. Element 38 is in Group IIA in the fifth period. Write the outer configuration of element 36 and state its group number.

11. Consult a periodic table, but not Figure 8.1, and write electronic configurations, with noble gas cores, for these.

a. F
b. Ba
c. S
d. Zn
e. Sc
f. Al
g. Pb
h. Xe

12. Write Lewis structures, using the symbol "X," for elements having these electronic configurations.
    a. s^2
    b. s^2p^2
    c. s^2p^5
13. Without looking at Figure 8.4, state which group would be represented by each of these general Lewis structures.
    a. $:\overset{\cdot}{\underset{\cdot}{X}}:$
    b. $:\overset{\cdot\cdot}{\underset{\cdot\cdot}{X}}:$
    c. $\overset{\cdot}{X}$
    d. $:\overset{\cdot\cdot}{\underset{\cdot}{X}}:$
    e. $\cdot\overset{\cdot}{\underset{\cdot}{X}}\cdot$
    f. $\overset{\cdot}{X}\cdot$
    g. $\cdot\overset{\cdot}{X}\cdot$
    h. $:\overset{\cdot}{\underset{\cdot}{X}}\cdot$
14. Which in these pairs would be the less metallic?
    a. Rb or Cs
    b. Cu or Ga
    c. Sb or Te

*15. Write the ion, with its charge, that would be formed if these transition metal atoms lost the specified electrons.
    a. Au: one 6*s* electron and two 5*d* electrons
    b. Pd: four 4*d* electrons
    c. Cr: one 4*s* electron and five 3*d* electrons
    d. Cu: one 4*s* electron

16. Which in these pairs would be the more reactive?
    a. Br_2 or Ga
    b. Ca or Mn
    c. Se or Cu
17. Which in these pairs would have the larger atomic radius?
    a. S or Si
    b. Zn or Hg
    c. Sr or I
18. Consider the undiscovered element having atomic number 107. Assume that it obeys all the rules.
    a. Which group would it be in?
    b. What would its core configuration be?
    c. What other elements would it resemble?
    d. What period would it be in?

*19. Suppose a hitherto undiscovered element has been discovered, and named "obscurium" (Ob). It has these properties:
    a. atomic weight 287 g/mole
    b. metallic
    c. forms oxide Ob_2O_3 and chloride $ObCl_3$

    What is its atomic number, its group, and its outer configuration?

CHAPTER 9

Chemical Bonding

LEARNING OBJECTIVES

After studying this chapter, you should be able to:

1. Supply a correct definition, explanation, or example for each of these:

chemical bond	coordinate covalent bond
valence electron	molecular orbital
octet rule	molecular orbital representation
crystal lattice	bond length
ionic bond	sigma molecular orbital
electrovalent bond	pi molecular orbital
heat of formation	bond energy
electron transfer	bond strength
first and second ionization energy	central atom
molecule	electronegativity
covalent bond	nonpolar covalent bond
nonbonding electron pair	polar covalent bond
single bond	dipole
double bond	
triple bond	

2. Write electron transfer equations using Lewis structures to show reactions between metals and nonmetals.
3. Explain the changes that take place when an ionic compound is formed from its elements.
4. Explain how ionization energy is related to ionic charge, and account for trends in ionization energy in the periodic table.
5. Explain how simple molecules are formed, using Lewis structures.
6. Explain and make predictions with regard to the relationship of bond length to bond strength.
7. Write correct Lewis structures from the formulas of covalent compounds or polyatomic ions.
8. Given electronegativity values, predict whether a bond would be classified as ionic, polar covalent, or nonpolar covalent.

Elements attempt to achieve a stable noble gas electronic structure by reacting to form compounds. There are two ways that elements can do this. One is to lose or gain electrons and form positive and negative ions, which then combine to form an ionic crystal. Another is to share electrons and form covalent molecules.

Compounds, whether ionic or covalent, are held together by chemical bonds. We can expand our Chapter 2 definition of *chemical bond* now and say that it's an attractive force that holds atoms or ions together in a molecule or crystal.

chemical bond

An atom's outer electrons are involved in chemical bonding. In particular, *valence electrons* are those electrons which participate in chemical bonds. The valence electrons of the representative elements are their outer *s* and *p* electrons, or the electron dots in the Lewis structures of Figure 8.4. Transition elements' valence electrons are *s* and *d*.

valence electrons

In this chapter, we'll use a convenient aid to keep track of representative elements' bonding electrons. The *octet rule* states that an atom attempts to obtain a complete octet of electrons. Exceptions to the octet rule are hydrogen, lithium, and beryllium, which attempt to achieve the helium structure of two electrons. Another exception is boron, which sometimes has only six electrons.

octet rule

9.1 FORMATION OF IONIC COMPOUNDS

Ionic compounds are formed when ions of opposite charge come together to form a crystal. The ions themselves are created by transferring electrons from metal atoms to nonmetal atoms. We'll look at a classic reaction in which an ionic compound is formed.

THE SODIUM-CHLORINE REACTION. When soft, silvery-white, reactive sodium metal and poisonous, green chlorine gas are brought together, they react violently and spontaneously. A *spontaneous reaction* is one that happens all by itself. The resulting white crystalline sodium chloride which we eat as table salt has its ions arranged in an orderly, geometric structure. Such a structure is called a *crystal lattice* (shown in Figure 9.1) and is typical of ionic compounds. The force of attraction that holds ions together in a crystal is an *ionic, or electrovalent, bond.* This regular arrangement happens because the unlike-charged ions get as close to each other as possible, and the like-charged ions get as far away from each other as possible. The geometric arrangement of an ionic compound depends on the sizes and charges of the individual ions. The structure shown for NaCl is called cubic because its particles resemble cubes. A piece of salt with a volume of about 10^{-2} mm^3 looks like a little cube under a microscope and contains about 2×10^{17} each of sodium and chloride ions. Because there can be arbitrary numbers of ions, depending on the crystal size, we don't speak of a NaCl molecule. But we understand that any sample always contains the same number of sodium ions as chloride ions.

spontaneous reaction

crystal lattice

ionic bond

electrovalent bond

FIGURE 9.1
An ionic crystal is formed when a metal reacts with a nonmetal

This is the equation for the reaction:

$$2\ Na(s) + Cl_2(g) \longrightarrow 2\ NaCl(s) + 196.4\ kcal$$

The heat of reaction, 196.4 kcal, is released when 2 moles of NaCl are formed. We could cut this in half and say that 98.2 kcal/mole of energy is released when NaCl is formed. The amount of energy released when one mole of a substance is formed from the individual elements is the compound's *heat of formation*. This is also the energy we'd have to supply to break the compound apart into its individual elements.

heat of formation

The reactants, sodium and chlorine, are very reactive as elements, but the product, sodium chloride, is very stable and unreactive. Going from a state of high energy (the reactants) to a state of low energy (the product) takes place with a release of energy. During this reaction, valence electrons have been transferred from sodium atoms to chlorine atoms in an exchange called *electron transfer.* Sodium ions and chloride ions are the results of this electron transfer. **electron transfer**

ELECTRON TRANSFER EQUATIONS. We can use Lewis structures to illustrate the electron transfer between sodium and chlorine.

$$\text{Na}\cdot + :\ddot{\text{Cl}}: \longrightarrow \text{Na}^+ :\ddot{\text{Cl}}:^-$$

Here, the Lewis structure of a sodium ion is the same as its ionic formula. And the Lewis structure of a chloride ion is a complete octet of electrons: chlorine's Lewis structure plus one electron. Both ions' Lewis structures must show their charges with plus and minus signs.

Electron transfer equations are useful to show electron transfers between any metal and any nonmetal.

EXAMPLE 9.1: Write an equation, using Lewis structures, that shows electron transfer between calcium and fluorine.

Solution: We look for the Lewis structures of calcium and of fluorine in Table 8.4 (these structures are $\dot{\text{Ca}}\cdot$ and $:\ddot{\text{F}}:$). We know from Chapter 4 that two fluorines are needed for every calcium. Now we see that calcium has two valence electrons to donate, and fluorine can accept only one to complete its octet. We write one Lewis structure for calcium and two for fluorine on the left side of the equation:

$$\text{Ca}: + \begin{matrix} :\dot{\text{F}}: \\ :\dot{\text{F}}: \end{matrix} \longrightarrow$$

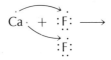

For the right side of the equation, we write the calcium ion without its two electrons, and we write the two fluoride ions with a complete octet of electrons and one minus charge each.

Answer:

$$\text{Ca}: + \begin{matrix} :\dot{\text{F}}: \\ :\dot{\text{F}}: \end{matrix} \longrightarrow \text{Ca}^{2+} \begin{matrix} :\ddot{\text{F}}:^- \\ :\ddot{\text{F}}:^- \end{matrix}$$

EXAMPLE 9.2: Write an equation, using Lewis structures, that shows electron transfer between aluminum and sulfur.

Solution: $\dot{\text{Al}}\cdot$ and $:\dot{\text{S}}:$ are the Lewis structures. Aluminum has three valence electrons; sulfur has six. We know from Chapter 4 that we need two aluminum atoms and three sulfur atoms for aluminum sulfide.

Answer:

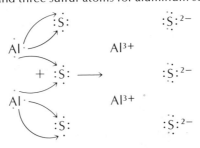

$$\begin{matrix} \dot{\text{Al}}\cdot \\ \dot{\text{Al}}\cdot \end{matrix} + \begin{matrix} :\dot{\text{S}}: \\ :\dot{\text{S}}: \\ :\dot{\text{S}}: \end{matrix} \longrightarrow \begin{matrix} \text{Al}^{3+} \\ \text{Al}^{3+} \end{matrix} \begin{matrix} :\ddot{\text{S}}:^{2-} \\ :\ddot{\text{S}}:^{2-} \\ :\ddot{\text{S}}:^{2-} \end{matrix}$$

In Example 9.2, we've drawn arrows from electrons in Al to places on S. But this is only bookkeeping: electrons aren't necessarily transferred in this way. We can't tell which electrons come from which aluminum atom or go to which sulfur atom.

From these two examples, we notice that the fluoride and sulfide ions both obey the octet rule. The positive calcium and aluminum ions obey the octet rule too, but we don't show their electrons. (Remember from Chapter 8 that a noble gas structure can be eight electrons or none.)

Electron transfer equations using the helium structure are written in the same way.

EXAMPLE 9.3: Write an equation, using Lewis structures, that shows electron transfer between lithium and hydrogen.

Solution: Li· and H· are the Lewis structures. Each has one electron. The electron transfer will go from lithium to hydrogen, because in that way each can achieve the helium structure of two electrons. Another way of looking at it is that lithium is a metal (and therefore loses electrons) and hydrogen is a nonmetal (and therefore gains electrons).

Answer:

$$Li + H\cdot \longrightarrow Li^{+} H\cdot^{-}$$

■

9.2 DESCRIBING IONIC COMPOUNDS

We've seen that ionic compounds are crystals composed of positive and negative ions. Now we'll take a closer look at these ions themselves.

IONIC RADIUS. The sodium and chloride ions that form during the sodium-chlorine reaction have changed in size (radius) from their atoms. A sodium ion is smaller (radius 0.095 nm) than a sodium atom (0.156 nm), and a chloride ion is larger (0.181 nm) than a chlorine atom (0.099 nm). In general, positive ions are smaller than the atoms from which they're formed. They are smaller for two reasons. First, by removing all the valence electrons, a shell has been stripped off. The ion is smaller in the same way as a coconut it smaller without its husk. Second, the positive ion then has more protons than electrons. The greater attraction for the electron cloud contracts it and makes the ion even smaller.

Negative ions are larger than the atoms from which they're formed. Electrons are added to the outer shell. The electrons repel each other and make the electron cloud spread out and occupy more volume. This increases the size of the negative ion beyond that of the neutral atom.

IONIC CHARGE. Sodium and the other alkali metals form ions of 1+ but not 2+ charge. The alkaline earth metals form ions of 2+ but not 3+ charge. Aluminum forms an ion of 3+ but not 4+ charge. In all these, valence electrons are removed up to, but not beyond, an atom's noble gas structure. It takes some energy to remove electrons to form noble gas structures, but it takes much, much more energy to remove electrons from noble gas structures.

The amount of energy, per mole, that it takes to remove one electron from an atom is an element's *ionization energy,* sometimes called *first ionization energy.* Taking a second electron off requires an element's *second ionization energy.* The first ionization energy of sodium is 118 kcal/mole; the second is 1091 kcal/mole, nearly ten times the first. The second ionization energy is much larger than the first because removing two electrons from a sodium atom means removing one electron from a stable noble gas structure. That takes so much energy that compounds of Na^{2+} don't form.

ionization energy

first ionization energy

second ionization energy

Figure 9.2 shows first, second, and subsequent ionization energies for a few metals. The dark bars represent energy required to remove an electron from a noble gas structure. This energy is much larger than the others before it. For the Group IIA elements, the first and second ionization energies are relatively small, but the third—removing an electron from a noble gas structure—is much larger. These elements can afford to lose two, but not three, electrons. Aluminum can afford to lose three, but not four.

IONIZATION ENERGY AND THE PERIODIC TABLE.

Figure 9.3 shows first ionization energies for the representative elements. Some definite trends appear across the periods and down the groups.

Proceeding down a group, we see that the first ionization energy de-

FIGURE 9.2
First, second, and subsequent ionization energies for selected metals

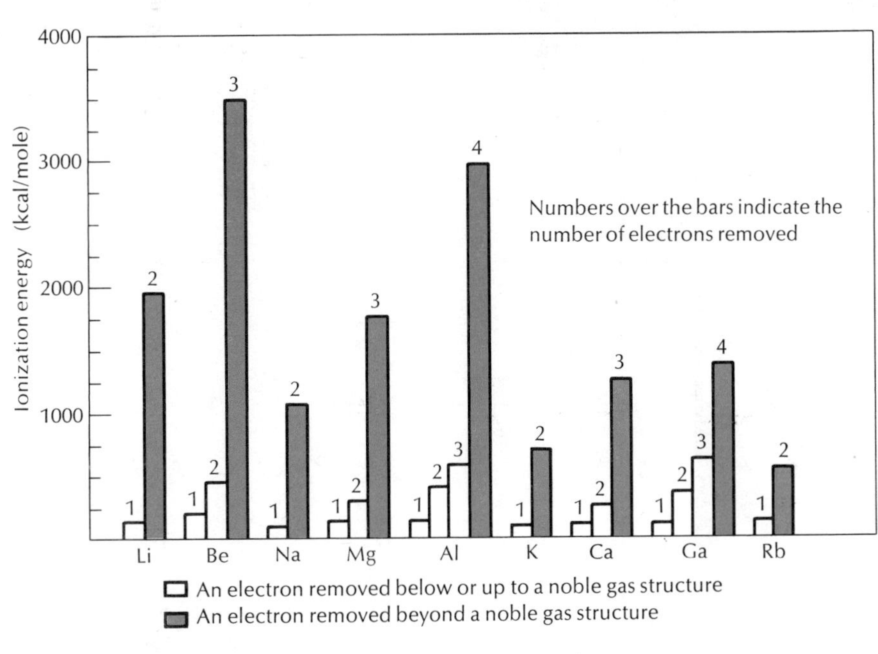

FIGURE 9.3
Ionization energies in kcal/mole for some representative elements

	IA	IIA	IIIA	IVA	VA	VIA	VIIA	VIIIA
1	H 312						H 312	He 567
2	Li 124	Be 214	B 190	C 259	N 334	O 312	F 400	Ne 494
3	Na 117.6	Mg 175	Al 137	Si 187	P 251	S 237	Cl 299	Ar 314
4	K 100	Ca 140	Ga 138	Ge 187	As 242	Se 224	Br 272	Kr 321
5	Rb 95.9	Sr 131	In 133	Sn 168	Sb 138	Te 207	I 244	Xe 279
6	Cs 89.2	Ba 120	Tl 140	Pb 170	Bi 184			

creases. Two factors are involved in this trend: (1) The valence electrons are farther from the nucleus toward the bottom of the groups because more filled shells separate the nucleus from the valence electrons. Greater distance means less attraction between unlike charged particles, and therefore less energy is required to remove an electron. (2) More electron shells act as shields or barriers between the nucleus and the valence electrons, lessening still more the attraction of the nucleus for the valence electrons.

Ionization energy increases from left to right across a period. Moving from left to right within a period means adding more nuclear and more electronic charge. The greater the charge, the greater the attraction, and the more energy needed to overcome the attraction and remove an electron.

It becomes more and more difficult to remove electrons as we go from the metals on the left to the nonmetals on the right. Thus a nonmetal cannot lose electrons, but can only accept them to achieve its noble gas structure. When a nonmetal reacts with a metal, this is easy. The nonmetal accepts electrons from the metal, and an ionic compound is formed.

When a nonmetal reacts with another nonmetal, a covalent compound is formed. Nonmetals have no choice when they're reacting with each other. Their ionization energies are too high to form positive ions, and though they can form negative ions, there's no atom around to donate electrons. Sharing electrons is the answer, and we'll look at this kind of bonding next.

9.3 MOLECULES AND COVALENT BOND FORMATION

We can expand our definition of a molecule now. A *molecule* is a particle of matter made of two or more atoms joined by covalent bonds. A *covalent bond* is a shared electron pair. Two atoms can cooperate to achieve noble gas structures, by forming a covalent bond. Each atom contributes a single electron to the formation of an electron pair. Then the two atoms share the pair, and each atom counts both electrons in the pair toward its noble gas structure. Some atoms can share more than one pair, if they need to, in order to have an octet of electrons.

molecule

covalent bond

DIATOMIC MOLECULES IN ELEMENTS. We saw in Chapter 2 that some elements exist as diatomic molecules. Now we'll find out why.

The halogens all have seven valence electrons. They each need one more electron to achieve a noble gas structure. Fluorine, for instance, needs one more electron for the neon structure. Let's look at the Lewis structure of a fluorine atom. When two electrons are together on a side, we say that these electrons are paired, or that they form an electron pair. If an electron is alone on a side, we say that it's an unpaired electron.

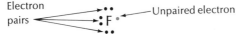

Electron pairs → $:\ddot{\underset{..}{F}}\cdot$ ← Unpaired electron

In covalent bond formation, fluorine can get the electron it needs by bonding to another element or by bonding to another fluorine. Two fluorine atoms can contribute their unpaired electrons to form a pair, and then they share the pair. The pair then belongs to them both equally, and not to one or the other. Then each atom has eight electrons, or four pairs, just like neon. We can use Lewis structures to see how this might happen.

$$:\ddot{\underset{\cdot\cdot}{F}}\cdot \quad \cdot\ddot{\underset{\cdot\cdot}{F}}: \longrightarrow :\ddot{\underset{\cdot\cdot}{F}}:\ddot{\underset{\cdot\cdot}{F}}:$$

The two dots between the two fluorine atoms now represent a shared electron pair, or covalent bond. The other pairs that aren't involved in bonding are called *nonbonding electron pairs.* The structure on the right is the Lewis structure for the F_2 molecule. Lewis structures for Cl_2, Br_2, and I_2 are the same as for F_2.

nonbonding electron pair

Let's look at hydrogen. Since it has only one electron, hydrogen's nearest noble gas configuration is that of helium, with two electrons. To achieve this structure by covalent bond formation, hydrogen can bond to another element, or it can form a diatomic molecule by bonding with another hydrogen atom, like this:

$$H\cdot \quad \cdot H \longrightarrow H:H$$

Notice that a diatomic hydrogen molecule has no nonbonding electron pairs.

F_2 and H_2 form *single* covalent *bonds,* because only one electron pair is involved in the bond. Oxygen, which has six valence electrons, has two unpaired electrons. To achieve its octet when bonding with another oxygen atom, it forms a *double bond,* like this:

single bond

double bond

$$:\ddot{O}\cdot \quad \cdot\ddot{O}: \longrightarrow :\ddot{O}::\ddot{O}:$$

The four dots between the two oxygen atoms stand for two single bonds, or one double bond. Notice that the Lewis structure of the O_2 molecule no longer has two electrons on a side, as atoms do. Instead, the four dots that represent two covalent bonds between the two atoms are owned equally by both atoms, so that each has an octet. The molecule has four nonbonding electron pairs, two on each O.

Nitrogen is an element whose diatomic molecule, N_2, has a *triple bond* between atoms. It does this because it has only five valence electrons, with three unpaired. It must pair three electrons to achieve the noble gas structure.

triple bond

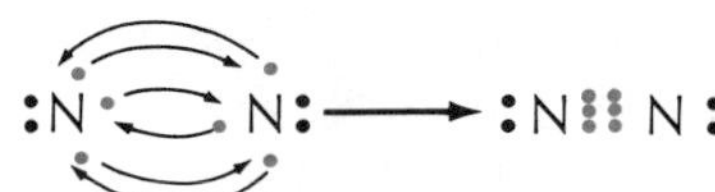

$$\longrightarrow :N\vdots\vdots N:$$

The resulting nitrogen molecule has a triple bond, as shown by the six dots between the nitrogen atoms, and two nonbonding electron pairs (one on each N).

MOLECULES FORMED FROM TWO ELEMENTS.

Table 9.1 shows Lewis structure of various elements and of compounds they form. We see that the number of atoms bonding together depends on the number of unpaired electrons each has to share. Hydrogen and fluorine have one unpaired electron each, so they share them to form one covalent bond in the molecule HF. Oxygen has two unpaired electrons, so it bonds to two hydrogen atoms, each of which has only one unpaired electron. Nitrogen, with three unpaired electrons, can bond to three hydrogen atoms. And so on.

These Lewis structures aren't meant to represent the shape or geometry of these molecules. All the following structures are correct for water, and all mean the same thing.

```
                    H      H
  ..      ..        ..     ..
H:O:H   H:O:      :O:H   :O:
  ..      ..        ..     ..
          H                H
```

Now that we know that two dots means a covalent bond or nonbonding electron pair, we can switch to the simpler bar notation for Lewis structures. In this notation, we use a straight line between atoms to represent a bond, and lines parallel to the edges of the letter used for the atom represent nonbonding electron pairs. We still use single dots to show unpaired electrons. Here are all the structures we've talked about so far, written in bar notation.

```
|F̄—F̄|  |O=O|  |N≡N|  H—F̄|  H—Ō—H  H—N̄—H
 ‾  ‾    ‾ ‾                 ‾        |
                                      H

                 H
                 |
        H—H   H—C—H
                 |
                 H
```

The number of single bonds that an element can form is equal to the number of unpaired electrons it has. Elements that have two unpaired elec-

TABLE 9.1
Lewis structures of some elements and their compounds

Elements	*Compounds*	
H. ·F: (with two dot pairs)	H:F: (with two dot pairs)	Hydrogen fluoride, HF (HCl, HBr, and HI are similar)
H. ·O· (with two dot pairs)	H:O:H (with two dot pairs)	Water, H_2O (H_2S, H_2Se, and H_2Te are similar)
H. ·N· (with one dot pair)	H:N:H with :H below	Ammonia, NH_3 (PH_3, AsH_3, and SbH_3 are similar)
H. ·C· (four single dots)	H:C:H with H above and below	Methane, CH_4 (SiH_4 and GeH_4 are similar)

trons can form double bonds; elements with more than two can form double or triple bonds (there's no such thing as a quadruple bond). Here are some of the possibilities:

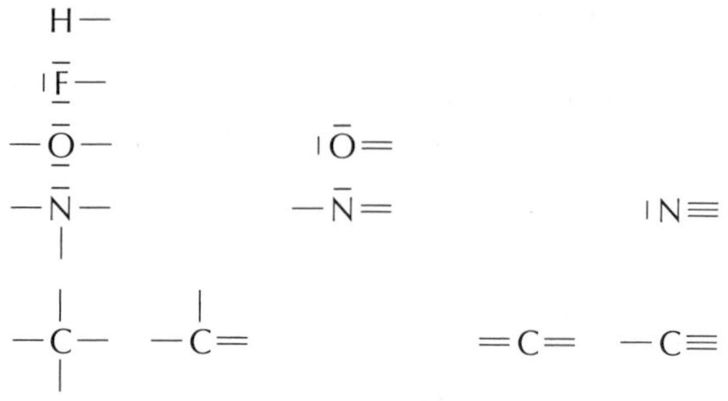

```
 H—

 |F̄—

—Ō—            |Ō=

—N̄—           —N̄=                  |N≡
  |

  |      |
—C—    —C=                =C=    —C≡
  |
```

COORDINATE COVALENT BONDING. We've seen that, in ordinary covalent bonding, each atom involved contributes one electron to a pair, and then each atom shares the pair.

Coordinate covalent bonding, on the other hand, allows an atom that's missing an entire electron pair to share the nonbonding electron pair on another atom. A *coordinate covalent bond* is a shared electron pair where both electrons have been contributed by one of the atoms.

coordinate covalent bond

Coordinate covalent bonding is best illustrated by example. Here are the Lewis structures of fluorine and of boron:

```
 ..      .
:F·      B·
 ..      .
```

Each of boron's three unpaired electrons can bond to one fluorine atom, like this:

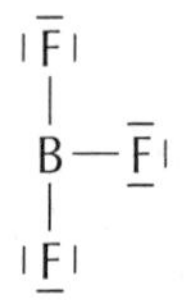

```
|F̄|
 |
 B—F̲|
 |
|F̲|
```

Notice that in this compound, boron trifluoride, boron doesn't have an octet. Nevertheless, BF_3 is a fairly stable compound by itself. If it is allowed to come in contact with a molecule that has nonbonding electron pairs, though, it can react with it to form a coordinate covalent bond and complete its octet.

Ammonia is a molecule that reacts with BF_3. Ammonia can donate its pair of nonbonding electrons to boron, which in turn shares the pair back with ammonia. The resulting molecule, called boron trifluoride-ammonia adduct, can be written like this, with the coordinate covalent bond shown in color:

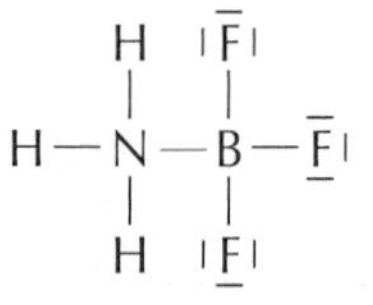

```
   H  |F̄|
   |   |
H—N—B—F̲|
   |   |
   H  |F̲|
```

We've used boron to illustrate coordinate covalent bonding, but other atoms can bond this way too. This kind of bonding is important in oxyacids and polyatomic ions.

9.4
DESCRIBING COVALENT BONDS

It's easy to see that ionic bonds form because of attraction between oppositely charged ions. But what about covalent bonds? Why should a shared electron pair be more stable than unpaired electrons on separate atoms? In this section, we'll look at a modern way of picturing covalent bonds.

MOLECULAR ORBITAL REPRESENTATION. According to current theory, two separate atomic orbitals, each containing one unpaired electron, can overlap to form a combined orbital, called a *molecular orbital.* The two electrons then occupy the molecular orbital as a pair. Describing a covalent bond as overlapping atomic orbitals is called *molecular orbital representation.*

molecular orbital

molecular orbital representation

For example, hydrogen's single unpaired electron is in an *s* orbital. The molecular orbital representation of two hydrogen atoms bonding together to form a diatomic H_2 molecule shows the two *s* orbitals overlapping like this.

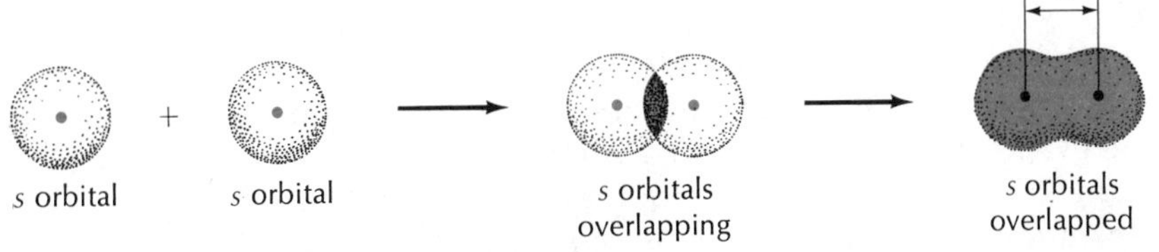

The colored dots represent the hydrogen nuclei. The distance (d) between these nuclei in the covalent bond is called the *bond length* and is 0.074 nm for the H—H bond in the H_2 molecule.

bond length

The molecular orbital, like an atomic orbital, is shown by a shaded area representing the likelihood of finding the electron pair. We see that the likelihood is high that the electron pair is somewhere between the two nuclei. Each nucleus can now be near two electrons instead of one, which results in a more stable arrangement.

The type of molecular orbital that results when two atomic orbitals overlap end-to-end is called a *sigma (σ) molecular orbital.* A sigma molecular orbital results when two *s* atomic orbitals overlap.

sigma (σ) molecular orbital

In a fluorine atom, the unpaired electron is in a *p* orbital. The molecular orbital representation of the F_2 molecule's formation shows two *p* orbitals overlapping end to end.

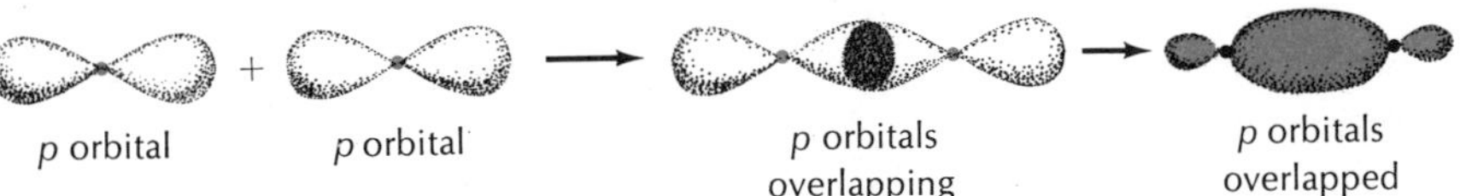

This molecular orbital is also called a sigma type, even though it looks different from the sigma molecular orbital formed from two *s* atomic orbitals. The molecular orbital above has three lobes: a large lobe between the nuclei and two small lobes on either end. These small lobes show that there is some likelihood of finding the electron pair in these locations, but much less than between the nuclei.

When double or triple bonds occur, one of the bonds is of the sigma type, and the second and third are of the pi type. A *pi* (π) *molecular orbital* results when two *p* orbitals overlap side by side instead of end to end, as shown below.

pi (π) molecular orbital

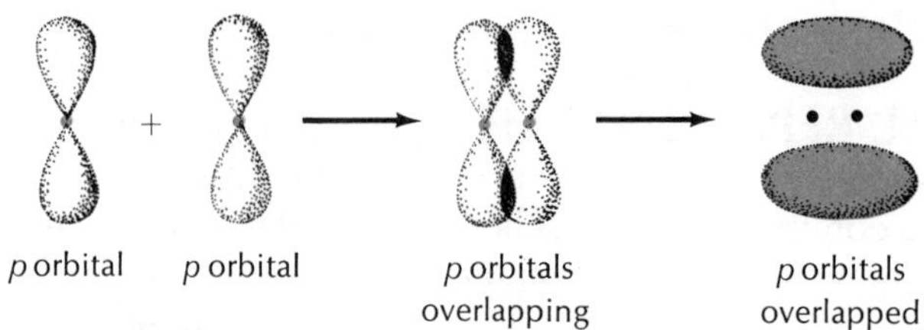

Like the *p* atomic orbitals, the pi molecular orbital has two lobes. An O_2 molecule contains one sigma and one pi molecular orbital; an N_2 molecule contains one sigma and two pi molecular orbitals.

BOND ENERGY. Like ionic compounds, covalent compounds have heats of formation. Unlike ionic compounds, covalent compounds contain individual molecules that, in turn, contain individual bonds. We can define *bond energy* as the energy, per mole, that is released when a covalent bond is formed. This is also the energy that would be required to break or dissociate the bond into separate atoms, so we sometimes see it called *bond dissociation energy,* or even *bond strength.* Translating this into an equation,

bond energy

bond dissociation energy

bond strength

$$HCl(g) + 102\ kcal \longrightarrow H(g) + Cl(g)$$

The bond energy of the H—Cl bond in the HCl molecule is 102 kcal/mole.

Bond strength and bond length are related; usually, shorter bonds are stronger than longer bonds. We see this effect here.

Bond	*Bond Energy (kcal/mole)*	*Bond Length (nm)*
H—F	135	0.0917
H—Cl	102	0.127
H—Br	86.5	0.141
H—I	70.5	0.161

Double and triple bonds increase the bond energy and decrease the bond length, as we see in this series.

Bond	Bond Energy (kcal/mole)	Bond Length (nm)
F—F	33	0.142
O=O	118	0.121
N≡N	225	0.110

9.5 PRACTICE IN WRITING LEWIS STRUCTURES

Oxyacids, polyatomic ions, and other compounds contain combinations of regular and coordinate covalent bonds. To write the Lewis structure of a molecule or ion, we don't have to know which bonds are which or which element contributed which electron to which bond. We'll take an example and go through the steps in writing the Lewis structure from a formula.

We want to write the Lewis structure of HNO_3.

Step 1. *Decide how the atoms are bonded to each other.* We start by figuring out what the *central atom* is—that is, the atom that has the most other atoms bonded to it. For binary covalent compounds and polyatomic ions the central atom is the one that appears only once in the formula. For oxyacids, the central atom is the atom other than hydrogen or oxygen. In an oxyacid, the oxygens are bonded to the central atom, and the hydrogens are bonded to the oxygens.

central atom

Here, we decide that N is the central atom; each of the O's is bonded to N; and the H is bonded to one of the O's.

Step 2. *Write the central atom and the others that are bonded to it. Make single bonds between them.*

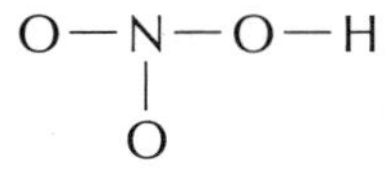

```
O—N—O—H
  |
  O
```

(Here, it doesn't matter which O we decide to attach the H to.)

Step 3. *Place nonbonding electron pairs around all atoms except H, so that each atom has a complete octet.*

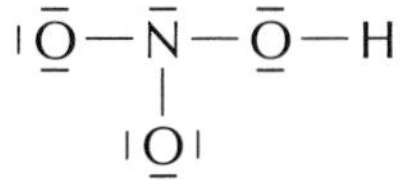

```
 _   _   _
|O—N—O—H
 ‾       ‾
    |
   |O|
    ‾
```

Step 4. *Count the electron pairs.* The structure above has thirteen.

Step 5. *Calculate the number of electron pairs that the structure should contain.* We find this number by adding up the valence electrons contributed by each atom. For negative ions, we also add in the amount of negative charge; for positive ions, we subtract the amount of positive charge. Then we divide the number of electrons by two to get the number of electron pairs.

N: 5 valence electrons
3(O): 18 valence electrons
H: 1 valence electron

24 electrons, or 12 electron pairs

(*Note:* For the great majority of cases, the number of electrons will be even, because pairing of electrons results in lower energy. There are a few cases where odd numbers of electrons occur, but we won't consider them in this book.)

Step 6. *If necessary, replace nonbonding electron pairs with double or triple bonds.* We will have to replace them if the numbers in Steps 4 and 5 aren't the same. Electrons must be conserved. We can't use more electron pairs in a Lewis structure than the molecule or ion had in the first place. To get rid of one electron pair, we replace two nonbonding pairs with one double bond. To get rid of two electron pairs, we replace four nonbonding pairs with two double bonds or one triple bond.

In our example, we need to get rid of one electron pair. We replace two nonbonding electron pairs (one from N and one from O) with a double bond between N and either of the two O's that's not bonded to H.

```
|O=N—O—H
   |
  |O|
```

Step 7. *Check the number of electron pairs again, and make sure that the octet rule is satisfied.*

Electron count: 12 pairs (same number as in Step 5)

Octet rule: O has 8, N has 8, O has 8, O has 8, H has 2

Our Lewis structure is correct.

Sometimes, as in writing chemical equations, we might have to repeat some of these steps until we get the right structure.

Now we can look at more examples.

EXAMPLE 9.4: Write the Lewis structure of PO_4^{3-}.

Solution:

Step 1: P is the central atom; each of the 4 O's is bonded to it.

Step 2:

```
    O
    |
O—P—O
    |
    O
```

Step 3:

```
    |O|
     |
|O—P—O|
     |
    |O|
```

Step 4: There are sixteen pairs.

Step 5: P: 5 valence electrons
4(O): 24 valence electrons
3−: 3 valence electrons
32 electrons, or 16 pairs

Step 6: No double or triple bonds are necessary. We have used the correct number of electron pairs.

Step 7: Electron count: 16 pairs
Octet rule: P has 8, all four O's have 8

Our structure is correct.

Answer:

$$\left[\begin{array}{ccc} & |\bar{\underline{O}}| & \\ & | & \\ |\bar{\underline{O}}- & P & -\bar{\underline{O}}| \\ & | & \\ & |\bar{\underline{O}}| & \end{array}\right]^{3-}$$

Notice: We enclose the Lewis structure of a polyatomic ion in brackets, with its charge written outside the brackets. ■

EXAMPLE 9.5: Write the Lewis structure of HCN. Carbon is the central atom; H and N are bonded to it.

Solution:

Step 1: When a situation arises that isn't covered by our rules, then the problem must give us additional information. Here, we couldn't have figured out that C is the central atom.

Step 2: H—C—N

Step 3: H—$\bar{\underline{C}}$—$\bar{\underline{N}}|$

Step 4: There are seven electron pairs.

Step 5: H: 1 valence electron
C: 4 valence electrons
N: 5 valence electrons
10 electrons, or 5 pairs

Step 6: We need to get rid of two electron pairs. Our choice is to make two double bonds or one triple bond. We can't make a double bond to H, because that would give it too many electrons. Our only choice is to remove two pairs each from C and N and to make a triple bond between them.

H—C≡N|

Step 7: Electron count: 5 pairs
Octet rule: H has 2, C has 8, N has 8

Our structure is correct.

Answer: H—C≡N| ■

Notice that polyatomic ions have the same number of electron pairs as their corresponding acids:

H—C≡N| and [|C≡N|]⁻; $|\bar{\underline{O}}$=N—$\bar{\underline{O}}$—H (with $|\bar{\underline{O}}|$ bonded to N) and $[|\bar{\underline{O}}$=N—$\bar{\underline{O}}|]^-$ (with $|\bar{\underline{O}}|$ bonded to N)

In each case the bond to H in the acid becomes a nonbonding pair in the ion.

9.6
BOND POLARITY

So far, we've talked about compounds as being either ionic or covalent. We've used the staircase-shaped line on the periodic table as a guide in deciding whether a compound is one or the other. This method works fine for general cases and for naming compounds, but actually elements don't let us classify them so easily. In reality, a great many bonds are neither purely ionic nor purely covalent, but somewhere between.

ELECTRONEGATIVITY. The chemist Linus Pauling proposed electronegativity to explain inequalities of bonding. *Electronegativity* is the tendency of an element to hold, or attract, the bonding electrons. For maximum attraction of the negative electrons in a bond, an element needs a large positive charge and a small distance between it and the electrons. Therefore electronegativity depends on nuclear charge and atomic radius (or number of shells between the nucleus and the valence electrons): in short, the same things that influence ionization energy. We might predict that elements having the highest ionization energy would also be the most electronegative, and they usually are.

electronegativity

Pauling calculated numerical values for electronegativities, which are shown in Figure 9.4. Elements having high electronegativities have the greatest attraction for bonding electrons. Fluorine is the most electronegative element, because of the combination of high nuclear charge and small radius. Oxygen is next.

Hydrogen's electronegativity is 2.1; fluorine's is 4.0. When they bond, fluorine tries to take the electron completely away from hydrogen. It can't, though, because hydrogen's ionization energy is too high. The result is a tug-of-war between hydrogen and fluorine, with the electron pair between—and fluorine almost winning, as shown in Figure 9.5.

POLAR BONDS. In a covalent bond, atoms share electron pairs "equally," but some atoms are more equal than others.[1] For instance, in a chlorine molecule, the electron pair is exactly shared between the two chlorine atoms. Neither one has any more control over it than the other. There is no doubt here about the sharing, because the bond is between atoms of the same element.

We can use the electronegativity values to decide how the electron-pair bond would be shared between various elements. The electrons forming a bond between two elements with about the same electronegativity values, such as nitrogen and chlorine, will be shared about equally by the elements. Such a bond is called *nonpolar covalent.* We consider bonds between atoms whose electronegativity difference is between 0 and 0.5 to be nonpolar covalent.

nonpolar covalent bond

[1]A paraphrase from George Orwell's *Animal Farm.*

FIGURE 9.4
Electronegativity values for the elements (exclusive of inner transition elements)

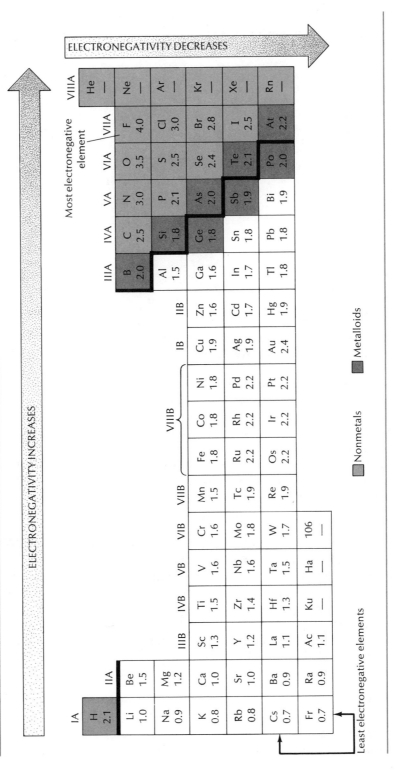

ELECTRONEGATIVITY INCREASES →

ELECTRONEGATIVITY DECREASES ↓

IA	IIA	IIIB	IVB	VB	VIB	VIIB	VIIIB	VIIIB	VIIIB	IB	IIB	IIIA	IVA	VA	VIA	VIIA	VIIIA
H 2.1																	He —
Li 1.0	Be 1.5											B 2.0	C 2.5	N 3.0	O 3.5	F 4.0	Ne —
Na 0.9	Mg 1.2											Al 1.5	Si 1.8	P 2.1	S 2.5	Cl 3.0	Ar —
K 0.8	Ca 1.0	Sc 1.3	Ti 1.5	V 1.6	Cr 1.6	Mn 1.5	Fe 1.8	Co 1.8	Ni 1.8	Cu 1.9	Zn 1.6	Ga 1.6	Ge 1.8	As 2.0	Se 2.4	Br 2.8	Kr —
Rb 0.8	Sr 1.0	Y 1.2	Zr 1.4	Nb 1.6	Mo 1.8	Tc 1.9	Ru 2.2	Rh 2.2	Pd 2.2	Ag 1.9	Cd 1.7	In 1.7	Sn 1.8	Sb 1.9	Te 2.1	I 2.5	Xe —
Cs 0.7	Ba 0.9	La 1.1	Hf 1.3	Ta 1.5	W 1.7	Re 1.9	Os 2.2	Ir 2.2	Pt 2.2	Au 2.4	Hg 1.9	Tl 1.8	Pb 1.8	Bi 1.9	Po 2.0	At 2.2	Rn —
Fr 0.7	Ra 0.9	Ac 1.1	Ku —	Ha —	106 —												

Most electronegative element (F)

Least electronegative elements (Cs, Fr)

Nonmetals

Metalloids

FIGURE 9.5
Elements with higher electronegativity tend to attract the bonding electrons

How do you like that? Why, that poor hydrogen atom doesn't have a *chance* at that electron pair against fluorine. You or I would have the same problem if we had to share a small blanket with a football tackle on a cold night! I guess they're still "sharing" it, but fluorine has the lion's share.

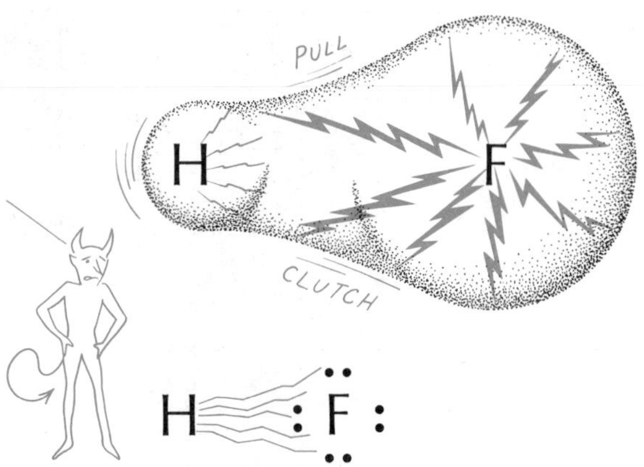

Most nonmetals have high electronegativity values; metals have low electronegativity values. We usually consider ionic bonds to be those where the difference in electronegativity values between the two elements is greater than 2.0.

A bond between two elements where the electronegativity difference is between 0.5 and 2.0 is considered to be a polar covalent bond. A *polar covalent bond* is one in which the bonding electrons are shared unequally. For instance, the electronegativity difference between hydrogen and fluorine is 1.9. This means that fluorine will control the electron pair much more than hydrogen, as we saw it doing in Figure 9.5. When discussing polarity, we sometimes use a crossed arrow to indicate a polar covalent bond. The arrow points to the more electronegative element.

polar covalent bond

$$\mathrm{H} \;{+}\!\!\!\longrightarrow \mathrm{F}$$

This kind of bond is called polar because it has two poles: a positive and a negative. Such a charge separation is called a *dipole*.

dipole

The greater the electronegativity difference, the more polar the bond, until, at the extreme, the electron pair comes off completely and we have an ionic bond. An ionic bond can be thought of as the extreme of bond polarity. (See Figure 9.6.)

Fluorine, oxygen, nitrogen, and chlorine are the "big four" electronegative elements. Bonds between these and the more electropositive nonmetals will usually be polar.

EXAMPLE 9.6: Classify bonds between these elements as ionic, polar covalent, or nonpolar covalent: a. N and H; b. K and Br; c. S and Cl.

Solution: We look at Figure 9.4 and get the electronegativity values of the two elements. If the difference between these values is less than 0.5, we

FIGURE 9.6
Greater electronegativity differences cause greater bond polarity

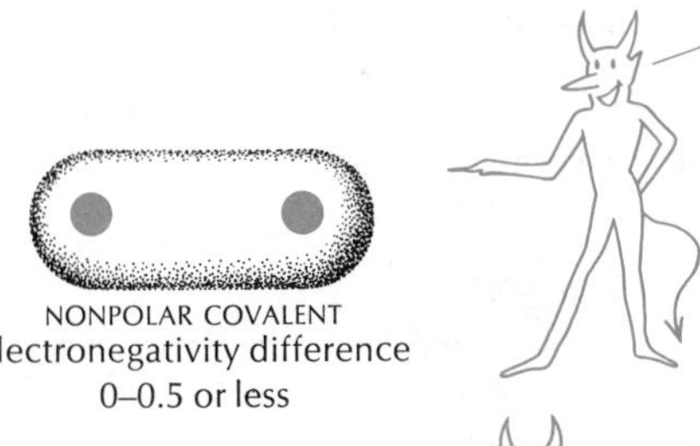

0–0.5 or less
Now, that's what I call sharing! These two atoms have about the same electronegativities, so each has a fair share of the electron pair. In these pictures, I've used circles to mean nuclei of atoms, and the shaded cloud to mean the electron pair.

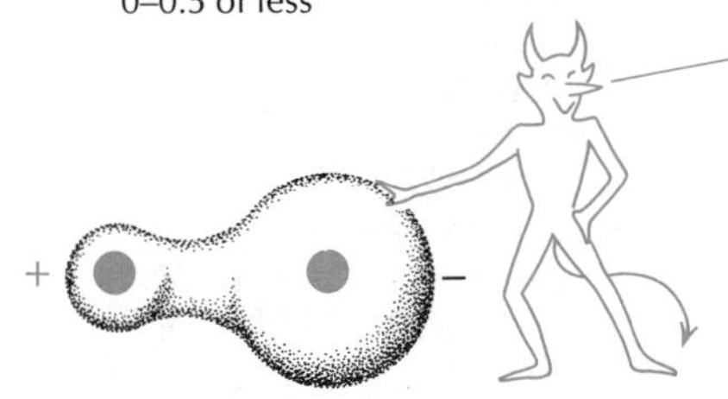

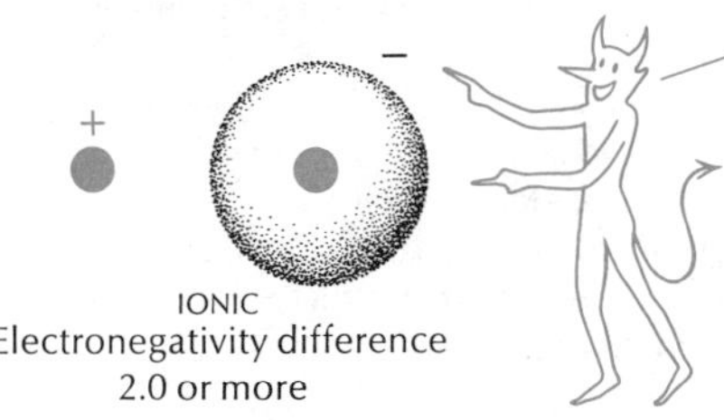

2.0 or more

classify the bond as nonpolar covalent. If it's between 0.5 and 2.0, it's polar covalent. 2.0 and above is ionic. We get these differences: a. N and H: 0.9; b. K and Br: 2.0; c. S and Cl: 0.5.

Answer: a. polar covalent; b. ionic; c. nonpolar covalent. ■

REVIEW QUESTIONS

1. How do we now define *chemical bond?* What are the *valence electrons* of representative and of transition elements?
2. What is the *octet rule,* and what are its exceptions?

Formation of Ionic Compounds

3. What is a *crystal lattice?* What holds it together?
4. What is a compound's *heat of formation?* How is it related to heat of reaction?
5. Is sodium chloride in a state of higher or lower energy than the elements sodium and chlorine?
6. What is *electron transfer?* Explain how to write electron transfer equations.

Describing Ionic Compounds

7. How are the sizes of ions different from the sizes of the atoms from which they are formed? Explain why they are different.

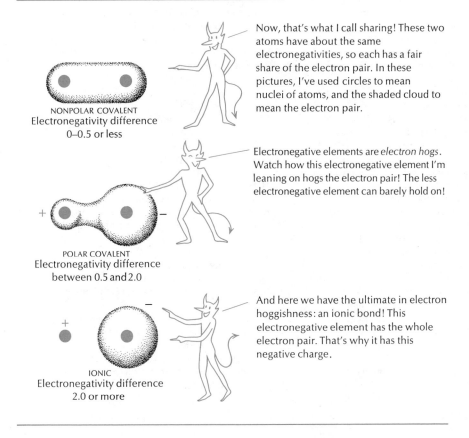

8. What are *first* and *second ionization energies?* Explain the relationship between these and ionic charge.
9. Describe and explain the trends in ionization energy in the periodic table.

Molecules and Covalent Bond Formation

10. How do we now define *molecule?* What is a *covalent bond?*
11. How is it possible for an atom to achieve an octet by covalent bond formation?
12. Why do hydrogen and the halogens exist as diatomic molecules? Draw Lewis structures for a diatomic hydrogen molecule and for one of the halogens.
13. What are double and triple covalent bonds? Give an example of each.
14. Write Lewis structures for compounds formed between hydrogen and the following: fluorine, oxygen, nitrogen, carbon.
15. What are *nonbonding electron pairs?* How many nonbonding electron pairs does each of the molecules in Question 14 have?
16. What is *bar notation?* Write Lewis structures for the compounds in Question 14 in bar notation.
17. How do the valence electrons of an element determine how many bonds it forms? Give the number of bonds for H, F, O, N, and C.
18. What is a *coordinate covalent bond?* Give an example. What other substances form coordinate covalent bonds? Explain why boron may have an incomplete octet.

Describing Covalent Bonds

19. What is a *molecular orbital?* Sketch the *molecular orbital representation* of the formation of a diatomic H_2 molecule. Show the *bond length.*
20. Show, with sketches, the difference between a *sigma molecular orbital* and a *pi molecular orbital.*
21. What is *bond energy?* What are some other words for it? What is the relationship between bond energy and bond length?

Practice in Writing Lewis Structures

22. What is a *central atom?* How do we decide which is the central atom in a binary covalent compound, polyatomic ion, or oxyacid?
23. Describe the steps in writing a correct Lewis structure from a formula.
24. How do we decide whether a Lewis structure should have double or triple bonds?
25. How does the Lewis structure of a polyatomic ion differ from the Lewis structure of a neutral molecule?

Bond Polarity

26. What is *electronegativity?* What does it depend on?
27. What is the most electronegative element? The least?
28. Describe a bond between a more electronegative element and a less electronegative element.
29. What are the electronegativity differences among *nonpolar covalent, polar covalent,* and *ionic bonds?*
30. What is a *dipole?* What is the symbol for a polar covalent bond?

EXERCISES

Set A (Answers at back of book. Asterisks mark the more difficult exercises.)

1. Write equations, using Lewis structures, that show electron transfer between these elements.
   a. Li and O   c. Al and H
   b. Mg and N   d. Rb and I
2. Predict which of these would have the larger radius.
   a. Ca or Ca^{2+} c. K^+ or Br^-
   b. O or O^{2-}
3. Without consulting Figure 9.2, predict which of these would be larger.
   a. the second ionization energy of Ca, or the second ionization energy of K
   b. the third ionization energy of Be or the third ionization energy of Al
   c. the first ionization energy of Na or the second ionization energy of Li
4. State the group in the periodic table that has, in general,
   a. the highest first ionization energy
   b. the lowest first ionization energy
5. Looking at the periodic table inside the covers of this book but without looking at Figure 9.3, predict which of these would have the higher first ionization energy.
   a. Na or Rb   c. Ba or Ca
   b. Al or Ar   d. N or Bi
6. Write Lewis structures for Cl_2, Br_2, and I_2.
7. The diatomic molecule S_2 exists, but it's not very stable. Write its Lewis structure.
8. Write Lewis structures for the compounds most likely to be formed between these pairs.

a. H and P
b. N and Cl
c. Br and O
d. Se and F

9. BF_3 can react with water to form a coordinate covalent bond. Write the Lewis structure for such a compound.

*10. The bond dissociation energy of an N—H bond is 93.4 kcal/mole. How much energy is required to take apart a mole of NH_3 molecules, containing 3 N—H bonds each?

11. Predict which of these would have the greater bond strength.
   a. H—H (bond length 0.0741 nm) or F—F (bond length 0.142 nm)
   b. C=N or C≡N

12. What is wrong with these Lewis structures?

   a. 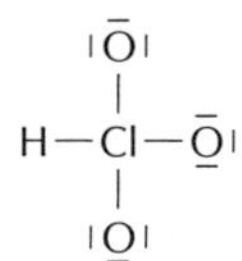 H—Cl—O (with two additional O atoms bonded to Cl above and below) for $HClO_3$

   b. 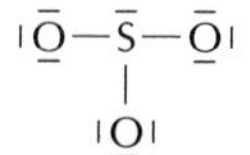 O—S—O (with an additional O atom bonded to S below) for SO_3

   c. 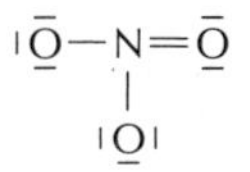 O—N=O (with an additional O atom bonded to N below) for NO_3^-

13. Write correct Lewis structures for the substances in Exercise 12.

*14. Boric acid, used diluted as an eyewash, has the formula H_3BO_3. Write its Lewis structure. (Boron does not have a complete octet.)

15. Write Lewis structures for H_2SO_4, HSO_4^-, and SO_4^{2-}. Show their similarities.

16. Write Lewis structures for SO_4^{2-}, PO_4^{3-}, ClO_4^-, and CCl_4. Point out their similarities.

17. Teflon is made from tetrafluoroethylene, C_2F_4. The two carbon atoms are bonded to each other, and two fluorine atoms are bonded to each carbon atom. Write its Lewis structure.

18. Write Lewis structures for these polyatomic ions.
   a. OH^-
   b. CN^-
   c. SH^-
   d. O_2^{2-}
   e. HCO_3^-
   f. ClO^-

19. Without looking at Figure 9.4, arrange each of these sets in order of increasing electronegativity.
   a. S, Cl, Se, F
   b. Rb, I, F, Sr

20. Using the values in Figure 9.4, classify the bonds that would be formed between these elements as nonpolar, polar covalent, or ionic.
   a. Al and Si
   b. Ba and Br
   c. C and Cl
   d. Be and Se

21. For the polar covalent bonds in Exercise 20, write out the bonds with the crossed arrow pointing at the more electronegative element.

22. We've seen that the distinctions between ionic and covalent compounds are not cut and dried; they are only guidelines. To prove this, select a pair of atoms in each of these categories (use Figure 9.4).
   a. a metal and a nonmetal, whose bond would be predicted to be polar covalent (by electronegativity difference) instead of ionic
   b. a nonmetal and a metalloid on the right-hand side of the staircase-shaped line, whose bond would be predicted to be ionic (by electronegativity difference) instead of covalent

---

Set B (Answers not given. Asterisks mark the more difficult exercises.)

1. Write equations, using Lewis structures, that show electron transfer between these elements.
   a. Al and N
   b. Ba and H
   c. K and S
   d. Be and O

2. Predict which of these would have the smaller radius.
   a. N or N^{3-}
   b. F^- or Li^+
   c. Al or Al^{3+}

3. Without consulting Figure 9.2, predict which of these energies would be smaller.
   a. the first or the second ionization energy of Rb
   b. the third ionization energy of Al or the third ionization energy of Mg
   c. the second ionization energy of Be, or the fourth ionization energy of Ga

4. Consider the element astatine, not shown in Figure 9.3.
   a. Would you expect its ionization energy to be larger or smaller than that of cesium?
   b. Would you expect its ionization energy to be larger or smaller than those of other members of its group?

5. Looking at the periodic table inside the covers of this book, but without looking at Figure 9.3, predict which of these would have the lower first ionization energy.
   a. Kr or He
   b. B or Al
   c. Mg or O
   d. Al or Cl

6. The halogens can form diatomic molecules with each other. These are called *interhalogen compounds*. Write Lewis structures for the interhalogen compounds ICl, BrF, ClF.

7. Write Lewis structures for compounds of hydrogen with Cl, S, P, and Si. Use both dot and bar notation.

8. Write Lewis structures for the most likely compounds formed between these pairs.
   a. O and F
   b. P and Br
   c. B and Cl
   d. C and S

9. BF_3 can react with a fluoride ion, F^-, and complete boron's octet by forming a polyatomic negative ion, BF_4^-. Write its Lewis structure.

*10. To take apart a mole of H_2O molecules requires 221 kcal. What is the bond dissociation energy of an O—H bond?

11. Predict which of these would have the greater bond strength.
    a. C—O or C=O
    b. Br—Br (bond length 0.228 nm) or Cl—Cl (bond length 0.199 nm)

12. What is wrong with these Lewis structures?

    a. 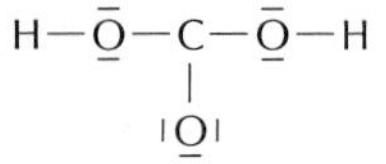 H—O—C—O—H for H_2CO_3, with |O| bonded below the C

    b. 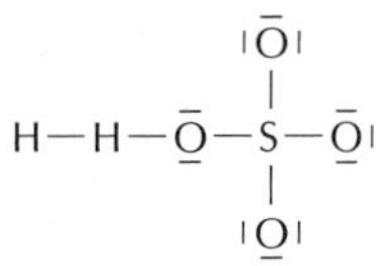 H—H—O—S—O| for H_2SO_4, with |O| bonded above and |O| bonded below the S

    c. 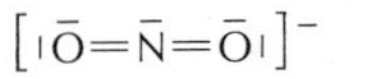 [|O=N=O|]⁻ for NO_2^-

13. Write correct Lewis structures for the substances in Exercise 12.

14. Write the Lewis structure of the borate ion, $B(OH)_4^-$.

15. Write Lewis structures of NO_3^-, CO_3^{2-}, and SO_3. Point out their similarities.

16. Write Lewis structures for NO_2^- and ClO_2^-. Point out their differences.

17. Hydrazine, N_2H_4, is used for rocket fuel. The two nitrogen atoms are bonded together, and two hydrogen atoms are bonded to each nitrogen. Write its Lewis structure.

18. Write Lewis structures for these compounds.
    a. formaldehyde, CH_2O (C is the central atom, and all other atoms are bonded to it.)
    b. ethane, C_2H_6 (The two carbon atoms are bonded together.)
    c. vinyl chloride, C_2H_3Cl (The two carbon atoms are bonded together. Two hydrogen atoms are bonded to one carbon atom, and a hydrogen and a chlorine are bonded to the other carbon atom.)

19. Write Lewis structures for these polyatomic ions.
    a. HPO_4^{2-} c. CO_3^{2-} e. SO_3^{2-}
    b. SO_4^{2-} d. ClO_3^- f. NO_2^-

20. Write Lewis structures for these oxyacids.
    a. HNO_3 c. $HClO$
    b. H_3PO_4 d. H_2SO_4

21. Without looking at Figure 9.4, arrange each of these sets in order of increasing electronegativity.
    a. B, Li, N, F b. H, Rb, O, Be

22. Using the values in Figure 9.4, classify the bonds that would be formed between these elements as nonpolar, polar covalent, or ionic.
    a. C and O c. S and I
    b. H and O d. Ca and O

23. For the polar covalent bonds in Exercise 22, write out the bonds with the crossed arrow pointing at the more electronegative element.

CHAPTER 10

Changes in States of Matter

LEARNING OBJECTIVES

After studying this chapter, you should be able to:

1. Supply a correct definition, explanation, or example for each of these:

atmospheric pressure	vapor pressure
atmosphere	dynamic equilibrium
equilibrium	boiling point
static equilibrium	sublimation
Le Chatelier's Principle	melting point
barometer	freezing point
torr	hydrogen bond
standard temperature and pressure	bond angle
vapor	hydrogen-bonded molecular crystal
	covalent crystal

2. Explain or interpret a heat-temperature graph such as the one in Figure 10.1.
3. Calculate the amount of heat in a temperature and state change from appropriate data.
4. Explain how a barometer works, and be able to answer questions about atmospheric pressure.
5. Explain the effects of temperature and pressure on a substance's physical state.
6. Explain the equilibrium between a substance and its vapor, and predict the effect of pressure, temperature, or other conditions on that equilibrium.
7. Explain what happens when a substance boils or freezes.
8. Explain the relationships among heat of fusion, heat of vaporization, melting and boiling points, molecular weight, and attractions among particles.

We use changes of state every day. When we boil water, dry clothes, or use a refrigerator, changes of state are working for us. We also make use of the fact that certain things are gases, liquids, and solids at room temperature. If water weren't a liquid, air weren't a gas, and rocks and metals weren't solids at room temperature, our life on earth (if any) would be far different from what it is.

In this chapter, we'll take a closer look at the states of matter themselves and at what happens when a substance changes from one state to another. We'll examine the interactions between the small particles—atoms or molecules or ions—that make up these substances and see what interactions cause their states to be what they are at room temperature.

No matter what state a substance is in at room temperature—gas or liquid or solid—we can change it to one of the others by changing the conditions around it. Two of these conditions are temperature and pressure.

10.1 TEMPERATURE AND HEAT

Figure 10.1 shows what happens to H_2O when we start at temperatures near absolute zero and increasingly add energy. (Throughout these discussions, we'll use the formula "H_2O" to cover all states of the compound water; we'll use the word "water" for the liquid state, "ice" for the solid state, and "steam" or "water vapor" for the gas state.) Three concepts we saw briefly in Chapter 3—specific heat, heat of fusion, and heat of vaporization—are illustrated again here.

At 0 K, or absolute zero, all motion has stopped and everything is a solid. The first sloping portion of the graph indicates the specific heat of ice (0.5 cal/[g $\times$ K]). Next, the first flat portion of the graph occurs at the melting point of ice. Here, the temperature remains constant while enough energy is supplied to melt *all* the ice. The energy needed to change H_2O from its solid state to its liquid state at its melting point is the heat of fusion (79.9 cal/g). The next sloping part of the graph indicates the specific heat of water (1 cal/[g $\times$ K]). Then we come to the boiling point of water, where the graph flattens out again. The temperature stays constant while enough energy is supplied to boil *all* the water. This energy is the heat of vaporization (540 cal/g). After that, the graph slopes up again, indicating the specific heat of steam (about 0.5 cal/[g $\times$ K]).

We see from Figure 10.1 that the heat of vaporization of water is much larger than the heat of fusion of ice. This imbalance is typical of most substances, because the difference between a liquid and a gas is much greater than that between a solid and a liquid. A liquid is like a rather disorganized solid. The particles aren't held rigidly together in the liquid, but the attractive forces between them are approximately like those in the solid. In a gas, though, the particles are very far apart and the attractions are slight. It takes much more energy to change a liquid to a gas than it does to change a solid to a liquid.

FIGURE 10.1
Temperature-heat curve for water

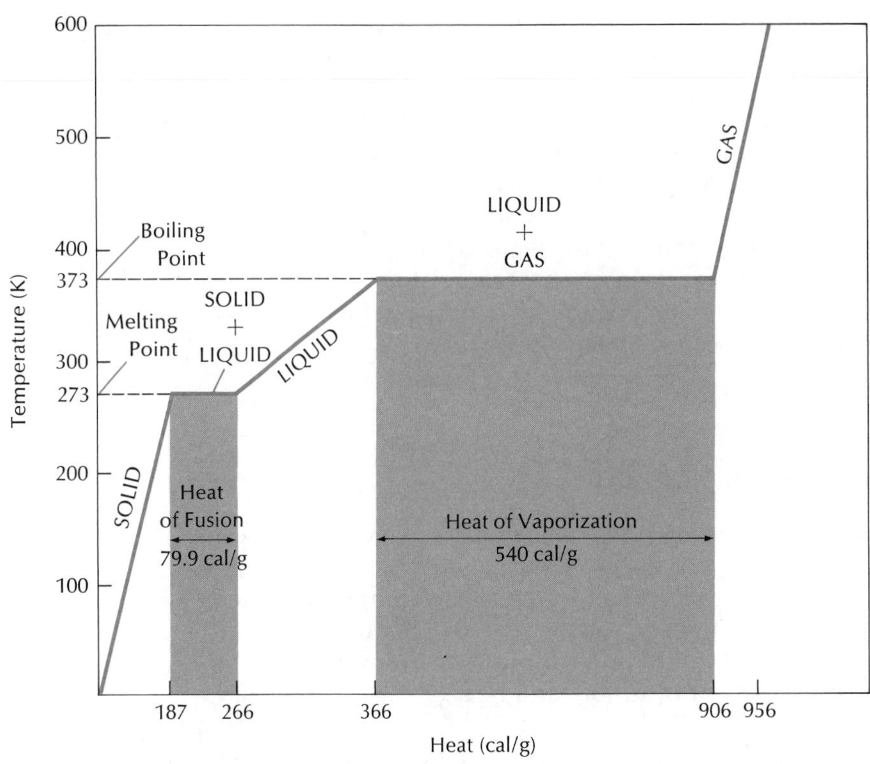

We can easily calculate how much heat is involved in taking H_2O through a series of changes in physical state.

EXAMPLE 10.1: A freezer normally operates at about −15°C (258 K). If 5 kg of ice is removed from a freezer and placed in a picnic cooler, how much heat (in kilocalories) will be absorbed as the ice melts and the water changes to room temperature (295 K)?

Solution: We want to convert 5 kg of ice to kilocalories over a temperature range of 258 K to 295 K. Figure 10.1 shows us that three things happen over this temperature range: the ice warms from 258 to 273, then it melts at 273, then water warms from 273 to 295. We must do this problem in three stages, using three conversion factors: the specific heat of ice (0.5 kcal/[kg × K]), the heat of fusion of ice (79.9 kcal/kg), and the specific heat of water (1 kcal/[kg × K]).

$$\text{To heat ice: } 5\,\cancel{\text{kg}} \times 0.5\,\frac{\text{kcal}}{\cancel{\text{kg}} \times \text{K}} \times (273 - 258)\cancel{\text{K}} = 37.5\text{ kcal}$$

$$\text{To melt ice: } 5\,\cancel{\text{kg}} \times 79.9\,\frac{\text{kcal}}{\cancel{\text{kg}}} = 399.5\text{ kcal}$$

$$\text{To heat water: } 5\ \cancel{kg} \times 1\,\frac{kcal}{\cancel{kg} \times \cancel{K}} \times (295 - 273)\cancel{K} = 110 \text{ kcal}$$

$$\text{Total } \quad 547 \text{ kcal}$$

Answer: 500 kcal. ■

Most of the heat that's absorbed is taken up in melting the ice. By comparison, heating the ice and the water requires much less heat. That's why ice is a better cooling agent than water at 273 K.

We can reverse this process by starting at the right of the graph and moving to the left. There's no upper limit of temperature, so let's start with steam at 600 K. As the steam cools, it gives off energy according to its specific heat until it gets to the condensation point (the same temperature as the boiling point). Then the temperature stays constant while all the steam condenses to liquid water. In this interval, the heat of condensation (the same as the heat of vaporization) is given off. Then as more energy is removed, liquid water cools according to its specific heat. At the freezing point (the same temperature as the melting point), the temperature stays constant until all the water has frozen. The heat of crystallization (the same as the heat of fusion) is given off. Then the ice cools according to its specific heat. We see that the same amount of energy is involved whether we heat or cool a substance. In the first case, the energy is supplied; in the second, it's released.

EXAMPLE 10.2: A heating plant might release steam at 110°C (383 K). Calculate the heat given off when 10 grams of steam at that temperature are condensed on a person's hand and then cooled to body temperature (310 K).

Solution: We want to convert 10 grams of steam to calories. Again, Figure 10.1 shows that three stages occur in the temperature range of 383 K to 310 K. First, the steam cools from 383 K to 373 K. Then the steam condenses at 373 K. Then the water cools from 373 K to 310 K. This gives us three conversion factors: the specific heat of steam (0.5 cal/[g × K]), the heat of condensation of steam (540 cal/g), and the specific heat of water (1 cal/[g × K]).

$$\text{To cool steam: } 10\ \cancel{g} \times 0.5\,\frac{cal}{\cancel{g} \times \cancel{K}} \times (383 - 373)\cancel{K} = 50 \text{ cal}$$

$$\text{To condense steam: } 10\ \cancel{g} \times 540\,\frac{cal}{\cancel{g}} = 5400 \text{ cal}$$

$$\text{To cool water: } 10 \text{ g} \times 1\,\frac{cal}{\cancel{g} \times \cancel{K}} \times (373 - 310)\cancel{K} = 630 \text{ cal}$$

$$\text{Total } \quad 6080 \text{ cal}$$

Answer: 6000 cal, or 6 kcal. ■

We see why a steam burn is so much worse than a boiling water burn. The heat that the water gave off in cooling from 373 K to 310 K is much less than the heat the steam gave off by condensing to water at 373 K. This high heat of condensation of steam makes it good for heating buildings, though.

Room temperature (about 295 K) occurs in Figure 10.1 where water is in its liquid state. Most substances have temperature-heat curves similar to water's. Of course, the numbers can differ, which is why some things are gases, others liquids, and still others solids at room temperature.

10.2 PRESSURE

Whenever we push on something, we exert a pressure on it. Anything that has weight can exert pressure. The air we live in is a mixture of gases that have volume and mass and weight. Thus air can exert pressure.

ATMOSPHERIC PRESSURE. The earth has air around it because it has enough gravity to hold onto the air. (The moon has gravity, too, but not enough to hold an atmosphere.) Air extends to about 500 miles above the earth, but it becomes less dense the further away it is. This is because the force of gravity grows weaker as the distance from the center of the earth increases. Figure 10.2 shows the various layers of our atmosphere and their relative densities.

In a game called "pig pile," people lie on top of each other in a pile of bodies. The person on the bottom feels more pressure than someone in the middle, who feels more than someone near the top. If we cut an imaginary column of air one meter square from where the atmosphere meets the earth to the top of the atmosphere, we'd be creating a kind of "pig pile" of air molecules. The more molecules that are present, the greater the pressure. The molecules on the bottom of the column (and the earth itself) feel a lot of pressure from all those molecules above. This is called *atmospheric pressure*. Figure 10.3 shows that atmospheric pressure is greater at sea level than it is on a mountain top, because the column of air extending from sea level contains more molecules than the one extending from a mountain top.

atmospheric pressure

We measure pressure by how much force is exerted on how much surface. This translates as force per unit of area. We fill our tires with air according to the number of pounds per square inch. In the English system, the atmospheric pressure at sea level is 14.7 pounds per square inch, because 14.7 pounds of air push down on a square inch of earth. In chemistry and other sciences, this pressure is called 1 *atmosphere* (*atm*). One atmosphere can be expressed in metric units as 1.04 kilograms per square centimeter, or in the SI system as 9.9×10^{-6} pascals. (See Appendix B for the definition of a pascal.)

atmosphere (atm)

Notice that a *pressure of 1 atmosphere* means the atmospheric pressure at sea level. *Atmospheric pressure* means simply the pressure of the atmosphere wherever it's measured. Unless the measurement is taken below sea level, atmospheric pressure is usually less than 1 atmosphere.

Since we live at atmospheric pressure, it might not seem as if any pressure is pushing down on us. After all, we don't feel anything. This is because our bodies are adjusted so that our internal pressure pushes *out* just to equalize the atmospheric pressure that pushes *in*. That is, our bodies are in a state of equilibrium with the atmospheric pressure. *Equilibrium* means

equilibrium

FIGURE 10.2
Density of air relative to distance above sea level

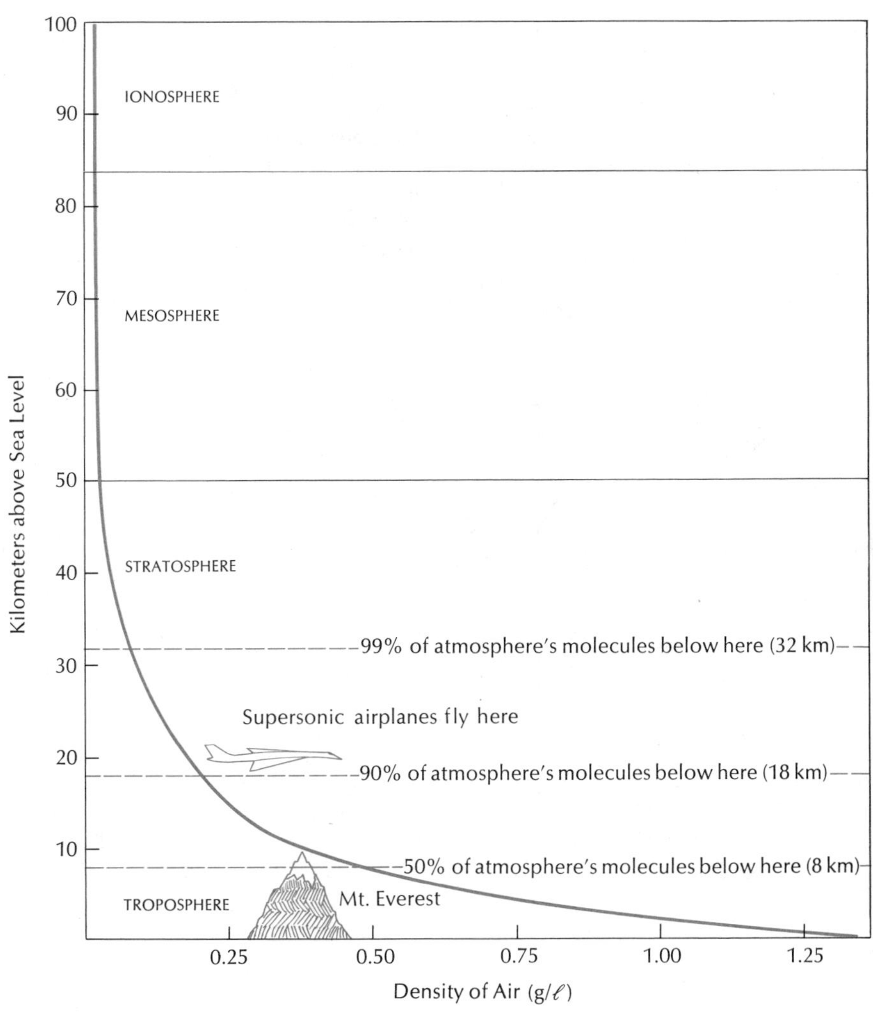

a static or dynamic state of balance between opposing forces. ("Static" means "standing still"; "dynamic" means "in motion.") Our bodies and atmospheric pressure are in a state of *static equilibrium* since the equal pressures don't change unless something disturbs the equilibrium. We'll see many examples of dynamic equilibrium a little later, and the contrast will become clear.

static equilibrium

States of equilibrium are often disturbed. Henri Le Chatelier stated a principle to explain what happens when an equilibrium is disturbed. Figure 10.4 illustrates *Le Chatelier's Principle,* which says that an equilibrium system, when disturbed, adjusts itself so as to restore equilibrium.

Le Chatelier's Principle

FIGURE 10.3
Pressure varies with altitude, because different amounts of molecules push down

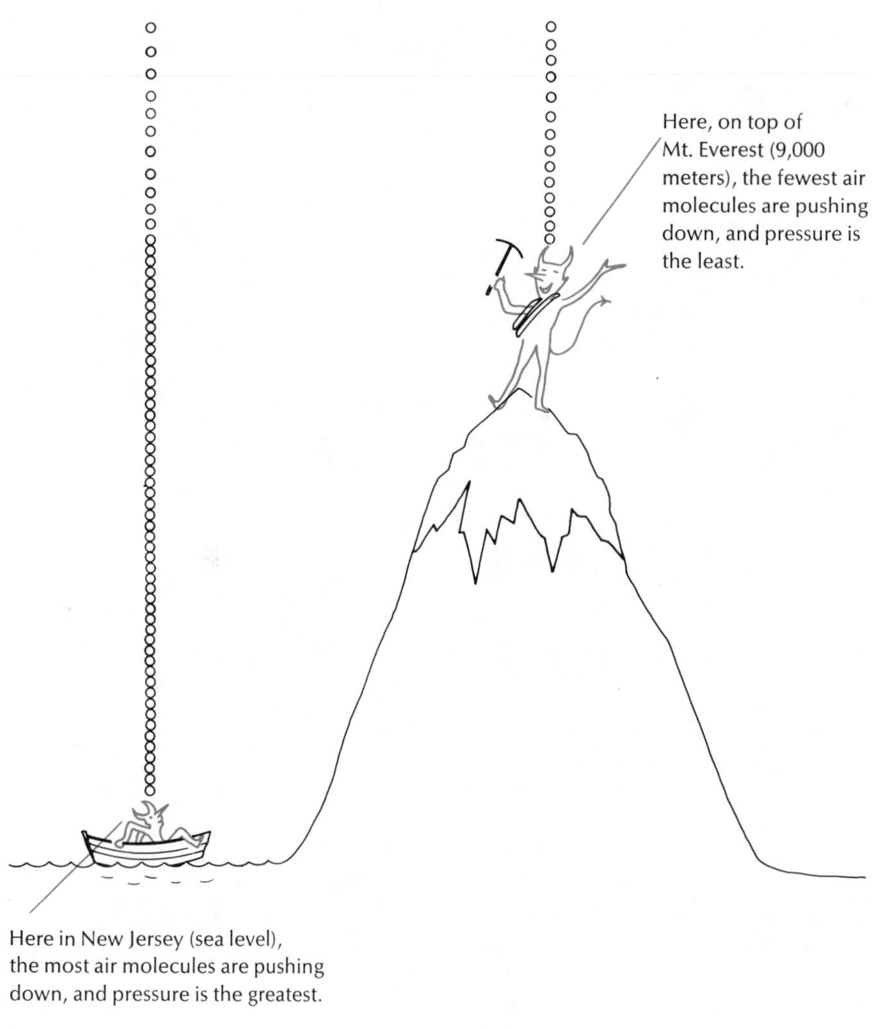

Here, on top of Mt. Everest (9,000 meters), the fewest air molecules are pushing down, and pressure is the least.
Here in New Jersey (sea level), the most air molecules are pushing down, and pressure is the greatest.

When we put our bodies into situations where the external pressure is different from the internal pressure, we're placing a stress on the pressure equilibrium between our bodies and their environment. If we climb a 2000-meter mountain, where the atmospheric pressure is lower than at sea level, then the equilibrium will be disturbed. However, given a little time, our bodies will adjust and the equilibrium will shift toward lower pressure. People with high blood pressure, especially, have to take it easy after experiencing large altitude changes, since a sudden release of external pressure can cause blood vessels to rupture. If this happens in the brain, it's a stroke.

FIGURE 10.4
An illustration of Le Chatelier's Principle

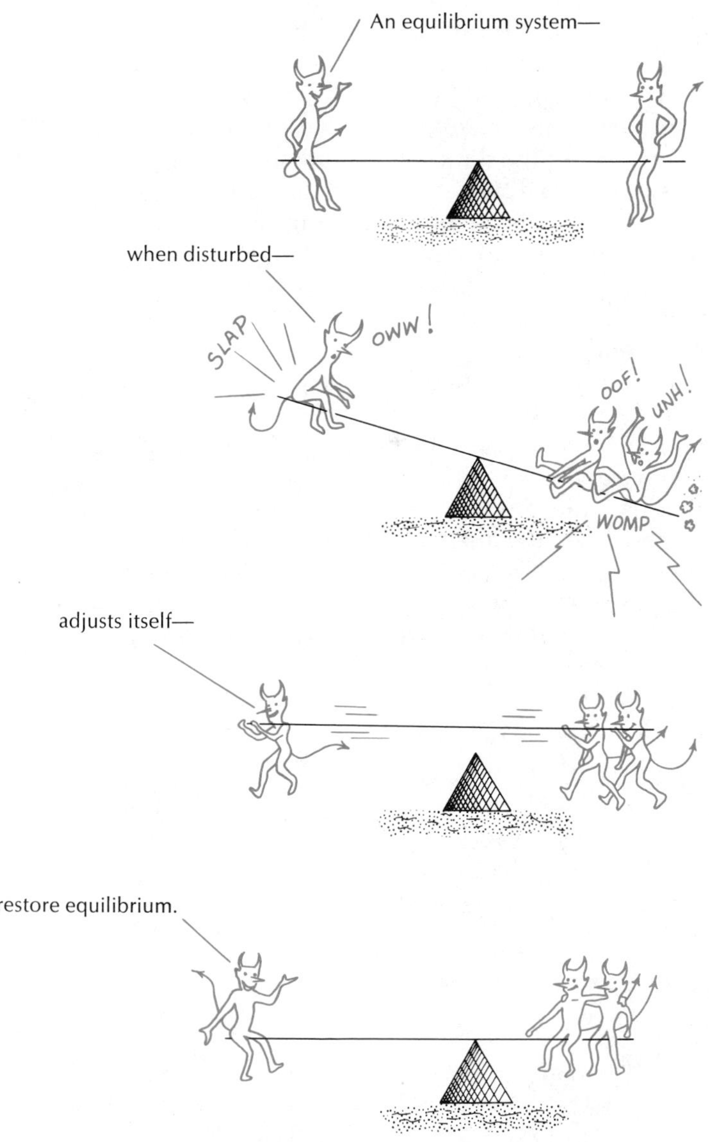
An equilibrium system—
when disturbed—
SLAP
OWW!
OOF!
UNH!
WOMP
adjusts itself—
so as to restore equilibrium.

A system succeeds in reestablishing equilibrium only if it hasn't been disturbed too much. How much is "too much" depends on the system itself. Although our bodies can adjust to relatively small pressure differences, they would explode in outer space, where the pressure inside our bodies would be pushing out and nothing would be pushing in. Astronauts wear pressurized space suits, which create a human-sized environment of atmospheric pressure. Airplane cabins are also pressurized to compensate for the lower pressure at high altitudes.

barometer

We measure pressure with a *barometer.* We can make a barometer with a glass tube, sealed at one end, and a flat-bottomed bowl. We fill the tube with a liquid like mercury and invert the tube carefully in the bowl. The liquid will drop to some level, but it won't all run out. The liquid stops running out when an equilibrium is established between the pressure caused by its weight in the tube and the atmosphere pressure on the mercury in the dish. If the liquid is mercury and the atmospheric pressure is 1 atmosphere, then the mercury will drop to a height of 760 millimeters. A millimeter of mercury in pressure is also called a *torr* after Evangelista Torricelli (1608–1647), who first devised the mercury barometer. (One torr = 133.3 pascals.)

torr

If we cut off the sealed end of the tube, then the mercury would all run out, because the same atmospheric pressure would be pushing down on the mercury in the tube *and* on the surface of the mercury in the dish. We could make the mercury rise in the tube again by sucking on it with a vacuum pump. The vacuum pump decreases the pressure inside the tube. When the pressure inside the tube is less than atmospheric pressure, the atmospheric pressure pushes the mercury up the tube. But no matter how hard we pump, we can never get the mercury above 760 millimeters, because that's as far as a pressure of 1 atmosphere can push the mercury. Figure 10.5 illustrates how a barometer works.

Mercury is a better barometer liquid than water. Since water is 1/13.6 as

FIGURE 10.5
A barometer measures the height of a liquid pushed up a sealed tube by atmospheric pressure

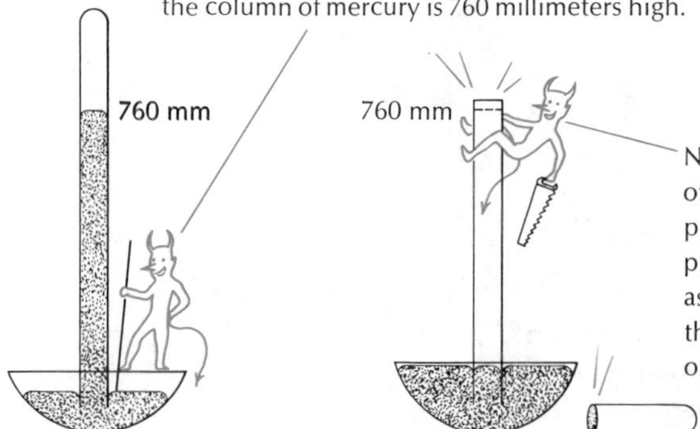

The atmospheric pressure in my lab today happens to be exactly 1 atmosphere. This pressure pushes down on the surface of the mercury in the dish. Since the other end of the tube is sealed off, the only thing that can push back is the weight of the column of mercury. This system is at equilibrium when the column of mercury is 760 millimeters high.
760 mm
760 mm
Now, I've sawed off the sealed end of the tube, to let the atmospheric pressure in. The same pressure is pushing down from inside the tube as outside, so there's no reason for the mercury to stay up. It all runs out.

dense as mercury, 1 atmosphere would push a column of water up 13.6 times higher than mercury, or 10.3 meters. We'd need a tall ladder to read this barometer. For this reason also, a water well can't be any deeper than 10.3 meters if the air pressure is supposed to push the water out. Pumping water out of a well is like sucking on a straw. Sucking on a straw decreases the pressure inside the straw, and the atmospheric pressure pushes the liquid up the straw. The atmospheric pressure has no trouble pushing liquids up a straw, because the straw is very short. Pumping water out of a well decreases the pressure at the top of the well, and the atmospheric pressure pushes the water up to the top. But if the well were deeper than 10.3 meters, the pump couldn't get the water all the way up, no matter how hard it pumped. In that case, the well would have to be sealed off, and a pressure greater than atmospheric pressure would have to be applied to the water in it.

PRESSURE AND A CONFINED GAS. Eighteen grams of water occupy only 18 milliliters as a liquid, but 22.4 *liters* as a vapor. This means that less than 0.08 percent of the volume of water vapor is actually occupied by the water molecules. The rest is empty space, which the gas molecules create among themselves by moving fast and bumping into each other frequently.

To measure a solid, it's most convenient to weigh it. We can also weigh a liquid, but since we have to put it in a container, we may as well put it in a marked container and measure its volume instead. Neither weighing nor measuring volume is the best way to measure a gas, because a gas takes up so much space and occupies all of the volume of any container. Instead, we measure the pressure of a gas. A confined gas creates pressure by bumping into the sides of its container. The more molecules there are to bump, the greater the pressure. Thus pressure is a measure of how much of a gas we have.

We can apply pressure to a confined gas, just as we could apply heat to a solid, and change the substance's state. If we start near zero pressure and increase the pressure continuously, we'll push the gas molecules closer together. At some point they will be pushed closely enough to form a liquid or even a solid. So-called bottled gases—like the butane used for cigarette lighters, or the propane used for camping stoves—are liquids when they are in their tanks under pressure. Turning a valve releases the pressure, and they come out as gases.

EFFECTS OF TEMPERATURE AND PRESSURE ON STATE.

From the discussion so far, we can see that both temperature *and* pressure affect states of matter. The temperature-heat curve for water in Figure 10.1 was really for a pressure of 1 atmosphere. In fact, the boiling point of water is not always 373 K. In the mountains, it is lower than that, as Table 10.1 shows. This is because boiling point decreases as atmospheric pressure decreases.

We see all around us the combined effects of temperature and pressure

TABLE 10.1
Variation of water's boiling point with atmospheric pressure

Location	Altitude (meters)	Pressure (atmospheres)	Boiling Point of Water (K)
Death Valley, California	−85	1.04	374
Sea level	0	1.00	373
Mt. Carmel, Israel	550	0.930	371
Mt. Vesuvius, Italy	1200	0.865	369
Mt. Olympus, Greece	3000	0.695	363
Pikes Peak, Colorado	4300	0.593	359
Mt. Kilimanjaro, Tanzania	5900	0.487	354
Mt. K2, Kashmir	8600	0.336	345

Brr! Here on Mt. K2 in Kashmir, it's taking me 24 minutes to cook a 3-minute egg. I wouldn't even be able to cook beans. Water doesn't get hot enough.

on states of matter. When bottled propane or butane is released from its tank, the tank nozzle might cool noticeably. In changing from a liquid to a gas, because of released pressure, the substance absorbs its heat of vaporization from the closest thing around. In a refrigerator, substances that are normally gases, such as ammonia, sulfur dioxide, or Freon (CF_2Cl_2), are compressed to liquids and pumped through the refrigerator coils. As the liquid vaporizes, it takes its heat of vaporization from its surroundings (the coils), cooling the coils and the inside of the refrigerator. Then the compressor changes the gas back into a liquid, releasing heat into the room, and the cycle repeats.

Because temperature and pressure affect a substance's state, we can't specify a state without also noting what the temperature and pressure are. A substance at 273 K and a pressure of 1 atmosphere is said to be at *standard temperature and pressure (STP)*.

standard temperature and pressure (STP)

10.3 EQUILIBRIA AMONG STATES

Water left in an uncovered glass will eventually evaporate. We can smell mothballs because some of the molecules have escaped into the air and found their way to our noses. Mothballs left out in the open will also eventually disappear.

Some of a liquid or solid always sneaks into the gas state, even though the substance is below its melting point or boiling point. The molecules that escape from a solid or a liquid make up its *vapor*. The pressure exerted by this vapor is the substance's *vapor pressure*. Everything has a vapor pressure, but some substances have more vapor pressure than others. (Tungsten, which doesn't vaporize much at all, is sometimes described as having a vapor pressure of "one atom per universe.")

vapor

vapor pressure

VAPOR PRESSURE. Although water in an uncovered glass will evaporate, water in a covered glass will not. Both, however, have vapor pressure. Molecules *are* escaping to the vapor in the covered glass, but molecules are also returning to the liquid state.

Here's our first example of *dynamic equilibrium,* which we run into a lot in chemistry. Two opposing processes are happening at once and at the same rate, so that the net result stays the same. Imagine an old silent film comedy, the action taking place in a grocery store. One of the comedians thinks he is supposed to be stacking cans. The other thinks he is supposed to be taking the cans down and putting them in their boxes. Both scurry around doing their jobs (with the camera going double-speed), oblivious to the other's activities. The result is that the pile of cans stays the same height, even though cans are being moved around at a furious rate. This is an example of dynamic equilibrium: something is being *done* at the same rate it's being *undone.*

dynamic equilibrium

Now, let's look again at the water in the sealed glass. Two processes are going on: (1) molecules are escaping from the liquid and going into the vapor; and (2) molecules are leaving the vapor and going into the liquid. Although the rates of these processes will differ at first, the vapor molecules will soon return to the liquid at the same rate the liquid molecules escape into the vapor. When this happens, the system is at equilibrium. At equilibrium, the net number of molecules in the vapor and the net number in the liquid always stay the same, even though both processes are still going on.

Of course, equilibrium is established only if the container is covered. This is one of many examples of Le Chatelier's Principle applied to dynamic equilibrium. We can write the equilibrium between water and water vapor like this:

$$H_2O(l) \rightleftharpoons H_2O(g)$$

Here, the double arrows mean a state of dynamic equilibrium. If the container is uncovered, molecules from the right (the vapor) side will escape to the outside world, disturbing the equilibrium. To attempt to reestablish equilibrium, more molecules from the liquid will go into the vapor. The equilibrium won't be reestablished until either the glass is covered or the liquid is evaporated. We could disturb the equilibrium in the opposite direction, too, by adding more vapor molecules (say, by piping steam into the glass). Then some of the excess vapor would condense.

Disturbing an equilibrium system shifts the equilibrium. Removing water vapor shifts the equilibrium to the vapor side. Adding water vapor shifts the equilibrium to the liquid side. Figure 10.6 shows the position of equilibrium shifting towards water vapor.

The air around us always contains some water vapor. There is a limit, though, to how much water vapor the air can hold at each temperature. In humid weather, our perspiration doesn't evaporate, our clothes don't dry, and crackers and potato chips get soggy. This is another example of Le Chatelier's Principle. If the air has as much, or nearly as much, water vapor as it can hold, then the vapor molecules have nowhere to go but back into

FIGURE 10.6
When an uncovered liquid evaporates, the equilibrium shifts toward the vapor

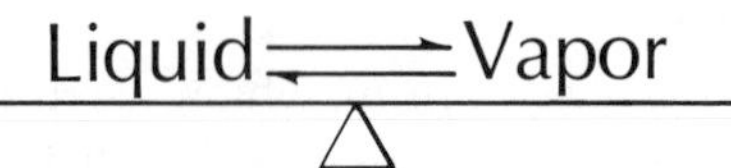

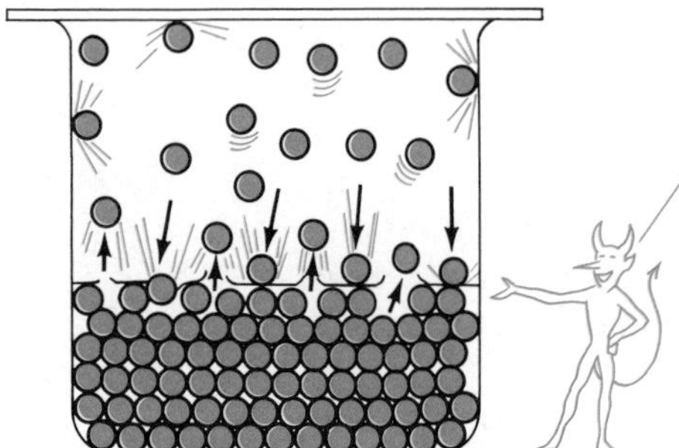

See this molecule entering the liquid phase? There are three others doing that in this picture. There are also four molecules going from the liquid to the vapor in this sealed container. We have a state of *equilibrium*.

Liquid
Vapor

Somebody left the cover off this beaker of water! This molecule I'm pointing to is the only one I can see that's going *into* the liquid phase. We still have four molecules leaving, though, just as before.

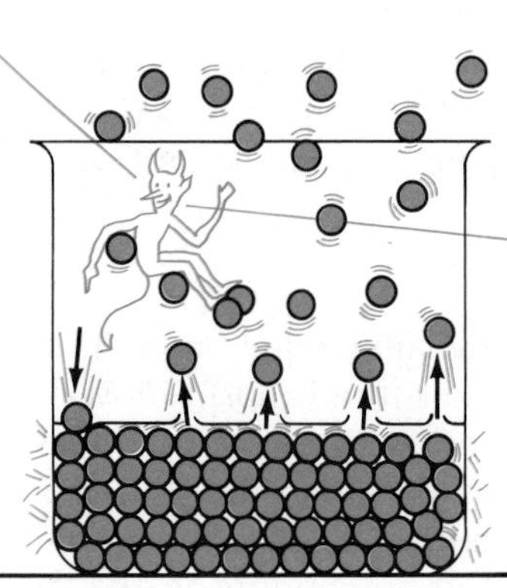

This system is *not* at equilibrium. It's trying to be, though. It won't succeed until all of the liquid is gone.

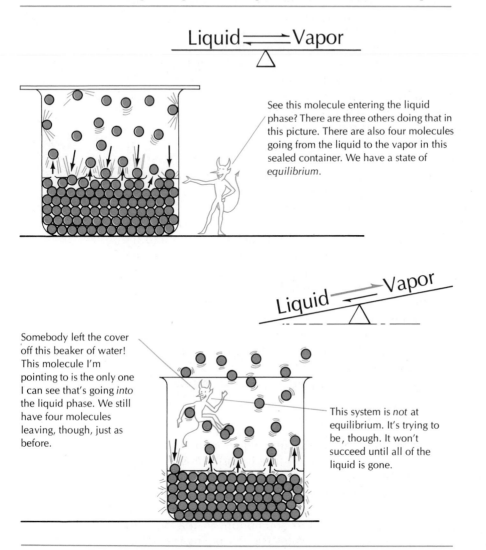

the liquid state. The equilibrium shifts toward the liquid. Warm air can hold more water vapor than cold air, so high humidity and hot weather often go together.

The air is much drier in cold climates than in warm climates. If the air is dry, the equilibrium between water and water vapor shifts toward the vapor, and substances lose moisture. Furniture, musical instruments, and our lips and nasal passages are accustomed to a certain amount of water vapor. If that amount drops drastically, they may develop cracks as they dry out. A humidifier can be used to supply the air with water vapor.

VAPOR PRESSURE AND TEMPERATURE. Why should a liquid or a solid have a vapor pressure? Why don't all of the molecules just stay in their proper states? To answer these questions, we need to think some more about temperature. We said in Chapter 3 that substances are at higher temperatures because their molecules are moving faster. But the temperature of a substance represents only an average of the energies of all the molecules in a substance.

In any group of shoppers, a few will be moving very fast, a few will be moving very slowly, and most will be moving at a speed somewhere in between. When we look at molecules, we see that they behave the same way. Some molecules in a substance move very fast, some move slowly, and most move at a speed somewhere in between. We can plot the percentage of molecules having certain energies against the energies themselves, and we always get a Maxwell-Boltzmann Distribution Curve, shown in Figure 10.7. At a given temperature, some molecules will have enough energy to escape from the liquid or solid and go into the vapor. As the temperature is increased, more molecules will acquire more energy and enter the vapor. Then the vapor pressure increases.

BOILING POINT. Up to the boiling point, the atmospheric pressure pushes down on the liquid, making most of the molecules stay in the liquid. If we steadily increase the temperature, the vapor pressure increases, too. When we reach a temperature where the vapor pressure of the liquid equals the external pressure that's pushing down on it, then vapor can form anywhere where the liquid is touching the container. The bubbles in a pan of boiling water are water vapor that has formed on the bottom and sides of the pan. Figure 10.8 shows that the temperature at which vapor pressure and external pressure are equal is a liquid's *boiling point.* If the external pressure is less than one atmosphere, then the liquid will boil at a temperature lower than its normal boiling point.

boiling point

We can cause water to boil at a temperature higher than its normal boiling point by going below sea level or by using a pressure cooker. In a pressure cooker, a valve maintains the inside pressure at 5 pounds per square inch (psi), 10 psi, or 15 psi *more* than atmospheric pressure. (The pressure becomes higher than atmospheric pressure since the steam molecules can't escape.) For water to boil, the vapor pressure of water must also increase, and that means increasing the temperature. At 5 psi (1.3 atm), water boils at 381 K; at 10 psi (1.7 atm), 389 K. For every 10 K increase in temperature, we cut the cooking time about in half. Cooking is a chemical reaction, and we'll see in Chapter 16 that increasing the temperature by 10 K doubles the speed of most chemical reactions. In the same way, we can compute the approximate time it takes to cook things in the mountains. For a 10-K decrease in the boiling point of water, cooking will take twice as long as it would at sea level.

SUBLIMATION. There is an equilibrium between solid and vapor, just as there is between liquid and vapor. Going directly from the solid state to the gas state is called *sublimation.* Almost any solid will sublime

sublimation

FIGURE 10.7
Maxwell-Boltzmann Distribution Curves

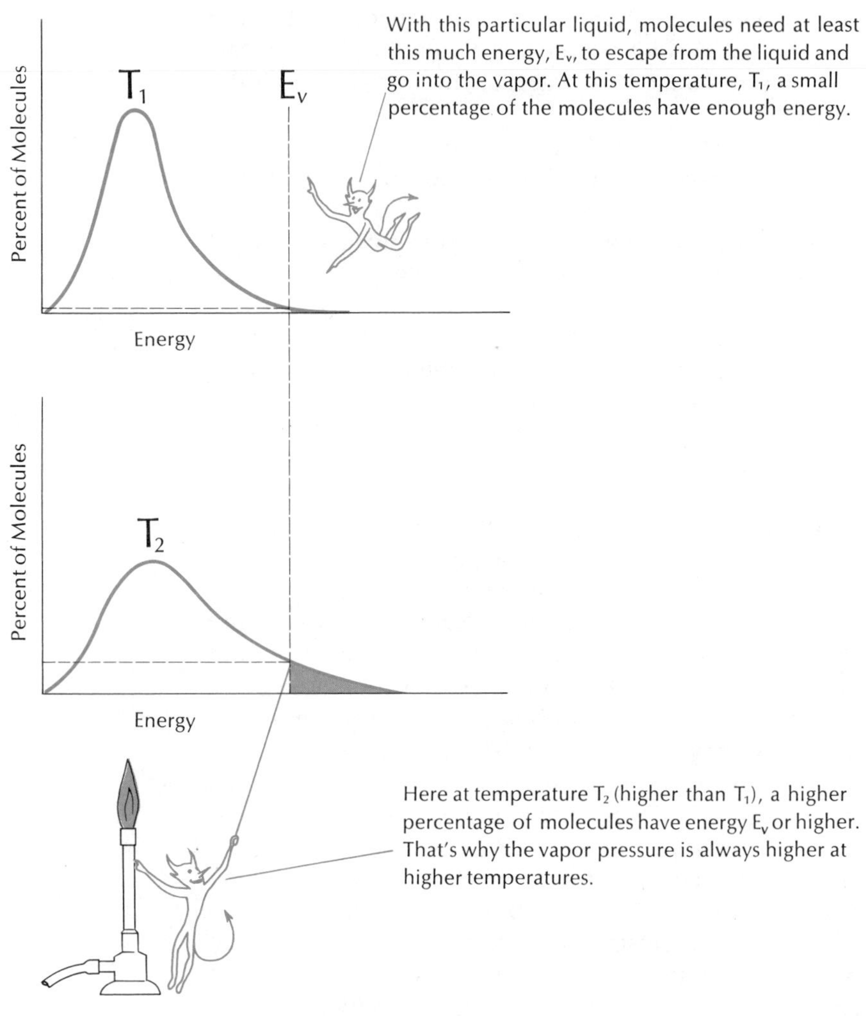

if it is subjected to low enough pressures, because the vapor molecules are removed and the equilibrium shifts toward the vapor. Some things sublime at atmospheric pressure. Dry ice (solid CO_2) sublimes instead of melting. So do mothballs.

With this particular liquid, molecules need at least this much energy, E_v, to escape from the liquid and go into the vapor. At this temperature, T_1, a small percentage of the molecules have enough energy.
T_1
E_v
Percent of Molecules
Energy
T_2
Percent of Molecules
Energy
Here at temperature T_2 (higher than T_1), a higher percentage of molecules have energy E_v or higher. That's why the vapor pressure is always higher at higher temperatures.

Ice has a vapor pressure lower than that of dry ice, but it can sublime because of Le Chatelier's Principle. Frozen clothes on a line will dry, especially if it's windy, because the changing air keeps removing the water vapor and drives the equilibrium toward the vapor. Like evaporation, sublimation works best if the humidity is low rather than high.

FIGURE 10.8
The vapor pressure of water increases with temperature up to the boiling point

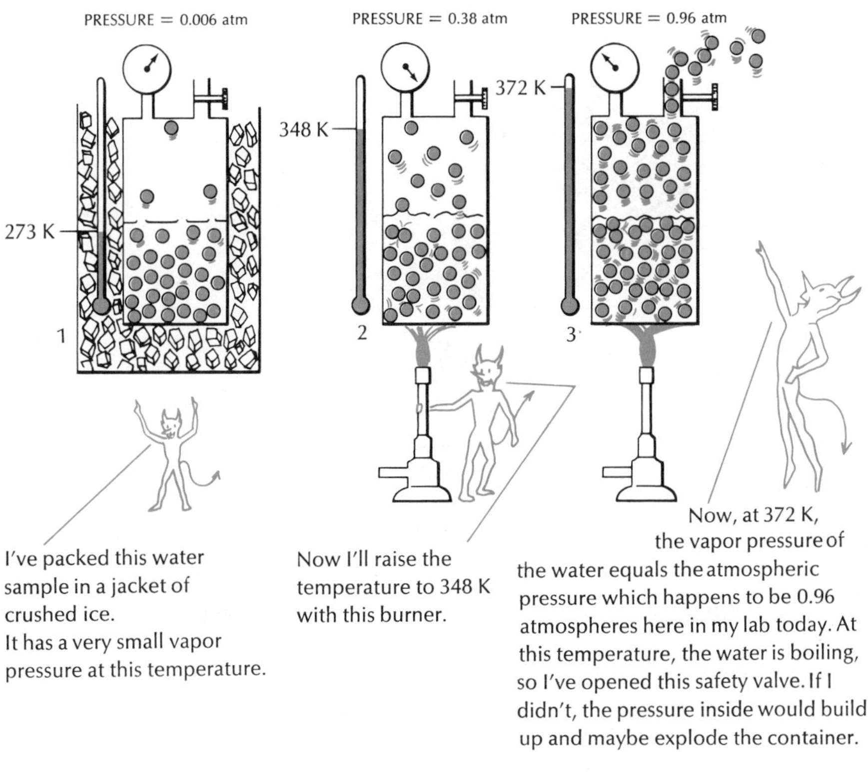

PRESSURE = 0.006 atm
PRESSURE = 0.38 atm
PRESSURE = 0.96 atm
273 K
348 K
372 K
1
2
3
I've packed this water sample in a jacket of crushed ice.
It has a very small vapor pressure at this temperature.
Now I'll raise the temperature to 348 K with this burner.
Now, at 372 K, the vapor pressure of the water equals the atmospheric pressure which happens to be 0.96 atmospheres here in my lab today. At this temperature, the water is boiling, so I've opened this safety valve. If I didn't, the pressure inside would build up and maybe explode the container.

Freeze-drying uses Le Chatelier's Principle, too. A substance is frozen, and the water vapor is continually removed with a pump. In a frost-free freezer, the circulating air is kept very dry, either with a drying agent or with a pump. The water molecules escaping from the ice are constantly being removed. As a result, the ice cubes in a frost-free freezer become smaller the longer they sit unused.

FREEZING (MELTING). The melting point can be defined as the temperature at which the solid is at equilibrium with the liquid. But it's really not quite that simple, because both the solid and the liquid have vapor pressures that we can't ignore. Solid and liquid are at equilibrium with vapor as well as with each other. Figure 10.9 shows that solid and liquid are at equilibrium with the same vapor. Thus *melting point* (*freezing point*) is the temperature at which the solid and liquid have the same vapor pressure. In Chapter 12, we'll see that equal vapor pressure for solid and liquid causes antifreeze to work and salt to melt ice.

melting point

freezing point

FIGURE 10.9
At the melting point, the solid and liquid have the same vapor pressure

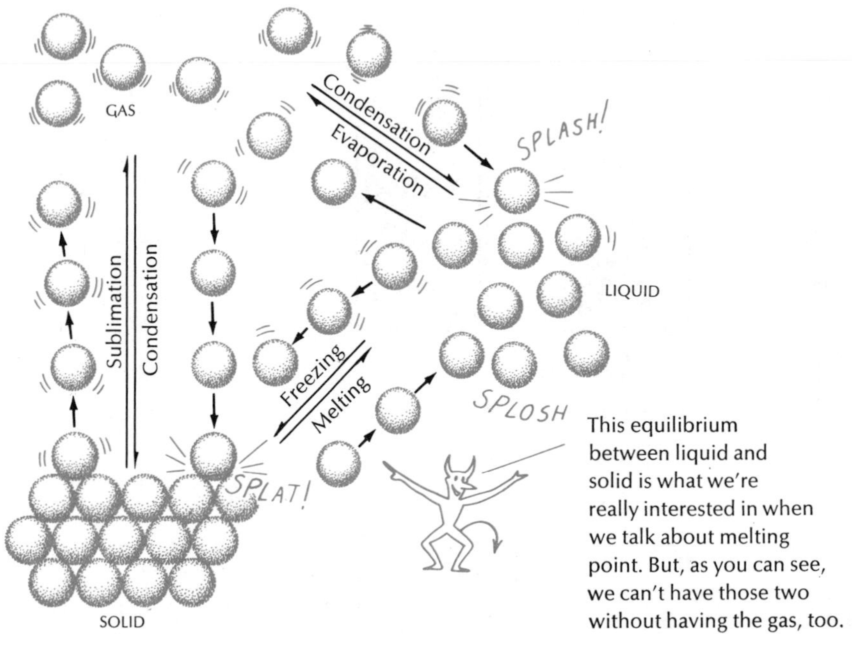

GAS
Condensation
Evaporation
SPLASH!
Sublimation
Condensation
LIQUID
Freezing
Melting
SPLOSH
SPLAT!
SOLID
This equilibrium between liquid and solid is what we're really interested in when we talk about melting point. But, as you can see, we can't have those two without having the gas, too.

10.4 ATTRACTIONS AMONG PARTICLES

At atmospheric pressure, substances melt and boil at different temperatures because the attractions that hold them together have different strengths. Here, we'll look at a few—but not all—of the interactions between particles that hold them together in solids and liquids.

HYDROGEN BONDING. For some substances, melting points and boiling points increase roughly in proportion to formula weight. Figure 10.10 shows the graphs of the melting points and boiling points for the hydrogen compounds of the Group VA, VIA, and VIIA elements. The melting points and boiling points for the last three members of each series increase regularly with formula weight. But the first member of each series seems out of place. For these three compounds, HF, H_2O, and NH_3, a strong attraction—hydrogen bonding—has been added.

We know from Chapter 9 (pp. 190–193) that the H—F bond is polar and is written like this:

$$H \mapsto F$$

FIGURE 10.10
Hydrogen bonding causes higher melting and boiling points

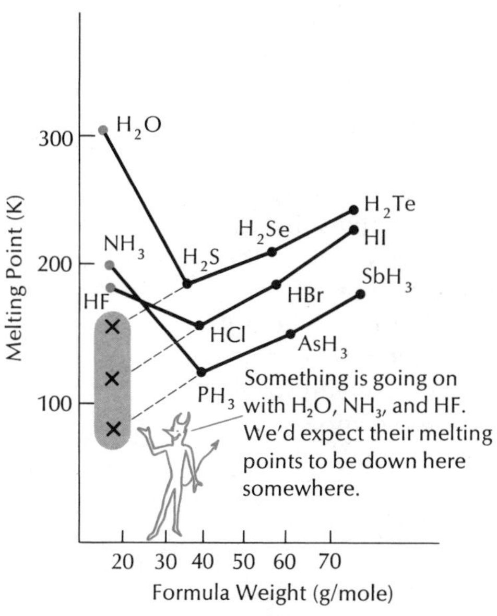

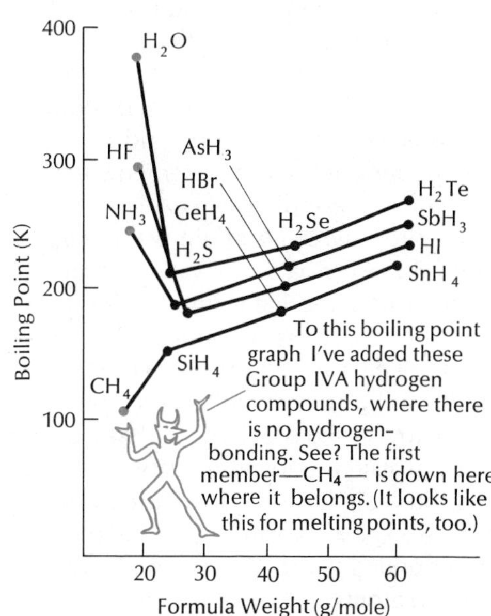

The fluorine has hogged the electron pair so much that the hydrogen is almost a naked proton hanging out on one side. Sometimes we call it an "unshielded proton." Part of its positive charge is exposed and is ready and able to attract something negative. If there are only other HF molecules around, it finds a nonbonding electron pair on another HF molecule and is attracted to it. We define a *hydrogen bond* as an attraction between a hydrogen atom that is bonded to fluorine, oxygen, or nitrogen and the nonbonding electron pair on a fluorine, oxygen, or nitrogen of another molecule. Only these three elements are electronegative enough to cause hydrogen bonding to any great extent. Here's how hydrogen bonding occurs in HF, H_2O, and NH_3:

hydrogen bond

```
  ..      ..             H         H              N          H
 :F—H----:F—H            |         |              |          |
  ..      ..            :O—H----:O—H          H—N:----H—N:
                         ..       ..              |          |
                                                  H          H
```

These are the most common examples, but any compound that contains H—F, H—O, or H—N bonds will form hydrogen bonds. Other examples are hydrogen peroxide (H_2O_2), hydrazine ($H_2N—NH_2$), and many carbon compounds derived from water and from ammonia. If these compounds are mixed together, they'll hydrogen-bond to one another.

To melt a hydrogen-bonded crystal, we have to break the hydrogen bonds but not the covalent bonds in the molecules. Although a hydrogen bond is not as strong as a regular covalent bond, it's still strong enough to cause the high melting points and boiling points of HF, H_2O, and NH_3. Fluorine, oxygen, and nitrogen are the only elements in their respective groups that are electronegative enough to cause their hydrogen compounds to have hydrogen bonding. Hydrogen bonding causes HF, H_2O and NH_3 to have unexpectedly high melting and boiling points compared to the hydrogen compounds of the other elements in their groups.

Hydrogen bonding also explains the fact that ice is less dense than water. If we drop the solid of almost any substance into its liquid, the solid will sink. But ice floats on water, because it's less dense than water. If this weren't true, the ice formed in winter would sink to the bottom of lakes, rivers, and oceans. In spring, the ice would be at the bottom, where it couldn't be warmed and melted. The next winter, more water would freeze, and it still wouldn't melt in the spring. Pretty soon the whole world would be nearly a solid block of ice that would not support life as we know it.

To explain the lower density of ice than liquid water, we need to look at the three-dimensional structure of the H_2O molecule. From its Lewis structure, we know that an H_2O molecule consists of a central oxygen atom with two bonded hydrogen atoms and two nonbonding electron pairs. The hydrogen atoms and nonbonding pairs are arranged about the central oxygen atom so that each occupies the corner of a tetrahedron, as shown in Figure 10.11. The angle that the two hydrogen atoms form with each other, called the *bond angle,* is 105°.

bond angle

When water freezes, its molecules spread apart as they form a continuous crystal structure, shown in Figure 10.11. This crystal has a lot of empty space in it—enough space for a stream of gaseous helium atoms to pass through and come out the other side. Because its particles are packed less tightly into space, ice is less dense than liquid water. But why do the H_2O molecules spread out when water freezes?

A hydrogen bond has some of the restrictions of a regular covalent bond. These restrictions are:

1. Only one hydrogen atom can hydrogen-bond to one electron pair.
2. Hydrogen bonding has to occur in a straight line. That is, each hydrogen bond must be in a straight line with the hydrogen nucleus and the nuclei of the two nonhydrogen atoms involved.

Ice is less dense than water because of these two rules. Since H_2O has two unshielded protons and two nonbonding electron pairs, a continuous hydrogen-bonded structure can be formed. When water freezes, this structure is locked in place. To keep the hydrogen bonds in straight lines and to preserve H_2O's bond angle (105°), the molecules have to spread apart.

Even though HF and NH_3 can hydrogen-bond, too, their solid states are more dense than their liquid states. HF has three nonbonding electron pairs and only one unshielded proton, so there aren't enough unshielded protons to bond with every electron pair. NH_3 has three unshielded protons and one nonbonding electron pair, so there will always be leftover protons.

FIGURE 10.11
Structures of H_2O and of an ice crystal

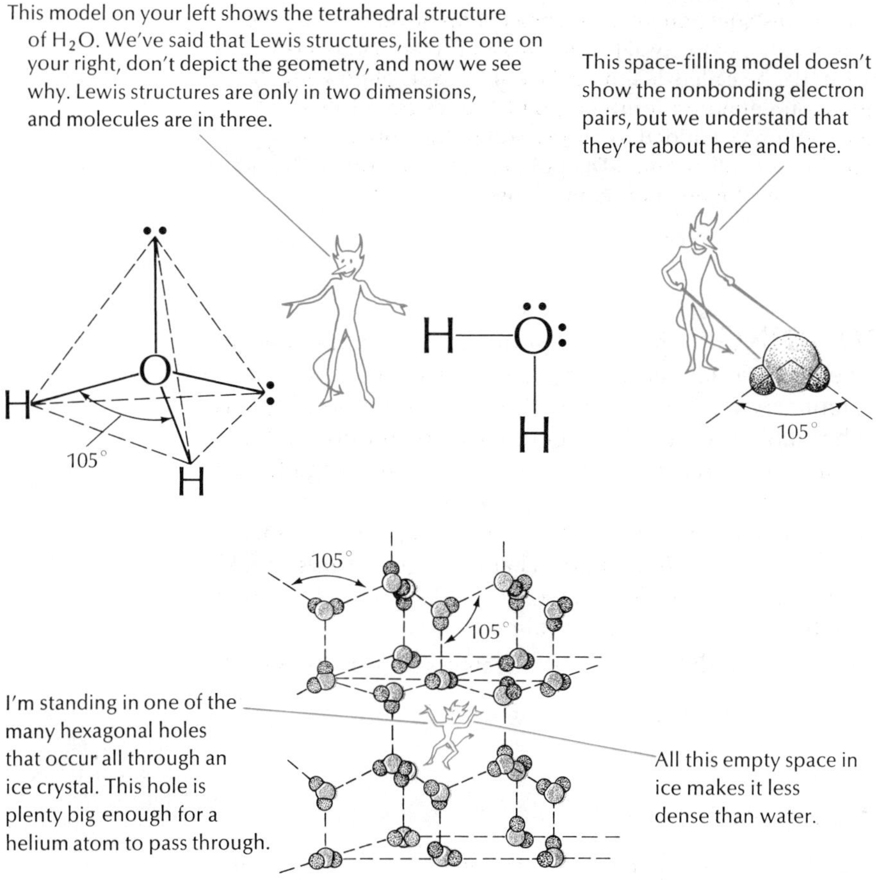

This model on your left shows the tetrahedral structure of H_2O. We've said that Lewis structures, like the one on your right, don't depict the geometry, and now we see why. Lewis structures are only in two dimensions, and molecules are in three.

This space-filling model doesn't show the nonbonding electron pairs, but we understand that they're about here and here.

I'm standing in one of the many hexagonal holes that occur all through an ice crystal. This hole is plenty big enough for a helium atom to pass through.

All this empty space in ice makes it less dense than water.

Thus neither has the possibility of a continuous hydrogen-bonded structure like that of water, which matches unshielded protons and nonbonding electron pairs exactly.

Hydrogen bonding occurs in liquid water, too, but the bonds are continuously formed and broken as the water molecules move around. However, it is the extensive hydrogen bonding in liquid water that gives water the high specific heat we saw in Chapter 3. Water can absorb a lot of energy, because the energy goes into breaking the many hydrogen bonds.

hydrogen-bonded molecular crystal

Ice is an example of a *hydrogen-bonded molecular crystal.* In these crystals, the single units are molecules and the attraction is hydrogen bonding.

IONIC BONDING. From Chapter 9 (pp. 175–176), we know that the units in ionic crystals are positive and negative ions, and the attraction is ionic bonding. Ionic crystals are all solids at room temperature because of the continuity of the crystal and the strength of the ionic bond.

ICE-NINE

What if ice had a melting point above room temperature? Kurt Vonnegut, Jr., asked and answered this question in his science fiction novel *Cat's Cradle*. In it, a scientist experimented with crystal structures and created a special crystalline form of ice that he called "ice-nine." Ice-nine "was blue-white. It had a melting point of one-hundred-fourteen-point-four degrees Fahrenheit" (45°C). When ice-nine was added to normal water, the normal water crystallized in the form of ice-nine. The political and environmental results of this situation make for an entertaining story.

COVALENT BONDING. The strongest of all are the attractions in *covalent crystals*. Their individual particles are atoms. They have a continuous structure, like ionic crystals, but the attraction—covalent bonding—is stronger than ionic bonding. We saw the structure of a diamond crystal in Chapter 2 (p. 27). Diamond won't melt. At 3773 K, it will rearrange to the graphite structure.

covalent crystal

The atoms in a covalent crystal don't have to be the same. Silicon dioxide is made of two kinds of particles: silicon atoms and oxygen atoms. Rocks are hard because they're made mostly of silicon dioxide. To melt or cut a covalent crystal, we actually have to break covalent bonds.

Table 10.2 summarizes the different kinds of crystals and the attractions that hold them together.

TABLE 10.2
Summary of interactions in solids

Kind of Crystal	*Units*	*Attraction*	*Examples*	*Comments*
Covalent Crystal	Atoms (same or different)	Covalent bonds	Diamond Silicon Boron SiO_2 Carborundum (SiC)	These are all very hard and have high melting points, because a covalent bond is the strongest interaction there is. All are solids at room temperature.
Ionic Crystal	Positive and negative ions	Ionic bonds	NaCl KNO_3, and all ionic compounds	Cover a wide range of melting points. All are solids at room temperature
Hydrogen-bonded Molecular Crystal	Hydrogen-bonding molecules	Hydrogen bonds	H_2O (ice) NH_3 HF	Melting points far below those of ionic crystals, because the interactions are much weaker. These substances are liquids or gases at room temperature.

REVIEW QUESTIONS

Temperature and Heat

1. Without looking at the text explanation, describe what is happening physically during each portion of Figure 10.1.
2. Why is ice a better cooling agent than water, when the two are at the same temperature? Why is steam a better heating agent than water, under the same conditions?
3. How can we tell from Figure 10.1 that water is a liquid at room temperature?
4. Which is usually larger, a substance's heat of fusion or its heat of vaporization? Explain.

Pressure

5. What is *pressure?* How do we measure it?
6. What is *atmospheric pressure?* What causes it? Why is it greater at sea level than on a mountain top?
7. What is 1 *atmosphere?* How can it be measured? What is the SI unit of pressure?
8. What is *equilibrium?* Give an example of *static equilibrium.*
9. Explain *Le Chatelier's Principle,* and give an example.
10. Explain how a *barometer* works. What is a *torr?*
11. How does a gas create pressure?
12. Why do we need to measure the pressure of a gas?
13. Explain how pressure can change a substance's physical state.
14. Are bottled gases liquid or gas? Explain.
15. Is the boiling point of water always 373 K? Explain.
16. Explain how a refrigerator works. How is heat of vaporization used?
17. What is *standard temperature and pressure (STP)?*

Equilibria Among States

18. What is *vapor? Vapor pressure?*
19. What is *dynamic equilibrium?* Give an example.
20. Explain how Le Chatelier's Principle is involved in the evaporation of water from an uncovered glass.
21. Why do objects dry out in cold winters?
22. Explain why solids and liquids have vapor pressure. Why does it increase with temperature?
23. What is our new definition of *boiling point?* Explain what happens when a liquid boils.
24. Why do liquids boil at lower temperatures in the mountains?
25. Why does food cook faster in a pressure cooker?
26. What is *sublimation?* Give some familiar examples.
27. How is Le Chatelier's Principle involved in freeze-drying?
28. How do we now define *melting point?*

Attractions Among Particles

29. Why do substances melt and boil at different temperatures?
30. What is a *hydrogen bond?* What evidence do we have for its existence?
31. What substances can hydrogen-bond? Give some examples.
32. What are the restrictions on hydrogen-bonding?
33. What is a *bond angle?* What is water's bond angle?
34. Why is ice less dense than water? How do hydrogen bonding and water's bond angle enter in?
35. HF and NH_3 can hydrogen-bond. Explain why their liquids are not denser than their solids as in H_2O.
36. Why does water have such a high specific heat?
37. How is a covalent crystal like an ionic crystal? How is it different?
38. Without looking at Table 10.2, summarize the kinds of interactions discussed in this section.

EXERCISES

Set A (Answers at back of book. Asterisks mark the more difficult exercises.)

1. An ice cube weighing 20. grams is taken from a refrigerator at −10°C. How much heat (in cal) would be required to change the ice cube to steam at 100°C?
2. For a ski resort to make snow, the temperature must be −5°C or below. As the snow is being made, heat is generated by the freezing of water. Calculate the amount of heat, in kcal, that would be produced if 15,000 liters of water at +5°C is sprayed into air at −5°C by a snowmaking machine.
3. Ammonia has a heat of vaporization of 327 cal/g at its boiling point, −33°C. If 1 kg of liquid ammonia is allowed to change to a gas at its boiling point, how many kilocalories of heat will it absorb?

4. Imagine an inhabitable planet whose atmospheric pressure is half that of earth.
   a. What would be the reading on an earth barometer?
   b. Would drinking through a straw be easier or more difficult than on earth?
   c. Would it take a longer or shorter time to boil an egg in water than it would on earth?
   d. Would an individual with high blood pressure be better or worse off than one with normal blood pressure?

*5. Calculate the pressure, in atmospheres, that a diver would experience at a depth of 5.0 meters in fresh water. (Remember that the pressure at the surface is already 1 atm.)

6. Here is a temperature-heat graph for an unknown substance.

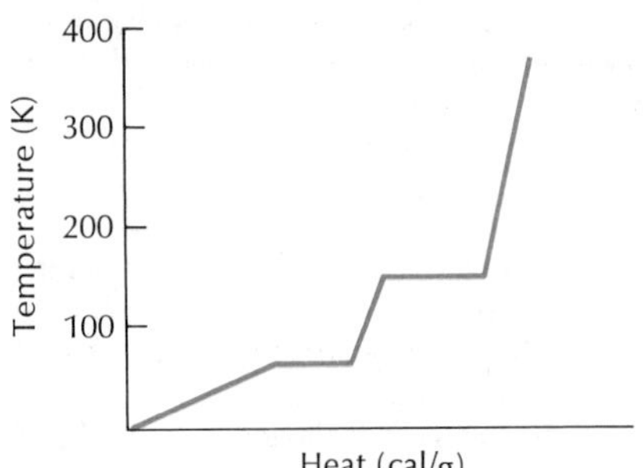

   a. What is its physical state at room temperature?
   b. Indicate which portions of the graph are its heat of fusion and heat of vaporization.
   c. Is the specific heat highest for gas, liquid, or solid?
   d. Are the attractions between its particles stronger or weaker than the attractions between particles in H_2O?

7. When your skin is cleansed with alcohol, it feels cold, even though the bottle of alcohol is at room temperature. Explain.

8. Which of these would help dry your clothes indoors on a rainy day? Explain.
   a. increasing the temperature of the room
   b. decreasing the temperature of the room
   c. putting a pan of water on the radiator
   d. blowing an electric fan on the clothes
   e. opening the window
   f. closing the window

9. Is it possible to cause water to boil at room temperature? How?

10. An efflorescent substance is a hydrate that loses its water of hydration at room temperature, without additional heating. (See Chapter 5, p. 98.) Would such a substance lose its water of hydration more readily in moist or dry air?

11. At the melting point, the solid and liquid must have the same vapor pressure to coexist. What would happen if a piece of ice were placed in liquid water at a temperature where the vapor pressure of water is greater than that of ice?

12. The following table shows heats of fusion and heats of vaporization for the hydrogen compounds of the Group VIIA elements.

Substance	*Heat of Fusion (cal/g)*	*Heat of Vaporization (cal/g)*
H_2O	79.9	540
H_2S	16.7	131
H_2Se	7.4	57
H_2Te	—	42

   a. Explain why the heats of vaporization are larger than the heats of fusion for each compound.
   b. Explain why both the heats of vaporization and the heats of fusion are largest for H_2O.

13. Explain these in terms of interactions between particles.
   a. H_2O has a higher boiling point than HF.
   b. Quartz (SiO_2) has a higher melting point than NaCl.
   c. The heat of fusion of NaCl is greater than that of H_2O.
   d. The heat of vaporization of NH_3 is greater than that of SO_2.

Set B (Answers not given. Asterisks mark the more difficult exercises.)

1. An isolated outpost in Siberia wishes to generate its electricity with steam made from snow. How many kilocalories must be supplied to convert 1 kg of snow at −40°C to steam at 100°C?

2. How many grams of steam at 373 K must be condensed to water at 373 K to provide 5 kcal of heat?

3. NH_3 and SO_2 were once used as refrigerating gases more than they are today. Both are toxic and corrosive, and therefore unpleasant to work with. Freon-12, CF_2Cl_2, is one substance that has replaced them because it is inert, nontoxic, and nonexplosive. The heats of vaporization of these substances follow:

NH_3:	327 cal/g
SO_2:	92.7 cal/g
CF_2Cl_2:	40 cal/g

Calculate the number of grams of each that would be required to absorb 1.00 kcal of heat in a refrigerator. Which substance is the best refrigerant?

4. In salt water, a depth of 33 feet subjects a diver to a total of 2 atmospheres pressure.
   a. Would a diver with high blood pressure be better or worse off at this depth than at the surface? Explain.
   b. What would be the barometer reading, in torr, at that depth?
   c. Would the diver's mask fit more or less tightly than it did at the surface?

*5. Calculate the depth at which a scuba diver would be subjected to 3 atmospheres total pressure in fresh water. (Remember that the pressure at the surface is 1 atm.)

6. Here is a temperature-heat graph for an unknown substance.

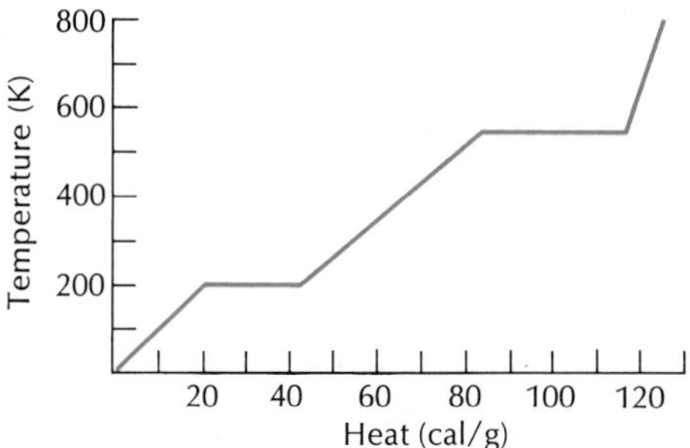

   a. Estimate, to two significant figures, the melting point, boiling point, heat of fusion, and heat of vaporization.
   b. Is the substance a solid, liquid, or gas at room temperature?
   c. Indicate, on the graph, which states are present for each portion of the graph.

7. Alcohol will evaporate faster from an uncovered glass at room temperature than water will. Which has the higher vapor pressure at room temperature?

8. Which of the following would help to keep your house plants from drying out while you went away for a few days? Explain.
   a. increasing the temperature of the room
   b. decreasing the temperature of the room
   c. opening the window
   d. closing the window
   e. placing a pan of water on the radiator
   f. blowing an electric fan on the plants
   g. putting the plants in sealed plastic bags

9. Hikers find that carrying water in a porous canvas container will keep the water cool even on a hot day. Explain how it works.

10. Dry ice sublimes at atmospheric pressure instead of melting. Yet a puddle of liquid often remains after a piece of dry ice has disappeared. What is the liquid, and where did it come from?

11. What would happen if a solid were placed in a sample of its liquid form at a temperature at which the vapor pressure of the liquid is less than the vapor pressure of the solid?

12. What is the expected order of increasing heats of fusion and of vaporization for these compounds: H_2O, NaCl, CH_4? Which would be greater, the heat of vaporization or the heat of fusion for each? Explain.

13. Explain why solid H_2S is more dense than liquid H_2S even though its molecular shape is similar to that of H_2O.

CHAPTER 11

Gases

LEARNING OBJECTIVES

After studying this chapter, you should be able to:

1. Supply a correct definition, explanation, or example for these:

photosynthesis	Boyle's Law
nitrogen fixation	Charles' Law
catalyst	Gay-Lussac's Law
visible light	combined gas laws
ultraviolet radiation	Dalton's Law
greenhouse effect	partial pressure
infrared radiation	kinetic-molecular theory
part per million (ppm),	ideal gas
photochemical smog	pressure
Avogadro's Law	ideal gas equation
molar volume	gas constant (*R*)

2. List air's important gases and explain their origin and importance, using appropriate equations.
3. List the major air pollutants and their sources, and explain their undesirable effects and ways of removal, illustrating with appropriate equations.
4. Use the molar mass of a gas to determine the mass, volume, density, or molecular weight at STP.
5. Perform mole-to-volume, volume-to-mole, volume-to-volume, mass-to-volume, and volume-to-mass conversions with chemical equations involving gases.
6. Use Boyle's, Charles', and Gay-Lussac's Laws to calculate and predict the new condition of a gas from appropriate data.
7. Use the combined gas laws to calculate the new volume of a gas when both temperature and pressure are changed.
8. Use Dalton's Law to calculate the partial or total pressure from appropriate data on mixtures of gases, and solve problems dealing with the collection of gases by displacement of water.
9. Describe a gas according to the kinetic-molecular theory, or make predictions using this theory.
10. Use the ideal gas equation to calculate the number of moles or the volume of a gas from appropriate data.
11. Distinguish between an ideal and a nonideal gas, and describe properties of each.

Gases are very interesting and useful, because of the huge changes they can undergo. In the last chapter, we saw some of the ways that changes of state can work for us, and that changes involving gases are usually the most significant. In this chapter, we'll look more closely at the gas state itself, starting with air, our most important gas. We'll see some chemical reactions that involve gases, and we'll calculate exactly how changes in pressure, temperature, and volume affect them.

11.1 SOME CHEMISTRY OF AIR

Air is our favorite gas. As we know, though, air isn't just one gas, but a mixture, as shown in Table 11.1. Each of air's major components is important to our life on earth. The gases that are present in smaller amounts affect us very little. Some of these, though, become air pollutants when present in larger amounts. We'll look at some of the important properties of normal air and see how air pollutants are introduced and how they can and should be minimized.

AIR'S IMPORTANT GASES. Of course, oxygen is number one in importance, because without it we couldn't survive. The hemoglobin in our blood combines with the oxygen that we breathe into our lungs and transports it to all parts of our bodies. When oxygen arrives at its various destinations, it combines with hydrogen (which results from other bodily processes) and forms water, liberating energy.

$$O_2 + 4\,H\text{ (in compounds)} \longrightarrow 2\,H_2O + \text{energy}$$

In effect, we use oxygen to "burn" our food, which makes the energy to run our bodies. We also use oxygen to burn fuels like coal and gasoline, which provide us with energy to heat our homes and drive our cars.

Oxygen wasn't always a part of our atmosphere. It didn't exist in any significant amount until about 1.8 billion years ago, when green plants were first evolving. Green plants use carbon dioxide plus the energy from the sun to make sugar, which they then use for energy. This process is called *photosynthesis,* and it produces oxygen.

photosynthesis

Living organisms need nitrogen in their bodily structures. Nitrogen gas is very inert, though, so plants and animals can't use it straight from the air. Animals get nitrogen they can use by eating plants and other animals. Plants obtain it by *nitrogen fixation,* which means that nitrogen in the air combines with other elements and enters the soil. The plants then take the compound up in their roots. Natural nitrogen fixation happens when lightning causes the nitrogen and oxygen in the air to react together.

nitrogen fixation

$$N_2(g) + O_2(g) \xrightarrow{\text{lightning}} 2\,NO(g)$$

NO_2 is formed when NO reacts with oxygen or ozone.

TABLE 11.1
Composition of clean dry air at ground level

Name	*Formula*	*Volume Percent*	*Comments*
Nitrogen	N_2	78.09	Rather unreactive gas.
Oxygen	O_2	20.95	Essential for air-breathing animals. Very reactive. Produced by plants.
Argon	Ar	0.93	Very unreactive gas.
Carbon dioxide	CO_2	0.032	Product of burning and breathing. Used by green plants to make sugar.
Neon	Ne	1.8×10^{-3}	Very unreactive gas.
Helium	He	5.2×10^{-4}	Very unreactive gas.
Methane	CH_4	1.5×10^{-4}	Produced by volcanos and some microorganisms.
Krypton	Kr	1.0×10^{-5}	Very unreactive gas.
Hydrogen	H_2	5.0×10^{-6}	Reactive; produced by volcanos and some microorganisms.
Dinitrogen oxide	N_2O	2.0×10^{-6}	Produced by lightning.
Carbon monoxide	CO	1.0×10^{-6}	Product of burning and of some marine organisms.
Xenon	Xe	8.0×10^{-7}	Very unreactive gas.
Ozone	O_3	2.0×10^{-8}	Formed by ultraviolet light on O_2, and by lightning.
Ammonia	NH_3	6.0×10^{-9}	Product of volcanos and microorganisms.
Nitrogen dioxide	NO_2	1.0×10^{-9}	Formed by lightning and hot fires.
Nitrogen monoxide	NO	6.0×10^{-10}	Formed by lightning and hot fires.
Sulfur dioxide	SO_2	2.0×10^{-10}	Produced by volcanos and burning S.
Hydrogen sulfide	H_2S	2.0×10^{-10}	Produced by volcanos and some microorganisms.

$$2\,NO(g) + O_2(g) \longrightarrow 2\,NO_2(g)$$

$$NO(g) + O_3(g) \longrightarrow NO_2(g) + O_2(g)$$

Rain washes the NO_2 down as nitrites and nitrates.

$$2\,NO_2(g) + H_2O(l) \longrightarrow HNO_2(aq) + HNO_3(aq)$$

Having reached the ground in a compound, the nitrogen can then be absorbed by the plants. Another source of nitrogen for plants is decaying animal waste and dead animal and plant matter. Also, some microorganisms that live in the soil and in the roots of peas, beans, or related plants can fix nitrogen.

With our tremendous agricultural demands on the soil, we can't depend on natural nitrogen fixation. Humans have developed industrial ways of fixing nitrogen. Nitrogen from the air is made to react to form a compound, which is then used as a fertilizer. This is more difficult than it sounds, because nitrogen is very unreactive. Some of these processes need the help of a *catalyst,* a substance that makes a reaction go much faster without itself being changed in the end. A very important nitrogen fixation process is the Haber process.

catalyst

$$N_2(g) + 3\,H_2(g) \xrightarrow[\text{catalyst}]{\text{pressure, }\Delta} 2\,NH_3(g)$$

The ammonia is either made into ammonium salts or used directly as a fertilizer.

Still other parts of the air regulate the way we receive and store the sun's energy. Sunlight has many components. The part that we see is *visible light.* We don't see ultraviolet and infrared radiation. *Ultraviolet radiation* has the highest energy. It gives some people sunburn. We'd burn up if we received all the ultraviolet radiation that the sun sent out. Before the oxygen in our atmosphere was formed, life had been confined to the water for protection from ultraviolet radiation. Then a protective screen of ozone (O_3) was also formed in the upper atmosphere, and life began to exist out of water. Ozone absorbs some ultraviolet radiation.

visible light

ultraviolet radiation

Ozone is produced by the action of ultraviolet radiation on oxygen. When this high-energy radiation hits an oxygen molecule, it breaks the bond and creates two oxygen atoms.

$$O_2(g) \xrightarrow{\text{uv}} 2\,O(g)$$

Single oxygen atoms will react immediately with the first thing they meet. If the first thing is an oxygen molecule, then ozone is formed.

$$O(g) + O_2(g) \longrightarrow O_3(g)$$

Ultraviolet light also destroys the ozone it helps create. When it does, though, energy is absorbed in breaking one of ozone's bonds.

$$O_3(g) \xrightarrow{\text{uv}} O_2(g) + O(g)$$

Of course, the new oxygen atom can react right away with another oxygen molecule and form ozone again. But meanwhile some harmful ultraviolet radiation will be absorbed and prevented from reaching us. The formation and decomposition of ozone are in a state of equilibrium, so that the amount of ozone in the upper atmosphere remains fairly constant. While this is being written, there is controversy over whether some gases we use—such as the Freon in aerosol cans—might be disturbing that equilibrium and reducing the amount of ozone. If this is happening, it could put us in danger of being overexposed to ultraviolet radiation.

Water vapor isn't listed as a component of air in Table 11.1, because its amount varies widely with location. Air can contain as much as 5 percent water vapor in tropical climates, and as little as 0.01 percent at the North Pole or South Pole. Water vapor is an important part of the air, though. Apart from its obvious role in making rain and snow, it and carbon dioxide help maintain a fairly even climate on earth. In what we call the *greenhouse effect,* carbon dioxide and water vapor form a blanket around the earth that keeps in some of the sun's warmth during the night. (See Figure 11.1.) The ultraviolet radiation from the sun changes into *infrared radiation* (which is heat) when it hits the earth. The water vapor and carbon dioxide

greenhouse effect

infrared radiation

FIGURE 11.1
The greenhouse effect

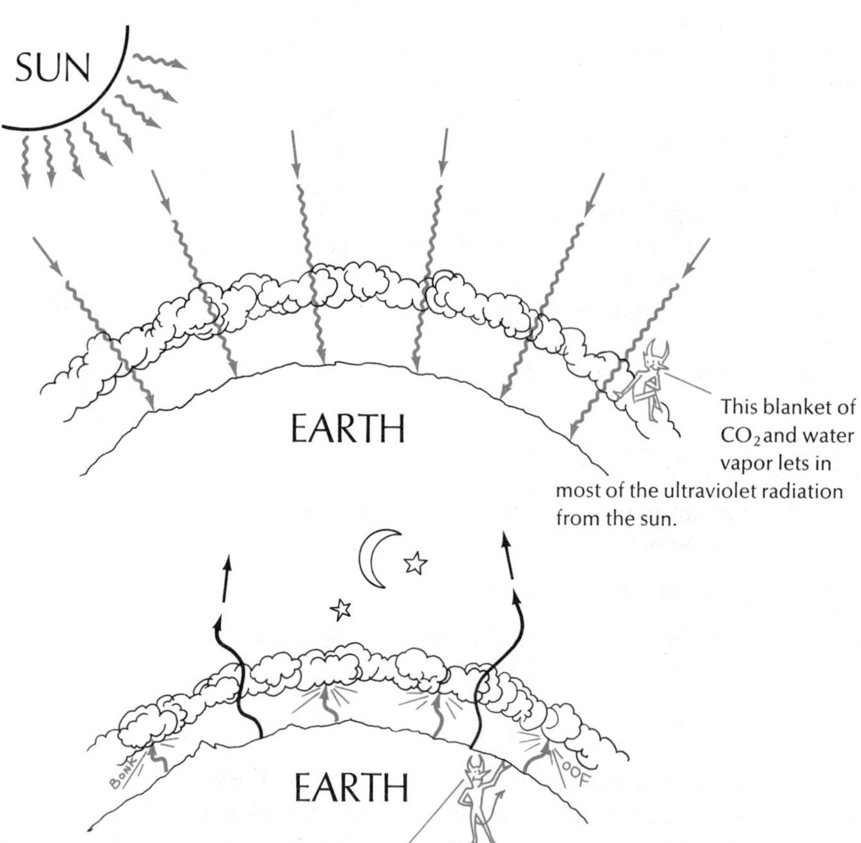
SUN
EARTH
This blanket of CO_2 and water vapor lets in most of the ultraviolet radiation from the sun.
BONK
EARTH
OOF

But it doesn't let all of the infrared radiation escape. Earth keeps some of it overnight. This is called the greenhouse effect because the glass of a greenhouse serves the same purpose as the blanket of CO_2 and water vapor.

TABLE 11.2
Some desirable reactions involving gases in air

Gas	*Reaction*	*Effect*
O_2	$Hb + O_2 \longrightarrow HbO_2$ hemoglobin	Hemoglobin transports oxygen to cells.
	$O_2 \longrightarrow 2\ O^{2-} + \text{energy}$ (in cells)	Cells use oxygen to obtain energy.
	$C + O_2 \longrightarrow CO_2 + \text{energy}$ (in fuels)	Souce of heat and energy.
O_3	$O_3 \xrightarrow{\text{uv}} O_2 + O$	Absorbs some harmful ultraviolet radiation in outer atmosphere.
CO_2	$6\ CO_2 + 6\ H_2O \xrightarrow{\text{light}} C_6H_{12}O_6 + 6\ O_2$ (photosynthesis) glucose	Origin of atmospheric oxygen. Vital source of energy for green plants and animals.
N_2	$N_2 + O_2 \xrightarrow{\text{lightning}} 2\ NO$	Natural nitrogen fixation.
	$N_2 + 3\ H_2 \xrightarrow[\text{catalyst}]{\text{pressure, }\Delta} 2\ NH_3$	Industrial Haber process for nitrogen fixation.

let the ultraviolet radiation through to the earth during the daytime, but they keep some of the infrared radiation from escaping at night. If the earth couldn't keep some heat this way, we'd be far too cold at night.

Table 11.2 summarizes some of the important reactions that involve gases in the air.

AIR'S UNDESIRABLE GASES. We saw some of the major air pollutants in Chapter 4. Now we know from Table 11.1 that traces of some of them are in the lower atmosphere: CO, NO, NO_2, O_3, and SO_2. Although these are pretty harmless in small amounts, our advanced technology is putting far more of them into the atmosphere than it can handle. Carbon monoxide comes from the incomplete burning of fuel, particularly in automobile engines. Nitrogen monoxide is formed whenever air is raised to high temperatures, so that the nitrogen and the oxygen in air react with each other. Automobiles and electric power plants are the primary sources of this pollutant. Nitrogen dioxide is formed from nitrogen monoxide, and ozone is formed from nitrogen dioxide and oxygen. Sulfur oxides form when anything containing sulfur is burned. The primary sources are smelting sulfur ores and burning coal and petroleum that contain sulfur.

part per million (ppm)

We measure air pollutants in *parts per million* (*ppm*). One part per million means one measure (volume, weight, molecule) of something in a million measures of a mixture. One molecule of oxygen in a million molecules of a mixture is one part per million.

Table 11.3 lists some air pollutants and their primary and secondary effects. Sometimes, secondary effects are worse than primary effects. This is especially true of nitrogen monoxide, which leads to nitrogen dioxide and ultimately to the formation of toxic ozone in the lower atmosphere. Although ozone forms a lifesaving screen in the upper atmosphere, it's a dangerous poison to breathe. Ozone, NO_2, and unburned gasoline in automobile exhaust produce a great many irritants when acted upon by ultraviolet radiation. We know the mixture as *photochemical smog.*

photochemical smog

Nature has some methods for removing air pollutants. Some organisms in the soil use up carbon monoxide. Rain will wash the oxides of nitrogen and sulfur out of the air and into the soil, for plants to use. But natural processes can handle only small amounts of these gases. It's important for us to start at the root of the pollution problem, by controlling the amounts of various gases that we release into our atmosphere. This is much easier said than done, however. Table 11.4 shows the sources of some air pollut-

TABLE 11.3
Some air pollutants and their effects

Pollutant	*Primary Effect*[a]	*Secondary Effect*[a]	*Allowed Exposure by Industries*
CO_2	None; nontoxic to animals.	Increased amounts may overdo the greenhouse effect, so that climate may become warmer. Eventual effects unknown.	
CO	Poisonous to oxygen-breathing organisms. Reacts with hemoglobin and prevents it from carrying O_2: $Hb + CO \longrightarrow HbCO$		100 ppm (but ill effects noticed at 5 ppm).
NO	Reacts with hemoglobin as CO does. Not as serious as secondary effect.	Reacts with ozone to produce NO_2: $NO + O_3 \longrightarrow NO_2 + O_2$	25 ppm
NO_2	Toxic. Causes respiratory irritation.	Initiator of "smog." Worst effect is as source of atomic oxygen and then ozone: $NO_2 \xrightarrow{uv} NO + O$; $O + O_2 \longrightarrow O_3$	5 ppm
O_3	Toxic. Destroys bronchial passages and prevents breathing. Attacks rubber and other substances.	Reacts with other smog components to form eye and throat irritants.	0.1 ppm
SO_2, SO_3	Suffocating and toxic to animals and plants.	Cause "acid rain," which corrodes statues and buildings.	5 ppm

[a]Primary effects have direct bearing on animals, plants, or the environment. Secondary effects indirectly cause other environmental problems.

TABLE 11.4
Sources of and possible ways to eliminate air pollutants

Pollutant	*Source*	*Way to Eliminate*	*Comments*
CO	Incomplete burning of fuel (automobiles primarily). Some industries.	Burn fuel completely. Supply more oxygen and burn at higher temperature. Or burn CO as it emerges from stack: $2\,CO + O_2 \longrightarrow 2\,CO_2$	This will lead to increased NO formation (see under NO).
O_3	$NO_2 \longrightarrow NO + O$ $O + O_2 \longrightarrow O_3$	Eliminate NO_2.	
NO_2	$NO + O \longrightarrow NO_2$	Eliminate NO.	
NO	Raising air to high temperatures: $N_2 + O_2 \longrightarrow 2\,NO$ Automobiles and electric power plants.	Burn at lower temperatures. Decrease O_2 supply.	This will lead to increased CO formation.
		Remove from smokestack. Some suggested ways: $2\,CO + 2\,NO \longrightarrow N_2 + 2\,CO_2$ $4\,NH_3 + 6\,NO \longrightarrow 5\,N_2 + 6\,H_2O$ $CH_4 + 4\,NO \longrightarrow 2\,N_2 + CO_2 + 2\,H_2O$ Convert to NO_2, then: $4\,NO_2 + Ca(OH)_2 \longrightarrow Ca(NO_3)_2 + 2\,HNO_2$	All substances are gases, which are hard to handle. Temperatures, pressures, and catalysts have to be found which will work and won't produce unwanted side effects. $Ca(NO_3)_2$ could be used as a fertilizer.
SO_2, SO_3	Smelting ore and burning sulfur-containing fuel (coal is the worst). $S + O_2 \longrightarrow SO_2$ (free or combined)	Don't burn any fuels that contain sulfur. Remove S from fuel before burning. Remove S from ore without burning. One way: $4\,CuFeS_2 + 3\,SO_2 + 12\,HCl \longrightarrow 4\,CuCl + 4\,FeCl_2 + 11\,S + 6\,H_2O$	Almost impossible, since all do. This is being attempted. Also would use up SO_2 from another source.
	$SO_2 + \frac{1}{2}O_2 \longrightarrow SO_3$	Remove SO_2 from stack gases: $CaCO_3 + SO_2 + \frac{1}{2}O_2 \longrightarrow CaSO_4 + CO_2$ $MnO(s) + SO_2(g) \longrightarrow MnSO_3(s)$ $NaOH(aq) + SO_2(g) \longrightarrow NaHSO_3(aq)$ $2\,H_2S(g) + SO_2(g) \longrightarrow 3\,S(s) + 2\,H_2O(g)$ $2\,CO(g) + SO_2(g) \longrightarrow S(s) + 2\,CO_2(g)$	Limestone-dolomite process. Solid products must be disposed of or used. Where all reactants are gases, handling is a problem.

ants and some of the complex problems involved in trying to remove them.

For instance, to minimize carbon monoxide formation, we supply more oxygen and burn fuel at a higher temperature. But when we do this, we increase the nitrogen monoxide formation. To cut down on the nitrogen monoxide, we supply less oxygen and burn at a lower temperature—but that increases carbon monoxide formation and also causes the release of unburned gasoline. It's hard to win. Engineers are now struggling with these and other conflicting demands.

Most of our fuel, especially coal, contains sulfur. We must either remove the sulfur before burning the fuel—which is very difficult—or remove the SO_2 from the smokestack as it's formed. There are many ways to do the latter, but most involve large quantities of solid by-products. These by-products have to be removed in some way, and often in themselves pose a pollution problem. If the SO_2 could be used to make sulfuric acid, this would be a practical use; unfortunately, it mostly comes out too dilute to be used for this purpose.

Removal of NO is a very difficult problem. Most of the reactions that would get rid of it involve other gases, which are hard to handle, and special temperatures, pressures, and catalysts must be found to make the reactions go in the desired way. We'll see next exactly how gases are affected by temperature and pressure, and how we measure them.

11.2 MEASURING GASES AT STP

In Chapter 10, we talked about some of the difficulties of measuring gases. The volume of a gas is a useful measurement only if we also measure its temperature and pressure. If we know the temperature and pressure and volume and know what the gas is, we can also know its mass.

Avogadro's Law

MOLAR VOLUME OF A GAS. *Avogadro's Law,* proposed in 1811 by Amadeo Avogadro (1776–1856), states that equal volumes of gases at the same temperature and pressure contain equal numbers of particles. This is true because the gas particles themselves occupy almost no volume compared with the space between them. From Avogadro's Law, we know that equal numbers of gas particles—or equal numbers of moles—occupy the same volume at the same temperature and pressure. In fact, the volume occupied by one mole of any gas at standard temperature and pressure (STP)—called the *molar volume* of a gas—is 22.4 liters (22,400 cm^3). It doesn't matter what the gas is. If we know the volume of a gas at STP, we automatically know how many moles—or molecules, or atoms—it contains. This is a unique property of gases. We can't say the same for liquids or solids.

molar volume

Equal volumes of different gases at STP don't have the same masses, though, as Figure 11.2 demonstrates. This is because the particles themselves have different masses.

We can use the molar volume, 22.4 ℓ/mole, as a conversion factor to find the mass or the volume of a gas at STP.

FIGURE 11.2
One mole of any gas occupies 22.4 liters at STP

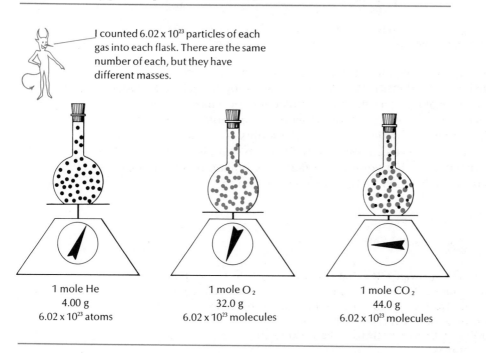

I counted 6.02 x 10^23 particles of each gas into each flask. There are the same number of each, but they have different masses.

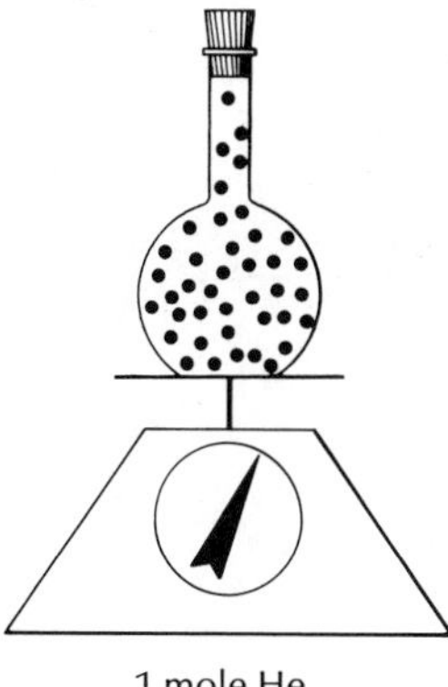

1 mole He
4.00 g
6.02×10^{23} atoms

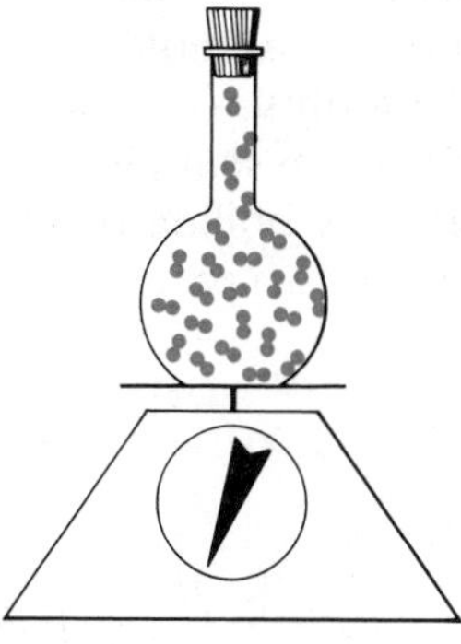

1 mole O_2
32.0 g
6.02×10^{23} molecules

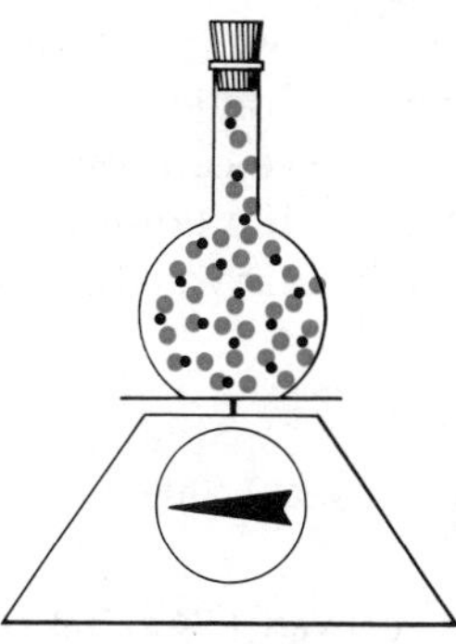

1 mole CO_2
44.0 g
6.02×10^{23} molecules

EXAMPLE 11.1: Burning a certain coal sample yields 455 grams of CO_2. What volume, in liters, would this occupy at STP?

Solution:
Step 1: We want to convert 455 g CO_2 to liters at STP.
Step 2: Our conversion factors are 44.0 g/mole and 22.4 ℓ/mole.
Step 3: Our setup is:

$$455 \text{ g } CO_2 \times \frac{1 \text{ mole}}{44.0 \text{ g } CO_2} \times 22.4 \frac{\ell}{\text{mole}} = \underline{\quad} \ell$$

Answer: 232 liters. ■

We can also calculate a gas's density at STP if we know what the gas is and have an atomic weight table. We don't need any other information. This is another thing we can't do with liquids or solids.

EXAMPLE 11.2: Calculate the density of SO_2 at STP, and compare it with that of air (1.29 g/ℓ).

Solution:
Step 1: Here, we seem to have no given quantity. However, our given quantity is built-in, because we know that the molecular weight of SO_2 is 64.1 g/mole. We want to convert 64.1 g/mole to grams/liter at STP.
Step 2: Our conversion factor is 22.4 ℓ/mole.

Step 3: Our setup is:

$$64.1 \frac{\text{g}}{\cancel{\text{mole}}} \times \frac{1 \cancel{\text{mole}}}{22.4\,\ell} = \_\_\_\_ \frac{\text{g}}{\ell}$$

Answer: 2.86 g/ℓ, more than twice the density of air. (We see why this air pollutant hangs heavily over a city instead of being dispersed.) ■

Of course, we can also do the reverse of this and calculate the molecular weight if we know the density. We might do this to identify an unknown gas, or to decide what a given substance's molecular formula is.

In Chapter 6, we saw how to determine empirical formulas from percent compositions, and how to determine molecular formulas from a substance's molecular weight. Now we can find out one way these molecular weights are actually determined. If a liquid has a fairly low boiling point, then we can determine its molecular weight by changing it to a gas.

EXAMPLE 11.3: A liquid whose empirical formula is CH is changed to a vapor by heating. Its vapor occupies a volume of 500. cm³ at STP and weighs 1.74 grams. What is its molecular weight?

Solution:

Step 1: We want to convert 1.74 g/500. cm³ to grams/mole. Notice that we could divide this out to get density in grams/cubic centimeter, but we don't need to. The division will be performed in the operation.

Step 2: Our conversion factors are 22.4 ℓ/mole and 10^{-3} ℓ/cm³. We may change 1.74 g/500. cm³ directly to 1.74 g/0.500 ℓ.

Step 3: Our setup is:

$$\frac{1.74 \text{ g}}{0.500 \cancel{\ell}} \times \frac{22.4 \cancel{\ell}}{\text{mole}} = \_\_\_\_ \frac{\text{g}}{\text{mole}}$$

Answer: 78.0 g/mole. In Example 6.17 (p. 126), we used this molecular weight to decide that the molecular formula of the substance (benzene) is C_6H_6. ■

In Chapter 6, we also did calculations with chemical equations that involved moles and masses. Now we'll see how to do similar calculations with gas volumes.

VOLUME-TO-VOLUME CONVERSIONS. In the reactions where both substances of interest are gases, Avogadro's Law allows us to substitute a liter ratio for a mole ratio. For gases, a liter ratio is the same as a mole ratio since equal volumes of gas contain equal numbers of molecules and therefore equal numbers of moles.

EXAMPLE 11.4: In the Haber process for fixing nitrogen, how many liters of NH_3 could be made from 2500 liters of N_2? The equation is:

$$N_2(g) + 3\,H_2(g) \longrightarrow 2\,NH_3(g)$$

Solution:

Step 1: We want to convert 2500 ℓ N_2 to liters of NH_3.

Step 2: The equation is balanced.

Step 3: Our conversion factor is 2 moles NH_3/mole N_2, which we may change to a liter ratio: 2 ℓ NH_3/ℓ N_2.

Step 4: Our setup is:

$$2500 \cancel{\ell\ N_2} \times \frac{2\ \ell\ NH_3}{\cancel{\ell\ N_2}} = \_\_\_\_ \ell\ NH_3$$

Answer: $5000\ \ell\ NH_3$. ■

Notice that this reaction also causes a change in the total volume occupied by the gases at STP. Whereas the reaction starts with 4 volumes total of gas—1 volume of N_2 and 3 volumes of H_2—it ends with only 2 volumes total of gas (NH_3). The total gas volume has been cut in half. This is another interesting feature of gases. Conservation of mass still prevails, but not conservation of volume.

EXAMPLE 11.5: The equation for photosynthesis is as follows:

$$6\ CO_2(g) + 6\ H_2O(l) \longrightarrow C_6H_{12}O_6(aq) + 6\ O_2(g)$$

If a plant were placed in a plastic bag, maintained at STP, with 5.00 liters of CO_2, how many liters of oxygen could it make? Would we notice any change in the volume of the plastic bag?

Solution:

Step 1: We want to convert $5.00\ \ell\ CO_2$ to liters of O_2.

Step 2: The equation is balanced.

Step 3: Although there is a substance in the equation that isn't a gas ($C_6H_{12}O_6$), the two substances we're interested in are gases. Our conversion factor is thus $6\ \ell\ CO_2/6\ \ell\ O_2$.

Step 4: Our setup is:

$$5.00 \cancel{\ell\ CO_2} \times \frac{6\ \ell\ O_2}{6 \cancel{\ell\ CO_2}} = \_\_\_\_ \ell\ O_2$$

Answer: $5.00\ \ell\ O_2$. We wouldn't notice any change in the volume, because there isn't any. (We could have gotten the answer by inspection, too, when we saw that equal numbers of moles of O_2 and CO_2 are involved.) ■

MASS-TO-VOLUME AND VOLUME-TO-MASS CONVERSIONS.

Here, we'll work problems converting grams of a solid to liters of a gas, and vice versa.

Sometimes we can make use of chemical reactions in which solids or liquids react to produce gases. A solid or liquid takes up much less room than a gas and is easier to handle, and the released gas can do work or perform a specific function for us because of its much larger volume. Baking powder, gasoline, jet and rocket fuels, and gunpowder are examples of such substances.

EXAMPLE 11.6: Gunpowder is a mixture of powdered carbon, potassium nitrate, and sulfur. When struck, this substance reacts to form a mixture of gases. Because of their much greater volume at STP, the gases create enough pressure to propel a bullet. The equation is

$$2\ C(s) + 2\ KNO_3(s) + S(s) \longrightarrow CO(g) + CO_2(g) + SO_2(g) + N_2(g) + K_2O(s)$$

If a 0.1-cm^3 sample of gunpowder contains 0.2 grams of KNO_3, and enough C and S to completely react with it, calculate the volume of all the gases produced, at STP, and compare this volume with that of solid gunpowder.

Solution:

Step 1: We want to convert 0.2 grams of KNO_3 to liters of CO, CO_2, N_2, and SO_2.

Step 2: The equation is balanced.

Step 3: Because one mole of each of the four gases is produced for every two moles of KNO_3 used, we may use 4 moles gas/2 moles KNO_3 as our conversion factor, as well as 22.4 ℓ gas/mole gas and the formula weight of KNO_3.

Step 4: Our setup is:

$$0.2\ \text{g}\ \cancel{KNO_3} \times \frac{1\ \text{mole}\ \cancel{KNO_3}}{101\ \text{g}\ \cancel{KNO_3}} \times \frac{4\ \text{moles gas}}{2\ \text{moles}\ \cancel{KNO_3}} \times \frac{22.4\ \ell\ \text{gas}}{1\ \text{mole gas}} = \_\_\_\ \ell\ \text{gas}$$

Answer: 0.09 liters, or 90 cm^3. Compared with the volume of 0.1 cm^3 for solid gunpowder, this represents a volume increase of about 900 times. ■

Sometimes, on the other hand, we want to get rid of a gas by letting it react to form a solid. The absorption of SO_2 gas by limestone to form solid calcium sulfate (Example 6.11, p. 121) is an example. In the chemistry laboratory, we often want to prepare a sample of a gas in order to study it. We usually do so by letting solids and/or liquids react to form gases.

EXAMPLE 11.7: A common way to prepare N_2 in the laboratory is to gently heat ammonium nitrite, according to this equation:

$$NH_4NO_2(s) \xrightarrow{\Delta} N_2(g) + 2\ H_2O(g)$$

How many grams of ammonium nitrite would have to be used to get 50.0 cm^3 of nitrogen, measured at STP?

Solution:

Step 1: We want to convert 50.0 cm^3 N_2 to grams of NH_4NO_2.

Step 2: The equation is balanced.

Step 3: Our conversion factors are 64.0 g NH_4NO_2/mole NH_4NO_2, 1 mole NH_4NO_2/mole N_2, and 22.4 ℓ N_2/mole N_2. We can change 50.0 cm^3 to 0.0500 liters.

Step 4: Our setup is:

$$0.0500\ \cancel{\ell\ N_2} \times \frac{1\ \text{mole}\ \cancel{N_2}}{22.4\ \cancel{\ell\ N_2}} \times \frac{1\ \text{mole}\ \cancel{NH_4NO_2}}{\text{mole}\ \cancel{N_2}} \times \frac{64.0\ \text{g}\ NH_4NO_2}{\text{mole}\ \cancel{NH_4NO_2}} = \_\_\_\ \text{g}\ NH_4NO_2$$

Answer: 0.143 g NH_4NO_2. ■

11.3 THE GAS LAWS

All the problems so far have involved gases at STP. In practice, though, gases are seldom—if ever—measured at STP. How do we handle that? We use some laws that let us calculate exactly how the pressure, temperature, and volume of a gas affect one another.

THE EFFECT OF PRESSURE ON VOLUME:

BOYLE'S LAW. At constant temperature, the change in volume of a gas is inversely proportional to the change in pressure, according to *Boyle's Law,* formulated in 1660 by Robert Boyle (1627–1691). This means that the pressure decreases if the volume increases, and vice versa. Figure 11.3 illus-

Boyle's Law

FIGURE 11.3
Boyle's Law: At constant temperature, change in gas volume is inversely proportional to change in pressure

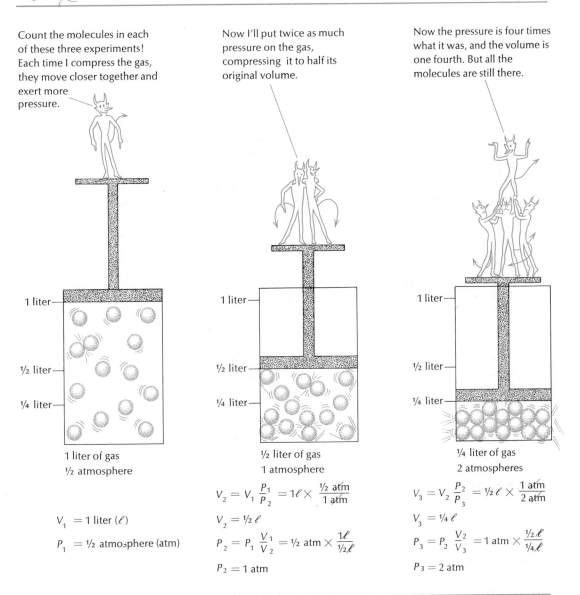

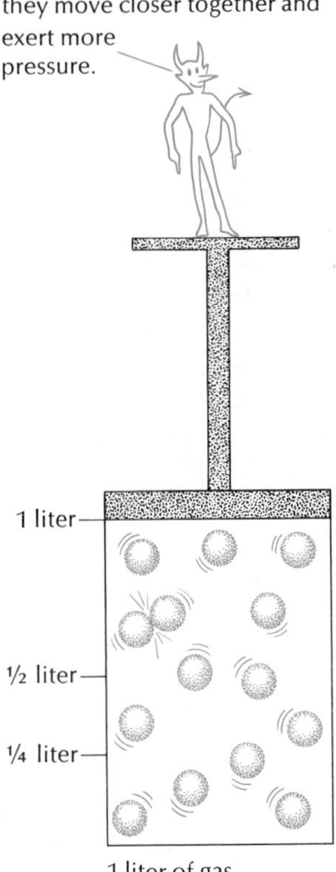

Count the molecules in each of these three experiments! Each time I compress the gas, they move closer together and exert more pressure.
1 liter
½ liter
¼ liter

1 liter of gas
½ atmosphere

$V_1 = 1$ liter (ℓ)

$P_1 = ½$ atmosphere (atm)

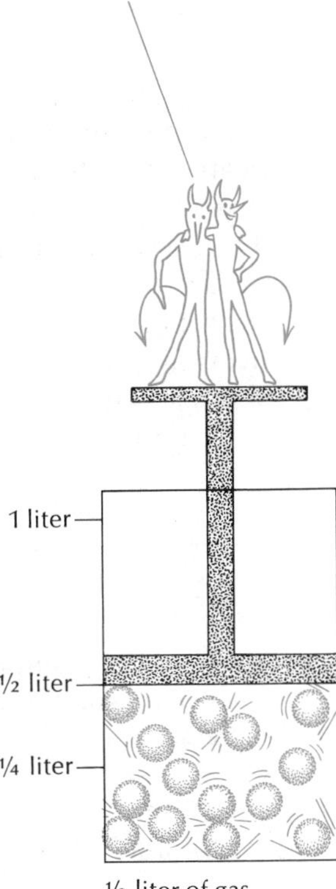

½ liter of gas
1 atmosphere

$$V_2 = V_1 \frac{P_1}{P_2} = 1\ell \times \frac{½ \text{ atm}}{1 \text{ atm}}$$

$$V_2 = ½\,\ell$$

$$P_2 = P_1 \frac{V_1}{V_2} = ½ \text{ atm} \times \frac{1\ell}{½\ell}$$

$$P_2 = 1 \text{ atm}$$

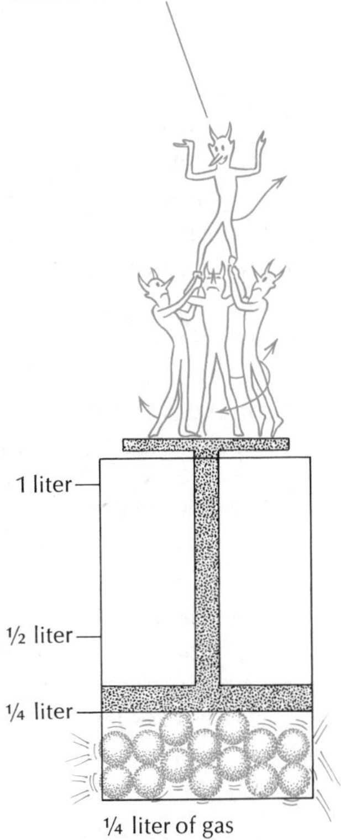

Now the pressure is four times what it was, and the volume is one fourth. But all the molecules are still there.
1 liter
½ liter
¼ liter

¼ liter of gas
2 atmospheres

$$V_3 = V_2 \frac{P_2}{P_3} = ½\,\ell \times \frac{1 \text{ atm}}{2 \text{ atm}}$$

$$V_3 = ¼\,\ell$$

$$P_3 = P_2 \frac{V_2}{V_3} = 1 \text{ atm} \times \frac{½\ell}{¼\ell}$$

$$P_3 = 2 \text{ atm}$$

trates Boyle's Law with a sliding piston, which changes the pressure on the gas in the container. At first, 1 liter of gas is at $\frac{1}{2}$ atmosphere. When the pressure is increased to 1 atmosphere, the volume drops to $\frac{1}{2}$ liter. The number of gas molecules remains the same, of course, but the molecules are half as far apart. In the second case, to maintain equilibrium with the increased outside pressure, molecules exert twice as much pressure on the walls of their container. When the pressure is increased even further to 2 atmospheres, the volume is reduced again, to $\frac{1}{4}$ liter.

Mathematically, this is what Boyle's Law says:

$$V_2 = V_1 \frac{P_1}{P_2}$$

That is, if we start with a gas having initial volume V_1 and initial pressure P_1 and change the pressure to P_2, we can calculate the new volume V_2. Boyle's Law can also be stated like this:

$$P_2 = P_1 \frac{V_1}{V_2}$$

That is, if we start with a gas having initial pressure P_1 and initial volume V_1, and change the volume to V_2, we can calculate the new pressure P_2. Calculations with these formulas are shown for the examples in Figure 11.3.

In working problems with these and other gas law formulas, we should always decide first what direction our answer will take. That is, for an increase in pressure, the volume will decrease; for an increase in volume, the pressure will decrease. That's what "inversely proportional" means. If we have some feeling at the beginning for what sort of an answer to expect, we'll avoid the errors that might result from a wrong calculator answer or an upside-down fraction.

EXAMPLE 11.8: In Example 11.1, we calculated the volume of a sample of CO_2 to be 232 liters at STP. Calculate the volume of this sample if it were collected at standard temperature and 0.876 atm.

Solution: First of all, we see that the pressure is changing from 1 atm (standard pressure) to 0.876 atm: a decrease in pressure. Thus we expect an increase in volume. We're given two pressures and a starting volume ($V_1 = 232\ \ell$). Our setup is:

$$V_2 = V_1 \frac{P_1}{P_2} = 232\ \ell \times \frac{1.00\ \cancel{\text{atm}}}{0.876\ \cancel{\text{atm}}} = \underline{\quad}\ \ell$$

Answer: 265 ℓ. We got the larger volume we expected. ■

EXAMPLE 11.9: A standard air tank for scuba diving has a capacity of 0.423 cubic feet. The tank is filled with air under pressure so that it will release the equivalent of 65.2 cubic feet of air at 1 atm when opened. What is the pressure of the tank before use?

Solution: Here, the units are in the English system, but that's all right as long as we stick to them throughout the problem. We're given two volumes and a finishing pressure ($P_2 = 14.7\ \text{lb/in}^2$ (1 atm); $V_1 = 0.423\ \text{ft}^3$; $V_2 = 65.2\ \text{ft}^3$), and we're asked to solve for starting pressure. We expect a larger starting pressure, because the starting volume is smaller, and we know we're dealing with a compressed gas. Our setup:

$$P_1 = P_2 \frac{V_2}{V_1} = 14.7 \text{ lb/in}^2 \times \frac{65.2 \cancel{\text{ft}^3}}{0.423 \cancel{\text{ft}^3}} = \underline{\quad} \text{ lb/in}^2$$

Answer: 2270 lb/in², a much larger pressure, as predicted. ■

THE EFFECT OF TEMPERATURE ON VOLUME: CHARLES' LAW. Jacques Charles (1746–1823) noticed that the volume of a gas, at constant pressure, increases in proportion to the Celsius temperature. To show this, blow up a balloon partway, and then put it close to a light bulb that's been on awhile. The heat from the bulb will cause the balloon to expand visibly, even though it's still subject to the same atmospheric pressure. By the same token, refrigerating a blown-up balloon will cause it to shrink, because a cooled gas contracts (at constant pressure).

Later, Joseph Gay-Lussac (1778–1850) discovered that no matter what its volume, a gas will shrink by $\frac{1}{273}$ of its volume for every degree below zero degrees Celsius it is cooled. In fact, this discovery led Lord Kelvin (William Thomson, 1824–1907) to the development of the Kelvin temperature scale in 1848. We've already seen that zero Kelvin is the temperature at which all motion stops. It's also the temperature at which a gas theoretically would shrink to zero volume, something we know is impossible.

The result of all this is what we now call Charles' Law, even though Gay-Lussac also contributed to it. *Charles' Law* states that the change in volume of a gas at constant pressure is directly proportional to the change in the Kelvin temperature. Or, stated mathematically: Charles' Law

$$V_2 = V_1 \frac{T_2(\text{K})}{T_1(\text{K})} \quad \text{and} \quad T_2(\text{K}) = T_1(\text{K}) \frac{V_2}{V_1}$$

Figure 11.4 shows the relationship between volume and temperature.

Hot-air balloons can float in air because the density of hot air is less than the density of the colder air near the earth. However, when the balloon reaches an altitude where the density of its air is equal to the density of the surrounding air, it will stop rising.

EXAMPLE 11.10: Calculate the density of air at 40°C and standard pressure. (The density of air at STP is 1.29 g/ℓ.) Then, from Figure 10.2 (p. 203), estimate the altitude at which a balloon filled with this 40°C air would stop rising.

Solution: The density of air tells us that 1.29 grams occupy 1.00 liter at STP. To calculate the density at the new temperature ([40. + 273] K = 313 K), we first find the volume that 1.29 grams would occupy at the same pressure and 313 K. Since we're increasing the temperature, we expect a volume greater than 1.00 liter. Thus we're given a starting volume (V_1 = 1.00 ℓ) and two temperatures (T_1 = 273 K, standard temperature; and T_2 = 313 K). Our setup is:

$$V_2 = V_1 \frac{T_2(\text{K})}{T_1(\text{K})} = 1.00\, \ell \times \frac{313 \cancel{\text{K}}}{273 \cancel{\text{K}}} = 1.15\, \ell \text{ (larger, as expected)}$$

Now we can calculate the density:

$$\text{density} = \frac{\text{weight}}{\text{volume}} = \frac{1.29\ \text{g}}{1.15\ \ell} = \underline{\quad}\ \frac{\text{g}}{\ell}$$

Answer: 1.12 g/ℓ less than the density of colder air, as we expected. Figure 10.2 shows the altitude to be about 2 km. ■

FIGURE 11.4
Charles' Law: At constant pressure, change in gas volume is directly proportional to change in Kelvin temperature

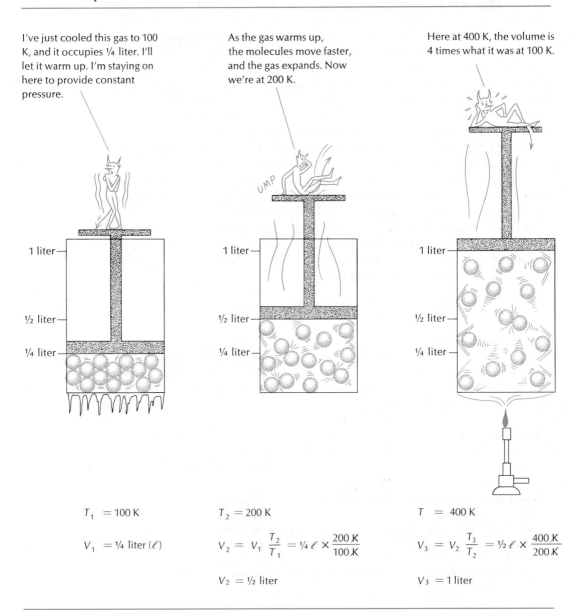

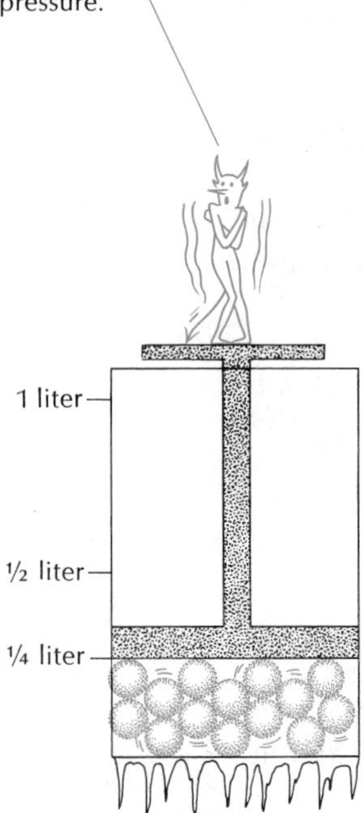
I've just cooled this gas to 100 K, and it occupies ¼ liter. I'll let it warm up. I'm staying on here to provide constant pressure.
1 liter
½ liter
¼ liter

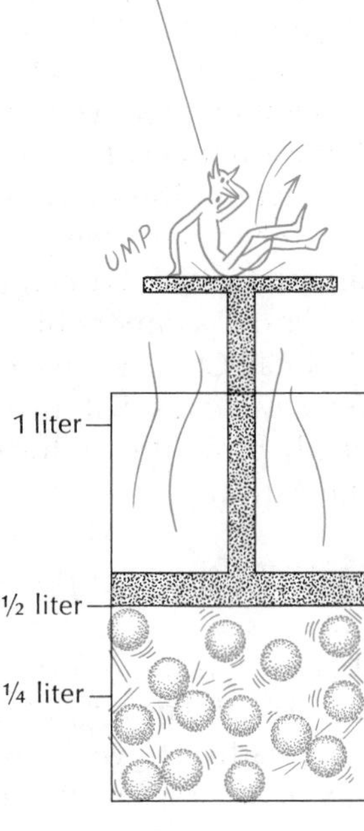
As the gas warms up, the molecules move faster, and the gas expands. Now we're at 200 K.
UMP
1 liter
½ liter
¼ liter

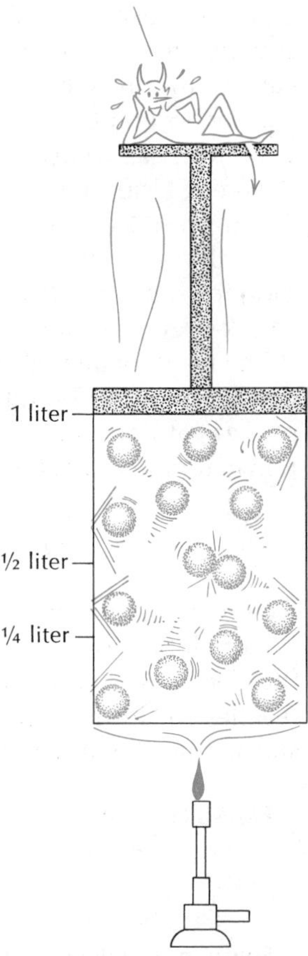
Here at 400 K, the volume is 4 times what it was at 100 K.
1 liter
½ liter
¼ liter

$T_1 = 100\ \text{K}$

$V_1 = ¼\ \text{liter}\ (\ell)$

$T_2 = 200\ \text{K}$

$V_2 = V_1 \frac{T_2}{T_1} = ¼\ \ell \times \frac{200\ \text{K}}{100\ \text{K}}$

$V_2 = ½\ \text{liter}$

$T = 400\ \text{K}$

$V_3 = V_2 \frac{T_3}{T_2} = ½\ \ell \times \frac{400\ \text{K}}{200\ \text{K}}$

$V_3 = 1\ \text{liter}$

THE EFFECT OF TEMPERATURE ON PRESSURE: GAY-LUSSAC'S LAW.

On a hot day, pressure will build up inside tires as a car is driven. This is an illustration of *Gay-Lussac's Law,* which states that the pressure of a gas, at constant volume, changes in direct proportion to the change in Kelvin temperature. Or, stated mathematically:

Gay-Lussac's Law

$$P_2 = P_1 \frac{T_2}{T_1} \quad \text{and} \quad T_2 = T_1 \frac{P_2}{P_1}$$

Figure 11.5 illustrates the relationship between temperature and pressure of a gas.

EXAMPLE 11.11: Calculate the pressure inside a used aerosol can if it is accidentally incinerated at 350°C. Assume that the gas inside the spent can before heating was at 1 atmosphere and 25°C.

Solution: We expect the pressure to be greater at a higher temperature. Here, $P_1 = 1$ atm; $T_1 = 298$ K; and $T_2 = 623$ K. Our setup is:

$$P_2 = 1\ \text{atm} \times \frac{623\ \cancel{\text{K}}}{298\ \cancel{\text{K}}} = \underline{\quad}\ \text{atm}$$

Answer: 2 atm. Greater, as expected. ■

Gay-Lussac's Law explains why we shouldn't throw aerosol cans into the fire. The can keeps the volume constant, but the increased temperature increases the pressure of the confined gas in the can to the point that the can may explode.

THE COMBINED GAS LAWS.

We usually measure gases at both temperatures and pressures that aren't STP. In the laboratory, we often generate gases under laboratory conditions of temperature and pressure and then must convert them to STP. To do this, we use the *combined gas laws,* which take into account changes of pressure, temperature, and volume all at once. Since we'll usually want to know how volume changes with temperature and pressure, we'll use the combined gas laws in this form:

combined gas laws

$$V_2 = V_1 \left(\frac{T_2}{T_1}\right)\left(\frac{P_1}{P_2}\right)$$

This equation, a combination of all the previous gas laws, says that the new volume of gas (V_2) is going to be the old volume (V_1) times two fractions. One of the fractions represents the temperature change, and one represents the pressure change. We won't always be able to predict the direction of the change when we're using this equation, but we can predict the effect of each fraction on the volume.

EXAMPLE 11.12: In Example 11.7, we calculated that it would take 0.143 g of NH_4NO_2 to generate 50.0 cm^3 of N_2 at STP. What volume would be generated under the laboratory conditions of 23°C and 0.943 atm?

FIGURE 11.5
Gay-Lussac's Law: At constant volume, change in gas pressure is directly proportional to change in Kelvin temperature

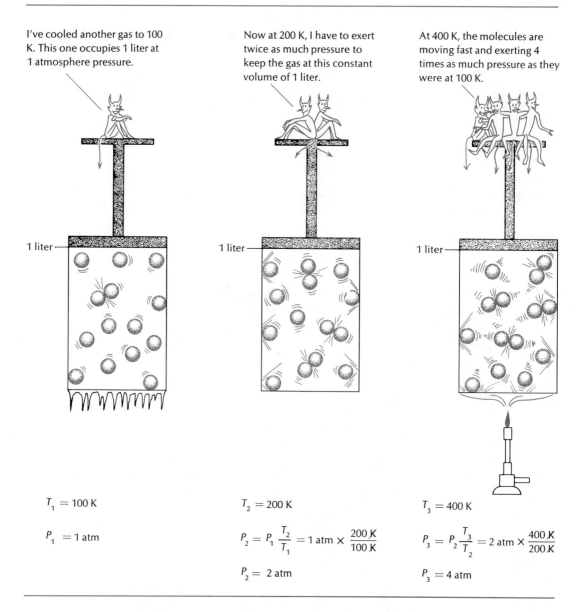

I've cooled another gas to 100 K. This one occupies 1 liter at 1 atmosphere pressure.

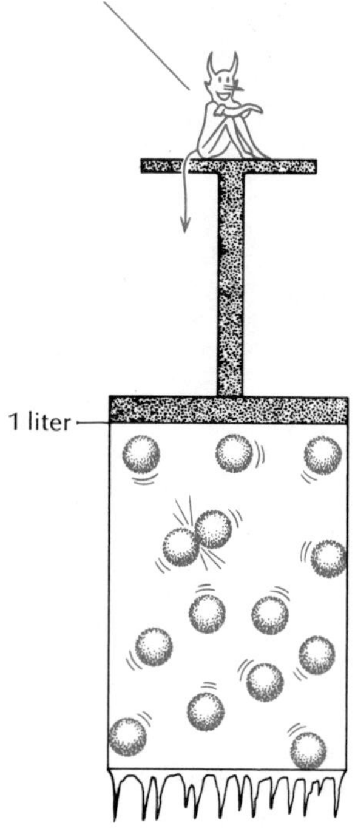

$T_1 = 100\text{ K}$

$P_1 = 1\text{ atm}$

Now at 200 K, I have to exert twice as much pressure to keep the gas at this constant volume of 1 liter.

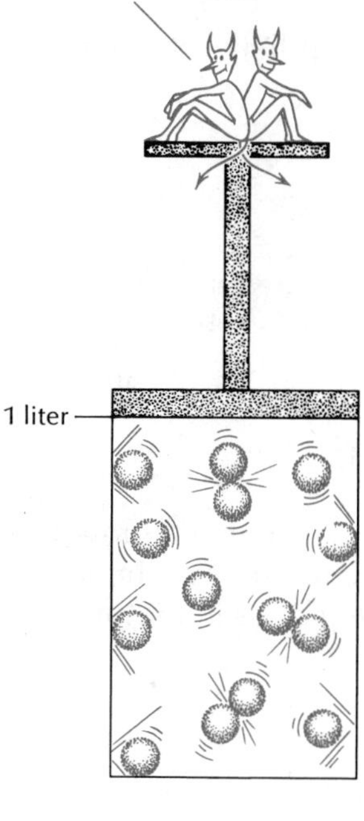

$T_2 = 200\text{ K}$

$P_2 = P_1 \dfrac{T_2}{T_1} = 1\text{ atm} \times \dfrac{200\text{ K}}{100\text{ K}}$

$P_2 = 2\text{ atm}$

At 400 K, the molecules are moving fast and exerting 4 times as much pressure as they were at 100 K.

1 liter

$T_3 = 400\text{ K}$

$P_3 = P_2 \dfrac{T_3}{T_2} = 2\text{ atm} \times \dfrac{400\text{ K}}{200\text{ K}}$

$P_3 = 4\text{ atm}$

Solution: We're changing our initial volume ($V_1 = 50.0\text{ cm}^3$) from standard temperature ($T_1 = 273\text{ K}$) to a laboratory temperature ($T_2 = [23 + 273]\text{ K} = 296\text{ K}$). This is an increase of temperature, so we expect this fraction to cause an increase in volume, and the larger number goes on top: 296 K/273 K. The pressure change is from standard pressure ($P_1 = 1\text{ atm}$) to a laboratory pressure ($P_2 = 0.943\text{ atm}$). This is a decrease in pressure, and we know that this too will increase the volume: our pressure fraction is

1 atm/0.943 atm. Both fractions increase the volume, so we expect V_2 to be larger than V_1. Our setup is:

$$V_2 = 50.0\ \text{cm}^3 \times \frac{296\ \cancel{\text{K}}}{273\ \cancel{\text{K}}} \times \frac{1\ \cancel{\text{atm}}}{0.943\ \cancel{\text{atm}}} = \_\_\_\ \text{cm}^3$$

Answer: 57.5 cm³. Larger, as we expected. ■

EXAMPLE 11.13: The Haber process (see Example 11.4) isn't really performed at STP. If 1500 liters of NH_3 are produced at 30 atm and 450°C, what volume would this be at STP?

Solution: Our initial volume (V_1) is 1500 liters. We're changing temperature from initial (T_1 = [450 + 273] K = 723 K) to standard (T_2 = 273). Since we're cooling the gas, this fraction will decrease its volume. The pressure change goes from P_1 = 30 atm to P_2 = 1 atm (standard pressure). A decrease in pressure means an increase in volume. Here, one fraction decreases the volume and one increases it. We can't tell which will win, but we can use our reasoning to make sure the temperature fraction is less than 1, and the pressure fraction is greater than 1. Our setup is:

$$V_2 = 1500\ \ell \times \frac{273\ \cancel{\text{K}}}{723\ \cancel{\text{K}}} \times \frac{30\ \cancel{\text{atm}}}{1\ \cancel{\text{atm}}} = \_\_\_\ \ell$$

Answer: 1.7×10^4 liters. (The pressure fraction wins.) ■

THE PRESSURE OF A MIXTURE OF GASES: DALTON'S LAW.

Air is a mixture of gases. Sometimes, a gas we collect in the laboratory is a mixture of gases. If two or more gases are mixed in a container, each one will exert its own pressure, without regard for the others. This fact is stated in *Dalton's Law,* named after John Dalton (1766–1844): In a mixture of gases at constant temperature and volume, the total pressure is the sum of the partial pressures. The *partial pressure* means the pressure the gas would exert if it were by itself.

Dalton's Law

partial pressure

Most often, laboratory mixtures contain water vapor as one of the gases. We know that liquids have vapor pressure. Therefore, a gas in a container above water will always have some water vapor in it. When we generate gases in the laboratory and collect them for study, we usually find it convenient to collect them by letting them displace water in a container, such as a bottle or a test tube. When we do this, we always end up with a mixture of the generated gas and water vapor. The pressure of the mixture is the sum of the pressure of the gas and the vapor pressure of water at that temperature. We can write this as an equation.

$$P_{\text{total}} = P_{\text{gas}} + P_{\text{water vapor}}$$

The total pressure is the barometric pressure. The vapor pressure of water is usually provided by a table, such as Table 11.5. To find the pressure of a gas we've collected by displacement of water, we read the barometer, take the temperature of the water, and look in Table 11.5 for the vapor pressure of water at that temperature. Then, we use this equation:

$$P_{\text{gas}} = P_{\text{total}} - P_{\text{water vapor}}$$

TABLE 11.5
Vapor pressure of water

Temperature (°C)	Pressure torr	Pressure atm	Temperature (°C)	Pressure torr	Pressure atm
−15.0	1.44	0.00189	16.0	13.6	0.0179
−14.0	1.56	0.00205	17.0	14.5	0.0191
−13.0	1.69	0.00222	18.0	15.5	0.0204
−12.0	1.83	0.00241	19.0	16.5	0.0217
−11.0	1.99	0.00261	20.0	17.5	0.0231
−10.0	2.15	0.00283	21.0	18.7	0.0245
−9.00	2.36	0.00306	22.0	19.8	0.0261
−8.00	2.51	0.00331	23.0	21.1	0.0277
−7.00	2.72	0.00357	24.0	22.4	0.0294
−6.00	2.93	0.00386	25.0	23.8	0.0313
−5.00	3.16	0.00416	26.0	25.2	0.0332
−4.00	3.41	0.00449	27.0	26.7	0.0352
−3.00	3.67	0.00483	28.0	28.3	0.0373
−2.00	3.96	0.00520	29.0	30.0	0.0395
−1.00	4.26	0.00596	30.0	31.8	0.0419
0.000	4.58	0.00602	35.0	40.2	0.0555
1.00	4.93	0.00648	40.0	55.3	0.0728
2.00	5.29	0.00697	45.0	71.9	0.0946
3.00	5.68	0.00748	50.0	92.5	0.122
4.00	6.10	0.00803	55.0	118.0	0.155
5.00	6.54	0.00861	60.0	149.4	0.197
6.00	7.01	0.00923	65.0	187.5	0.247
7.00	7.51	0.00988	70.0	233.7	0.307
8.00	8.04	0.0106	75.0	289.1	0.380
9.00	8.61	0.0113	80.0	355.1	0.467
10.0	9.21	0.0121	85.0	433.6	0.570
11.0	9.84	0.0130	90.0	525.8	0.692
12.0	10.5	0.0138	95.0	633.9	0.834
13.0	11.2	0.0148	100.	760.0	1.00
14.0	12.0	0.0158	105.	906.1	1.19
15.0	12.8	0.0168	110.	1075	1.41

Figure 11.6 illustrates a gas being collected by displacement of water.

EXAMPLE 11.14: In Example 11.12, we calculated that a nitrogen sample would occupy 57.5 cm³ at 23°C and 0.943 atm. If this sample were collected by displacement of water, calculate (A) the pressure of the nitrogen, and (B) the corrected volume occupied by the nitrogen.

Solution A: We use the equation $P_{gas} = P_{total} - P_{water\ vapor}$. $P_{total} = 0.943$ atm (the barometric pressure). $P_{water\ vapor} = 0.0277$ atm (at 23°C, from Table 11.5). Then:

$$P_{gas} = 0.943 \text{ atm} - 0.0277 \text{ atm} = \_\_\_\_ \text{ atm}$$

Answer A: 0.915 atm.

FIGURE 11.6
A gas collected by displacement of water always contains water vapor

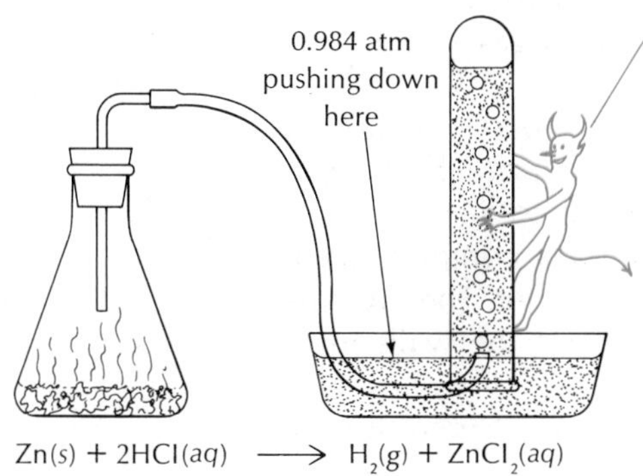

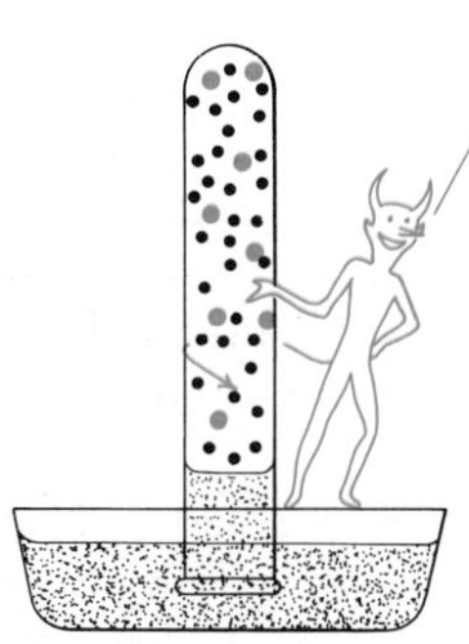

Solution B: The volume of the nitrogen sample was 57.5 cm³ at 0.915 atm. We want to find the volume at the same temperature but at 0.943 atm. This is an increase in pressure, so we expect the volume to decrease.

$$V_2 = V_1 \frac{P_1}{P_2} = 57.5 \text{ cm}^3 \times \frac{0.915 \text{ atm}}{0.943 \text{ atm}} = \_\_\_\_ \text{ cm}^3$$

Answer B: 55.8 cm³. Smaller, as predicted. ■

11.4 THE IDEAL GAS EQUATION

The quantities that we've used to calculate with gases can all be tied together into one equation: the ideal gas equation. It contains volume, moles, temperature, and pressure. It's based on the Kinetic-Molecular Theory.

THE KINETIC-MOLECULAR THEORY. To explain the properties of gases, and to develop the ideal gas equation, the *Kinetic-Molecular Theory* makes these assumptions about gases.

Kinetic-Molecular Theory

1. *Gases are composed of particles that are very far apart. The particles themselves occupy negligible volume.* This explains gases' ability to expand and contract, and their low density.
2. *Gas molecules are always moving rapidly, and they move more rapidly as the temperature increases. They collide with each other and with the walls of the container.* This explains gases' ability to mix with each other. It also explains that the pressure exerted by a confined gas is caused by the particles hitting the walls of the container. The pressure increases with temperature because the particles hit the walls more often.
3. *The particles of a gas have no attraction for each other. When they do collide with each other, they bounce right off.*

An *ideal gas* is one that obeys these three assumptions. The energy of an ideal gas at a certain temperature is given by:

ideal gas

$$\text{energy} = nRT$$

Here, n is the number of moles of the gas, R is a conversion factor called the gas constant, and T is the Kelvin temperature. We can see two things right off. First, at 0 K ($T = 0$), there is zero energy, which we already know is true. Also, if there are no moles ($n = 0$), there is no gas, and there can be no energy.

We can redefine *pressure* as energy per unit of volume, instead of force per unit of area. Pressure is thus a measure of how hard a gas is trying to get out of its container, or how hard an outside pressure is trying to keep the gas in its container. Stated mathematically:

pressure

$$P = \frac{\text{energy}}{V}$$

If we substitute nRT for energy, then we have this equation:

$$P = \frac{nRT}{V}$$

This gives us the usual form of the *ideal gas equation.*

ideal gas equation

$$PV = nRT$$

In this equation, P is in atmospheres, V is in liters, n is in moles, and T is in Kelvins. To make the units come out right, we use the *gas constant* (R) as a conversion factor. R has the value of:

gas constant (R)

$$0.0821 \frac{(\ell \times \text{atm})}{(\text{mole} \times \text{K})} \quad \text{or} \quad 62.4 \frac{(\ell \times \text{torr})}{(\text{mole} \times \text{K})}$$

which we read, "liter-atmospheres per mole-Kelvin" and "liter-torr per mole-Kelvin." Which of these we use depends on the units given to us in the problem.

WORKING PROBLEMS WITH THE IDEAL GAS EQUATION. The ideal gas equation is most convenient for solving problems with only one set of conditions. Most commonly, we'll want to solve either for volume or for moles. To solve for volume:

$$V = \frac{nRT}{P}$$

To solve for moles, we must convert the equation to:

$$n = \frac{PV}{RT}$$

EXAMPLE 11.15: A gas balloon is charged with 12.5 pounds of helium. What will the volume of the balloon be when it rises to an altitude of 6 miles, where the pressure is 210 torr and the temperature is −40°C?

Solution: We want to solve for volume. Before we can use the equation

$$V = \frac{nRT}{P},$$

though, we have to convert 12.5 lb helium to moles:

$$12.5\ \cancel{\text{lb}} \times \frac{454\ \cancel{\text{g}}}{\cancel{\text{lb}}} \times \frac{1.00\ \text{mole}}{4.00\ \cancel{\text{g}}} = 1420\ \text{moles}$$

Next,

$$V = \frac{[1420\ \cancel{\text{moles}}]\left[62.4\,\frac{(\ell \times \cancel{\text{torr}})}{(\cancel{\text{mole}} \times \cancel{\text{K}})}\right][233\ \cancel{\text{K}}]}{210\ \cancel{\text{torr}}} = \underline{\qquad}\ \ell$$

Answer: 98,300 liters. ■

EXAMPLE 11.16: A chemical process has generated 525 liters of NO_2. We want to convert the NO_2 to $Ca(NO_3)_2(s)$, a fertilizer, according to this equation:

$$4\ NO_2(g) + Ca(OH)_2(s) \longrightarrow Ca(NO_3)_2(s) + 2\ HNO_2(l)$$

A. If the NO_2 is at 4.2 atm and 350 K, how many moles are there?
B. How many grams of $Ca(OH)_2$ will it take to convert the NO_2?

Solution A: We want to solve for moles. We substitute into this equation:

$$n = \frac{PV}{RT} = \frac{(4.2\ \cancel{\text{atm}})(525\ \cancel{\ell})}{\left[0.0821\,\frac{(\cancel{\ell} \times \cancel{\text{atm}})}{(\text{mole} \times \cancel{\text{K}})}\right](350\ \cancel{\text{K}})} = \underline{\qquad}\ \text{moles}$$

Notice that:

$$\frac{1}{\left(\frac{1}{\text{mole}}\right)} = \text{mole}$$

Inverting reciprocals like this is discussed in Appendix B, p. 514.

Answer A: 77 moles.

Solution B: We want to convert moles of NO_2 to grams of $Ca(OH)_2$. The equation is balanced, and our conversion factors are 1 mole $Ca(OH)_2$/4 moles NO_2 and 74.1 g $Ca(OH)_2$/mole $Ca(OH)_2$. Our setup is:

$$77 \text{ moles } NO_2 \times \frac{1 \text{ mole } Ca(OH)_2}{4 \text{ moles } NO_2} \times \frac{74.1 \text{ g } Ca(OH)_2}{\text{mole } Ca(OH)_2} = \_\_\_ \text{ g } Ca(OH)_2$$

Answer B: 1400 g $Ca(OH)_2$, or 1.4 kg $Ca(OH)_2$.

Notice that we chose *R* in liter-atmospheres per mole-Kelvin when the pressure was given in atmospheres, and in liter-torr per mole-Kelvin when the pressure was given in torr. ■

NON-IDEAL BEHAVIOR. No gas behaves ideally. The noble gases come closest to ideal behavior, because the attraction between their particles is the least. In general, gases will behave less ideally if we put them under high pressure or at low temperature. When their particles are close enough together, they will attract each other, and assumption 3 of the Kinetic-Molecular Theory will no longer be true. Water molecules, when they're in the gas state, don't bounce off one another as they're supposed to. Instead, they attract each other, because of hydrogen bonding. Molecules with this kind of attraction usually don't obey the ideal gas equation. We need a table of water vapor pressure at each temperature (Table 11.5), because we have to measure it instead of calculating it.

REVIEW QUESTIONS

Some Chemistry of Air

1. What is *photosynthesis?* How does it explain why our atmosphere contains oxygen?
2. What is *nitrogen fixation?* List some natural and artificial ways of fixing nitrogen.
3. What is a *catalyst?* Why is a catalyst sometimes needed in nitrogen fixation?
4. What part of the sunlight gives us sunburn? How does our ozone layer protect us from this part of sunlight?
5. Explain the *greenhouse effect.* What parts of sunlight are involved, and how?
6. Review the important reactions of gases in air.
7. What are some of the major air pollutants? Where do they come from and why are they hazardous?
8. What are the main ingredients of *photochemical smog?*
9. Why is it dangerous to have ozone in our lower atmosphere?
10. How does nature get rid of some air pollutants? Can we depend on this?
11. Why is it difficult to minimize CO and NO at the same time?
12. What are some of the problems connected with SO_2 removal?

Measuring Gases at STP

13. State *Avogadro's Law.* What is the *molar volume* of a gas? How are these two concepts related?
14. Why do molar volumes of different gases have different masses?
15. How can we use molar volume as a conversion factor to calculate a gas's density at STP?
16. How can we use gas density to determine the molecular weight of a liquid?
17. What are some examples of chemical reactions that produce gases?

The Gas Laws

18. State *Boyle's Law,* and explain it in your own words. What formulas do we use in calculations with Boyle's Law?
19. What law relates temperature and volume of a gas? State the law, and explain how it is related to the Kelvin temperature scale.
20. What is *Gay-Lussac's Law?* How does it explain the behavior of tires on a hot day?
21. How can we predict the direction of an answer when doing gas law calculations?
22. Write the equation for calculating the volume of a gas using the *combined gas laws,* and explain what each fraction means.

23. What is *Dalton's Law?* Why might we need to use it in the laboratory?
24. Explain how we correct the pressure of a gas for the presence of water vapor.

The Ideal Gas Equation

25. State the assumptions of the Kinetic-Molecular Theory. How do these explain all of the previous gas laws?
26. What is an *ideal gas?* What is the energy of an ideal gas?
27. What is the *gas constant?* What are its units?
28. Write the *ideal gas equation* in three forms. How do we tell which one to use in calculations?
29. Do all gases behave ideally? When do they deviate the most from ideal behavior?

EXERCISES

Set A (Answers at back of book. Asterisks mark the the more difficult exercises.)

1. An average automobile, without pollution control devices, produces about 5 grams of NO for each mile it is driven. How many liters (measured at STP) would be produced by such an automobile on a 100-mile trip?
2. In 1974, about 2×10^{10} liters (STP) of ammonia were produced. How many metric tons is this?
3. Compute and compare the densities of NO and NO_2 at STP.
4. An unknown air pollutant is analyzed and found to have a density of 1.35 g/ℓ at STP.
   a. Calculate its molecular weight.
   b. Which air pollutant (from Table 11.3) is it likely to be?
5. A sample of an unknown liquid weighing 0.469 grams when converted to its vapor, is found to occupy a volume of 125 cm³ at STP.
   a. Calculate its molecular weight.
   b. If the substance's empirical formula is CH_2, what is its molecular formula?
6. In the natural fixation of nitrogen by lightning (see Table 11.2), how many liters of NO would be produced from 500 liters of nitrogen (measured at STP)?
7. One way to get rid of NO is to let it react with ammonia (see Table 11.4 for the equation). How many liters of ammonia would it take to get rid of 12,000 liters of NO? (All measurements are at STP.)
8. Natural gas contains mostly CH_4, with some sulfur in the form of H_2S. The H_2S can be removed by controlled oxidation, according to this equation:

   $$6\,H_2S(g) + 3\,O_2(g) \longrightarrow 6\,S(s) + 6\,H_2O(g)$$

   How many grams of sulfur will be obtained by treating 250 liters (at STP) of H_2S in this manner?
9. In the process of photosynthesis (Table 11.2), how many liters of O_2 (STP) would be obtained for every 2.5 moles of glucose produced?

*10. The Solvay process for making sodium bicarbonate uses this reaction:

   $$H_2O(l) + NH_3(g) + CO_2(g) + NaCl(aq) \longrightarrow NH_4Cl(aq) + NaHCO_3(aq)$$

   How many liters of NH_3 and of CO_2 (both at STP) are needed to make 1.00 kg of sodium bicarbonate, $NaHCO_3$ if the reaction proceeds at 46.4% yield?

11. Each person breathes about 20,000 liters of air (at STP) daily. How many liters would a person have to breathe at 0.72 atmospheres to get the same amount of air? (Assume no change in temperature.)
12. A compressor can reduce a gas from 1000 liters at 1 atmosphere to 0.5 liters. What pressure can such a compressor provide?
13. Gas turbines burn liquid fuel to provide high temperatures, producing gases, which then drive the turbine. These gases have a temperature of about 650°C. What volume would 1 liter of such gas have at standard temperature and the same pressure?
14. Compare the density of air on a hot day in Needles, California (43°C), with the density of air on a cold day in Bozeman, Montana (−29°C). Assume a pressure of 1 atmosphere. The density of air at STP is 1.29 g/ℓ.
15. In an explosive like blasting gelatin, a solid reacts to form gases. When confined, the gases build up tremendous pressure and explode. If gases are produced at an initial pressure of 713 atm (at 25°C) and finally reach a pressure of 13,000 atm because of the temperature change involved in the reactions, calculate the final temperature.
16. A gas sample is taken outdoors in the winter at −10°C and 0.972 atm. What will be its new pressure if it warms to room temperature (27°C)?
17. A sample of oxygen gas is collected in the laboratory. It has a volume of 89.2 cm³ at a temperature of 22°C and a pressure of 0.978 atm. Calculate its volume at STP.

18. A chemical reaction predicts a volume of 533 cm^3 of hydrogen at STP. What volume would be obtained at 745 torr and 25°C?
19. What is the total pressure of a mixture of 0.234 atm H_2, 0.438 atm N_2, and 0.199 atm He?
20. Using displacement of water, 44.0 cm^3 of nitrogen gas is collected at 24.0°C and 0.957 atm.
    a. What is the pressure of the nitrogen?
    b. Calculate the volume of the nitrogen at STP.
21. A chemical reaction predicts 97.2 cm^3 of hydrogen gas at STP. What volume would be obtained, if the hydrogen were collected by displacement of water at 22.0°C and 0.954 atm?
22. The manufacture of acetylene gas, C_2H_2, involves letting calcium carbide react with water according to this reaction:

    $$CaC_2(s) + 2\,H_2O(l) \longrightarrow Ca(OH)_2(aq) + C_2H_2(g)$$

    a. How many liters of C_2H_2 (at STP) would be obtained from 550 kg of CaC_2?
    b. How many liters would be obtained at 300°C and 2.00 atm?
23. A person can breathe 0.05 ppm of NO a day without harm. At this level, what volume of NO (at STP) would a person breathe in a daily total of 20,000 ℓ of air?
24. One way to remove SO_2 from the air is to let it react with CO:

    $$SO_2(g) + 2\,CO(g) \longrightarrow S(s) + 2\,CO_2(g)$$

    Using this formula, compute how many moles of sulfur would be produced if 6×10^5 liters of SO_2 were removed from the air at 0.976 atmospheres and 200°C.
25. SO_2 from ore smelters can be used to make sulfuric acid, according to this equation:

    $$2\,SO_2(g) + O_2(g) + 2\,H_2O(l) \longrightarrow 2\,H_2SO_4(aq)$$

    Using the SO_2 from Exercise 24, (a) how many moles, and (b) how many kilograms of H_2SO_4 could be produced?

*26. Air is about 0.031 percent CO_2, by volume. How many liters of air does a sugar cane plant need to produce 1.00 lbs of glucose (molecular weight 180.) according to the equation in Table 11.2?

Set B (Answers not given. Asterisks mark the more difficult exercises.)

1. Carbon monoxide is the air pollutant emitted in the largest quantity in the United States. In 1968, it amounted to 100 million metric tons. How many liters (STP) is this?
2. Sulfur dioxide is used in the wine industry to keep white grapes from browning. If 5 liters of SO_2 (STP) from a tank are bubbled through a vat of crushed grapes, what weight is lost from the SO_2 tank?
3. A railroad car containing a shipment of fluorine tanks has overturned and the tanks have broken open. Will the escaping fluorine rise into the atmosphere or stay near the ground? Prove your answer by calculating the density of fluorine gas at STP and comparing it with that of air (1.29 g/ℓ).
4. The density of an unknown gas is found to be 3.17 g/ℓ (STP). Calculate its molecular weight. It is a gaseous element: which one?
5. A 0.562-gram sample of a liquid, when converted to vapor, occupies 369 cm^3 at STP.
    a. Calculate its molecular weight.
    b. Its empirical formula is HO. What is its molecular formula?
6. How many liters (STP) of methane, CH_4, will it take to remove 200,000 liters (STP) of NO emitted from an electric power plant? The equation (from Table 11.4) is:

    $$CH_4 + 4\,NO \longrightarrow 2\,N_2 + CO_2 + 2\,H_2O$$

7. How many liters (STP) of nitrogen will be produced by the reaction in Exercise 6?
8. How many grams of MnO per liter of SO_2 (STP) would be needed to remove SO_2 from stack gases, using the equation in Table 11.4?
9. How many liters of NO_2 (STP) can be absorbed in 2.5 kg of $Ca(OH)_2$ according to the equation in Table 11.4?

*10. A power plant that burns 10,000 tons of coal containing 2.5 percent sulfur will emit about 1.8×10^5 liters (STP) of SO_2 each day. If this were all absorbed with calcium carbonate in the limestone-dolomite process (see Table 11.4), how many metric tons of $CaSO_4$ would be produced daily?

*11. The burning of charcoal yields mostly carbon monoxide, according to this equation:

$$2\,C(s) + O_2(g) \longrightarrow 2\,CO(g)$$

a. How many liters of CO, measured at STP, would result from burning 1 kg of charcoal in a grill?
b. In a sealed room having a volume of 2×10^5 liters, how many parts per million is that? How does it compare with the maximum tolerance level in Table 11.3?

12. Using the Haber process, 525 liters of NH_3 are produced at 40 atmospheres. What volume would this occupy at standard pressure and the same temperature?
13. In a steam turbine, the energy in steam under pressure is used to make mechanical energy, which in turn generates electric power. The steam

at about 100 atmospheres is allowed to flow through a nozzle and reach 1 atmosphere, and the resulting expansion turns the turbine. Calculate the volume that 15 liters of high-pressure steam would occupy after such expansion.

14. Air to fill a hot-air balloon is heated to 50°C. What is the density of air at this temperature (same pressure)?
15. On a cool morning (23°C), you fill your tires with air to a pressure of 2.5 atmospheres. The temperature of the tire later reaches 66°C. Calculate the new pressure inside the tire.
16. A gas initially at 1.24 atm and 120°C is cooled until its pressure is 0.85 atm. Calculate the final temperature.
17. A chemical equation predicts that 5.40 liters of CO_2 will be obtained at STP. What volume would actually be obtained if the gas were collected under the laboratory conditions of 25°C and 0.902 atmospheres?
18. 54.7 cm^3 of HCl gas was generated in the laboratory at 754 torr and 23°C. Calculate its volume at STP.
19. A mixture of oxygen and helium sampled at 745 torr was found to have a partial oxygen pressure of 350 torr. Calculate the partial pressure of the helium.
20. Methane gas, CH_4, is bubbled through water into a measuring tube, and the measured volume is 24.3 cm^3 at 739 torr and 24°C. Calculate the pressure of the methane and its corrected volume at STP.
21. From a reaction, 53.6 cm^3 of oxygen are expected. Calculate the actual volume if the oxygen is collected by displacement of water at 29°C and 0.878 atm.

*22. Baking powders are mostly made of sodium bicarbonate and calcium hydrogen phosphate. When baking powder is wet, this reaction produces carbon dioxide:

$$NaHCO_3(aq) + CaHPO_4(aq) \longrightarrow NaCaPO_4(aq) + CO_2(g) + H_2O(l)$$

a. If a teaspoon of baking powder contains 2.00 grams of sodium bicarbonate, how many cubic centimeters of carbon dioxide (at STP) will be obtained from this reaction?

b. The baking powder is used to make biscuits in a 400°F oven. What will be the volume at this temperature and 1 atm?

c. What would be the volume if the same biscuits were baked at an altitude of 0.860 atm? What would you notice about the biscuits?

23. One way to detect very small amounts of ozone is with this reaction:

$$Hg(l) + O_3(g) \longrightarrow HgO(s) + O_2(g)$$

The HgO forms a dirty-looking scum on top of the shiny mercury. If 0.012 moles of HgO are obtained, what volume of O_3 at 24°C and 0.989 atm was present?

*24. Use the data of Example 11.6, and assume that the gases are confined to a volume of 0.1 cm^3 and reach a temperature of 1500 K. The bullet is propelled by the pressure the gases create when they're confined to this volume at this temperature. Calculate the pressure. (Hint: solve for P in the ideal gas equation.)

25. A supersonic transport plane (SST) releases about 3 metric tons of NO into the atmosphere for every hour it flies. It flies at an altitude of about 20 kilometers; at this altitude the pressure is about 0.13 atmospheres and the temperature is about −56°C.

a. What volume would 3 metric tons of NO occupy under those conditions?

b. What volume of O_3 would the NO be capable of destroying, under the same conditions? Here is the equation:

$$NO + O_3 \longrightarrow NO_2 + O_2$$

*26. The scuba tank of Example 11.9 is to be used by a diver at a depth of 33 m in seawater, where the pressure is 2 atm. The diver breathes at a rate of 1 ft^3/min. If a reserve of 8.2 ft^3 (STP) is to be left in the tank as a safety margin, how long can the diver stay under water with this tank?

CHAPTER 12

Solutions

LEARNING OBJECTIVES

After studying this chapter, you should be able to:

1. Supply a correct definition, explanation, or example for each of these:

solution	solvation energy
homogeneous	hydration energy
solvent	parts per million
solute	water fluoridation
concentration	water chlorination
grams per liter	colloid
weight/volume concentration term	emulsion
weight/volume percent	emulsifying agent
percent by weight	Tyndall Effect
percent by volume	Brownian Movement
molarity	detergent
saturated solution	hydrophilic
soluble	hydrophobic
solubility	Cottrell Precipitator
unsaturated	flocculation
insoluble	semipermeable membrane
sparingly soluble	dialysis
immiscible	Henry's Law
miscible	supersaturated solution
polar molecule	eutrophication
center of symmetry	colligative properties
ion-dipole attraction	osmotic pressure
distillation	reverse osmosis
molality	
freezing-point constant	

2. Calculate the concentration of a solution in grams per liter, weight/volume percent, percent by weight, percent by volume, molarity, and molality, and use these as conversion factors in appropriate calculations.
3. Describe how water dissolves various types of solutes.
4. Describe "hard" water, what it contains, and various ways of dealing with water hardness.
5. Describe the formation and destruction of emulsions and colloids, with appropriate examples.
6. Calculate the effect of a change in pressure on the solubility of a gas.
7. Describe the effects of pressure and temperature on different types of solutions.
8. Calculate the freezing point of a solution from appropriate data.
9. Describe how a mixture may be separated by selective freezing or by distillation.
10. Use osmosis or osmotic pressure to explain various natural occurrences.

Many familiar substances are solutions. The liquid part of blood is a solution of various salts, gases, and biological molecules in water. Gasoline, vodka, tincture of iodine, and ocean water are also solutions. Even the water we drink is a solution.

We usually think of solutions as being liquids, but they don't have to be. For instance, air is a gaseous solution, and brass is a solid solution. However, most solutions of interest to us here are liquids. Liquid solutions are very useful in chemistry because many chemical reactions take place easily in liquid solutions, which are easy to measure and convenient to handle.

12.1 DESCRIBING SOLUTIONS

A solution is a homogeneous mixture of two or more substances. *Homogenous* means alike throughout. No matter where we take a sample from a solution, or how large the sample is, it will have the same composition and properties as any other sample taken from the same solution. In a solution, the substance present in the largest amount is called the *solvent.* The substances dissolved in the solvent are called *solutes.* Table 12.1 lists some solutions with their solutes and solvents.

solution

homogeneous

solvent

solute

TABLE 12.1
Some common solutions

Name	*Description*	*Solute*	*Solvent*
Air	Various gases dissolved in N_2	Gas	Gas
Soda water	CO_2 dissolved in water	Gas	Liquid
Vodka	Alcohol dissolved in water	Liquid	Liquid
Antifreeze	Ethylene glycol dissolved in water	Liquid	Liquid
Dental fillings	Mercury dissolved in silver	Liquid	Solid
Blood plasma	CO_2, O_2, salts, and biological molecules dissolved in water	Gases, solids	Liquid
Tincture of iodine	Iodine dissolved in alcohol	Solid	Liquid
Household ammonia	NH_3 dissolved in water	Gas	Liquid
Gasoline	Various carbon-containing liquids dissolved in each other	Liquids	Liquid
Seawater	Various salts dissolved in water	Solids	Liquid
Syrup	Sugar dissolved in water	Solid	Liquid
Brass	Zinc dissolved in copper	Solid	Solid
Solder	Tin dissolved in lead	Solid	Solid
Vinegar	Acetic acid dissolved in water	Liquid	Liquid

Besides being homogeneous, a solution has these characteristics:

1. The solute is mixed throughout the solvent as ions or molecules, and not as larger particles.
2. A solution is usually transparent, although it may be colored.
3. Usually a physical means, such as evaporation, is needed to separate a solute from a solvent in a solution.

THE CONCENTRATION OF SOLUTIONS. The amount of solute that is present in a specified amount of solution or of solvent is called the *concentration*. We can express concentration in any number of ways, depending on what we want to use the solution for. These are the concentration terms that we'll explain in this chapter:

concentration

Grams per liter (g/ℓ)
Weight/volume percent (wt/vol %)
Weight/weight percent (% by wt)
Volume/volume percent (% by vol)
Molarity (M), or moles per liter (m/ℓ)
Molality (m), or moles per 1000 g solvent

First, we'll introduce the concentration term *grams per liter* (g/ℓ), which means the number of grams of solute in one liter of solution. We can express the concentration of any kind of solute in this way, even if we don't know what the solute is.

grams per liter

EXAMPLE 12.1: 100.-$m\ell$ samples of water are taken from the Pacific Ocean and from the Great Salt Lake, and the water is allowed to evaporate from them. The salts that remain (mostly NaCl) are 3.85 grams from the Pacific Ocean and 31.9 grams from the Great Salt Lake. Calculate the original concentration of each in grams per liter.

Solution: It doesn't matter that we don't know the exact composition of these salt residues. We know how many grams are in 100. $m\ell$, or 0.100 ℓ, so we can easily find the number of grams per liter.

$$\frac{3.85\text{ g}}{0.100\ \ell} = \underline{\qquad}\frac{\text{g}}{\ell}$$

$$\frac{31.9\text{ g}}{0.100\ \ell} = \underline{\qquad}\frac{\text{g}}{\ell}$$

Answer: 38.5 g/ℓ for the Pacific Ocean; 319 g/ℓ for the Great Salt Lake. (The Great Salt Lake has a salt concentration nearly ten times that of the Pacific Ocean.) ■

Grams per liter is a measure of *weight/volume* concentration. That is, the solution is described according to the weight of solute that is dissolved in a given volume of solution. Another weight/volume term, used widely in medicine, is *weight/volume percent* (wt/vol %), which is the weight in grams of solute in 100 $m\ell$ of solution. This term is very convenient to use with physiological fluids such as blood and urine. A "0.9% saline solution,"

weight/volume

weight/volume percent

used for intravenous feeding, is in weight/volume percent, and means 0.9 grams of NaCl per 100 mℓ of solution. Here, the 100 mℓ is a pure number. (For a discussion of pure numbers, see Appendix A, p. 492.)

EXAMPLE 12.2: A blood sample taken from a patient is analyzed for glucose; 5.00 mℓ of blood contains 0.00812 grams of glucose. What is the weight/volume percent of glucose in the blood sample?

Solution: We use this relationship:

$$\text{wt/vol \%} = \frac{\text{wt. solute (in grams)}}{\text{vol. solution (in m}\ell)} \times 100$$

Then,

$$\frac{8.12 \times 10^{-3}\ \text{g glucose}}{5.00\ \text{m}\ell\ \text{solution (blood)}} \times 100 = \_\_\_\%\ \text{(wt/vol)}$$

Answer: 0.162% (wt/vol).
(A normal person who has eaten nothing for eight to ten hours has about 0.08% to 0.1% glucose in the blood. This patient may have just eaten or may be a diabetic.) ■

In clinical chemistry, milligram percent (mg %), meaning mg of solute in mℓ of solution, is used. The answer to Example 12.2 could be expressed as 162 mg%.

Converting between grams per liter and weight/volume percent is easy if we remember that grams per liter is the same as grams per 1000 mℓ. We multiply by ten to convert weight/volume percent to grams per liter, and divide by ten to convert grams per liter to weight/volume percent.

Another concentration term that uses weight of solute is weight/weight percent. More commonly called *percent by weight* (wt%), it is the number of grams of solute in 100 grams of solution. Again, the 100 grams of solution is a pure number.

percent by weight

EXAMPLE 12.3: Concentrated sulfuric acid is 96% H_2SO_4 by weight and has a density of 1.84 g/mℓ. How many grams of H_2SO_4 are in (a) 10.0 g and (b) 10.0 mℓ of concentrated sulfuric acid?

Solution: (a) We use the weight percent as a conversion factor:

$$10.0\ \text{g}\ \cancel{\text{conc. acid}} \times \frac{96\ \text{g}\ H_2SO_4}{100\ \text{g}\ \cancel{\text{conc. acid}}} = \_\_\_\text{g}\ H_2SO_4$$

(b) We use density and weight percent as conversion factors:

$$10.0\ \cancel{\text{m}\ell\ \text{conc. acid}} \times \frac{1.84\ \cancel{\text{g conc. acid}}}{\cancel{\text{m}\ell\ \text{conc. acid}}} \times \frac{96\ \text{g}\ H_2SO_4}{100\ \cancel{\text{g conc. acid}}} = \_\_\_\text{g}\ H_2SO_4$$

Answer: (a) 9.6 g H_2SO_4; (b) 18 g H_2SO_4. ■

When a liquid is dissolved in another liquid, it's often convenient to use *percent by volume* (vol%), which is the number of mℓ of solute per 100 mℓ of solution. The label on a bottle of rubbing alcohol may say "70% by volume," which means that 70 mℓ of alcohol is present for every 100 mℓ of the solution.

percent by volume

EXAMPLE 12.4: A bottle of hydrogen peroxide in a drugstore is labeled "3% by volume." The bottle contains 250. mℓ of solution. How many mℓ of pure hydrogen peroxide are contained in the bottle?

Solution: We want to find mℓ of hydrogen peroxide. The percentage by volume is our conversion factor.

$$250.\ \cancel{\text{m}\ell\ \text{solution}} \times \frac{3\ \text{m}\ell\ \text{hydrogen peroxide}}{100.\ \cancel{\text{m}\ell\ \text{solution}}} = \_\_\_\_\ \text{m}\ell\ \text{hydrogen peroxide}$$

Answer: 8 mℓ hydrogen peroxide. ■

In the chemistry laboratory, solutions are generally used in chemical reactions. We remember from Chapter 6 that equations describing chemical reactions are expressed in moles. So knowing how many moles of a chemical are in the solution is useful. Moles per liter (m/ℓ) or *molarity* (*M*) means the number of moles of solute in one liter of solution. (One liter is a pure number.) To find molarity, we have to know the solute's formula weight. We didn't need this information with any of the earlier concentration terms. A 1.00 *M* (molar) solution of NaCl is a solution of 58.5 g (1 mole) of NaCl in 1 liter of solution.

molarity

EXAMPLE 12.5: A 100.-mℓ sample of a brine solution containing only NaCl and water was allowed to evaporate to dryness. The residue, NaCl, weighed 2.58 g. What was the molarity of the brine solution?

Solution: We use the relationship

$$M = \frac{\text{m}}{\ell}\ (\text{molarity} = \text{moles/liters}).$$

Before we can use this expression, we have to find the number of moles of NaCl, which we do by dividing the grams by the formula weight of NaCl (58.5 g/mole).

$$\text{Moles NaCl} = 2.58\ \cancel{\text{g}} \times \frac{1\ \text{mole NaCl}}{58.5\ \cancel{\text{g}}} = 0.0440\ \text{moles NaCl}$$

$$\text{Molarity}\ (M) = \frac{0.0440\ \text{moles}}{0.100\ \ell} = \_\_\_\_\ M$$

Answer: 0.440 *M* (molar). ■

The remaining concentration term, molality, will be explained later in the chapter (p. 275).

SOLUBILITY.

If we keep adding salt to a solution of salt in water, we eventually reach a point where no more salt will dissolve, because the water has dissolved all the salt it can. When this happens, we have a *saturated solution*. It means the maximum amount of solute that is *soluble* (will dissolve) to form a given amount of solution at a certain temperature. We define *solubility* as the concentration of a saturated solution. If the concentration of any solution is less than that of a saturated solution, then the solution is *unsaturated*. Solubilities of some common substances in grams/liter are shown in Table 12.2.

saturated solution

soluble

solubility

unsaturated

TABLE 12.2
Solubilities of some common substances in water at 20°C

Substance	*Solubility (g/ℓ)*	*Comments*
O_2	4.5×10^{-2}	Sparingly soluble.
CO_2	0.145	Soluble enough to be carried mostly by water portion of blood.
N_2	2.4×10^{-2}	Sparingly soluble.
HCl	719	Soluble. Water solutions are hydrochloric acid.
NH_3	320	Soluble. Solutions are aqueous ammonia, or household ammonia.
Cl_2	6.3	Rather soluble. Solutions are chlorine bleaches.
He	2×10^{-3}	Sparingly soluble.
Gasoline	0.4	Sparingly soluble.
Ethyl alcohol	Infinitely soluble	Mixes with water in all proportions.
Ethylene glycol (antifreeze)	Infinitely soluble	Mixes with water in all proportions.
NaCl	358	Soluble.
AgCl	1.43×10^{-3}	Sparingly soluble.
Diamond	0	Insoluble.
$CaCO_3$	1.4×10^{-2}	Sparingly soluble.
CuS	3.3×10^{-6}	Rather insoluble.
HgS	1×10^{-6}	Rather insoluble.
NH_4NO_3	118	Soluble.

A diamond is insoluble in water, which means that it doesn't dissolve in water. There is no doubt about diamond, but most things we call "insoluble" in water do dissolve to a small extent. By *insoluble,* then, we usually mean "not very soluble." AgCl is "insoluble" in water. If we stir some around in water, we don't notice any dissolving. Actually, though, some does dissolve: the solubility of AgCl in water is 0.00143 grams/liter at 25°C. This is slight compared with the solubility of NaCl. Sometimes we say that a substance is *sparingly soluble* instead of "insoluble" when it dissolves to a small extent.

insoluble

sparingly soluble

EXAMPLE 12.6: The solubility of oxygen in water is about 4.5×10^{-2} grams/liter. The water portion of an adult's total blood supply is about 5 liters. How many grams of oxygen could dissolve in 5 liters of water?

Solution: We want to convert 5 liters of solution to grams of O_2. Our conversion factor is the solubility. Our setup is:

$$5 \not{\ell} \times 4.5 \times 10^{-2} \frac{\text{g } O_2}{\not{\ell}} = \underline{\qquad} \text{g } O_2$$

Answer: 2×10^{-1} g O_2. Actually, whole blood contains about 7×10^{-1} O_2/ℓ. Because of the low solubility of oxygen in water, most of the oxygen has to be carried by the hemoglobin in the red blood cells. Otherwise, blood couldn't carry enough oxygen from our lungs to our tissues. ■

12.2 WATER AS A SOLVENT

Water is the most important chemical on earth. Some organisms can live without oxygen, but none can live without water. All organisms use water to carry salts, sugars, and various biological molecules from one place to another. Human blood is 92 percent water.

But some substances are relatively insoluble in water. We've already seen that O_2 is sparingly soluble and must be carried in the blood by hemoglobin. Table 12.2 shows that the gases N_2 and He are also only sparingly soluble. We've seen a solid—diamond—that won't dissolve in water at all, and another solid—AgCl—that is sparingly soluble. Some liquids are insoluble or sparingly soluble in water—oil and grease, for instance, and gasoline. When liquids won't mix with each other, we say they are *immiscible*. (Liquids that *do* mix with each other are *miscible*. Table 12.2 shows that water and ethyl alcohol are miscible.)

immiscible

miscible

Why does water dissolve some things better than others? Why does it dissolve CO_2 better than O_2, and HCl better still? Why does it dissolve ethyl alcohol and not gasoline? Why does it dissolve NaCl much better than AgCl?

For a solute to dissolve in a solvent, the particles in both the pure solute and the pure solvent must separate from each other and mix together. In a solution, the attractions among particles in the pure state are replaced by attractions between solute and solvent particles. If the attractions between solute and solvent particles aren't as strong (or stronger), then the solute won't dissolve well in the solvent. Let's look at a characteristic of water that makes it a particularly good solvent.

THE POLAR NATURE OF WATER. A consequence of polar bonds is that entire molecules can be polar, too. We define a *polar molecule* as a molecule whose centers of positive and negative charge aren't in the same place. Although a molecule is electrically neutral overall, within it there can be such a separation of charge. We often represent a polar molecule with a bar that has these different centers of charge:

polar molecule

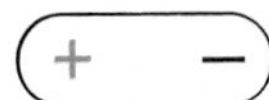

It takes more than just polar bonds to cause a molecule to be polar. HF, CF_4, H_2O, and CO_2 all contain polar bonds, but only HF and H_2O are polar molecules. What makes the difference is the shape of the molecule. Symmetrical molecules can't be polar, even though they might contain polar bonds. This is the case with CF_4 and CO_2. In both cases, the polar bonds pulling against one another exactly cancel, and there is no *net* dipole, as shown in Figure 12.1.

Both CO_2 and CF_4 have *centers of symmetry*. This means that if we start at the central atom and move out any one of the bonds, we will always encounter the same structure. CO_2 is a linear molecule with carbon in the center of two oxygens; CF_4 is tetrahedral, with carbon in the middle of the tetrahedron and each of the four fluorines at a corner. center of symmetry

It's easy to see why HF is polar. It contains only one bond, and that bond is polar. It lacks a center of symmetry. Water also lacks a center of symmetry, because the molecule is bent so that both polar bonds are on the same side. Figure 12.2 shows symmetrical molecules and unsymmetrical molecules.

To summarize, a polar molecule must have *both* of these things:

1. polar bond or bonds
2. no center of symmetry

FIGURE 12.1
Polar bonds acting unsymmetrically cause polar molecules; polar bonds acting symmetrically cause nonpolar molecules

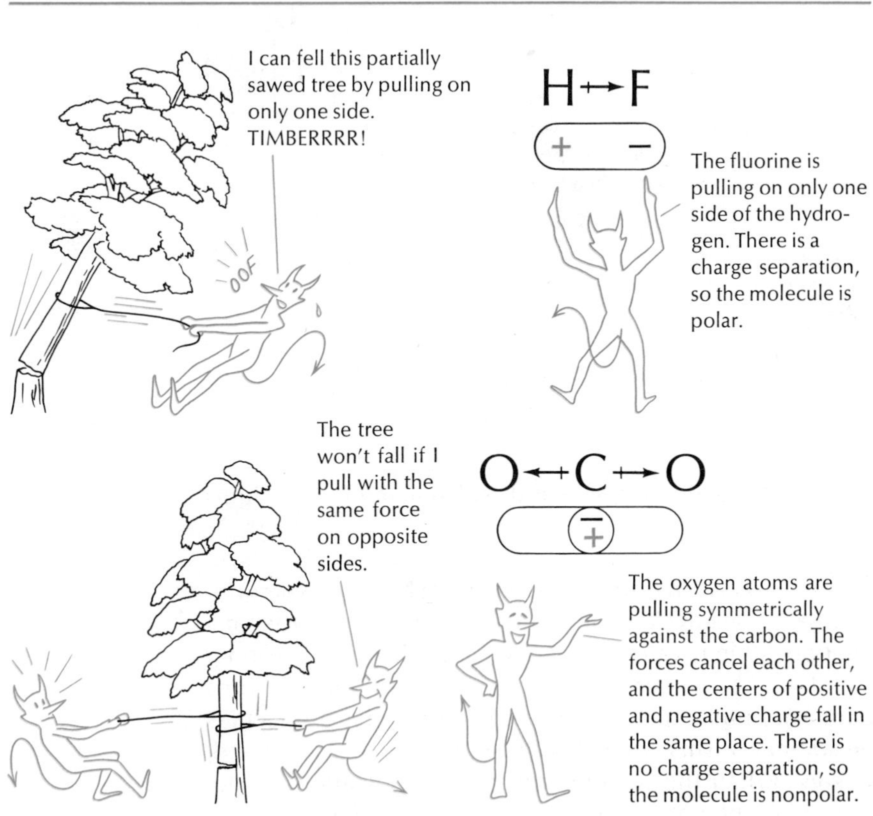

FIGURE 12.2
CF_4 has a center of symmetry; H_2O has none

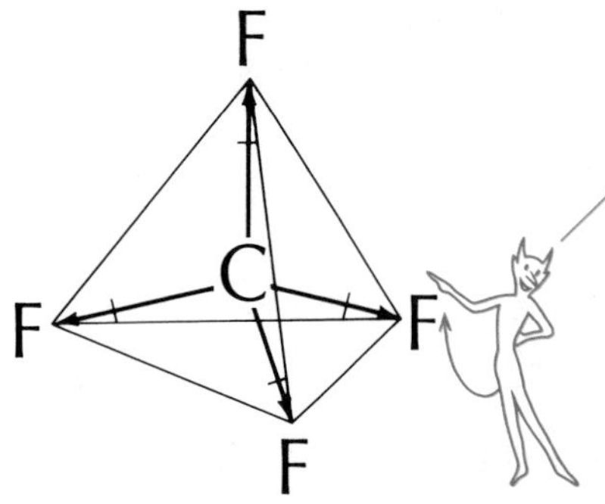

This CF_4 molecule has a center of symmetry. If I climbed inside to the carbon atom and crawled along each bond, I'd find a fluorine atom the same distance from the center in each direction.

This water molecule has no center of symmetry. If I crawled along each line starting with the oxygen atom in the center, sometimes I'd find a hydrogen atom, and sometimes I'd find a nonbonding electron pair.

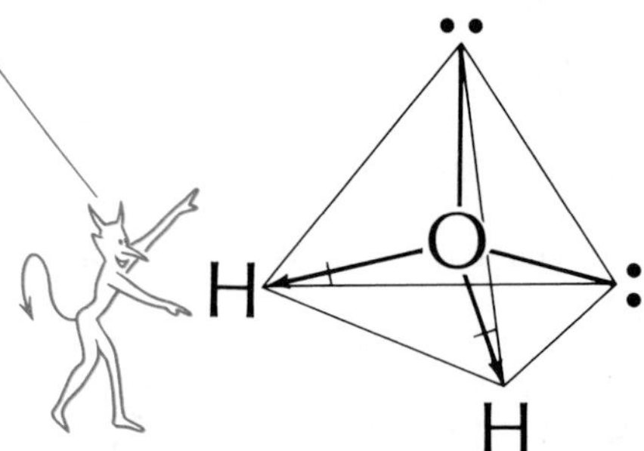

If we consider just the atoms, we see that water is bent. Its polar bonds pulling unsymmetrically cause a net dipole in this direction.

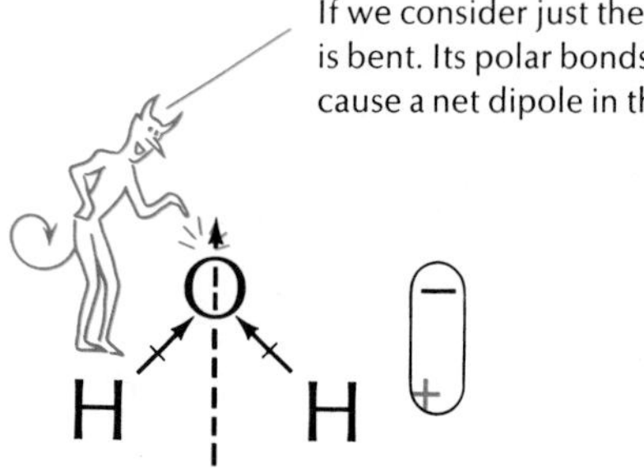

Next, we'll see how water's polar nature lets it dissolve many ionic solutes easily.

IONIC SOLUTES. To dissolve an ionic crystal, we have to overcome the attractions between the positive and negative ions and disperse the ions throughout the solution. When water, a polar solvent, dissolves an ionic crystal, the polar ends of the water molecules cluster around the ions and pull them apart. The interaction between ions and polar molecules is called *ion-dipole attraction.* This attraction causes energy to be released, called *solvation energy.* When the solvent is water, the solvation energy is called *hydration energy.* The net energy released when a mole of a substance dissolves in water is that substance's *heat of solution.*

ion-dipole attraction

solvation energy

hydration energy

heat of solution

If the solvation energy is enough to pay for the energy needed to break apart the ionic crystal, then the crystal will dissolve. If it isn't enough, the crystal won't dissolve. Figure 12.3 shows the behavior of NaCl and AgCl in water. Here, the attractive forces between ions in the AgCl crystal are stronger than those in NaCl—too large to be overcome by the hydration energy.

But why do so-called insoluble ionic compounds like AgCl dissolve at all?

FIGURE 12.3
The weaker attractions in NaCl make it more soluble than AgCl

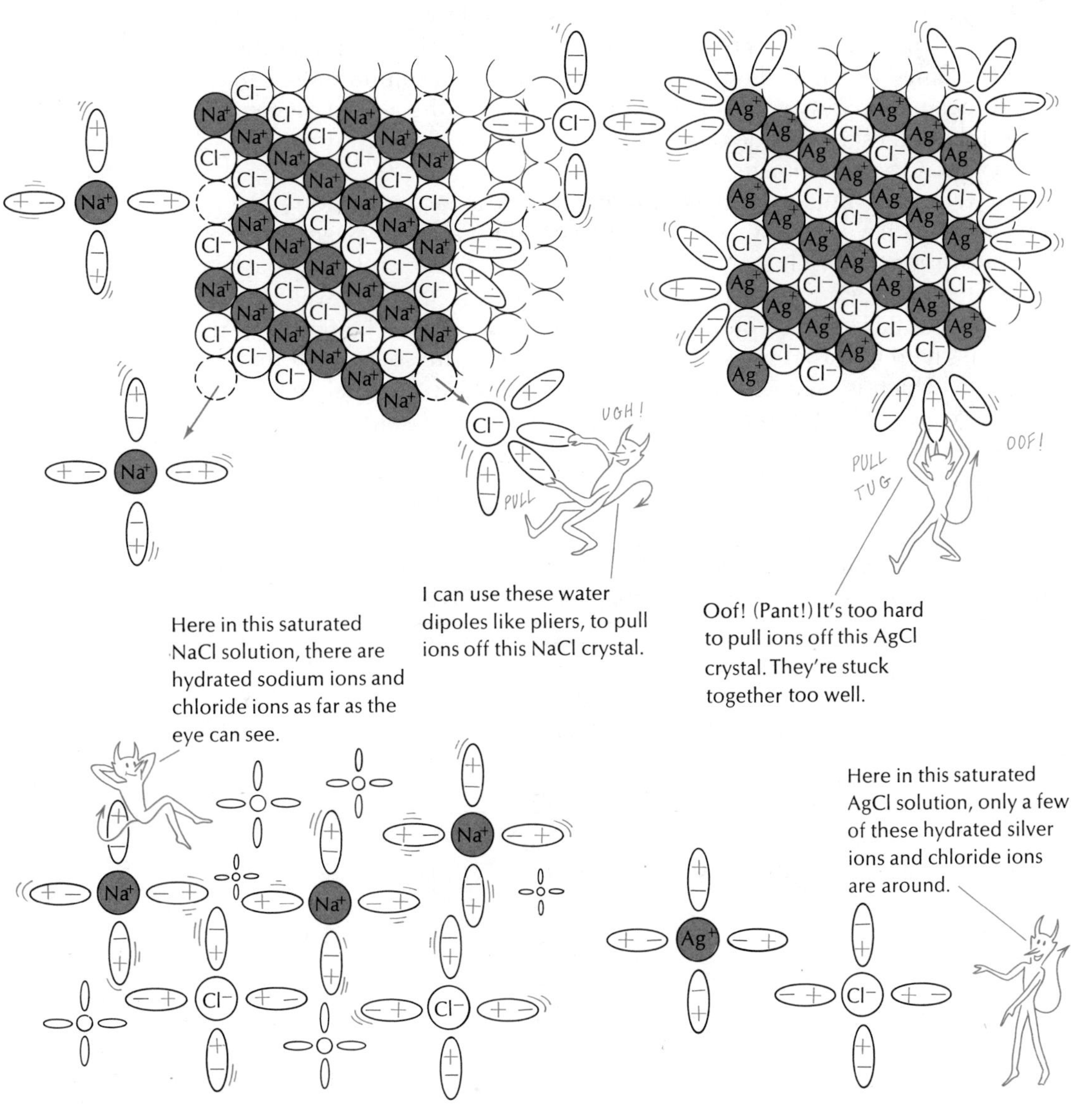

UGH!
PULL
PULL
TUG
OOF!
I can use these water dipoles like pliers, to pull ions off this NaCl crystal.
Here in this saturated NaCl solution, there are hydrated sodium ions and chloride ions as far as the eye can see.
Oof! (Pant!) It's too hard to pull ions off this AgCl crystal. They're stuck together too well.
Here in this saturated AgCl solution, only a few of these hydrated silver ions and chloride ions are around.

For the answer, we should recall from Chapter 10 that some of a substance's molecules always have enough energy to escape to the vapor state, no matter what the temperature, because of the Maxwell-Boltzmann distribution. In the same way, a few ions in a crystal always have more energy than others. The additional hydration energy is enough to dissolve these few ions.

Oceans and lakes without outlets (such as the Great Salt Lake) contain a lot of dissolved salts. The moving water of rivers and streams constantly dissolves parts of the rocks and earth it runs through. These dissolved particles collect in their final destinations—oceans or lakes. We express concentrations in natural water as parts per million (ppm) by weight. This is similar to parts per million by volume, which we used for gases in the last chapter. One part per million by weight means one weight measure of the solute contained in one million weight measures of total solution. We can get parts per million from weight percent by multiplying the weight percent by 10^4. Table 12.3 shows some of the main ionic solutes in seawater. Note that Na^+ and Cl^- predominate.

Fresh water also contains many ions as impurities. In fact, these ions give spring water its flavor. Spring water usually contains calcium, magnesium, and iron(III) ions, as well as sulfate, hydrogen carbonate, and carbonate ions. Water without any ions at all is flat and tasteless. If spring water, or well water, contains quite a bit of Ca^{2+} and Mg^{2+}, it's called "hard water," because it's hard to wash things in it. These ions react with soap and form a precipitate, which is "soap scum." When hard water is completely boiled away, the CO_3^{2-} that it also contains reacts with the Mg^{2+} and Ca^{2+}. This

TABLE 12.3
Main ingredients in seawater

Positive Ions	*Ppm (by weight)*	*Negative Ions*	*Ppm (by weight)*	*Gases*	*Ppm (by weight)*
Na^+	10,760	Cl^-	19,353	N_2	10
Mg^{2+}	1,294	SO_4^{2-}	2,712	O_2	7
Ca^{2+}	413	HCO_3^-	142	CO_2	600
K^+	387	Br^-	67		
Sr^{2+}	8	F^-	1.4		
Ba^{2+}	0.05	I^-	0.06		

Elements	*Ppm (by weight)*	*Elements*	*Ppm (by weight)*	*Elements*	*Ppm (by weight)*
B	5	Rb	0.12	Fe	0.01
Si	3	P	0.072	Al	0.01
N	0.52	In	0.02	Mo	0.01
Li	0.17	Zn	0.01		

Note: About twenty-nine other elements are present in traces.

reaction leaves a residue of $CaCO_3$ and $MgCO_3$, sometimes called "boiler scale." Synthetic detergents work better in hard water than soaps do, because synthetic detergents don't form precipitates (soap scum) with these ions. However, even synthetic detergents don't work as well in hard water as in soft water.

One way of dealing with these ions in hard water is to add a substance to the detergent that will react with the metal ions and get them out of the way. "Phosphate" detergents contain sodium tripolyphosphate (STPP), $Na_5P_3O_{10}$. The tripolyphosphate ion has this Lewis structure:

```
[      |Ō|      |Ō|      |Ō|      ]5−
        |        |        |
 |Ō̲ — P — Ō̲ — P — Ō̲ — P — Ō̲|
        |        |        |
       |O̲|      |O̲|      |O̲|
]
```

These ions can surround a metal ion and bond to it, so that the metal ion stays in solution and doesn't react with anything else. Substances that can tie up metal ions in this way are called "complexing agents," or "sequestering agents."

Another way of handling Ca^{2+} and Mg^{2+} in household water is to use a water softener. Most water softeners contain ion-exchange resins, in the form of small brown beads resembling sand. The beads contain sodium ions attached to large resin molecules. When water containing Ca^{2+}, Mg^{2+}, or other positive ions is poured through the water softener, which contains a column of this resin, the sodium ions trade places with the metal ions in the water.

$$2\ Na(Res) + Ca^{2+} \longrightarrow Ca(Res)_2 + 2\ Na^+$$

Now the water contains sodium ions, which don't interfere with detergent action. However, people who must limit their sodium intake should not continuously drink water softened by this method.

Along with natural impurities, municipal water that has been treated for household use may have some substances added to it. In some areas, drinking water is *fluoridated* by adding sodium fluoride, NaF, to help prevent tooth decay. To kill harmful bacteria, water is usually *chlorinated* by adding chlorine to it. When chlorine is added to water, this reaction takes place:

fluoridation

chlorination

$$Cl_2 + H_2O \longrightarrow HClO + HCl$$

Hypochlorous acid (HClO) is probably the active agent when it comes to disinfecting with chlorine. HClO decomposes to form atomic oxygen:

$$HClO \longrightarrow HCl + \underset{\text{Atomic oxygen}}{[O]}$$

Atomic oxygen reacts quickly with nearby chemicals, including those in bacteria and viruses. Salts of hypochlorous acid, such as NaClO or $Ca(ClO)_2$, are equally effective. Often these solid salts, instead of chlorine gas, are added to water to chlorinate it, because solids are safer and easier to handle.

POLAR HYDROGEN-BONDED SOLUTES. Dissolving a polar hydrogen-bonded solute presents the same problem as dissolving an ionic solute, but the interactions between solute particles aren't as strong. Also, ionic solutes are only solids, but polar hydrogen-bonded solutes can be solids, liquids, or gases. In Table 12.1, alcohol, ethylene glycol (antifreeze), and ammonia are examples. When these solutes dissolve, their own hydrogen bonds are replaced by similar attractions with water molecules. Figure 12.4 shows some solutes forming hydrogen bonds with water.

FIGURE 12.4
Hydrogen bonds among solute molecules are replaced by hydrogen bonds between solute and water

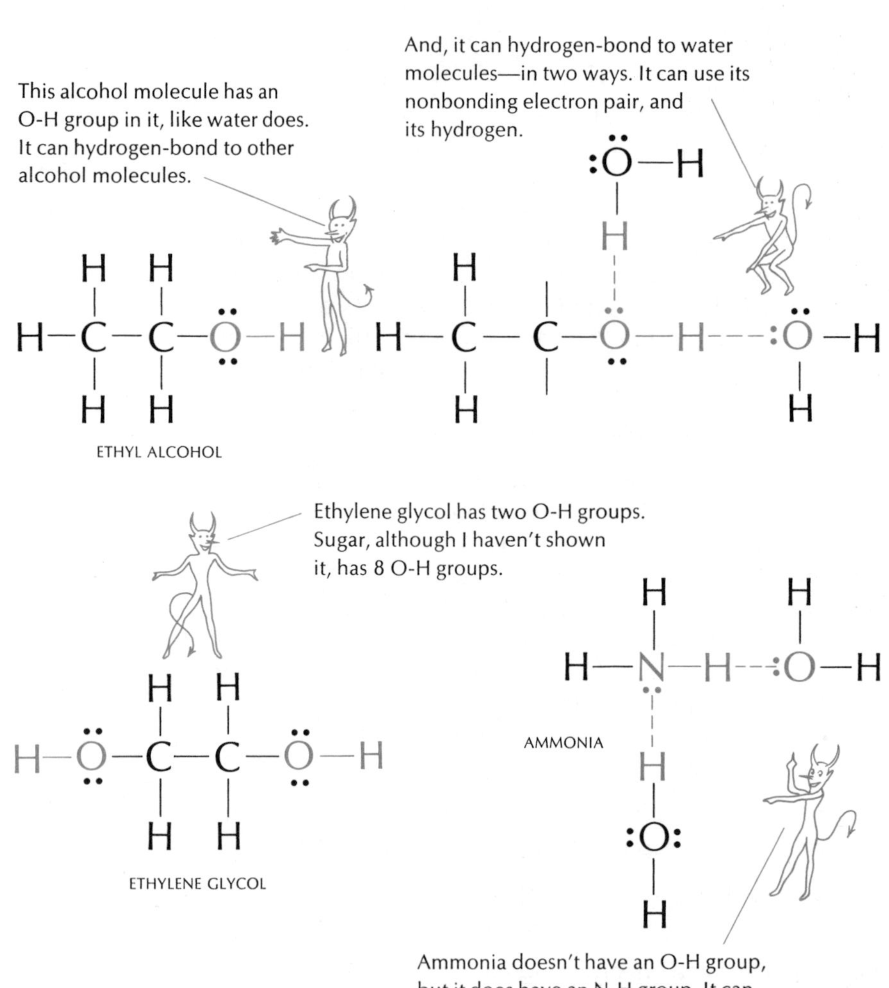

This alcohol molecule has an O-H group in it, like water does. It can hydrogen-bond to other alcohol molecules.
And, it can hydrogen-bond to water molecules—in two ways. It can use its nonbonding electron pair, and its hydrogen.
ETHYL ALCOHOL
Ethylene glycol has two O-H groups. Sugar, although I haven't shown it, has 8 O-H groups.
ETHYLENE GLYCOL
AMMONIA
Ammonia doesn't have an O-H group, but it does have an N-H group. It can also hydrogen-bond in two ways.

Carbon dioxide is a nonpolar molecule, but it does contain polar bonds. The difference in electronegativity between carbon and oxygen makes the oxygen slightly negative, so that water can hydrogen-bond to it.

$$|\overline{O}=\underline{\overline{C}}=\overline{O}| \cdots H-\underline{\overline{O}}-H$$

Since CO_2 has no hydrogens, water provides all the hydrogens. This makes CO_2 less soluble in water than NH_3 or HF, both of which do have hydrogens that can hydrogen-bond to water. But the fact that water can hydrogen-bond to it, even one-way, makes CO_2 more soluble than O_2.

NONPOLAR SOLUTES. "Oil and water don't mix." This old saying is true. A chemist would put it this way: "Nonpolar solutes don't dissolve well in polar solvents." Some examples of nonpolar solutes, besides oil, are gasoline, oxygen, iodine, wax, and turpentine.

Water doesn't dissolve nonpolar solutes well, because the attractions among water molecules are greater than the attractions that the nonpolar solute has to offer. The nonpolar molecules are just squeezed out. Oil is less dense than water, so it floats on water. This lets us clean up oil spills on oceans, with difficulty, by scooping the oil off the top. When a nonpolar liquid is more dense than water, as CCl_4 is, then the water floats on top. In either case, the liquids are immiscible, and two layers are formed.

A little bit of nonpolar solute will dissolve in water, for the same reasons "insoluble" ionic compounds will. We do find O_2 in natural fresh water to the extent of 10 ppm, although it dissolves much better in gasoline (800 ppm). Water dissolves in gasoline just enough to plug up a car's gas line when it freezes on a cold morning. In general, though, we have to use nonpolar solvents to dissolve large quantities of nonpolar solutes. Turpentine or paint thinner dissolves oil-based paints. Cleaning fluids dissolve grease on clothes. But we can get water to dissolve grease by adding soap or detergent. We'll see how that works next.

12.3 EMULSIONS AND COLLOIDS

GENERAL PROPERTIES. Many familiar substances are emulsions or colloids, as shown in Table 12.4. The more general term is *colloid*, a system in which the particles of one substance are dispersed throughout another substance without bonding to solvent molecules. Colloids have properties that are different from those of solutions:

colloid

1. The dispersed particles are between 1 nanometer and 200 nanometers in diameter (solute particles are smaller). This is the property that's used to distinguish a colloid from a solution. Sometimes these particles are clusters of many ions or molecules, but they can also consist of just one huge molecule (such as starch or protein).
2. Colloids are usually cloudy rather than transparent because the particles

TABLE 12.4
Some familiar colloidal systems

Substance	*Type of Colloid*
Shaving foam	Gas dispersed in liquid
Styrofoam	Gas dispersed in solid
Smoke, dust	Solid dispersed in gas
Muddy water	Solid dispersed in liquid
Gelatin	Liquid dispersed in solid
Fog, mist, clouds	Liquid dispersed in gas
Milk	Liquid (butterfat) dispersed in liquid (water) —emulsion
Mayonnaise, butter, cold cream	Liquid dispersed in liquid—emulsion
Latex paint	Solid dispersed in liquid

are large enough to scatter light. A beam of light can be seen going through a colloid because the dispersed particles reflect the light; this is called the *Tyndall Effect.* For the same reason, sunlight can be seen passing through air that contains dust particles, and car headlights can be seen shining through fog.

Tyndall Effect

3. The random motion of colloidal particles in a colloid can be seen with a special microscope that has a strong light source focused on the colloid. This motion is called *Brownian Movement* after Robert Brown (1773–1858), who discovered it. Solute particles are in motion too, but are too small to reflect light, so we can't see their motion in a solution.

Brownian Movement

emulsion

An *emulsion* is a special kind of colloid, where two immiscible liquids are held in suspension by another substance, called an *emulsifying agent.*

emulsifying agent

FORMING EMULSIONS AND COLLOIDS. Emulsions or colloids can form by themselves, or we can form them on purpose. We form an emulsion when we use a detergent to help water dissolve grease. A *detergent* is an emulsifying agent that lets oil mix with water.

detergent

A detergent molecule is long and has two ends. One end is ionic and *hydrophilic* (water-loving); the other is nonpolar and *hydrophobic* (water-hating). The nonpolar end sticks into a globule of grease to avoid the water. The polar end sticks into the water, which suspends the grease particles in the water so they can be washed away. Figure 12.5 shows how a soap (detergent) works. The suspended grease particles are negatively charged, since the negative ends of the soap molecules stick out into the water. Typically, colloidal particles are electrically charged. They can be positively or negatively charged, but they're all the same within a given colloid. The like charges on these particles keep them from coming together and forming bigger particles, which could then settle out and destroy the colloid.

hydrophilic

hydrophobic

FIGURE 12.5
Detergents can emulsify grease with their hydrophilic and hydrophobic ends

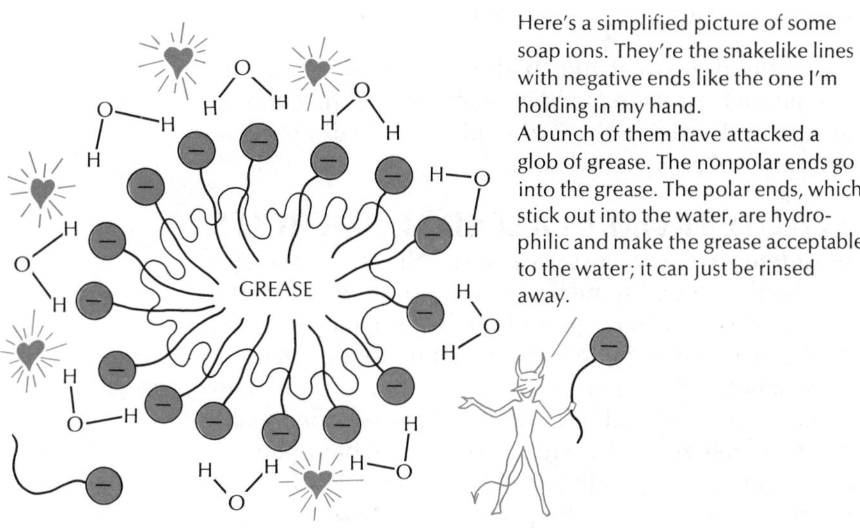

GREASE
Here's a simplified picture of some soap ions. They're the snakelike lines with negative ends like the one I'm holding in my hand.
A bunch of them have attacked a glob of grease. The nonpolar ends go into the grease. The polar ends, which stick out into the water, are hydrophilic and make the grease acceptable to the water; it can just be rinsed away.

DESTROYING EMULSIONS AND COLLOIDS.

In general, any means that will overcome the repulsion of the charged colloidal particles for each other, and cause them to form larger particles, will break up a colloid. Sometimes a colloid can be destroyed by heat, which gives the particles enough energy to overcome the repulsion and collide with each other. Salt will also destroy a colloid, because the positive and negative ions mask the colloidal particles' electrical repulsion. We see a destroyed colloid in the large sand or silt bars at the mouths of rivers. These are created when suspended dirt particles in a river hit the salt in ocean water.

Rain clouds build up tremendous amounts of electrical charge because of the accumulation of charges on the individual water droplets suspended in air. A rain cloud gets rid of this electrical charge by discharging to the ground or to another cloud. This is lightning. The water droplets can come together when the clouds get rid of this charge. Then it rains.

Cottrell Precipitator

A *Cottrell Precipitator* uses this electrical effect to remove dust or smoke particles from air or smokestack gases. The particles are given an intense electrical charge in a special "ionizing chamber." Then they are collected on metal plates having the opposite electrical charge, which removes them from the air.

A colloid can also be destroyed by giving the particles a large surface area to stick onto. A cloudy soup (cloudy because colloidal protein particles are suspended in water) can often be cleared up ("clarified") by adding egg white, boiling, and straining. As the egg white coagulates in the hot water, the colloidal particles stick to it and can be removed with it. In water purification, sediments suspended in water are often removed by forming a precipitate with a large surface area and letting the particles adhere to it. One such reaction is this one:

$$2\ Al_2(SO_4)_3(aq) + 6\ Ca(OH)_2(aq) \longrightarrow 4\ Al(OH)_3(s) + 6\ CaSO_4(s)$$

The aluminum sulfate (alum) and calcium hydroxide are added to the cloudy water. The aluminum hydroxide precipitate forms, and the colloidal dirt particles stick to it. Both the aluminum hydroxide and the dirt particles can then be removed. This process is called *flocculation*. flocculation

SEPARATING SOLUTE FROM COLLOIDAL PARTICLES. We've said that the difference between solute and colloidal particles is their size. To separate them, we need some sort of sieve or strainer that will let the smaller solute particles go through and keep the larger colloidal particles behind. A *semipermeable membrane* is a membrane that has pores. Particles that are smaller than the pores can pass through the membrane, but larger particles are kept behind. We can buy semipermeable membranes with specific pore sizes to let through or keep behind particles of the size we want in or out. The separating of solute from colloidal particles with a semipermeable membrane is called *dialysis*.

semipermeable membrane

dialysis

Dialysis is used in artificial kidney machines, needed by patients whose kidneys do not function. The job of the kidneys is to constantly filter blood to remove waste products while leaving behind the blood cells and dissolved protein molecules. The kidney machine does it by circulating the blood through a semipermeable membrane. The waste products are smaller and pass through the membrane but the larger cells and protein molecules stay behind.

12.4 CONDITIONS AFFECTING SOLUTIONS

The amount of a solute that will dissolve in a given solvent depends on various conditions. These conditions, in turn, depend on the physical state of the solute.

SOLUTIONS OF GASES IN LIQUIDS. Gases become less soluble in liquids as the temperature increases. An open bottle of soda water will lose its carbonation more rapidly as it warms up. Oxygen is less soluble in warm water than in cold water. The thermal pollution caused by industrial use of water for cooling can decrease the amount of dissolved oxygen in natural water and damage aquatic life.

Pressure is also important to the solubility of gases. A gas is more soluble in a liquid at high pressure than at low pressure. When we uncap a bottle of a carbonated beverage, we release the pressure and the carbon dioxide comes out of solution. Rivers and lakes at high altitudes have less dissolved oxygen in them than those at low altitudes.

A quantitative statement about the effect of pressure on solubility of a gas is given by *Henry's Law,* named after William Henry (1774–1836). It states that the solubility of a relatively insoluble gas in a liquid is directly proportional to the pressure of the gas above the liquid. Or, stated mathematically,

Henry's Law

$$S_2 = S_1 \frac{P_2}{P_1}$$

That is, if we start with a gas having solubility S_1 at pressure P_1 and we change the pressure to P_2, we can calculate the new solubility S_2 with the equation above.

EXAMPLE 12.7: The solubility of nitrogen in water is 2.4×10^{-2} g/ℓ at 1 atmosphere pressure. Assuming the same solubility in blood as in water, calculate the solubility of nitrogen in the blood of a diver who has descended to a depth of 66 ft below the surface of the ocean.

Solution: From Chapter 10, we know that every 33 ft of ocean depth bring an additional pressure of 1.0 atm. Then 66 ft give us 2.0 atm in addition to the 1.0 atm at the surface, for a total pressure of 3.0 atm.

$$S_2 = (2.4 \times 10^{-2}\ \text{g}/\ell) \times \frac{(3.0\ \cancel{\text{atm}})}{(1.0\ \cancel{\text{atm}})} = \_\_\_\ \text{g}/\ell$$

Answer: 7.2×10^{-2} g/ℓ. ■

Deep-sea divers sometimes experience a painful and dangerous condition called the "bends." Breathing air under the high pressures deep in the ocean causes the nitrogen in the air to be much more soluble in blood than it is at atmospheric pressure. As the divers come up to the surface and the pressure decreases, the nitrogen comes out of solution quickly, with damage to blood vessels. This can be overcome by giving the divers a mixture of helium and oxygen to breathe. Helium isn't very soluble in blood at any pressure.

SOLUTIONS OF SOLIDS IN LIQUIDS. Pressure doesn't affect the solubility of a solid in any noticeable way. Temperature often does, though. Heating may cause a solid to be more soluble in a liquid. Stirring and decreasing the particle size of a solid will usually help it to dissolve faster.

We can sometimes prepare a *supersaturated solution*—one that contains more solute than a saturated solution at the same temperature. If a solute is more soluble at a higher temperature, we can prepare a saturated solution at a higher temperature and let the solution cool undisturbed. If no solute precipitates out as the solution cools, then we have a supersaturated solution. Such solutions are usually very unstable. They will release their excess solute if we add a small crystal of solute, or if a piece of dust falls into the solution, or if we scratch the sides of the container with a glass rod. All these things provide a surface for the solute to crystallize on.

supersaturated solution

DISSOLVED OXYGEN IN NATURAL WATER. Oxygen dissolves in rivers and streams when the water's constant tumbling over rocks brings it in contact with air. Once the water feeds into a lake or ocean, though, it can obtain oxygen from the air only at the surface. Thus it's difficult to replenish dissolved oxygen in lakes or oceans if the amount is decreased in any way.

Natural fresh water at sea level contains only 10 ppm of oxygen. Seawater contains even less (7 ppm), because oxygen is less soluble in salt water. These small quantities of dissolved oxygen are easily removed by water pollutants in various ways.

All living organisms need phosphate. The small amount of phosphate in water usually limits the amount of algae that can grow. But as our sewers carry phosphate detergents to water bodies, large amounts of algae form. As the algae grow and die, they use oxygen. They also clog the water surface, which prevents more oxygen from replenishing what's used up. Eventually, fish and plants die because of lack of oxygen. This process—called *eutrophication* (overnourishment)—can occur naturally over many years. But the addition of large amounts of phosphate to natural waters can greatly speed up this natural process. Clearly, it's important to keep excess phosphates out of our natural waters.

eutrophication

Excess phosphate can be removed by flocculation. Adding alum ($Al_2[SO_4]_3$) to lake water precipitates $Al(OH)_3$ and brings the phosphate down with it. In 1978, "floccing" Medical Lake, Washington, reduced its phosphate content by 90 percent.

Although oxygen-forming water plants can help replenish the oxygen in water, excess sediment will shut out the sunlight these plants need for photosynthesis. Excessive amounts of organic matter in natural water also decrease the oxygen supply, because decaying animal matter and plant matter use oxygen.

12.5 THE BEHAVIOR OF SOLUTIONS

Solutions behave differently from the pure solvents they contain in a number of interesting ways.

VAPOR PRESSURE. The vapor pressure of a solution is always lower than the vapor pressure of the pure solvent at the same temperature. To escape into the vapor state, solvent molecules have to be at the surface of the liquid. If a solute is dissolved in the liquid, some solute particles take up space at the surface. Fewer solvent particles can escape, and the rate of escape is slower. The solvent vapor, though, deposits particles into the liquid at the same rate it would if no solute were there. Thus when equilibrium is reached, fewer molecules are in the vapor than would be there if the solvent were pure, as Figure 12.6 shows.

We can infer from the figure that the more solute there is in the solution, the fewer solvent particles will be at the surface of the solution and the

FIGURE 12.6
A solute lowers the vapor pressure of a solvent

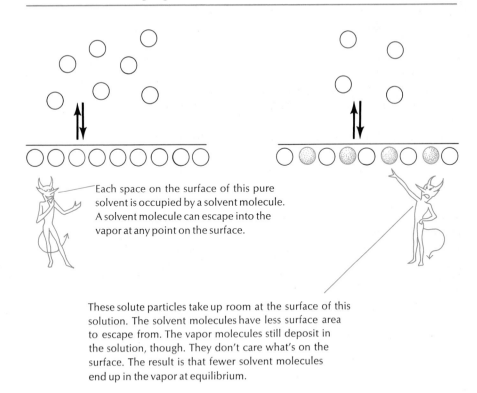

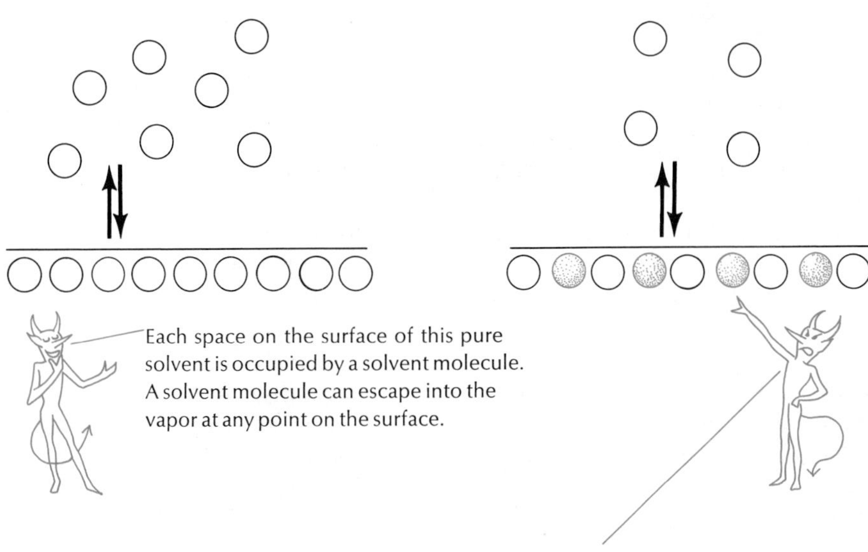

Each space on the surface of this pure solvent is occupied by a solvent molecule. A solvent molecule can escape into the vapor at any point on the surface.
These solute particles take up room at the surface of this solution. The solvent molecules have less surface area to escape from. The vapor molecules still deposit in the solution, though. They don't care what's on the surface. The result is that fewer solvent molecules end up in the vapor at equilibrium.

lower the vapor pressure will be. Properties of a solution that depend on the number of solute particles it contains are called *colligative properties*. Some colligative properties, besides lowering of vapor pressure, are elevation of boiling point, depression of freezing point, and osmotic pressure.

colligative properties

BOILING POINT. Since the vapor pressure of a solvent is lowered by the addition of solute, we have to raise the temperature so that the vapor pressure will equal the atmospheric pressure. This results in an elevation of the boiling point. A solvent's boiling point will be raised a certain amount by the addition of a given amount of solute. For water, the boiling point will be raised 0.512°C for every mole of solute particles that's dissolved in 1000. grams of water. Salt water takes longer to boil than pure water, because the salt raises the boiling point of water.

When a solution containing a solid solute like salt evaporates, only the pure solvent enters the vapor; the solid solute is left behind. This explains why the oceans are salty and keep getting saltier. In the "water cycle" of natural waters, rain falls on land and finds its way to rivers and streams. They dissolve parts of rocks and ground as they move toward the oceans. Water evaporates from the oceans, leaving behind the dissolved salts. The water vapor becomes rain, and the cycle starts over. More salts are washed into the ocean as the water is recycled, but the total amount of water stays the same. This natural "water cycle" is an example of *distillation*, a process by which a liquid is converted to vapor and then the vapor is condensed to a liquid.

distillation

We can use distillation to obtain pure water or any pure liquid. We boil water that contains a solute that doesn't evaporate itself, let the steam condense on something cold, and collect the condensed water (called "distilled water"). The impurities will be left behind. Figure 12.7 shows a setup for chemical distillation.

When a mixture of liquids is distilled, the liquids with the lower boiling point will usually boil off first. Distillation is used to make alcoholic beverages stronger, because alcohol has a different boiling point than water.

FREEZING POINT. Adding a solute to a solvent lowers the solvent's freezing point, because it lowers its vapor pressure. When a solution freezes, the pure solid solvent freezes out at first. At the freezing point, then, we have pure solid solvent, solution, and solvent vapor—all at equilibrium.

FIGURE 12.7
A simple distillation experiment

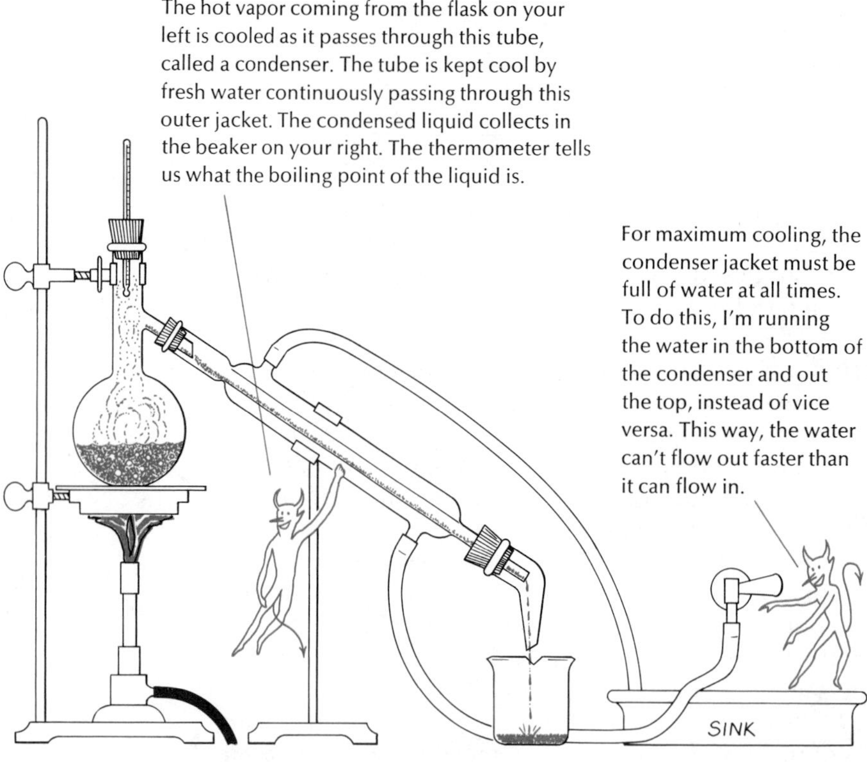

The hot vapor coming from the flask on your left is cooled as it passes through this tube, called a condenser. The tube is kept cool by fresh water continuously passing through this outer jacket. The condensed liquid collects in the beaker on your right. The thermometer tells us what the boiling point of the liquid is.
For maximum cooling, the condenser jacket must be full of water at all times. To do this, I'm running the water in the bottom of the condenser and out the top, instead of vice versa. This way, the water can't flow out faster than it can flow in.
SINK

As we saw before, the vapor pressures of the liquid and the solid have to be the same at the freezing point. Since the vapor pressure of the solvent in the solution is now lower, the temperature at which the pure solid solvent has this same vapor pressure will also have to be lower. This is the new freezing point, lower than the freezing point of the pure solvent.

Adding antifreeze to the water in a car radiator lowers the freezing point of water by lowering its vapor pressure. One mole of solute particles added to 1 kilogram of water will lower the freezing point of water by 1.84°C. Any solute can act as an antifreeze as long as it's soluble enough. Ethylene glycol is used because it's very soluble, doesn't damage the cooling system, doesn't boil off when the radiator gets hot, and is itself a fairly good coolant.

In experiments that involve changes in temperature and changes of state, the concentration term is not grams per liter. We use *molality* (m), which is moles of solute per kilogram of solvent. We use moles instead of grams because the lowering of the freezing point depends on moles and not grams. We use 1 kilogram of solvent instead of 1 liter of solution because mass doesn't change with temperature, whereas volume does. If we prepared a molar solution at 25°C and used it at 0°C, we'd introduce some error into our experiment.

molality

EXAMPLE 12.8: Calculate the molality of a radiator solution that is prepared by dissolving 3.5 kg of ethylene glycol (molecular weight 62.0 g/mole) in 6.5 kg of water.

Solution: We want to convert 3.5 kg of ethylene glycol per 6.5 kg of water to moles of ethylene glycol per kilogram of water. Our conversion factor is 62.0 g/mole, the formula weight of ethylene glycol. We may convert 3.5 kg directly to 3500 g.

$$\frac{3500\ \cancel{\text{g solute}}}{6.5\ \text{kg solvent}} \times \frac{1\ \text{mole solute}}{62.0\ \cancel{\text{g solute}}} = \_\_\_\_ m$$

Answer: 8.7 m (8.7 moles solute/kg solvent). ■

To calculate what the freezing point of a water solution will be, we use this formula:

$$T_f = 0.00°\text{C} - \left(1.84\frac{°\text{C}}{m}\right)m$$

Here, T_f is the freezing point of the solution (the *T*emperature at which ice *f*reezes from the solution); 0.00°C is the freezing point of pure water; 1.84°C/m (degrees Celsius per molal) is the *freezing-point constant* of water; and m is the molality of the solution.

freezing-point constant

EXAMPLE 12.9: Calculate the freezing point of the antifreeze solution in Example 12.8.

Solution: The molality of the solution is 8.7 m. Thus:

$$\begin{aligned} T_f &= 0.00°\text{C} - \left(1.84\frac{°\text{C}}{\cancel{m}}\right)(8.7\ \cancel{m}) \\ &= 0.00°\text{C} - 16.0°\text{C} \end{aligned}$$

Answer: $-16°C$. (The radiator would be protected from freezing down to this temperature.) ■

We use the lowered freezing point of solutions to make ice cream and to melt the ice on streets and sidewalks. We've seen that any solute will lower water's freezing point. Salt is especially useful for this because it's plentiful and dissolves well in water. To make ice cream, we pack the outer jacket of the ice cream freezer with a mixture of ice and salt. The ice will melt somewhat, and the temperature will go down. The ice cream in the middle container freezes when the temperature is low enough (about $-10°C$). The same thing happens when we throw rock salt on a frozen sidewalk: the ice melts, and the temperature goes down. The salt on the ice makes a solution. The ice—now trying to be at equilibrium with a solution of lower freezing point—melts. As it melts, it takes its heat of fusion from its surroundings, lowering the temperature.

Selective freezing, like distillation, can be used to separate some mixtures. Figure 12.8 shows that when a mixture of liquids is cooled, the liquid having the highest freezing point will freeze out of the solution first, leaving the other one behind.

OSMOTIC PRESSURE. When placed in water, a dried prune will swell and become plump. A cucumber placed in brine will wrinkle and become smaller. Our skin becomes wrinkled after a time in soapy water. Salt and sugar solutions can be used to preserve food. All these things happen because of a solution behavior called osmosis. *Osmosis* means the flow of pure solvent through a semipermeable membrane from a dilute solution to a more concentrated solution. Here, the semipermeable membrane lets solvent but not solute particles pass through its pores. In any living organism, the solvent is always water, and the water has many solutes dissolved in it. Skin and cell membranes are semipermeable. When a prune is placed in water, the water goes through the membrane (the skin) from the more dilute solution (pure water) to the more concentrated solution (the juice inside the prune). A brine solution has more solute in it than the juice inside a cucumber, so the water will pass from the inside of the cucumber into the brine solution, and the cucumber will shrink. Sugar and salt solutions can be used as preservatives because bacteria that cause spoilage lose water by osmosis through their cell walls, and die.

osmosis

Osmosis causes a pressure to be exerted, as we can tell when the prune's skin is pushed outward. *Osmotic pressure* is what would have to be applied to prevent osmosis from happening, as shown in Figure 12.9. If we apply more than osmotic pressure to a solution, the solvent molecules will go in the opposite direction. This is called *reverse osmosis*. It's used to remove salt from water and to help industries recycle their waste products. An industrial waste solution might contain small concentrations of a usable substance, formerly too small to bother trying to recover. Reverse osmosis lets the industries concentrate their waste solutions and recover the materials instead of throwing them away. This cuts down on water pollution and also saves money.

osmotic pressure

reverse osmosis

FIGURE 12.8
Separating a mixture by selective freezing

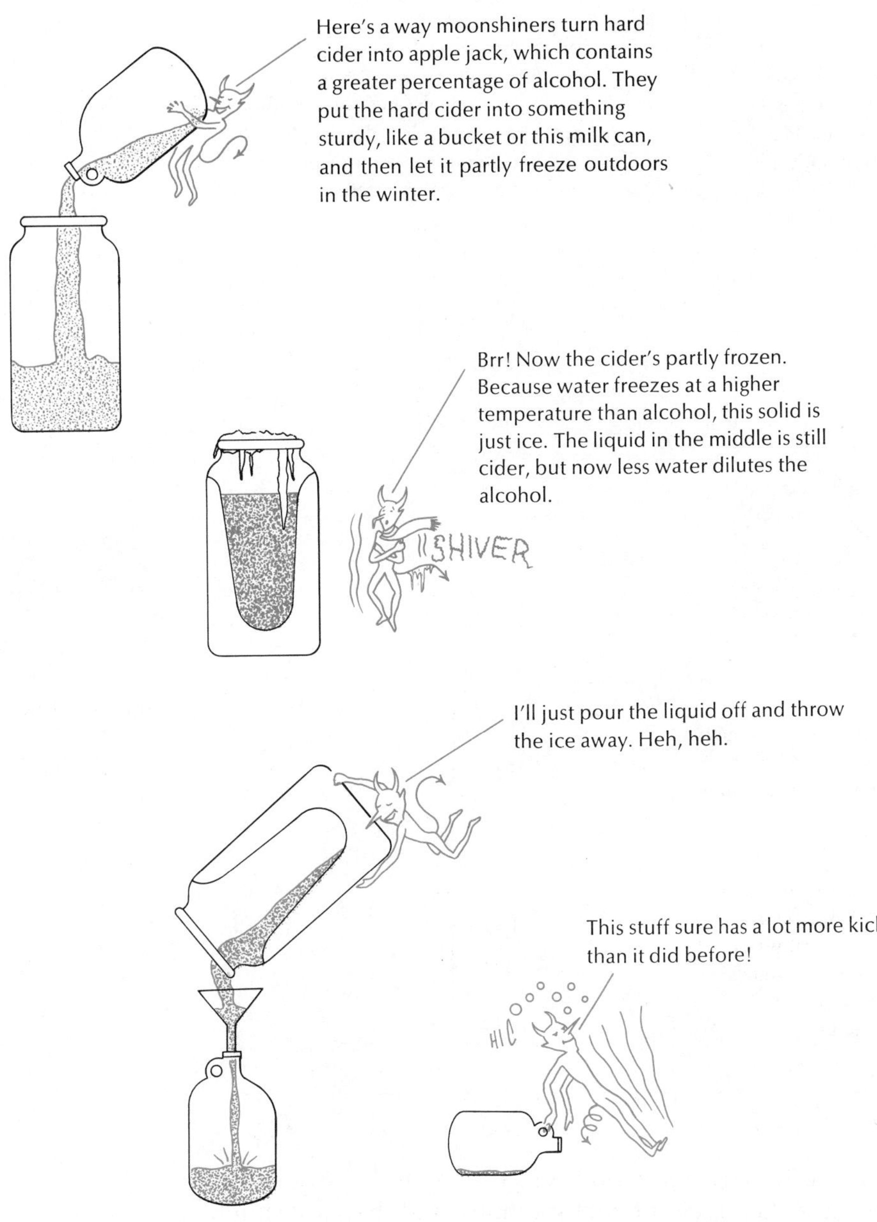

Here's a way moonshiners turn hard cider into apple jack, which contains a greater percentage of alcohol. They put the hard cider into something sturdy, like a bucket or this milk can, and then let it partly freeze outdoors in the winter.
Brr! Now the cider's partly frozen. Because water freezes at a higher temperature than alcohol, this solid is just ice. The liquid in the middle is still cider, but now less water dilutes the alcohol.
SHIVER
I'll just pour the liquid off and throw the ice away. Heh, heh.
This stuff sure has a lot more kick than it did before!
HIC

Organisms are very sensitive to changes in the osmotic pressure of their environment or their bodily fluids. For this reason, many saltwater fish can't survive in fresh water, and vice versa. When a hospital patient receives a solution intravenously, the solution must have the same osmotic pressure

FIGURE 12.9
Osmosis occurs when solvent molecules pass through a semipermeable membrane into a more concentrated solution

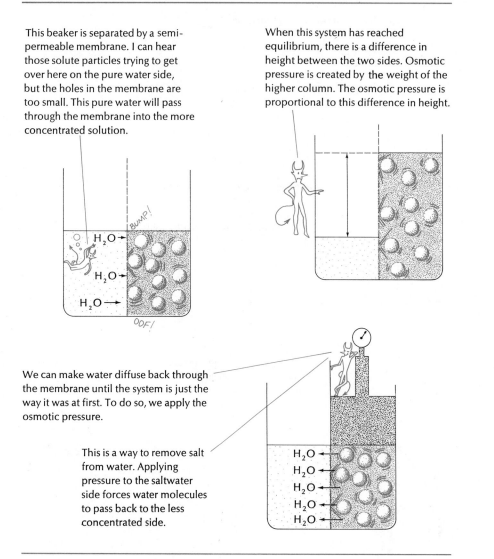

This beaker is separated by a semipermeable membrane. I can hear those solute particles trying to get over here on the pure water side, but the holes in the membrane are too small. This pure water will pass through the membrane into the more concentrated solution.

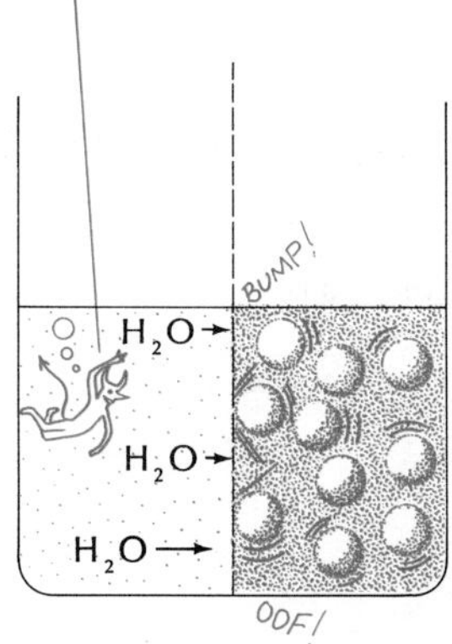

When this system has reached equilibrium, there is a difference in height between the two sides. Osmotic pressure is created by the weight of the higher column. The osmotic pressure is proportional to this difference in height.

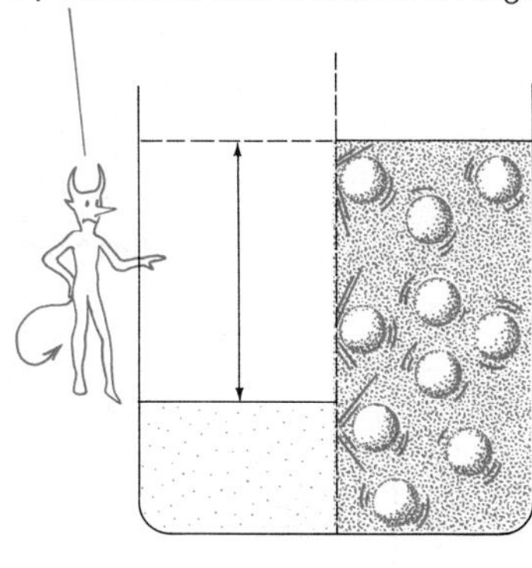

We can make water diffuse back through the membrane until the system is just the way it was at first. To do so, we apply the osmotic pressure.

This is a way to remove salt from water. Applying pressure to the saltwater side forces water molecules to pass back to the less concentrated side.

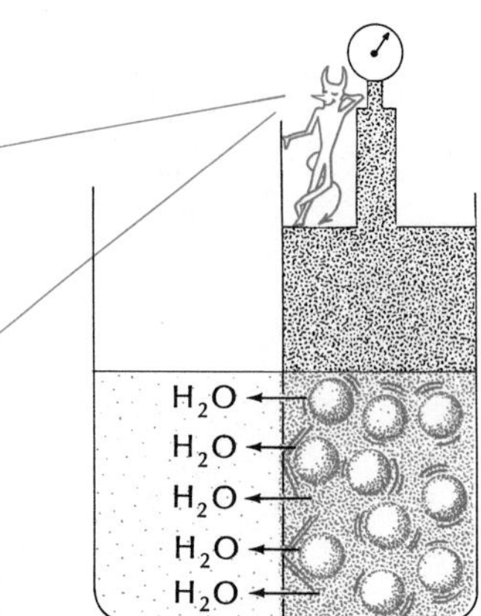

as blood plasma. This is because red blood cells will expand or even burst in solutions more dilute than their internal solutions, and they will shrink in more concentrated solutions. In either case, the red blood cell won't function properly.

Solutions that have the same osmotic pressure are said to be *isotonic*. The 0.9 percent saline solution, mentioned in Section 12.1, is called "isotonic saline" because it is made to have the same osmotic pressure as that of blood.

In this chapter, we've looked at the behavior and some properties of solutions. In the next two, we'll write equations and do calculations involving reactions in solution.

REVIEW QUESTIONS

Describing Solutions

1. What is a *solution?* What is a *solvent,* and what is a *solute?*
2. Give some examples of solutions. For each, state which is the solvent and which is the solute.
3. List and describe four characteristics of a solution.
4. What do we mean by the *concentration* of a solution? List and describe five concentration terms.
5. What is a *saturated solution?* How do we express its concentration?
6. What do the words *soluble* and *insoluble* mean? What does *sparingly soluble* mean?

Water as a Solvent

7. List some substances that are soluble and insoluble in water.
8. Name two liquids that are *miscible*, and two that are *immiscible.*
9. What is a *polar molecule?* Why are H_2O and HF polar but CO_2 and CF_4 nonpolar? What must a molecule have to be polar?
10. What energy must be overcome in dissolving an ionic solute? How does water overcome this energy? Why are some ionic solutes insoluble in water?
11. Explain why some ions are always present in solution, even when the compound is insoluble.
12. How do we express the concentration of natural water? What are some of the chief ions present in seawater?
13. What is hard water? Why does it make washing difficult? What are some ways of dealing with it?
14. What are some hydrogen-bonded solutes? Explain how water can dissolve these.
15. Give an example of a nonpolar solute to which water can hydrogen-bond.
16. Why doesn't water dissolve nonpolar solutes? What solvents do we use to dissolve nonpolar solutes?
17. What substances are sometimes added to municipal water, and why?

Emulsions and Colloids

18. Explain the relationship between *colloids, emulsions,* and *emulsifying agents.* List some familiar colloids.
19. List three characteristics of a colloid.
20. How do *detergents* work?
21. What keeps colloidal particles from coming together and settling out?
22. How can we destroy a colloid? Why might we want to?
23. What is a *Cottrell Precipitator?* How does it work?
24. What is a *semipermeable membrane?* What is *dialysis?* Explain how these are used in an artificial kidney machine.

Conditions Affecting Solutions

25. How do temperature and pressure affect the solubility of a gas in a liquid?
26. What is *Henry's Law?* Write the formula used to do calculations using Henry's Law.
27. What is a *supersaturated solution?* How can we prepare one?
28. How does natural water get its dissolved oxygen? How much is there?
29. Why is phosphate a water pollutant? What is *eutrophication?*

The Behavior of Solutions

30. Why is the vapor pressure of a solution lower than the vapor pressure of the pure solvent?
31. What are *colligative properties?* List four.
32. How does the lower vapor pressure of the solvent in a solution affect the solvent's boiling point in that solution ?
33. What is *distillation?* Give an example.
34. How does the lower vapor pressure of a solution affect its freezing point?
35. Why does antifreeze work?
36. What is *molality?* Why must we use it in experiments with freezing point?
37. How can we calculate the freezing point of a water solution?
38. What is the *freezing-point constant* of water? What does it mean?
39. How does salt melt ice and help make ice cream?
40. What happens when the temperature of a mixture of liquids gradually decreases?
41. What is *osmosis*? Give some examples.
42. What is *osmotic pressure?* What causes it? How do we measure it?
43. Explain how *reverse osmosis* can desalinate water and concentrate solutions.
44. What does *isotonic* mean? Explain what is meant by an "isotonic saline solution."

EXERCISES

Set A (Answers at back of book. Asterisks mark the more difficult exercises.)

1. A 50.0-mℓ sample of water from the Colorado River is found to contain 0.352 grams of dissolved solids. Calculate the concentration in grams per liter.
2. A certain preparation of corn syrup contains 1.1 kg of corn sugar per liter of corn syrup. A recipe calls for 150 g of corn sugar, to be dissolved in water. How many milliliters of corn syrup must be used to get 150 g of corn sugar?
3. How many grams of NaCl must be used to prepare 250 mℓ of a 0.900 percent (wt/vol) saline solution?
4. Express the concentration in Exercise 3 in grams per liter.
5. If 25.0 g of a glucose solution contain 1.44 g of glucose, what is the weight percent of glucose in this solution?

*6. Concentrated nitric acid is 72% by weight and has a density of 1.42 g/mℓ. How many mℓ of concentrated nitric acid must be used to get 15.0 grams of HNO_3?

7. A solution of hydrogen peroxide is prepared by diluting 5.0 mℓ of H_2O_2 to 500. mℓ of solution. Calculate the volume percent.
8. The word "proof" on alcoholic beverages is defined as twice the percentage by volume. How many mℓ of alcohol are in a 1-liter bottle of 80-proof gin?
9. Calculate the molarity of a solution made by dissolving 45.0 g of KOH in water to a final volume of 250. mℓ.
10. How many mℓ of a 0.30 *M* HCl solution are needed to get 5.0 grams of HCl?
11. Solubility may be expressed in moles per liter (*M*) as well as in grams per liter. Ag_2S has a solubility of about 2×10^{-17} *M*. How many grams of Ag_2S are present in 500. mℓ of a saturated solution?
12. Using the data in Table 12.2, calculate the solubilities of O_2, CO_2, and NH_3 in moles per liter.
13. It is desired to separate the much more valuable AgCl from the NaCl in a solid mixture of the two. Explain how this might be done.

*14. Calculate the molarities of Na^+ and of Mg^{2+} in seawater, using the data of Table 12.3. The density of seawater is 1.03 g/mℓ.

15. The "safe" limit of mercury in fish for us to eat is 0.5 ppm. At this level, how much mercury would be contained in a 15-kg swordfish?

*16. Bromine chloride, BrCl, has been suggested as an alternative to chlorine in water purification, because it's twelve times as soluble as chlorine and reacts with water immediately to form hypobromous acid, HBrO.
   a. Calculate the solubility of BrCl in moles per liter.
   b. Write a complete equation for the reaction of BrCl with water.

17. Which compound in each of the following pairs would probably be the more soluble in water? Explain for each.
    a. HF or F_2
    b. N_2 or NO_2
    c. H_2S or H_2SO_4
    d. CH_4 or CH_3OH
18. An old saying is, "Don't try to make mayonnaise during an electrical storm." Does this have any basis in fact? Explain.
19. Is milk of magnesia a solution or a colloid? How do you know?
20. Biologists often free protein solutions of unwanted salts by putting the protein solution into a semipermeable bag and suspending it in a large amount of pure water. Explain how this works.
21. A cookbook offers this advice to prevent cloudy ice when making an ice mold for a punch bowl: boil the water first, then pour into mold and freeze. Explain why this works.
22. The solubilities in Table 12.2 are for 1 atm pressure. Calculate
    a. the solubility of CO_2 at 350 torr
    b. the solubility of Cl_2 at 2.5 atm

*23. What pressure would be needed to obtain a 0.200 g/ℓ solution of CO_2 in water?

24. Why does honey often contain sugar crystals after it has been standing for some time?
25. In 1974, oxygen gas was the third most important chemical produced. It is obtained from air by distillation: air is liquefied, and its components are allowed to boil off. The three chief components of air have these boiling points: O_2, $-183°C$; N_2, $-196°C$; Ar, $-186°C$. List these in the order they would boil off.
26. How many grams of ethyl alcohol (M.W. 46.1) are needed to make a 1.00 molal solution using 250. grams of water?
27. What is the molality of a solution that contains 5.54 grams of glycerin (M.W. 92.1) in 125 g of water?
28. Methyl alcohol (molecular weight 32.0) has been used as an antifreeze. What would be the freezing point of a solution containing 27.0 grams of methyl alcohol and 73.0 grams of water?

*29. How many grams of glucose, $C_6H_{12}O_6$, would have to be dissolved in 100. g of water to give a solution that freezes at $-1.00°C$?

$$(m = [-T_f]/[1.84°C/m])$$

30. Could antifreeze (ethylene glycol) be used instead of salt to freeze ice cream? Why, or why not?
31. Explain the following, on the basis of osmosis or osmotic pressure.
    a. An effective way to kill a snail or slug in your garden is to sprinkle it with salt.
    b. Drinking salt water actually dehydrates (removes water from) our tissues.
    c. Celery that has gone limp can often be revived by cutting the ends off and placing it in water for a time.
32. Describe what might happen to red blood cells if they were placed in each of these solutions:
    a. 0.154 *M* NaCl
    b. 0.154 percent (wt/vol) NaCl
    c. 0.154 g/ℓ NaCl

Set B (Answers not given. Asterisks mark the more difficult exercises.)

1. It has been suggested that gold might be recovered from seawater. If seawater contains about 4×10^{-9} g/ℓ of dissolved gold, how many liters of seawater would have to be processed to get 1 gram of gold? Does it seem feasible to process this much seawater?
2. How many grams of NaCl must be dissolved in 250. mℓ of solution to make a solution that is 0.542 g/ℓ?
3. The average adult has about 6 grams total of glucose in the blood. Assuming that the same average adult has about 5 liters of blood, calculate the concentration of blood glucose in weight/volume percent.
4. Express the blood glucose of Exercise 3 in grams per liter.
5. Concentrated hydrochloric acid is 36 percent by weight and has a density of 1.18 g/mℓ.
    a. How many grams of HCl are in 25.0 grams of concentrated hydrochloric acid?
    b. How many grams of HCl are in 50.0 mℓ of concentrated hydrochloric acid?
6. How many grams of solid boric acid must be used to make 155 g of a solution that is 3.0 percent by weight?
7. How many mℓ of alcohol are contained in a 1.5-liter bottle of rubbing alcohol that is 70 percent by volume?
*8. In one state, an alcohol concentration of 0.05 percent by volume in blood indicates legal drunkenness for operating a motor vehicle. Assuming that wine is 12 percent alcohol by volume, that the average human being has 5 liters of blood, and that 25 percent of the alcohol from the wine goes directly into the blood:
    a. Calculate the concentration in volume percent that one would achieve after drinking 1 liter of wine.
    b. Would this person be legally drunk?
    c. How much wine can be consumed without exceeding this limit?
9. How many grams of $AgNO_3$ should be weighed out to prepare 500. mℓ of a 0.100 *M* solution?
10. How many mℓ of 2.0 *M* $BaCl_2$ can we get by dissolving 100. g of $BaCl_2$ in water?
11. A 20.0-liter saturated solution of iodine in water contains 5.80 grams of iodine. A 10.0-milliliter saturated solution of iodine in alcohol contains 2.05 grams of iodine. Calculate the solubility of iodine in each solvent. In which is it more soluble? Why do you think alcohol is used as the solvent for tincture of iodine? (See Table 12.1.)
12. Using the data in Table 12.2, calculate the solubilities of NaCl, $CaCO_3$, and HgS in moles per liter.
13. Is it possible to prepare a 5.0 *M* solution of CuS in water? Explain.
*14. Calculate the concentration of Ca^{2+} and of HCO_3^- in seawater in moles per liter, using the data in Table 12.3. The density of seawater is 1.03 g/mℓ.
15. If the concentration of mercury in a recalled batch of canned tuna is 0.75 ppm, calculate the number of grams of mercury in a $6\frac{1}{2}$-ounce can.
16. Using the data in Table 12.3, calculate the total weight of dissolved materials in one kg of seawater.
17. Which compound in each of the following pairs would probably be the more soluble in water? Explain for each.
    a. $CHCl_3$ or HCl
    b. NH_3 or NCl_3
    c. CCl_4 or $CaCl_2$
    d. SO_3 or O_3
18. Would a detergent work better or worse in salt water than it does in fresh water? Explain.
19. When it's past time to clean a fish tank, we can often see the beam of light from a flashlight that's shining through it. Explain.
20. A detergent is suspected of being "padded" with an inert filler, sodium sulfate. Suggest a way of separating the large detergent molecules from the smaller sodium and sulfate ions.
21. When opened, a refrigerated bottle of soda water will fizz more at 10,000 feet than at sea level. Explain.
22. The solubility of oxygen is 4.5×10^{-2} g/ℓ at 1.00 atm. Calculate its solubility in a mountain lake at 0.85 atm.
23. Calculate the solubility of CO_2 gas at 978 torr. (Data in Table 12.2 is for 1 atm.)
24. Which of these would slow down the eutrophication of a lake:
    a. circulating a stream of air through the water constantly

b. adding phosphate to the lake
c. planting water lilies
d. raking dead leaves out of the lake

25. The boiling point of water is 100.°C; of ether, 34.5°C; and of ethyl alcohol, 78.3°C. Put these in order of increasing vapor pressure at room temperature.
26. What is the molality of a solution that contains 10.0 grams of NaCl in 500. grams of water?
27. Propylene glycol (molecular weight 76.1 g/mole) can be used as antifreeze. How many kilograms of propylene glycol would be needed to make a solution of 9.0 *m*, using 6.5 kilograms of water?
28. What is the freezing point of a solution made by dissolving 50.0 grams of ethyl alcohol (M.W. 46.1) in 100. grams of water?

*29. Maple syrup is mostly a solution of sucrose in water. Calculate the molality of the syrup if a sample freezes at −0.50°C. Assume that the syrup contains no solutes other than sucrose.

$$(m = [-T_f]/[1.84°C/m])$$

30. A little water dissolved in gasoline can plug a gas line as it freezes in cold weather. If a mixture of alcohols called "dry gas" is added directly to the gasoline, the water will not freeze. Explain how this works.
31. Explain the following, on the basis of osmosis or osmotic pressure.
    a. When sprinkled with sugar, a dish of sliced fruit will form its own juice.
    b. Meat that is salted before cooking tends to dry out.
    c. Although trees have no pump (heart) to circulate their fluids, water is drawn from the soil up into the branches and leaves.
32. A leading candy maker uses reverse osmosis to concentrate its waste sugar solutions so that the sugar can be recovered and used again. Explain how this could be done.

CHAPTER 13

Acids and Bases

LEARNING OBJECTIVES

After studying this chapter, you should be able to:

1. Supply a correct definition, explanation, or example for each of these:

hydronium ion	Brønsted-Lowry acid and base
strong acid	proton
strong base	proton transfer
ionic equation	conjugate pair
deliquescence	conjugate base
weak acid	conjugate acid
weak base	amphoteric
pH	Lewis acid and base
indicator	
Arrhenius acid and base	

2. Name the strong acids and bases and write ionic equations showing their reaction with water.
3. Name a common weak acid and weak base and write ionic equations showing their reaction with water.
4. Explain the relationship between hydronium-ion and hydroxide-ion concentrations.
5. Explain qualitatively the relationship between pH and hydronium-ion concentration, and state whether a given solution is acidic, basic, or neutral, based on its pH.
6. Given the concentration of a strong acid or strong base, state its hydronium-ion and hydroxide-ion concentration.
7. Calculate among pH, hydronium-ion, and hydroxide-ion concentrations, using a calculator and a log table.
8. State some applications of pH.
9. Write ionic equations for proton transfer with water for any of the species given in this chapter and predict, where possible, whether their solutions would be acidic or basic.
10. Explain the relationship of acid and base structure to strength, and predict whether a given acid or base would be stronger or weaker than another.

As early as the seventeenth century, a class of compounds was recognized whose water solutions tasted sour, turned litmus (a vegetable dye) from blue to red, and could dissolve many things. These substances were named acids. Another class of compounds was known whose water solutions were slippery or soapy to the touch, turned litmus blue, and could dissolve grease. These were called alkalis, or bases. It was observed that acids react with bases, or neutralize them, to form salts.

Many familiar substances are acids or bases. Vinegar and citrus juices are acids. Stomach acid helps begin the digestive process and kills many microorganisms that we ingest. Household ammonia, lye, and drain or oven cleaner are common bases, as are most laundry or dishwasher detergents. Some of these, like household ammonia, are solutions.

Maintaining just the right amount of acid or base in a solution is often extremely important. For instance, if blood isn't kept at just the right acidity, the hemoglobin can't take on oxygen and release carbon dioxide when it's supposed to. We'll learn how to measure acidity or basicity with pH, and what the term means. But first, we need to find out what acids and bases are and what they do.

13.1 STRONG AND WEAK ACIDS AND BASES

We learned the names of acids and bases, and saw some of their reactions, in Chapters 4 and 5. Now we'll see that there's a difference between strong and weak acids and bases. This difference shows up in their water solutions, which are the convenient substances to work with. Both strong acids and strong bases have a high affinity for water.

WATER SOLUTIONS OF STRONG ACIDS. A solution of hydrochloric acid can be prepared by bubbling hydrogen chloride gas through water. What happens is shown by this equation:

$$HCl(g) + H_2O(l) \longrightarrow Cl^-(aq) + H_3O^+(aq)$$

The ion H_3O^+ is called *hydronium ion*, and it is always formed when a strong acid is dissolved in water. In fact, a *strong acid* is defined as one that forms as many hydronium ions when dissolved in water as there were acid molecules at the start. According to one theory, an acid contains a "loose" hydrogen, which breaks away from its bonding electron pair and fastens itself onto one of water's available electron pairs to form H_3O^+, as illustrated in Figure 13.1.

hydronium ion

strong acid

The equation shown above is an *ionic equation,* in which substances that exist as ions in aqueous solution are written as ions. The acids H_2SO_4, HNO_3, $HClO_4$, HBr, and HI are all strong acids that convert completely to hydronium ions in water. Table 13.1 shows ionic equations for these conversions. These strong acids are all covalent compounds, but they ionize (form ions) in water. We see the dibasic acid H_2SO_4 using only one of its hydrogens as a strong acid. The other hydrogen acts as a weak acid.

ionic equation

FIGURE 13.1
In a strong acid, molecules are converted to hydronium ions and negative ions

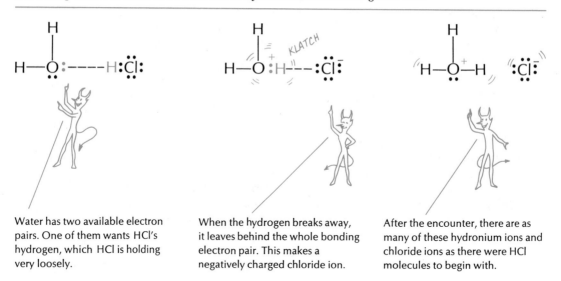

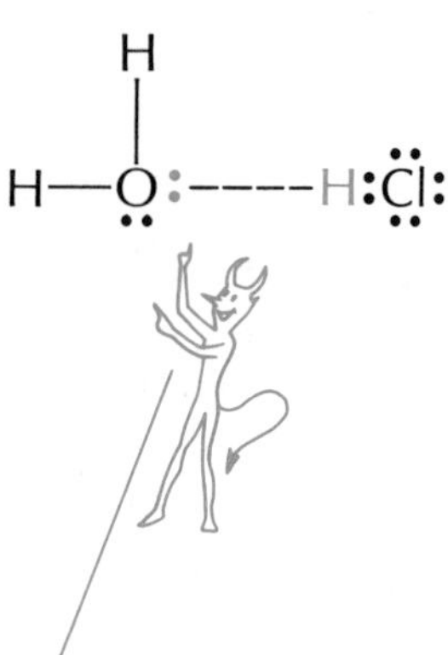

Water has two available electron pairs. One of them wants HCl's hydrogen, which HCl is holding very loosely.

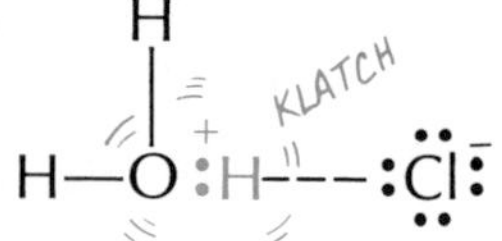

When the hydrogen breaks away, it leaves behind the whole bonding electron pair. This makes a negatively charged chloride ion.

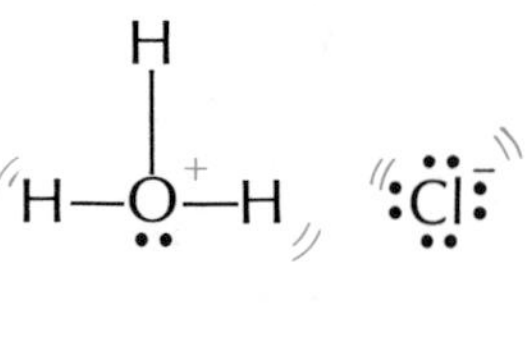

After the encounter, there are as many of these hydronium ions and chloride ions as there were HCl molecules to begin with.

The strong acids commonly used in the chemistry laboratory are sulfuric, nitric, and hydrochloric. Concentrated sulfuric acid is 96 percent H_2SO_4 by weight and about 18 molar. It can cause serious damage to human tissues if spilled on the skin and should always be handled with care. Concentrated nitric acid is about 72 percent HNO_3 by weight and about 16 molar, and it is also hazardous. Concentrated hydrochloric acid is about 36 percent HCl by weight and about 12 molar. It's not quite as dangerous as the other two, but it should still be used with caution. Solutions of these acids that are labeled "dilute" are usually 6 molar for hydrochloric and nitric acids and 3 molar for sulfuric acid.

WATER SOLUTIONS OF STRONG BASES. Whereas the strong acids are covalent compounds, the common strong bases are all ionic compounds. We already know from Chapter 12 how an ionic compound dissolves in water. Metal hydroxides are the most common strong bases. Here's an equation showing how one dissolves in water.

$$NaOH(s) \longrightarrow Na^+(aq) + OH^-(aq)$$

Here ions are not being created but only separated and mixed through the solution. Hydroxide ions are responsible for the properties of a base solution. When a *strong base* dissolves in water, the amount that dissolves yields exactly as many hydroxide ions in solution as its formula would predict. For instance, every mole of dissolved NaOH produces a mole of hydroxide ions; and every mole of dissolved $Ba(OH)_2$ produces two moles of hydroxide ions. But Appendix D (p. 547) shows that few hydroxides are very

strong base

TABLE 13.1
Equations showing the strong acids dissolving in water

$HCl(g)$	$+ H_2O(l) \longrightarrow$	$Cl^-(aq)$	$+ H_3O^+(aq)$
$H_2SO_4(l)$	$+ H_2O(l) \longrightarrow$	$HSO_4^-(aq)$	$+ H_3O^+(aq)$
$HNO_3(l)$	$+ H_2O(l) \longrightarrow$	$NO_3^-(aq)$	$+ H_3O^+(aq)$
$HClO_4(l)$	$+ H_2O(l) \longrightarrow$	$ClO_4^-(aq)$	$+ H_3O^+(aq)$
$HBr(g)$	$+ H_2O(l) \longrightarrow$	$Br^-(aq)$	$+ H_3O^+(aq)$
$HI(g)$	$+ H_2O(l) \longrightarrow$	$I^-(aq)$	$+ H_3O^+(aq)$

soluble. Table 13.2 shows some soluble strong bases and how they form hydroxide ions in solution. These and other soluble bases are the only ones useful in making base solutions. NaOH and KOH are more soluble than the other three.

The strong base used most often in the chemistry laboratory is sodium hydroxide (lye), although potassium hydroxide is sometimes used, too. Both NaOH and KOH are available in solid form as pellets. The pellets have such a high affinity for water that they absorb water vapor from the air and form a solution. Some NaOH or KOH pellets spilled on a lab bench will soon form a puddle of strong base solution, which is hazardous to anyone resting an arm on the bench top. Solids that absorb water vapor from the air and form solutions are said to be *deliquescent*. deliquescence

Concentrated solutions of these bases are usually 9 molar; dilute solutions are 6 molar. Both of these bases will burn the skin and dissolve wool and other fabrics, and they should be handled carefully. They also etch glass slowly and are usually kept in plastic or wax-lined bottles for this reason.

Just as the strong acids are completely converted to hydronium ion in water, some strong bases besides hydroxide ion are completely converted to hydroxide ion in solution. One of these is the oxide ion. In Chapter 5, we learned that metal oxides are basic anhydrides; that is, that they form hydroxides in water. Now we can describe this reaction as an acid-base reaction between water and oxide ion:

$$O^{2-} + H_2O \longrightarrow OH^- + OH^-$$

Metal oxides are all solids. If we took a sample of, say, Na_2O, and dissolved it in water, the above reaction would take place immediately. We would find no oxide ions in the solution, because they'd have been completely converted to hydroxide ions. We see from the equation that two

TABLE 13.2
Some common soluble strong bases

Soluble	$NaOH(s)$	$\longrightarrow$	$Na^+(aq)$	$+ OH^-(aq)$
	$KOH(s)$	$\longrightarrow$	$K^+(aq)$	$+ OH^-(aq)$
Sparingly soluble	$Ba(OH)_2(s)$	$\longrightarrow$	$Ba^{2+}(aq)$	$+ 2\,OH^-(aq)$
	$Ca(OH)_2(s)$	$\longrightarrow$	$Ca^{2+}(aq)$	$+ 2\,OH^-(aq)$
	$Sr(OH)_2(s)$	$\longrightarrow$	$Sr^{2+}(aq)$	$+ 2\,OH^-(aq)$

moles of hydroxide ions are produced for every mole of oxide ion. One of the hydroxide ions came from the oxide ion, and the other from the water, but they are identical. Here's an equation showing solid Na_2O dissolving in water:

$$Na_2O(s) + H_2O(l) \longrightarrow 2\ Na^+(aq) + 2\ OH^-(aq)$$

Weak acids and *weak bases* don't ionize much in water, and that's why they're considered weak. Their water solutions don't contain very many hydronium or hydroxide ions.

weak acid

weak base

WEAK ACIDS IN WATER. For a moment let's consider a solution of water in water. The following reaction happens to a very small extent:

$$H_2O + H_2O \longrightarrow H_3O^+ + OH^-$$

Both hydronium ions and hydroxide ions are produced, which means that water is both a weak acid and a weak base. In a 1-liter sample of normal water at room temperature, we'd find only 10^{-7} moles each of hydronium and hydroxide ions.

Other weak acids dissolved in water form some hydronium ions—more than water alone does. For instance, a 1-liter solution containing one mole of acetic acid ($HC_2H_3O_2$) will have about 4×10^{-3} moles of hydronium ions, as shown in Figure 13.2. Weak acids include all acids that we haven't specified previously as strong.

Acetic acid is the most common weak acid used in the chemistry labora-

FIGURE 13.2
Weak acids give only a few hydronium ions in water

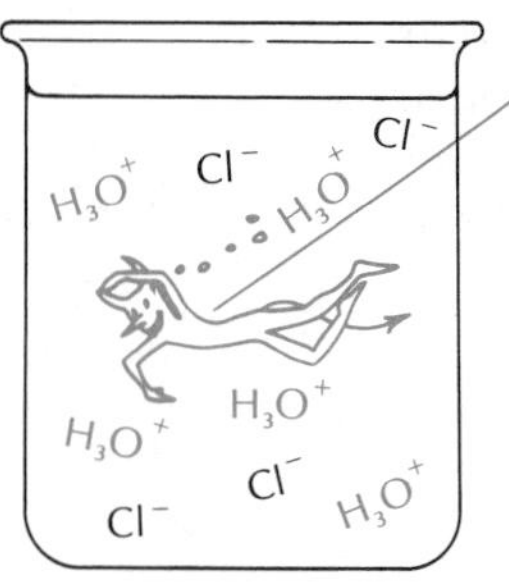

Swimming around in this beaker of hydrochloric acid, I can't find *any* HCl molecules. *None!* That's because HCl is a strong acid that is 100 percent ionized in water.

$$HCl + H_2O \rightarrow Cl^- + H_3O^+$$
100 %

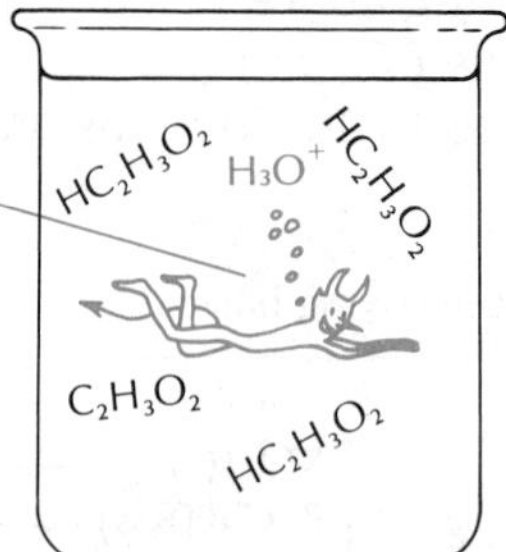

Now here in this beaker of acetic acid, it's a different story! I have a hard time finding any hydronium ions and acetate ions, although there are a few here. Acetic acid is a weak acid and is only about 1 to 2 percent ionized.

$$HC_2H_3O_2 + H_2O \rightarrow C_2H_3O_2^- + H_3O^+$$
1%–2%

tory. We already know that vinegar is a dilute solution of acetic acid in water. Pure acetic acid, without any water in it, is a clear, colorless liquid, like water. Its melting point is 16.7°C, so in a cold room acetic acid usually has some solid crystals in it. Because these crystals look like ice, pure acetic acid is also called "glacial" acetic acid. It has an overpowering vinegar smell and, like any concentrated acid, can burn skin. Dilute acetic acid is usually 6 molar. Vinegar is about 5 percent acetic acid and about 0.8 molar.

WEAK BASES IN WATER. As we saw in the previous section, water itself is a weak base. When dissolved in water, other weak bases also form a few hydroxide ions—more than plain water does but fewer than a strong base does.

Ammonia is the weak base used most often in the chemistry laboratory. When ammonia gas is bubbled through water to make aqueous ammonia, this reaction happens to a small extent.

$$NH_3(g) + H_2O(l) \longrightarrow NH_4^+(aq) + OH^-(aq)$$

Ammonia is a covalent compound that ionizes in solution—in contrast with the strong bases, which are ionic compounds. In a 1-liter solution containing one mole of ammonia, we'd find only about 4×10^{-3} moles of hydroxide ions. Concentrated aqueous ammonia (sometimes incorrectly labeled "ammonium hydroxide, NH_4OH") is about 27 percent NH_3, and about 16 molar. Dilute aqueous ammonia is about 10 percent and 6 molar.

Another weak base sometimes used in solution is the carbonate ion (CO_3^{2-}). Any soluble carbonate, such as Na_2CO_3, is a source of carbonate ions. Carbonate produces some hydroxide ions in water, according to this reaction:

$$CO_3^{2-} + H_2O \longrightarrow HCO_3^- + OH^-$$

Other negative ions can also be weak bases, as we'll see later.

The equation for the reaction that happens when solid Na_2CO_3 is dissolved in water is:

$$Na_2CO_3(s) + H_2O(l) \longrightarrow 2\ Na^+(aq) + OH^-(aq) + HCO_3^-(aq)$$

Since the carbonate ion is a weak base, this reaction happens only to a limited extent.

13.2
MEASURING ACIDS AND BASES

We know that acids and bases have different strengths. Spilling vinegar on the skin does no harm, but we'd feel it if we spilled sulfuric acid of the same concentration. Our stomachs contain about 0.02 molar HCl, but if we drank 6 molar HCl we'd be very sick. Milk of magnesia, $Mg(OH)_2$, is a base that we take for an upset stomach, but the same number of moles of NaOH would burn our insides. The difference in these reactions is in how many hydronium or hydroxide ions are present in solution.

HYDRONIUM-ION AND HYDROXIDE-ION CONCENTRATION. We can easily predict the hydronium-ion concentration of a strong acid and the hydroxide-ion concentration of a strong base. Each comes from its molarity. A 0.01-molar HCl solution has a hydronium-ion concentration of 0.01 *M*, because HCl ionizes completely. A 0.05 *M* NaOH solution has a hydroxide-ion concentration of 0.05 *M*. We write square brackets to mean "molar concentration." For example, "$[H_3O^+] = 0.01\ M$" means "a hydronium-ion concentration of 0.01 molar." Similarly, "$[OH^-] = 0.05\ M$" means "a hydroxide-ion concentration of 0.05 molar."

As we saw earlier, pure water contains 10^{-7} moles per liter each of H_3O^+ and OH^-. Or, for pure water:

$$[H_3O^+] = 1.0 \times 10^{-7}\ M$$
$$[OH^-] = 1.0 \times 10^{-7}\ M$$

In pure water at 25°C, multiplying the hydronium-ion concentration by the hydroxide-ion concentration gives us a value of 1.0×10^{-14}.

$$[H_3O^+][OH^-] = (1.0 \times 10^{-7})(1.0 \times 10^{-7}) = 1.0 \times 10^{-14}$$

Adding an acid to water will increase the hydronium-ion concentration and decrease the hydroxide-ion concentration. Adding a base will do the opposite. Thus in acid or base solutions, the concentrations of hydronium ions and hydroxide ions aren't 1.0×10^{-7}, but their product is still 1.0×10^{-14}. That is:

$$[H_3O^+][OH^-] = 1.0 \times 10^{-14}$$

The hydronium-ion concentration of a strong acid and the hydroxide-ion concentration of a strong base depend only on how much acid or base is present in solution. But the concentrations can't be predicted that easily for weak acids and bases, because they depend on the relative strengths of the weak acids or bases. We won't go into the ways hydronium- and hydroxide-ion concentrations can be calculated for weak acids and bases. An easier and more accurate way of finding out these concentrations is just to measure them. We'll soon see how this is done.

THE pH OF SOLUTIONS. The term "pH" is a familiar one in the advertising and the environmental vocabularies. An antiblemish soap or a shampoo is supposed to be "pH-adjusted." By law, a laundry detergent that gives a pH greater than 11 when dissolved in a washerful of water must be so labeled. pH is a simple and widely used way of expressing hydronium-ion concentration.

As a kind of shorthand notation, pH makes it easier to express acidity and basicity since it does not use exponential numbers. For instance, a pH of 7.00 means a hydronium-ion concentration of $1.0 \times 10^{-7}\ M$—the pH of pure water at 25°C. The lower the pH (below 7, down to zero), the more acidic the solution; the higher the pH (above 7, up to 14), the more basic.

TABLE 13.3
pH values of some common solutions

Solution	pH	
Battery acid	0	INCREASING ACIDITY
Stomach acid	1.4–1.8	
Lemon juice	2.1	
Vinegar	2.9	
Soda water	3	
Wine	3.5	
Tomato juice	4	
Black coffee	5	
Urine, sour milk	6	
Rainwater	6.5	
Pure water, 25°C	7.00	Neutral
Blood	7.35–7.45	INCREASING BASICITY
Seawater	8	
Soaps, shampoos	8–9	
Detergents	9–10	
Milk of magnesia	10	
Household ammonia	11.9	
Liquid bleach	12	
Household lye	14	

A pH of 7.00 is neutral and is the dividing line between acids and bases. Table 13.3 lists pH values for some common solutions.

If the coefficient of the hydronium-ion concentration is 1.0, it's easy to find the pH: just use the exponent without the minus sign. A hydronium-ion concentration of 1.0×10^{-2} *M* would have a pH of 2.0. Figure 13.3 is a scale that can be used to find either hydronium-ion concentration, hydroxide-ion concentration, or pH if we know just one of these, and if the concentrations have coefficients of 1.0, then the pH will be an even number. From the scale, we see that a pH of 3.0 means a hydronium-ion concentration of 1.0×10^{-3} *M* and a hydroxide-ion concentration of 1.0×10^{-11} *M*. On this scale, multiplying the two concentrations together (that is, adding their exponents) always gives 1.0×10^{-14}. Each pH unit represents a tenfold increase in $[H_3O^+]$.

FIGURE 13.3
Scale for conversion between pH and concentrations

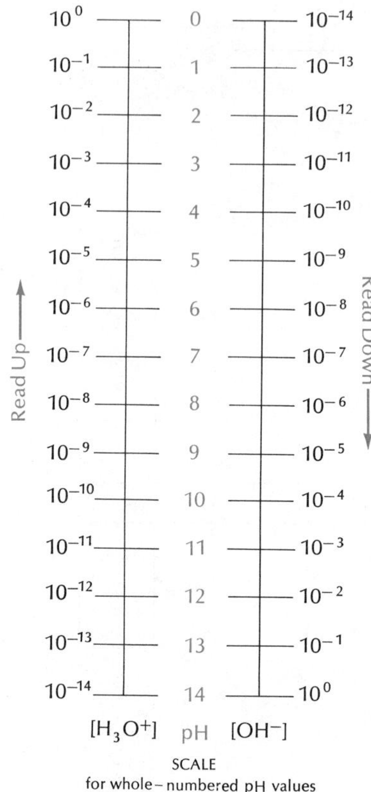

EXAMPLE 13.1: What are the hydronium- and hydroxide-ion concentrations of tomato juice?

Solution: Looking at Table 13.3, we see that tomato juice has a pH of 4 (acidic, because it's less than 7). We find a pH of 4 in Figure 13.3 and read across to find the hydronium-ion concentration (1.0×10^{-4} *M*) and hydroxide-ion concentration (1.0×10^{-10} *M*).

Answer: $[H_3O^+] = 1.0 \times 10^{-4}\ M$; $[OH^-] = 1.0 \times 10^{-10}\ M$. ■

EXAMPLE 13.2: What is the pH of: (a) 0.010 *M* HCl; and (b) 0.050 *M* $Ca(OH)_2$?

Solution: First, we express the hydronium- and hydroxide-ion concentrations in exponential notation. Because HCl is a strong acid and $Ca(OH)_2$ is a strong base, we get these from the molarities.
In HCl,

$$[H_3O^+] = 1.0 \times 10^{-2}\ M$$

In $Ca(OH)_2$,

$$[OH^-] = 2 \times (0.050\ M) = 1.0 \times 10^{-1}\ M$$

(We multiply the molarity by 2 to get the hydroxide-ion concentration, because each mole of $Ca(OH)_2$ gives 2 moles of hydroxide ion.) Next, we get the pH from Figure 13.3.

Answer: (a) 2.0; (b) 13.0. ■

FRACTIONAL pH. Nature isn't always so obliging as to give us solutions that have whole-numbered pH values. Most of the really interesting solutions have values somewhere between. Blood with a pH of 7.35 would have a hydronium-ion concentration somewhere between 1.0×10^{-7} *M* and 1.0×10^{-8} *M*. The exponential term is 10^{-8}, but the coefficient is some number larger than 1.0. To see how to handle fractional pH, we'll define pH by using logarithms. (For a discussion of logarithms, see Appendix A, p. 499.) We know that a hydronium-ion concentration of 1.0×10^{-7} *M* means a pH of 7.0. Another way of saying this is that

$$\text{pH} = -\log(1.0 \times 10^{-7}) = -(0 + [-7]) = 7$$

This equation illustrates the formal way of defining pH: *pH is the negative logarithm of the molar hydronium-ion concentration*. That is, pH

$$\text{pH} = -\log[H_3O^+]$$

To find pH, we take the logarithm of the hydronium-ion concentration and put a minus sign in front of it. If you have a calculator with a logarithm function, this technique is extremely easy. Many calculators have both natural (base e) and common (base 10) logarithms; be sure to use the common log function. (Consult the instructions that accompany your calculator.)

EXAMPLE 13.3: What is the pH of a 0.055 *M* HCl solution?

Solution: HCl is a strong acid, so $[H_3O^+] = 5.5 \times 10^{-2}$ *M*. The pH will be somewhere between 1 and 2. To find it, punch either 0.055 or 5.5×10^{-2} on your calculator (consult the instructions to find out if either of those is preferable). Then press the "log" key. The display probably reads "−1.2596373," which we round off to −1.26.

The calculator has told us that $\log(5.5 \times 10^{-2}) = -1.26$. Now, because $\text{pH} = -\log(5.5 \times 10^{-2})$, all we have to do is put a minus sign in front of our result: $-(-1.26) = 1.26$.

Answer: pH = 1.26. (It's less than 7, as we expect for an acid solution.) ■

If the solution whose pH we're trying to determine is basic, we get the hydronium-ion concentration by first finding the hydroxide-ion concentration. Then we use this relationship, introduced earlier:

$$[H_3O^+][OH^-] = 1.0 \times 10^{-14}$$

Rearranging this equation, we can get H_3O^+ if we know OH^-:

$$[H_3O^+] = \frac{1.0 \times 10^{-14}}{[OH^-]}$$

EXAMPLE 13.4: What is the pH of a 0.055 *M* NaOH solution?

Solution: Here, $[OH^-] = 5.5 \times 10^{-2}$ *M*. We substitute it into the above equation:

$$[H_3O^+] = \frac{1.0 \times 10^{-14}}{5.5 \times 10^{-2}} = ?$$

If you're doing this on a calculator, the display probably shows "1.8182 – 13," meaning, of course, 1.8182×10^{-13}, the value of the hydronium-ion concentration. We could write this value down, but it's just as easy to press the "log" button right away.

Press "log." The display probably reads "–1.274 01," which means –12.74. Now our pH is 12.74. Reasonable, because the solution is basic; we expect a pH above 7.

Answer: pH = 12.74. ■

Alternatively, if your calculator doesn't have logs on it, you can still calculate fractional pH with the log table (Appendix A). Example 13.5 illustrates how. (You'll find a discussion of logarithms in Appendix A, p. 499.)

EXAMPLE 13.5: Find the pH of a 0.055 *M* HCl solution by using the log table.

Solution:

Step 1: Express $[H_3O^+]$ in exponential notation.

$$[H_3O^+] = 5.5 \times 10^{-2}\ M$$

Step 2: Take logs of both sides, separating the coefficient and the exponential parts. Affix a minus sign to both sides.

$$-\log[H_3O^+] = -(\log 5.5 + \log 10^{-2})$$

Step 3: Evaluate the log of the coefficient, using the log table.

$$\log(5.5) = 0.74 \quad \text{(from table)}$$

Step 4: Substitute this value into the equation from Step 2. Evaluate the log of the exponential and substitute it also.

$$-\log[H_3O^+] = -(0.74 - 2) = 1.26$$

Since $-\log[H_3O^+] = \text{pH}$, this is our answer.

Answer: pH = 1.26. (The same answer we got using a calculator.) ■

To get $[H_3O^+]$ from pH, we use this form of the pH equation:

$$[H_3O^+] = 10^{-\text{pH}}$$

Converting from pH to hydronium- and hydroxide-ion concentration can be done by using the INV and log functions on a calculator.

EXAMPLE 13.6: A blood sample has a pH of 7.51. Find its hydronium-ion and hydroxide-ion concentration with a calculator.

Solution: Punch in –7.51 on the calculator. Then press INV and then log. The display reads 3.0903–08, which, rounded off, means a hydronium-ion concentration of 3.1×10^{-8} *M*.

To get hydroxide-ion concentration:

$$[OH^-] = \frac{1.0 \times 10^{-14}}{3.1 \times 10^{-8}} = 3.2 \times 10^{-7}\ M.$$

Answer: $[H_3O^+] = 3.1 \times 10^{-8}$ *M*; $[OH^-] = 3.2 \times 10^{-7}$ *M*. ■

EXAMPLE 13.7: Do Example 13.6 using a log table.

Solution: We reverse the procedure of Example 13.5.

Step 1: Decide what the exponent of 10 will be.

7.51 is between 7 and 8, so the exponent will be −8.

Step 2: Subtract the pH from the absolute value of the exponent.

$$8.00 - 7.51 = 0.49$$

Step 3: Find 0.49 in the log table, and find its antilog.

From the table, the antilog of 0.49 is 3.1

Step 4: Put the coefficient and the exponential together. This is the hydronium-ion concentration.

$$[H_3O^+] = 3.1 \times 10^{-8}\ M$$

Step 5: Calculate the hydroxide-ion concentration in the usual way.

$$[OH^-] = \frac{1.0 \times 10^{-14}}{3.1 \times 10^{-8}} = 3.2 \times 10^{-7}\ M$$

Answer: $[H_3O^+] = 3.1 \times 10^{-8}$ *M*; $[OH^-] = 3.2 \times 10^{-7}$ *M*. ■

APPLICATIONS AND MEASUREMENT OF pH. Maintaining a certain pH value or range is often important, especially in biological systems. For example, our blood maintains itself at a pH of between 7.35 and 7.45. If the pH falls slightly below 7.35, the condition is called acidosis; if it rises slightly above 7.45, the condition is called alkalosis. (In Example 13.6, the donor of the blood with pH 7.51 has alkalosis.) Both conditions must be treated. If the blood changes more than a few tenths of a pH unit from the normal range, the results are usually fatal. Hemoglobin itself is a weak acid, and too much acid or base interferes with its function of picking up, carrying, and releasing oxygen.

Most plants grow best in soil with a pH between 6 and 7. Higher or lower values prevent them from absorbing nutrients from the soil. For instance, a too-acidic soil can prevent plants from absorbing phosphate. Most plants absorb this nutrient as $H_2PO_4^-$. A too-acidic soil would convert it to H_3PO_4, and a too-basic soil would convert it to HPO_4^{2-}. In either case, plants would not be able to absorb as much phosphate.

Bacteria have an optimum pH range, too. Pickling foods in vinegar is an effective way to preserve them, because most bacteria that cause food spoilage won't grow in solutions having the low pH of vinegar.

Shampoos normally have a pH of about 8. Scalp, on the other hand, is on the acid side, with a pH of around 6. Rinsing the hair with vinegar after shampooing neutralizes the base left by the shampoo and restores the acid condition of the scalp. "Alkaline scalp" is a dry, flaky condition resembling dandruff. A shampoo advertised as "pH-adjusted" has a pH of around 6. Its makers have added various acids to lower the pH.

Rainwater is naturally slightly acidic (pH of 6.5), as we saw in Table 13.3. This is because some atmospheric carbon dioxide is always dissolved in it, which makes carbonic acid. The air pollutants NO_2, SO_2, and SO_3 increase the acidity of rainwater even further. In 1966 in Europe, the pH of rainwater was measured to be about 4. In Pasadena in 1976 and 1977, rainfall had an average pH of 3.9. That's frightening when we remember that a pH decrease of 2.5 means that acid concentrations increase more than 100 times. We saw in Chapter 5 that such acid rain can dissolve buildings and statues.

Mine drainage also adds to the "pH pollution" of our environment. Any kind of mining exposes buried sulfur compounds to the oxygen of the air. The oxygen attacks the sulfur and converts it to sulfuric acid, which then washes into streams, rivers, and lakes. Salmon, trout, and other kinds of fish can't live in water with a pH below 5.5.

Maintaining a certain pH, or correcting a too-high or a too-low pH, is thus very important. To do it, we have to be able to measure pH in the first place. One way to measure pH is with a pH meter. This instrument has a glass probe that is inserted into the solution to be measured. The pH can be read on a scale or digital display.

indicator

Another way of measuring pH is with indicators or pH paper. *Indicators* are dyes that change color at a certain pH value. The first indicators were vegetable dyes and we still use some of these, such as litmus. Impregnating this dye onto absorbent paper and cutting it into strips gives us litmus paper, commonly used in the chemistry laboratory. Litmus paper turns red in acid and blue in base, but it doesn't tell what the pH is. We can use vegetable dyes to observe the behavior of acids and bases in a simple kitchen experiment (see Box below).

Now a great variety of synthetic dyes exist which change color over a wide range of pH values. A mixture of these dyes, covering a desired pH

THE EFFECT OF pH ON A VEGETABLE DYE

This interesting and colorful kitchen experiment will demonstrate the effects of varying pH on vegetable dyes. Grind up part of a red cabbage in about two cups of water (use a blender if you have one). Strain the pulp out with a sieve or some cheesecloth. Divide the purple liquid into a number of portions. Test foods and household substances like vinegar, lemon juice, detergent, ammonia, drain cleaner (dangerous, so use a small amount), wine, and coffee by adding them to separate portions of the purple liquid. The dye goes through a number of delightful color changes, as shown below. This is one way you can estimate the pH values of some of these substances. If enough things are tried, the set of samples will show very gradual color changes. The dye in red cabbage has these colors for approximate pH values:

pH	2	3 4 5	6	7		8 9	10 11	12 13 14
Color	Red	Red-purple	Purple	Blue-purple	Blue	Blue-green	Green	Yellow

range, impregnated onto paper and cut in strips gives us pH paper. A strip of the paper is dipped into the test solution and compared with a color chart provided with the pH paper. pH paper is less accurate but sometimes more convenient—and certainly less expensive—than a pH meter.

Soil can be tested with pH paper. Acidity is a more common problem than basicity; if a soil is found to be too acidic, lime (CaO) or ground limestone ($CaCO_3$) is usually added to neutralize the acid. A pH meter would be used for more critical measurements, such as blood pH. Chemistry laboratories use both methods, depending on the nature of the experiment.

13.3 THE STRUCTURES OF ACIDS AND BASES

We've seen how acids and bases behave in water, and we've seen how to measure them. Now we'll find out what makes an acid an acid and a base a base.

BRØNSTED-LOWRY ACIDS AND BASES. Up to now, we've seen acids as substances that release H_3O^+ in water, and bases as substances that release OH^- in water. This is formally called the *Arrhenius* concept of acids and bases. A more recent and more general description of acids and bases was proposed independently by J. N. Brønsted and T. M. Lowry in 1923. This description is as follows: An *acid* is a proton donor, and a *base* is a proton acceptor. A *proton* is a hydrogen atom without its electron, or a hydrogen ion. (Since a hydrogen atom consists only of a proton and an electron, removing the electron leaves the hydrogen nucleus, which is a proton.) When a proton is removed from its electron pair on an acid and bonded to an available electron pair on a base, this process is called *proton transfer.* All the acid-base reactions we've seen so far have been proton transfer reactions. The process is shown in Figure 13.4. Looking back at Figure 13.1, we see that the ionization of HCl in water is proton transfer, with the HCl as the acid and water as the base.

Arrhenius acid

Arrhenius base

Brønsted-Lowry acid

Brønsted-Lowry base

proton

proton transfer

At the very least, an acid must contain a hydrogen, and a base must have an available electron pair. However, the converse isn't always true: containing a hydrogen or an available electron pair doesn't guarantee that a substance will act as an acid or base.

We see from Figure 13.4 that proton transfer is like a game of "musical protons," with electron pairs on an acid and base competing for a single proton. The "winner" that holds the proton more strongly is the base in that relationship. When the reaction is over, a new acid and base have been created: HB^+ now has a hydrogen, and A^- has an available electron pair.

We can illustrate proton transfer with this general equation:

$$HA + B \longrightarrow A^- + BH^+$$

Here, HA represents a Brønsted-Lowry acid, and B a Brønsted-Lowry base.

FIGURE 13.4
Illustration of proton transfer

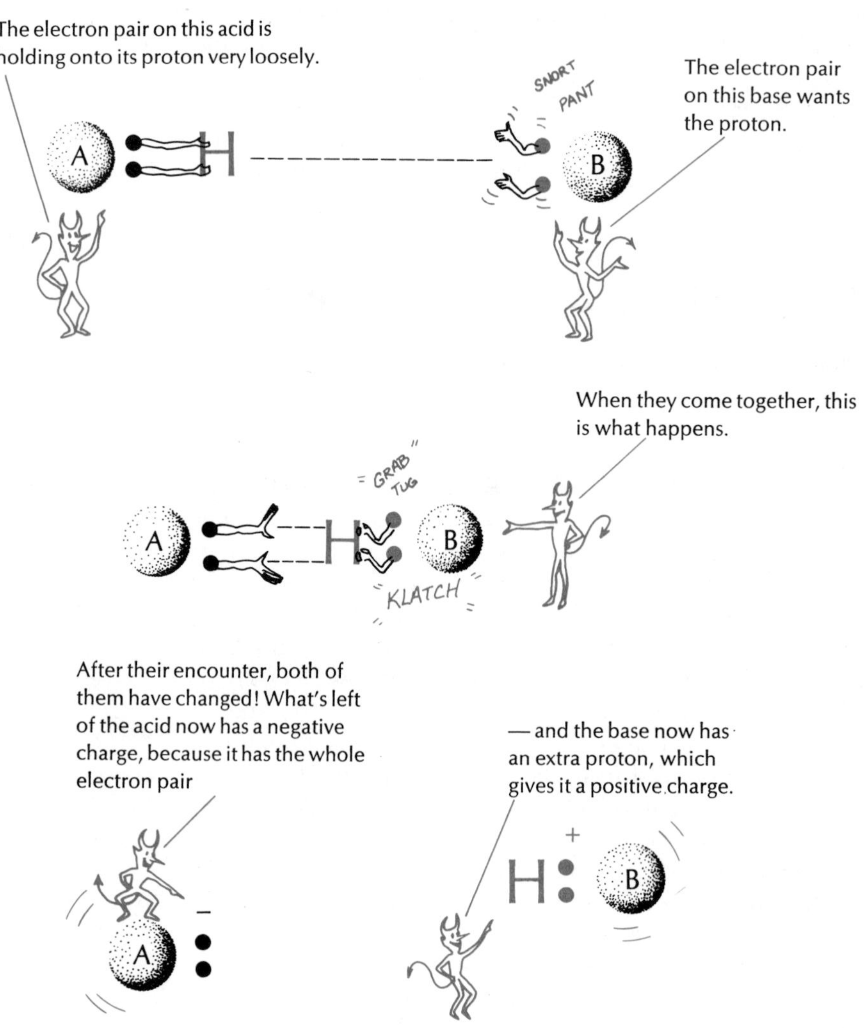

The electron pair on this acid is holding onto its proton very loosely.
SNORT
PANT
The electron pair on this base wants the proton.
A
H
B
When they come together, this is what happens.
GRAB
TUG
KLATCH
A
H
B
After their encounter, both of them have changed! What's left of the acid now has a negative charge, because it has the whole electron pair
— and the base now has an extra proton, which gives it a positive charge.
A
H
B

The acid, HA, forms its *conjugate base,* A^-, by donating a proton. And the base, B, forms its *conjugate acid,* BH^+, by accepting a proton. An acid and its conjugate base (or a base and its conjugate acid) constitute a *conjugate pair.* For the above conjugate pairs, it's equally correct to say that HA is the conjugate acid of A^- and that B is the conjugate base of BH^+.

conjugate base

conjugate acid

conjugate pair

Table 13.4 shows some common conjugate pairs of acids and bases, arranged in order of strength. The stronger the acid, the weaker its conjugate base, and vice versa.

At the top of the table, we have the four strong acids, $HClO_4$, HCl, HNO_3, and H_2SO_4. The first three lose their protons completely to form hydronium

TABLE 13.4
Some common conjugate acid-base pairs

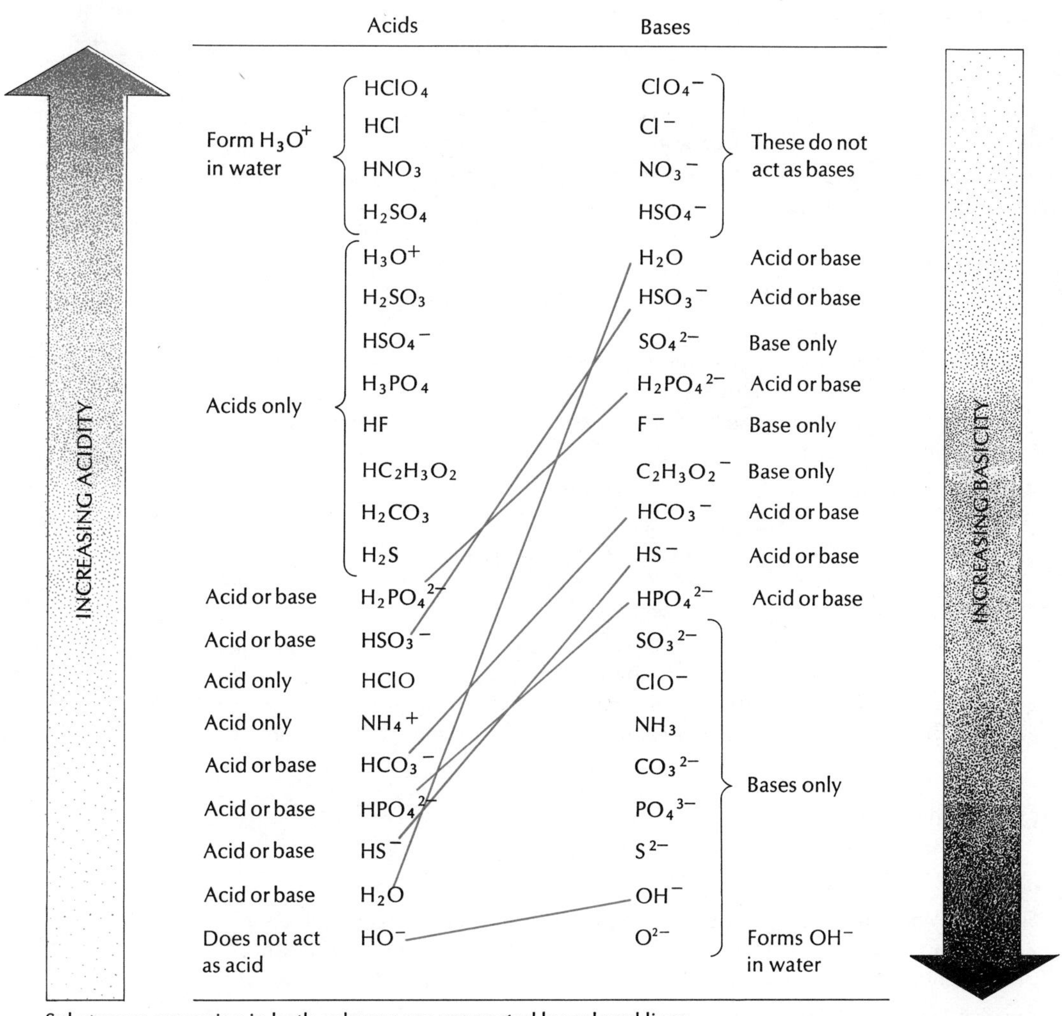

	Acids	Bases	
Form H_3O^+ in water	$HClO_4$	ClO_4^-	These do not act as bases
	HCl	Cl^-	
	HNO_3	NO_3^-	
	H_2SO_4	HSO_4^-	
Acids only	H_3O^+	H_2O	Acid or base
	H_2SO_3	HSO_3^-	Acid or base
	HSO_4^-	SO_4^{2-}	Base only
	H_3PO_4	$H_2PO_4^{2-}$	Acid or base
	HF	F^-	Base only
	$HC_2H_3O_2$	$C_2H_3O_2^-$	Base only
	H_2CO_3	HCO_3^-	Acid or base
	H_2S	HS^-	Acid or base
Acid or base	$H_2PO_4^{2-}$	HPO_4^{2-}	Acid or base
Acid or base	HSO_3^-	SO_3^{2-}	Bases only
Acid only	$HClO$	ClO^-	
Acid only	NH_4^+	NH_3	
Acid or base	HCO_3^-	CO_3^{2-}	
Acid or base	HPO_4^{2-}	PO_4^{3-}	
Acid or base	HS^-	S^{2-}	
Acid or base	H_2O	OH^-	
Does not act as acid	HO^-	O^{2-}	Forms OH^- in water

Substances appearing in both columns are connected by colored lines.

ions and their conjugate bases as shown in Table 13.4. Notice that H_2SO_4, a diprotic acid, loses one hydrogen completely to form its conjugate base, HSO_4^- (hydrogen sulfate ion), which is a weaker acid than H_2SO_4. Here is the equation for the proton transfer between HSO_4^- and water:

$$\underset{\text{Acid}}{HSO_4^-} + \underset{\text{Base}}{H_2O} \longrightarrow \underset{\text{Conj. base}}{SO_4^{2-}} + \underset{\text{Conj. acid}}{H_3O^+}$$

This reaction happens only to a limited extent.

We see that the conjugate bases of the four strong acids—ClO_4^-, Cl^-, NO_3^-, and HSO_4^-—are such weak bases that they don't act as bases at all. This illustrates the concept that, the stronger the acid, the weaker its conjugate base, and vice versa. At the other end of the list, H_2O, the weakest acid on the list, has OH^- as its conjugate base, which is the strongest base on the list.

We can use this table to predict whether a water solution of a substance will be acidic or basic. The strong acids, and anything marked "acid only," will give water solutions that are acidic. Anything marked "base only" will give a basic solution.

EXAMPLE 13.8: Would you expect a solution of NH_4Cl in water to be acidic, basic, or neutral? Write the equation for proton transfer with water, if any, and label conjugate pairs.

Solution: Dissolving NH_4Cl in water produces a solution containing NH_4^+ and Cl^-. We look for these two ions in Table 13.4. NH_4^+ appears in the acids column; Cl^- is the conjugate base of HCl, a strong acid, and is therefore not itself going to act as a base.

Answer: The solution will be acidic. Equation:

$$\underset{\text{Acid}}{NH_4^+} + \underset{\text{Base}}{H_2O} \longrightarrow \underset{\text{Conj. base}}{NH_3} + \underset{\text{Conj. acid}}{H_3O^+}$$

We see that when an acid reacts with water, a proton is transferred from the acid to water, forming the conjugate base of the acid, and hydronium ion. For any of the weak acids, this proton transfer won't take place to a large extent; that is, we'd find some hydronium ions, but not many, in a water solution of a weak acid. ■

EXAMPLE 13.9: State whether solutions of these would be acidic, basic, or neutral. Write equations for the proton transfer, if any, and label conjugate pairs: (a) Na_2CO_3; (b) NaCl.

Solution: We don't find Na^+ anywhere on the list, so that we assume it's neither acidic nor basic; CO_3^{2-} we find among the bases; and Cl^- is neither an acid nor a base.

Answer: (a) basic. (When a base reacts with water, a proton is transferred from water to the base, forming hydroxide ion and the conjugate acid of the base.)

$$\underset{\text{Base}}{CO_3^{2-}} + \underset{\text{Acid}}{H_2O} \longrightarrow \underset{\text{Conj. acid}}{HCO_3^-} + \underset{\text{Conj. base}}{OH^-}$$

(b) neutral. ■

We can predict the acidity or basicity of solutions where there is only one acid or base, or none. With only the information given in this book, we wouldn't be able to predict acidity or basicity of a species marked "acid or base" (for example, $H_2PO_4^-$). Also, for a compound such as NH_4F, where NH_4^+ is a weak acid and F^- is a weak base, we'd need more information to predict whether the acid or base would win. But we can still write all the proton transfer reactions with water.

EXAMPLE 13.10: Write proton transfer equations that show HPO_4^{2-} acting as an acid and as a base with water.

Solution: When acting as an acid, it loses a proton to water; when a base, it gains a proton from water.

Answer: Acid: $HPO_4^{2-} + H_2O \longrightarrow PO_4^{3-} + H_3O^+$
Base: $HPO_4^{2-} + H_2O \longrightarrow H_2PO_4^- + OH^-$ ■

Substances such as $H_2PO_4^-$ and HPO_4^{2-} that can act either as an acid or as a base are called *amphoteric* substances.

amphoteric

LEWIS ACIDS AND BASES. An even more general description of acids and bases is that proposed by G. N. Lewis, of Lewis-structure fame. According to the Lewis concept, an *acid* is an electron-pair acceptor, and a *base* is an electron-pair donor. All acids and bases covered by the Brønsted-Lowry theory are also covered by the Lewis theory. A proton can certainly accept an electron pair, so a proton donor is also an electron-pair acceptor. By the same token, a proton acceptor must have an available electron pair and therefore is also an electron-pair donor.

Lewis acid

Lewis base

There are many substances that are defined as acids in the Lewis sense but not in the Brønsted-Lowry sense. One of the most common of these is boric acid, often written H_3BO_3. Its Lewis structure is:

```
        H
        |
       |O|
        |
  H—Ō—B—Ō—H
    ‾     ‾
```

The structure of this acid is better described with the formula $B(OH)_3$.

We saw in Chapter 9 that boron and other members of Group IIIA can participate in coordinate covalent bonds, because they have room to accept a whole electron pair. Boric acid reacts with a base, such as hydroxide ion, not by donating a proton, but by accepting an electron pair to form $B(OH)_4^-$.

```
┌           H          ┐ −
│           |          │
│          |O|         │
│           |          │
│  H—Ō—B—Ō—H  │
│    ‾   |   ‾         │
│          |O|         │
│           |          │
└           H          ┘
```

The equation for this reaction is:

$$\underset{\text{Lewis acid}}{B(OH)_3} + \underset{\text{Lewis base}}{OH^-} \longrightarrow \underset{\text{Borate ion}}{B(OH)_4^-}$$

As we might expect, $Al(OH)_3$ also reacts as a Lewis acid. We already know that it acts as a Brønsted-Lowry base. $Al(OH)_3$ is amphoteric. Another amphoteric hyroxide is $Zn(OH)_2$. These two hydroxides react with base as follows:

$$Al(OH)_3(s) + OH^-(aq) \longrightarrow \underset{\text{Aluminate ion}}{Al(OH)_4^-(aq)}$$

$$Zn(OH)_2(s) + 2\,OH^-(aq) \longrightarrow \underset{\text{Zincate ion}}{Zn(OH)_4^{2-}(aq)}$$

Although the Lewis acid-base theory is more general than the Brønsted-Lowry theory, we'll find the latter more convenient to use in this chapter.

ACID STRUCTURE RELATED TO STRENGTH. As we know, HCl is an acid; H_2O is a weaker acid; NH_3 is a base; and CH_4 is neither an acid nor a base. To be an acid, a compound's hydrogen must be "loose." Although other energies are involved, one measure of "looseness" is bond energy. The acidity of the halogen acids increases in the order HF, HCl, HBr, HI. This is the order of decreasing bond energy. For oxyacids, where the acidic hydrogen is always bonded to oxygen, anything that will weaken the O—H bond will make the acid stronger. Let's consider the series of chlorine oxyacids.

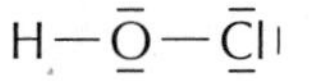

H—O—Cl   Hypochlorous

H—O—Cl—O   Chlorous

H—O—Cl(—O)—O   Chloric

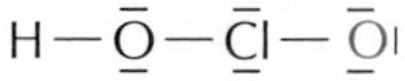

H—O—Cl(—O)(—O)—O   Perchloric

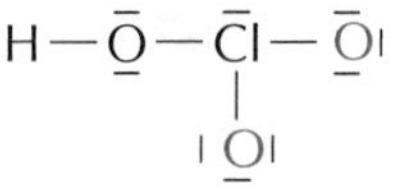

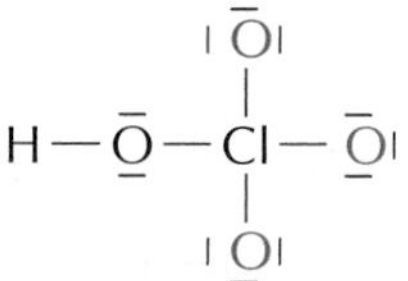

The acid strength increases in this order: HClO, $HClO_2$, $HClO_3$, $HClO_4$. The acid strength increases with the number of oxygens (in color) that aren't also bonded to hydrogen. $HClO_4$ has three such oxygens; HClO has none. Now we can see why H_2SO_4 is strong, whereas H_2SO_3 and H_3PO_4 are weak.

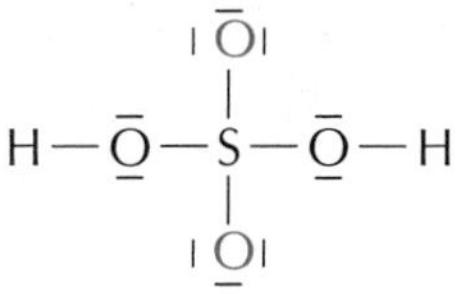

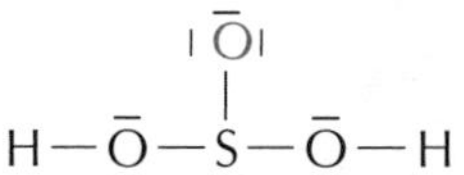

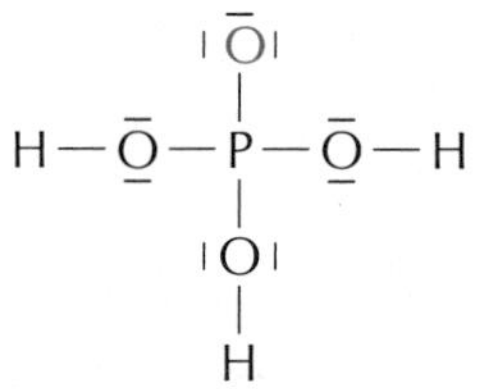

H—O—S(—O)(—O)—O—H

H—O—S(—O)—O—H

H—O—P(—O)(—O—H)—O—H

Sulfuric acid has two oxygens that aren't bonded to hydrogen; sulfurous and phosphoric acids each have only one. Extremely strong oxyacids have three; strong oxyacids have two; weak oxyacids have one; and extremely weak oxyacids have none.

It might seem as if we're neglecting the structure of bases and their strengths in this discussion. This isn't so if we remember that the stronger the acid, the weaker its conjugate base. Using this reasoning, we see that the base strength of polyatomic ions *decreases* with the number of oxygens that aren't bonded to hydrogen in their conjugate acids. That is, the base strength decreases in the series ClO^-, ClO_2^-, ClO_3^-, ClO_4^-. By the same token, HSO_3^- and $H_2PO_4^-$ are stronger bases than HSO_4^-.

Oxygen is the second most electronegative element, which makes it more electronegative than the central atom in any oxyacid. The more oxygens that are attached to the central atom, the more polar bonds there are. The more polar bonds there are, the more positive the central atom becomes. The more positive the central atom becomes, the weaker the O—H bond, and the stronger the acid.

If an acid has more than one proton to lose, the first is always much stronger than the second. H_2SO_4 is a strong acid, but HSO_4^- is weak. H_2SO_3 is weak, but HSO_3^- is weaker still. This is because it takes more energy to remove a proton from a negatively charged ion, as we'd expect.

Organic acids—that is, carbon-based acids—have this characteristic grouping:

```
   |O̅
    ‖
 —C—O̅—H
```

(This is often abbreviated —COOH.) Most organic acids are weak, because they have only one oxygen that isn't bonded to hydrogen. Here are the structures of three common acids with the —COOH grouping.

```
     |O̅                   H  O̅|
      ‖                   |  ‖
  H—C—O̅—H             H—C—C—O̅—H
                          |
                          H
```

Formic acid (the simplest)

Acetic acid

```
               H
               |
       |O̅  H  |O|  H  O̅|
        ‖  |   |   |  ‖
   H—O̅—C—C—C—C—C—O̅—H
           |   |   |
           H   |   H
               |
           |O=C
               |
              |O—H
```

Citric acid

In these, the acidic hydrogens are marked in color. The —OH group where carbon is not bonded to another oxygen isn't appreciably acidic, as we'd expect. Although organic acids may have seemingly complicated structures, the important part is the —COOH group.

In Chapter 14, we'll see that most reactions involving acids and bases are proton transfer reactions.

REVIEW QUESTIONS

Strong and Weak Acids and Bases

1. What is *hydronium ion?* When does it form?
2. What are the *strong acids?* Write ionic equations for their ionizations in water.
3. What are some *strong bases?* What do they have in common?
4. Besides the fact that they are bases, how are the strong bases different from the strong acids?
5. What do we mean by *deliquescent?* What strong bases are deliquescent?
6. Do we find any oxide ions in a water solution of a metal oxide? Explain, and write an appropriate ionic equation.
7. Write the equation for water ionizing in water.
8. How much hydronium ion and hydroxide ion are present in normal water at room temperature?
9. What is a *weak acid?* What is the most commonly used weak acid?
10. What is a *weak base?* Write an ionic equation for the most common weak base ionizing in water.
11. Write an ionic equation showing the carbonate ion acting as a weak base in water.

Measuring Acids and Bases

12. How can we find $[H_3O^+]$ for a strong acid and $[OH^-]$ for a strong base?
13. What is the relationship between $[H_3O^+]$ and $[OH^-]$ for any acid or base solution?
14. What is pH? What does a pH of 7.00 mean? How does pH tell us whether a solution is acidic or basic?
15. How do we find pH when the coefficient of the hydronium-ion concentration is 1?
16. How do we use Figure 13.3 to find pH, hydronium-ion concentration, or hydroxide-ion concentration? When can we use this scale?
17. Why do we multiply the molarity of $Ca(OH)_2$ by 2 to get the hydroxide-ion concentration?
18. What should be the product of the hydronium-ion and hydroxide-ion concentrations of the same solution?
19. Define pH in terms of logarithms. How do we find pH with a calculator?
20. How do we find the pH of a basic solution from the hydroxide-ion concentration?
21. Explain how to find pH using a log table.
22. How do we convert from pH to hydronium- and hydroxide-ion concentration using a calculator? How do we do it with a log table?
23. What happens if blood falls outside the pH range of 7.35 to 7.45?
24. Why do we need to measure pH? What are two ways of measuring pH?
25. What is an *indicator?* Give an example.
26. How can a too-acidic soil be corrected?

The Structures of Acids and Bases

27. Give the Brønsted-Lowry definitions of acids and bases, and illustrate.
28. What is meant by *proton* in the Brønsted-Lowry concept? What is *proton transfer?* Give an example.
29. What is a *conjugate pair?* Give an example, and label each half.
30. How are the strengths of acids and their conjugate bases related? Give examples of this from Table 13.4.
31. Does sulfuric acid lose both protons completely? Explain.
32. How can we use Table 13.4 to predict whether a water solution of a substance will be acidic or basic?
33. Can we predict acidity or basicity of all solutions using Table 13.4? Explain.
34. What is an *amphoteric* substance? Write equations showing an amphoteric substance acting as an acid and as a base with water.
35. What is a *Lewis acid?* Give an example of a substance that is a Lewis acid and not a Brønsted acid.
36. What is a *Lewis base?* Give an example.
37. What energies are involved in determining the strength of an acid?
38. How does the Lewis structure of an oxyacid give us a rough idea of how strong it is?
39. Which is the stronger base, ClO_4^- or ClO^-? How do we know?
40. When an acid has two or more protons to lose, what are the relative acidities of each successive proton? Why?
41. Why are most organic acids weak?

EXERCISES

Set A (Answers at back of book. Asterisks mark the more difficult exercises.)

Note: In all exercises having to do with fractional pH, use either the calculator or the log table method, or both, as specified by your instructor.

1. Write ionic equations for the dissolving of each of these in water. After each, state whether the reaction would take place completely or only partially.
   a. $HNO_3(l)$ c. $HF(g)$ e. $NaOH(s)$
   b. $K_2O(s)$ d. $NH_3(g)$ f. $K_2CO_3(s)$
2. Assume that S^{2-} behaves the same way as O^{2-}, and write an ionic equation for the reaction of Na_2S with water.
3. Give the ion concentration requested for each of the following.
   a. 0.035 *M* HNO_3:$[H_3O^+]$
   b. 0.15 *M* KOH:$[OH^-]$
   c. 0.0021 *M* HCl:$[H_3O^+]$
   d. 6.8×10^{-14} *M* $Ca(OH)_2$:$[OH^-]$
4. From the pH values in Table 13.3, and using the scale in Figure 13.3, give $[H_3O^+]$ and $[OH^-]$ for these:
   a. sour milk c. household lye e. pure water, 25°C
   b. black coffee d. seawater
5. Using Figure 13.3, find the pH of these solutions, and classify each as acidic or basic:
   a. 0.100 *M* NaOH c. 0.0050 *M* $Ca(OH)_2$
   b. 1.0 *M* HCl d. 0.00010 *M* HNO_3
6. Find the pH of the following, and classify each as acidic or basic.
   a. 0.036 *M* $HClO_4$
   b. 0.15 *M* KOH
   c. a solution whose $[H_3O^+]$ is 7.5×10^{-5} *M*
   d. a solution whose $[OH^-]$ is 4.3×10^{-8} *M*
   e. a solution whose $[H_3O^+]$ is 8.7×10^{-9} *M*
   f. a solution whose $[OH^-]$ is 0.20 *M*
7. Fill in the following table.

$[H_3O^+]$	$[OH^-]$	$[H_3O^+]$	$[OH^-]$
10^{-7}	______	2.3×10^{-4}	______
______	10^{-1}	______	9×10^{-8}
______	0.045	0.12	______

8. Find the hydronium-ion and hydroxide-ion concentrations of solution having these pH values:
   a. 1.26 c. 8.53
   b. 6.54 d. 10.49
9. A blood sample is found to have a hydronium-ion concentration of 5.0×10^{-8} *M*.
   a. Calculate its pH and hydroxide-ion concentration.
   b. Is the pH within the normal range?

*10. Forest soil has a hydronium-ion concentration of about 3×10^{-5} *M*; desert soil, about 1×10^{-10} *M*. Suggest a reason for this difference.

11. State the effect of litmus paper on each of these (red, blue, no effect).
    a. a solution of HCl
    b. a solution having a pH of 9.3
    c. rainwater
    d. a solution of dishwasher detergent
12. Which of these could be used to correct a solution that has too high a pH? (More than one may be correct.)
    a. vinegar d. limestone
    b. washing soda (Na_2CO_3) e. HCl
    c. drain cleaner
13. Fill in the following table without looking at Table 13.4.

Conjugate Acid	*Conjugate Base*	*Conjugate Acid*	*Conjugate Base*
$HC_2H_3O_2$	______	______	SO_4^{2-}
______	H_2O	HCO_3^-	______
NH_4^+	______	______	OH^-
______	F^-	HPO_4^{2-}	______
H_3O^+	______	HNO_3	______
______	CO_3^{2-}	______	NH_3

14. Write ionic equations for the proton transfer reactions of each of these with water. Label the acid and base and identify the conjugate pairs.
    a. CO_3^{2-} b. HSO_4^- c. HF d. S^{2-}
15. Consult Table 13.4, and arrange these in order of increasing acid strength: H_3PO_4, HClO, H_2O, HSO_4^-.
16. State whether a water solution of each of these would be acidic, basic, or neutral, or whether no prediction can be made. For the acidic or basic solutions, write ionic equations and identify the conjugate pairs.
    a. NH_4NO_3 c. $NaHCO_3$ e. $CaCl_2$
    b. Na_3PO_4 d. $KC_2H_3O_2$ f. Li_2O
17. Write proton transfer reactions for each of these amphoteric substances, showing it reacting as an acid and as a base, with water. Identify the conjugate pairs.
    a. $H_2PO_4^{2-}$ b. HSO_3^-

*18. Carbide ion (C^{2-}) is like oxide ion in that it is a strong base that's completely converted to hydroxide ions in water. This happens in the commercial preparation of acetylene (C_2H_2), where calcium carbide (CaC_2) is added to water. Write an equation showing the proton transfer between water and carbide ion.

*19. The odor of hydrogen cyanide, HCN, can usually be detected over a water solution of KCN. (WARNING: Don't try it!) Explain. Write an appropriate equation.

20. Give two examples of:
   a. an acid weaker than H_2S
   b. an oxide whose water solution is basic
   c. a substance whose water solution is neutral

21. Classify these naturally occurring substances as acids, bases, or neither.
   a. lime, CaO
   b. bauxite ore, Al_2O_3
   c. phosphate rock, $Ca_3(PO_4)_2$

22. Classify each of these as a Brønsted acid, Brønsted base, Lewis acid, or Lewis base. (More than one classification may apply.)
   a. BF_3 b. $Al(OH)_3$ c. OH^-

*23. Identify the Lewis acid and Lewis base in this reaction.

$$AlCl_3 + Cl^- \longrightarrow AlCl_4^-$$

24. On the basis of Lewis structures, predict which acid in these pairs is the stronger.
   a. $HClO_3$ or H_2SO_3
   b. $HC_2H_3O_2$ or HClO
   c. H_3PO_3 or H_2SO_3
   d. HBrO or $HBrO_2$
   e. $HClO_4$ or H_2SO_4
   f. HNO_3 or H_2SO_3

Set B (Answers not given. Asterisks mark the more difficult exercises.)

Note: In all exercises having to do with fractional pH, use either the calculator or the log table method, or both, as specified by your instructor.

1. Write ionic equations for the reactions of these with water. After each, state whether the reaction would take place completely or only partially.
   a. CaO
   b. H_2SO_4
   c. HBr
   d. NaOH
   e. $HC_2H_3O_2$
   f. HClO

*2. H^- is a strong base that is completely converted to OH^- when added to water. Write an ionic equation for this reaction.

3. Give the ion concentration requested for each of these.
   a. 0.223 *M* NaOH:$[OH^-]$
   b. 2.0 *M* $HClO_4$:$[H_3O^+]$
   c. 0.00102 *M* $Ba(OH)_2$:$[OH^-]$
   d. 4.5 *M* HBr:$[H_3O^+]$

4. From the pH values given in Table 13.3, and using the scale in Figure 13.3, give $[H_3O^+]$ and $[OH^-]$ for these.
   a. battery acid
   b. milk of magnesia
   c. liquid bleach
   d. urine
   e. soda water

5. Using Figure 13.3, find the pH of these solutions, and classify each as acidic or basic.
   a. 0.0100 *M* HNO_3
   b. 0.100 *M* $HClO_4$
   c. 0.00005 *M* K_2O
   d. 0.050 *M* $Ba(OH)_2$

6. Find the pH of these, and classify each as acidic or basic.
   a. 0.00450 *M* HCl
   b. 0.0250 *M* NaOH
   c. a solution whose $[OH^-]$ is 3.4×10^{-2} *M*
   d. a solution whose $[H_3O^+]$ is 0.500 *M*
   e. a solution whose $[OH^-]$ is 0.125 *M*
   f. a solution whose $[H_3O^+]$ is 0.0036 *M*

7. Fill in this table.

$[H_3O^+]$	$[OH^-]$	$[H_3O^+]$	$[OH^-]$
______	1.0×10^{-3}	7.5×10^{-6}	______
10^{-2}	______	______	4×10^{-7}
6.3×10^{-9}	______	0.1	______

8. Find the hydronium- and hydroxide-ion concentrations of solutions having these pH values, and classify each as acidic or basic.
   a. 11.69
   b. 3.8
   c. 0.22
   d. 9.5

9. The pH of urine may span a range of 5.5 to 7.0, depending on what we've been eating. What range of hydronium-ion concentration does this represent?

*10. A soil sample is found to have a hydronium-ion concentration of 4.4×10^{-4} *M*.
   a. Calculate the pH of the soil.
   b. Is the soil too acidic, too basic, or just right?
   c. If not just right, what could be done to correct it?

11. State the effect on litmus paper of each of these (red, blue, no effect).
   a. lemon juice
   b. household ammonia
   c. a solution of table salt
   d. a solution having a pH of 4.3

12. Which could be used to correct a too-low pH? (More than one may be correct.)
   a. club soda
   b. household ammonia
   c. vinegar
   d. milk of magnesia
   e. battery acid

13. Fill in this table without looking at Table 13.3:

Conjugate Acid	*Conjugate Base*	*Conjugate Acid*	*Conjugate Base*
OH^-	______	______	HSO_3^-
______	Cl^-	$H_2PO_4^-$	______
______	HS^-	______	ClO^-
H_2CO_3	______	HSO_4^-	______

14. Write ionic equations for the proton transfer reactions of each of these with water. Label the acid and base and identify the conjugate pairs.
   a. $HClO_3$ c. PO_4^{3-}
   b. H_2S d. SO_3^{2-}
15. Consult Table 13.4, and arrange these in order of increasing base strength: SO_4^{2-}, HCO_3^-, S^{2-}, O^{2-}.
16. State whether a water solution of each of these would be acidic, basic, or neutral, or whether no prediction can be made. For the acidic or basic solutions, write ionic equations and identify the conjugate pairs.
   a. NaF c. $(NH_4)_2CO_3$ e. CaO
   b. $KClO_4$ d. $KHSO_4$ f. $LiC_2H_3O_2$
17. Write proton transfer reactions for each of these amphoteric substances, showing each reacting as an acid and as a base, with water. Identify the conjugate pairs.
   a. HCO_3^- b. HS^-

*18. Chlorine bleach, a solution of sodium hypochlorite in water, feels slippery. Explain.
19. Give two examples of each.
   a. a base stronger than Na_2CO_3
   b. a salt whose water solution is acidic
   c. an acid stronger than $HC_2H_3O_2$
*20. Potash, K_2CO_3, is present in wood ashes. Caustic potash, KOH, can be made by adding water to potash. Write an appropriate equation showing a proton transfer with water.
21. Classify these naturally occurring substances as acids, bases, or neither.
   a. zinc ore, ZnS
   b. Chile saltpeter, KNO_3
   c. boiler scale, $CaCO_3$
22. Classify each of these as a Brønsted acid, Brønsted base, Lewis acid, or Lewis base. (More than one classification may apply.)
   a. H_3PO_4 b. H_3BO_3 c. H_2O
*23. Identify the Lewis acid and Lewis base in this reaction.

$$ZnCl_2 + 2\,Cl^- \longrightarrow ZnCl_4^{2-}$$

24. On the basis of Lewis structures, predict which acid in the following pairs is the stronger of the two.
   a. HClO or $HClO_3$ d. $HClO_2$ or H_2SO_4
   b. H_3PO_4 or H_3PO_3 e. $HBrO_2$ or $HBrO_4$
   c. HNO_2 or HNO_3 f. H_2SO_3 or HClO

CHAPTER 14

Reactions in Solution

LEARNING OBJECTIVES

After studying this chapter, you should be able to:

1. Supply a correct definition, explanation, or example for each of these:

net ionic equation	titration
complete ionic equation	endpoint
	standardization
spectator ion	normality
filtration	equivalent
centrifugation	equivalent weight
volumetric flask	

2. Write complete and net ionic equations given the reactants, and translate molecular equations into ionic equations.
3. Choose appropriate solutions to precipitate or dissolve a given substance, or predict the outcome of a reaction, using the solubility rules and the list of acids and their conjugate bases.
4. Explain how acids and bases may dissolve insoluble substances, including gases. Write complete and net ionic equations for examples of these.
5. Describe how to prepare a solution, and calculate the mass of a given solute, the concentration of a solution, or its volume, given the other two quantities.
6. Perform dilution calculations where either molarity or volume of the dilute or concentrated solution is unknown, given the other three quantities.
7. Perform titration calculations: given any three of concentration and volume of acid and base solution, calculate the other. Or given two out of three of mass of solid acid or base, volume or concentration of base or acid solution, calculate the other.
8. Given an acid or base and how it will react, calculate its equivalent weight or convert between molarity and normality of its solutions.

A solution is the chemist's favorite medium for carrying out reactions. Solutions are convenient to prepare, handle, and measure. And in solutions, the reactant ions and molecules can move around and come into contact with each other, which they must do in order to react. Learning how to prepare solutions, understanding how reactions take place in them, and doing calculations with them is important in chemistry.

In this chapter, we'll look at acid-base and precipitation reactions and find out how to write equations describing them. We'll also learn how to calculate the concentrations of solutions and how to prepare a solution of a specific concentration.

14.1 ACID-BASE REACTIONS IN SOLUTION

We learned in Chapter 13 how acids and bases ionize in water, and in Chapter 6, we wrote equations that described reactions of acids and bases. Now we'll look at acid-base reactions in more detail.

REACTIONS BETWEEN STRONG BASES AND STRONG ACIDS. We know that a strong acid solution contains hydronium ions and a strong base solution contains hydroxide ions. When these solutions are mixed, this reaction takes place:

$$H_3O^+ + OH^- \longrightarrow 2\ H_2O$$

(In this chapter, we'll often omit the physical states (*aq*) and (*l*), because we assume these substances to be in aqueous solution.) In Chapter 13, we learned how to write ionic equations. The equation above is a *net ionic equation,* which shows only the ions that are going into or out of solution. Here, hydronium and hydroxide ions go out of solution as they react to form water.

net ionic equation

The complete ionic equation for this reaction depends on which acid and which base are used. A *complete ionic equation* shows all substances in the form that they predominantly have in solution. The complete ionic equation for the reaction between solutions of HCl and of NaOH is written like this:

complete ionic equation

$$H_3O^+ + Cl^- + Na^+ + OH^- \longrightarrow 2\ H_2O + Na^+ + Cl^-$$

Na^+ and Cl^- appear on the right as well as the left of the equation, because they haven't changed. Ions that are present during a reaction but that don't themselves react are called *spectator ions.* If we remove these spectator ions from the equation, then we're back to our original net ionic equation.

spectator ion

We'll introduce a shorthand notation that will make writing ionic equations a bit simpler. It's convenient to abbreviate the hydronium ion, H_3O^+, as H^+. Sometimes this is called "hydrogen ion," even though we know that it's not really a naked hydrogen ion. We use this notation in Table 14.1, which shows complete and net ionic equations for some strong acid-base reactions.

When acids and bases react, there isn't much to see. If, however, we mix 6-molar hydrochloric acid and 6-molar sodium hydroxide together, we can feel the beaker getting warm. Acid-base reactions are exothermic. Mixing concentrated acids and bases together is dangerous, because so much heat is generated that the solution usually splatters and can splash out of its container.

REACTIONS OF WEAK ACIDS AND WEAK BASES.

Weak acids and bases exist predominantly in the molecular form, so we write them that way in complete and net ionic equations. Here's the equation for the reaction between aqueous ammonia and acetic acid:

$$NH_3 + HC_2H_3O_2 \longrightarrow NH_4^+ + C_2H_3O_2^-$$

Here, ammonium and acetate ions are formed as a proton is transferred from the acetic acid molecule to the ammonia molecule. This equation doesn't have any spectator ions. The net ionic equation is the same as the complete ionic equation.

Table 14.2 shows the ionic equations for some reactions between weak acids and weak bases, strong acids and weak bases, and weak acids and strong bases. In the latter two cases, we write the strong acids or bases in the ionic form.

A strong acid donates all its protons to a weak base, and a strong base takes all the protons away from a weak acid. These reactions take place completely, but reactions between weak acids and weak bases don't. Some of the protons, but not all, are transferred from a weak acid to a weak base. We saw an example of this type of reaction in Chapter 13, where very few hydronium ions are produced in the proton transfer between acetic acid, a weak acid, and water, a weak base.

TABLE 14.1
Complete and net ionic equations for some strong acid-base reactions

Complete Ionic Equation	*Net Ionic Equation*
$HSO_4^- + H^+ + Na^+ + OH^- \longrightarrow H_2O + Na^+ + HSO_4^-$	$H^+ + OH^- \longrightarrow H_2O$[a]
$H^+ + NO_3^- + K^+ + OH^- \longrightarrow H_2O + K^+ + NO_3^-$	$H^+ + OH^- \longrightarrow H_2O$
$2\,H^+ + 2\,ClO_4^- + Ba^{2+} + 2\,OH^- \longrightarrow 2\,H_2O + Ba^{2+} + 2\,ClO_4^-$	$2\,H^+ + 2\,OH^- \longrightarrow 2\,H_2O$
$2\,H^+ + 2\,Cl^- + Ca^{2+} + 2\,OH^- \longrightarrow 2\,H_2O + Ca^{2+} + 2\,Cl^-$	$2\,H^+ + 2\,OH^- \longrightarrow 2\,H_2O$

[a]Only the first proton of H_2SO_4 acts as a strong acid. The second is a weak acid.

TABLE 14.2
Ionic equations showing reactions of weak acids and weak bases

Complete Ionic Equation	*Net Ionic Equation*
Weak Acids and Weak Bases	
$HF + 2\ Na^+ + CO_3^{2-} \longrightarrow HCO_3^- + 2\ Na^+ + F^-$	$HF + CO_3^{2-} \longrightarrow HCO_3^- + F^-$
$H_3PO_4 + NH_3 \longrightarrow NH_4^+ + H_2PO_4^-$	Same as complete ionic equation
$HC_2H_3O_2 + 3\ Na^+ + PO_4^{3-} \longrightarrow C_2H_3O_2^- + 3\ Na^+ + HPO_4^{2-}$	$HC_2H_3O_2 + PO_4^{3-} \longrightarrow C_2H_3O_2^- + HPO_4^{2-}$
$HC_2H_3O_2 + 2\ Na^+ + CO_3^{2-} \longrightarrow C_2H_3O_2^- + 2\ Na^+ + HCO_3^-$	$HC_2H_3O_2 + CO_3^{2-} \longrightarrow C_2H_3O_2^- + HCO_3^-$
$HC_2H_3O_2 + Na^+ HCO_3^- \longrightarrow C_2H_3O_2^- + Na^+ + H_2CO_3$ $(CO_2 + H_2O)$	$HC_2H_3O_2 + HCO_3^- \longrightarrow C_2H_3O_2^- + H_2CO_3$ $(CO_2 + H_2O)$
Strong Acids and Weak Bases	
$H^+ + Cl^- + 2\ Na^+ + CO_3^{2-} \longrightarrow Cl^- + 2\ Na^+ + HCO_3^-$	$H^+ + CO_3^{2-} \longrightarrow HCO_3^-$
$H^+ + Cl^- + Na^+ + HCO_3^- \longrightarrow Cl^- + Na^+ + H_2CO_3$ $(CO_2 + H_2O)$	$H^+ + HCO_3^- \longrightarrow H_2CO_3$ $(CO_2 + H_2O)$
$H^+ + HSO_4^- + NH_3 \longrightarrow HSO_4^- + NH_4^+$	$H^+ + NH_3 \longrightarrow NH_4^+$
$H^+ + NO_3^- + PO_4^{3-} \longrightarrow NO_3^- + HPO_4^{2-}$	$H^+ + PO_4^{3-} \longrightarrow HPO_4^{2-}$
Weak Acids and Strong Bases	
$Na^+ + HSO_4^- + Na^+ + OH^- \longrightarrow 2\ Na^+ + SO_4^{2-} + H_2O$	$HSO_4^- + OH^- \longrightarrow SO_4^{2-} + H_2O$
$HC_2H_3O_2 + K^+ + OH^- \longrightarrow C_2H_3O_2^- + K^+ + H_2O$	$HC_2H_3O_2 + OH^- \longrightarrow C_2H_3O_2^- + H_2O$
$Na^+ + HCO_3^- + Na^+ + OH^- \longrightarrow 2\ Na^+ + CO_3^{2-} + H_2O$	$HCO_3^- + OH^- \longrightarrow CO_3^{2-} + H_2O$
$2\ HF + Ba^{2+} + 2\ OH^- \longrightarrow 2\ F^- + Ba^{2+} + 2\ H_2O$	$2\ HF + 2\ OH^- \longrightarrow 2\ F^- + 2\ H_2O$

14.2
FORMING AND DISSOLVING PRECIPITATES

Many practical problems that industrial chemists face involve getting solids into and out of solution. To extract metals from their ores, the ores are usually dissolved first, that is, put into solution. Then the pure metal must be taken out of solution. Cleaning up polluted water means taking unwanted substances out of solution. To clean a rusty tool, we must dissolve the rust without dissolving the tool. In this section, we'll learn how to approach problems like these.

REACTIONS THAT FORM PRECIPITATES. We've seen that an unwanted substance, such as phosphate ion, can be removed from water by precipitating it as part of an insoluble compound. Wanted substances, too, can be removed from solution by precipitation. The ocean contains many ions, which makes it a potential source of chemicals. Magnesium, for one, can be recovered from seawater by precipitating Mg^{2+} as insoluble $Mg(OH)_2$. The net ionic equation for this reaction is:

$$Mg^{2+} + 2\,OH^- \longrightarrow Mg(OH)_2(s)$$

This equation shows magnesium ions and hydroxide ions going out of solution as solid magnesium hydroxide is formed. But how do we know what to add to get magnesium hydroxide to precipitate?

Table 14.3 gives some solubility rules. We need to know which salts are soluble, because these will be sources of the positive or negative ions that we might need to form a precipitate. We need to know which salts are insoluble, so that we will know what precipitates form.

TABLE 14.3
Solubility rules

1. Nitrates are soluble.
2. Compounds of the alkali metals and ammonium ion are soluble.
3. Chlorides, bromides, and iodides are soluble except those of Ag^+, Hg_2^{2+}, and Pb^{2+}. Also HgI_2 is insoluble.
4. Sulfates are soluble except those of Ca^{2+}, Ba^{2+}, Sr^{2+}, Pb^{2+}, and Ag^+.
5. Carbonates and phosphates are insoluble except those of Na^+, K^+, and NH_4^+.
6. Hydroxides are insoluble except those of the alkali metals; $Ca(OH)_2$, $Ba(OH)_2$, and $Sr(OH)_2$ are sparingly soluble.
7. Sulfides are insoluble except those of the alkali metals and ammonium. The sulfides of Ca^{2+}, Mg^{2+}, Ba^{2+}, and Al^{3+} decompose in water.
8. Chromates are insoluble, except those of the alkali metals, Ca^{2+}, and Mg^{2+}.

Note: Also see Appendix D for solubilities of specific compounds.

EXAMPLE 14.1: Suggest a solution of an ionic compound that could be used to precipitate $Mg(OH)_2$ from a solution of $MgCl_2$. Write complete and net ionic equations.

Solution: We need a source of hydroxide ions, so we look in Table 14.3 for a soluble hydroxide. Both Na and K are possibilities (Rule 6); we'll choose NaOH. It and $MgCl_2$ are written in the ionic form because both are soluble.

The complete ionic equation:

$$Mg^{2+} + 2\,Cl^- + 2\,Na^+ + 2\,OH^- \longrightarrow Mg(OH)_2(s) + 2\,Cl^- + 2\,Na^+$$

The ions that are left over after $Mg(OH)_2$ is formed are Na^+ and Cl^-. We don't write "NaCl" in this ionic equation, because NaCl is a soluble salt. Because Na^+ and Cl^- haven't changed during the reaction, they're spectator ions. Removing the spectator ions gives us this net ionic equation:

$$Mg^{2+} + 2\,OH^- \longrightarrow Mg(OH)_2(s)$$

Answer: $Mg^{2+} + 2\,Cl^- + 2\,Na^+ + 2\,OH^- \longrightarrow Mg(OH)_2(s) + 2\,Na^+ + Cl^-$

$Mg^{2+} + 2\,OH^- \longrightarrow Mg(OH)_2(s)$ ■

EXAMPLE 14.2: Suggest a solution of an ionic compound that could be used to remove Hg^{2+} from mercury-polluted water. Write the complete and net ionic equations.

Solution: Here, we're not given a specific compound to precipitate. We need to find an insoluble compound of Hg^{2+}. We find it in Rule 3: HgI_2. The polluted water provides the Hg^{2+}, but we need a source of iodide ions. Rule 3 gives us quite a lot of soluble iodides to choose from; we'll choose KI.

Answer: A KI solution. Complete ionic equation:

$$Hg^{2+} + 2\,K^+ + 2\,I^- \longrightarrow HgI_2(s) + 2\,K^+$$

Net ionic equation:

$$Hg^{2+} + 2\,I^- \longrightarrow HgI_2(s)$$ ■

In this example, there's no negative ion to go with the Hg^{2+} in the polluted water. There had to be negative ions there, but we didn't care about them or even know what they were. This will often be true.

In cases like the one above, we can choose one of many compounds to precipitate an ion. In the laboratory, the choice might be simplified by the compounds that are available or the solutions already prepared. We chose KI in Example 14.2. When chemists want a negative ion, they use sodium, potassium, and sometimes ammonium compounds of the desired negative ion most often. When they want a positive ion, chemists use a nitrate or sometimes a chloride compound of the desired positive ion. In industry, cost is the primary reason for choosing a compound, followed by convenience. To remove phosphate from large amounts of polluted water $Ca(OH)_2$ is used because it's easily obtained from limestone and therefore cheaper than other sources of calcium ions. A chemist would probably choose $CaCl_2$ or $Ca(NO_3)_2$ as a source of calcium ions, because they're more soluble than $Ca(OH)_2$.

Unlike acid-base reactions, precipitation reactions give us something to see. When a precipitate forms, the solution becomes cloudy. Most precipitates are white, but some are colored, and many sulfides are even black. To remove the precipitate, we have to separate it from the liquid part of the solution. We have two ways to do this in the laboratory. One is *filtration*. We pour the liquid through a funnel with a piece of filter paper in it. The solution goes through the paper and the precipitate is trapped on it. Another way is *centrifugation*. A test tube containing the cloudy solution is whirled around very fast in a centrifuge, which packs the small solid particles together at the bottom of the tube. Then the liquid part can be poured off. Precipitation and centrifugation are demonstrated in Figure 14.1.

filtration

centrifugation

In industry, where large amounts of precipitate have to be removed,

FIGURE 14.1
Precipitation of mercury(II) iodide

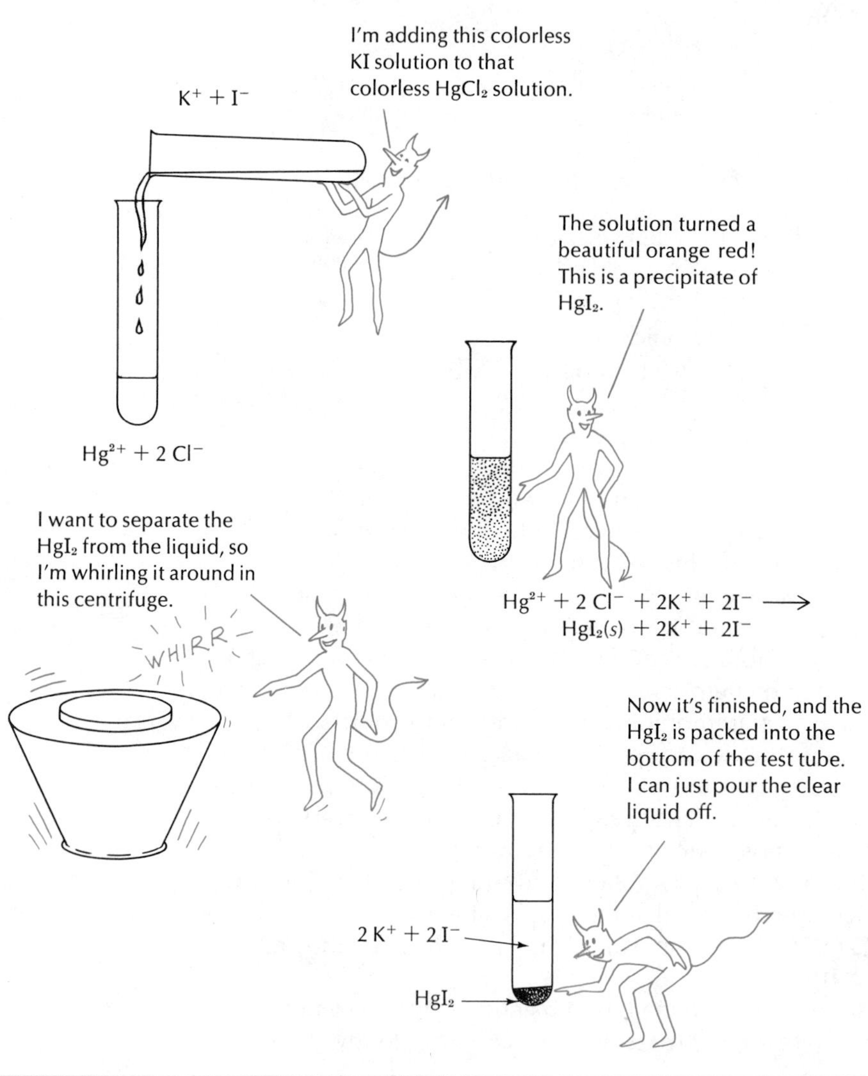
I'm adding this colorless KI solution to that colorless $HgCl_2$ solution.
$K^+ + I^-$
$Hg^{2+} + 2\,Cl^-$
The solution turned a beautiful orange red! This is a precipitate of HgI_2.
$Hg^{2+} + 2\,Cl^- + 2K^+ + 2I^- \longrightarrow HgI_2(s) + 2K^+ + 2I^-$
I want to separate the HgI_2 from the liquid, so I'm whirling it around in this centrifuge.
WHIRR
Now it's finished, and the HgI_2 is packed into the bottom of the test tube. I can just pour the clear liquid off.
$2\,K^+ + 2\,I^-$
HgI_2

different methods are used. Sometimes, the solution is allowed to stand for a time in large tanks, and the precipitate simply settles out.

We can use the solubility rules to predict reactions.

EXAMPLE 14.3: Would anything happen if solutions of $Pb(NO_3)_2$ and K_2CrO_4 were mixed? Write the net ionic equation, if any.

Solution: First, we write all soluble salts in their ionic forms. Rules 1 and 2 tell us that both compounds are soluble, so we write:

$$Pb^{2+} + 2\ NO_3^- + 2\ K^+ + CrO_4^{2-}$$

Next, we see whether an insoluble compound is formed by switching the ions around and recombining them. KNO_3 is soluble (Rules 1 and 2), but $PbCrO_4$ is insoluble (Rule 8).

Finally, we write the net ionic equation for the formation of the insoluble substance, leaving out the spectator ions (K^+ and NO_3^-, in this case).

Answer: Yes, $PbCrO_4$ will precipitate.

$$Pb^{2+} + CrO_4^{2-} \longrightarrow PbCrO_4(s)$$ ■

EXAMPLE 14.4: Would anything happen if solutions of $Zn(NO_3)_2$ and $BaBr_2$ were mixed? Write the net ionic equation, if any.

Solution: We write these separate ions: $Zn^{2+} + 2\ NO_3^- + Ba^{2+} + 2\ Br^-$. No insoluble compounds can be formed from these ions, according to the solubility rules.

Answer: No, nothing would happen. We show this by writing N.R. (no reaction).

$$Zn^{2+} + 2\ NO_3^- + Ba^{2+} + 2\ Br^- \longrightarrow \text{N.R.}$$ ■

DISSOLVING PRECIPITATES. If an insoluble substance contains a negative ion that's a base, we can often dissolve it by treating it with a strong acid. As it happens, a large number of insoluble substances do contain negative ions that are bases. Sulfides, carbonates, hydroxides, and phosphates are mostly insoluble and are bases. Also, many oxides are insoluble. Dissolving the FeO coating on steel with H_2SO_4 (called "pickling")—a reaction we saw in Chapter 5, p. 100—is an acid-base reaction. The ionic equation is:

$$\underset{\text{Acid}}{H^+} + \underset{\text{Acid}}{HSO_4^-} + \underset{\text{Base}}{FeO(s)} \longrightarrow Fe^{2+} + SO_4^{2-} + H_2O$$

Here, there are no spectator ions, so the complete and net ionic equations are the same.

EXAMPLE 14.5: Could H_2SO_4 be used to dissolve CuS? Write the ionic equation for the reaction, if any.

Solution: We find the sulfide ion in the list of bases in Table 13.4, so this reaction is possible. The ionic equation is written in the usual way.

Answer: H_2SO_4 could be used to dissolve CuS.

$$H^+ + HSO_4^- + CuS(s) \longrightarrow SO_4^{2-} + Cu^{2+} + H_2S(g)$$

(The sulfide ion, when treated with strong acid, will always change to H_2S by accepting two protons.) ■

EXAMPLE 14.6: Could a strong acid be used to dissolve AgCl? Choose any strong acid and write the ionic equation for the reaction, if any.

Solution: The chloride ion is at the top of the list of bases in Table 13.4, and it's labeled as not acting as a base. Therefore AgCl is neither acidic nor basic.

Answer: A strong acid would be no help in dissolving AgCl. Insoluble substances whose negative ions aren't bases can't be dissolved by acids.

$$AgCl(s) + H^+ \longrightarrow N.R.$$

■

Acids can dissolve some metals, too, as we saw in Chapter 5. Here's how we'd write such a reaction as complete and net ionic equations:

$$\text{Complete: } 2\,H^+ + 2\,Cl^- + Zn(s) \longrightarrow H_2(g) + Cl^- + Zn^{2+}$$

$$\text{Net: } 2\,H^+ + Zn(s) \longrightarrow H_2(g) + Zn^{2+}$$

Because Cl^- is a spectator ion, it doesn't matter which strong acid we use. It's the hydronium ion that dissolves metal. The net ionic equation shows H^+ going out of solution as H_2 is formed, and Zn^{2+} going into solution as zinc metal dissolves.

Acids can also dissolve gases, if the gases are basic. The only common basic gas is ammonia. Ammonia dissolves very well in solutions of strong acids, because of this reaction:

$$H^+ + NH_3(g) \longrightarrow NH_4^+$$

If ammonia is an unwanted product of a reaction, it can be kept from escaping into the air by passing it through an acid solution.

Most gases are acidic, though. Basic solutions are most useful for dissolving gases such as SO_2, CO_2, and NO_2. For example,

$$CO_2(g) + OH^- \longrightarrow HCO_3^-$$

$$SO_2(g) + OH^- \longrightarrow HSO_3^-$$

$$SO_3(g) + OH^- \longrightarrow HSO_4^-$$

$$2\,NO_2(g) + 2\,OH^- \longrightarrow NO_2^- + NO_3^- + H_2O$$

Bases can also dissolve insoluble substances that are acids. Many greases contain heavy organic acids. We can write $-COOH$ for any organic acid, because it's the $-COOH$ part that's the acid.

$$\underset{\text{Grease}}{-COOH} + OH^- \longrightarrow \underset{\text{Dissolved grease}}{-COO^-} + H_2O$$

This is one reason detergents should be basic: to help dissolve the grease.

Drain and oven cleaners contain NaOH because of its ability to dissolve grease. But bases, like acids, will also dissolve some metals. For this reason, makers of drain and oven cleaners warn us not to use their products on aluminum pans. Strong bases dissolve aluminum in this way:

$$Al(s) + 2\,OH^- + 2\,H_2O \longrightarrow \underset{\text{Aluminate ion}}{Al(OH)_4^-} + H_2(g)$$

PRACTICE IN WRITING IONIC EQUATIONS. Table 14.4 summarizes the rules for writing ionic equations. We might also want to consult the table of conjugate acid-base pairs, in Chapter 13 (p. 299) or the solubility rules, p. 314, or Appendix D.

We've already seen how to predict reactions with ionic equations. We can also translate any molecular equation into an ionic equation, using the rules in Table 14.4.

EXAMPLE 14.7: If hydrogen sulfide gas is bubbled through a solution of copper(II) nitrate, a black precipitate of CuS is formed, according to this molecular equation:

$$H_2S(g) + Cu(NO_3)_2(aq) \longrightarrow CuS(s) + 2\ HNO_3(aq)$$

Translate this into a net ionic equation.

Solution: H_2S is a gas, so it's written in the molecular form. $Cu(NO_3)_2$ is a soluble salt, so it's written in ionic form; CuS is insoluble and written in molecular form; HNO_3 is a strong acid, so it's written in ionic form. The complete ionic equation is:

$$H_2S(g) + Cu^{2+} + 2\ NO_3^- \longrightarrow CuS(s) + 2\ H^+ + 2\ NO_3^-$$

Nitrate ions are spectators, so we eliminate them.

$$H_2S(g) + Cu^{2+} \longrightarrow CuS(s) + 2\ H^+$$

The equation is balanced for atoms. To check the charge balance, we write the charge of each species under it:

$$\underset{0}{H_2S(g)} + \underset{2+}{Cu^{2+}} \longrightarrow \underset{0}{CuS(s)} + \underset{2+}{2\ H^+}$$

There is a charge of 2+ on each side of the equation, so the equation is balanced for charge.

Answer: $H_2S(g) + Cu^{2+} \longrightarrow CuS(s) + 2\ H^+$ ■

In Example 14.7, we were given the information that CuS was insoluble. Otherwise, we could have found that out by consulting Table 14.3, the solubility rules (p. 314).

EXAMPLE 14.8: One type of baking powder contains $NaHCO_3$ and KH_2PO_4. When water is added, this reaction takes place:

$$NaHCO_3 + KH_2PO_4 \longrightarrow NaKHPO_4 + H_2O + CO_2$$

Write the net ionic equation.

Solution: Both substances on the left are soluble salts, so we write them as

TABLE 14.4
Rules for writing ionic equations

1. Completely ionized substances (soluble salts, strong acids and bases) are written in their ionic forms.
2. Incompletely ionized substances (insoluble salts, gases, weak acids and bases) are written in their molecular forms.
3. The net ionic equation should exclude spectator ions.
4. The net ionic equation must be balanced by atoms and charge.

ions. Also $NaKHPO_4$ is a soluble salt. H_2O and CO_2 we write in the molecular form because they are incompletely ionized substances.

$$\cancel{Na^+} + \underset{1-}{HCO_3^-} + \cancel{K^+} + \underset{1-}{H_2PO_4^-} \longrightarrow \cancel{Na^+} + \cancel{K^+} + \underset{2-}{HPO_4^{2-}} + \underset{0}{H_2O} + \underset{0}{CO_2}$$

Above, we crossed out the spectator ions (Na^+ and K^+) and wrote charges beneath the remaining species. The atoms are balanced, and each side of the equation has a charge of 2−, so the charge is balanced too.

Answer: $HCO_3^- + H_2PO_4^- \longrightarrow HPO_4^{2-} + H_2O + CO_2(g)$ ■

This is an example of a reaction in solution that doesn't happen out of solution. Both $NaHCO_3$ and KH_2PO_4 are also salts and thus occur as solid compounds. We can mix them in the solid form and they won't react. Baking powder contains a drying agent so that it won't react on the shelf. When we want it to react, we add water, and that lets the ions move around and come in contact with each other.

EXAMPLE 14.9: Write the net ionic equation, if any, for the reaction that takes place when solutions of NaOH and NH_4Cl are mixed.

Solution: We don't know what to look for here. But if a reaction does occur, it will be either acid-base or precipitate formation. First, we write the substances in their correct forms in solution:

$$Na^+ + OH^- + NH_4^+ + Cl^-$$

Looking at the solubility rules, we can't find any insoluble combination of these ions, so no precipitate will form. Next, we look at the table of conjugate acid-base pairs (Table 13.4, p. 299). OH^- appears in the bases column, and NH_4^+ appears in the acids column. There could thus be an acid-base reaction, between NH_4^+ and OH^-. Na^+ and Cl^- aren't involved, so we might as well go right to the net ionic equation.

$$\underset{1+}{NH_4^+} + \underset{1-}{OH^-} \longrightarrow \underset{0}{NH_3(g)} + \underset{0}{H_2O}$$

The equation is balanced for atoms and for charge.

Answer: $NH_4^+ + OH^- \longrightarrow NH_3(g) + H_2O$ ■

In fact, adding a strong base to an ammonium salt is a common laboratory way of preparing ammonia gas.

14.3 QUANTITATIVE PREPARATION OF SOLUTIONS

In the chemistry laboratory, we'll often need to use solutions of known concentration. To prepare these, we measure out a given amount of solute and dilute it to a given volume with water. If the solute is a solid, we measure it by weighing; if it's already in solution, such as one of the common laboratory acids, we measure it by volume.

MEASUREMENT BY WEIGHING. To find out how much solute we need to make up a known volume of a solution of known concentration, we use the concentration and formula weight as conversion factors.

EXAMPLE 14.10: An experiment calls for 500. mℓ of a 0.250 *M* NaCl solution. How many grams of NaCl are needed to make this solution?

Solution: We want to convert 500. mℓ of solution to grams of NaCl. The molarity (0.250 moles/ℓ) and the formula weight of NaCl (58.5 g/mole) are the conversion factors.

$$500.\ \cancel{m\ell} \times \frac{0.250\ \cancel{\text{moles}}}{1000\ \cancel{m\ell}} \times \frac{58.5\ \text{g}}{\cancel{\text{mole}}} = \underline{\quad}\ \text{g}$$

Answer: 7.31 g NaCl. ■

After weighing out the solute, we would dilute it to a volume of 500. mℓ. We could use a *volumetric flask,* which is made to contain a specified volume. Some common sizes for volumetric flasks are 100.-mℓ, 250.-mℓ, 500.-mℓ, and 1000.-mℓ.

volumetric flask

Here's the procedure for making up 500. mℓ of 0.250 *M* NaCl:

1. Weigh out 7.31 grams of NaCl.
2. Put the NaCl into a 500.-mℓ volumetric flask. (Don't spill any.)
3. Add water to a little below the neck, and swirl the flask until the solute is dissolved. (Filling it above the neck makes it harder to swirl.)
4. Add water carefully to the mark. Stopper the flask and mix by inverting and swirling.
5. Set upright for a few moments; check the mark. Add a few additional drops of water if necessary to bring the volume back up to the mark. Mix again.

Figure 14.2 illustrates this procedure.

MEASUREMENT BY VOLUME. To prepare a solution of a common laboratory acid, we usually start with the concentrated solutions, whose concentrations are given in Table 14.5. The problem is to dilute a solution that's too concentrated for our purposes. To do so, we take a known amount of the concentrated solution and, adding more solvent, bring it up to a known new volume. In any dilution, the total amount of solute stays the same, but the concentration changes, as Figure 14.3 shows.

TABLE 14.5
Concentrations of concentrated laboratory acids

Acid	*Molarity*	*Weight Percent*
Hydrochloric acid, HCl	12	36
Nitric acid, HNO_3	16	72
Sulfuric acid, H_2SO_4	18	96
Acetic acid, $HC_2H_3O_2$	17.5	100

FIGURE 14.2
Preparation of a 0.250 *M* NaCl solution

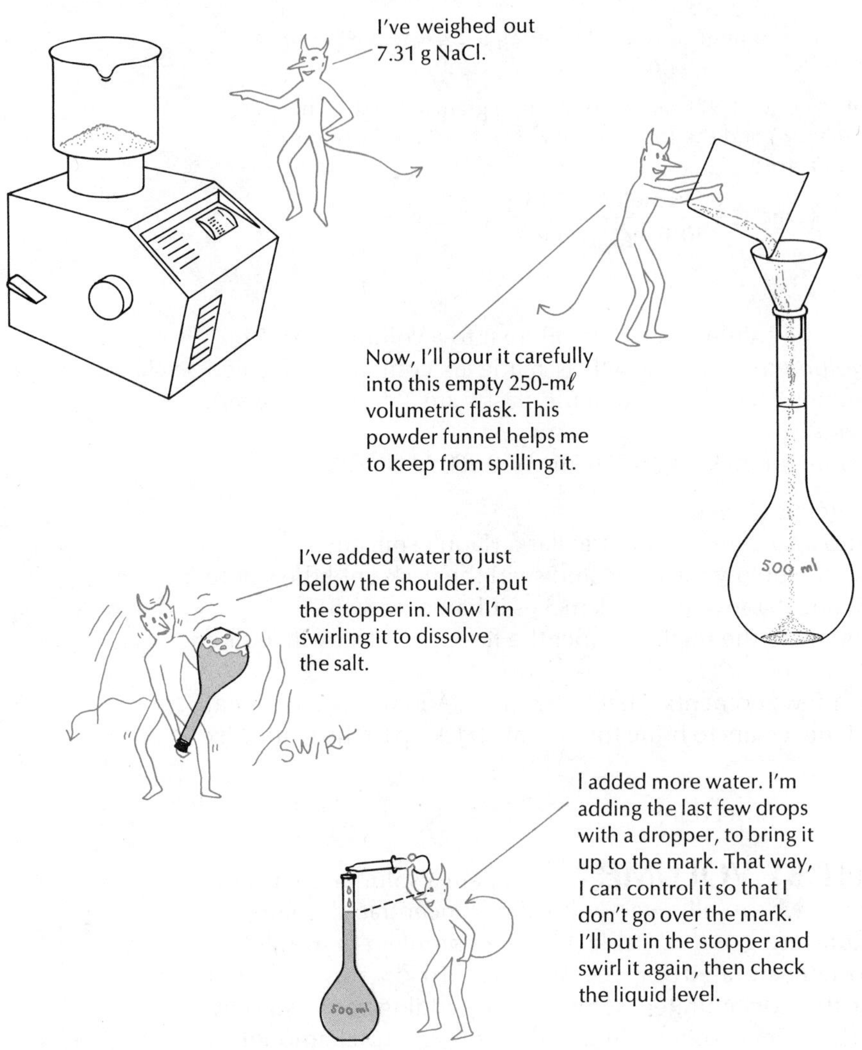

I've weighed out 7.31 g NaCl.
Now, I'll pour it carefully into this empty 250-mℓ volumetric flask. This powder funnel helps me to keep from spilling it.
500 ml
I've added water to just below the shoulder. I put the stopper in. Now I'm swirling it to dissolve the salt.
SWIRL
I added more water. I'm adding the last few drops with a dropper, to bring it up to the mark. That way, I can control it so that I don't go over the mark. I'll put in the stopper and swirl it again, then check the liquid level.

This equation is useful in dilution problems:

$$V_1C_1 = V_2C_2$$

It says that the old volume times the old concentration equals the new volume times the new concentration. We can use any units for volume and concentration as long as we use the same units on both sides of the equation. We can rearrange the equation to solve for whatever quantity we want to find.

To find the volume of a concentrated solution that must be diluted to get

FIGURE 14.3
Diluting a concentrated solution

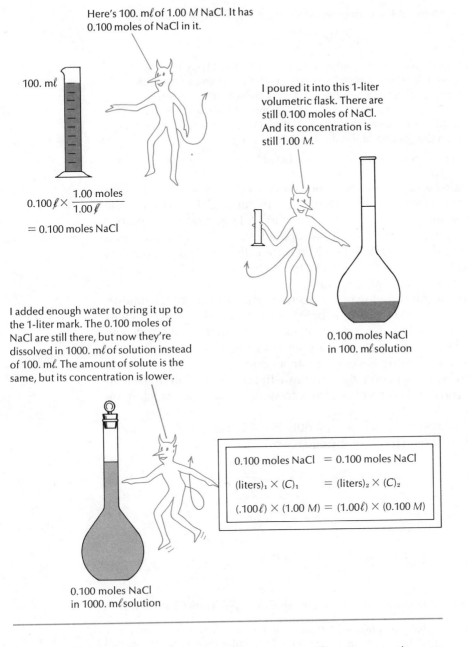

Here's 100. mℓ of 1.00 *M* NaCl. It has 0.100 moles of NaCl in it.

100. mℓ

$0.100 \not{\ell} \times \dfrac{1.00 \text{ moles}}{1.00 \not{\ell}}$

$= 0.100$ moles NaCl

I poured it into this 1-liter volumetric flask. There are still 0.100 moles of NaCl. And its concentration is still 1.00 *M*.

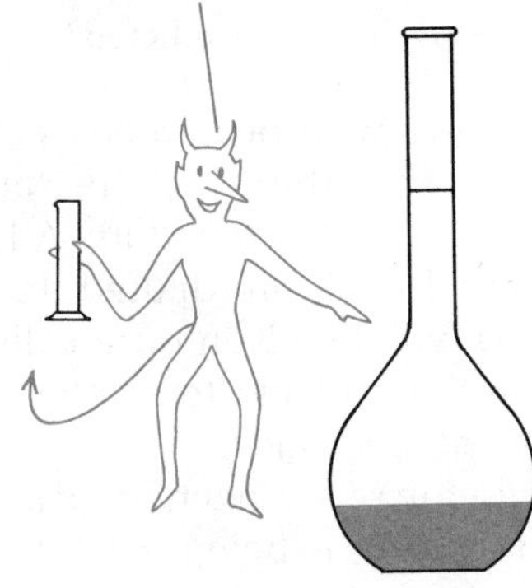

0.100 moles NaCl
in 100. mℓ solution

I added enough water to bring it up to the 1-liter mark. The 0.100 moles of NaCl are still there, but now they're dissolved in 1000. mℓ of solution instead of 100. mℓ. The amount of solute is the same, but its concentration is lower.

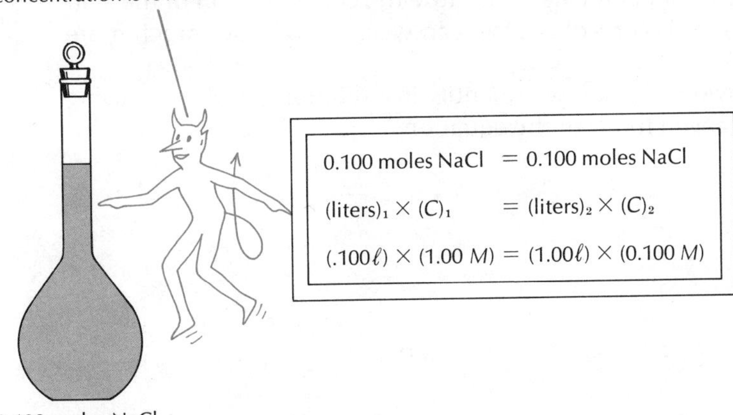

0.100 moles NaCl = 0.100 moles NaCl
(liters)₁ × (C)₁ = (liters)₂ × (C)₂
(.100ℓ) × (1.00 M) = (1.00ℓ) × (0.100 M)

0.100 moles NaCl
in 1000. mℓ solution

a specific volume of a dilute solution having a specific concentration, we use this form of the equation:

$$V_1 = V_2 \frac{C_2}{C_1}$$

EXAMPLE 14.11: A laboratory experiment calls for 0.60 *M* HCl, but the only solution in the lab is 12 *M* (concentrated) HCl. What volume of 12 *M* HCl must be diluted to make 0.50 liters of 0.60 *M* HCl?

Solution: The volume of the concentrated solution (V_1) is the unknown here.

$$V_1 = 0.50\,\ell \times \frac{0.60\,\cancel{M}}{12\,\cancel{M}} = \_\_\_\ell$$

Here, as with the "plug-in" formulas we used in Chapter 11, we'll be better off knowing what kind of an answer to expect. Then we can catch errors caused by putting the wrong numbers on top or bottom. Here, we're diluting a concentrated solution to get a much more dilute solution, so we expect the volume of the concentrated solution (V_1, in this case) to be much less than the volume of the diluted solution (V_2).

Answer: 0.025 ℓ, or 25 mℓ. Much less, as predicted. ■

A concentrated acid should always be diluted by adding acid slowly, with stirring, to some of the water. Then the rest of the water is added to make the desired final volume. This is a safety procedure. A large heat of solution is given off when concentrated acids are diluted. If water is poured into a concentrated acid, the heat will be given off locally, where the water hits the acid, which can cause the hot acid to spatter. It's all right to add the water directly to less concentrated acids.

Figure 14.4 shows the dilution of a concentrated acid. Here, an Erlenmeyer flask, and not a volumetric flask, is being used. Whether we use a volumetric or Erlenmeyer flask depends on the desired accuracy of the solution. For two significant figures, we can use an Erlenmeyer flask. For three or more, we use a volumetric flask. To dilute a concentrated acid, it doesn't make sense to use a volumetric flask, because the concentration of the acid isn't that accurate in the first place. We can weigh a solid out much more accurately.

We can find any other unknown quantity in a dilution problem by using one of these rearranged forms of the equation:

$$V_2 = V_1\frac{C_1}{C_2} \qquad C_1 = C_2\frac{V_2}{V_1}$$

$$C_2 = C_1\frac{V_1}{V_2}$$

EXAMPLE 14.12: 25 mℓ concentrated NH_3 is diluted to 1.0 liter. What is the molarity of the diluted solution?

Solution: We want to know molarity, so we use $C_2 = C_1\frac{V_1}{V_2}$. From Chapter 13, p. 289, we know that concentrated NH_3 is 16 *M*. Because we are diluting the solution, we expect the new concentration to be smaller than 16 *M*.

$$M_2 = 16\,M \times \frac{0.025\,\cancel{\ell}}{1.0\,\cancel{\ell}} = \_\_\_M$$

Answer: 0.40 *M*. Smaller, as expected. ■

FIGURE 14.4
Diluting a concentrated acid

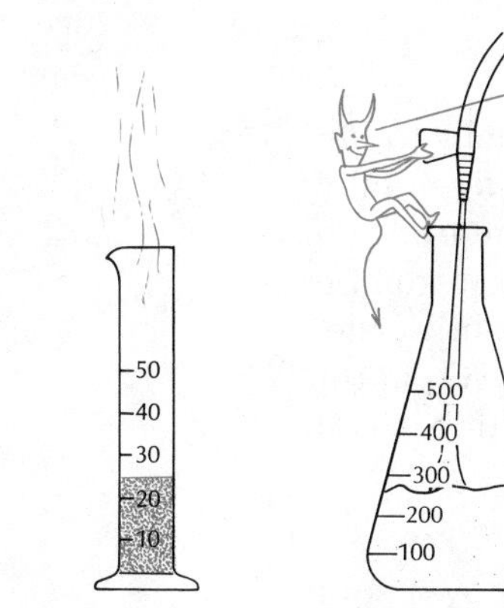

I've measured out 25 mℓ (0.025ℓ) of concentrated (12*M*) HCl. Now I'm putting some water into this Erlenmeyer flask. I'll make sure that the total won't go over the final volume I want, which is 500 mℓ (0.5 ℓ).

25 mℓ of 12 *M* HCl

About 300 mℓ of water

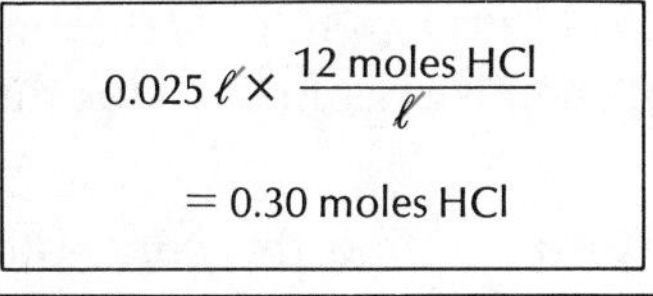

$$0.025\ \ell \times \frac{12 \text{ moles HCl}}{\ell} = 0.30 \text{ moles HCl}$$

Now I'm adding the concentrated HCl very carefully to the water, swirling this flask as I go, to keep them constantly mixing.

Of course, I'm doing all of this under a hood, so that I won't get HCl fumes all over my lab.

Now, I add more water to bring it up to the 500-mℓ mark. There.

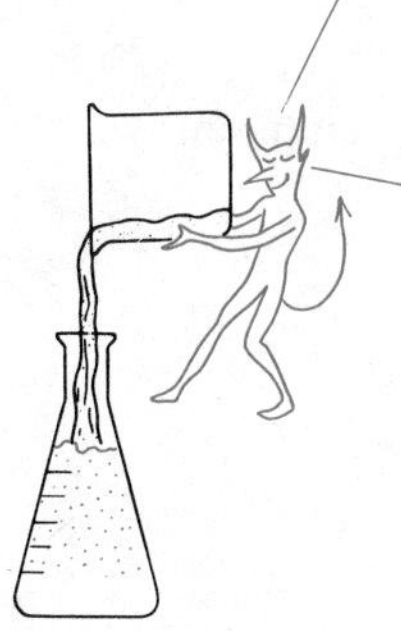

I'll mix it again before I use it.

$$0.50\ \ell \times \frac{0.60 \text{ moles HCl}}{\ell} = 0.30 \text{ moles HCl}$$

This calculation shows that the number of moles of solute hasn't changed.

14.4 QUANTITATIVE REACTIONS IN SOLUTION

In Chapter 6, we did calculations with chemical equations, usually to find out how many grams of something it took to react with a specific number of grams of something else. We're going to do the same thing here, except that our chemicals are now in solution, and we'll be looking for the number of milliliters instead of the number of grams. Chemists are frequently called upon to answer questions like "What's the concentration of H_2SO_4 in this sample of mine drainage?" Determining the concentration of solutions is an important aspect of chemistry, and we'll look at this aspect next.

TITRATION. To determine the concentration of an unknown solution, we let a specific volume of it react with another solution whose concentration we do know (a "known" solution). When the reaction has taken place completely, we can calculate the concentration of the unknown solution from the amount of known solution used.

Titration is the gradual adding of one solution to another until the solute in the first solution has reacted completely with the solute in the second solution. For a titration to work, two things must be true: (1) we must choose the solute in the known solution so that it will indeed react completely with the solute in the unknown solution; (2) we must have some way of knowing at what point the two solutions have reacted completely. Titration can be used for a variety of solutes, but we'll confine our discussions to acids and bases. titration

If we want to determine the concentration of a strong or weak acid solution, we place a known volume of it in an Erlenmeyer flask. Then we add a few drops of an indicator called phenolphthalein, which is colorless in acid and pink in base. We add a strong base solution of known concentration gradually from a buret, as shown in Figure 14.5. When the mixture shows the slightest tinge of pink, we know a very slight excess of base is present and that our acid has been neutralized. The buret indicates the amount of base solution we added. The point of neutralization, as shown by the indicator, is called the *endpoint*. Knowing the original volume of the acid, the concentration of the base, and the volume of the base needed to reach the endpoint, we can calculate the concentration of the acid. endpoint

EXAMPLE 14.13: A technician in a quality-control laboratory of a vinegar plant titrated a 25.0 mℓ sample of vinegar with a 0.512 *M* NaOH solution. It took 35.2 mℓ of the NaOH solution to neutralize the vinegar. What is the molarity of the acetic acid in the vinegar sample?

Solution: First, we write the equation:

$$NaOH(aq) + HC_2H_3O_2(aq) \longrightarrow NaC_2H_3O_2(aq) + H_2O(l)$$

Next, we want to convert 35.2 mℓ (0.0352 ℓ) of NaOH to moles of $HC_2H_3O_2$. Our conversation factors are the molarity of the NaOH and the mole ratio.

FIGURE 14.5
Titration of an acid against a base

I've put 25.0 mℓ of vinegar into this flask. It contains a certain number of moles of acetic acid, but I don't know how many yet.

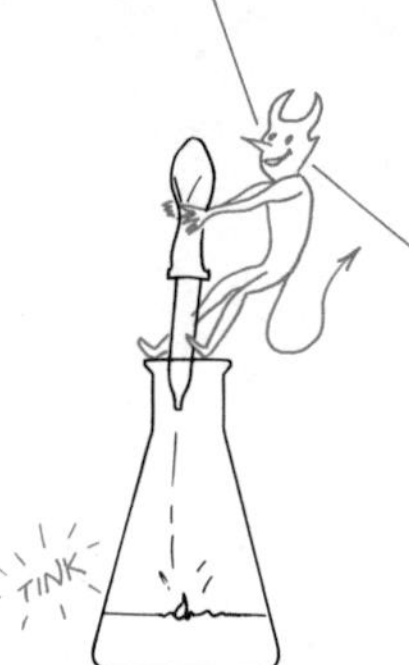

Now, I'll add a drop of phenolphthalein indicator. I can't see it, because it's colorless in acid.

And, I've filled this buret up to the 50-mℓ mark with 0.512 *M* NaOH. The buret has markings on it so that I know how much base I've added.

This stopcock lets me add the base to the acid solution as fast or as slowly as I want to.

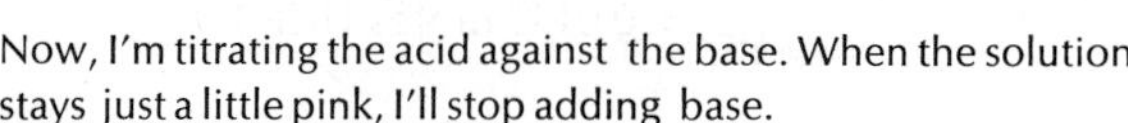

Now, I'm titrating the acid against the base. When the solution stays just a little pink, I'll stop adding base.

$$NaOH\,(aq) + HC_2H_3O_2\,(aq) \longrightarrow NaC_2H_3O_2\,(aq) + H_2O(l)$$

Now the solution is slightly pink, and it took 35.2 mℓ (0.0352ℓ) of base solution to do it. According to these calculations, I used 0.0180 moles of base to neutralize the acid, so there must have been 0.0180 moles of acid there in the first place—right?

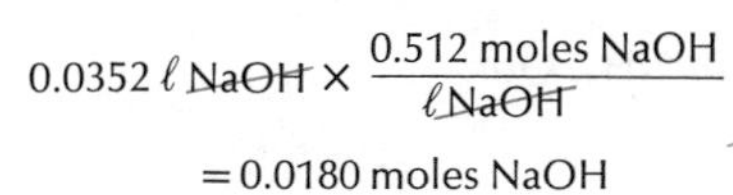

$$0.0352\ \ell\ \cancel{NaOH} \times \frac{0.512\text{ moles NaOH}}{\ell\ \cancel{NaOH}} = 0.0180\text{ moles NaOH}$$

$$0.0352\ \cancel{\ell\ NaOH} \times \frac{0.512\ \cancel{moles\ NaOH}}{\cancel{\ell\ NaOH}} \times \frac{1\ mole\ HC_2H_3O_2}{\cancel{mole\ NaOH}}$$
$$= 0.0180\ moles\ HC_2H_3O_2$$

Now that we know that 0.0180 moles of acetic acid are contained in 25.0 mℓ, we can easily calculate molarity.

$$molarity = \frac{moles}{liter} = \frac{0.0180\ moles}{0.0250\ \ell} = \_\_\_ M\ HC_2H_3O_2$$

Answer: 0.721 M $HC_2H_3O_2$. ■

Earlier, we said that it's not possible to make a solution accurate to more than two significant figures by diluting a concentrated acid. But there might be times when we would want to know the concentration more accurately. We can determine the concentration of an acid solution by titrating it against a solution whose concentration we do know accurately (a standard solution), or against a solid base that we've weighed accurately. This process is called *standardization*.

standardization

EXAMPLE 14.14: The HCl solution prepared in Example 14.11 was standardized by titrating against solid Na_2CO_3. A sample of Na_2CO_3 weighing 0.792 grams required 24.9 mℓ of the HCl solution to neutralize it. Calculate the molarity of the HCl solution.

Solution: First, we write the equation:

$$Na_2CO_3(s) + 2\ HCl(aq) \longrightarrow NaCl(aq) + 2\ H_2O(l) + CO_2(g)$$

Next, we want to convert grams of Na_2CO_3 to moles of HCl. Our conversion factors are the formula weight of Na_2CO_3, and the mole ratio.

$$0.792\ \cancel{g\ Na_2CO_3} \times \frac{1\ \cancel{mole\ Na_2CO_3}}{106\ \cancel{g\ Na_2CO_3}} \times \frac{2\ moles\ HCl}{\cancel{mole\ Na_2CO_3}} = 0.0149\ moles\ HCl$$

Then,

$$molarity = \frac{moles}{liter} = \frac{0.0149\ moles\ HCl}{0.0246\ \ell} = \_\_\_ M$$

Answer: 0.606 M.

Notice: Because the solution was made to be approximately 0.60 M, we see that the answer is reasonable. If the answer were far from that value, we might want to recheck the experiment or the calculations. ■

EXAMPLE 14.15: A solution made by diluting 25.0 mℓ of concentrated (16 M) NH_3 to 1 liter was standardized against the HCl solution in Example 14.14. 25.0 mℓ of the diluted NH_3 solution required 17.1 mℓ of the HCl solution to neutralize it. Calculate the molarity of the NH_3 solution.

Solution: First, we write the equation:

$$NH_3(aq) + HCl(aq) \longrightarrow NH_4Cl(aq)$$

Next, we want to convert milliliters of HCl solution to moles of NH_3. Our conversion factors are the molarity of the HCl and the mole ratio.

$$0.0171\ \cancel{\ell\ HCl} \times \frac{0.606\ \cancel{moles\ HCl}}{\cancel{\ell\ HCl}} \times \frac{1\ mole\ NH_3}{\cancel{mole\ HCl}} = 0.0104\ moles\ NH_3$$

And $$molarity = \frac{0.0104\ moles\ NH_3}{0.0250\ \ell} = 0.416\ M$$

We can check whether this answer is reasonable by using the dilution information.

$$M_2 = M_1\frac{V_1}{V_2} = 16\ M \times \frac{0.0250\ \ell}{1\ \ell} = 0.4\ M$$

Our answer is reasonable. The solution was diluted to be about 0.4 *M*.

Answer: 0.416 *M*. ■

NORMALITY. Normality is a concentration term useful in analytical chemistry. To find the normality of a solution, we need to know not only the formula weight of the solute, but also *how* it will react.

To illustrate normality, let's first look at two acids: HCl and H_2SO_4, each 6 *M*, and compare the equations for each to neutralize 6 *M* NaOH:

$$HCl(aq) + NaOH(aq) \longrightarrow NaCl(aq) + H_2O(l)$$

$$H_2SO_4(aq) + 2\ NaOH(aq) \longrightarrow Na_2SO_4(aq) + 2\ H_2O(l)$$

To neutralize 50 mℓ of 6 *M* NaOH, we'd need 50 mℓ of 6 *M* HCl. But we'd need only 25 mℓ of 6 *M* H_2SO_4 to neutralize the same amount of NaOH, because H_2SO_4 has two protons to donate. One mole of H_2SO_4 can take care of two moles of NaOH.

The concentration term normality is useful to compensate for this difference. *Normality (N)* means the number of equivalents of solute per liter of solution. An *equivalent* of an acid or base is the number of grams that contains or can neutralize one mole of acidic hydrogens. We get the normality from the molarity like this:

normality

equivalent

normality = molarity × no. of acidic hydrogens (for an acid)

or,

normality = molarity × no. of protons accepted (for a base)

EXAMPLE 14.16: What is the normality of (a) 6*M* HCl; (b) 3 *M* H_2SO_4; (c) 6 *M* NH_3; (d) 1 *M* $Ca(OH)_2$?

Solution: We multiply the molarity by 1 for HCl; by 2 for H_2SO_4; by 1 for NH_3 (although NH_3 is a weak base, it is capable of reacting with one mole of H^+); and by 2 for $Ca(OH)_2$.

Answer: (a) 6 *N* HCl; (b) 6 *N* H_2SO_4; (c) 6 *N* NH_3; (d) 2 *N* $Ca(OH)_2$. ■

In chemistry laboratories, often dilute acids and bases are labeled "6 *N*" or "6 Normal." We see that for monoprotic acids like HCl, the molarity and normality are the same, but for H_2SO_4, the molarity is half the normality (3 *M*, in this case).

To calculate the normality of a solution from the weight of acid or base, we divide that weight by the substance's equivalent weight. The *equivalent weight (g/eq)* of an acid or base is the weight in grams that contains or neutralizes one mole of acidic hydrogen. We get the equivalent weight by dividing the formula weight by the number of acidic hydrogens contained or neutralized per formula.

equivalent weight

EXAMPLE 14.17: What is the normality of a solution that contains 163 grams of phosphoric acid per liter, assuming that all three protons will be neutralized?

Solution: First, determine the equivalent weight by dividing the formula weight by 3 (because H_3PO_4 has 3 acidic hydrogens).

$$\text{equivalent wt.} = \frac{98.0\text{ g}/\cancel{\text{mole}}}{3\ \cancel{\text{moles}}/\text{eq}} = 32.7\text{ g/eq}$$

Next, we want to convert grams per liter to equivalents per liter (normality). The equivalent weight is our conversion factor.

$$163\ \cancel{\text{g H}_3\text{PO}_4}/\ell \times \frac{1.00\text{ eq}}{32.7\ \cancel{\text{g H}_3\text{PO}_4}} = \_\_\_N$$

Answer: 4.98 *N*. ■

The usefulness of normality comes from the fact that it automatically incorporates the mole ratio into its value. Then, in titration calculations, we can use this simple relationship, similar to the one we used for dilution:

$$V_1N_1 = V_2N_2$$

Here, the volume of solution 1 times its normality equals the volume of solution 2 times its normality.

From this relationship, we can get either the volume or the normality of an unknown solution in a titration problem, if we know the other three values. We use these versions of the equation above, similar to those we had earlier for dilution:

$$V_2 = V_1\frac{N_1}{N_2} \qquad N_2 = N_1\frac{V_1}{V_2}$$

Also useful in solving normality problems are these:

liters times normality equals equivalents:

$$\ell \times N = \text{eq}$$

equivalents of reactant 1 equals equivalents of reactant 2:

$$\text{eq}_1 = \text{eq}_2$$

EXAMPLE 14.18: A sample of too-alkaline water is found to have a hydroxide-ion concentration of 0.015 *N*. How many milliliters of a strong acid solution that's 2.0 *N* will it take to neutralize 2.5 liters of the water?

Solution: We don't know what the acid and base are, but it doesn't matter because we know their normalities. We want to know volume, so we use

$$V_2 = V_1\frac{N_1}{N_2} = 2.5\,\ell \times \frac{0.015\,\cancel{N}}{2.0\,\cancel{N}} = \_\_\_\ell$$

As before, we try to have some feeling for the kind of answer to expect. Here, the acid solution is much more concentrated than the base solution, so we expect its volume will be much smaller than that of the base solution.

Answer: 0.019 ℓ, or 19 mℓ. Much smaller, as expected. ■

The last problem couldn't have been solved using molarities unless we knew what the acid and base were, or at least whether they were mono-, di-, or triprotic.

EXAMPLE 14.19: Calculate the normality of a base solution if 25.0 mℓ of it is neutralized by 32.7 mℓ of a 0.110 *N* standard acid solution.

Solution: We want to know normality, so we use $N_2 = N_1 \frac{V_1}{V_2}$. The volumes of the two reactants are similar in size, so we expect the normalities to be similar too. It took more of the acid than it did of the base, though, and we expect the base solution to be slightly more concentrated than the acid solution.

$$N_2 = 0.110\ N \times \frac{32.7\ \cancel{m\ell}}{25.0\ \cancel{m\ell}} = \_\_\_ N$$

Answer: 0.144 *N*. Slightly more concentrated, as expected. ■

EXAMPLE 14.20: An acid solution is standardized by titrating against 0.312 g of Na_2CO_3. It took 44.8 mℓ of the acid to neutralize the Na_2CO_3. Calculate the normality of the acid.

Solution: Here, we use $eq_1 = eq_2$. First, we find eq_1, the number of equivalents of the Na_2CO_3.

$$\text{eq. wt. } Na_2CO_3 = \text{form. wt.}/2 = 106.0/2 = 53.0\ \text{g/eq}$$

$$0.312\ \cancel{g\ Na_2CO_3} \times \frac{1\ \text{eq}}{53.0\ \cancel{g\ Na_2CO_3}} = 5.89 \times 10^{-3}\ \text{eq}\ Na_2CO_3 = \text{eq acid}$$

$$N = \text{eq}/\ell = \frac{5.89 \times 10^{-3}\ \text{eq}}{0.0448\ \ell} = \_\_\_ N\ \text{acid}$$

Answer: 0.131 *N* acid. ■

We've explained normality in terms of acids and bases, but it can be used for any type of compound if we know how it's going to react. In this book, we won't go beyond the discussion of acids and bases.

REVIEW QUESTIONS

Acid-Base Reactions in Solution

1. What do we mean by *net ionic* and *complete ionic equations?* Give an example of each.
2. What are *spectator ions?* What role do they play in reactions in solution?
3. What reaction occurs between any strong acid and any strong base? Write the net ionic equation.
4. Write a complete ionic equation for the reaction of a strong acid with a strong base. Label the spectator ions.
5. Why is it dangerous to mix concentrated acids and bases together?
6. Write ionic equations that show: (a) a weak acid reacting with a weak base; (b) a strong acid reacting with a weak base; (c) a weak acid reacting with a strong base.
7. Give examples of: (a) a base that can gain two hydrogens; and (b) an acid that can lose two hydrogens.

Forming and Dissolving Precipitates

8. Explain how precipitation is used to remove ions from solutions.
9. How can we decide which of several possible

precipitate-forming ions to use in the laboratory?

10. How do we know what substances are soluble and which are insoluble?
11. How can we remove a precipitate from solution after it's formed?
12. How can we predict whether a precipitate will be formed if solutions of two compounds are mixed?
13. What kind of insoluble substances can strong acids dissolve? Give an example, and write the equation.
14. Why does ammonia gas dissolve well in strong acid? Write the equation.
15. What kinds of solids do bases dissolve well? Write an equation.
16. Why are bases better at dissolving gases than acids are? Write an equation as an example.
17. Why do some reactions take place more readily in solution than out of it? Give an example.

Quantitative Preparation of Solutions

18. Explain how to prepare a solution of known concentration from a solid solute. Include the calculations in your explanation.
19. Explain how to prepare a dilute solution of a common laboratory acid, including the calculations.
20. Why must a concentrated acid be added to water, rather than vice versa?
21. What equation do we use in dilution calculations? Write it in all its forms, and explain how to use it.
22. Why can't we prepare a dilute solution of a laboratory acid to the same accuracy as we can a solution of a solid solute?

Quantitative Reactions in Solutions

23. What is *titration?* How is it used to find the concentration of an acid or base solution? What is an *endpoint?*
24. Explain how *standardization* is used to determine accurately the concentration of a strong acid solution.
25. What is *normality?* What is an *equivalent?* How do we calculate normality from molarity?
26. What is *equivalent weight?* How do we find a substance's equivalent weight?
27. Why is normality useful? What equation can we use to solve titration problems with normality?

EXERCISES

Set A (Answers at back of book. Asterisks mark the more difficult exercises.)

1. Write complete and net ionic equations for reactions between these acids and bases in water solution. Classify each acid and base as weak or strong.
   a. $HClO_4$ and KOH
   b. $HC_2H_3O_2$ and $Ca(OH)_2$
   c. NH_3 and $HC_2H_3O_2$
   d. HCl and Na_3PO_4
   e. HNO and $Sr(OH)_2$
   f. $NaHCO_3$ and HF
2. One type of fire extinguisher contains sulfuric acid and sodium hydrogen carbonate in separate containers. When the tank is tipped upside down, these mix and form carbon dioxide, which puts out the fire. Write complete and net ionic equations for this reaction, and identify the acid and the base.
3. Suggest a compound whose solution could be used to remove each of these ions from solution. Write complete and net ionic equations for each.
   a. Pb^{2+}
   b. CO_3^{2-}
   c. Cu^{2+}
   d. F^-
   e. Cr^{3+}
   f. OH^-
   g. Co^{2+}
   h. Cl^-
   i. Al^{3+}
4. Would a precipitate form if water solutions of these pairs were mixed? If so, write the complete and net ionic equations.
   a. K_2CO_3 and $MgCl_2$
   b. $Pb(NO_3)_2$ and SrI_2
   c. $ZnBr_2$ and $NiSO_4$
   d. CoF_2 and $Ca(OH)_2$
   e. NaOH and KCl
   f. $HgCl_2$ and $(NH_4)_2S$

*5. Suggest a compound whose water solution could be used to tell whether you had one or the other solution in each of these pairs. Write net ionic equations for the reactions, and tell your reasoning.
   a. NaOH or Na_2SO_4
   b. $CoCl_2$ or $CaCl_2$
   c. KCl or K_2CrO_4
   d. $Mg(NO_3)_2$ or $Ca(NO_3)_2$

6. A commercial source of phosphoric acid is the treatment of phosphate rock, $Ca_3(PO_4)_2$, with sulfuric acid. Write the complete ionic equation and explain why sulfuric acid will dissolve phosphate rock.
7. Write net ionic equations for dissolving these substances in strong acid. If a substance cannot be dissolved in strong acid, explain why.
   a. zinc ore, ZnS
   b. Chile saltpeter, KNO_3
   c. boiler scale, $CaCO_3$
   d. calomel, Hg_2Cl_2
   e. lime, CaO
   f. $Al(OH)_3$
8. Write net ionic equations for dissolving these substances in strong base. If a substance cannot be dissolved in strong base, explain why.
   a. copper ore, CuS
   b. butyric acid (rancid butter), C_3H_7COOH
   c. SO_3

d. NH_3
e. phosphate rock, $Ca_3(PO_4)_2$
f. HCl gas

*9. Whenever a solution is made from NaOH pellets, it is always contaminated with Na_2CO_3. Suggest an explanation, and write appropriate equations.

10. Translate these molecular equations into net ionic equations.
   a. $Na_2SO_4(aq) + Ba(NO_3)_2(aq) \longrightarrow BaSO_4(s) + 2\ NaNO_3(aq)$
   b. $ZnS(s) + H_2SO_4(aq) \longrightarrow ZnSO_4(aq) + H_2S(g)$
   c. $2\ NaOH(aq) + SO_3(g) \longrightarrow Na_2SO_4(aq) + H_2O(l)$
   d. $(NH_4)_2S(aq) + 2\ NaOH(aq) \longrightarrow 2\ NH_3(g) + Na_2S(aq) + 2\ H_2O(l)$
   e. $CaCO_3(s) + H_2SO_4(aq) \longrightarrow CaSO_4(aq) + CO_2(g) + H_2O(l)$

11. Sodium bicarbonate and solid citric acid ($H_2C_5H_4O_5COOH$) are the ingredients in Alka Seltzer that make it fizz in water. (a) Write the net ionic equation for this reaction. (b) Explain why the reaction does not occur until the tablets hit the water.

12. Calculate the number of grams of solute needed to make each of these solutions.
   a. 1.00 ℓ of 0.500 *M* Na_2SO_4
   b. 0.500 ℓ of 1.00 *M* Na_2SO_4
   c. 100. mℓ of 0.100 *M* $AgNO_3$
   d. 250. mℓ of 0.0250 *M* KCl

13. Calculate the number of milliliters of concentrated acid needed to make each of these solutions.
   a. 500. mℓ of 0.25 *M* HCl
   b. 1.00 ℓ of 0.10 *M* $HC_2H_3O_2$
   c. 250. mℓ of 2.0 *M* HNO_3
   d. 100. mℓ of 6.0 *M* HCl

14. Sometimes, people whose stomachs don't manufacture enough HCl must take synthetic HCl as medicine. How many milliliters of 12 *M* HCl would be needed to make 1.00 ℓ of HCl in the same concentration as stomach acid (0.016 *M*)?

15. Determine the concentrations of these solutions.
   a. 10.0 mℓ of 0.15 *M* HCl diluted to 100. mℓ
   b. 25.0 mℓ of 0.50 *M* H_2SO_4 diluted to 250. mℓ
   c. 10.0 mℓ of 0.112 *M* NaCl diluted to 1.00 ℓ
   d. 25.0 mℓ of 6.0 *M* NaOH diluted to 1.00 ℓ

16. What volume of the specified new solutions will be produced by diluting each of these?
   a. 10.0 mℓ of 1.00 *M* Na_2CO_3 to 0.100 *M*
   b. 25.0 mℓ of 0.250 *M* NH_3 to 0.100 *M*
   c. 5.0 mℓ of 6.0 *M* HCl to 1.0 *M*
   d. 50.0 mℓ of 3.0 *M* H_2SO_4 to 0.50 *M*

17. Calculate the volume of 0.122 *M* NaOH that would be needed to neutralize each of these.
   a. 25.0 mℓ of 0.233 *M* HCl
   b. 0.566 grams of $NaHCO_3$

18. Palmitic acid, an ingredient in palm oil, is used in making soap:

$$C_{15}H_{31}COOH(l) + NaOH(aq) \longrightarrow C_{15}H_{31}COONa(aq) + H_2O(l)$$

Soap

How many milliliters of 1.00 *M* NaOH would it take to neutralize 50.0 grams of palmitic acid? The molecular weight of palmitic acid is 256 g/mole.

19. What is the molarity of a 25.0-mℓ vinegar solution that is neutralized by 28.1 mℓ of 0.122 *M* NaOH?

20. A sample of mine drainage containing H_2SO_4 was titrated against 0.246 *M* Na_2CO_3. 50.0 mℓ of the mine drainage required 15.3 mℓ of the Na_2CO_3 solution. Calculate the concentration of the H_2SO_4 in the mine drainage, assuming it's the only acid present.

21. A student is analyzing a sample of household ammonia by titrating it against 0.0500 *M* HCl. A 10.0-mℓ sample of the ammonia requires 24.1 mℓ of the HCl. Calculate the molarity of the ammonia.

22. 25.0 mℓ of a solution of NaOH required 45.7 mℓ of 0.206 *M* H_2SO_4 to neutralize it. What is the molarity of the NaOH?

*23. A solution of a rat poison containing F^- is allowed to react with $CaCl_2$. It is found that all the F^- in 10.0 mℓ of rat poison solution can be precipitated as CaF_2 when 28.2 mℓ of 1.0 *M* $CaCl_2$ are added.
   a. Write the complete and net ionic equations.
   b. What is the concentration of fluoride ion in the rat poison?
   c. How many grams of fluoride would be contained in a 6-oz bottle of the rat poison?

24. What is the normality of each?
   a. 0.5 *M* NaH_2PO_4 (as a diprotic acid)
   b. 1.00 *M* HCl
   c. 0.200 *M* Na_3PO_4 (as a triprotic base)
   d. 0.5 *M* $Ca(OH)_2$ (as a diprotic base)

25. Determine the equivalent weights, assuming complete neutralization.
   a. H_3PO_4
   b. NaOH
   c. $KC_2H_3O_2$
   d. H_2SO_3

26. Determine the normalities of these solutions, assuming complete neutralization.
   a. 98.0 g of H_3PO_4 per 500. mℓ
   b. 4.01 g $Ca(OH)_2$ per liter

27. Determine the normalities of these.
   a. a base solution, 25.0 mℓ of which required 37.8 mℓ of 0.315 *N* H_2SO_4 to neutralize it
   b. an acid solution, 10.0 mℓ of which required 15.2 mℓ of 0.175 *N* NaOH to neutralize it

28. What are the normalities of these laboratory solutions?
   a. concentrated H_2SO_4
   b. concentrated (16 *M*) NH_3

29. An NaOH solution is prepared by diluting 90 mℓ of 9 *M* NaOH to 1.00 ℓ. To standardize it, 2.25 grams of $H_2C_2O_4 \cdot 2\ H_2O$ (oxalic acid dihydrate) are titrated against it, and 42.9 mℓ of the NaOH solution is

required to neutralize the oxalic acid. Calculate the normality of the ammonia.

*30. A chemist in a crime laboratory is analyzing a substance believed to be strychnine ($C_{21}H_{22}N_2O_2$). A 5.34-gram sample of the substance required 28.7 mℓ of 0.448 *N* HCl to neutralize it. Calculate the equivalent weight of the substance. Could it be strychnine? (Strychnine is a diprotic base and reacts completely with HCl.)

Set B (Answers not given. Asterisks mark the more difficult exercises.)

1. Write complete and net ionic equations for reactions between these acids and bases in water solution. Classify each acid and base as weak or strong.
   a. $NaHSO_4$ and KOH
   b. NH_3 and H_2SO_4
   c. HCl and NaOH
   d. $NaHCO_3$ and HNO_3
   e. $HC_2H_3O_2$ and KOH
   f. Na_2CO_3 and KH_2PO_4
2. Solutions of $Al_2(SO_4)_3$ and $Fe_2(SO_4)_3$ are used to remove phosphate from water. Write complete and net ionic equations.
3. Suggest a compound whose solution could be used to remove each of these ions from solution. Write complete and net ionic equations for each.
   a. Ag^+
   b. Sn^{2+}
   c. Br^-
   d. S^{2-}
   e. SO_4^{2-}
   f. CrO_4^{2-}
   g. Cd^{2+}
   h. Ba^{2+}
   i. Zn^{2+}
4. Would a precipitate form if water solutions of these pairs were mixed? If so, write the complete and net ionic equations.
   a. $Hg_2(NO_3)_2$ and K_2CrO_4
   b. $AlCl_3$ and $Ca(C_2H_3O_2)$
   c. $(NH_4)_2SO_4$ and $AgNO_3$
   d. KOH and H_2SO_4
   e. Na_3PO_4 and K_2S
   f. $CrCl_3$ and NaOH

*5. Suggest a compound whose water solution could be used to tell whether you had one or the other solution in each of these pairs. Write net ionic equations for the reactions, and tell your reasoning.
   a. $MgBr_2$ or $MnBr_2$
   b. $CuCl_2$ or $BaCl_2$
   c. NaCl or $NaNO_3$
   d. NH_4Cl or NH_4F
   e. $Mg(C_2H_3O_2)$ or K_3PO_4
   f. $FeSO_4$ or FeI_2
6. The second step in recovering magnesium from seawater is to dissolve the $Mg(OH)_2$ precipitate in HCl. Write the complete and net ionic equations, and explain why HCl dissolves $Mg(OH)_2$.
7. Write net ionic equations for dissolving these substances in strong acid. If a substance cannot be dissolved in strong acid, explain why.
   a. phosphate rock, $Ca_3(PO_4)_2$
   b. silver tarnish, Ag_2S
   c. limestone, $CaCO_3$
   d. bauxite ore, Al_2O_3
   e. saltpeter, $NaNO_3$
   f. NH_3 gas
   g. Cu metal
   h. PbI_2
8. Write net ionic equations for dissolving these substances in strong base. If a substance cannot be dissolved in strong base, explain why.
   a. calomel, Hg_2Cl_2
   b. $Ca(ClO)_2$
   c. HF gas
   d. CO_2 gas
   e. lactic acid (sour milk), C_2H_5OCOOH
   f. cinnabar, HgS

*9. Suggest a way in which helium gas, contaminated with SO_2, could be purified.

10. Translate these molecular equations into balanced net ionic equations.
   a. $Na_2CO_3(aq) + CuSO_4(aq) \longrightarrow CuCO_3(s) + Na_2SO_4(aq)$
   b. $Al(s) + HC_2H_3O_2(aq) \longrightarrow Al(C_2H_3O_2)_3(aq) + H_2(g)$
   c. $NaCl(aq) + H_2SO_4(aq) \longrightarrow HCl(g) + Na_2SO_4(aq)$
   d. $Cr(OH)_3(s) + HNO_3(aq) \longrightarrow Cr(NO_3)_3(aq) + H_2O(l)$
   e. $Pb(NO_3)_2(aq) + KCl(aq) \longrightarrow PbCl_2(s) + KNO_3(aq)$
11. Jars for collecting insect specimens often have a cap that is impregnated with, for example, a mixture of solid KCN and NaH_2PO_4. When water is sprinkled on the cap, HCN gas is released, killing the insect quickly. Write the net ionic equation for the reaction.
12. Calculate the number of grams of solute needed to make each of these solutions.
   a. 1.50 ℓ of 0.200 *M* K_2CrO_4
   b. 100. mℓ of 0.10 *M* NaOH
   c. 500. mℓ of a 0.500 *M* solution using oxalic acid dihydrate, $H_2C_2O_4 \cdot 2\,H_2O$, as the solute
   d. 1.00 ℓ of 0.485 *M* $Ca(NO_3)_2$
13. Calculate the number of milliters needed to make each of these solutions.
   a. 100. mℓ of 0.60 *M* H_2SO_4 from 18 *M* H_2SO_4
   b. 0.50 ℓ of 3.0 *M* NH_3 from 16 *M* NH_3
   c. 150 mℓ of 0.10 *M* NaOH from 12 *M* NaOH
   d. 200. mℓ of 5.0 *M* HNO_3 from 16 *M* HNO_3
14. A chemist wants to use some acetic acid to make salad dressing. How many milliliters of 6 *M* acetic acid must be diluted to get 0.50 ℓ of acetic acid having about the same concentration as vinegar (0.8 *M*)?
15. Determine the concentrations of these solutions.
   a. 5.0 mℓ of 12 *M* HCl diluted to 250. mℓ

b. 10.0 mℓ of 6.0 M NH_3 diluted to 100. mℓ
c. 25.0 mℓ of 0.116 M KNO_3 diluted to 1.00 ℓ
d. 10.0 mℓ of 10.0 M H_2SO_4 diluted to 500. mℓ

16. What volume of the specified new solutions will be produced by diluting each of these?
a. 25.0 mℓ of 12 M HCl to 6.0 M
b. 10.0 mℓ of 9 M NaOH to 1.0 M
c. 5.0 mℓ of 0.405 M NaCl to 0.200 M
d. 10.0 mℓ of 6.00 M HNO_3 to 0.300 M

17. Calculate the volume of 0.245 M HCl that would be needed to neutralize each.
a. 50.0 mℓ of 0.179 M KOH
b. 10.4 grams of Na_2CO_3

18. Lactic acid, C_2H_5OCOOH, is the sour ingredient in sour milk and buttermilk. A recipe that calls for sour milk or buttermilk usually also calls for baking soda ($NaHCO_3$). Write the equation for the reaction between lactic acid and baking soda, and calculate the number of grams of baking soda needed to neutralize 0.0510 M lactic acid in 1 cup (about 250 mℓ) of sour milk.

19. What is the molarity of a HCl solution if 25.0 mℓ of it required 28.4 mℓ of 0.122 M NaOH to neutralize it?

*20. A sample of rainwater collected near a copper smelter is analyzed for acid content. It turns out that a 100.-mℓ sample of the rainwater is neutralized by 22.4 mℓ of 0.0122 M NaOH. Assuming that the acid present is sulfurous acid that resulted from the reaction of SO_2 with water, what is the molarity of acid in the rainwater?

21. A sodium carbonate solution is standardized against potassium hydrogen phthalate, $KHC_8H_4O_4$ (M.W. = 204 g/mole) according to this reaction.

$$2\,KHC_8H_4O_4(aq) + Na_2CO_3(aq) \longrightarrow 2\,NaKC_8H_4O_4(aq) + CO_2(aq) + CO_2(g) + H_2O(l)$$

Then, 5.25 grams of potassium hydrogen phthalate required 35.2 mℓ of sodium carbonate solution to neutralize it. What is the molarity of the sodium carbonate solution?

22. A solution made by diluting 10.0 mℓ of concentrated (12 M) HCl to 1.00 liters was standardized against 0.105 M NaOH. A 25.0-mℓ sample of HCl required 35.2 mℓ of the base solution to neutralize it. Calculate the molarity of the HCl solution.

*23. The U.S. Department of Public Health limit on the amount of cyanide ion allowed in drinking water is 8×10^{-6} M. An industrial waste solution containing CN^- is allowed to react with 0.10 M $AgNO_3$ solution. It is found that 45.2 mℓ of silver nitrate solution will precipitate all the cyanide ion in 10.0 liters of waste.
a. Write the complete and net ionic equations.
b. What is the concentration of CN^- in the waste?
c. Will the government allow the waste to be used for drinking water?

24. What is the normality of each?
a. 1.0 M $NaHCO_3$ (as a monoprotic acid)
b. 0.500 M H_3PO_4 (as a triprotic acid)
c. 0.100 M NaOH
d. 2.0 M H_2SO_4 (as a diprotic acid)

25. Determine the equivalent weights, assuming complete neutralization.
a. H_2SO_4
b. $Ca(OH)_2$
c. HNO_3
d. Na_3PO_4

26. Determine the normalities of these solutions, assuming complete neutralization.
a. 98.0 g H_2SO_4 per liter
b. 10.5 g NaOH per 100. mℓ

27. Determine the normalities.
a. an acid solution, 25.0 mℓ of which required 21.2 mℓ of 0.133 N NaOH to neutralize it
b. a base solution, 10.0 mℓ of which required 28.9 mℓ of 0.225 N H_2SO_4 to neutralize it

28. What are the normalities of these laboratory solutions: (a) concentrated HCl; (b) concentrated HNO_3

29. A sulfuric acid solution is prepared by diluting 10.0 mℓ concentrated H_2SO_4 to 1.00 liter. It is then standardized by titrating against solid $NaHCO_3$. A 0.620 g sample of $NaHCO_3$ required 21.1 mℓ H_2SO_4. Calculate the normality of the H_2SO_4.

*30. An unknown acid is titrated against 0.195 N NaOH. A 0.250-g sample of the acid required 26.2 mℓ of the NaOH solution. Calculate the equivalent weight of the acid.

CHAPTER 15

Electrochemistry

LEARNING OBJECTIVES

After studying this chapter, you should be able to:

1. Supply a correct definition, explanation, or example for each of these:

electron donor	strong electrolyte
electron acceptor	weak electrolyte
oxidation	nonelectrolyte
reduction	electrolysis
reducing agent	electrolytic cell
oxidizing agent	cation
redox reaction	anion
galvanic cell	electroplating
battery	standard electrode potential
electrode	fuel cell
cathode	voltage
anode	oxidation number
electrolytic reaction	
electrolyte	

2. Write correct electron transfer half-reactions, state whether oxidation or reduction is occurring, identify the oxidizing and reducing agent, and identify the electrode at which each reaction would take place.
3. Sketch a galvanic cell and explain how it works.
4. Sketch an electrolytic cell, explain how it works, and identify the half-reaction that occurs at each electrode.
5. State whether a given substance is a strong, weak, or nonelectrolyte.
6. Given a table of oxidation potentials and half-reactions, predict whether a given reaction would occur spontaneously or not, and write a balanced ionic equation.
7. Given a table of oxidation potentials, calculate the voltage and/or energy for a given cell or electrochemical reaction.
8. Determine the oxidation number of any element in any compound.
9. Use oxidation numbers to balance redox reactions where half-reactions are not available.

So far, the energy we've seen in chemical reactions has been only in the form of heat. But heat isn't the only form of energy. Electricity is another one, and it can take part in chemical reactions just as heat can. Electrochemistry covers a class of chemical reactions that involve electrical energy. When we turn on a flashlight or start a car, we're using electrochemistry. The chemical activity in a battery is like an exothermic reaction: the battery releases energy in the form of electricity. A car bumper's chromium plating comes from the equivalent of an endothermic reaction: electricity must be supplied. Electrochemical processes that extract metals and nonmetals from their ores or from seawater are also like endothermic reactions.

We've seen precipitation reactions in which positive and negative ions are traded, and we've seen acid-base reactions involving proton transfer (Chapter 13). In this chapter, we'll look at electrochemical reactions, where electrons are transferred.

15.1 ELECTRONS IN CHEMICAL REACTIONS

Electrons are involved in all chemical reactions, but in electron transfer reactions they're involved as actual reactants and products.

SPONTANEOUS ELECTRON TRANSFER REACTIONS.

Many familiar reactions and applications involve electron transfer. Table 15.1 lists some typical spontaneous electron transfer reactions. Whereas precipitation and acid-base reactions involve only double-replacement, electron transfer reactions can be a combination of elements to form compounds, decomposition where elements are formed, or single replacement, as well as other more complex types.

In electron transfer, the substance that gives up electrons is the *electron donor,* and the substance that receives electrons is the *electron acceptor.* Another term for electron transfer is "oxidation-reduction." *Oxidation* means loss of electrons, and *reduction* means gain of electrons. ("Oxidation" used to mean only combining with oxygen. Combining with oxygen does cause a substance to lose electrons, so now oxidation means more generally any process in which electrons are lost.) The electron donor, since it causes another substance to gain electrons, is called the *reducing agent.* The electron acceptor, since it causes another substance to lose electrons, is called the *oxidizing agent.* In oxidation-reduction reactions, also called *redox* (*red*uction + *ox*idation) *reactions,* the reducing agent is oxidized and the oxidizing agent is reduced. Figure 15.1 illustrates these relationships.

electron donor

electron acceptor

oxidation

reduction

reducing agent

oxidizing agent

redox reaction

The substances on Table 15.1 that are reacting as oxidizing agents are shown in color. These equations don't show any electrons being reactants or products. To show this, we break each reaction into two parts.

TABLE 15.1
Some spontaneous electron transfer reactions

Reaction	*Type*	*Comments*
$2\,Na(s) + Cl_2(g) \longrightarrow 2\,NaCl(s)$	Combination (elements)	When an ionic compound forms, electrons are transferred from the metal to the nonmetal.
$CuSO_4(aq) + Zn(s) \longrightarrow ZnSO_4(aq) + Cu(s)$ Net ionic: $Cu^{2+} + Zn(s) \longrightarrow Zn^{2+} + Cu(s)$ This is the reaction that happens in one kind of galvanic cell.	Single replacement	When a metal reacts with a metal ion, electrons are transferred from the metal atom to the metal ion.
$6\,H^+ + Al(s) \longrightarrow Al^{3+} + 3\,H_2(g) + 3\,H_2O(l)$ Strong acids dissolve many metals.	Single replacement	When a metal reacts with a strong acid, electrons are transferred from the metal atoms to the acid's protons.
$2\,Al(s) + 6\,H_2O(l) + 2\,OH^-(aq) \longrightarrow 2\,Al(OH)_4^- + 3\,H_2(g)$ Strong bases dissolve many metals.	Single replacement	When a metal reacts with a strong base, electrons are transferred from the metal atoms to hydroxide ion's hydrogens.
$2\,I^- + Cl_2(g) \longrightarrow I_2(s) + 2\,Cl^-$ Extracting iodine from seawater with chlorine.	Single replacement	When a nonmetal reacts with a nonmetal ion, electrons are transferred from the nonmetal ion to the nonmetal atom.
$4\,Fe(s) + 3\,O_2(g) + H_2O(l) \longrightarrow 2\,Fe_2O_3(s) + H_2O$ Iron rusting in moist air.	Combination	Electrons are transferred from the iron to the oxygen.
$(NH_4)_2Cr_2O_7 \longrightarrow Cr_2O_3(s) + N_2(g) + 4\,H_2O(l)$ "Volcano" reaction.	Decomposition	Some electron transfer reactions are very complex.

ELECTRON TRANSFER HALF-REACTIONS. We can see that electron transfer is happening in this net ionic equation taken from Table 15.1:

$$Cu^{2+}(aq) + Zn(s) \longrightarrow Cu(s) + Zn^{2+}(aq)$$

Clearly, the zinc atom is giving its electrons to the copper(II) ion. Copper(II) ion is the oxidizing agent and zinc atom is the reducing agent. At the end of the reaction, copper(II) ion has been reduced and zinc has been oxidized. As we've done with several other processes before, we can break this down into partial reactions.

$$Zn(s) \longrightarrow Zn^{2+}(aq) + 2e^- \qquad \textit{Oxidation}$$
$$Cu^{2+} + 2e^- \longrightarrow Cu(s) \qquad \textit{Reduction}$$

FIGURE 15.1
Reducing agents are electron donors, and oxidizing agents are electron acceptors

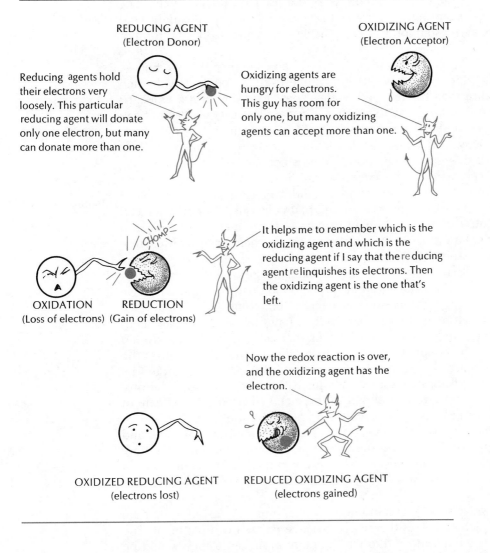

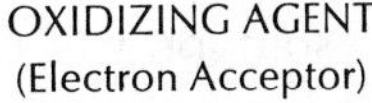
OXIDIZING AGENT
(Electron Acceptor)

REDUCING AGENT
(Electron Donor)
Reducing agents hold their electrons very loosely. This particular reducing agent will donate only one electron, but many can donate more than one.

Oxidizing agents are hungry for electrons. This guy has room for only one, but many oxidizing agents can accept more than one.

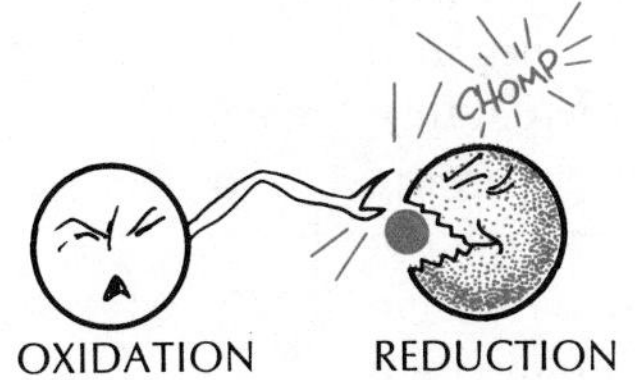
CHOMP
OXIDATION
(Loss of electrons)
REDUCTION
(Gain of electrons)
It helps me to remember which is the oxidizing agent and which is the reducing agent if I say that the reducing agent relinquishes its electrons. Then the oxidizing agent is the one that's left.

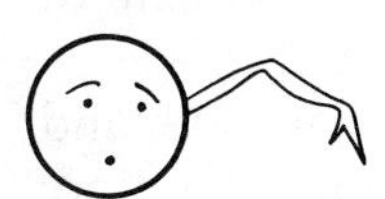
OXIDIZED REDUCING AGENT
(electrons lost)

Now the redox reaction is over, and the oxidizing agent has the electron.
REDUCED OXIDIZING AGENT
(electrons gained)

Each of these represents a *half-reaction*. Adding the half-reactions gives the total reaction. In electron transfer reactions, one half-reaction must always show electrons being given up and one must show electrons being accepted. As with chemical substances, electrons can be neither destroyed nor created, so the number of electrons lost is always the same as the number gained. Although electrons don't appear anywhere in the complete equation, they do appear as either reactants or products in the half-reactions. These half-reactions happen simultaneously, and not one at a time.

half-reaction

If we put a piece of zinc metal into a solution containing copper(II) ions, we'd see something happen. The solution, originally blue because of the presence of the blue copper(II) ion, becomes colorless as the copper(II) ions are replaced by colorless zinc ions. The surface of the zinc metal changes from shiny silver grey to dark as copper metal is deposited (plated) on it. The solution warms up as energy is released in the form of heat.

Because this reaction releases energy, we can use it in a battery. But we can't let the zinc and copper(II) ions come in direct contact with each other, because then the released energy doesn't do any useful work. To make it do useful work, and to get the energy as electricity instead of heat, we have to make the electron transfer happen through an external circuit and not directly. This means separating the zinc from the copper(II) ions.

Figure 15.2 shows zinc and copper(II) ions connected in a galvanic cell. A *galvanic cell,* named after Luigi Galvani, is any cell in which a spontaneous redox reaction is harnessed to produce electricity. *Batteries* contain various types of galvanic cells. In this particular one, a bar of zinc metal is sticking into the zinc solution and a bar of copper metal is sticking into the copper solution. These bars are *electrodes,* where each half-reaction will take place. The partition is there to keep the copper(II) ion from coming into contact with the zinc electrode and taking the zinc's electrons directly.

galvanic cell

battery

electrode

When the electrodes aren't connected, nothing happens. But when we connect them through an external circuit, something like a relay race begins. A zinc atom on the zinc electrode donates two electrons, and goes into solution as a zinc ion. The electrons that zinc donates whiz around the circuit to the copper electrode, causing the light bulb to light on the way. As the electrons reach the copper electrode, they are accepted by waiting copper(II) ions. As soon as a copper(II) ion accepts two electrons, it becomes a copper atom and part of the metal electrode. The stream of electrons flowing around the circuit is an electric current.

In this cell, positive zinc ions are constantly formed on one side, and positive copper(II) ions are constantly lost on the other side. But a solution, like a chemical formula, must be electrically neutral. The reaction would stop as soon as there was a slight excess of either positive or negative ions in either compartment. To keep the reaction going, we make the partition porous. Negative sulfate ions and positive zinc ions can move through the partition, which equalizes the charge balance in each compartment. The copper(II) ions don't move through the partition because they would be moving into a compartment that already had too many positive ions in it.

In any cell, the electrode where reduction takes place is called the *cathode,* and the electrode where oxidation takes place is called the *anode.* In this cell, the copper electrode is the cathode, where the reduction half-reaction takes place. The zinc ions won't be reduced, even though they can come in contact with the cathode, because they are much worse electron acceptors than copper(II) ions.

cathode

anode

The zinc electrode is the anode, where the oxidation half-reaction takes place. The sulfate ions can come in contact with the anode, but nothing happens to them. They are neither oxidizing nor reducing agents, but spectator ions. If a copper(II) ion touched the zinc electrode, though, it

FIGURE 15.2
A galvanic cell harnesses spontaneous electron transfer by passing electrons around a circuit

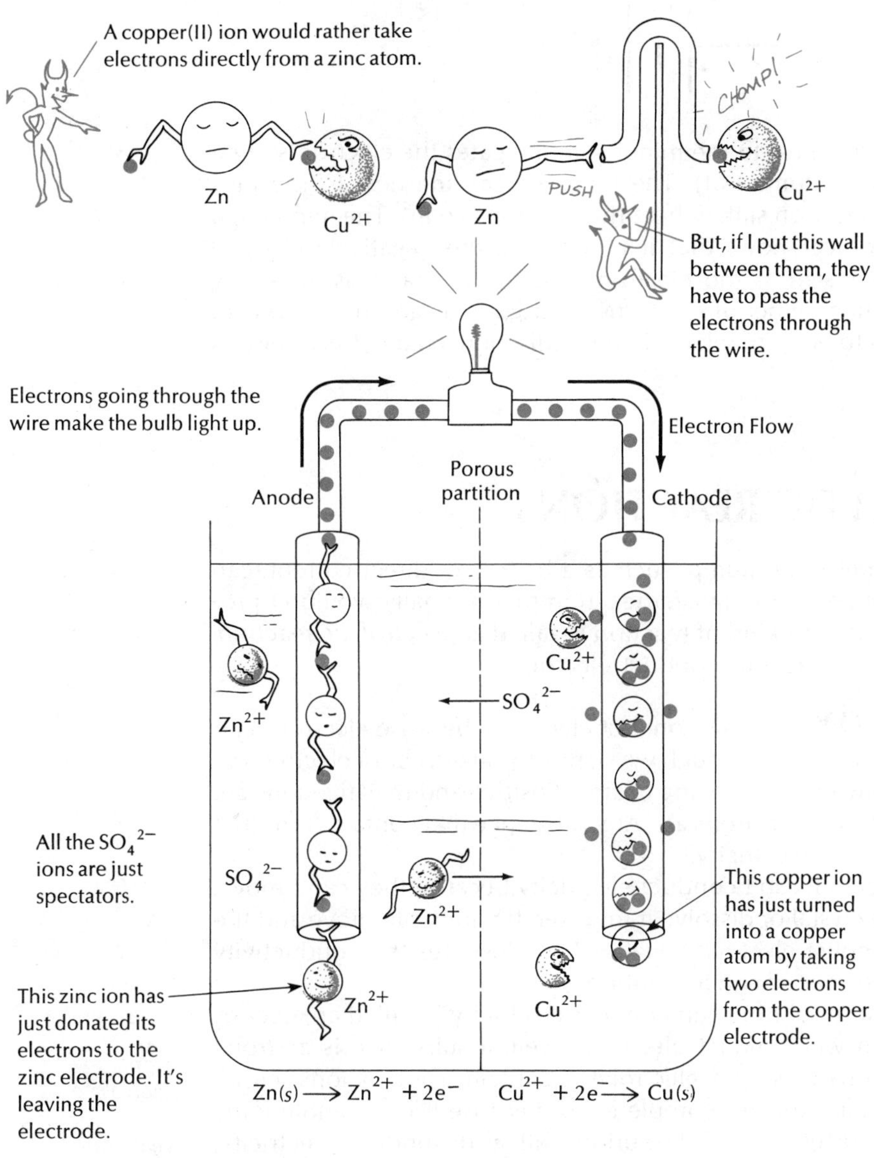

A copper(II) ion would rather take electrons directly from a zinc atom.
Zn
Cu2+
Zn
PUSH
CHOMP!
Cu2+
But, if I put this wall between them, they have to pass the electrons through the wire.
Electrons going through the wire make the bulb light up.
Electron Flow
Anode
Porous partition
Cathode
Zn2+
SO4 2−
Cu2+
All the SO4 2− ions are just spectators.
SO4 2−
Zn2+
This copper ion has just turned into a copper atom by taking two electrons from the copper electrode.
This zinc ion has just donated its electrons to the zinc electrode. It's leaving the electrode.
Zn2+
Cu2+
Zn(s) ⟶ Zn2+ + 2e−
Cu2+ + 2e− ⟶ Cu(s)

would take electrons directly from the zinc atoms and stop the current flow through the circuit. But the porous partition helps keep the copper(II) ions on their own side.

When the zinc has all been oxidized, or when the copper(II) ions have all been reduced, the battery is dead.

This particular battery is only one of many possible ones. We see, though, that a battery is an electron pump. When we connect it, it pumps electrons around a circuit. We use the following general symbol to mean a battery.

The short vertical line next to the minus sign indicates the electron source (where the electrons come out). The long vertical line next to the plus sign indicates the electron sink (where the electrons go in). The minus sign at the electron source shows that it is repelling the negatively charged electrons. The plus sign at the electron sink shows that it is attracting electrons. The same number of electrons must go in as go out. For this to happen, there has to be a complete circuit. A disconnected battery delivers no current.

15.2 ELECTROLYTIC REACTIONS

Electrons from an electron pump, such as a battery or direct current, can be used to drive an electron transfer reaction that normally wouldn't proceed spontaneously. This kind of reaction is called an *electrolytic reaction.* It occurs only if the reactants conduct electricity.

electrolytic reaction

CONDUCTIVITY. Metals conduct electricity because electrons can move easily through them. Although we've mostly talked about electrons so far, electricity is any kind of moving charge. Positive and negative ions are charged particles too, as electrons are. Any substance that contains ions that can move will conduct electricity.

Ions in a solid crystal can't conduct electricity because they can't move. But if we melt the crystal or dissolve it in water, the ions can move and the substance will conduct electricity. Figure 15.3 illustrates the conductivity of sodium chloride, melted and in solution.

A substance whose solution can conduct electricity is called an *electrolyte.* If its solution will conduct electricity well, a substance is a *strong electrolyte.* Solutions of strong electrolytes contain a lot of ions, either because the substance ionizes completely or because it is a soluble ionic compound. A *weak electrolyte's* solution will also conduct electricity, but not nearly as well; this is because it doesn't contain many ions. Weak electrolytes either don't ionize completely—like weak acids and bases—or are relatively insoluble ionic compounds. A *nonelectrolyte* is a substance whose solution won't conduct electricity at all, because it contains no ions. Table 15.2 gives examples of each kind of electrolyte.

electrolyte

strong electrolyte

weak electrolyte

nonelectrolyte

Water itself is listed as a nonelectrolyte, even though it does form a few ions. The concentration of hydronium and hydroxide ions in water—

FIGURE 15.3
Moving ions conduct electricity

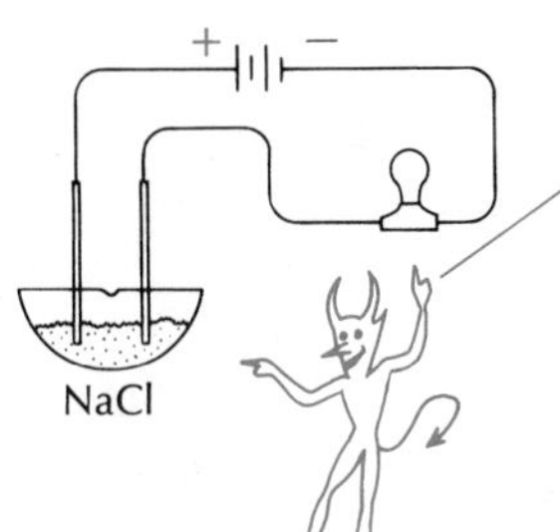

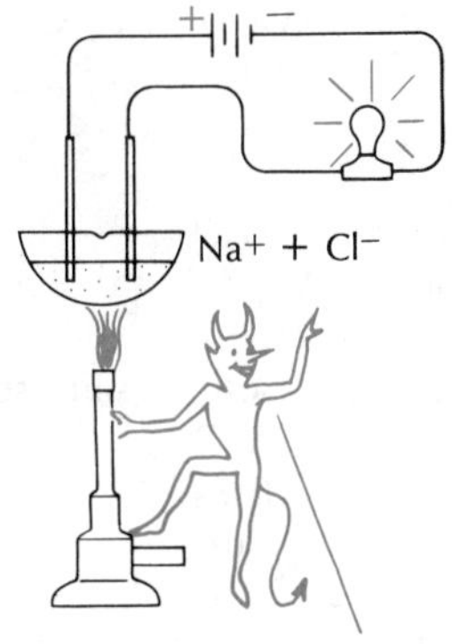

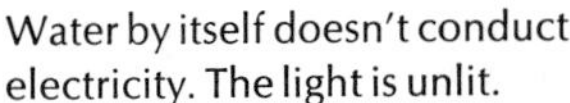

NaCl
When I put these electrodes into a dish of salt crystals, nothing happens. The light doesn't light. No conductivity. The ions in sodium chloride are fastened tightly in their crystal structure.

Na+ + Cl−
If I melt the salt, though, the light lights up! There is conductivity. The ions can move and conduct current.

Water by itself doesn't conduct electricity. The light is unlit.

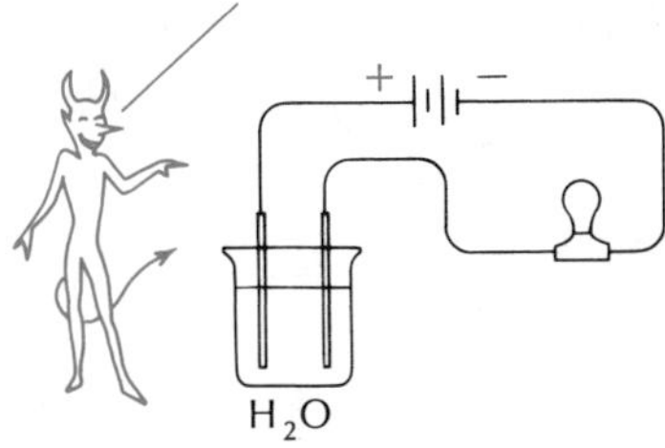

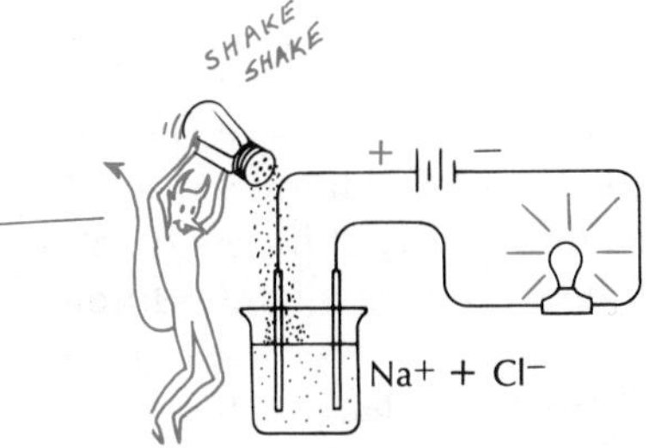

H₂O

SHAKE
SHAKE

Na+ + Cl−
But now watch what happens when I shake some salt into the water. When the ions are dissolved, they can move and conduct current. The light lights up.

10^{-7} *M* for each—isn't enough for water to conduct electricity. Even so, we shouldn't be in contact with water when we're using electricity. Dissolved ions in tap water, plus salt that's on our skin, can make water conduct enough to be dangerous.

Cars that spend a lot of time around oceans or where winter roads must be salted tend to rust rapidly. The rusting of iron is an electrochemical process, as we saw in Table 15.1. The salt makes an electrolytic solution that speeds up the rusting reaction.

ELECTROLYSIS. Conductivity and electrolysis go together. A battery or direct current source conducts electricity when connected to a solution or a melted ionic compound; electrolysis happens then, too. *Electrolysis* is the use of electrical energy to drive a chemical reaction.

electrolysis

TABLE 15.2
Examples of strong, weak, and nonelectrolytes

Strong Electrolytes	*Weak Electrolytes*	*Nonelectrolytes*
Strong acids, such as:	Weak acids, such as:	Any covalent compound that isn't an acid or base, such as:
$H^+ + HSO_4^-$	$HC_2H_3O_2$	ethyl alcohol
$H^+ + Cl^-$	H_3PO_4	sugar
$H^+ + NO_3^-$	HF	glycerin
$H^+ + ClO_4^-$	H_2SO_3	antifreeze
Strong bases, such as:	Weak bases, such as:	water
$Na^+ + OH^-$	NH_3	
$Ca^{2+} + 2\,OH^-$	Insoluble salts, such as:	
Soluble salts, such as:	AgCl	
$Na^+ + Cl^-$	$BaSO_4$	
$K^+ + NO_3^-$	CuO	
$2\,NH_4^+ + SO_4^{2-}$		

We know this reaction occurs spontaneously:

$$2\,Na(s) + Cl_2(g) \longrightarrow 2\,NaCl(s) + \text{energy}$$

Sodium chloride won't decompose spontaneously to form sodium and chlorine. But we can force it to decompose by supplying electrical energy. This is the reaction that takes place in the electrolysis of melted NaCl:

$$2\,NaCl(l) + \text{energy} \longrightarrow 2\,Na(l) + Cl_2(g)$$

This reaction isn't quite the reverse of the formation of sodium chloride, because it happens at a higher temperature, where both sodium chloride and sodium are liquids. This difference, plus some other distinctions, cause the two energies to have slightly different values.

Electrolysis is a way to recover sodium and chlorine from seawater. First, the water is evaporated, leaving the solid salt. Then the salt is melted and electrolyzed. Chlorine gas comes off at the anode and liquid sodium drips from the cathode.

electrolytic cell

An *electrolytic cell* is a cell in which electrical energy is pumped in to cause a chemical reaction. An electrolytic cell is the opposite of a galvanic cell, where a chemical reaction is used to provide electrical energy. Figure 15.4 gives us a detailed look at the sodium chloride electrolytic cell. The circuit must be complete for electrical conductivity to occur. The circuit is completed by the movement of the ions, which pick up and deposit electrons. The positive ions move to the cathode, where they accept electrons pumped in by the battery.

Cathode: $Na^+(l) + e^- \longrightarrow Na(l)$ *Reduction*

At the same time, negative ions move to the anode and donate their electrons to the battery's sink.

Anode: $2\,Cl^-(l) \longrightarrow Cl_2(g) + 2e^-$ *Oxidation*

FIGURE 15.4
The electrolysis of melted NaCl

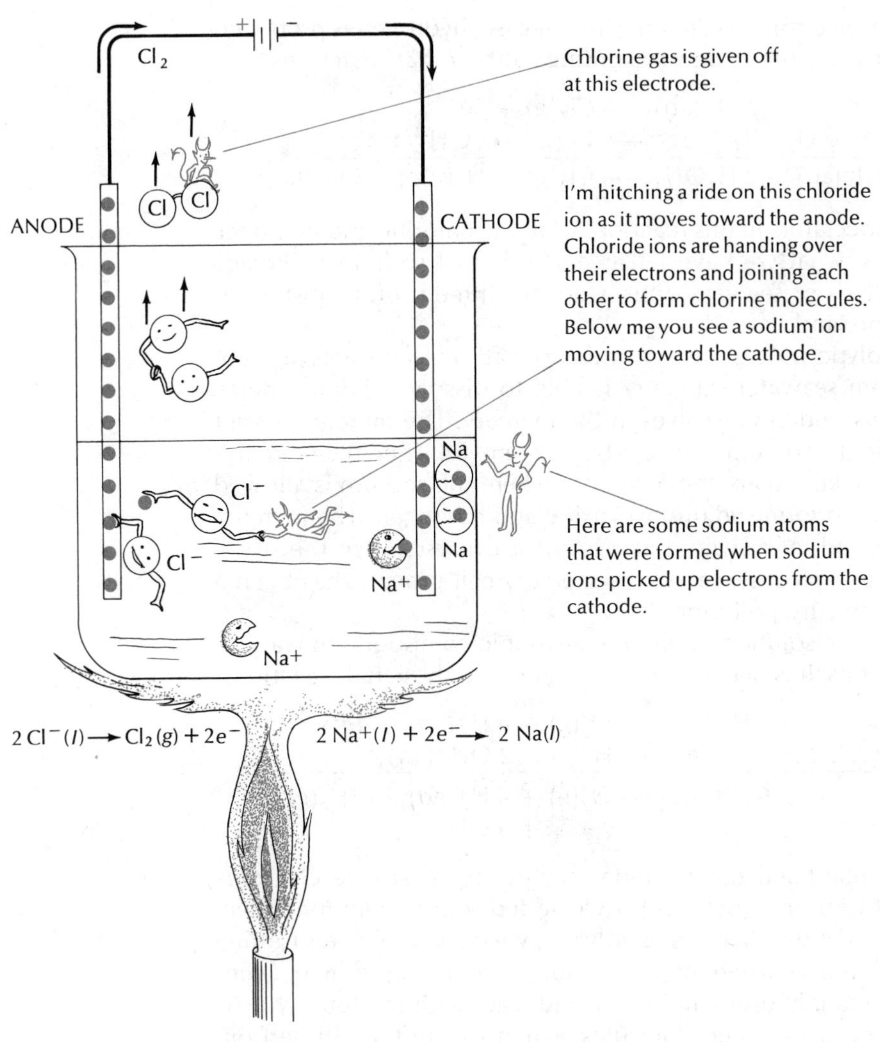
Cl_2
Chlorine gas is given off at this electrode.
ANODE
CATHODE
I'm hitching a ride on this chloride ion as it moves toward the anode. Chloride ions are handing over their electrons and joining each other to form chlorine molecules. Below me you see a sodium ion moving toward the cathode.
Cl
Cl
Cl^-
Cl^-
Na
Na
Na^+
Na^+
Here are some sodium atoms that were formed when sodium ions picked up electrons from the cathode.
$2\,Cl^-(l) \rightarrow Cl_2(g) + 2e^-$
$2\,Na^+(l) + 2e^- \rightarrow 2\,Na(l)$

Because they move toward the cathode, positive ions are also called *cations* (read, "cat-ions"). Negative ions are called *anions* (read, "an-ions"), because they move toward the anode.

cation

anion

We said that the number of electrons going in and out of a battery must be the same. To show this, we make sure that the electrons are balanced by multiplying one or both of the half-reactions by appropriate numbers. In this case, we multiply the cathode half-reaction by 2. Then we can add the half-reactions to get the overall reaction, and the electrons will cancel.

Cathode: $2\,Na^+(l) + \cancel{2e^-} \longrightarrow 2\,Na(l)$
Anode: $2\,Cl^-(l) \longrightarrow Cl_2(g) + \cancel{2e^-}$
Overall: $2\,Na^+(l) + 2\,Cl^-(l) \longrightarrow 2\,Na(l) + Cl_2(g)$

Seawater can be electrolyzed directly. In that case, hydrogen is produced at the cathode, instead of sodium, according to these half-reactions:

Anode: $2\,Cl^-(aq) \longrightarrow Cl_2(g) + \cancel{2e^-}$
Cathode: $2\,H_2O(l) + \cancel{2e^-} \longrightarrow H_2(g) + 2\,OH^-(aq)$
Overall: $2\,Cl^-(aq) + 2\,H_2O(l) \longrightarrow Cl_2(g) + H_2(g) + 2\,OH^-(aq)$

Sodium ions are spectators in this reaction. With the chlorine and hydrogen going off as gases, what we have left is a solution of sodium hydroxide (lye). The electrolysis of seawater thus produces three useful substances: hydrogen, chlorine, and lye.

Another electrolytic process uses a mercury cathode and gets sodium metal directly from seawater. Mercury is able to dissolve sodium metal. As soon as it forms, sodium dissolves in the mercury. The mercury is kept flowing through so the sodium is always being removed. Then the mercury-sodium mixture is taken to another chamber where the sodium is allowed to react with water to form sodium hydroxide and hydrogen. The mercury can be used over and over again, but a lot of it can escape in the waste water. This particular process, called the mercury cell process, has been a major source of mercury pollution.

When a more dilute solution of NaCl or any ionic compound in water is electrolyzed, water itself is decomposed, according to these half-reactions:

Anode: $2\,H_2O(l) \longrightarrow O_2(g) + 4\,H^+(aq) + \cancel{4e^-}$
Cathode: $2[2\,H_2O + \cancel{2e^-} \longrightarrow H_2(g) + 2\,OH^-(aq)]$
Overall: $6\,H_2O(l) \longrightarrow O_2(g) + 4\,H^+(aq) + 2\,H_2(g) + 4\,OH^-(aq)$

(Notice that we multiplied the cathode reaction by 2, so the electrons would cancel out of both equations.) If we added some phenolphthalein indicator to water that was being electrolyzed, we'd see the water turning pink around the cathode, where hydroxide ions were being formed. Eventually, though, the four hydronium ions would react with the four hydroxide ions to produce eight water molecules. When we subtract these from the ten on the left, our net reaction is:

$$2\,H_2O(l) \longrightarrow 2\,H_2(g) + O_2(g)$$

Here, both the sodium and chloride ions are spectators. They serve only to carry the current and to make the solution electrically neutral. This is another way that hydrogen is produced commercially, with oxygen as a by-product.

Many metals are produced by electrolysis of their compounds. Magnesium is prepared this way. In Chapter 14, we saw seawater being treated with slaked lime, $Ca(OH)_2$, to precipitate the magnesium as $Mg(OH)_2$.

$$Mg^{2+}(aq) + 2\ OH^-(aq) \longrightarrow Mg(OH)_2(s)$$

Next, the $Mg(OH)_2$ is filtered out and dissolved in HCl.

$$Mg(OH)_2(s) + 2\ H^+(aq) + 2\ Cl^-(aq) \longrightarrow Mg^{2+}(aq) + 2\ Cl^-(aq) + 2\ H_2O(l)$$

Now the water is removed by evaporation, leaving solid $MgCl_2$, which is then melted and electrolyzed.

$$Mg^{2+}(l) + 2\ Cl^-(l) \longrightarrow Mg(s) + Cl_2(g)$$

ELECTROPLATING. "Plating" means depositing a thin coating of one metal over another metal. When this is done in an electrolytic cell, it's called *electroplating*. Silver plate, a thin layer of silver over an object made of another metal (usually nickel), is an example of electroplating.

electroplating

In silver-plating an object like a nickel spoon, the spoon itself is used as the cathode in a solution of silver nitrate. The silver ions move to the cathode, accept electrons, and plate out on the spoon.

Cathode: $Ag^+(aq) + e^- \longrightarrow Ag(s)$

To replace the silver ions, a bar of silver metal is used for the anode.

Anode: $Ag(s) \longrightarrow Ag^+(aq) + e^-$

The silver atoms, giving up their electrons to the anode, dissolve into the solution as silver ions. The nitrate ions are spectators. There is no net reaction here, but silver has been moved from the anode to the spoon.

Copper and chromium plating are also done this way. "Tin" cans are actually steel cans that have been tin-plated because steel rusts and tin doesn't. In plating, the object to be plated is the cathode, the solution contains ions of the metal to be plated out, and the anode is a bar of the solid metal that is being plated out.

15.3 OXIDIZING AND REDUCING AGENTS

We've seen that some redox reactions proceed spontaneously and others require energy. Now we'll see how to tell which is which.

STANDARD ELECTRODE POTENTIALS. Table 15.3 shows oxidizing and reducing agents, with their half-reactions, arranged in order of strength. We can use this table to decide whether a given oxidizing agent will react with a given reducing agent or not. Every oxidizing agent has a reduced form, and every reducing agent has an oxidized form, just as every acid has a conjugate base and every base has a conjugate acid. The stronger an oxidizing agent, the weaker its reduced form is as a reducing agent, and

TABLE 15.3
Oxidizing and reducing agents

Oxidation Half-Reaction	Reducing Agents	Electrode Potential, E^0 (volts)
$2\,F^-(aq) \longrightarrow F_2(g) + 2e^-$	F^-	−2.65
$2\,H_2O(l) \longrightarrow H_2O_2(aq)\ 2\,H^+(aq) + 2e^-$	H_2O	−1.77
$PbSO_4(s) + 2\,H_2O(l) \longrightarrow PbO_2(s) + SO_4^{2-}(aq) + 4\,H^+(aq) + 2e^-$	$PbSO_4$	−1.68
$Au(s) \longrightarrow Au^{3+}(aq) + 3e^-$	Au	−1.50
$Mn^{2+}(aq) + 4\,H_2O(l) \longrightarrow MnO_4^-(aq) + 8\,H^+(aq) + 5e^-$	Mn^{2+}	−1.50
$2\,Cl^-(aq) \longrightarrow Cl_2(g) + 2e^-$	Cl^-	−1.36
$2\,Cr^{3+}(aq) + 7\,H_2O(l) \longrightarrow Cr_2O_7^{2-}(aq) + 14\,H^+ + 6e^-$	Cr^{3+}	−1.33
$Mn^{2+}(aq) + 2\,H_2O(l) \longrightarrow MnO_2(s) + 4\,H^+(aq) + 2e^-$	Mn^{2+}	−1.23
$2\,H_2O(l) \longrightarrow O_2(g) + 4\,H^+(aq) + 4e^-$	H_2O	−1.23
$2\,Br^-(aq) \longrightarrow Br_2(l) + 2e^-$	Br^-	−1.06
$NO(g) + 2\,H_2O(l) \longrightarrow NO_3^-(aq) + 4\,H^+(aq) + 3e^-$	NO	−0.96
$Ag(s) \longrightarrow Ag^+(aq) + e^-$	Ag	−0.80
$Fe^{2+}(aq) \longrightarrow Fe^{3+}(aq) + e^-$	Fe^{2+}	−0.77
$Mn(OH)_3(s) + NH_3(aq) \longrightarrow MnO_2(s) + H_2O(l) + NH_4^+(aq) + e^-$	$Mn(OH)_3$	−0.74
$H_2O_2(aq) \longrightarrow O_2(g) + 2\,H^+(aq) + 2e^-$	H_2O_2	−0.68
$2\,I^-(aq) \longrightarrow I_2(s) + 2e^-$	I^-	−0.54
$Ni(OH)_2(s) + 2\,OH^-(aq) \longrightarrow NiO_2(s) + 2\,H_2O(l) + 2e^-$	$Ni(OH)_2$	−0.49
$Cu(s) \longrightarrow Cu^{2+}(aq) + 2e^-$	Cu	−0.34
$H_2S(aq) \longrightarrow S(s) + 2\,H^+(aq) + 2e^-$	H_2S	−0.14
$Hg(l) + 2\,OH^-(aq) \longrightarrow HgO(s) + H_2O(l) + 2e^-$	Hg	−0.10
$H_2(g) \longrightarrow 2\,H^+(aq) + 2e^-$	H_2	0.00
$Pb(s) \longrightarrow Pb^{2+}(aq) + 2e^-$	Pb	+0.13
$Sn(s) \longrightarrow Sn^{2+}(aq) + 2e^-$	Sn	+0.14
$Ni(s) \longrightarrow Ni^{2+}(aq) + 2e^-$	Ni	+0.25
$Pb(s) + SO_4^{2-}(aq) \longrightarrow PbSO_4(s) + 2e^-$	Pb	+0.36
$Fe(s) \longrightarrow Fe^{2+}(aq) + 2e^-$	Fe	+0.44
$Zn(s) \longrightarrow Zn^{2+}(aq) + 2e^-$	Zn	+0.76
$Cd(s) + 2\,OH^-(aq) \longrightarrow Cd(OH)_2(s) + 2e^-$	Cd	+0.81
$H_2(g) + 2\,OH^-(aq) \longrightarrow 2\,H_2O(l) + 2e^-$	H_2	+0.83
$Cr(s) \longrightarrow Cr^{2+}(aq) + 2e^-$	Cr	+0.91
$Zn(s) + 2\,OH^-(aq) \longrightarrow Zn(OH)_2(s) + 2e^-$	Zn	+1.25
$Al(s) \longrightarrow Al^{3+}(aq) + 3e^-$	Al	+1.66
$Mg(s) \longrightarrow Mg^{2+}(aq) + 2e^-$	Mg	+2.37
$Na(s) \longrightarrow Na^+(aq) + e^-$	Na	+2.71
$Li(s) \longrightarrow Li^+(aq) + e^-$	Li	+3.01

INCREASING STRENGTH OF REDUCING AGENTS

Electrode Potential, E^0 (volts)	*Oxidizing Agents*	*Reduction Half-Reaction*
+2.65	F_2	$F_2(g) + 2e^- \longrightarrow 2\,F^-(aq)$
+1.77	H_2O_2	$H_2O_2(aq) + 2\,H^+(aq) + 2e^- \longrightarrow 2\,H_2O(l)$
+1.68	PbO_2	$PbO_2(s) + SO_4^{2-}(aq) + 4\,H^+ + 2e^- \longrightarrow PbSO_4(s) + 2\,H_2O(l)$
+1.50	Au^{3+}	$Au^{3+}(aq) + 3e^- \longrightarrow Au(s)$
+1.50	MnO_4^-	$MnO_4^-(aq) + 8\,H^+(aq) + 5e^- \longrightarrow Mn^{2+}(aq) + 4\,H_2O(l)$
+1.36	Cl_2	$Cl_2(g) + 2e^- \longrightarrow 2\,Cl^-(aq)$
+1.33	$Cr_2O_7^{2-}$	$Cr_2O_7^{2-}(aq) + 14\,H^+(aq) + 6e^- \longrightarrow 2\,Cr^{3+}(aq) + 7\,H_2O(l)$
+1.23	MnO_2	$MnO_2(s) + 4\,H^+(aq) + 2e^- \longrightarrow Mn^{2+}(aq) + 2\,H_2O(l)$
+1.23	O_2	$O_2(g) + 4\,H^+(aq) + 4e^- \longrightarrow 2\,H_2O(l)$
+1.06	Br_2	$Br_2(l) + 2e^- \longrightarrow 2\,Br^-(aq)$
+0.96	NO_3^-	$NO_3^-(aq) + 4\,H^+(aq) + 3e^- \longrightarrow NO(g) + 2\,H_2O(l)$
+0.80	Ag^+	$Ag^+(aq) + e^- \longrightarrow Ag(s)$
+0.77	Fe^{3+}	$Fe^{3+}(aq) + e^- \longrightarrow Fe^{2+}(aq)$
+0.74	MnO_2	$MnO_2(s) + H_2O(l) + NH_4^+(aq) + e^- \longrightarrow Mn(OH)_3(s) + NH_3(aq)$
+0.68	O_2	$O_2(g) + 2\,H^+(aq) + 2e^- \longrightarrow H_2O_2(aq)$
+0.54	I_2	$I_2(s) + 2e^- \longrightarrow 2\,I^-(aq)$
+0.49	NiO_2	$NiO_2(s) + 2\,H_2O(l) + 2e^- \longrightarrow Ni(OH)_2(s) + 2\,OH^-(aq)$
+0.34	Cu^{2+}	$Cu^{2+}(aq) + 2e^- \longrightarrow Cu(s)$
+0.14	S	$S(s) + 2\,H^+(aq) + 2e^- \longrightarrow H_2S(aq)$
+0.10	HgO	$HgO(s) + H_2O(l) + 2e^- \longrightarrow Hg(l) + 2\,OH^-(aq)$
0.00	H^+	$2\,H^+(aq) + 2e^- \longrightarrow H_2(g)$
−0.13	Pb^{2+}	$Pb^{2+}(aq) + 2e^- \longrightarrow Pb(s)$
−0.14	Sn^{2+}	$Sn^{2+}(aq) + 2e^- \longrightarrow Sn(s)$
−0.25	Ni^{2+}	$Ni2+(aq) + 2e^- \longrightarrow Ni(s)$
−0.36	$PbSO_4$	$PbSO_4(s) + 2e^- \longrightarrow Pb(s) + SO_4^{2-}$
−0.44	Fe^{2+}	$Fe^{2+}(aq) + 2e^- \longrightarrow Fe(s)$
−0.76	Zn^{2+}	$Zn^{2+}(aq) + 2e^- \longrightarrow Zn(s)$
−0.81	$Cd(OH)_2$	$Cd(OH)_2(s) + 2e^- \longrightarrow Cd(s) + 2\,OH^-(aq)$
−0.83	H_2O	$2\,H_2O(l) + 2e^- \longrightarrow H_2(g) + 2\,OH^-(aq)$
−0.91	Cr^{2+}	$Cr^{2+}(aq) + 2e^- \longrightarrow Cr(s)$
−1.25	$Zn(OH)_2$	$Zn(OH)_2(s) + 2e^- \longrightarrow Zn(s) + 2\,OH^-(aq)$
−1.66	Al^{3+}	$Al^{3+}(aq) + 3e^- \longrightarrow Al(s)$
−2.37	Mg^{2+}	$Mg^{2+}(aq) + 2e^- \longrightarrow Mg(s)$
−2.71	Na^+	$Na^+(aq) + e^- \longrightarrow Na(s)$
−3.01	Li^+	$Li^+(aq) + e^- \longrightarrow Li(s)$

INCREASING STRENGTH OF OXIDIZING AGENTS

vice versa. Like proton transfer and protons, electron transfer is a contest between two substances for one or more electrons.

Unlike acid-base reactions, redox reactions are usually visible: a gas or solid is formed, a solid is dissolved, or there is a color change. Heat is usually involved, too. We see in the table that changes between ions in solution, and solids or gases out of solution, are frequent. For example, when iron metal is oxidized, it goes into solution as Fe^{2+}; when H^+ is reduced, H_2 goes out of solution as a gas. Also, many half-reactions contain substances other than the oxidizing or reducing agent, such as H_2O, OH^-, H^+, and others. These are present to provide atom balance for the half-reactions. And finally, the number of electrons being transferred is often more than one, while in proton transfer reactions one is the usual number.

Accompanying each half-reaction is a number called the *standard electrode potential* (E^0). This potential describes the tendency of a given half-reaction to proceed in the direction it's written. The more positive the potential, the more the half-reaction would tend to proceed. The more negative the potential, the less it would tend to proceed. These potentials provide the basis for the order of oxidizing and reducing agents in the table. We had a partial, simplified version of this table in the Activity Series in Chapter 5. The hydrogen/hydronium-ion electrode—with a potential of zero—is the dividing line between positive and negative. We'll be explaining and working with these potentials in more detail. Right now, though, we'll see how to use the table to predict reactions.

standard electrode potential (E^0)

PREDICTING REDOX REACTIONS. An oxidizing agent will react only with reducing agents that are below it in the table. Generally speaking, alkali metal ions and alkaline earth metal ions in solution will not be oxidizing agents, but they usually will be spectator ions. Similarly, the negative ions sulfate and phosphate are usually neither oxidizing nor reducing agents but spectator ions. This information, plus the table, can be used to predict reactions. Let's see how by finding out if a reaction will take place when aluminum metal is placed in a solution of nickel(II) sulfate. Our goal is to write the net ionic equation for the reaction, if there is one.

Step 1. *Write all substances in their correct forms.*
In our example, they are $Al(s)$, $Ni^{2+}(aq)$, and $SO_4^{2-}(aq)$.

Step 2. *Decide whether any of these are oxidizing or reducing agents.*
We've been told that SO_4^{2-} is neither, but we do find Ni^{2+} in the oxidation column above $Al(s)$ in the reduction column. Thus $Ni^{2+}(aq)$ will oxidize $Al(s)$.

Step 3. *Write the two half-reactions, balance the electrons, and add the two equations.*

$$Al(s) \longrightarrow Al^{3+}(aq) + 3e^-$$
$$Ni^{2+}(aq) + 2e^- \longrightarrow Ni(s)$$

Here, we have a loss of three electrons but a gain of only two. To balance the electrons, we multiply the top equation by 2 and the bottom equation by 3.

$$
\begin{array}{r}
2\,Al(s) \longrightarrow 2\,Al^{3+}(aq) + \cancel{6e^-} \\
3\,Ni^{2+}(aq) + \cancel{6e^-} \longrightarrow 3\,Ni(s) \\
\hline
2\,Al(s) + 3\,Ni^{2+}(aq) \longrightarrow 2\,Al^{3+}(aq) + 3\,Ni(s)
\end{array}
$$

Step 4. *Check the atom and charge balance.*
The atoms balance, and the charges are 6+ on each side. Yes, a reaction will occur, and the equation we've written is correct.

This reaction would allow us to nickel-plate an aluminum object without supplying any energy, because the reaction proceeds spontaneously. While it was happening, we'd see the mixture bubbling, and the green $Ni^{2+}(aq)$ would be replaced by the colorless $Al^{3+}(aq)$. A thin layer of nickel would coat the aluminum object.

EXAMPLE 15.1: Write the net ionic equation for the reaction, if any, that would occur if a piece of iodine were placed in a solution of NaCl.

Solution:
Step 1: The substances are $I_2(s)$, $Na^+(aq)$, and $Cl^-(aq)$.
Step 2: We find I_2 as an oxidizing agent *below* Cl^- as a reducing agent.

Answer: No reaction would occur. ■

In fact, if we look back at Table 15.1, we can see that the reverse of this reaction is used to remove iodine from seawater. If we bubbled chlorine gas through a solution containing iodide ions, we'd see fine dark crystals of iodine precipitating. Water solutions of chlorine are often used to test for the presence of iodide ion in an unknown solution.

EXAMPLE 15.2: Could $KMnO_4$ be used to remove H_2S from waste water? (That is, will $KMnO_4$ react with H_2S?) Write the equation for the reaction, if any.

Solution:
Step 1: $K^+(aq)$, $MnO_4^-(aq)$, and $H_2S(aq)$.
Step 2: We find MnO_4^- as an oxidizing agent above H_2S as a reducing agent, so a reaction will take place.
Step 3: The half-reactions are:

$$
\begin{array}{r}
2\,[MnO_4^-(aq) + 8\,H^+(aq) + \cancel{5e^-} \longrightarrow Mn^{2+}(aq) + 4\,H_2O(l)] \\
5\,[H_2S(aq) \longrightarrow S(s) + 2\,H^+(aq) + \cancel{2e^-}] \\
\hline
2\,MnO_4^-(aq) + 5\,H_2S(aq) + 6\,H^+(aq) \longrightarrow 2\,Mn^{2+}(aq) + 5\,S(s) + 8\,H_2O(l)
\end{array}
$$

(Notice that some H_2O and H^+ canceled on each side.)
Step 4: Atom balance checks. The charge on the left is $2(1-) + 6(1+) = 4+$; the charge on the right is $2(2+) = 4+$. The equation is correct.

Answer: Yes, $KMnO_4$ could be used to remove H_2S. The equation is given above. ■

Permanganate ion is a good oxidizing agent, as we see from its position in the table. It also provides its own built-in indicator for seeing when the reaction is complete. MnO_4^- is a deep purple, which changes to the almost colorless Mn^{2+} as the reaction proceeds.

EXAMPLE 15.3: Will a copper penny dissolve in nitric acid? Write the net ionic equation for the reaction, if any.

Solution: Here, we're asked not only whether a reaction will take place, but also to interpret the question. The only way a metal can dissolve in an aqueous solution is if it's converted to its positive ion. Therefore, if a metal is to dissolve, it must be oxidized. Thus our task is to find out if nitric acid will oxidize copper metal.

Step 1: $Cu(s)$, $H^+(aq)$, and $NO_3^-(aq)$.

Step 2: H^+ is below Cu, but NO_3^- is above Cu. Cu can be oxidized by NO_3^-.

Step 3: The half-reactions are:

$$3\,[Cu(s) \longrightarrow Cu^{2+}(aq) + \cancel{2e^-}]$$
$$2\,[NO_3^-(aq) + 4\,H^+(aq) + \cancel{3e^-} \longrightarrow NO(g) + 2\,H_2O(l)]$$
$$\overline{3\,Cu(s) + 2\,NO_3^-(aq) + 8\,H^+(aq) \longrightarrow 3\,Cu^{2+}(aq) + 2\,NO(g) + 4\,H_2O(l)}$$

Step 4: Atom balance checks. The charge on the left is $2(1-) + 8(1+) = 6+$; on the right, $3(2+) = 6+$. The equation is correct.

Answer: A penny will dissolve in nitric acid. The equation is above. ■

This reaction is often demonstrated in chemistry classes. As the solution turns blue from the blue Cu^{2+}, a brown gas bubbles out of it. This gas is NO_2, which forms as the NO reacts with the oxygen of the air.

EXAMPLE 15.4: Would there be any reaction if hydrogen and oxygen gas were bubbled through water at the same time? Write the net ionic equation for the reaction, if any.

Solution:

Step 1: $H_2(g)$, $O_2(g)$, and $H_2O(l)$.

Step 2: O_2 is in the column of oxidizing agents. Both H_2 and H_2O are in the column of reducing agents, but only H_2 is below O_2.

Step 3:

$$2\,[H_2(g) \longrightarrow 2\,H^+(aq) + \cancel{2e^-}]$$
$$O_2(g) + 4\,H^+(aq) + \cancel{4e^-} \longrightarrow 2\,H_2O(l)$$
$$\overline{2\,H_2(g) + O_2(g) \longrightarrow 2\,H_2O(l)}$$

Step 4: Atoms and charges balance.

Answer: A reaction will occur, as in the equation above. ■

This answer shouldn't surprise us, because we know that hydrogen and oxygen react spontaneously to form water. Our bodies use this reaction indirectly to provide us with energy. And the reaction is also used in *fuel cells,* which are batteries that use substances normally burned as fuel. When we make electricity by burning hydrogen, we have to convert the heat energy to mechanical energy by making steam turn a turbine. Then we convert mechanical energy to electricity. Every time one kind of energy is changed to another kind, some of the energy is lost. If we can go directly from the chemical reaction to electrical energy, we won't waste as much. A hydrogen-oxygen fuel cell does that, at a high temperature and with catalysts. At the same time, it produces water that is suitable for drinking. Fuel cells are efficient, but they are expensive. The fuels have to be very

fuel cell

pure, and expensive catalysts must be used. So far, fuel cells have been used extensively only in space missions.

Where more than one reaction is possible, the reaction involving the oxidizing and reducing agents furthest apart will be preferred in exothermic reactions, because the most energy can be released. For endothermic reactions, the oxidizing and reducing agents closest together will be preferred, because less energy input is required. We see now why, in the electrolysis of a moderately concentrated NaCl solution, hydrogen is obtained at the anode instead of sodium. Chloride ion, in the column of reducing agents, is closer to H_2O as an oxidizing agent than it is to sodium.

15.4 CALCULATIONS WITH ELECTRODE POTENTIALS

Combinations of standard electrode potentials determine how much voltage can be obtained from a certain reaction, or what voltage will have to be supplied to make a nonspontaneous reaction proceed. We'll take a look at voltage itself and how it's related to the energy of reaction.

VOLTAGE CALCULATIONS. A river is flowing water. We saw in Chapter 7 that we can make it do work for us by damming it up and then letting it fall from higher potential energy to lower potential energy. The dam is a barrier—a potential barrier. The water on top of the barrier will fall spontaneously to the bottom and release energy. The water at the bottom of the barrier will not rise spontaneously to the top. If we want the water at the bottom to go to the top, we have to supply energy and pump it up there.

Electricity is flowing charge. To do work for us, it too must fall from higher potential to lower potential. Here, *voltage* is the potential barrier. voltage
A positive electrode potential means that the electrons are on top of the barrier and will fall spontaneously to the bottom. A negative electrode potential means that the electrons are at the bottom of the barrier and won't rise spontaneously to the top. If we want electrons to go from the bottom to the top, we have to supply energy and pump them up there with a battery or some other voltage source.

To find the voltage of any reaction, we combine the individual electrode potentials of its two half-reactions by adding them algebraically. (In electrolysis of melted salts, we'll use the electrode potentials of Table 15.3 even though these are for ions in aqueous solution and not for melted salts. Although this usage is not strictly correct, the values are close enough for our purposes.) We find that the standard electrode potential is -2.71 volts for the reduction of sodium ion, and -1.36 volts for the oxidation of chloride ion. The fact that the oxidation of the chloride ion involves two electrons and the reduction of the sodium ion involves only one makes no

TABLE 15.4
Voltage calculations for some reactions

Reaction with Half-Reactions	*E^0 and Total Potential*
$2\,I^-(aq) + Cl_2(g) \longrightarrow I_2(s) + 2\,Cl^-(aq)$	
$2\,I^- \longrightarrow I_2(s) + 2e^-$	-0.54 V
$Cl_2(g) + 2e^- \longrightarrow 2\,Cl^-(aq)$	$+1.36$ V
Extraction of iodine from seawater with chlorine.	$+0.82$ V
$Cu^{2+}(aq) + Zn(s) \longrightarrow Cu(s) + Zn^{2+}(aq)$	
$Cu^{2+}(aq) + 2e^- \longrightarrow Cu(s)$	$+0.34$ V
$Zn(s) \longrightarrow Zn^{2+}(aq) + 2e^-$	$+0.76$ V
A simple galvanic cell.	$+1.10$ V
$Mg^{2+}(l) + 2\,Cl^-(l) \longrightarrow Mg(s) + Cl_2(g)$	
$Mg^{2+}(aq) + 2e^- \longrightarrow Mg(s)$	-2.37 V
$2\,Cl^-(aq) \longrightarrow Cl_2(g) + 2e^-$	-1.36 V
Electrolysis of $MgCl_2$.	-3.73 V

difference to the potential. The potential barrier is the same height no matter how many electrons go over it, just as a dam stays the same height no matter how much water goes over it. The number of electrons does affect the energy, though, as we'll see a little later.

The total potential for the electrolysis of sodium chloride is thus $(-2.71) + (-1.36) = -4.07$ volts. This means that we'd have to provide at least 4.07 volts to make the reaction go, as illustrated in Figure 15.5. A positive total potential for a reaction means that the reaction will go spontaneously and provide as much voltage as the value of the potential. To make a battery, a reaction must have a positive voltage. Table 15.4 shows voltage calculations for some of the reactions we've already discussed.

Standard electrode potentials are measured relative to one another. We couldn't measure an oxidation or a reduction potential by itself, because these half-reactions don't go by themselves. There has to be one of each. The values of the electrode potentials were arrived at by assigning the hydronium-ion/hydrogen electrode a value of 0.00 and then measuring the others relative to that one.

EXAMPLE 15.5: Calculate the voltage provided by a flashlight battery. The total reaction is:

$$Zn(s) + 6\,NH_4^+(aq) + 2\,MnO_2(s) \longrightarrow Zn^{2+}(aq) + 2\,Mn(OH)_3(s) + 6\,NH_3(aq)$$

Solution: First, select the appropriate half-reactions from Table 15.3. We find $Zn(s)$ in the column of reducing agents but MnO_2 appears twice in the column of oxidizing agents. We choose the equation that contains NH_4^+. Next, add up the half-reactions to make sure that they give the correct overall equation.

FIGURE 15.5
Electrolysis proceeds only when the driving voltage exceeds the negative potential

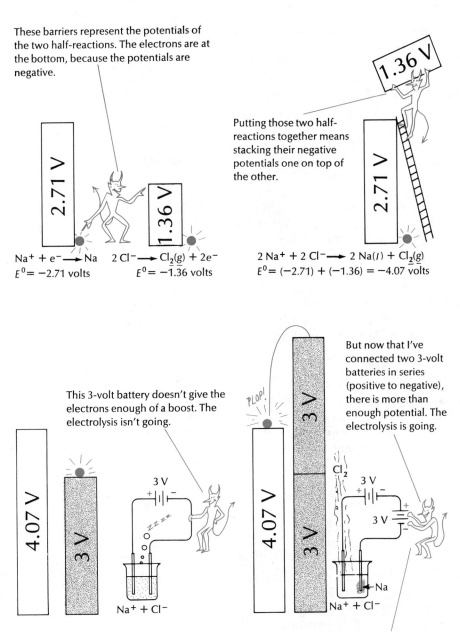

These barriers represent the potentials of the two half-reactions. The electrons are at the bottom, because the potentials are negative.

2.71 V

1.36 V

$Na^+ + e^- \longrightarrow Na$
$E^0 = -2.71$ volts

$2\,Cl^- \longrightarrow Cl_2(g) + 2e^-$
$E^0 = -1.36$ volts

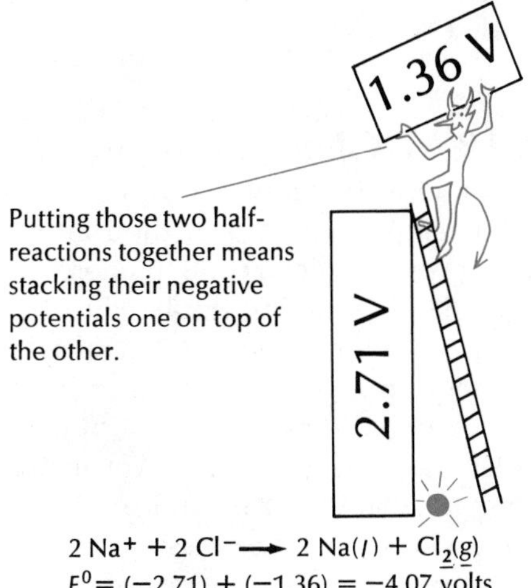

1.36 V
Putting those two half-reactions together means stacking their negative potentials one on top of the other.
2.71 V

$2\,Na^+ + 2\,Cl^- \longrightarrow 2\,Na(l) + Cl_2(g)$
$E^0 = (-2.71) + (-1.36) = -4.07$ volts

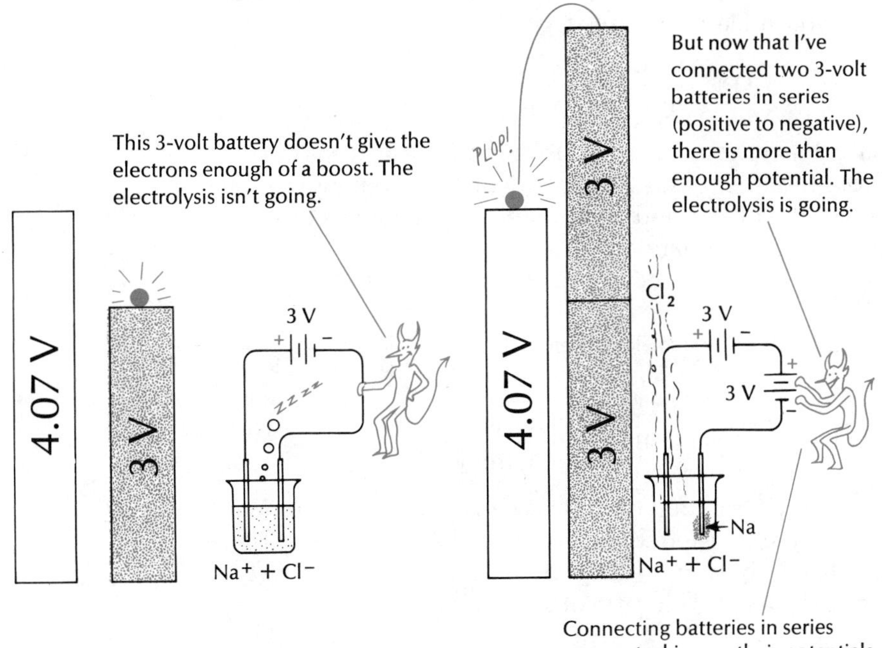

This 3-volt battery doesn't give the electrons enough of a boost. The electrolysis isn't going.
4.07 V
3 V
3 V
+ −
zzzz
Na+ + Cl−
PLOP!
3 V
But now that I've connected two 3-volt batteries in series (positive to negative), there is more than enough potential. The electrolysis is going.
4.07 V
3 V
Cl2
3 V
+ −
+
3 V
−
Na
Na+ + Cl−
Connecting batteries in series means stacking up their potentials.

$$
\begin{array}{rll}
 & & E^0 \\
Zn(s) \longrightarrow & Zn^{2+}(aq) + \cancel{2e^-} & +0.76\ V \\
2\,[MnO_2(s) + H_2O(l) + NH_4^+(aq) + \cancel{e^-} \longrightarrow & Mn(OH)_3(s) + NH_3(aq)] & +0.74\ V \\
\hline
Zn(s) + 2\,MnO_2(s) + 2\,H_2O(l) + 2\,NH_4(aq) \longrightarrow & Zn^{2+}(aq) + 2\,Mn(OH)_3(s) & \\
 & + 2\,NH_3(aq) & +1.50\ V
\end{array}
$$

The half-reactions are correct, so we can go ahead and combine their electrode potentials: $(+0.76\ V) + (+0.74\ V) =$ ____V.

Answer: 1.50 volts. ■

ENERGY CALCULATIONS. The amount of energy released when water falls over a dam depends on two things: the height of the dam, and the amount of water falling over it. Ten thousand liters of water falling over a 50-meter dam gives us twice as much energy as 5,000 liters of water falling over the same dam. Doubling the height of the dam also gives us twice as much energy with the same amount of water.

The amount of energy released when electrons fall from a potential barrier also depends on two things: the height of the potential barrier, and the number of moles of electrons falling over it. One mole of electrons falling through 1 volt will provide 23.1 kilocalories of energy. Two moles falling through 1 volt will provide twice as much energy: 46.2 kilocalories. One mole falling through 2 volts also provides 46.2 kilocalories. Figure 15.6 illustrates these concepts.

We've introduced a new conversion factor. If 23.1 kilocalories of energy are released when 1 mole of electrons falls through 1 volt, then we have 23.1 kilocalories per volt-mole of electrons, or:

$$23.1\,\frac{\text{kilocalories}}{(\text{volt})\,(\text{mole e}^-)}$$

This is also the energy that it would take to push 1 mole of electrons through a potential of 1 volt. In either case, we can calculate the energy for any reaction if we know its voltage and how many moles of electrons are being transferred. This is the equation:

Energy in kilocalories

$$= (\text{potential in volts})\,(\text{moles of electrons})\left[23.1\,\frac{\text{kilocalories}}{(\text{volt})\,(\text{mole e}^-)}\right]$$

or

$$E = (\text{volts})\,(\text{moles e}^-)\left[23.1\,\frac{\text{kilocalories}}{(\text{volt})\,(\text{mole e}^-)}\right]$$

To calculate the energy for a total reaction, we use the total number of moles of electrons that are transferred, as shown in the examples in Table 15.5. If, on the other hand, we want to know the energy based on a specific reactant or product, we use the mole ratio of substance to electrons.

EXAMPLE 15.6: How much energy would be provided if 250 grams of lead were used up in a lead storage (car) battery? The two half-reactions are:

$$
\begin{array}{rll}
Pb(s) + SO_4^{2-} \longrightarrow & PbSO_4(s) + 2e^- & +0.36\ V \\
PbO_2(s) + 4\,H^+(aq) + SO_4^{2-}(aq) + 2e^- \longrightarrow & PbSO_4(s) + 2\,H_2O(l) & +1.68\ V
\end{array}
$$

FIGURE 15.6
Energy depends on quantity and potential

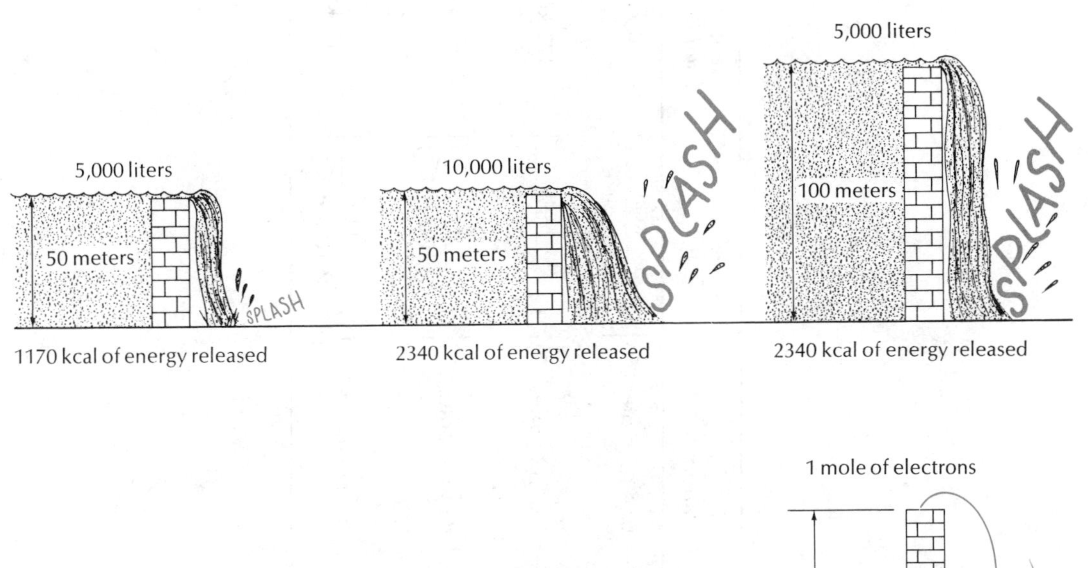

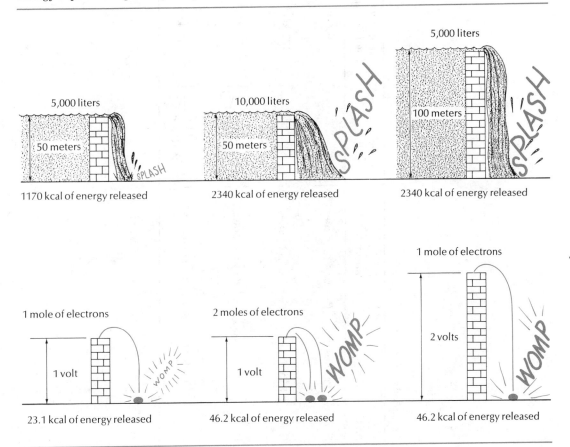

Solution: First, let's make sure we have all the quantities needed to solve the energy equation. Adding the voltages of the half-reactions gives us the volts: $(+1.68) + (+0.36) = 2.04$ volts. We get the moles of electrons from the grams of lead given, by using these two conversion factors: the atomic weight of lead (207 g/mole); and the mole ratio of electrons to lead, using the total number of electrons transferred (2 moles e^-/1 mole pb). Our setup is:

$$250 \text{ g } \cancel{\text{Pb}} \times \frac{1 \text{ } \cancel{\text{mole Pb}}}{207 \text{ g } \cancel{\text{Pb}}} \times \frac{2 \text{ moles } e^-}{\cancel{\text{mole Pb}}} = 2.4 \text{ moles } e^-$$

Next, we put all the values into the energy equation and solve.

$$E = (+2.04 \text{ volts})(2.4 \text{ } \cancel{\text{moles } e^-})\left[23.1 \frac{\text{kilocalories}}{(\cancel{\text{volts}})(\cancel{\text{moles } e^-})}\right] = \text{\_\_\_\_ kilocalories}$$

Answer: 110 kilocalories. ■

Figure 15.7 shows a diagram of a lead storage battery. The anode is coated with lead and the cathode is coated with lead(IV) oxide. The electrodes

TABLE 15.5
Total energy calculations for some electrochemical reactions

Reaction	*Potential*	*Moles* e^-	*Calculation*	*Energy for Total Reaction*
$2\,Na^+(l) + 2\,Cl^-(l) \longrightarrow 2\,Na(l) + Cl_2(g)$ Electrolysis of NaCl.	−4.07 V	2 moles e^-	$(-4.07\,\cancel{V})(2\,\cancel{\text{moles e}^-})\left[23.1\,\frac{\text{kcal}}{(\cancel{V})(\cancel{\text{mole e}^-})}\right]$	−188 kcal
$Mg^{2+}(l) + 2\,Cl^-(l) \longrightarrow Mg(l) + Cl_2$ Electrolysis of $MgCl_2$.	−3.73 V	2 moles e^-	$(-3.73\,\cancel{V})(2\,\cancel{\text{moles e}^-})\left[23.1\,\frac{\text{kcal}}{(\cancel{V})(\cancel{\text{mole e}^-})}\right]$	−172 kcal
$2\,I^-(aq) + Cl_2(g) \longrightarrow I_2(s) + 2\,Cl^-(aq)$ Iodine from seawater.	+0.83 V	2 moles e^-	$(+0.83\,\cancel{V})(2\,\cancel{\text{moles e}^-})\left[23.1\,\frac{\text{kcal}}{(\cancel{V})(\cancel{\text{mole e}^-})}\right]$	+38.3 kcal
$3\,Cu(s) + 2\,NO_3^-(aq) + 8\,H^+(aq) \longrightarrow 3\,Cu^{2+}(aq) + 2\,NO(g) + 4\,H_2O(l)$ Copper penny dissolving in HNO_3.	+0.62 V	6 moles e^-	$(+0.62\,\cancel{V})\,(6\,\cancel{\text{moles e}^-})\left[23.1\,\frac{\text{kcal}}{(\cancel{V})(\cancel{\text{mole e}^-})}\right]$	+85.9 kcal

FIGURE 15.7
One cell of a lead storage battery

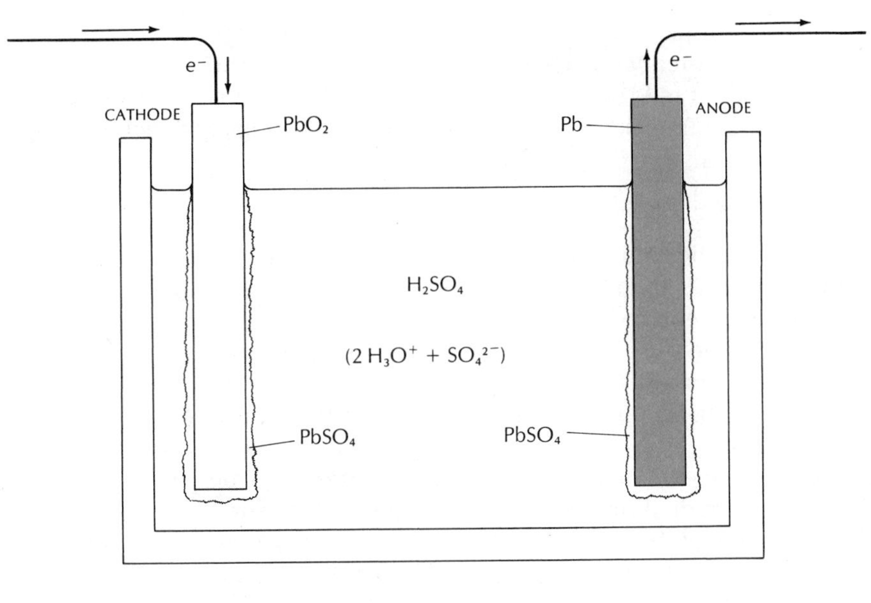

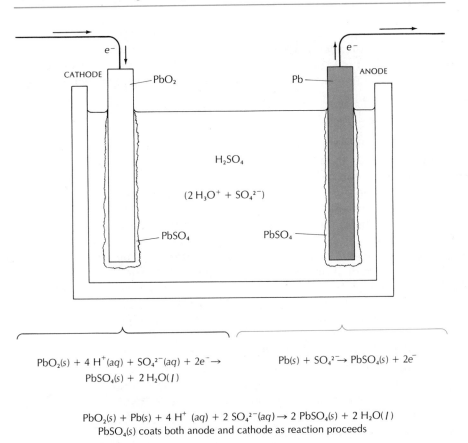

$PbO_2(s) + 4\,H^+(aq) + SO_4{}^{2-}(aq) + 2e^- \rightarrow$
$PbSO_4(s) + 2\,H_2O(l)$

$Pb(s) + SO_4{}^{2-} \rightarrow PbSO_4(s) + 2e^-$

$$PbO_2(s) + Pb(s) + 4\,H^+\ (aq) + 2\,SO_4{}^{2-}(aq) \rightarrow 2\,PbSO_4(s) + 2\,H_2O(l)$$

$PbSO_4(s)$ coats both anode and cathode as reaction proceeds

are held apart with plastic spacers, and the space inside is filled with 3-*M* sulfuric acid. As we just saw, this cell provides about 2 volts. For a 6- or a 12-volt battery, three or six of these cells are connected in series.

Some of the water in a lead storage battery evaporates, and some is lost by electrolysis. If a battery's water weren't replenished now and then, the battery would dry out and the ions could no longer conduct current. The lead(II) sulfate produced in both half-reactions sticks to the electrode where it's formed. When all the lead and lead(IV) oxide have been changed to lead(II) sulfate, the battery is dead. But it can be recharged, by pumping electrons into the battery in the opposite direction. This reverses the reactions, and turns the lead(II) sulfate on the electrodes back into lead and lead(IV) oxide. Since a car recharges its battery while it runs, a battery should last forever. But after a while the plastic spacers begin to loosen, and some lead(II) sulfate falls to the bottom of the battery and isn't available for recharging. Eventually, the battery won't hold enough charge to start a car.

15.5 OXIDATION NUMBERS

The number of electrons lost or gained in any half-reaction is determined by changes in the oxidizing and reducing agents' oxidation numbers. The *oxidation number* of an element tells how many more or fewer electrons it has than it would have as the free element. For an example, let's look at this half-reaction:

oxidation number

$$Zn^{2+} + 2e^- \longrightarrow Zn(s)$$

The reactant Zn^{2+} has an oxidation number of 2+, which means that the element has two fewer electrons than it would have as a free element. The product Zn is the free element and has an oxidation number of zero. Two electrons are transferred, because zinc's oxidation number goes from 2+ to zero. If an oxidation number is negative (as it is with Cl^-, whose oxidation number is 1−), then the element has more electrons than it would have as a free element.

In oxidation, the oxidizing agent's oxidation number is decreased (the element gains electrons). In reduction, the reducing agent's oxidation number is increased (the element loses electrons). Once we learn how to determine oxidation numbers, we can use them to decide how many electrons are being transferred and finally to write complete half-reactions.

DETERMINING OXIDATION NUMBERS. For monatomic ions, the oxidation number is the same as the ionic charge. Zinc ion, whose ionic charge is 2+, has an oxidation number of 2+. And chloride ion, whose ionic charge is 1−, has an oxidation number of 1−. For covalent compounds and polyatomic ions, the oxidation numbers of the atoms involved are based on electronegativity differences. For these, oxidation number is an electron bookkeeping device. Each electron pair shared between two atoms is assigned to the more electronegative atom, and then the valence electrons are counted for each element. For example:

$$|\overline{\underline{Cl}}-\overline{\underline{Cl}}| \qquad H-\overline{\underline{Cl}}| \qquad \left[|\overline{\underline{O}}-\overline{\underline{Cl}}-\overline{\underline{O}}|\right]^-$$

In Cl_2, there is no electronegativity difference. Neither owns the pair, and the oxidation number is 0. In HCl, Cl is more electronegative than H, so the bonding electron pair is assigned to Cl. This gives Cl eight electrons, one more than it would have as a free element, so its oxidation number is 1−. H has no electrons, one fewer than it would have as a free element, so its oxidation number is 1+. In ClO_2^-, O is more electronegative than Cl, so each O gets a bonding electron pair. This gives each O eight electrons—two more than it would have as a free element—so its oxidation number is 2−. Cl now only has four electrons—three less than it would have as a free element—so its oxidation number is 3+. All oxidation numbers in a compound must total zero, and in a polyatomic ion, they must add up to the ion's total charge.

EXAMPLE 15.7: Find the oxidation number of each element in $PbSO_4$ and PbO_2.

Solution: In $PbSO_4$, the oxidation number of Pb^{2+} is 2+. In the sulfate ion, O is more electronegative than S, so the four O's get all four electron pairs. Each O then has eight electrons—two more than it has as the free element—so its oxidation number is 2−. S has no electrons—six less than it has as the free element—so its oxidation number is 6+. Then the charge of the sulfate ion is (6+) + (8−) = 2−. In PbO_2, the oxidation numbers are the ionic charges.

Answer: $PbSO_4$: Pb = 2+, S = 6+, O = 2−. (2+) + (6+) + 4(2−) = 0
PbO_2: Pb = 4+, O = 2−. (4+) + 2(2−) = 0 ■

Determining oxidation numbers for elements in many compounds would lead us to see that the oxidation numbers of O and H usually remain the same in most of their compounds. We can use their oxidation numbers as references to find other oxidation numbers, according to the rules in Table 15.6.

EXAMPLE 15.8: What are the oxidation numbers of Mn in $KMnO_4$ and in MnO_2?

Solution: In $KMnO_4$, we use K (1+) and O (2−) as references. Then we must balance Mn with their total: (1+) + (4 × −2) = 7−. For MnO_2, Mn must balance oxygen's total of 4−.

Answer: 7+ in $KMnO_4$; 4+ in MnO_2. ■

EXAMPLE 15.9: What are the oxidation numbers of N in NO_3^- and in NH_4^+?

Solution: In NO_3^-, oxygen totals 6−. For a total charge of 1−, N must be 5+. In NH_4^+, hydrogen totals 4+. For a total charge of 1+, N must be 3−.

Answer: 5+ in NO_3^-; 3− in NH_4^+. ■

USING OXIDATION NUMBERS. Sometimes we might need to balance an oxidation-reduction reaction when we don't have access to the necessary half-reactions. Redox reactions are often complex, and attempting to balance them by the trial-and-error method of Chapter 6 frequently leads only to frustration. Also, to be balanced properly, a redox equation must have its electrons balanced. It's possible to have an equation whose atoms balance correctly but whose electrons don't.

The first problem in balancing a redox equation is deciding what elements have changed oxidation numbers. One approach is to check every single atom's oxidation number on both sides. But we can narrow the choice a little with these guidelines: (1) If possible, write the equation in net ionic form (this will eliminate the spectator ions); and (2) if hydrogen and oxygen do not appear as H_2, O_2, or a peroxide, hydrogen and oxygen have not changed.

TABLE 15.6
Rules for determining oxidation numbers, with examples

Rules	*Examples* Compound	Element	Oxidation Number
1. The oxidation number of a monatomic ion is equal to its charge. Metals always have zero or positive oxidation numbers.	FeO	Fe	2+
		O	2−
2. The oxidation number of a free element is zero.		H_2	0
		Cu	0
3. The usual oxidation number of hydrogen in a compound is 1+, except in metal hydrides, where it is 1−.	HCl	H	1+
		Cl	1−
	NaH	Na	1+
		H	1−
4. The usual oxidation number of oxygen in a compound is 2−, except in peroxides, where it's 1−.	H_2O	H	1+
		O	2−
	H_2O_2	H	1+
		O	1−
5. All other oxidation numbers are assigned to produce the total charge on the substance. Rules 1 through 4 take precedence. The total of the oxidation numbers in a covalent compound must add up to zero; the total for a positive or negative polyatomic ion must add up to the ion's total charge. Usually the central atom in a polyatomic ion or covalent compound is the one whose oxidation number is in question.	CO_2	O	2− (Rule 4)
		C	4+ (needed for total of zero)
	SO_4^{2-}	O	2− (Rule 4)
		S	6+ (needed for total of 2−)
	HCO_3^-	H	1+ (Rule 3)
		O	2− (Rule 4)
		C	4+ (needed for total of 1−)

A test for sulfurous acid in rainwater is to add an acid solution of $K_2Cr_2O_7$. If sulfurous acid is present, the yellow color of the dichromate ion changes to the green color of the chromium(III) ion, according to this unbalanced equation:

$$K_2Cr_2O_7(aq) + H_2SO_3(aq) + H_2SO_4(aq) \not\longrightarrow Cr_2(SO_4)_3(aq) + K_2SO_4(aq) + H_2O(l)$$

Our goal is to balance the equation.

Step 1. *Write the equation in net ionic form.*

$$Cr_2O_7^{2-} + H_2SO_3 + 2\,H^+ \not\longrightarrow 2\,Cr^{3+} + 3\,SO_4^{2-} + H_2O$$

Step 2. *Decide which substances have changed oxidation numbers.*
Since the equation contains no H_2, O_2, or peroxides, we know hydrogen and oxygen haven't changed. That leaves only Cr and S. It's a

good bet that both have changed, since one substance must lose and one must gain electrons. A check of the oxidation numbers verifies this. Cr is 6+ on the left and 3+ on the right. S is 4+ on the left and 6+ on the right.

Step 3. *Determine how many electrons have been lost and gained, and write incomplete half-reactions.*
In our example, Cr went from 6+ to 3+, so each Cr gained 3 electrons. However, there are two Cr in $Cr_2O_7^{2-}$, so that makes 6 electrons gained altogether.

$$Cr_2O_7^{2-} + 6e^- \not\rightarrow 2\ Cr^{3+}$$

S went from 4+ on the left to 6+ on the right, so each S lost 2 electrons.

$$H_2SO_3 \not\rightarrow SO_4^{2-} + 2e^-$$

Step 4. *Balance the oxygens and hydrogens in the half-reactions by adding H_2O, OH^-, or H^+.*
What combinations of these to add depends on whether the reaction occurs in acidic or basic solution, and on which side of the half-reaction lacks oxygen or hydrogen. We use these guidelines:

	To side where O needed, add, for each O needed:	*To other side, add for each O:*
Acidic solution	1 H_2O	2 H^+
Basic solution	2 OH^-	1 H_2O
	To side where H needed, add, for each H needed:	*To other side, add for each H:*
Acidic solution	1 H^+	---
Basic solution	1 H_2O	1 OH^-

(When it's not apparent whether the solution is acidic or basic, this information would be given.) Now let's balance the hydrogens and oxygens in our example. We know the solution is acidic because of H_2SO_4. In the first half-reaction, we need to balance $Cr_2O_7^{2-}$ by adding seven oxygen atoms on the right. We add 7 H_2O to the right and 14 H^+ to the left. We check our atom and charge balances and find that they're correct.

$$Cr_2O_7^{2-} + 6e^- + 14\ H^+ \longrightarrow 2\ Cr^{3+} + 7\ H_2O$$

In the second half-reaction, SO_4^{2-} on the right has one more oxygen than H_2SO_3 on the left, so we need one oxygen on the left. We add 1 H_2O to the left and 2 H^+ to the right. We also need two hydrogens on the right, so we add 2 H^+ to the right. We write all this down so that we can simplify it.

$$H_2SO_3 + H_2O \longrightarrow SO_4^{2-} + 2\ H^+ + 2\ H^+ + 2e^-$$

Then we combine the waters and hydronium ions, and check the atom and charge balances, which are correct.

$$H_2SO_3 + H_2O \longrightarrow SO_4^{2-} + 4\,H^+ + 2e^-$$

Step 5. *Balance the electrons and add the half-reactions.*

To balance the electrons in our example, we must multiply the second half-reaction by 3.

$$3(H_2SO_3 + H_2O \longrightarrow SO_4^{2-} + 4\,H^+ + 2e^-)$$

Then we can add the half-reactions.

$$\begin{array}{r} Cr_2O_7^{2-} + 14\,H^+ + \cancel{6e^-} \longrightarrow 2\,Cr^{3+} + 7\,H_2O \\ 3\,H_2SO_3 + 3\,H_2O \longrightarrow 3\,SO_4^{2-} + 12\,H^+ + \cancel{6e^-} \\ \hline Cr_2O_7^{2-} + 3\,H_2SO_3 + 3\,H_2O + 14\,H^+ \longrightarrow \\ 2\,Cr^{3+} + 3\,SO_4^{2-} + 7\,H_2O + 12\,H^+ \end{array}$$

Step 6. *Consolidate the equation and check the atom and charge balances.*

In our example we consolidate the waters and hydronium ions. The balanced equation is:

$$Cr_2O_7^{2-} + 3\,H_2SO_3 + 2\,H^+ \longrightarrow 2\,Cr^{3+} + 3\,SO_4^{2-} + 4\,H_2O$$

EXAMPLE 15.10: Nitric acid dissolves many substances more effectively than hydrochloric or sulfuric acid does. This is because nitric acid contains nitrate ion, a powerful oxidizing agent, in addition to the oxidizing agent H_3O^+. The other two strong acids have only H_3O^+. CuS dissolves with difficulty in HCl but easily in HNO_3, according to this unbalanced equation:

$$CuS(s) + HNO_3(aq) \not\longrightarrow Cu(NO_3)_2(aq) + S(s) + NO(g) + H_2O(l)$$

Balance the equation.

Solution:

Step 1:

$$CuS(s) + H^+ + NO_3^- \not\longrightarrow Cu^{2+} + S(s) + NO(g) + H_2O(l)$$

Step 2: S has changed from 2− in CuS to zero in S. N has changed from 5+ in NO_3^- to 2+ in NO.

Step 3:

$$\begin{array}{r} CuS(s) \longrightarrow Cu^{2+}(aq) + S(s) + 2e^- \\ NO_3^-(aq) + 3e^- \not\longrightarrow NO(g) \end{array}$$

Step 4: The first half-reaction is balanced. Now for the second.

$$4\,H^+(aq) + NO_3^-(aq) + 3e^- \longrightarrow NO(g) + 2\,H_2O(l)$$

Step 5:

$$\begin{array}{r} 3\,CuS(s) \longrightarrow 3\,Cu^{2+}(aq) + 3\,S(s) + \cancel{6e^-} \\ 8\,H^+(aq) + 2\,NO_3^-(aq) + \cancel{6e^-} \longrightarrow 2\,NO(g) + 4\,H_2O(l) \\ \hline 3\,CuS(s) + 8\,H^+(aq) + 2\,NO_3^-(aq) \longrightarrow 3\,Cu^{2+}(aq) + 3\,S(s) + 2\,NO(g) \\ + 4\,H_2O(l) \end{array}$$

Step 6: The equation is consolidated. Atoms and charges balance.

Answer:

$$3\,CuS(s) + 8\,H^+(aq) + 2\,NO_3^-(aq) \longrightarrow 3\,Cu^{2+}(aq) + 3\,S(s) + 2\,NO(g) + 4\,H_2O(l)$$ ■

In this example, we see that sulfur undergoes a change in oxidation number different from that of the previous example (2− to zero, instead of 4+ to 6+). We also see from the initial unbalanced equation that some of the nitrate ions remained intact at the end of the reaction; they provide a negative ion to balance the Cu^{2+} in solution. Thus some of the nitrate ions acted as oxidizing agents and some acted as spectator ions.

REVIEW QUESTIONS

Electrons in Chemical Reactions

1. What reaction types are usually electron transfer? Give examples.
2. Define the following words, and explain how they relate to one another: *oxidation, reduction, oxidizing agent, reducing agent, redox.*
3. What is a *half-reaction?* Write two examples, one showing electrons as reactants and one showing electrons as products.
4. Describe what happens when a piece of zinc metal is placed into a solution of copper(II) sulfate.
5. What is a *galvanic cell?* Sketch the zinc-copper galvanic cell, and describe what happens when it is connected.
6. Define *anode* and *cathode.* Write the half-reactions that occur at each electrode in the zinc-copper galvanic cell.
7. What is the symbol for a battery? Where do the electrons come out, and where do they go in?

Electrolytic Reactions

8. What is an *electrolytic reaction?*
9. Explain how ions can conduct electricity. Define *strong electrolyte, weak electrolyte,* and *non-electrolyte,* and give examples of each.
10. Why is it dangerous to use electricity when in contact with water?
11. Describe the *electrolysis* of sodium chloride. What are *cations* and *anions,* and how did they get their names?
12. How do we add half-reactions to get a total reaction?
13. How is electrolysis of seawater different from electrolysis of melted NaCl? What are the products of each?
14. What are the products of electrolysis of a very dilute NaCl solution?
15. Explain how magnesium metal is extracted from seawater.
16. What is *electroplating?* How does it work?

Oxidizing and Reducing Agents

17. How do we use Table 15.3 to decide whether a redox reaction will occur spontaneously?
18. What is a *standard electrode potential?* What is the significance of its sign?
19. List the steps involved in predicting a redox reaction. Illustrate each with an example.
20. What happens to a metal when it dissolves in an aqueous solution?
21. If more than one oxidizing or reducing agent is present, how do we know which will be preferred?

Calculations with Electrode Potentials

22. What is *voltage?* What does its sign mean? How do we find the voltage of a redox reaction?
23. What determines the amount of energy involved in a redox reaction?
24. How many kilocalories of energy do we get when one mole of electrons falls through one volt?

Oxidation Numbers

25. What is the *oxidation number* of an element? What happens to the oxidation number during oxidation? During reduction?
26. Explain how oxidation numbers are determined, and give an example.
27. In balancing redox equations, how do we use oxidation numbers to decide which substances have been oxidized and which reduced?
28. How do we balance hydrogen and oxygen in half-reactions?
29. Summarized the steps involved in balancing a redox equation when the half-reactions are not given.

EXERCISES

Set A (Answers at back of book.)

1. Identify the oxidizing agent and the reducing agent in each of these electron transfer reactions.
   a. $Cu(s) + S(s) \longrightarrow CuS(s)$
   b. $N_2(g) + O_2(g) \longrightarrow 2\ NO(g)$
   c. $4\ H^+(aq) + Fe(s) \longrightarrow 2\ H_2(g) + Fe^{2+}(aq)$
   d. $2\ Cl^-(aq) + F_2(g) \longrightarrow Cl_2(g) + 2\ F^-(aq)$
   e. $Zn(s) + 2\ OH^-(aq) + 2\ H_2O(l) \longrightarrow Zn(OH)_4{}^{2-}(aq) + H_2(g)$
   f. $2\ AgNO_3(aq) + Mg(s) \longrightarrow 2\ Ag(s) + Mg(NO_3)_2(aq)$
2. In Exercise 1, identify the oxidized substances and the reduced substances.
3. In a galvanic cell that consists of a lead electrode and an aluminum electrode, these half-reactions take place:

   $$Al(s) \longrightarrow Al^{3+}(aq) + 3e^-$$
   $$Pb^{2+}(aq) + 2e^- \longrightarrow Pb(s)$$

   a. Which is the anode, and which is the cathode? How do you know?
   b. Sketch such a cell and indicate the direction of electron flow.
4. Classify each of the following as a weak electrolyte, a strong electrolyte, or a nonelectrolyte (assume aqueous solution).

a. $MgCl_2$	e. $PbCl_2$	i. H_2O
b. H_2CO_3	f. $BaSO_4$	j. KOH
c. HNO_3	g. NH_3	k. $Mg(OH)_2$
d. $NaHCO_3$	h. H_3PO_4	

5. Aluminum is prepared commercially by electrolyzing melted Al_2O_3.
   a. Write half-reactions for this electrolysis, balance the electrons, and add the half-reactions to obtain a balanced equation.
   b. Sketch this electrolytic cell. Indicate the cathode and anode and battery terminals, and show the direction of electron flow.
   c. Describe what happens at each electrode.
6. Use Table 15.3 to predict whether a reaction would occur spontaneously when each of the folowing pairs was mixed. If a reaction would occur, write the balanced ionic equation.
   a. $Pb(s) + Cr(NO_3)_2(aq)$
   b. $KMnO_4(aq) + KI(aq)$
   c. $Ag(s) + HCl(aq)$
   d. $Ag(s) + HNO_3(aq)$
   e. $FeSO_4(aq) + MnO_2(s)$
   f. $CuSO_4(aq) + HNO_3(aq)$
   g. $Na_2Cr_2O_7(aq) + H_2O_2(aq)$
   h. $PbO_2(s) + KBr(aq) + H_2SO_4(aq)$
7. A student used an aluminum weighing cup to weight out some solid copper(II) sulfate. Afterward, he found that the weighing cup had holes in it where individual copper(II) sulfate crystals had come in contact with it. Explain what happened and write a balanced equation for the reaction.
8. Could H_2SO_4 clean a gold ring without damaging it? How do you know?
9. Calculate the voltage produced by a hydrogen-oxygen fuel cell (see Example 15.4). How many fuel cells would be needed to electrolyze NaCl?
10. Alkaline storage batteries contain hydroxide ions. One kind is a nickel-cadmium (Nicad) battery, used in calculators and in other portable electronic equipment. The overall reaction is as follows:

    $$Cd(s) + NiO_2(s) + 2\ H_2O(l) \longrightarrow Cd(OH)_2(s) + Ni(OH)_2$$

    Write the two half-reactions, chosen from Table 15.3, and calculate the voltage of this cell. (Notice the unusual oxidation number of nickel.)
11. Calculate the voltage of this cell:

    $$Ni(s) + 2\ Ag^+(aq) \longrightarrow Ni^{2+}(aq) + 2\ Ag(s)$$

12. If a mixture of NaCl and LiBr is melted and electrolyzed, what product will first be obtained at each electrode? Explain.
13. A student wishes to identify the terminals of a battery whose markings have been removed. She uses the battery to electrolyze water containing a little Na_2SO_4. What would be observed, and what conclusions could be drawn?
14. Aluminum metal is added to a solution containing $Fe(NO_3)_2$ and $Pb(NO_3)_2$.
    a. Two redox reactions are possible. Write their net ionic equations.
    b. Which reaction would be preferred? Explain.
15. Calculate the energy, in kilocalories, needed to electrolyze one mole of molten $ZnCl_2$.
16. For the copper-zinc galvanic cell, (a) calculate the energy provided by the total reaction, and (b) find how many of these cells would provide enough voltage and energy to electrolyze one mole of NaCl.
17. Which takes more energy to produce by electrolysis, sodium metal or magnesium metal? (Answer this question by calculating the energy required to produce a mole of each one.)
18. Determine the oxidation number of each element in these compounds.

a. H_2O_2	d. Hg_2Cl_2	f. $MnSO_3$
b. $K_2Cr_2O_7$	e. SO_2	g. NiO_2
c. $NaClO_3$		

19. Silver-zinc batteries are efficient but expensive.

They use an unusual oxidation number of silver. The total reaction is:

$$Zn(s) + AgO(s) + H_2O(l) \not\rightarrow Zn(OH)_2(s) + Ag_2O(s)$$

Use oxidation numbers to write balanced half-reactions, and write the balanced ionic equation.

20. Write balanced ionic equations for the following reactions by writing balanced half-reactions and combining them.
   a. $MnO_2(s) + HCl(aq) \not\rightarrow MnCl_2(aq) + Cl_2(g) + H_2O(l)$
   b. $I_2 + Cl_2 + H_2O \not\rightarrow HIO_3 + HCl$
   c. $BaO_2(s) + HCl(aq) \not\rightarrow BaCl_2(aq) + H_2O(l) + Cl_2(g)$
   d. $H_2O_2(aq) + KMnO_4(aq) + H_2SO_4(aq) \not\rightarrow O_2(g) + MnSO_4(aq) + K_2SO_4(aq) + H_2O(l)$
   e. $KIO_3(aq) + SO_2(g) + H_2O(l) \not\rightarrow I_2(s) + K_2SO_4(aq)$ (acid solution)

Set B (Answers not given.)

1. Identify the oxidizing agent and the reducing agent in each of these electron transfer reactions.
   a. $3\ Ag_2S(s) + 8\ H^+(aq) + 2\ NO_3^- \longrightarrow 6\ Ag^+(aq) + 3\ S(s) + 2\ NO(g) + 4\ H_2O(l)$
   b. $2\ Fe(s) + 3\ Cl_2(g) \longrightarrow 2\ FeCl_3(aq)$
   c. $H_2(g) + F_2(g) \longrightarrow 2\ HF(g)$
   d. $2\ Al(s) + 3\ Ni^{2+}(aq) \longrightarrow 2\ Al^{3+}(aq) + 3\ Ni(s)$
   e. $CuO(s) + H_2(g) \longrightarrow Cu(s) + H_2O(g)$
   f. $2\ Al(s) + 2\ OH^-(aq) + 6\ H_2O(l) \longrightarrow 2\ Al(OH)_4^-(aq) + 3\ H_2(g)$
2. In Exercise 1, identify the oxidized substances and the reduced substances.
3. In a galvanic cell that consists of a zinc electrode and a cadmium electrode, these half-reactions take place.

$$Zn(s) \longrightarrow Zn2^+(aq) + 2e^-$$
$$Cd2^+(aq) + 2e^- \longrightarrow Cd(s)$$

   a. Which is the anode, and which is the cathode? How do you know?
   b. Sketch such a cell and indicate the direction of electron flow.
4. Classify each as a weak electrolyte, a strong electrolyte, or a nonelectrolyte (assume aqueous solution).

a. $NaOH$	e. HNO_2	i. C_2H_5OH
b. $HClO_4$	f. $AgCl$	j. $HClO$
c. $BaCl_2$	g. NaH_2PO_4	k. $CaCO_3$
d. HgI_2	h. $C_6H_{12}O_6$	

5. Impure copper metal is refined by an electrolytic process similar to electroplating. A bar of impure copper is used as the anode, and a bar of pure copper is used as the cathode. Dilute sulfuric acid is used as the electrolyte. Sketch such a cell and describe each electrode reaction.
6. Use Table 15.3 to predict whether a reaction would occur spontaneously if each of these pairs were mixed. If a reaction would occur, write the balanced ionic equation.
   a. $MnO_2(s) + NaF(aq)$
   b. $H_2O_2(aq) + KI(aq)$
   c. $H_2S(aq) + I_2(aq)$
   d. $Sn(s) + H_2SO_4(aq)$
   e. $Br_2(aq) + HNO_3(aq)$
   f. $Au(s) + HCl(aq)$
   g. $K_2Cr_2O_7(aq) + NaBr(aq)$
   h. $FeCl_2(aq) + PbSO_4(s)$
7. Should HCl be used to remove an oxide coating from aluminum? Why?
8. In Table 12.3, we saw that the amount of fluoride ion in seawater is between that of bromide and iodide ions. Could the method used to extract Br_2 and I_2 from seawater (addition of chlorine) also be used to extract F_2? Explain, with reference to appropriate electrode potentials.
9. Calculate the voltage of this cell:

$$F_2(aq) + 2\ I^-(aq) \longrightarrow 2\ F^-(aq) + I_2(aq)$$

10. If a tin can is scratched, the iron underneath the tin will corrode much faster than it would have if the tin hadn't been there at all. This is because the iron acts as the anode and the tin acts as the cathode. These half-reactions take place:

Cathode:

$$Sn^{2+}(aq) + 2e^- \longrightarrow Sn(s)$$

Anode:

$$Fe(s) \longrightarrow Fe^{2+}(aq) + 2e^-$$

The tin acts as the cathode because iron is a better reducing agent than tin.
   a. Calculate the voltage of this cell.
   b. Would the iron corrode as quickly if zinc were used as a coating instead? Explain.
11. Calculate the voltage of this cell:

$$3\ Cu(s) + 2\ Au^{3+}(aq) \longrightarrow 3\ Cu^{2+}(aq) + 2\ Au(s)$$

How many of these cells would be needed to electrolyze NaBr?
12. If a water solution of NaI is electrolyzed, what product will be obtained at each electrode? Explain.
13. Iron pipes can be prevented from corroding by connecting them to a block of magnesium and burying the "mixture." The magnesium, called a "sacrificial metal," eventually disappears and must be replaced. Explain how this works.
14. FeI_3 does not exist. After referring to Table 15.3, give a possible reason why.
15. Calculate the energy, in kilocalories, provided by

the total reaction that takes place in a flashlight battery. (See Example 15.5.)

16. An alkaline storage battery used in artificial heart pacemakers uses this overall equation:

$$HgO(s) + Zn(s) + H_2O(l) \longrightarrow Zn(OH)_2(s) + Hg(l)$$

Calculate the energy provided by the total reaction.

17. How much energy would be required to electrolyze a box (454 grams) of table salt?

18. Determine the oxidation number of each element in these compounds.

a. H_2SO_3	d. NO_2	g. $KClO_4$
b. AgO	e. Na_2O_2	h. $Fe(NO_3)_2$
c. CrO_3	f. NH_4NO_2 (N has two oxidation numbers here)	

19. This reaction is a test for Mn, with the purple MnO_4^- appearing as a positive test.

$$Mn^{2+} + H^+(aq) + BiO_3^-(aq) \not\rightarrow MnO_4^- + Bi^{3+}(aq)$$

Use oxidation numbers to write balanced half-reactions, and write the balanced ionic reaction.

20. Write balanced ionic equations for these reactions by writing balanced half-reactions and combining them.

a. $H_2C_2O_4(aq) + KBrO_3(aq) \not\rightarrow CO_2(g) + KBr(aq) + H_2O(l)$

b. $NH_3(aq) + NaClO(aq) \not\rightarrow N_2H_4(g) + NaCl$ (basic solution)

c. $As(s) + HNO_3(aq) + H_2O(l) \not\rightarrow AsO_4^{3-}(aq) + NO(g)$

d. $H_2S(aq) + H_2O_2(aq) \not\rightarrow S(s) + H_2O(l)$

e. $CuO(s) + NH_3(g) \not\rightarrow Cu(s) + N_2(g) + H_2O(l)$

CHAPTER 16

Rates and Equilibria of Chemical Reactions

LEARNING OBJECTIVES

After studying this chapter, you should be able to:

1. Supply a correct definition, explanation, or example for each of these:

reaction rate	equilibrium constant
collision theory	ion product
activation energy	solubility product
effective collision	common ion effect
chemical equilibrium	buffer ratio
reversible chemical reaction	

2. Predict the effect of temperature, pressure, and catalyst on rate of reaction and position of equilibrium.
3. Draw or interpret energy-barrier diagrams in terms of activation energy, reaction rate, heat of reaction, and position of equilibrium.
4. Specify the conditions that would drive a particular reaction in the desired direction.
5. Write the equilibrium-constant expression for a given chemical equation.
6. Use solubility product to calculate the concentration of either ion from appropriate data.
7. Calculate a buffer ratio, hydronium-ion concentration, and pH of a buffer system from appropriate data.
8. Calculate the change in pH that adding a small amount of acid or base would cause in a given buffer system.

When we put food into a refrigerator, boil an egg, or plunge a burned finger into ice-cold water, we're using temperature to change the rates of chemical reactions. Putting food into the refrigerator slows down the chemical reactions that can cause food spoilage. Cooking an egg or anything else involves chemical reactions that are speeded up at higher temperatures. Ice-cold water will slow down the chemical reactions that damage tissue right after a burn. From this we see that chemical reactions can take place at different rates, or speeds. We'll look at some other ways besides temperature that we can use to slow down a reaction we don't want to happen or to speed up one we do want to happen.

Besides reaction rates, we'll also consider how far reactions go to completion. Some reactions go all the way; in Chapter 6, we assumed they all did. But in some reactions, the products react with each other to form reactants and reverse the reaction. When we inhale and exhale, our bodies are using just such reversible reactions. Hemoglobin reacts to bind oxygen, but what if it didn't let go of the oxygen when it got to the tissues? And what if our blood dissolved carbon dioxide but didn't release it at the lungs? Here, concentration changes are being used to drive a chemical reaction in a desired direction. We'll see some other ways to drive a reaction in the direction we want it to go.

16.1 RATES OF CHEMICAL REACTIONS

When we take sodium bicarbonate for an upset stomach, we should feel better right away. Acid-base reactions happen as soon as their solutions are mixed, and are over before you can say "phenolphthalein." Precipitation reactions happen just about as fast. If we were trying to remove phosphate from polluted water, we wouldn't have to wait for the precipitate of calcium phosphate after we'd added the calcium hydroxide. Both acid-base and precipitation reactions depend only on how fast the reacting ions or molecules can combine. Once they meet, it's all over.

Some kinds of reactions—mostly redox and some biological or organic reactions—take longer to happen. Iron takes a while to rust, fortunately. Many reactions between gases are quite slow. We'll look next at explanations of why some reactions go faster than others.

THE COLLISION THEORY OF REACTION RATES. When we define *reaction rate* as how fast the reaction goes, we're really talking about how many molecules, ions, or atoms react in a certain amount of time. We can measure the rate of a reaction by removing samples at various times to find out how much reactant (or how much product) there is. If we did this, we'd find that the rate depends on the number (or concentration) of reactant atoms, ions, or molecules. As the reaction progresses, the concentration of reactants decreases, because they're being used up in the reaction. Figure 16.1 illustrates the decrease of rate as concentration decreases.

reaction rate

FIGURE 16.1
Rate of reaction decreases as reactants disappear

13, 14, 15—I counted 15 sheep going over the fence in the last minute. That's a rate of 15 sheep per minute. This corral is really jammed full of sheep and they're all pushing and shoving and jostling each other. I don't blame them for wanting to get out of there.

Now most of the sheep have gone, and the rate has decreased a lot. Only one sheep escapes every minute. A chemical reaction is like this. The rate is fast at first, and grows slower and slower as more and more of the reactants are used up.

If we did further experiments, we'd find that the rate of a reaction increases with temperature as well as with concentration. An explanation of these observations lies in the *collision theory:* For molecules (or atoms or ions) to react, they have to bump into each other (collide), and they have to have enough energy to react. The reason the particles must bump each other is pretty obvious: How else can they react? But how much energy is "enough" for a reaction?

collision theory

Every reaction has an energy barrier that its reactants must overcome to become products. This energy barrier is like the potential barrier in Chapter 15; to push electrons over the potential barrier, we had to give them a boost of at least as much voltage as the barrier. To push reactants over the energy barrier, we have to give them a boost of at least as many kilocalories per mole as the energy barrier. This amount of energy is the *activation energy,* and it's different for every reaction.

activation energy

To cross the energy barrier, the reactants have to hit each other *and* they have to have the required activation energy after they've hit. When they do, an *effective collision* occurs, and the particles react. The rate of a reaction depends on the number of effective collisions that occur in a given amount of time. But where do the reactants get the energy?

effective collision

To answer that, we recall the Maxwell-Boltzmann Distribution Curve of Chapter 10 (p. 212): Some molecules always have enough energy to escape from the liquid and enter the vapor state. This can be applied to reactants, too: Some reactant molecules always have enough energy to fall over the barrier. If the energy barrier is high, then only a few will have enough energy; if it's low, then a lot will. The number of molecules that have enough energy to fall over the energy barrier determines the rate of a reaction, as shown in Figure 16.2.

When a reaction starts, the first particles to react are the ones with enough energy to fall over the barrier. The reaction keeps on going, though, even when these have reacted. The remaining particles hit each other and exchange energy, so that some of them are always getting more. The Maxwell-Boltzmann curve keeps redistributing itself, so that there are always particles that have high energy, as illustrated in Figure 16.3.

FACTORS AFFECTING REACTION RATE. If we want a reaction to go faster, we can increase the number of effective collisions (and thus the reaction rate) in these ways.

1. We can increase the concentrations of the reactants.
2. If one or more of the reactants is a gas, we can increase the pressure. Increasing the pressure of a gas is the same as increasing its concentration, as we learned in Chapter 11.
3. We can increase the temperature. A temperature increase of 10 K will make a reaction go roughly twice as fast.
4. We can use a catalyst, if we can find one that works. A catalyst speeds up a reaction by lowering the energy barrier, so that more particles can get across it. The catalyst usually remains unchanged at the end of the reaction. By using a catalyst, it is often possible to avoid the high temperatures that might otherwise be needed to make a reaction with high activation energy proceed. (See Box.)

16.2 CHEMICAL EQUILIBRIUM

Chemical equilibrium is a state in which two opposing chemical reactions are taking place at the same rate. This is another example of the dynamic equilibrium we saw in Chapter 10, where a process is being done as fast as it's being undone. Here, the process being done and undone is a reversible chemical reaction.

chemical equilibrium

FIGURE 16.2
Reactions having low activation energies go fast

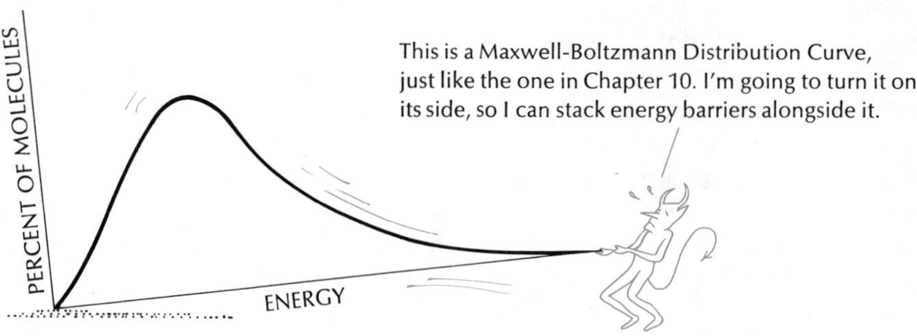

PERCENT OF MOLECULES
ENERGY
This is a Maxwell-Boltzmann Distribution Curve, just like the one in Chapter 10. I'm going to turn it on its side, so I can stack energy barriers alongside it.

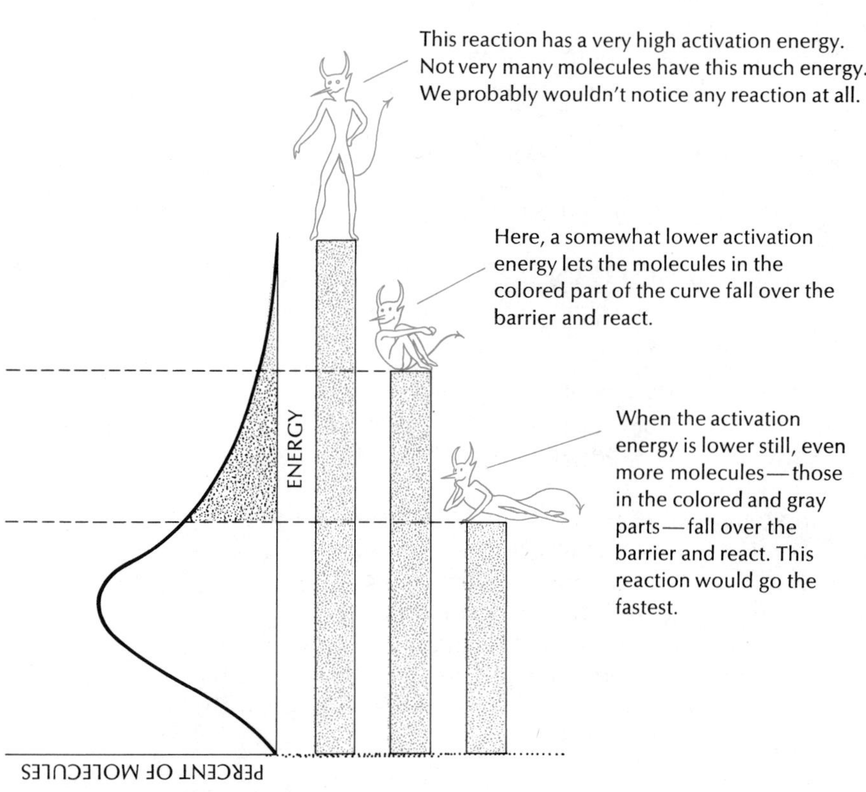

This reaction has a very high activation energy. Not very many molecules have this much energy. We probably wouldn't notice any reaction at all.
Here, a somewhat lower activation energy lets the molecules in the colored part of the curve fall over the barrier and react.
When the activation energy is lower still, even more molecules—those in the colored and gray parts—fall over the barrier and react. This reaction would go the fastest.
ENERGY
PERCENT OF MOLECULES

REVERSIBLE CHEMICAL REACTIONS. The reaction of oxygen with sulfur dioxide to make sulfur trioxide, an important industrial reaction, is an example of a *reversible chemical reaction,* one that can take place in either of two directions. To show that a reaction is reversible, we write it with double arrows, like this:

reversible chemical reaction

$$2\,SO_2(g) + O_2(g) \rightleftharpoons 2\,SO_3(g)$$

FIGURE 16.3
As high-energy molecules react, others take their places

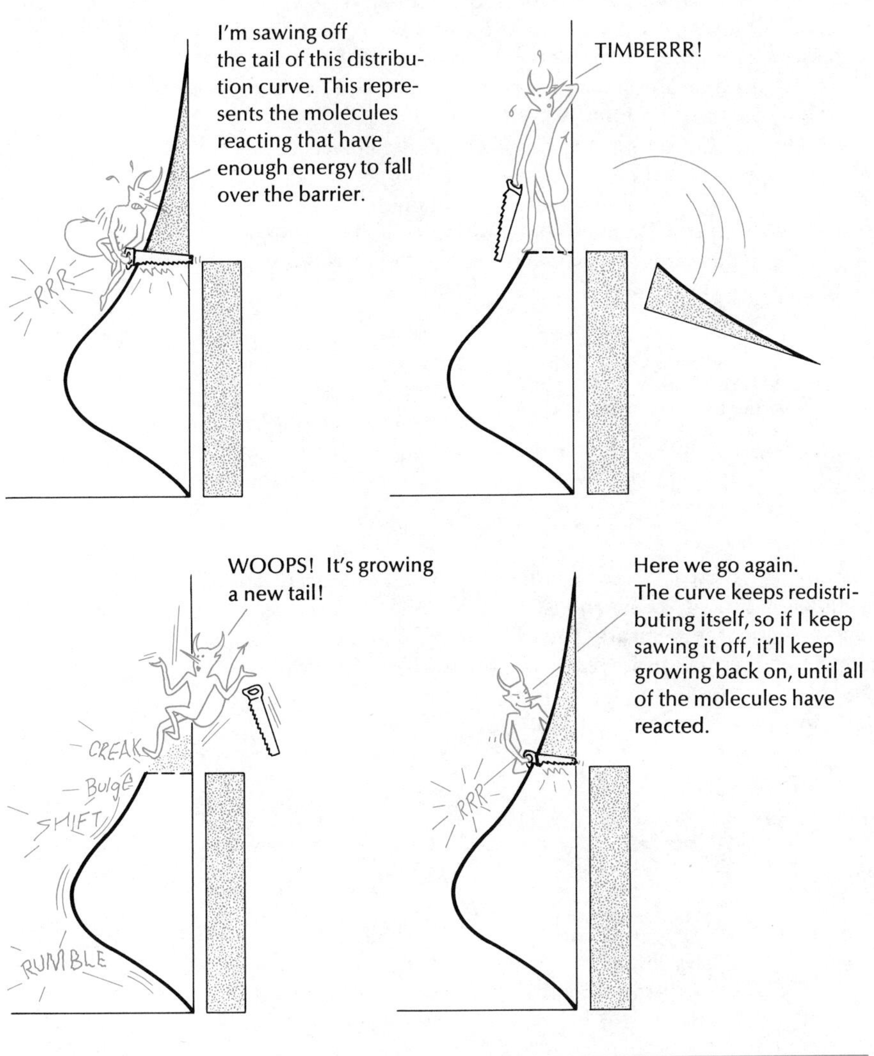

In this reaction, we might start with a mixture of sulfur dioxide and oxygen. The rate would be high at first, because the concentration of the reactants would be high. As the reactants disappeared, the rate of the forward reaction would decrease. Meanwhile, as sulfur trioxide was formed, it would become the reactant in the reverse reaction. At first, the rate would be low because there wouldn't be very much sulfur trioxide. As the forward reaction progressed, more sulfur trioxide would be available and the reverse rate would increase. The forward rate would decrease and the reverse rate would increase, until eventually the two rates would be the same. Then the system would be at equilibrium.

THE CATALYTIC CONVERTER

Up to now, the three biggest problems with automobile emissions have been carbon monoxide, unburned gasoline, and nitrogen monoxide. The CO and unburned gasoline are from incomplete combustion. One way to avoid them is to increase the amount of air and the temperature in the engine. But this also increases the amount of NO formed.

The reaction between N_2 and O_2 to form NO has a high energy barrier—too high for much NO to form at ordinary temperatures. This is lucky, because air is mostly N_2 and O_2, which mix freely, but which don't react to produce poisonous NO in the air. In a gasoline engine, though, the heat produced by the burning of gasoline gives some of the N_2 and O_2 enough energy to get over the barrier and form NO.

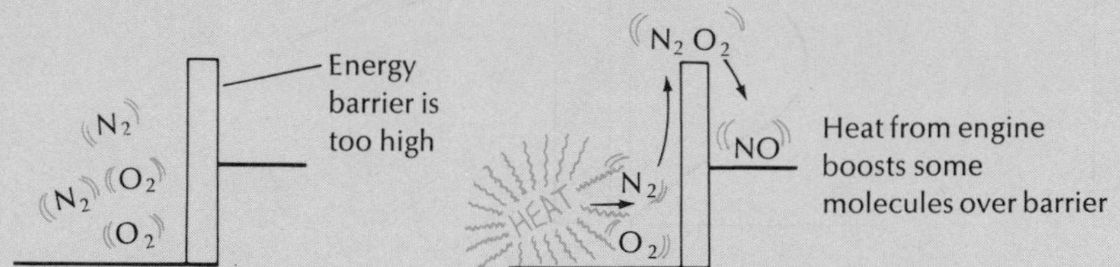

A relatively new device, called a "catalytic converter," takes the exhaust from the engine and allows the gasoline and CO to burn as completely as they would at a much higher temperature. The catalyst, a mixture of platinum and palladium, does this by lowering the energy barrier for the burning of CO and gasoline. The catalyst has no effect on the NO energy barrier and so does not cause more NO to form.

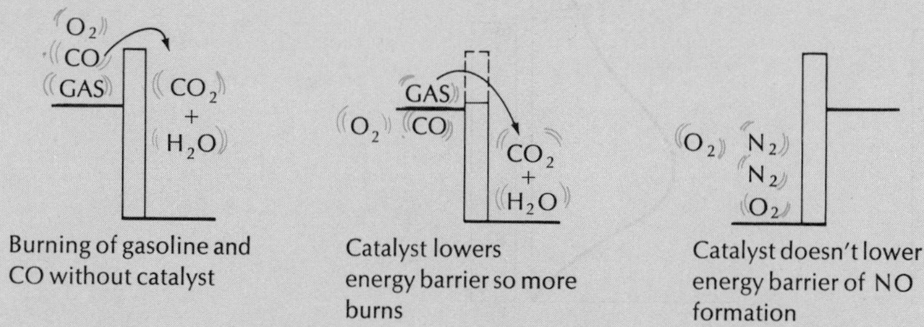

Burning of gasoline and CO without catalyst

Catalyst lowers energy barrier so more burns

Catalyst doesn't lower energy barrier of NO formation

Leaded gasoline can't be used in an automobile that has a catalytic converter. Lead "poisons" the catalyst so that it doesn't work.

An undesirable side effect of the catalytic converter is increased emissions of SO_3, formerly not a big problem in car exhaust. Small amounts of SO_2 are always produced by combustion of gasoline. But the catalytic converter lowers the energy barrier for SO_3 formation. Whereas a car without a converter releases about 1 milligram of SO_3 per mile, a car with a converter releases 10 milligrams to 30 milligrams per mile. And that's not all. H_2O—another product of combustion—can combine with SO_3 to produce sulfuric acid and, in time, destroy the converter.

We can always approach an equilibrium from either side. We could start with SO_3 and let it decompose to SO_2 and O_2. In that case, we'd write:

$$2\,SO_3(g) \rightleftharpoons 2\,SO_2(g) + O_2(g)$$

The two ways of writing the same reversible reaction mean exactly the same thing. At equilibrium, we can no longer tell the products from the reactants, because all substances are both. However, the way a reaction is written usually determines which we call the reactants and which we call the products. Often, the way we write the equation reflects what the equation is used for. The first of the above reactions is important industrially to make sulfur trioxide, so it would more commonly be written that way.

$$2\,SO_2(g) + O_2(g) \rightleftharpoons 2\,SO_3(g)$$

When we write the equation this way, SO_2 and O_2 are the reactants and the reaction between them is the forward reaction. SO_3 is the product and its decomposition is the reverse reaction.

FACTORS AFFECTING THE POSITION OF EQUILIBRIUM.

In a reversible reaction, the energy of the products can be—and usually is—different from the energy of the reactants. The energy barrier is approached from two different directions, and from different starting energies. The activation energies of the two reactions are different, so each reaction has a different distance to travel up the energy barrier. The difference between the activation energies of the reactants and the products is the heat of reaction. The reaction between sulfur dioxide and oxygen is exothermic.

$$2\,SO_2(g) + O_2(g) \rightleftharpoons 2\,SO_3(g) + 41.2\text{ kcal}$$

The energy barrier for the forward reaction is lower than that for the reverse reaction, and this favors the formation of product, as shown in Figure 16.4. In this case, we say that the position of the equilibrium lies to the right. As we see in the figure, though, an equilibrium doesn't necessarily lie all the way to the right or left. We'd find that at equilibrium at 1000 K, we'd actually get only about 60 percent of the theoretical yield of SO_3.

For endothermic reactions, the energy of the reactants is lower than the energy of the products, and the position of the equilibrium lies on the side of the reactants. We'd get less than 50 percent of the theoretical yield in that case. Figure 16.5 shows various possibilities of equilibrium positions.

In Chapter 10 (p. 203), we introduced Le Chatelier's Principle, which says that an equilibrium system, when disturbed, adjusts itself so as to restore equilibrium. This principle applies to chemical equilibrium, too—very much so. Disturbing a chemical equilibrium by changing its conditions causes the position to shift as equilibrium is reestablished. Which way the position is shifted depends on how the change affects the rates of the forward and reverse reactions. We'll look at these condition changes one by one, as we did for the reaction rate.

FIGURE 16.4
The position of equilibrium lies to the side of lower energy

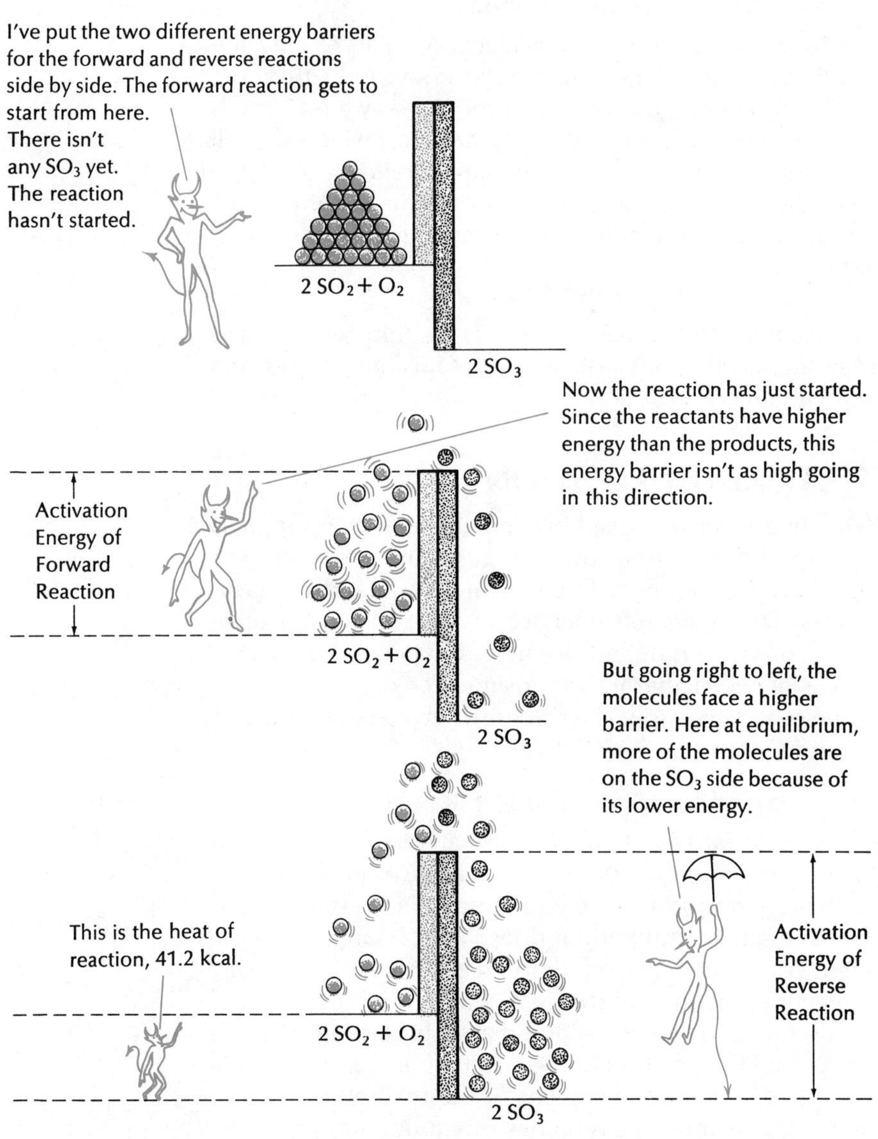

I've put the two different energy barriers for the forward and reverse reactions side by side. The forward reaction gets to start from here. There isn't any SO_3 yet. The reaction hasn't started.
2 SO_2 + O_2
2 SO_3
Now the reaction has just started. Since the reactants have higher energy than the products, this energy barrier isn't as high going in this direction.
Activation Energy of Forward Reaction
2 SO_2 + O_2
2 SO_3
But going right to left, the molecules face a higher barrier. Here at equilibrium, more of the molecules are on the SO_3 side because of its lower energy.
This is the heat of reaction, 41.2 kcal.
2 SO_2 + O_2
2 SO_3
Activation Energy of Reverse Reaction

Increasing the temperature will increase the rates of both the forward and reverse reactions. Whichever reaction has the higher activation energy will be affected more by temperature. If we know what the relative activation energies are, we can predict the direction of shift; if not, we can still predict that equilibrium will be established faster. Increasing the temperature also means supplying heat. If heat is a reactant or a product (which it

FIGURE 16.5
Equilibrium can lie to the right, to the left, or in between

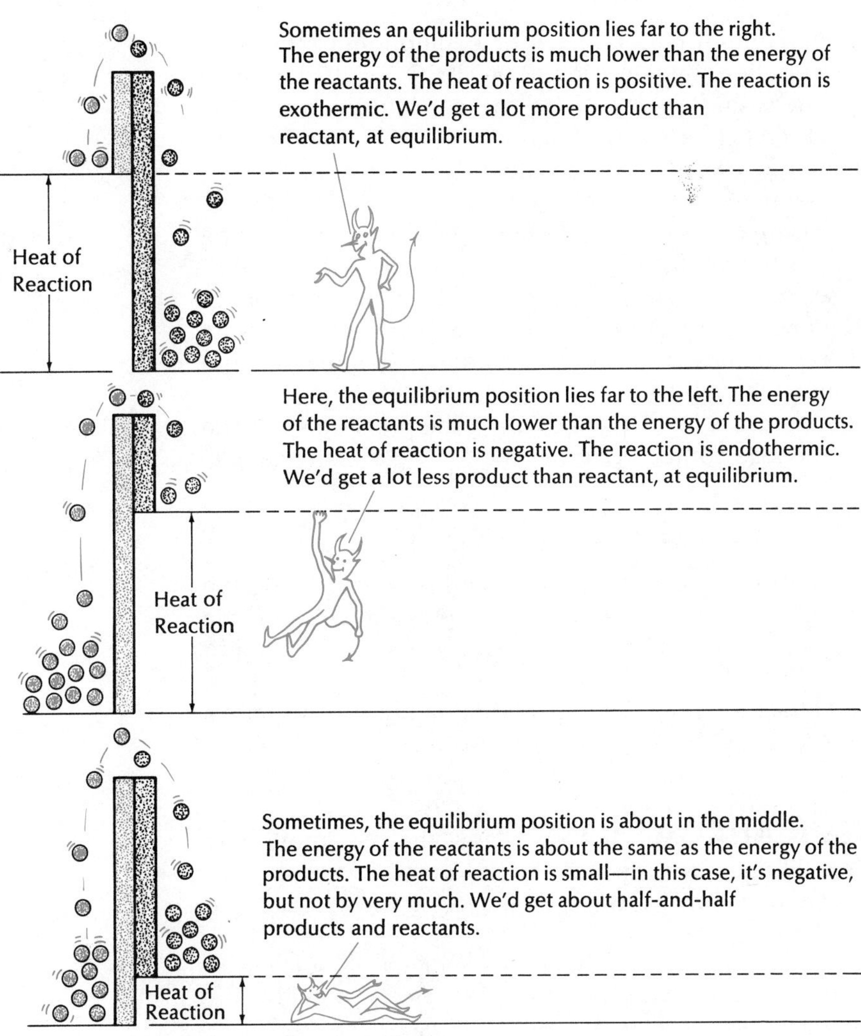

Sometimes an equilibrium position lies far to the right. The energy of the products is much lower than the energy of the reactants. The heat of reaction is positive. The reaction is exothermic. We'd get a lot more product than reactant, at equilibrium.
Heat of Reaction
Here, the equilibrium position lies far to the left. The energy of the reactants is much lower than the energy of the products. The heat of reaction is negative. The reaction is endothermic. We'd get a lot less product than reactant, at equilibrium.
Heat of Reaction
Sometimes, the equilibrium position is about in the middle. The energy of the reactants is about the same as the energy of the products. The heat of reaction is small—in this case, it's negative, but not by very much. We'd get about half-and-half products and reactants.
Heat of Reaction

always is), then this will increase the rate of the reaction in which heat appears as a reactant. The equilibrium will shift away from the side that contains heat. For example, in this reaction:

$$2\,SO_2(g) + O_2(g) \rightleftharpoons 2\,SO_3(g) + 41.2\text{ kcal}$$

adding heat shifts the equilibrium to the left, and taking heat away shifts the equilibrium to the right.

Increasing the concentration of any substance shifts the equilibrium away from that substance, and decreasing the concentration of any substance shifts the equilibrium toward that substance. For example, in the reaction:

$$H_2CO_3(aq) \rightleftharpoons H_2O(l) + CO_2(g)$$

increasing the concentration of H_2CO_3 shifts the equilibrium to the right. Decreasing the concentration (by removing some H_2CO_3) shifts the equilibrium to the left. Increasing or decreasing the concentration of H_2O won't have any effect. (Where a reaction happens in water solution and water is a reactant, there is always so much water that a change in concentration makes no difference.)

Increasing or decreasing the pressure of a gas is the same as increasing or decreasing its concentration. Increasing the pressure shifts the equilibrium away from a gas, and decreasing the pressure shifts the equilibrium toward a gas. In the above reaction, CO_2 is the only gas involved. Increasing the pressure shifts the equilibrium to the left, and decreasing the pressure shifts the equilibrium to the right.

Where more than one gas is involved, increasing the pressure shifts the equilibrium toward the side containing fewer moles of gas. For example, in the reaction:

$$2\,SO_2(g) + O_2(g) \rightleftharpoons 2\,SO_3(g) + 42.1 \text{ kcal}$$

the left side has three moles of gas—two of SO_2 and one of O_2—and the right side has only two moles of gas. Increasing the pressure will shift the equilibrium to the right, and decreasing the pressure will shift the equilibrium to the left.

A catalyst will not affect the position of equilibrium, but it will speed a reaction up and help equilibrium be established faster.

DRIVING CHEMICAL REACTIONS. Now that we know how conditions affect equilibrium, we can use this information to make an equilibrium shift in the direction we want.

EXAMPLE 16.1: We've seen that the equilibrium position of the reaction for the formation of SO_3 is such that we get only 60 percent of the theoretical yield at 1000 K. What conditions could we change to help us get more product?

Answer: Getting more product means shifting the equilibrium to the right. We can increase the pressure, add more SO_2 or O_2, or remove SO_3 or heat as they are formed. ■

Figure 16.6 shows how some of these conditions are actually used in the industrial preparation of SO_3.

We see that removing a product is a good way to shift an equilibrium to the right, and removing a reactant is a good way to shift it to the left. Sometimes another chemical reaction can be used to remove a product or a reactant. Some important reversible reactions that happen in blood are:

FIGURE 16.6
Industrial preparation of SO_3

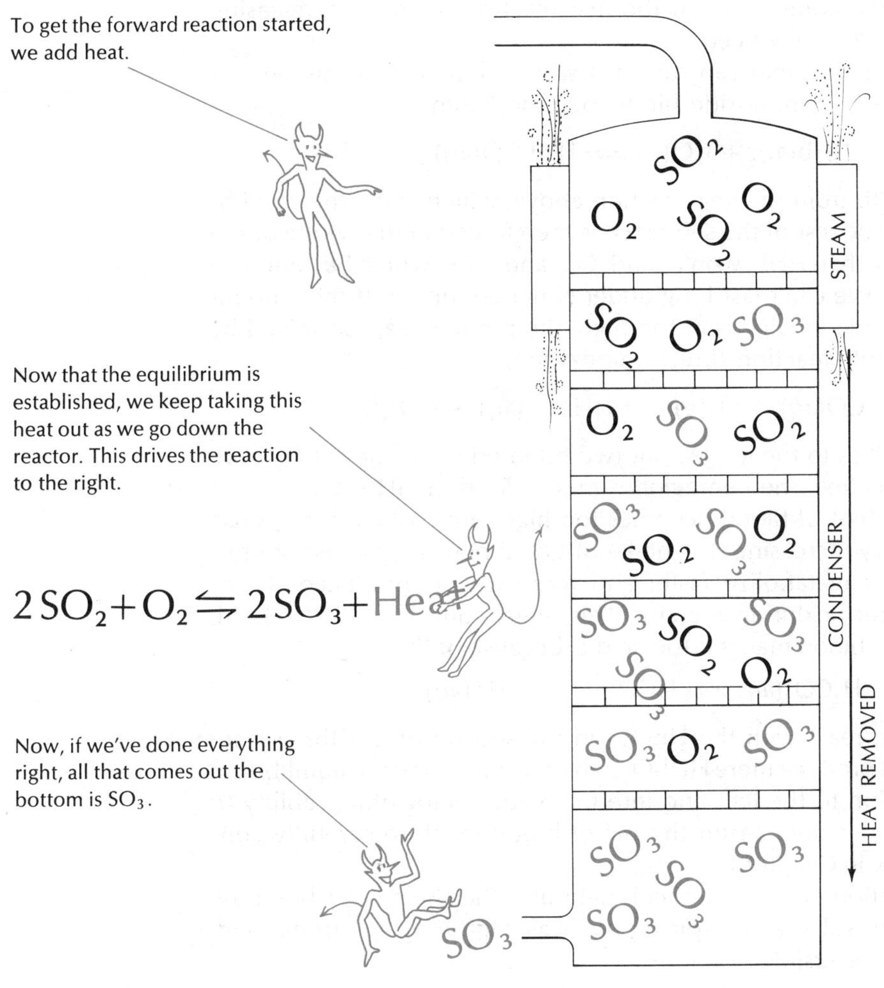

To get the forward reaction started, we add heat.
Now that the equilibrium is established, we keep taking this heat out as we go down the reactor. This drives the reaction to the right.
$2\,SO_2 + O_2 \rightleftharpoons 2\,SO_3 + Heat$
Now, if we've done everything right, all that comes out the bottom is SO_3.
STEAM
CONDENSER
HEAT REMOVED
SO_2
O_2
SO_3

$$Hb + O_2(g) \rightleftharpoons HHbO_2(aq) \quad \text{(hemoglobin bound to } O_2 \text{ is more acidic)}$$
$$HHbO_2(aq) + HCO_3^-(aq) \rightleftharpoons H_2CO_3(aq) + HbO_2^-(aq)$$
$$H_2CO_3(aq) \rightleftharpoons CO_2(g) + H_2O(l)$$

At the lungs, there is a high pressure (concentration) of O_2, which pushes the equilibrium of the first reaction to the right, forming more $HHbO_2$. The increased $HHbO_2$ pushes the equilibrium of the second reaction to the right, forming more H_2CO_3. And the increased H_2CO_3 pushes the equilibrium of the third reaction to the right, releasing from the blood the CO_2 we exhale.

At the tissues, there is a high pressure of CO_2 because it is a waste product of metabolism. The high pressure of CO_2 pushes the equilibrium of the third reaction to the left, forming more H_2CO_3, which pushes the second reaction's equilibrium to the left. The increased $HHbO_2$ that's formed pushes the equilibrium of the first reaction to the left, releasing to the tissues the O_2 they need.

A reversible reaction that can ruin this whole sequence is this one, in which inhaled carbon monoxide binds to hemoglobin:

$$HHb(aq) + CO(g) \rightleftharpoons HHbCO(aq)$$

This removes HHb from the first reaction above, which shifts the equilibrium to the left. The rest of the sequence is then forced in the wrong direction. This means that HHb won't bind O_2, and CO_2 won't be removed from the tissues. We can't last long under either condition. If the amount of carbon monoxide inhaled isn't too high, the problem can be solved by using this reversible reaction (breathing oxygen):

$$HHbCO(aq) + O_2(g) \rightleftharpoons HHbO_2(aq) + CO(g)$$

This equilibrium lies to the left. About two hundred molecules of O_2 have to be supplied to free one hemoglobin molecule from HHbCO.

If the CO_2 content of blood becomes too high, our biological response is to pant, thereby increasing the intake of O_2. When we exercise strenuously, our bodies metabolize faster and more CO_2 is produced. That's why we pant after or during exercise. This reflex controls the CO_2 level. Too much CO_2 in blood makes it too acidic, because of this reaction:

$$H_2CO_3(aq) \rightleftharpoons HCO_3^-(aq) + H^+(aq)$$

The H^+ can then react with the HbO_2^- in the second of our three reactions. This would produce more $HHbO_2$, which would force the equilibrium of the first reaction to the left, and interfere with hemoglobin's ability to bind oxygen. This is one reason the pH of blood must be carefully controlled, as we saw in Chapter 13.

Using one reaction to drive another is helpful in the chemistry laboratory, too. When we dissolve a precipitate, such as $Mg(OH)_2$, in strong acid, we're using these reversible reactions:

$$Mg(OH)_2(s) \rightleftharpoons Mg^{2+}(aq) + 2\,OH^-(aq)$$
$$2\,H^+(aq) + 2\,OH^-(aq) \rightleftharpoons 2\,H_2O(l)$$

The hydronium ion reacts with the hydroxide ion, removing hydroxide ion from the first reaction. The first equilibrium shifts to the right, and $Mg(OH)_2$ dissolves.

16.3 EQUILIBRIUM CALCULATIONS

So far we've seen what causes an equilibrium to shift and how to predict the direction of the shift. Now we'll see how to describe the positions of equilibria with numbers, which we can then use for more exact predictions.

THE EQUILIBRIUM CONSTANT. We could investigate an equilibrium system by analyzing samples of it to find out the concentration of each substance. If we took three samples from the reaction between SO_2 and O_2 to produce SO_3, at a constant temperature of 1000 K, we might obtain the three sets of data in Table 16.1. The concentrations themselves are different in each sample, but if we substitute them into this formula:

$$\frac{[SO_3][SO_3]}{[O_2][SO_2][SO_2]}$$

we come out with the same value for each sample: 284. This value is the *equilibrium constant* (K) for this reaction. It stays constant as long as the temperature stays the same, even though the individual concentrations may change.

equilibrium constant (K)

Every reversible reaction has such an equilibrium constant. Its value has to be determined experimentally, but we can write the equilibrium constant expression for any reversible reaction from its equation. The products are in the numerator, and the reactants are in the denominator. Each substance appears the same number of times that it appears in the chemical equation.

EXAMPLE 16.2: Write the equilibrium constant expression for this reaction:

$$N_2(g) + 3\ H_2(g) \rightleftharpoons 2\ NH_3(g)$$

Answer:

$$\frac{[NH_3][NH_3]}{[N_2][H_2][H_2][H_2]} = K$$

■

The value of the equilibrium constant tells us the position of an equilibrium. If the constant is greater than 1, as it is with 284, then the equilibrium lies to the right. If it's less than 1, then the equilibrium lies to the left. If it's around 1, then the equilibrium is about in the middle. The larger the equilibrium constant for a reaction, the smaller it is for the reaction written

TABLE 16.1
Possible concentrations for a reaction at equilibrium

$2\ SO_2(g) + O_2(g) \rightleftharpoons 2\ SO_3(g)$

Concentration				
SO_2	O_2	SO_3	*Calculations*	*Equilibrium Constant (K)*
0.100	0.225	0.800	$\frac{(0.800)(0.800)}{(0.100)(0.100)(0.225)} =$	284
1.00	0.352	10.0	$\frac{(10.0)(10.0)}{(1.00)(1.00)(0.352)} =$	284
0.400	0.507	4.80	$\frac{(4.80)(4.80)}{(0.400)(0.400)(0.507)} =$	284

in the opposite direction. When we write a reaction in the opposite direction, the new equilibrium constant is the reciprocal of the old one. For instance, when the SO_2/SO_3 reaction is written this way:

$$43.2 \text{ kcal} + 2\,SO_3(g) \rightleftharpoons 2\,SO_2(g) + O_2(g)$$

Equilibrium constants change with temperature. For an exothermic reaction, such as the formation of SO_3, the equilibrium constant decreases as the temperature increases. For an endothermic reaction, such as the decomposition of SO_3, the equilibrium constant increases as the temperature increases. Table 16.2 shows equilibrium constants for both the formation and decomposition of SO_3 at various temperatures. At room temperature (300 K), the equilibrium constant for the formation of SO_3 is extremely large, meaning that the equilibrium lies far on the SO_3 side. However, if we mix SO_2 and O_2 at room temperature, nothing happens because the energy barrier is high. Heat must be supplied to push the reactants over the energy barrier, but heat must be removed after equilibrium is established, so the equilibrium will shift in favor of SO_3. We already saw this happening in Figure 16.6.

At around 1400 K, the equilibrium constant for the formation of SO_3 drops below 1. At temperatures above this, the equilibrium lies further and further toward the SO_2 side.

SPECIAL KINDS OF EQUILIBRIUM CONSTANTS. Many equilibrium constant expressions aren't as complicated as those we've just seen. We introduced one simplified equilibrium constant in Chapter 13:

$$[H^+][OH^-] = 10^{-14}$$

This really states the value of the equilibrium constant for the ionization of water, which is a reversible reaction.

$$H_2O(l) \rightleftharpoons H^+(aq) + OH^-(aq)$$

For reversible reactions that happen in water solution, the concentration

TABLE 16.2
Equilibrium constants vary with temperature

K for Formation of SO_3 $2\,SO_2(g) + O_2(g) \rightleftharpoons 2\,SO_3(g)$	*Temperature (K)*	*K for Decomposition of SO_3* $2\,SO_3(g) \rightleftharpoons 2\,SO_2(g) + O_2(g)$
1.1×10^{22}	300	9.1×10^{-23}
3.2×10^{6}	700	3.12×10^{-7}
8.1×10^{4}	800	1.2×10^{-5}
3.2×10^{3}	900	3.1×10^{-4}
2.8×10^{2}	1000	3.6×10^{-3}
3.9×10^{1}	1100	2.6×10^{-2}
7.5	1200	1.3×10^{-1}
1.8	1300	5.6×10^{-1}
5.6×10^{-1}	1400	1.8

of water stays pretty much the same even though small amounts of water might be appearing or disappearing. The water concentration doesn't appear in the denominator of the equilibrium constant expression. The value of the water concentration need not be known, and is automatically included in the equilibrium constant. This particular equilibrium constant is called K_w, the *ion product* of water (10^{-14}).

ion product, K_w

Another equilibrium constant whose expression doesn't have a denominator is the *solubility product,* K_{sp}. In a saturated solution of a relatively insoluble compound, dissolving and crystal formation are reversible processes at equilibrium.

solubility product, K_{sp}

$$AgCl(s) \rightleftharpoons Ag^+(aq) + Cl^-(aq)$$
$$K_{sp} = [Ag^+][Cl^-] = 1.8 \times 10^{-10}$$

The value of K_{sp} tells us that the equilibrium lies far to the left, which we know because silver chloride doesn't dissolve much. Sometimes we might want to decrease its solubility even further, to avoid losing too much expensive silver ion. By adding a large amount of chloride ion, we can shift the equilibrium even further to the left, removing more Ag^+ from solution. This is the *common ion effect:* adding one of the ions of an insoluble substance decreases the solubility of that substance. We can use K_{sp} to illustrate the common ion effect.

common ion effect

EXAMPLE 16.3: A precipitate of AgCl has been prepared in the laboratory and is about to be washed free of impurities. The concentration of Ag^+ in a saturated solution of AgCl is 1.35×10^{-5} *M*. To avoid losing this much Ag^+ in the washing, the experimenter washes the precipitate with 1 *M* NaCl. How many moles per liter of Ag^+ will be lost in this case?

Solution: Since:

$$[Ag^+][Cl^-] = 1.8 \times 10^{-10}$$

then

$$[Ag^+] = \frac{1.8 \times 10^{-10}}{[Cl^-]}$$

The NaCl solution is the source of the common ion, Cl^-. Therefore:

$$[Cl^-] = 1\ M$$
$$[Ag^+] = \frac{1.8 \times 10^{-10}}{1} = \_\_\_\_\ M\ Ag^+$$

Answer: 1.8×10^{-10} moles Ag^+ per liter. The Ag^+ loss has been cut by 10^5. ■

EXAMPLE 16.4: Silver acetate is quite a bit more soluble than silver chloride. The K_{sp} of silver acetate is 2.3×10^{-3}. What should be the concentration of a $NaC_2H_3O_2$ solution to keep the silver ion concentration down to 1.0×10^{-3}?

Solution: From the K_{sp} expression:

$$[Ag^+][C_2H_3O_2^-] = 2.3 \times 10^{-3}$$

we may write:

$$[C_2H_3O_2^-] = \frac{2.3 \times 10^{-3}}{[Ag^+]} = \frac{2.3 \times 10^{-3}}{1.0 \times 10^{-3}} = \_\_\_\_\ M\ NaC_2H_3O_2$$

Answer: 2.3 *M* $NaC_2H_3O_2$. Since silver acetate is so much more soluble than silver chloride, a lot more Ag^+ would be lost. ■

Another kind of equilibrium expression is K_a, which is the *acid dissociation constant,* or an equilibrium constant for the ionization of an acid. For example:

acid dissociation constant, K_a

$$H_2CO_3(aq) \rightleftharpoons HCO_3^-(aq) + H^+(aq)$$

$$K_a = \frac{[HCO_3^-][H^+]}{[H_2CO_3]} = 4.4 \times 10^{-7}$$

The equilibrium lies to the left, as we know because carbonic acid is a weak acid. The ordering of acid strengths in Table 13.4 (p. 299) was based on decreasing K_a values for the acids. The larger the K_a, the stronger the acid.

BUFFERS. We've looked at some cases where equilibrium positions have been changed by large amounts. In biological systems like blood, though, very small changes are constantly taking place. Blood must maintain a pH of between 7.35 and 7.45, even as small amounts of acid and base are constantly being added to it. A *buffer system* is a mixture of a weak acid and its conjugate base, that resists changes in pH. Two buffer systems especially help the blood maintain its pH. These buffer systems are H_2CO_3/HCO_3^- and $H_2PO_4^-/HPO_4^{2-}$. To see how a buffer really works, let's look first at a simple analogy.

buffer system

The five members of a neighborhood club take a vote. Three vote no and two vote yes, which is a 60 percent vote against. If one of the no voters switches to a yes, then the yeses have 60 percent of the vote. In such a small club, one person is enough to tip the scale, because one person is a large percentage of the total. But what if the club were much larger, with 5000 members? In a vote of 3000 noes and 2000 yeses, the noes still have it with 60 percent. If one no voter switches to yes, the totals become 2999 to 2001 and the no percentage becomes 59.98. Not much of a change. That one vote makes almost no difference at all, because the one voter is buffered by the large total number of participants.

Essentially, this is how a mixture of a weak acid and its conjugate base works as a buffer. To see how a buffer works, we'll first rearrange the K_a expression.

$$[H^+] = K_a \times \frac{[\text{acid}]}{[\text{conjugate base}]}$$

We see that the hydronium-ion concentration, and therefore the pH, depends on two things: the K_a of the particular acid, and the ratio of acid to conjugate base, called the *buffer ratio.* To keep a pH constant, a K_a should be fairly close (within about one power of ten) to the desired hydronium-ion concentration. Thus the H_2CO_3/HCO_3^- system, with K_a of 4.4×10^{-7}, qualifies to maintain a hydronium-ion concentration between 4.5×10^{-8}

buffer ratio

and 3.6×10^{-8}. If the hydronium-ion concentration we want to maintain happens to be exactly the same as the K_a, then the buffer ratio has to be exactly 1. If the desired hydronium-ion concentration is slightly different from the K_a, as it is in blood, then the buffer ratio will be some number other than 1.

EXAMPLE 16.5: A sample of blood contains 1.1×10^{-3} *M* HCO_3^- and 1.0×10^{-4} *M* H_2CO_3. Calculate the buffer ratio and the resulting pH.

Solution: First, the buffer ratio:

$$\frac{[H_2CO_3]}{[HCO_3^-]} = \frac{1.0 \times 10^{-4}}{1.1 \times 10^{-3}} = 9.1 \times 10^{-2}$$

Now for the hydronium-ion concentration:

$$[H^+] = (4.4 \times 10^{-7})(9.1 \times 10^{-2}) = 4.0 \times 10^{-8}$$

We convert it to pH.

Answer: Buffer ratio $= 9.1 \times 10^{-2}$, pH $= 7.40$. ■

We see that to achieve the hydronium-ion concentration we need, we must multiply the K_a of carbonic acid by 9.1×10^{-2}. If this buffer ratio changes, then the hydronium-ion concentration changes too. A good buffer is one that keeps its buffer ratio from changing very much, even though hydronium ion or hydroxide ion is added.

When hydronium ion is added to the buffer system, it reacts with HCO_3^- like this:

$$HCO_3^-(aq) + H^+(aq) \rightleftharpoons H_2CO_3(aq)$$

This is the reverse of the ionization of carbonic acid, so its equilibrium constant is the reciprocal of the K_a for carbonic acid: $K = 1/(4.4 \times 10^{-7}) = 2.3 \times 10^6$. The equilibrium lies far to the right. Essentially all the hydronium ion will react completely with HCO_3^- to form H_2CO_3. If we start with more HCO_3^- than H^+, there will be no H^+ at the end of the reaction; the number of moles of H_2CO_3 formed will exactly equal the number of moles of H^+ added; and the amount of HCO_3^- destroyed will exactly equal the amount of H^+ added.

EXAMPLE 16.6: For the blood sample of Example 16.5, calculate the concentrations of HCO_3^- and H_2CO_3 that would remain after adding H^+ equivalent to a concentration of 1.0×10^{-5} *M*.

Solution: The concentration of hydronium ion added is less than the concentration of HCO_3^-. For the 1.0×10^{-5} *M* H^+ that reacts, 1.0×10^{-5} *M* HCO_3^- will be destroyed. The concentration of HCO_3^- is then $(1.1 \times 10^{-3}) - (1.0 \times 10^{-5}) = 1.1 \times 10^{-3}$—no change, within the limits of our accuracy. The H_2CO_3 concentration becomes $(1.0 \times 10^{-4}) + (1.0 \times 10^{-5}) = 1.1 \times 10^{-4}$—a very small change.

Answer: $[HCO_3^-] = 1.1 \times 10^{-3}$ *M*; $[H_2CO_3] = 1.1 \times 10^{-4}$ *M*. ■

Now let's see how these concentration changes affect the buffer ratio and the pH.

EXAMPLE 16.7: Calculate the new buffer ratio and pH for the concentrations of Example 16.6.

Solution:

$$\text{buffer ratio} = \frac{1.1 \times 10^{-4}}{1.1 \times 10^{-3}} = 1.0 \times 10^{-1}$$

$$[H^+] = (4.4 \times 10^{-7})(1.0 \times 10^{-1}) = 4.4 \times 10^{-8}$$

Answer: Buffer ratio = 1.0×10^{-1}; pH = 7.36. ■

The pH is still within the allowed range. The buffer ratio can maintain the pH because its individual concentrations are very high compared with the added amounts of acid or base. The buffered H_2CO_3/HCO_3^- solution is like the club with many members. The 1.0×10^{-5} *M* H^+ makes hardly any difference to the large (by comparison) 1.1×10^{-3} *M* HCO_3^-. The buffer system in blood works because the changes in acidity are small. But if the amounts of added acid or base were as large or larger than the amount of buffer acid or conjugate base, then the buffer wouldn't work any more.

To show just how important this buffering system is, let's see what would happen if we added the same amount of acid to an unbuffered solution. A hydronium-ion concentration of 4.0×10^{-8} *M* means a hydroxide-ion concentration of 2.5×10^{-7} *M*. We can prepare a solution having the same pH as blood by making a 2.5×10^{-7} *M* NaOH solution.

EXAMPLE 16.8: Calculate the pH resulting when the equivalent of 1.0×10^{-5} *M* H^+ is added to a solution of 2.5×10^{-7} *M* NaOH.

Solution: We know that hydronium and hydroxide ions react with each other completely. Since there is more acid than base, all of the base will be used up, and there will be some acid left.

$$[H^+] = (1.0 \times 10^{-5}) - (2.5 \times 10^{-7}) = 1.0 \times 10^{-5}$$

The H^+ has completely overwhelmed the OH^-.

Answer: pH = 5.00. ■

The pH has changed by nearly 2. The unbuffered solution is like the club with only a few members. The amount of hydronium ion added was very large compared with the amount of hydroxide ion already there, and the addition was enough to change the solution from basic to acidic.

We could do the same kind of calculations for the addition of base instead of acid. When base is added to the buffer system, this reaction takes place:

$$H_2CO_3(aq) + OH^-(aq) \rightleftharpoons HCO_3^-(aq) + H_2O(l)$$

In the reactions of a weak acid with a strong base, the equilibrium lies far to the right. As with the addition of acid, we assume that a given amount of base produces the same amount of HCO_3^- and destroys the same amount of H_2CO_3.

REVIEW QUESTIONS

Rates of Chemical Reactions

1. What do we mean by *reaction rate?* How does it change as a reaction proceeds?
2. How does the *collision theory* explain changes in reaction rates?
3. What is *activation energy?* How is it like voltage?
4. What is an *effective collision?* How does the number of effective collisions influence the reaction rate?
5. When a chemical reaction starts, what determines which particles react first? What determines how many will react at a time?
6. What are the ways we can make a reaction go faster?
7. How does a catalyst increase the rate of a reaction?

Chemical Equilibrium

8. What is *chemical equilibrium?*
9. Give an example of a *reversible chemical reaction.* Tell what happens to the forward and reverse rates as the system approaches equilibrium.
10. How do we know which is the forward reaction and which the reverse reaction?
11. Sketch energy barriers for an endothermic reaction and for an exothermic reaction. Label the activation energies and the heat of reaction.
12. Explain how the position of equilibrium is affected by (a) temperature; (b) concentration; (c) pressure; and (d) a catalyst.
13. Under what conditions would changing the water concentration make no difference to the position of equilibrium? Explain.
14. Explain how a reaction can be driven in a desired direction by changing the conditions.
15. What are some reversible reactions that happen in blood? How do they depend on each other?
16. Why do we pant? Why is this an important biological response?

Equilibrium Calculations

17. What is an *equilibrium constant?* Give an example of an equilibrium constant and its expression.
18. What happens to an equilibrium constant when its reaction is written in reverse?
19. How does an equilibrium constant change with temperature?
20. Write the expression for K_w. Why doesn't it contain water concentration?
21. Write the expression for a *solubility product.*
22. What is the *common ion effect?* Explain how it can be used to decrease a salt's solubility.
23. Write the expression for the ionization of an acid.
24. What two quantities determine the pH of a buffer solution?
25. Explain how a buffer can resist changes in pH.
26. Under what conditions would the H_2CO_3/HCO_3^- buffer system in blood no longer be effective? Why?

EXERCISES

Set A (Answers at back of book. Asterisks mark the more difficult exercises.)

1. Some energy barriers are shown below. In each pair, which reaction would go faster, 1 or 2?

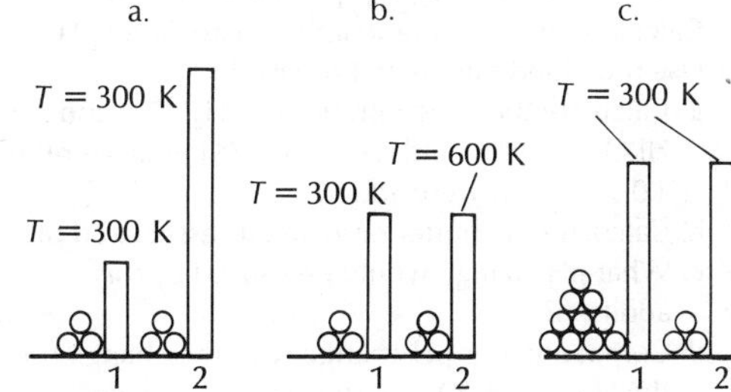

2. Sketch energy-barrier diagrams that show these (assume reversible reactions).
   a. an exothermic reaction
   b. an endothermic reaction
   c. zero activation energy for reverse reaction, nonzero activation energy for forward reaction
   d. a reaction with zero heat of reaction but nonzero activation energies
3. State where the equilibrium position lies for each part of Exercise 2.
4. For each part of Exercise 2, state whether the forward reaction or the reverse reaction would be more affected by changes in temperature.

5. TiO_2, a white paint pigment, is prepared industrially according to this reversible reaction:

$$TiCl_4(g) + O_2(g) \rightleftharpoons TiO_2(s) + 2\,Cl_2(g) + 17\text{ kcal}$$

State the effect each would have on the position of equilibrium.
   a. increasing the pressure
   b. decreasing the pressure
   c. removing solid TiO_2 as it is formed
   d. supplying heat
   e. removing heat
   f. adding a catalyst
6. Sketch an energy-barrier diagram for the reversible reaction in Exercise 5. Label the heat of reaction and the activation energies for the forward and reverse reactions.
7. In each of the following industrial processes, state what conditions should be used to obtain the maximum amount of product.
   a. preparation of lime from limestone:
   $$CaCO_3(s) + 44\text{ kcal} \rightleftharpoons CaO(s) + CO_2(g)$$
   b. first step in acetylene preparation (preparation of calcium carbide):
   $$CaO(s) + 3\,C(s) + 111\text{ kcal} \rightleftharpoons CaC_2(s) + CO(g)$$
   c. second step in acetylene preparation:
   $$CaC_2(s) + 2\,H_2O(l) \rightleftharpoons Ca(OH)_2(s) + \underset{\text{acetylene}}{C_2H_2(g)} + 149\text{ kcal}$$
   d. Haber process for nitrogen fixation:
   $$N_2(g) + 3\,H_2(g) \rightleftharpoons 2\,NH_3(g) + 22\text{ kcal}$$
   e. first step in Solvay process for making sodium bicarbonate:
   $$NH_3(g) + CO_2(g) + H_2O(l) \rightleftharpoons NH_4HCO_3(aq)$$
8. Sketch energy barrier diagrams for parts a, b, c, and d of Exercise 7. Label each activation energy and the heat of reaction.
9. Sketch energy barrier diagrams for each of the following cases, and state whether the position of equilibrium would lie to the left or to the right of the reaction.
   a. energy of products greater than energy of reactants; large activation energy
   b. energy of products greater than energy of reactants; no activation energy
   c. energy of products less than energy of reactants; small activation energy
10. With reference to the blood equilibria on pages 383–384, state whether each of the following would cause slower or more rapid breathing.
   a. breathing CO_2
   b. breathing a basic gas, such as NH_3
   c. breathing pure oxygen instead of air
11. Write an equilibrium-constant expression for each.
   a. $2\,NO(g) + O_2(g) \rightleftharpoons 2\,NO_2(g)$
   b. $4\,NH_3(g) + 5\,O_2(g) \rightleftharpoons 4\,NO(g) + 6\,H_2O(g)$
   c. $2\,H_2(g) + O_2(g) \rightleftharpoons 2\,H_2O(g)$
12. For each of the following, state whether the equilibrium favors reactants or products.
   a. $N_2(g) + 3\,H_2(g) \rightleftharpoons 2\,NH_3(g)$
   K (at 300 K) $= 6.85 \times 10^5$
   b. $N_2(g) + 3\,H_2(g) \rightleftharpoons 2\,NH_3(g)$
   K (at 720 K) $= 2.5 \times 10^{-5}$
   c. $2\,NO_2(g) \rightleftharpoons N_2O_4(g)$
   K (at 300 K) $= 9.1$
13. In the Haber process, the reaction between N_2 and H_2 to form NH_3 has such a high energy barrier that it must be carried out at 720 K rather than at room temperature (300 K). Refer to the equilibrium constants at these two temperatures in Exercise 12.
   a. Is the reaction endothermic or exothermic?
   b. What specific problem would be created by carrying out the reaction at 720 K?
   c. How could this problem be overcome so as to produce as much NH_3 as possible?
14. Write solubility product expressions for the following.
   a. $PbCrO_4(s) \rightleftharpoons Pb^{2+}(aq) + CrO_4^{2-}(aq)$
   b. $BaSO_4(s) \rightleftharpoons Ba^{2+}(aq) + SO_4^{2-}(aq)$
15. How could the common ion effect be used to remove as much Hg_2^{2+} as possible from polluted water by precipitation of Hg_2Cl_2? The equation is:
   $$Hg_2^{2+}(aq) + 2\,Cl^-(aq) \rightleftharpoons Hg_2Cl_2(s)$$
16. Calculate the amount of Ag^+ in a solution containing 1.0 M NaBr and solid AgBr. (The K_{sp} of AgBr is 5.2×10^{-13}.)
17. Solid $PbCrO_4$ is at equilibrium with its solution. What must be the concentration of Pb^{2+} to keep the CrO_4^{2-} concentration down to $1.0 \times 10^{-8}\ M$? (The K_{sp} of $PbCrO_4$ is 2.0×10^{-16}.)
18. For each of the following acids, write the equilibrium-constant expressions and arrange the acids in order of increasing acid strength.
   a. $HC_2H_3O_2$: $K_a = 1.8 \times 10^{-5}$
   b. HCO_3^-: $K_a = 4.7 \times 10^{-11}$
   c. H_3PO_4: $K_a = 7.1 \times 10^{-3}$
   d. HSO_4^-: $K_a = 1.0 \times 10^{-2}$
19. The blood buffer system $H_2PO_4^-/HPO_4^{2-}$ has a K_a of 6.3×10^{-8}. The following concentrations are found in a sample of blood: $[HPO_4^{2-}] = 2.6 \times 10^{-4}\ M$; $[H_2PO_4^-] = 1.8 \times 10^{-4}\ M$. Calculate the buffer ratio and its associated pH.
20. Use the blood sample in Exercise 19.
   a. Calculate the concentrations of $H_2PO_4^-$ and HPO_4^{2-} that would remain if $[H^+]$ equivalent to $1.0 \times 10^{-5}\ M$ were added.
   b. Calculate the buffer ratio and its associated pH.
   c. What pH change would be caused by this addition?
   d. Comparing this pH change with the change for the H_2CO_3/HCO_3^- buffer system on page 389, state which is the more effective buffer.

21. Calculate the pH change that would be produced on the buffer system of Exercise 19 if $[OH^-]$ were added equivalent to 1.0×10^{-5} *M*.

*22. Would you expect the $H_2PO_4^-/HPO_4^{2-}$ buffer in blood to be able to withstand an addition of $[H_3O^+]$ equivalent to 1.0×10^{-4} *M*? Explain, and prove your answer with appropriate calculations.

23. An industrial fermentation process requires that the pH be maintained at 5.40. Which of the following buffer systems would be the best choice? Explain.
    a. $HC_2H_3O_2(aq) \rightleftharpoons C_2H_3O_2^-(aq) + H^+(aq)$ $K_a = 1.8 \times 10^{-5}$
    b. $H_2PO_4^-(aq) \rightleftharpoons HPO_4^{2-}(aq) + H^+(aq)$ $K_a = 6.3 \times 10^{-8}$
    c. $NH_4^+(aq) \rightleftharpoons NH_3(aq) + H^+(aq)$ $K_a = 5.7 \times 10^{-10}$

Set B (Answers not given. Asterisks mark the more difficult exercises.)

1. Sketch energy-barrier diagrams that show:
    a. Reaction A and reaction B are carried out at the same temperature and concentration, but reaction A goes much faster than reaction B.
    b. Reaction A and reaction B have the same activation energy. At equal temperatures, reaction B can be made to go faster than reaction A by varying the concentrations of reactants.
    c. Reaction A and reaction B have the same activation energy. At the same initial concentrations of reactants, reaction A can be made to go faster than reaction B by varying the temperature.
2. State whether each of the following is endothermic or exothermic. Which has the higher activation energy, the forward reaction or the reverse reaction?

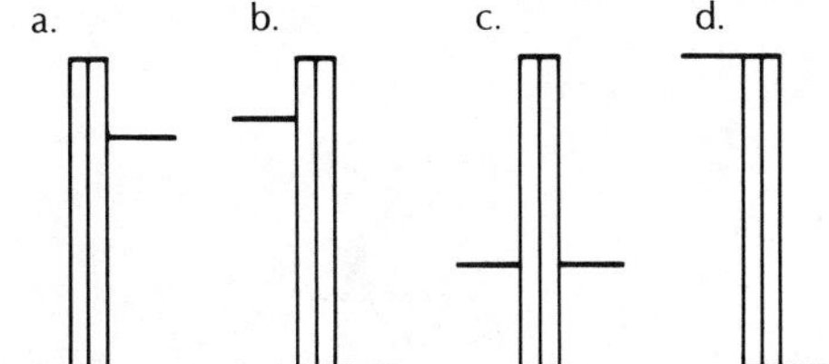

3. State where the equilibrium position would lie for each part of Exercise 2.
4. For each part of Exercise 2, state whether the forward reaction or the reverse reaction would be more affected by changes in temperature.
5. The formation of ozone can be described by this equation.
$$3\,O_2(g) + 64.8\text{ kcal} \rightleftharpoons 2\,O_3(g)$$
State the effect each would have on the position of equilibrium.
    a. increasing the pressure
    b. decreasing the pressure
    c. supplying heat
    d. removing heat
    e. adding a catalyst
    f. increasing the temperature
6. Sketch an energy-barrier diagram for the reversible reaction in Exercise 5. Label the heat of reaction and the activation energies for the forward and reverse reactions.
7. In each of these, state what conditions should be used to get the maximum amount of product.
    a. preparation of methyl chloride (CH_3Cl)
    $$CH_4(g) + Cl_2(g) \rightleftharpoons CH_3Cl(g) + HCl(g) + 26.4\text{ kcal}$$
    b. first step in preparation of nitric acid
    $$4\,NH_3(g) + 5\,O_2(g) \rightleftharpoons 4\,NO(g) + 6\,H_2O(g) + 216\text{ kcal}$$
    c. preparation of hydrogen iodide
    $$H_2(g) + I_2(g) + 6.2\text{ kcal} \rightleftharpoons 2\,HI(g)$$
    d. reduction of iron ore to obtain iron metal
    $$3\,C(s) + 2\,Fe_2O_3(s) + 111\text{ kcal} \rightleftharpoons 4\,Fe(s) + 3\,CO_2(g)$$
8. Sketch energy-barrier diagrams for each part of Exercise 7. Label each activation energy and the heat of reaction.
9. Sketch energy-barrier diagrams for each of these cases, and state whether the position of equilibrium would lie to the left or to the right.
    a. energy of products the same as energy of reactants; large activation energy
    b. energy of products greater than energy of reactants; small activation energy
    c. energy of products less than energy of reactants; large activation energy
10. Scuba divers occasionally suffer from carbon dioxide poisoning. Could it be caused by breathing either too rapidly or too slowly? Explain.
11. Write an equilibrium constant expression for each.
    a. $PCl_5(g) \rightleftharpoons PCl_3(g) + Cl_2(g)$
    b. $2\,NO(g) + 2\,H_2(g) \rightleftharpoons N_2(g) + 2\,H_2O(g)$
    c. $H_2(g) + Cl_2(g) \rightleftharpoons 2\,HCl(g)$
12. For each, state whether the equilibrium favors reactants or products.
    a. $HC_2H_3O_2(aq) \rightleftharpoons C_2H_3O_2^-(aq) + H^+(aq)$
    $K = 1.8 \times 10^{-5}$
    b. $H_2(g) + I_2(g) \rightleftharpoons 2\,HI(g)$
    K (at 973 K) $= 54.6$

c. $NH_4^+(aq) + OH^-(aq) \rightleftharpoons NH_3(aq) + H_2O(l)$
$K = 5.6 \times 10^4$

13. Explain how a reaction may be exothermic and still require a high temperature to get it started.
14. Write solubility product expressions for these.
   a. $CaCO_3(s) \rightleftharpoons Ca^{2+}(aq) + CO_3^{2-}(aq)$
   b. $PbSO_4(s) \rightleftharpoons Pb^{2+}(aq) + SO_4^{2-}(aq)$
15. What would happen if HCl gas were bubbled into a saturated solution of AgCl? Explain.
16. Calculate the concentration of Sr^{2+} in a solution that contains 0.50 *M* Na_2SO_4 and solid $SrSO_4$. (The K_{sp} of $SrSO_4$ is 7.6×10^{-7}.)
17. In the precipitation of AgI ($K_{sp} = 1.5 \times 10^{-16}$), what must be the iodide ion concentration so that the silver ion concentration is no greater than 1.0×10^{-10} *M*?
18. For each of these acids, write the equilibrium-constant expressions and arrange the acids in order of increasing strength.
   a. HNO_2: $K_a = 4.5 \times 10^{-4}$
   b. HClO: $K_a = 3.5 \times 10^{-8}$
   c. HCN: $K_a = 4.0 \times 10^{-10}$
   d. $HC_7H_5O_2$: $K_a = 6.3 \times 10^{-5}$
19. An acetic acid-acetate buffer is prepared by making a solution that is 1.00 *M* each in acetic acid and sodium acetate. (a) Calculate the buffer ratio and the associated pH. (b) Calculate the pH of a solution that is 0.100 *M* each in acetic acid and sodium acetate.
20. Use the two buffer solutions in Exercise 19.
   a. For each solution, calculate the concentrations of $C_2H_3O_2^-$ and $HC_2H_3O_2$ that would remain if $[H^+]$ equivalent to 0.01 *M* were added.
   b. Calculate the buffer ratio and its associated pH.
   c. What pH change would be caused by this addition?
   d. Which of these solutions is the more effective buffer? Explain why.
21. Calculate the pH change that would be produced on the buffer systems of Exercise 19 if $[OH^-]$ were added equivalent to 0.05 *M*.

*22. To maintain a pH value within 0.05 pH units, calculate the maximum amount of acid that could be added to each buffer system in Exercise 19.

23. Which of the acids of Exercise 18 would be the best choice to use in a buffer system where the pH was to be maintained at 4.70? Explain.

CHAPTER 17

Nuclear Reactions

LEARNING OBJECTIVES

After studying this chapter, you should be able to:

1. Supply a correct definition, explanation, or example for each of these:

nuclear reaction	Geiger-Müller counter
radioactivity	becquerel
radiation	scintillation counter
alpha particle	autoradiography
beta particle	half-life
gamma ray	carbon dating
alpha emitter	radioactive decay series
alpha emission	uranium dating
beta emitter	mass defect
beta emission	nuclear binding energy
gamma emitter	nuclear fusion
gamma emission	nuclear fission
decay	fissionable nuclei
electron capture	chain reaction
projectile	critical mass
transmutation	breeder reactor
artificial radioactivity	
radioisotope	
transuranium element	

2. Referring to a table of symbols and atomic numbers, write nuclear equations.
3. Write or recognize symbols, names, mass numbers, and charges of the nuclear particles in Table 17.1.
4. Describe and illustrate the ways in which a nucleus may be unstable and how it can stabilize itself.
5. Describe and work simple problems in carbon dating.
6. Explain why radioactive decay series exist and how they may be used in dating.
7. Write or recognize some uses of specific radioisotopes.
8. Perform appropriate mass defect calculations.
9. Given a graph of nuclear binding energy vs. mass number, predict relative amounts of energy released or absorbed by a fission or fusion process.
10. Given a partial equation for nuclear fusion or fission, find the mission products and complete the equation.

When the alchemists tried to change lead into gold, they were really trying to change one element into another. Little did they know that similar reactions were happening all around them. And they are happening all around us, too. In the ground, for instance, uranium is changing into thorium, thorium into protactinium, radium into radon, and radon into polonium. On the sun, hydrogen is changing into helium, releasing the energy that we need to live on earth. These changes are all spontaneous nuclear reactions.

Nuclear reactions are different from the chemical reactions we've been studying up to now. In chemical reactions, the elements keep their identities, even though they are bonded in different ways. Only the outer electrons are involved in chemical reactions. In *nuclear reactions,* the nuclei and sometimes the inner electrons are involved, and different elements are formed because the nuclei themselves change.

nuclear reaction

Humans can cause nuclear reactions to happen, too, and make one element change into another. The change that the alchemists wanted—lead into gold—isn't practical, but many other changes can be made to happen that are just as valuable. Humans can use nuclear reactions to make radioactive isotopes for medicine, industry, and atomic power plants.

The words "radioactivity," "nuclear reactor," "radioisotope," and "atomic power plant" are all part of our modern vocabulary. We'll take a closer look at the nuclear reactions that are behind them.

17.1 RADIOACTIVITY

All the spontaneous nuclear changes mentioned above happen with a release of energy. When something releases energy, sooner or later someone is going to notice it. In 1896, Henri Becquerel (1852–1908) noticed that a photographic plate on which he had laid a sample of a uranium salt had been exposed. The plate had been packaged to protect it from light, so he correctly concluded that the film had been exposed by some other kind of rays coming from the uranium. This spontaneous emission of radiation he named *radioactivity. Radiation* is energy traveling in a straight line.

radioactivity

radiation

This discovery opened the door to a whole world of nuclear reactions no one knew existed. Scientists started looking for and finding other radioactive substances. In 1898, Marie and Pierre Curie isolated two previously unknown elements, which they named polonium and radium. These are also radioactive. Since then, many other radioactive substances have been found in nature.

If we had a sample of a radioactive substance, we too would notice the energy it released. Some radioactive substances glow in the dark, some expose photographic film—and they can burn and damage tissue. Placed in a glass of water, a radioactive substance will cause the water to warm up. This energy is released as the nucleus of the substance changes into another nucleus.

THE PRODUCTS OF RADIOACTIVITY.

Scientists soon discovered that radioactive substances give off more than one kind of radiation. Some have more energy than others. Some radiations have mass and charge; some don't. Three kinds of radiation were recognized at first. They were named after the first three letters in the Greek alphabet—alpha, beta, and gamma—because no one knew exactly what they were. Later, when scientists found out what each kind of radiation was, the Greek names stuck. An *alpha particle* (symbolized with the Greek letter α) is a helium nucleus. A *beta particle* (β) is a high-energy electron. And a *gamma ray* (γ) is pure high-energy radiation (without mass or charge). These and other nuclear particles are shown in Table 17.1, where we use this shorthand notation:

alpha particle (α)

beta particle (β)

gamma ray (γ)

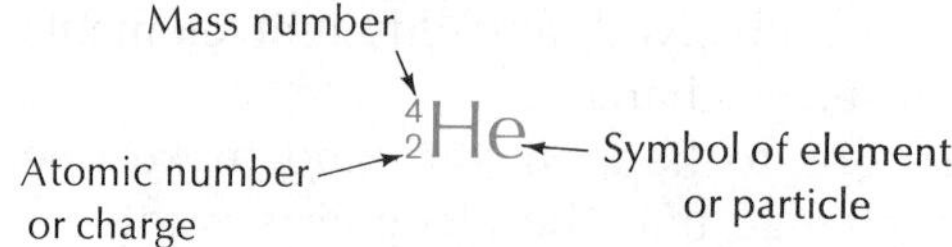

Different radiations have different abilities to penetrate skin and tissue. When it penetrates matter, nuclear radiation knocks electrons off the molecules it contacts, forming ions. These ions, too, have high energy and react with molecules near them. Radiation damage to tissue happens when the structures of important biological molecules are changed as a result of this ionization. Our bodies can repair small amounts of this kind of damage but not large amounts. The early nuclear scientists didn't realize the danger. All of them suffered from radiation burns, and some developed cancer as a result of exposure to radioactivity. Today, scientists protect themselves with lead shields and remote-control devices, but accidents causing exposure can still happen.

WRITING NUCLEAR EQUATIONS.

When a radioactive substance emits any of the particles in Table 17.1, except gamma rays, its nucleus changes. The new nucleus will contain whatever is left afterward. A substance that emits an alpha particle is called an *alpha emitter,* and the process is *alpha emission.*

alpha emitter

alpha emission

EXAMPLE 17.1: Uranium-238 is an alpha emitter. Write the equation for this reaction and identify the new nucleus.

Solution: First, we look in the table of atomic numbers on pages 16–17, where we find that the atomic number of uranium is 92. That lets us write U-238 in shorthand notation. Next, we write the incomplete equation.

$$^{238}_{92}U \not\longrightarrow {}^{4}_{2}He + ?$$

Nuclear equations, like chemical equations, must obey conservation of mass and charge. Mass is conserved, so the same number of atomic mass units must appear on each side. Our new nucleus must therefore have a mass number of 234. Charge is conserved, so the sum of the atomic numbers on each side must be equal. Our new nucleus must therefore have an atomic number of 90. Next, we look again in the table of atomic num-

TABLE 17.1
Characteristics of some nuclear particles

Name	*Symbols*	*Mass No.*	*Mass*[a] *(amu)*	*Charge*	*Identity*	*Comments*
Alpha particle	${}^{4}_{2}He$, α	4	4.00150	2+	Helium nucleus (He^{2+})	Can penetrate tissue only to a depth of 0.01 cm. Can be stopped by a sheet of paper.
Beta particle	${}^{0}_{-1}e$, β	0	0.000549	1−	High-energy electron	Can penetrate tissue to a depth of 1 cm. Goes through a sheet of paper but can be stopped by a 1-cm piece of aluminum.
Gamma ray	γ	0	0	0	High-energy radiation, like X rays	Can penetrate tissue to a depth of 100 cm, or pass completely through the body. Can go through paper and aluminum but can be stopped by a 5-cm sheet of lead.
Neutron	${}^{1}_{0}n$	1	1.00867	0	Neutron	Can penetrate tissue to a depth of 10 cm.
Positron	${}^{0}_{1}e$	0	0.000549	1+	Electron with positive charge (positive electron)	Same penetrating power as beta particle.
Proton	${}^{1}_{1}H$	1	1.00728	1+	Proton	Not usually a product of radioactivity.

[a]In Chapter 2, we used rounded amu values. Now we'll use exact values of amu.

bers to identify the new nucleus. We see that the element having an atomic number of 90 is thorium.

Answer: ${}^{238}_{92}U \longrightarrow {}^{4}_{2}He + {}^{234}_{90}Th$
Alpha emission decreases the mass number by 4 and the atomic number by 2. ■

A substance that emits a beta particle is called a *beta emitter,* and the process is *beta emission.*

beta emitter

beta emission

EXAMPLE 17.2: Carbon-14, a naturally occurring radioactive isotope of carbon, is a beta emitter. Write the equation for its beta emission and identify the new nucleus.

Solution: ${}^{14}_{6}C \not\longrightarrow {}^{0}_{-1}e + ?$
Our new nucleus must have a mass number of 14 and an atomic number of 7. We find that this element is nitrogen.

Answer: $^{14}_{6}C \longrightarrow {}^{0}_{-1}e + {}^{14}_{7}N$
Beta emission doesn't change the mass number, and it increases the atomic number by 1. ■

A substance that emits a gamma ray is a *gamma emitter,* and the process is *gamma emission.* Usually, gamma emission accompanies other kinds of nuclear reactions. Since a gamma ray has no charge or mass number, it doesn't change the nucleus, and it usually isn't included in the nuclear equation. Just as energy from chemical reactions is usually liberated in the form of heat, energy from nuclear reactions can be released in the form of gamma rays.

gamma emitter

gamma emission

From Table 17.1 we can deduce that positron emission will not affect the mass number but will decrease the atomic number by 1. Neutron emission will decrease the mass number by 1 but it will not affect the atomic number.

NUCLEAR STABILIZATION. All isotopes of all elements having atomic numbers higher than 83 (bismuth) are radioactive. In addition, some isotopes of other elements, such as carbon-14, are also radioactive. Twenty-two elements found in nature contain radioactive nuclei; of these, ten elements have atomic numbers less than 83.

When a nucleus emits radiation and changes to another nucleus, it *decays.* Nuclei decay because they are unstable, or have high potential energy, relative to their decay products. The new nucleus may also be unstable and decay even further. Decay will stop when a stable nucleus is reached. To see what causes certain nuclei to be unstable, we should first look at the stable nuclei.

decay

Figure 17.1 shows a plot of atomic number (number of protons) against the number of neutrons for the stable nuclei. The stable nuclei fall within a narrow band. Nuclei that aren't in this band are unstable. A nucleus is unstable either because it has an atomic number greater than 83, because it has too many neutrons per proton (left of the stable band), or because it has too many protons per neutron (right of the band).

If a nucleus is unstable because it has too high an atomic number, it often emits an alpha particle. By alpha emission, the nucleus can decrease its atomic number by 2. All alpha emitters have an atomic number greater than 83, but not all substances with an atomic number over 83 are alpha emitters.

If a nucleus is unstable because it has too many neutrons per proton, it can emit a neutron (rare) or it can change a neutron to a proton. To understand how this can happen, we can oversimplify a bit and think of a neutron as being made of a proton and an electron. Then, a neutron can change to a proton like this:

$$^{1}_{0}n \longrightarrow {}^{1}_{1}H + {}^{0}_{-1}e$$

(The masses and charges balance.) The new proton stays behind in the nucleus, and the electron leaves as a beta particle. Nuclei that have too

FIGURE 17.1
Neutron-proton relationship for stable nuclei

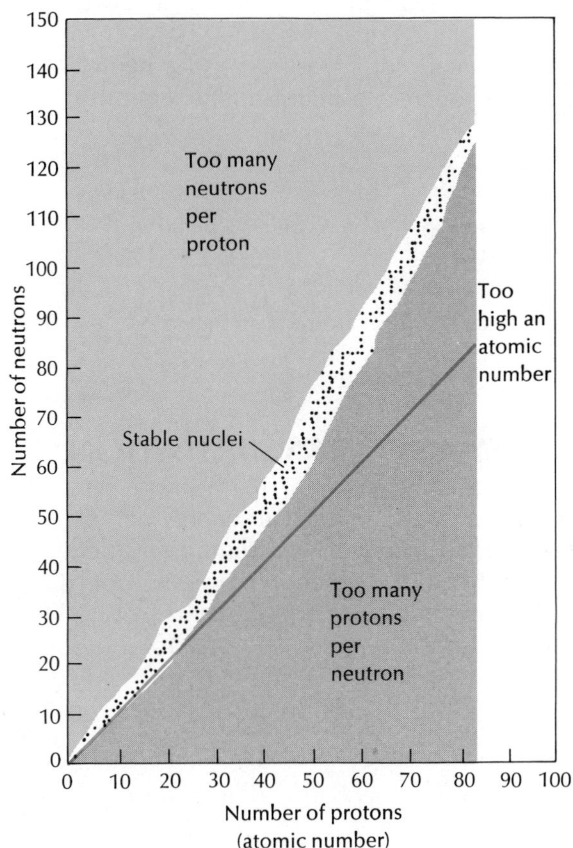

many neutrons per proton to be stable are usually beta emitters. C-14, a beta emitter, falls to the left of the band of stable nuclei.

If a nucleus is unstable because it has too many protons per neutron, it changes a proton to a neutron in one of two ways. In one way, a proton reacts with an electron in the reverse of the above equation. When the electron is one of the atom's own inner electrons, the process is called *electron capture*.

electron capture

EXAMPLE 17.3: Molybdenum-90 falls to the right of the band of stable nuclei and undergoes electron capture. Write the equation.

Solution: ${}^{90}_{42}\text{Mo} + {}^{0}_{-1}\text{e} \xrightarrow{\;/\;} ?$

Our new nucleus will have a mass number of 90 and an atomic number of 41: niobium. We write "electron capture" with the arrow to show that the electron came from within the atom itself.

Answer: ${}^{90}_{42}Mo + {}^{0}_{-1}e \xrightarrow[\text{capture}]{\text{electron}} {}^{90}_{41}Nb$ ■

In electron capture, no particles are emitted, but gamma rays are always released.

We can think of a proton as being made of a positron and a neutron. Then, the second way a proton can change to a neutron is by emitting a positron.

$${}^{1}_{1}H \longrightarrow {}^{1}_{0}n + {}^{0}_{1}e$$

EXAMPLE 17.4: C-11 has too many protons per neutron and is a positron emitter. Write the equation.

Solution: ${}^{11}_{6}C \not\longrightarrow {}^{0}_{1}e + ?$

Our new nucleus has a mass number of 11 and an atomic number of 5: boron.

Answer: ${}^{11}_{6}C \longrightarrow {}^{0}_{1}e + {}^{11}_{5}B$ ■

ARTIFICIAL RADIOACTIVITY AND TRANSMUTATION.

Besides the naturally occurring radioactive isotopes, many more have been made by humans. This is done by shooting a high-energy particle, called a *projectile,* at a large target nucleus. The word *transmutation* strictly means one element changing to another, but the current usage is applied mostly to changes caused by human beings. The first human-caused transmutation—but one that didn't result in a radioactive nucleus—was done by Ernest Rutherford in 1919. He bombarded nitrogen with alpha particles.

projectile

transmutation

$${}^{14}_{7}N + {}^{4}_{2}He \longrightarrow {}^{17}_{8}O + {}^{1}_{1}H$$

Then, in 1934, Irène Joliot-Curie did the first transmutation to result in a radioactive nucleus.

$${}^{27}_{13}Al + {}^{4}_{2}He \longrightarrow {}^{30}_{15}P + {}^{1}_{0}n$$

When aluminum-27 was bombarded with alpha particles, the resulting phosphorus-30 was found to be radioactive. Since a radioactive substance was produced from nonradioactive substances, this was called *artificial radioactivity.* Naturally occurring phosphorus is stable P-31, and so this process had created a new radioactive isotope (*radioisotope*). Today, many other radioisotopes are produced artificially by transmutation.

artificial radioactivity

radioisotope

Uranium has the highest atomic number (92) of any of the naturally occurring elements. The elements with atomic numbers greater than 92—up to 106 at the time of this writing—have been made artificially by similar transmutation reactions. These are called *transuranium elements.* Some reactions used to produce them are shown in Table 17.2. Of course, all these elements are radioactive.

transuranium element

Radioactive decay reactions are spontaneous and take place with a release of energy. Artificial transmutation reactions are not spontaneous, and great quantities of energy must be supplied to make them happen.

TABLE 17.2
Some transmutation reactions used to produce transuranium elements

Atomic Number	Name and Symbol	Target Nucleus		Projectile		
93	Neptunium, Np	$^{238}_{92}U$	+	$^{1}_{0}n$	$\longrightarrow$	$^{239}_{93}Np + {}^{0}_{-1}e$
94	Plutonium, Pu	$^{238}_{92}U$	+	$^{2}_{1}H$	$\longrightarrow$	$^{238}_{94}Pu + 2\,^{1}_{0}n + {}^{0}_{-1}e$
95	Americium, Am	$^{239}_{94}Pu$	+	$^{1}_{0}n$	$\longrightarrow$	$^{240}_{95}Am + {}^{0}_{-1}e$
96	Curium, Cm	$^{239}_{94}Pu$	+	$^{4}_{2}He$	$\longrightarrow$	$^{242}_{96}Cm + {}^{1}_{0}n$
97	Berkelium, Bk	$^{241}_{95}Am$	+	$^{4}_{2}He$	$\longrightarrow$	$^{243}_{97}Bk + 2\,^{1}_{0}n$
98	Californium, Cf	$^{242}_{96}Cm$	+	$^{4}_{2}He$	$\longrightarrow$	$^{245}_{98}Cf + {}^{1}_{0}n$
99	Einsteinium, Es	$^{238}_{92}U$	+	$15\,^{1}_{0}n$	$\longrightarrow$	$^{253}_{99}Es + 7\,{}^{0}_{-1}e$
100	Fermium, Fm	$^{238}_{92}U$	+	$17\,^{1}_{0}n$	$\longrightarrow$	$^{255}_{100}Fm + 8\,{}^{0}_{-1}e$
101	Mendelevium, Md	$^{253}_{99}Es$	+	$^{4}_{2}He$	$\longrightarrow$	$^{256}_{101}Md + {}^{1}_{0}n$
102	Nobelium, No	$^{246}_{96}Cm$	+	$^{12}_{6}C$	$\longrightarrow$	$^{254}_{102}No + 4\,^{1}_{0}n$
103	Lawrencium, Lr	$^{252}_{98}Cf$	+	$^{10}_{5}B$	$\longrightarrow$	$^{257}_{103}Lr + 5\,^{1}_{0}n$
104	Kurchatovium, Ku (tentative)	$^{242}_{94}Pu$	+	$^{22}_{10}Ne$	$\longrightarrow$	$^{260}_{104}Ku + 4\,^{1}_{0}n$
105	Hahnium, Ha (tentative)	$^{249}_{98}Cf$	+	$^{15}_{7}N$	$\longrightarrow$	$^{260}_{105}Ha + 4\,^{1}_{0}n$
106	Unnamed	$^{249}_{98}Cf$	+	$^{18}_{8}O$	$\longrightarrow$	$^{263}_{106}? + 4\,^{1}_{0}n$

17.2 SOME CONSEQUENCES OF RADIOACTIVE DECAY

Radioisotopes have many uses—and many hazards. They can be used as energy sources, as tracers in medicine, industry, or research, and as "clocks" to determine ages or events in natural history. The hazards, of course, have to do with biological damage that can occur from exposure. To use radioisotopes, and to guard against exposure, we need ways of detecting radiation.

RADIATION DETECTORS. The best-known radiation detector is the *Geiger-Müller counter,* sometimes called simply a Geiger counter. The part of the counter that receives the radiation is a tube with a window. The tube is filled with a gas and has a wire sticking into its center. A voltage source makes the wire an anode and the jacket of the tube a cathode. The same ionizing effect of radiation on matter that causes biological damage lets the radiation be detected in a Geiger counter. When radiation goes through the window of the tube, it hits the gas molecules inside the tube

Geiger-Müller counter

and knocks electrons off them. The electrons hit the anode and cause a current to flow, which registers as a click or a light flash. Geiger counters work best for beta and gamma emitters rather than for alpha emitters, because alpha particles don't have enough energy to get through the tube's window.

Each time a click or flash is registered, it means that a nucleus has decayed. Most radiation detectors count the number of such nuclear events and collect them on a digital readout. The usual way of measuring radioactivity is in counts per minute. The SI unit of radioactivity is the *becquerel* (*Bq*), which is one nuclear event per second (sixty counts per minute). Since an event may result in alpha, beta, or gamma radiation, all of which have different energies and produce different amounts of biological damage, the becquerel isn't an exact measure of how much damaging radiation is present. However, we can say roughly that exposure to between 10^4 and 10^5 becquerels will endanger the life of a human being.

becquerel

More sophisticated counters are used in research and medicine. One is the *scintillation counter.* The substance to be measured is placed in a container holding a solution of a radiation-sensitive dye. When radiation hits the dye, the dye gives a flash of light that is picked up by a light detector and translated into counts per minute.

scintillation counter

Light-sealed photographic film badges are worn by people who work around radiation. These badges are checked frequently to determine how much radiation their wearers have been exposed to. Photographic film is also used in *autoradiography,* in which a radioactive substance contained in a given sample takes its own picture, showing its location in the sample.

auto-radiography

HALF-LIVES OF RADIOACTIVE SUBSTANCES. Each radioisotope decays at a certain rate. Nothing will stop it from decaying or make it decay faster or slower. The time it takes for half of any specific amount of any radioactive substance to decay is its *half-life.* If an isotope's half-life is twenty-eight years and we start with one gram of it, then after twenty-eight years we'll have half a gram left. After another twenty-eight years, we'll have one-fourth of a gram left, as shown in Figure 17.2. Half-lives range from fractions of a second to billions of years. Some of the heavier artificial elements have very short half-lives (that of element 106 is 0.9 seconds).

half-life

Archaeologists use half-lives to determine how long once-living things have been dead. In a process called *carbon dating,* they measure the amount of radioactive carbon-14 a substance contains. Most naturally occurring carbon is C-12, but small amounts of C-14 are formed in the atmosphere by high-energy neutrons.

carbon dating

$$^{14}_{7}N + ^{1}_{0}n \longrightarrow ^{14}_{6}C + ^{1}_{1}H$$

The formation and decay of C-14 are at equilibrium in the atmosphere, so the ratio of C-12 to C-14 is always constant. Living plants and animals take in C-14, and the amount they contain is also at equilibrium with the amount

FIGURE 17.2
Strontium-90 has a half-life of 28 years

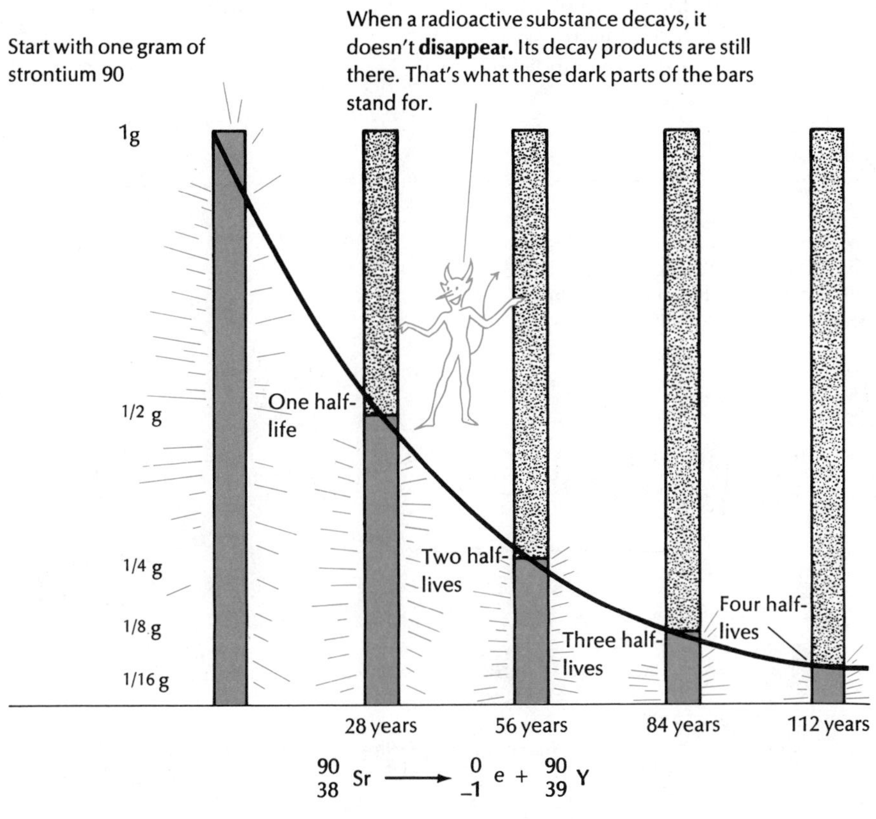

Start with one gram of strontium 90
When a radioactive substance decays, it doesn't **disappear.** Its decay products are still there. That's what these dark parts of the bars stand for.
1g
1/2 g
1/4 g
1/8 g
1/16 g
One half-life
Two half-lives
Three half-lives
Four half-lives
28 years
56 years
84 years
112 years

$$^{90}_{38}\text{Sr} \longrightarrow {}^{0}_{-1}\text{e} + {}^{90}_{39}\text{Y}$$

in the atmosphere. When a plant or animal dies, though, it stops taking in new C-14 while the C-14 it contains still decays.

One gram of carbon from anything alive contains about enough C-14 to release sixteen beta particles per minute. The half-life of C-14 is 5730 years. Thus if a gram of carbon from a dead substance releases only eight beta particles per minute, that means it's been dead for one half-life of C-14, or 5730 years.

EXAMPLE 17.5: A 0.50-gram sample of carbon isolated from a fossilized plant specimen measures two beta emissions per minute. How long has the plant been dead?

Solution: Two beta emissions per minute for 0.50 gram of carbon means four beta particles per minute for one gram of carbon. Since one-fourth of the original activity remains, two half-lives must have elapsed.

Answer: 11,460 years.

The alchemists believed that baser metals like lead "ripened" by changing first to silver and finally gold. Such a series of changes does occur among some metals, but it's nearly the reverse of what the alchemists wanted. One series starts with uranium-238, goes through fourteen changes, and finally ends with lead-206, instead of starting with it. This is called a *radioactive decay series*. The U-238 series is shown in Figure 17.3.

radioactive decay series

FIGURE 17.3
Uranium-238 decay series

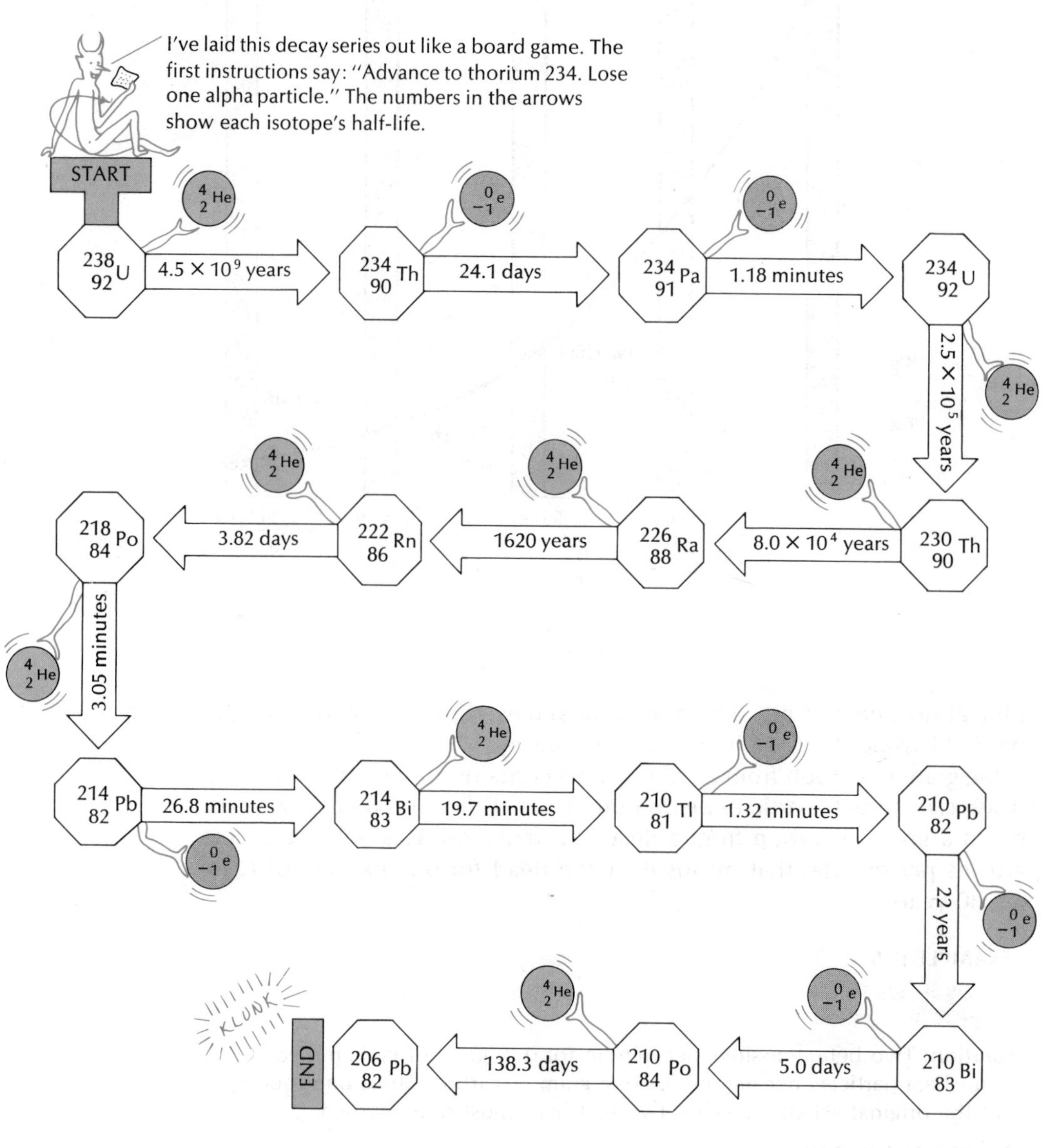

There are three naturally occurring decay series. Besides the U-238 decaying to Pb-206, Th-232 decays to Pb-208, and U-235 decays to Pb-207. For a decay series to exist, the parent isotope (for example, U-238) must have a long half-life, and the daughter (product) isotopes must be relatively short-lived. This explains why some naturally occurring isotopes have relatively short half-lives (for instance, Pa-234, with a half-life of only 1.18 minutes). They are being continuously replaced by their parent isotope. U-238 has a half-life of 4.5 billion years. The earth is believed to be about 4.6 billion years old, so it isn't surprising that there's still plenty of U-238 and its daughter isotopes around.

Lead-206 is not a natural isotope but is formed only through the U-238 decay series. If Pb-206 turns up in a rock, it must have come from the decay of U-238. Because we know U-238's half-life, we can tell how long it took that much uranium to decay to Pb-206. The larger the lead-to-uranium ratio, the older the rock is. Samples of moon rock have been analyzed by *uranium dating* and found to be about 4.6 billion years old, or about the same age as the earth.

uranium dating

SOME USES OF RADIOISOTOPES. Table 17.3 shows a few of the many useful radioisotopes. They can sterilize food, because the radiations kill microorganisms that cause spoilage. They can be used to control insect populations by sterilizing male insects. The males are then released to mate with females, whose eggs won't be fertilized. Once a female has mated, she doesn't mate again. Radioisotopes will also destroy cancerous tissue. Radiation from cobalt-60 is aimed at affected areas. Chromium-51 wires are implanted directly into tumors to destroy them. Iodine-131 concentrates in the thyroid gland and it can be used to destroy cancerous tissue there.

Radioisotopes are also used as tracers, in the same way that a detective might rig a suspicious vehicle with a radio transmitter so that the vehicle could always be located. Radioisotopes behave chemically in exactly the same way as their stable isotopes, but they send out signals telling where they are. Because of this, they can be substituted for their stable isotopes in compounds, and the compounds can be traced. Such labeled or tagged compounds are used in medicine to diagnose illnesses. For instance, blockages in the circulatory system can be located by injecting radioactive sodium-24 in the form of NaCl. In industry, pipe leaks can be detected by adding a radioisotope and detecting where it appears.

Radioisotopes that are placed into human beings must be carefully chosen. First, no alpha emitters are used. Even though external exposure to alpha radiation isn't serious because of its low penetrating power, alpha emission is the most harmful type of radiation if taken internally. The large massive particles can do a great deal of damage to nearby tissues or organs. Second, substances with relatively short half-lives are used, so that they will disappear from the body as soon as possible after their job is done.

TABLE 17.3
Some useful radioactive substances

Isotope	*Half-Life*	*Emission*	*Uses*
Cobalt-58	71.3 days	${}^{0}_{1}e$, γ	Used to determine intake of vitamin B-12, a cobalt-containing vitamin.
Cobalt-60	5.3 years	β, γ	A strong beta and gamma emitter Used in cancer radiation therapy.
Carbon-14	5730 years	β	Used in carbon dating and in chemical and biological research to determine reaction paths. Also used to check wear on tires.
Iodine-131	8 days	β, γ	Iodine concentrates in the thyroid gland. Used to diagnose thyroid malfunction and to treat thyroid cancer, since it concentrates in thyroid gland; its beta and gamma radiations destroy cancerous thyroid tissue without harming other cells.
Iron-59	45.6 days	β, γ	Used to study formation of red blood cells (hemoglobin contains iron).
Phosphorus-32	14.3 days	β	Used in biochemical research and to treat leukemia and skin lesions. Also used in industry to measure tire wear and thickness of films.
Radium-226	1602 years	α, γ	Like cobalt-60, used in cancer radiation therapy.
Sodium-24	15.0 hours	β, γ	Used as NaCl in water solution to check for proper function of circulatory system.
Hydrogen-3	12.3 years	β	Used in chemical and biochemical research in the form of various compounds. As H_2O, used to determine a person's total body water.

17.3 NUCLEAR ENERGY

Stable nuclei have lower energies than unstable ones. Stable nuclei also have lower energies than their separate neutrons and protons. If they didn't, the nuclei would fly apart. Just as energy is released when compounds are formed from atoms, energy is released when nuclei are formed from neutrons and protons.

ENERGY AND THE MASS DEFECT. If we add up the individual masses of the subatomic particles that make up a nucleus, we always find that the sum of these masses is greater than the actual mass of the nucleus itself. For instance, we can add up the masses of the ingredients of a helium nucleus.

2 protons, each 1.00728 amu =	2.01456 amu
2 neutrons, each 1.00867 amu =	2.01734 amu
Total	4.03190 amu

However, the actual mass of a helium nucleus is only 4.00150 amu. The difference of 0.03040 amu is called the *mass defect*. Every nucleus has a mass defect, and each value is different. Energy equivalent to the mass defect goes into holding the nucleus together.

mass defect

Mass-energy equivalence was discovered by Albert Einstein (1879–1955) and stated in his famous equation, $E = mc^2$. E stands for energy, m for mass, and c for the speed of light (3×10^8 meters/second). He said that it should be possible to convert energy to mass and vice versa, because they are different forms of the same thing. This relationship can be used to calculate exactly how much energy is equivalent to how much mass. A mass of 1.0 g is equivalent to about 2.2×10^{10} kcal of energy. We see that a little bit of mass produces a lot of energy, because the speed of light is a very large number.

$E = mc^2$

The energy equivalent to the mass defect is the energy released when the nucleus is formed from the neutrons and protons, or the energy required to take the nucleus apart into its separate particles. This is the *nuclear binding energy*. Figure 17.4 shows a plot of binding energy per neutron or proton against mass number. The binding energy decreases sharply with mass number at first, then levels off and increases slowly toward the elements of very high mass number. Changing from nuclei with larger binding energies to nuclei with smaller binding energies will release energy. We can do this by going downhill on the curve from left to right or from right to left. Iron is the dividing line.

nuclear binding energy

NUCLEAR FUSION. Putting two nuclei together to make a larger nucleus is called *nuclear fusion*. In Figure 17.4, putting hydrogens together to form helium means going downhill from left to right. Quite a lot of energy should be released in this process, and it is. One of the many fusion reactions that takes place on the sun is this one:

nuclear fusion

$$^{2}_{1}H + ^{2}_{1}H \longrightarrow ^{4}_{2}He$$

Fusion reactions have very high activation energies, because a lot of energy is needed to force the two positive nuclei close enough together to fuse. But much more energy is released once the reaction starts, as in

FIGURE 17.4
Nuclear binding energies of some of the elements

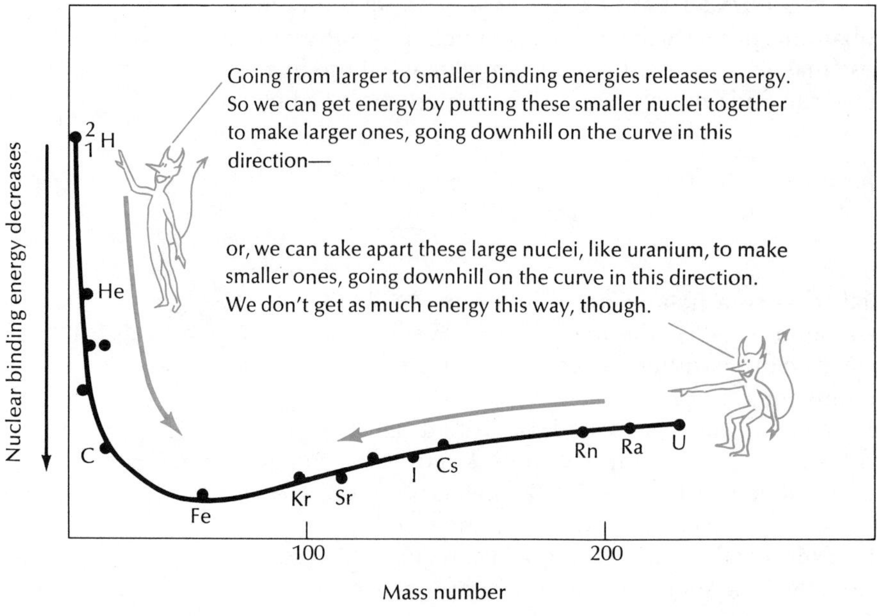
Going from larger to smaller binding energies releases energy. So we can get energy by putting these smaller nuclei together to make larger ones, going downhill on the curve in this direction—
or, we can take apart these large nuclei, like uranium, to make smaller ones, going downhill on the curve in this direction. We don't get as much energy this way, though.
Nuclear binding energy decreases
$^2_1$H
He
C
Fe
Kr
Sr
I
Cs
Rn
Ra
U
100
200
Mass number

many reactions we saw in Chapter 16. High temperatures like those found on the sun (about 40 million degrees) are needed to overcome the high activation energy. For this reason, using nuclear fusion as a controlled energy source is a long way in the future for us. Meanwhile, though, we do have a fusion reactor: the sun.

The hydrogen bomb is a fusion bomb. It also contains a fission bomb, which we'll talk about later. The fission bomb is needed to provide the high temperature for the fusion reaction.

NUCLEAR FISSION. Nuclear fusion occurs going downhill from left to right on the binding energy curve, and it releases a lot of energy. Going downhill from right to left also releases energy, but not as much. We can see about how much energy we'd get from breaking a uranium nucleus apart into a strontium nucleus (atomic number 38) and a xenon nucleus (atomic number 54). Breaking a nucleus apart into pieces of comparable size is *nuclear fission.*

nuclear fission

At the time of this writing, only three isotopes—U-235, U-233, and Pu-239—are known to have *fissionable nuclei* (nuclei capable of fission). Nuclear fission happens when a slow-moving neutron hits a fissionable nucleus. The nucleus, already unstable, is distorted by the impact of the neutron and elongated. This makes it easy for the nucleus to break apart, as shown in Figure 17.5, releasing energy and neutrons.

fissionable nuclei

FIGURE 17.5
Nuclear fission of U-235

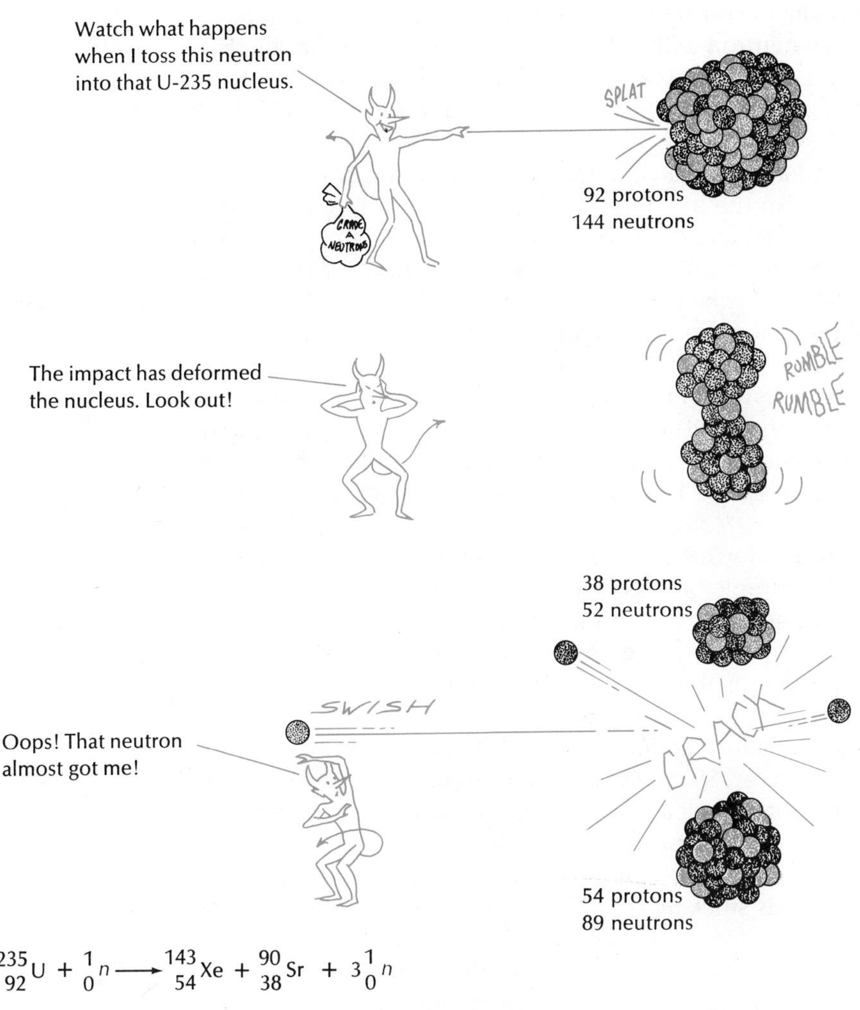

Many different products are obtained from nuclear fission, because each nucleus breaks apart randomly. If we hit a lot of marbles with a hammer, most of them would break more or less in half, but we'd get a variety of different-sized pieces. We'd also get some little splinters of glass, just as we get a few neutrons along with the big fragments. Some possible fission products of U-235 are these.

$$^{139}_{56}Ba + ^{95}_{36}Kr + 2\,^{1}_{0}n \qquad ^{103}_{42}Mo + ^{131}_{50}Sn + 2\,^{1}_{0}n$$

$$^{140}_{57}La + ^{94}_{35}Br + 2\,^{1}_{0}n \qquad ^{139}_{54}Xe + ^{95}_{38}Sr + 2\,^{1}_{0}n$$

$$^{135}_{53}I + ^{97}_{39}Y + 4\,^{1}_{0}n$$

Of the three fissionable nuclei, only U-235 occurs in nature; the other two are produced artificially by transmutation. U-235 itself is only about 1 percent of the total amount of uranium; the rest is U-238, which isn't fissionable. In naturally occurring uranium, fission of U-235 does happen. Now and then a stray neutron will strike a U-235 nucleus and that nucleus will undergo nuclear fission instead of radioactive decay. The neutrons produced by the fission will simply be lost to the surroundings. We wouldn't notice this isolated event.

However, if we had a great many U-235 nuclei close together, the neutrons (two or more) from the fission of a single nucleus could go out and hit other fissionable nuclei and cause more fission. The snowballing effect of more neutrons causing fission and producing more neutrons to cause still more fission is called a *chain reaction,* shown in Figure 17.6. In a chain reaction more than one neutron per nucleus is released, so that the fission keeps going and releases a lot of energy. For a chain reaction to happen, enough of the fissionable nuclei have to be in the same place, so that few neutrons can escape without hitting other nuclei. The amount of a fissionable material needed for a chain reaction is called the *critical mass.* The critical mass of U-235 is about 10 kilograms; for Pu-239, it's about 3 kilograms.

chain reaction

critical mass

If a chain reaction is allowed to proceed uncontrolled, the result is an atomic explosion. The atomic bombs of World War II were fission bombs, one using U-235 and the other Pu-239. In these bombs, several pieces of the fissionable material, equal to the critical mass when combined, were kept apart. To trigger the explosion, the parts were forced together with ordinary explosives.

In nuclear reactors, chain reactions are controlled. Neutrons from fission have rather high energy. Slow-moving neutrons are needed for fission; fast ones just bounce off. Nuclear reactors usually contain something to slow the neutrons down. The first nuclear reactor, built in 1942 by Enrico Fermi (1901–1954), contained a critical mass of U-235, with graphite blocks and cadmium rods stuck all through it. The graphite served to slow the neutrons down, and the cadmium absorbed neutrons. The rate of energy release could be controlled by pushing the cadmium rods in, to slow down the fission reaction, or by pulling them out, to speed it up.

Although nuclear reactors have become more sophisticated since that of Fermi, the principle is the same: the reaction is kept from getting out of hand by absorbing some of the neutrons when necessary. The first major nuclear power plant for producing electricity began operating in 1957 in Shippingport, Pennsylvania. Since then, over a hundred nuclear power plants have been or are being built in the United States. The method of power generation is the same as for conventional electrical power plants: heat is used to turn water into steam to drive a turbine to make electricity. The difference is that the source of heat is controlled fission and not burning fuel. Of course, the nuclear reactor part of the plant is also much more complex than the fuel-burning part of a conventional power plant.

The energy produced by the fission of 1 kilogram of U-235 is equivalent

FIGURE 17.6
Chain reaction involving uranium-235

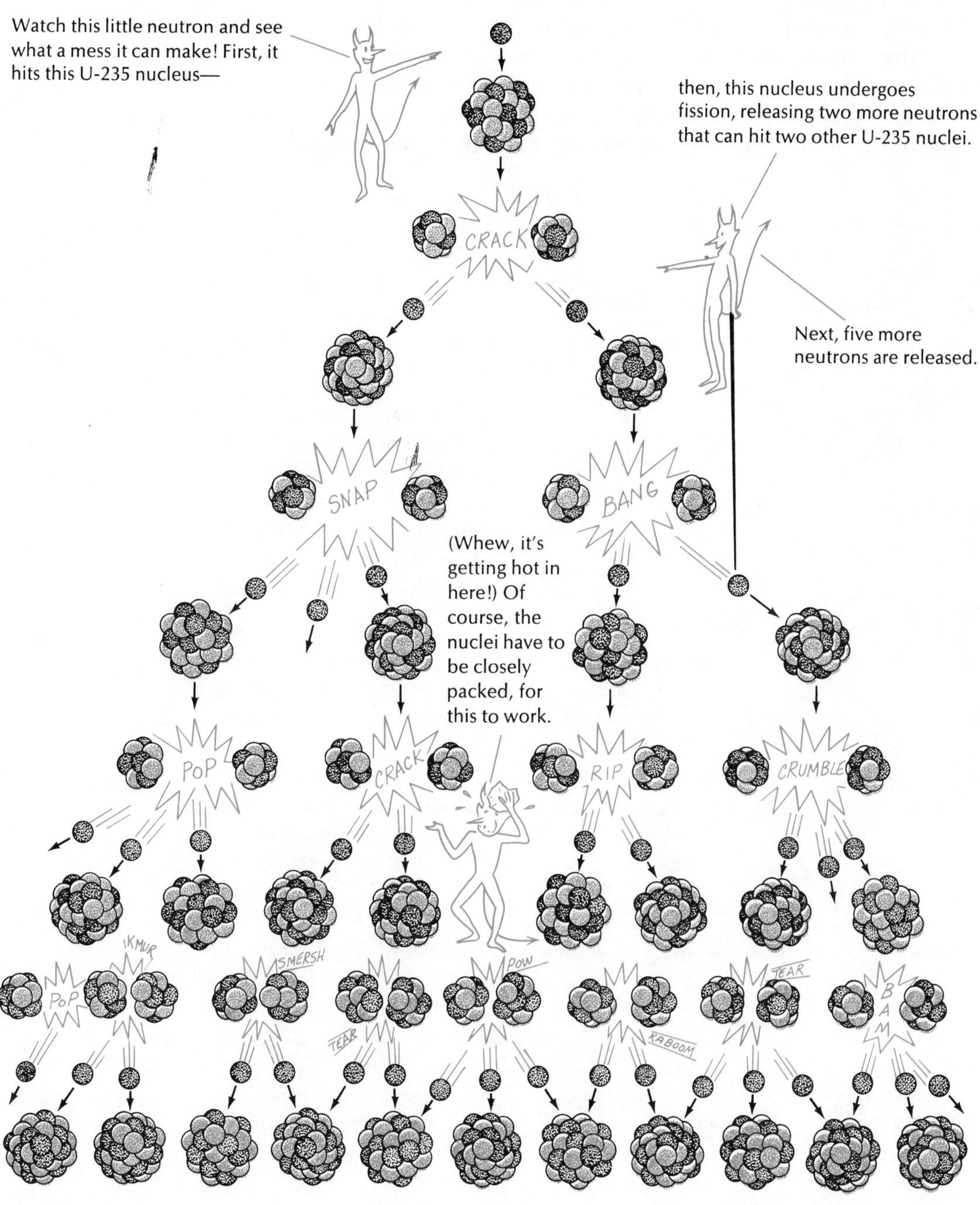

to that produced by burning about 2800 tons of coal. But there are problems with nuclear reactors. First, U-235 is scarce and hard to separate from U-238. Second, there is danger of releasing radioactive materials to the surroundings during fission. And third, the products of nuclear fission are themselves radioactive. That presents the problem of disposing safely of large amounts of radioactive waste products, some with very long half-lives.

breeder reactor

The *breeder reactor* would solve the first of these problems. A breeder reactor makes its own fuel by causing transmutation reactions that produce either U-233 from Th-232 or Pu-239 from U-239, using neutrons from fission.

$$^{232}_{90}\text{Th} + ^{1}_{0}n \longrightarrow ^{233}_{92}\text{U} + 2\ ^{0}_{-1}\text{e}$$

$$^{239}_{92}\text{U} + ^{1}_{0}n \longrightarrow ^{239}_{94}\text{Pu} + 2\ ^{0}_{-1}\text{e} + ^{1}_{0}n$$

TABLE 17.4
Some dangerous products of nuclear fission

Isotope	*Half-Life*	*Emission*	*Comments*
Krypton-85	10.8 years	β, γ	These noble gases escape directly from nuclear power plants. In the atmosphere, they expose skin and lungs to radiation.
Xenon-133	5.3 days	β, γ	
Xenon-135	9.14 hours	β, γ	
Strontium-90	27.7 years	β	A member of Group IIA that behaves like calcium, and collects in bones. Can cause bone cancer. Also absorbed by plants.
Radium-226	1,602 years	α, γ	Same as strontium-90.
Iodine-131	8 days	β, γ	A major contaminent from nuclear accidents. Concentrates in thyroid and causes thyroid cancer.
Hydrogen-3 (Tritium)	12.3 years	β	Escapes as H_2 or water vapor from reactors. Can be incorporated into most parts of the body. Not as serious as others because its radiation has relatively low energy.
Cesium-137	30 years	β, γ	A member of Group IA that imitates the behavior of sodium and potassium, finding its way into all animal and plant systems. Exposes the entire body to radiation.
Plutonium-239	24,390 years	α, γ	Also collects in bone. Extremely dangerous to ingest because of alpha emission.
Carbon-14	5,730 years	β	Collects in all parts of plants and animals.

Several prototype breeder reactors are now functioning in Europe, but as yet none are operating in the United States.

While the breeder reactor would partially solve the fuel problem, it wouldn't remove the considerable environmental threat posed by all nuclear reactors. Even during normal operation, nuclear power plants inevitably release radioactive substances to the atmosphere. Many radioactive products, such as xenon and krypton, are gases, and preventing their escape is difficult. In 1967, one power plant alone released about 4×10^{17} becquerels of radioactivity. The contamination would be even more serious if accidents occurred—leaks, ruptures, or runaway reactions that might cause melting and breakage. Some of the radioactive waste products, and their significance and half-lives, are shown in Table 17.4.

A lesser but still serious problem is that of thermal pollution. Nuclear power plants use huge amounts of cooling water, which is then released to natural water systems. We saw some of the consequences of thermal pollution in Chapter 12.

Even with all its problems, nuclear energy holds great promise as an energy source. Someday it may completely replace the fuels—coal, oil, natural gas—that we now use in large quantities.

REVIEW QUESTIONS

1. What are some differences between chemical reactions and nuclear reactions?

Radioactivity

2. What is *radioactivity*? What is *radiation*? How can we tell whether a substance is radioactive?
3. What are *alpha particles, beta particles,* and *gamma rays*? Why were they so named?
4. Write the symbols for six nuclear particles, and give their mass numbers and charges. Which type of external radiation exposure is most dangerous?
5. Give examples of *alpha emission* and *beta emission.* State how each type of emission changes the nucleus of the radioactive substance.
6. How are mass and charge conserved in a nuclear equation? How can the new nucleus be identified?
7. State how the following will affect the mass number and atomic number of a radioactive nucleus: *gamma emission,* positron emission, neutron emission.
8. Give three reasons why a nucleus might be unstable. How might a nucleus stabilize itself in each case?
9. Define and give an example of *electron capture.*
10. How are radioactive isotopes made by humans? What is meant by the word *transmutation?*
11. What is meant by *artificial radioactivity?* Give an example.
12. What are *transuranium elements?* Where do they come from? Why are they all radioactive?

Some Consequences of Radioactive Decay

13. Name and describe some radiation detectors. What is the usual way of measuring radioactivity?
14. What is the SI unit of radioactivity? What does it mean?
15. What is meant by *half-life?* Describe what happens to a certain amount of radioactive substance after one, two, and three half-lives.
16. Explain how *carbon dating* works. Why is there always a constant amount of C-14 in the atmosphere?
17. What is a *radioactive decay series?* What element is the end product of the three known radioactive decay series?
18. Protactinium-234 has a half-life of 1.18 minutes. Why is there still any of it around?
19. How can the ages of rocks be estimated by *uranium dating?*
20. Name some uses of radioisotopes.
21. Explain how radioactive tracers work. Give a few examples.

Nuclear Energy

22. How do we know that stable nuclei have lower energy than their separate neutrons and protons do?
23. What is the *mass defect*? How can it be calculated?
24. What is mass-energy equivalence? Why does a little bit of mass produce a lot of energy?
25. What is *nuclear binding energy*? How is it related to the mass defect?
26. What is *nuclear fusion*? Where does it happen naturally? Why does it require very high temperatures?
27. What is *nuclear fission*? How is it different from nuclear fusion?
28. What are the three *fissionable nuclei*? Describe what happens when one of these undergoes fission.
29. Are the same products always obtained from fission? Explain.
30. U-235 occurs in uranium ore. Why doesn't an atomic explosion happen in the ore?
31. What is a *chain reaction*? A *critical mass?*
32. How do nuclear reactors generate power? What are some problems connected with nuclear reactors?
33. What is a *breeder reactor*? Why might it be desirable?
34. Why is it impossible to prevent release of radioactivity from nuclear power plants?
35. Name a few hazardous fission products and say why they are dangerous.

EXERCISES

Set A (Answers at back of book. Asterisks mark the more difficult exercises.)

1. Write nuclear equations for alpha emission of these nuclei.
   a. U-235
   b. Th-232
   c. Pu-239
   d. Ra-223
2. Write nuclear equations for these beta emitters.
   a. P-32
   b. I-131
   c. Co-60
   d. Na-24
3. None of the naturally occurring radioactive substances emits positrons. However, the following artificial nuclei do. Write their nuclear equations.
   a. P-30
   b. Na-22
4. Helium is often found trapped in small holes in rocks. Where did it come from?
5. Potassium-40 is a naturally occurring radioactive isotope that occurs in about 30 percent of the existing potassium (a large percentage). It is interesting because it can undergo either beta emission or electron capture. Write nuclear equations for each reaction.
6. Fill in the missing numbers for these transmutation reactions.
   a. $^{130}_{52}? + ^{2}_{1}? \nrightarrow ^{?}_{53}\mathrm{I}$
   b. $^{?}_{?}? + ^{0}_{-1}\mathrm{e} \nrightarrow ^{32}_{15}\mathrm{P}$
   c. $^{239}_{94}? + ^{4}_{2}? \longrightarrow ^{?}_{?}\mathrm{Cm} + ^{?}_{0}n$
7. Which would be more dangerous to be exposed to internally?
   a. 1000 becquerels of an alpha emitter, or 1000 becquerels of a gamma emitter
   b. 500 becquerels of a beta emitter, or 1000 becquerels of a gamma emitter
8. How many counts per minute would be registered by each sample in Exercise 7?
9. Co-60 has a half-life of 5.3 years. If a 10.0-mg sample is purchased, approximately how much will be left after 15 years?
10. An archaeological sample has a carbon-14 content of 0.501 times that of a living plant. How old is the sample?
11. Phosphorus-32, in the form of phosphate, can be used to measure the amount of fertilizer plants take up. Explain how this might be done.
12. Small amounts of radioactive substances are added to automobile piston material to test how well lubricating oils are doing their job. If, after equal time of operation, oil sample A measures 42 counts per minute and oil sample B measures 15 counts per minute, which oil causes the least wear?
13. The mass of a Hg-200 nucleus is 199.9683 amu. Using the values for the masses of the neutron and proton in Table 17.1, calculate the mass defect of Hg-200.
14. Could we get energy from the fusion of krypton with strontium? Explain.
15. In Figure 17.4, which element has the smallest nuclear binding energy?
16. One of the reactions that takes place in the hydrogen bomb is this one:
    $$^{7}_{3}\mathrm{Li} + ^{1}_{1}\mathrm{H} \longrightarrow 2\,^{4}_{2}\mathrm{He}$$
    If a lithium-7 nucleus has a mass of 7.0160 amu calculate the mass difference between the reac-

tants and products. Would this reaction release or absorb energy?

*17. Calculate the energy, in kcal/mole, produced by the reaction of Exercise 16.

18. Zirconium-97, a radioactive fission product that can concentrate in bone tissue, results from fission of U-235. Two neutrons are also released. What is the other product? Write the equation.

19. Could a chain reaction occur from a fission reaction that produced only one neutron? Explain.

20. Bromine-90, a fission product, decays to the hazardous strontium-90. Write the equation.

Set B (Answers not given. Asterisks mark the more difficult exercises.)

1. Write nuclear equations for alpha emission of these nuclei.
   a. U-234 c. Po-210
   b. Bi-214 d. Th-230
2. Write nuclear equations for these beta emitters.
   a. Pb-214 c. Th-234
   b. Pa-234 d. C-14
3. Write nuclear equations for positron emission of these artificial nuclei.
   a. F-18 b. C-11
4. Nitrogen-14 is a stable nucleus; N-13 and N-16 are unstable. Give probable reasons for their instabilities and suggest possible methods of stabilization. Write nuclear equations for each.
5. Beryllium-7 undergoes electron capture.
   a. Write the equation.
   b. Explain why this happens with more difficulty when beryllium is in the form of Be^{2+}.
6. Some radioisotopes are made by transmutation. Fill in the missing numbers, symbols, or both for the following transmutation reactions.
   a. $^{59}_{26}Fe + ^{1}_{0}n \longrightarrow ^{60}_{?}Co + ^{?}_{?}?$
   b. $^{35}_{17}Cl + ^{?}_{?}? \not\longrightarrow ^{35}_{16}? + ^{1}_{1}H$
   c. $^{27}_{?}Al + ^{1}_{0}? \not\longrightarrow ^{24}_{11}Na + ^{?}_{?}?$
7. Which would be more dangerous to be exposed to externally?
   a. 500 becquerels of a beta emitter, or 500 becquerels of an alpha emitter
   b. 300 becquerels of a gamma emitter, or 600 becquerels of a beta emitter
8. How many counts per minute would be registered by each sample in Exercise7?
9. Element 106 has a half-life of 0.9 second. If 1,000,000 atoms of it were prepared, how many would remain after 4.5 seconds?
10. How old is a bone sample if 0.152 grams of carbon from the sample emit 1.2 counts per minute?
11. Explain how carbon-14 might be used to measure wear on tires.
12. Why do you think cobalt-58, and not cobalt-60, is used to determine intake of vitamin B-12 in human beings? (Refer to Table 17.3.)
13. The mass defect of a U-238 atom is 1.9353. Using the values of the masses of the neutron and proton in Table 17.1, calculate the mass of a U-238 atom.
14. Could we get energy by breaking apart a carbon atom into two lithium atoms? Explain.
15. In Figure 17.4, which element has the largest nuclear binding energy?
16. Calculate the mass difference between the reactants and products for this fusion reaction. The mass of an H-2 atom is 2.01355 amu.

$$^{?}_{1}H + ^{?}_{1}H \longrightarrow ^{4}_{2}He$$

*17. Calculate the energy, in kcal/g, of the reaction in Exercise 16.

18. What is the decay product of plutonium-239? Write the equation. (Consult Table 17.4.)

19. Is the amount of lead on earth increasing, decreasing, or staying the same? Explain.

20. Iodine-131 is formed by the loss of three beta particles from another fission product. What is that fission product? Write the equations.

CHAPTER 18

Introduction to Organic Chemistry

LEARNING OBJECTIVES

After studying this chapter, you should be able to:

1. Supply a correct definition, explanation, or example for each of these:

   organic compound, organic chemistry, inorganic chemistry, structure, carbon skeleton, hydrocarbon, structural formula, straight chain hydrocarbon, branched chain hydrocarbon, isomer, condensed structural formula, unsaturated, saturated, aromatic hydrocarbon, ring hydrocarbon, aliphatic hydrocarbon, alkyl group, coal gas, heat of combustion, octane rating, fraction, alkylation, cracking, functional group, alcohol, ether, dehydration, condensation, amine, primary amine, secondary amine, tertiary amine, substitution, chlorination, addition, hydrogenation, aldehyde, ketone, ester, hydrolysis, anhydride, polymer, addition polymer, monomer, vinyl polymer, copolymer, amide, polyamide, polyester, cross-link

2. Write condensed structural formulas from structural formulas, and vice versa.
3. Classify a hydrocarbon from its structure using these words: saturated, unsaturated, alkene, alkane, alkyne, aromatic, aliphatic, straight chain, branched chain, cyclic.
4. Name the simple alkyl groups and give their formulas (methyl, ethyl, propyl).
5. List the various fuels obtained from petroleum and from coal and tell how they are prepared.
6. Explain the effect of C-H bonds on heats of combustion of hydrocarbons.
7. Describe the refining of petroleum and its objectives.
8. Compare solubilities of alcohols and amines, using their formulas.
9. Write an equation for an amine acting as a base.
10. Recognize or predict products for these reactions, with equations: substitution, chlorination, addition, hydrogenation, dehydration, condensation, hydrolysis, oxidation.
11. Recognize the formula of any organic compound in the chapter, and give its name, origin, and function.

Photosynthesis began on the earth about three billion years ago. In photosynthesis, energy from the sun joins together atmospheric carbon dioxide and water molecules to form larger molecules. Energy stored in the bonds of these molecules was used by primitive organisms to carry out bodily functions. In the bodies of plants and animals, carbon atoms were arranged into complex chains and interlocking structures—some stiff enough to make a giant redwood or to help support the weight of a brontosaurus.

The dinosaurs and early plants died, decayed, and became the coal and petroleum deposits that we use today. These deposits are treasure chests of energy and of structure. Energy from the sun that went into making bonds is stored in them. Many molecules that existed in these early life forms still remain as chains or rings of carbon atoms in petroleum and coal. Now we harvest the stored energy as fuel and the molecular structure to make other structures like plastics, rubber, and fabrics.

Photosynthesis is still going on, and all organisms still use carbon to build their bodies. Carbon-based compounds are the foundation of living organisms, and for this reason were dubbed *organic compounds* in the late eighteenth century. *Organic chemistry* is the study of organic compounds. In addition to providing structure and energy, organic compounds provide function. Small organic molecules—vitamins, hormones, drugs, insecticides—perform certain functions because of their structures. Many of these are now made synthetically from raw materials found in petroleum and coal. Figure 18.1 shows just a few pathways from petroleum and coal to some of the products we use today. In all, about three million organic compounds are known. This number contrasts sharply with the approximately one hundred thousand *inorganic compounds,* which are not carbon based.

organic compound

organic chemistry

inorganic compound

Up to now, we've studied mostly inorganic compounds. But our bodies, the food we eat, the clothes that cover us, the drugs that keep us well, the fuel that gives us heat—all are organic compounds. We'll take a brief look at the broad spectrum of organic chemistry and, in the next chapter, at some special molecules of the life processes. Here, we'll consider the structure of organic compounds themselves; in the next chapter we'll see how they provide us with energy and function and are used to produce other structures.

18.1 THE STRUCTURES OF ORGANIC COMPOUNDS

So far, we've used the word "structure" in a broad sense. But compounds have structure, too, just as a tree or a bridge does. By *structure* of a compound, we mean the arrangement of the elements' atoms in the compound relative to each other, or the way the compound is put together. Structures of organic compounds determine how they behave.

structure

FIGURE 18.1
Consumer products from petroleum and coal

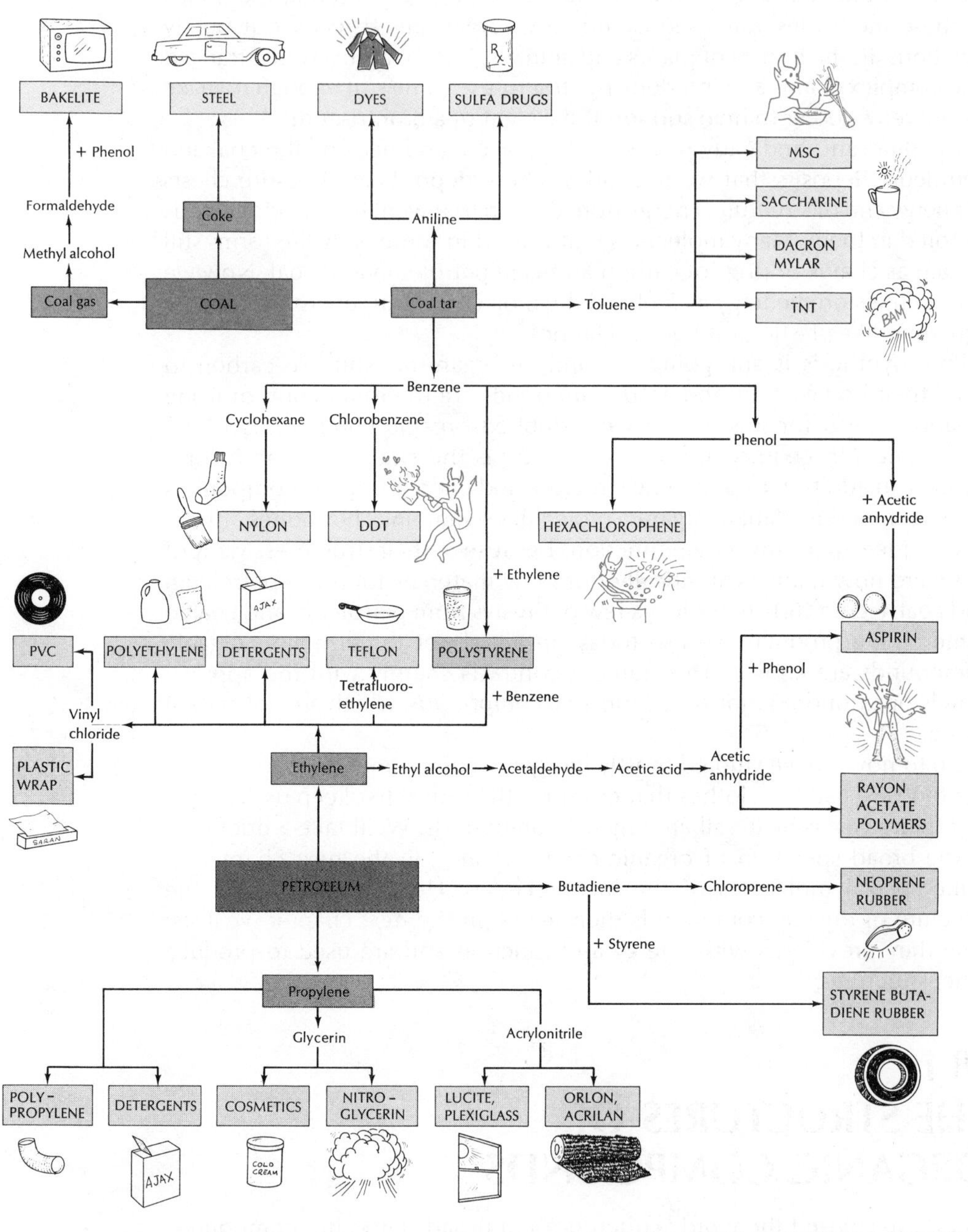

THE CARBON SKELETON. In the huge continuous structure of a diamond crystal, shown in Chapter 2 (Figure 2.7, p. 27), each single carbon atom is bonded to four others, each at the corner of a tetrahedron. In the graphite structure, also shown in Chapter 2, carbon atoms are joined together in flat connected six-membered rings, with carbon forming only three bonds. Most carbon compounds contain structures that are similar to those between carbon atoms in diamond and graphite, but often other elements are involved in addition to carbon.

The *carbon skeleton* of a compound shows only how the carbon atoms themselves are connected to each other. Figure 18.2 shows a few of the limitless possibilities. If we look back at Figure 1.1 in Chapter 1 (p. 3) we recognize these as skeletons of the organic compounds there. Styrofoam and rubber have carbon skeletons that are long and continuous and can provide structural material. Sugar has breaks in its carbon skeleton where there are oxygen atoms. Elements that interrupt the carbon skeleton provide a vulnerable spot where the molecule can be broken, and it's at the oxygen between the carbon rings that our bodies break down sugar to get energy.

carbon skeleton

The remaining bonds on the carbon atoms in each skeleton are connected to other elements. Hydrogen is usually present on most of the bonds. If hydrogen is the only other element in an organic compound, the compound is a *hydrocarbon*. The other elements besides carbon and hydrogen that are most often on the bonds are oxygen, nitrogen, a halogen, or sulfur. Some compounds in Figure 18.2—ethyl alcohol, vinegar, and PAN (a component of smog)—have the same carbon skeletons but vastly different properties. In these and most other cases, the elements other than carbon and hydrogen determine function.

hydrocarbon

These considerations will be important in all the organic compounds we'll study in this chapter. Now, though, we'll see how to represent the structures of organic compounds in the simplest way possible.

STRUCTURAL FORMULAS. All the atoms that are connected to the carbon skeleton and all the single, double, and triple bonds are shown in the *structural formulas*. Unlike Lewis structures, structural formulas don't show nonbonding electron pairs. Like Lewis structures, however, structural formulas are drawn on a two-dimensional sheet of paper and don't show the three-dimensional nature of the molecules. When carbon forms four single bonds, the geometry is tetrahedral.

structural formula

To illustrate structural formulas, let's take two ingredients of gasoline, whose structural formulas are shown below.

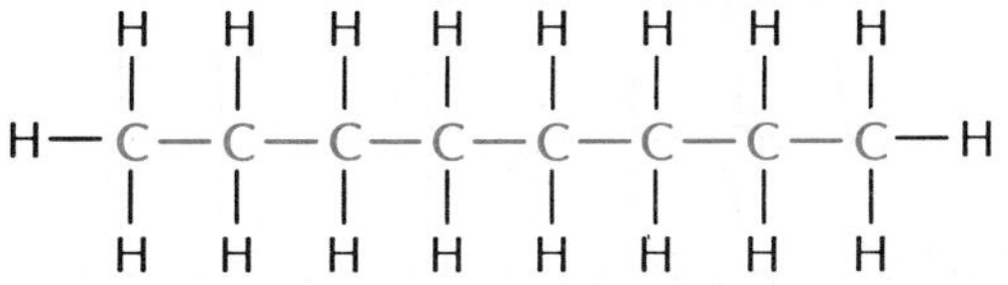

```
    H   H   H   H   H   H   H   H
    |   |   |   |   |   |   |   |
H — C — C — C — C — C — C — C — C — H
    |   |   |   |   |   |   |   |
    H   H   H   H   H   H   H   H
```

Normal octane

```
                        H
              H         |
              |     H — C — H
          H — C — H     |
          H   |   H     |   H
          |   |   |     |   |
      H — C — C — C — C — C — H
          |   |   |   |   |
          H   |   H   H   H
          H — C — H
              |
              H
```

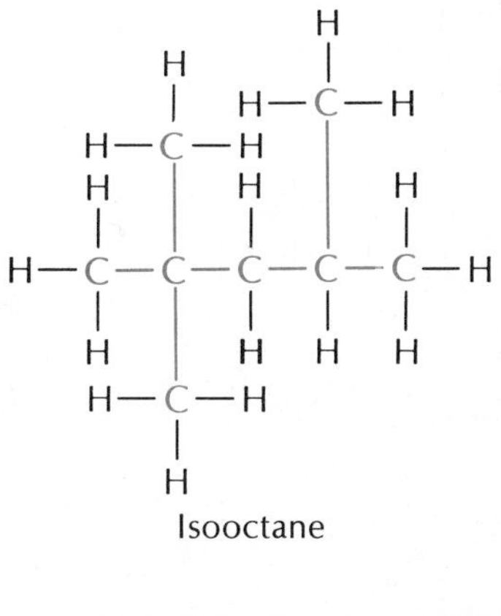

Isooctane

FIGURE 18.2
Carbon skeletons of some common substances

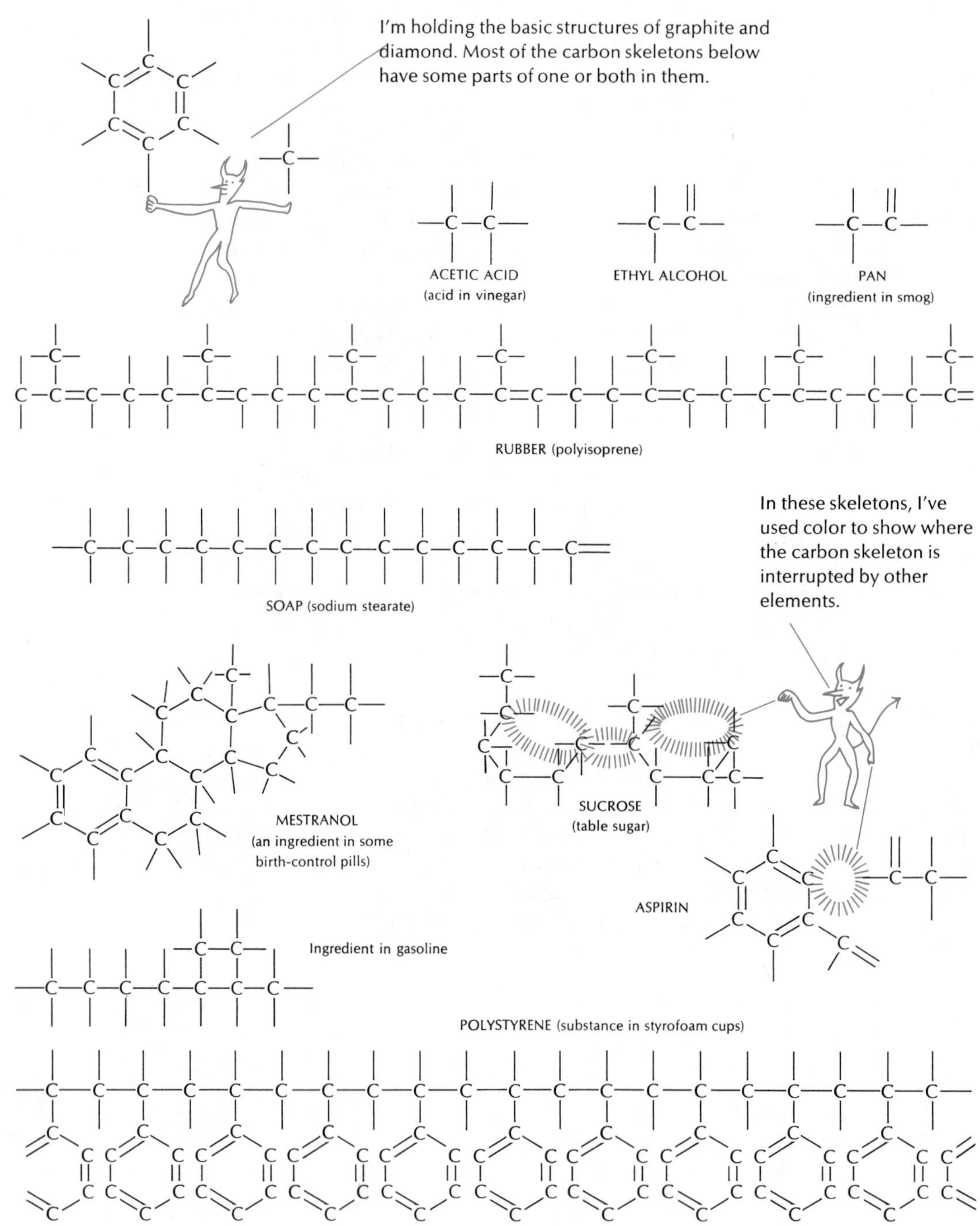

The two carbon skeletons, marked in color, are different from one another. Normal octane is an example of a *straight chain hydrocarbon,* while isooctane is a *branched chain hydrocarbon.* Even though these two compounds have the same molecular formula, C_8H_{18}, they have different structural formulas. Such compounds are called *isomers.* There are eighteen other isomers of these two compounds. We need the structural formula to know which isomer we're talking about.

straight chain hydrocarbon

branched chain hydrocarbon

isomer

Differences in structure cause isomers to behave differently from one another. The two substances above have different melting and boiling points and different densities because of differing attractions between adjacent molecules. Isooctane has an octane rating of 100, meaning that it's a verv good antiknock ingredient in gasoline. Normal octane has an octane rating of only −19, meaning that it's a terrible antiknock ingredient. We'll learn more about this later.

Structural formulas have their drawbacks. For large molecules, they're tedious to write and sometimes they tend to be cluttered. Instead, we use *condensed structural formulas,* which let us see the carbon skeleton more clearly. These formulas show each carbon atom and the noncarbon atoms bonded to it all on the same line. Everything that comes between two carbon atoms is bonded to the carbon atom to the left. The formulas of normal octane and isooctane, written as condensed structural formulas, are shown below.

condensed structural formula

$$CH_3-CH_2-CH_2-CH_2-CH_2-CH_2-CH_2-CH_3$$

or

$$CH_3(CH_2)_6CH_3$$

Normal octane

$$\begin{array}{ccccccc} & & CH_3 & & & & CH_3 & & \\ & & | & & & & | & & \\ CH_3 & - & C & - & CH_2 & - & CH & - & CH_3 \\ & & | & & & & & & \\ & & CH_3 & & & & & & \end{array}$$

Isooctane

For *ring hydrocarbons,* in which the carbon atoms are bonded to form a ring, we often leave the carbon and hydrogen atoms out altogether and just draw the rings. A carbon atom is understood to be where two straight lines intersect. If double bonds are present, they are shown. Each carbon atom is assumed to have as many bonded hydrogen atoms as it needs to satisfy its four single bonds, or two singles and a double. Table 18.1 illustrates condensed structural formulas involving carbon and hydrogen. We see that compounds with single bonds contain more hydrogen than those with double bonds. Compounds with double or triple bonds between carbons are *unsaturated;* compounds without double or triple bonds are *saturated.*

ring hydrocarbon

unsaturated

saturated

WHY CARBON? Of all the elements, carbon is the only one that bonds to itself and to other elements in such infinite variety. Forming a maximum of four bonds allows more possibilities than forming only three or two. This explains why elements like nitrogen, boron, and oxygen don't provide the backbone for organic compounds, whereas carbon does. But what about the other elements in Group IVA? Silicon, germanium, tin,

TABLE 18.1
Condensed structural formulas of hydrocarbons

Condensed Structural Formula	*Meaning*	*Example*	*Name*
$-CH_3$ or CH_3-	$-\overset{H}{\underset{H}{\overset{\vert}{\underset{\vert}{C}}}}-H$ or $H-\overset{H}{\underset{H}{\overset{\vert}{\underset{\vert}{C}}}}-$	CH_3-CH_2-OH	Ethyl alcohol
$-CH_2-$	$-\overset{H}{\underset{H}{\overset{\vert}{\underset{\vert}{C}}}}-$	$CH_3-CH_2-CH_3$	Propane
$-\overset{\vert}{C}H-$	$-\overset{\vert}{\underset{H}{\underset{\vert}{C}}}-$	$CH_3-\overset{OH}{\overset{\vert}{C}}H-CH_3$	Isopropyl alcohol (rubbing alcohol)
$CH_2=$	$H-\underset{H}{\underset{\vert}{C}}=$	$CH_2=CH_2$	Ethylene
$-CH=$	$-\underset{H}{\underset{\vert}{C}}=$	$CH_3-CH=CH_2$	Propylene
$-\overset{\vert}{C}=$	$-\overset{\vert}{C}=$	$CH_3-\overset{CH_3}{\overset{\vert}{C}}=CH_2$	Isobutylene (used to make synthetic rubber)
	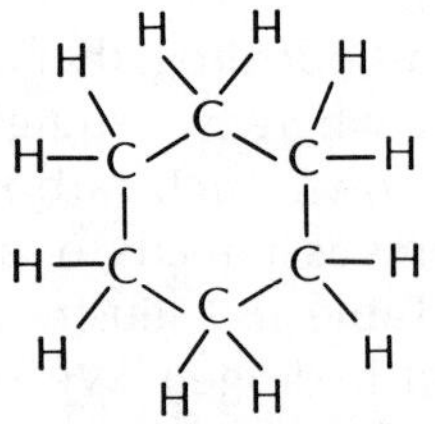		Cyclohexane (used to make nylon)
 or 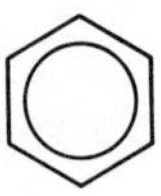	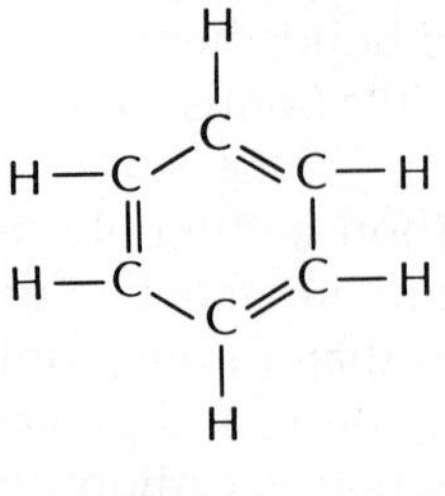 C_6H_6 (H—C ring, alternating double bonds) Benzene ring	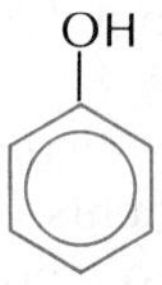 C_6H_5OH	Phenol (used to make aspirin)

and lead also have four valence electrons and can form four bonds. Why don't they also form the backbone of organic compounds?

The answer lies in the bond energies. We saw in Chapter 9 that bond energy means the energy needed to break a bond. Thus a higher bond energy means a stronger or more stable bond. Carbon bonds to itself and to other elements with high bond energy. Also, the strengths of the C—C bond and the C—O bond are about the same, giving a C—C bond about the same stability as a C—O bond. Silicon, on the other hand, forms a much weaker bond with itself than it does with oxygen. This means that an Si—Si bond is unstable relative to an Si—O bond, and if there is any oxygen around, an Si—O bond will form instead of an Si—Si bond.

Why should an Si—Si bond be weaker than a C—C bond? From Chapter 9, we know that the longer the bond, the lower the bond energy. Silicon, in the third period, has valence electrons one energy level further from the nucleus than carbon. When silicon bonds to itself, its atoms have to be further away from each other than carbon atoms are in a C—C bond. An Si—Si bond is longer and therefore weaker than a C—C bond. This trend continues with Ge—Ge, Sn—Sn, and Pb—Pb bonds becoming weaker and weaker. Carbon is thus the only element that forms strong bonds with itself and with other elements at the same time.

18.2 ENERGY FROM CARBON COMPOUNDS

Some fuels obtained from petroleum and coal are shown in Figure 18.3. All of these are hydrocarbons. When we burn these fuels, we're releasing the sun's energy stored in the bonds of their hydrocarbons. We'll be able to talk about hydrocarbons more easily once we have classified them.

THE CLASSIFICATION OF HYDROCARBONS.

aromatic hydrocarbon

An *aromatic hydrocarbon* is any hydrocarbon that contains a benzene ring (shown in Table 18.1) or has a benzene-like structure.

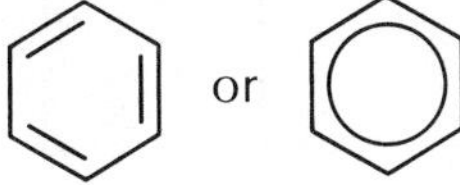

Benzene

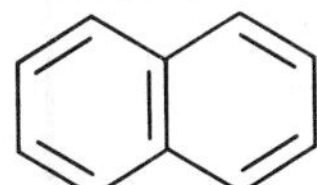

Naphthalene (moth balls)

(A circle inside the ring instead of the three alternating double bonds is a shorthand way of showing that the sequence of bonds is continuous.) The alternating double and single bonds around a ring are the characteristic feature of an aromatic hydrocarbon. Thus, neither of the two hydrocarbons on page 429 is aromatic even though they might look nearly the same as benzene. They lack the sequence of alternating single and double bonds.

FIGURE 18.3
Fuel products of petroleum and coal

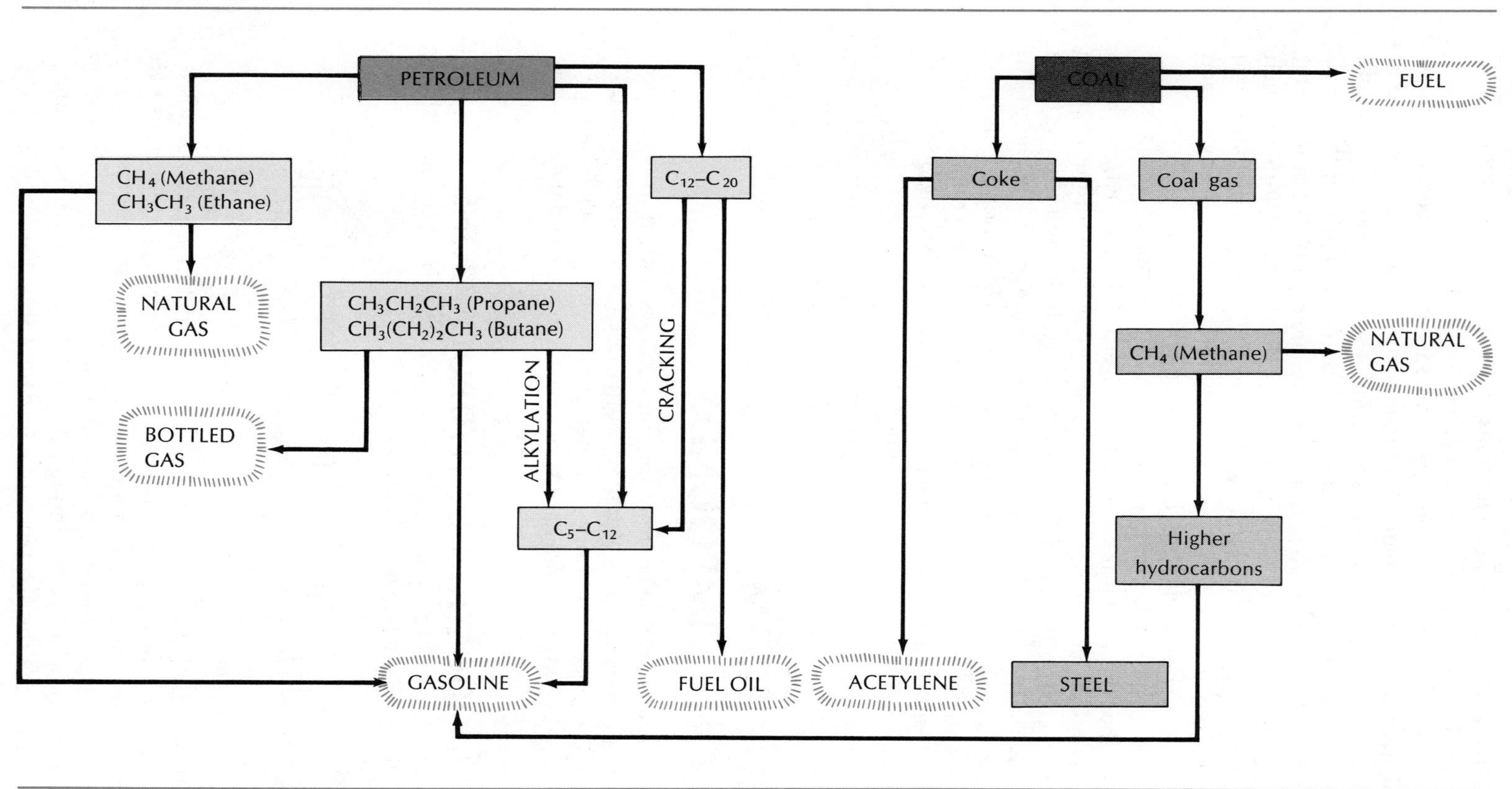

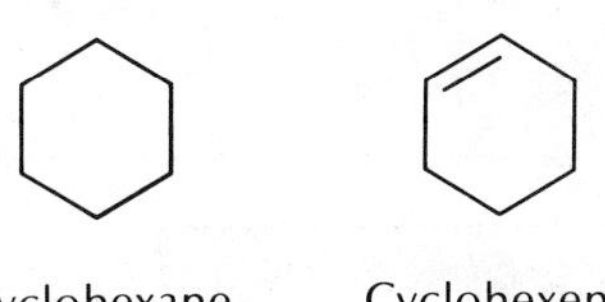

Cyclohexane Cyclohexene

Aromatic hydrocarbons come mostly from coal. They got their name because some early compounds isolated from coal and shown to contain benzene rings had characteristic aromas.

aliphatic hydrocarbon

An *aliphatic hydrocarbon* is any hydrocarbon that does not contain a benzene ring or benzenelike structure. In Table 18.1, all the examples except phenol are aliphatic. Aliphatic hydrocarbons come mostly from petroleum. They got their name from a Greek word meaning "oil." Even though coal contains mostly aromatic compounds and petroleum contains mostly aliphatic compounds, modern technology can synthesize (make) aromatic compounds from petroleum products and aliphatic compounds from coal products.

The classification of hydrocarbons appears in Figure 18.4. We learned in Chapter 4 that inorganic chemicals have common and systematic names. This is true of organic chemicals, too. So that chemists could discuss organic compounds, a system of naming was developed by the International Union of Pure and Applied Chemistry (IUPAC). In this system, the names of all alkanes end in *-ane,* alkenes in *-ene,* and alkynes in *-yne*. We won't go into this system here, although we should know that it exists. Usually we'll use the common name of a substance, sometimes indicating its IUPAC name in parentheses. If the common name is nonexistent—or unfamiliar—we'll use only the IUPAC name. The common names "ethylene," "propylene," and "acetylene" are so much more widely used than their IUPAC names "ethene," "propene," and "ethyne" that we'll use only the common names of these.

alkyl group

If a hydrogen atom is removed from an alkane, leaving a bonding position that can hold another atom or group, the rest of the alkane is an *alkyl group*. A particular alkyl group is named by dropping the *-ane* and adding *-yl*, as in these examples:

CH_4	$CH_3—$	$CH_3—OH$
Methane	Methyl group	Methyl alcohol
CH_3CH_3	$CH_3CH_2—$	$CH_3CH_2—OH$
Ethane	Ethyl group	Ethyl alcohol

COAL AS AN ENERGY SOURCE. Coal is mostly carbon with the graphite structure. Some other atoms—sulfur, nitrogen, and oxygen—are bonded chemically to carbon in coal's organic structure. When coal is burned, oxides of nitrogen and sulfur are products as well as carbon dioxide. This is why coal has been labeled a "dirty" fuel.

In Chapter 11, we discussed the possibilities of removing sulfur from

FIGURE 18.4
The classification of hydrocarbons

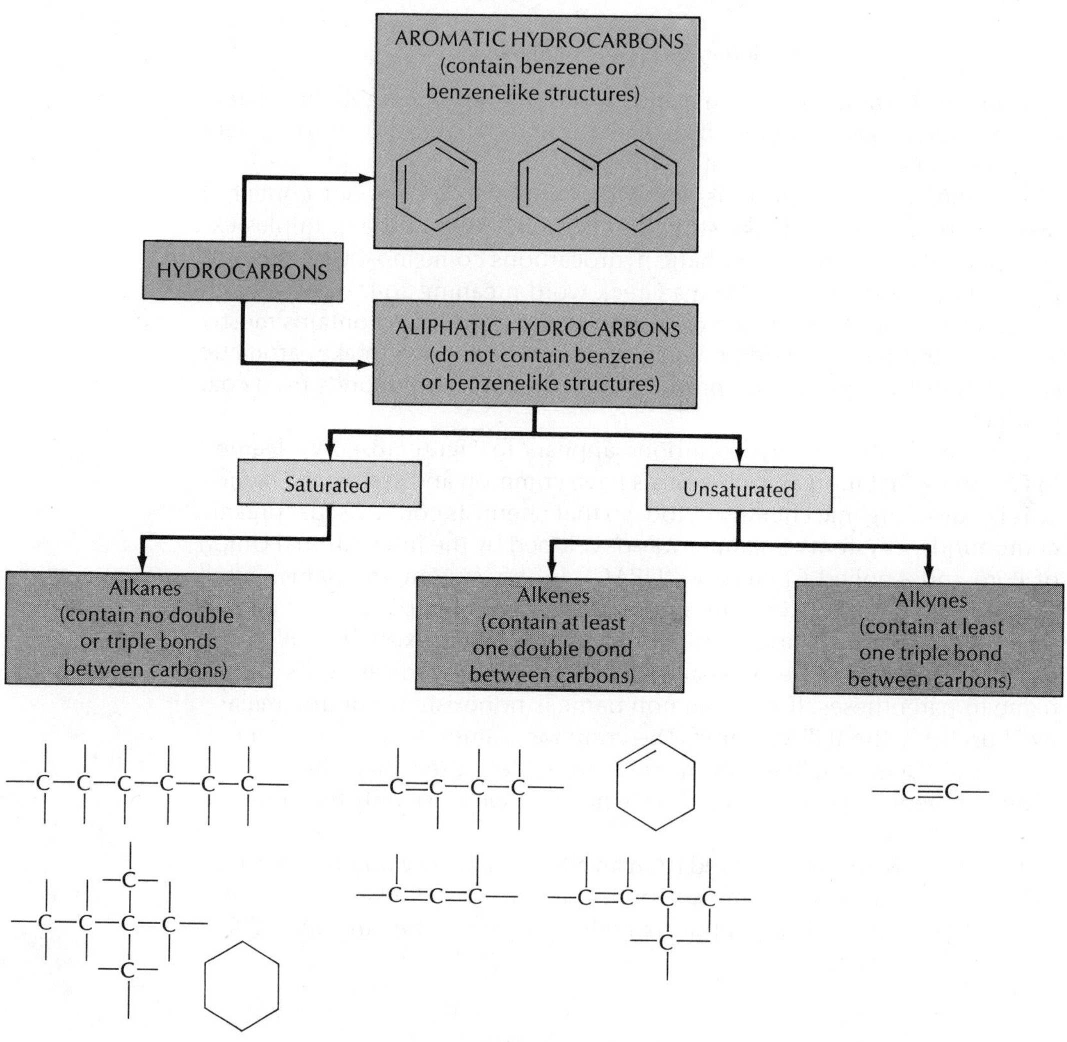

All categories may contain branched chain, straight chain, or cyclic hydrocarbons.

coal before it's burned, or removing SO_2 from the smokestack afterward. The sulfur dioxide from burning coal comes from two places: the sulfur that's bound chemically in the coal, and the sulfur from the pyrite (FeS_2) that's mixed in with the coal. The pyrite can be removed, with difficulty, by grinding up the coal and separating the coal particles from the pyrite particles using gravity. But the chemically bound sulfur stays behind; no good way has yet been found to remove it.

About 70 percent of the coal that is mined is burned to give heat, and the energy yield is about 7.8 kilocalories per gram of coal. The other 30 percent is used to make coke (charcoal) for the steel industry. In the coking process, coal is heated to about 1000°C without any air, as shown by this unbalanced equation:

$$\text{Coal(C, H, O, N, S)} \not\longrightarrow \text{Coke(C)} + \text{Coal tar} + CH_4 + H_2S + HCN$$
$$+ NH_3 + CO + H_2 + H_2C{=}CH_2 + CH_3{-}CH{=}CH_2$$

The mixture of gases produced as a by-product of the coking industry is called *coal gas*. This gas is usually passed through a water scrubber that removes the obnoxious gases—H_2S, HCN, and NH_3—leaving mostly CO and H_2, with some ethylene and propylene. This relatively clean gas can itself be burned as a fuel, or it can be made into other gases by using a nickel catalyst.

coal gas

$$CO(g) + 3\,H_2(g) \xrightarrow{\text{nickel}} CH_4(g) + H_2O(g)$$

Higher hydrocarbons (those having more carbon atoms) can be obtained by using an iron catalyst. These higher hydrocarbons can be made into gasoline.

PETROLEUM AS AN ENERGY SOURCE. Petroleum is a mixture of hydrocarbons having from one to about thirty carbon atoms. Most are straight chain saturated hydrocarbons, but some are branched chain and some are cyclic. To get some idea of the relative energies of these hydrocarbons, we can compare their heats of combustion. For example, in the burning of methane:

$$CH_4(g) + 2\,O_2(g) \longrightarrow CO_2(g) + 2\,H_2O(g) + 211\text{ kcal}$$

The *heat of combustion*, 211 kcal/mole in this case, is the energy released when a mole of any substance burns completely. Energy is released because the energy required to break the reactants' bonds is more than paid for by the energy released when the products' bonds are formed.

heat of combustion

Heats of combustion of some hydrocarbons are shown in Table 18.2. Of course, the total energy in kilocalories/mole increases with the number of carbon and hydrogen atoms, but it's more useful to compare energy released per gram. The heat of combustion in kilocalories/gram is greatest for methane, less for ethane, and levels off for the rest of that series. The amount of energy released compares roughly with the number of C—H bonds per carbon atom. This is because a C—H bond changing into an O—H bond releases more energy than does a C—C bond changing into a C=O bond. Unsaturated hydrocarbons have fewer C—H bonds and therefore lower heats of combustion than saturated hydrocarbons. All the heats of combustion are greater than that of coal (7.8 kcal/g), indicating that coal has even fewer C—H bonds per carbon atom than benzene. In fact, as we mentioned previously, coal resembles graphite in its structure, with many unsaturated rings fused together.

Heat of combustion is an important consideration when heat is the

TABLE 18.2
Heats of combustion of some hydrocarbons

Name	Condensed Structural Formula	Heat of Combustion (kcal/mole)	Heat of Combustion (kcal/g)	Number of C—H Bonds Per Carbon Atom
Methane	CH_4	211	13	4
Ethane	$CH_3—CH_3$	368	12	3
Propane	$CH_3—CH_2—CH_3$	526	11.9	2.7
Butane	$CH_3—(CH_2)_2—CH_3$	684	11.8	2.5
Pentane	$CH_3—(CH_2)_3—CH_3$	838	11.5	2.4
Hexane	$CH_3—(CH_2)_4—CH_3$	990	11.5	2.3
Heptane	$CH_3—(CH_2)_5—CH_3$	1150	11.5	2.3
Octane	$CH_3—(CH_2)_6—CH_3$	1303	11.5	2.3
Ethylene (ethene)	$CH_2═CH_2$	337	10.5	2
Acetylene (ethyne)	HC≡CH	302	10.1	1
Benzene	(benzene ring)	781	10	1

desired product of a fuel. As we saw in Chapter 6, though, we don't want all of gasoline's heat of combustion in a car engine. Instead, we want the fuel to burn evenly and produce gases that expand and push the pistons out when ignited with a spark. If the gasoline explodes when it's compressed, before the spark ignites it, the result is engine knock or preignition. Engine knock causes poor engine performance and low gasoline mileage. The octane rating of a gasoline tells how much engine knock it causes. The higher the octane rating, the less the engine knock. *Octane ratings* are based on a score of 100 for isooctane, and the others measured relative to that. Much of gasoline technology is aimed at improving the octane ratings of gasolines.

octane rating

We already saw the difference in octane rating between normal octane and isooctane. Straight chain hydrocarbons cause much more engine knock than branched chain or cyclic ones. Among the straight chain hydrocarbons, longer chains cause more engine knock than shorter chains. Compact molecules seem to burn more evenly and cause less engine knock than long, stringy molecules.

PETROLEUM REFINING. Crude petroleum is a thick black sludge. Distillation is used to separate it into natural gas, kerosene, gasoline, and so forth. The boiling points of these components are roughly proportional to molecular weight and therefore to the number of carbon atoms. When petroleum is heated and changed into a mixture of gases, the gases condense in different temperature ranges as the mixture cools. A mixture of substances that condenses within a given range is called a *fraction*. Because the number of carbon atoms in hydrocarbons is related to their molecular weights, each fraction will contain hydrocarbons having a certain range of numbers of carbon atoms. The lightest molecules—those having from one to four carbon atoms (C_1 to C_4)—do not condense at all and make up the lowest-boiling fraction, below 20°C. The next fraction, with a boiling range of 20°C to 60°C, includes hydrocarbons in the C_5 to C_6 range. And so forth. Figure 18.5 shows a diagram of an industrial distillation tower with its petroleum fractions.

fraction

The gasoline fraction presents two basic problems. First, there isn't enough of it. Petroleum is only about 30 percent to 40 percent gasoline, but the demand is for over 50 percent. Two things can be done to make more gasoline:

1. Combine smaller molecules in the C_1 to C_5 range to make molecules in the C_6 to C_{12} range. This is called *alkylation*.
2. Break apart larger molecules in the fuel oil range (C_{15} to C_{20}) to make molecules in the C_6 to C_{12} range. This is called *cracking*.

alkylation

cracking

Another problem with the gasoline fraction is that it's not good enough for modern high-compression car engines that need an octane rating of about 90. Various substances have to be added to gasoline to raise its octane rating. Tetraethyl lead [$Pb(C_2H_5)_4$] can be added, but the resulting leaded gasoline can't be used with catalytic converters, as we saw in Chapter 16. Also, leaded gasoline releases harmful lead compounds to the air. To produce unleaded gasoline with an increased octane rating, straight chain hydrocarbons have to be changed into more compact cyclic or branched chain hydrocarbons. This process is called *reforming*. Some unleaded gasolines have a characteristic smell because they contain benzene, a good antiknock agent. Unleaded gasolines are more expensive than leaded, because reforming is an expensive process.

reforming

18.3 FUNCTION FROM CARBON COMPOUNDS

We said earlier that elements other than carbon and hydrogen often determine function. Atoms or combinations of atoms that cause an organic compound to react in a certain way are called *functional groups*. We've already had a little experience with organic functional groups. In Chapter 13, we learned that the —COOH group causes an organic compound to behave like an acid. We've also seen the —OH, or alcohol, group in ethyl alcohol.

functional group

Some important functional groups, and a typical compound containing

FIGURE 18.5
A petroleum distillation tower separates light molecules from heavy molecules

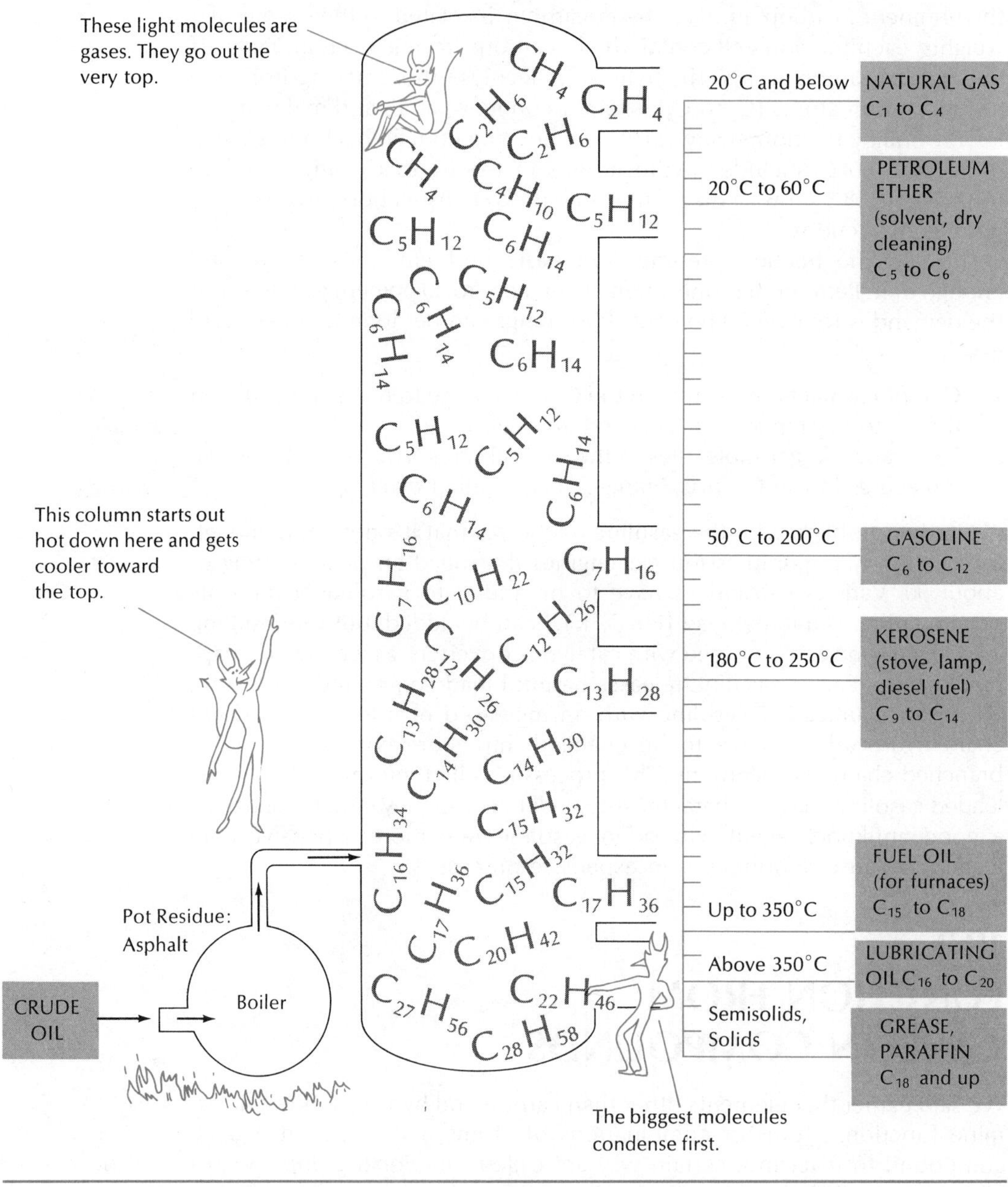

each one, are shown in Table 18.3. Textbooks of organic chemistry contain a discussion of each functional group, how it's made, how it reacts, and varying examples of compounds. In this brief introduction to organic chemistry, we can't cover every aspect of functional groups, any more than

TABLE 18.3
Structural formulas of some functional groups

Name of Functional Group	*Condensed Structural Formula*	*Meaning*	*Example*	
Chloro, Fluoro, Bromo, Iodo	$-Cl$ $-F$ $-Br$ $-I$	Same	CCl_2F_2	Freon
			$Cl-C_6H_4-CH(C_6H_4-Cl)-CCl_3$	DDT (dichlorodi-phenyltrichloro-ethane)
Alcohol	$-OH$	$-O-H$	CH_3-OH	Wood alcohol (methyl alcohol)
Ether	$-O-$	$-O-$	$CH_3CH_2-O-CH_2CH_3$	Ether (diethyl-ether)
Amino	$-NH_2$	$-N{<}^{H}_{H}$	$NH_2-(CH_2)_5-NH_2$	Cadaverine (odor of rotting flesh)
Nitro	$-NO_2$	$-N{<}^{\nearrow O}_{\searrow O}$	$C_6H_2(CH_3)(NO_2)_3$	TNT (trinitro-toluene)
Aldehyde	$-CHO$	$-C(=O)H$	$H-CHO$	Formaldehyde
Ketone	$-CO-$	$-C(=O)-$	$CH_3-CO-CH_3$	Acetone
Carboxylic acid	$-COOH$	$-C(=O)OH$	CH_3-COOH	Acetic acid
Ester	$-COO-$	$-C(=O)O-$	$CH_3-COO-C_6H_4-COOH$	Aspirin (acetylsalicylic acid)
Amide	$-CONH_2$	$-C(=O)NH_2$	NH_2-CONH_2	Urea (animal waste product)

we could see all of Europe on a five-day whirlwind tour. Instead, we're going to focus on a few important compounds and some industrial reactions or reaction types that convert raw materials from petroleum or coal to products that we use.

ORGANIC COMPOUNDS DERIVED FROM WATER AND FROM AMMONIA.

The substance that we usually think of as alcohol is ethyl alcohol. But any organic compound that contains the —OH functional group is an *alcohol*, and there are many of them. Here are the formulas of ethyl alcohol and water side by side: alcohol

$$CH_3{-}CH_2{-}O{-}H \qquad H{-}O{-}H$$

We see that ethyl alcohol is derived from water by substituting an ethyl group for one of water's hydrogen atoms. In the same way, we could substitute methyl, propyl, butyl, and so forth, and obtain a series of simple alcohols derived from alkanes. Besides these simple alcohols, more complex ones exist that contain more than one —OH group or are derived from aromatic or cyclic hydrocarbons. Table 18.4 shows a few common examples.

Simple alcohols are part water and part alkane, with properties somewhere in between. The —OH functional group can hydrogen-bond to itself and to water. Alcohols having one, two, and three carbon atoms (methyl, ethyl, and propyl) are miscible with water in all proportions. But as the alkyl group gets larger (four carbon atoms and higher), the alcohol becomes more like the alkane, less like water, and therefore less water soluble.

Some reactions of alcohols are like those of water. For instance, alcohols react with the Group IA metals to liberate hydrogen.

$$2\,CH_3CH_2OH(l) + 2\,Na(s) \longrightarrow H_2(g) + \underset{\text{Sodium ethoxide}}{2\,CH_3CH_2ONa(\text{solution})}$$

We obtain a solution of sodium ethoxide in ethyl alcohol in much the same way that we obtain a solution of sodium hydroxide in water when sodium reacts with water.

Other reactions involve both the alkyl group and the —OH group. If an alcohol is treated with sulfuric acid (which acts as a catalyst) at a fairly high temperature (180°C), the alcohol loses a water molecule to form an alkene.

$$CH_3CH_2OH\,(g) \xrightarrow[180°C]{H_2SO_4} CH_2{=}CH_2(g) + H_2O\,(g)$$

A reaction like this, where a single molecule loses the elements of water to form a more simple molecule, is called *dehydration*. dehydration

If, however, an alcohol is treated with sulfuric acid at a lower temperature, an ether is formed.

$$CH_3CH_2OH\,(l) + CH_3CH_2OH\,(l) \xrightarrow[\text{low temp.}]{H_2SO_4} \underset{\text{Diethyl ether}}{CH_3CH_2OCH_2CH_3(l)} + H_2O\,(l)$$

Here, the elements of water are taken from two molecules instead of one. A reaction where two or more molecules form a more complex molecule with the loss of water is called *condensation*. This meaning is somewhat different from the one we saw in Chapter 10, where condensation meant condensation

TABLE 18.4
Some common alcohols

Formula	*Name*	*Source*	*Comments*
$CH_3—OH$	Methyl alcohol (wood alcohol)	Older source: distillation of wood chips. Modern source: from coal gas: $CO(g) + 2\,H_2(g) \xrightarrow{\text{Cat.}} CH_3—OH$	Very poisonous; can cause blindness and death. Used to "denature" ethyl alcohol or make it unfit for drinking.
$CH_3—CH_2—OH$	Ethyl alcohol (grain alcohol)	For beverages: fermentation of sugar. Industrial source: from ethylene: $CH_2{=}CH_2(g) + H_2O(g) \xrightarrow{H_2SO_4} CH_3—CH_2—OH(l)$	The active ingredient in alcoholic drinks.
$CH_3—CH(OH)—CH_3$	Isopropyl alcohol	From propylene: $CH_3—CH{=}CH_2(g) + H_2O(g) \longrightarrow CH_3—CH(OH)—CH_3(l)$	Best known as "rubbing alcohol," but also important industrially.
$HO—CH_2—CH_2—OH$	Ethylene glycol	Derived from ethylene.	Used as antifreeze. Two —OH groups on one molecule make it very water soluble.
$HO—CH_2—CH(OH)—CH_2—OH$	Glycerin (glycerol)	By-product of soap manufacture. Occurs naturally in animal fats. Also made from propylene.	Very water soluble. Used to make cosmetics and nitroglycerin.

going from a less dense phase (such as a gas) to a more dense phase (such as a liquid). However, in the broad sense—going from something less dense to something more dense—the meaning of the word "condensation" still applies. We'll be seeing many more examples of condensation reactions throughout this and the next chapter.

Now let's look at the formulas of diethyl ether and of water side by side.

$$\underset{\text{Diethyl ether}}{CH_3CH_2—O—CH_2CH_3} \qquad \underset{\text{Water}}{H—O—H}$$

We see that diethyl ether is derived from water by substituting an ethyl group for each of water's hydrogen atoms. As with alcohols, we can substitute different alkyl groups to form different *ethers*. On a given ether molecule, the alkyl groups can be the same or they can be different.

ether

Ethers are much less like water than alcohols are. With two alkyl groups, only the oxygen of water is left, there is no hydrogen to hydrogen-bond with, and the ether behaves much more like an alkane than like water. Ethers are mostly low boiling and water insoluble. Because they lack a hydrogen bonded to oxygen, ethers don't participate in reactions similar to the ones we've mentioned for alcohols. (Ethers do burn and participate in some reactions, but we won't discuss them here.)

Amines are organic compounds derived from ammonia in the same way that alcohols and ethers are derived from water. amine

$$\underset{\text{Ammonia}}{-\underset{\displaystyle H}{\underset{|}{N}}-H} \qquad \underset{\text{Methylamine}}{CH_3-\underset{\displaystyle H}{\underset{|}{N}}-H} \qquad \underset{\text{Dimethylamine}}{CH_3-\underset{\displaystyle CH_3}{\underset{|}{N}}-H} \qquad \underset{\text{Trimethylamine}}{CH_3-\underset{\displaystyle CH_3}{\underset{|}{N}}-CH_3}$$

We see that there are three possibilities, since ammonia has three hydrogen atoms. All three types of compounds are amines: *primary* (one alkyl group), *secondary* (two alkyl groups), and *tertiary* (three alkyl groups). These compounds have some of the properties of alkanes and some of ammonia. The more alkyl groups substituted, the fewer hydrogens left to hydrogen-bond, the less water soluble the compounds become, and the lower their boiling points. As with alcohols, increasing the number of carbon atoms also decreases the water solubility. Most amines are foul smelling. The characteristic smell of putrid meat or rotten fish is caused by the amines they contain. primary amine secondary amine tertiary amine

Some reactions of amines are like those of ammonia. For instance, amines can act as weak bases.

$$CH_3NH_2(aq) + HCl(aq) \longrightarrow \underset{\text{Methylammonium ion}}{CH_3NH_3^+(aq)} + Cl^-(aq)$$

$$(CH_3)_3N(aq) + HCl(aq) \longrightarrow \underset{\text{Trimethylammonium ion}}{(CH_3)_3NH^+(aq)} + Cl^-(aq)$$

Since it is the nonbonding electron pair on nitrogen that makes ammonia a base, the primary, secondary, and tertiary amines alike can also be bases.

Primary and secondary, but not tertiary, amines can undergo condensation reactions as alcohols can. We'll see some important condensation reactions later.

REACTIONS OF HYDROCARBONS. In Chapter 10, we talked about the use of Freons—carbon compounds containing fluorine and chlorine—as refrigerants. These are made from the raw materials methane and ethane in petroleum. Saturated hydrocarbons don't have any functional groups, so their reactions are limited. We've already seen that they can burn and can be broken into smaller hydrocarbons (cracked). About the only reaction left is to trade one or more hydrogen atoms for one or more atoms of a different kind. This is called *substitution*. substitution

The first step in making one kind of Freon from methane is the substitution of a chlorine atom for one of methane's hydrogen atoms. The process is called *chlorination*. chlorination

$$CH_4(g) + Cl_2(g) \longrightarrow \underset{\text{Methyl chloride}}{CH_3Cl(g)} + HCl(g)$$

When this process is carried out three more times, the product is carbon tetrachloride (CCl_4), with four molecules of HCl as a by-product. Chlorination of hydrocarbons is an important commercial source of HCl. To make Freon 11 (trichlorofluoromethane), carbon tetrachloride is treated with HF, and one fluorine is substituted for one chlorine.

$$CCl_4(g) + HF(g) \longrightarrow \underset{\text{Freon-11}}{CCl_3F(g)} + HCl$$

Other Freons are these:

CCl_2F_2	$CClF_3$	CCl_2FCClF_2	$CClF_2CF_3$
Freon-12	Freon-13	Freon-113	Freon-115

Most organic compounds can be chlorinated. Bacteria and viruses will stop multiplying when their biological molecules are chlorinated, which is why chlorine is a good disinfectant. However, a side effect is that organic matter dissolved in water is also being chlorinated, putting chlorinated hydrocarbons into some public water supplies. Chlorinated hydrocarbons have been associated with various liver ailments, including cancer.

We see from Figure 18.1 that ethylene and propylene are very important raw materials from petroleum. They don't occur naturally in petroleum but are produced during the cracking process. These unsaturated hydrocarbons have a functional group: a double bond. Double bonds can serve as "handles" to get other functional groups onto hydrocarbons. The first step usually involves addition to the double bond. In *addition,* a two-part reactant such as Cl_2 (Cl—Cl) causes the double bond to open up and form two single-bonding positions, one for each part of the reactant. The resulting compound is saturated. Addition is different from substitution because atoms are added to the compound while none are taken away. addition

Addition is best illustrated by example. Looking at Figure 18.1, we see that the path from ethylene to PVC includes a compound called vinyl chloride. The first step in making PVC, then, is to change ethylene to vinyl chloride. This is done by addition of chlorine to ethylene.

$$CH_2{=}CH_2 + Cl_2 \xrightarrow{\text{low temp.}} \underset{\text{1,2-Dichloroethane}}{\underset{Cl\quad\; Cl}{CH_2{-}CH_2}} \xrightarrow{500°C} \underset{\text{Vinyl chloride}}{CH_2{=}CH{-}Cl} + HCl$$

The 1,2-dichloroethane decomposes at high temperature to vinyl chloride. We'll see how to get from vinyl chloride to PVC in Section 18.4.

In the pathway from ethylene to aspirin, the first step is changing ethylene to ethyl alcohol. As we saw in Table 18.4, this is done industrially by adding water, which we can think of as a two-part reactant H—OH.

$$CH_2{=}CH_2 + H_2O \xrightarrow{H_2SO_4} \underset{H\quad\; OH}{CH_2{-}CH_2} \quad \text{(written } CH_3CH_2OH\text{)}$$

Another reaction involving addition to a double bond is *hydrogenation,* where the two-part reactant is H_2 (H—H). Polyunsaturated fats and oils hydrogenation

FIGURE 18.6
Addition to a double bond involves a two-part reactant

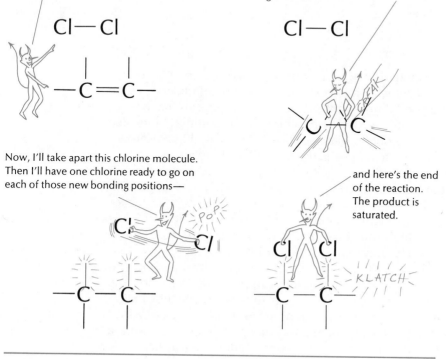

I'm going to make this two-part chlorine molecule add to the double bond.

Cl—Cl

—C=C—

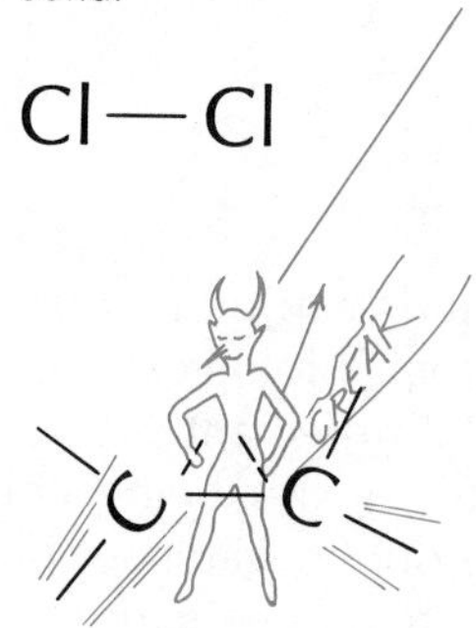
First, I'll open up this double bond. Then each carbon atom can form one more single bond.
Cl—Cl
CREAK
C
C

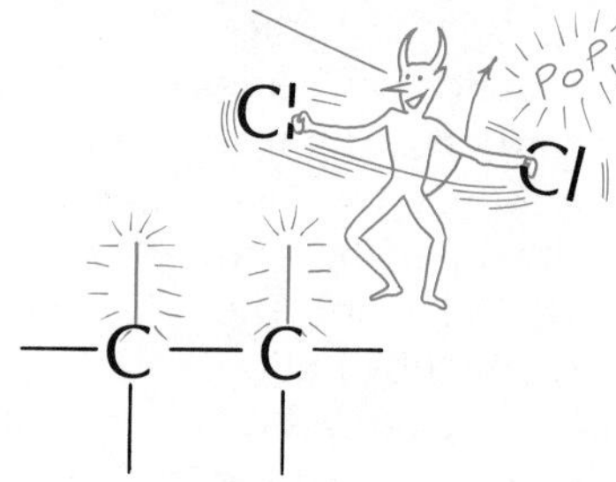
Now, I'll take apart this chlorine molecule. Then I'll have one chlorine ready to go on each of those new bonding positions—
POP
Cl
Cl
—C—C—

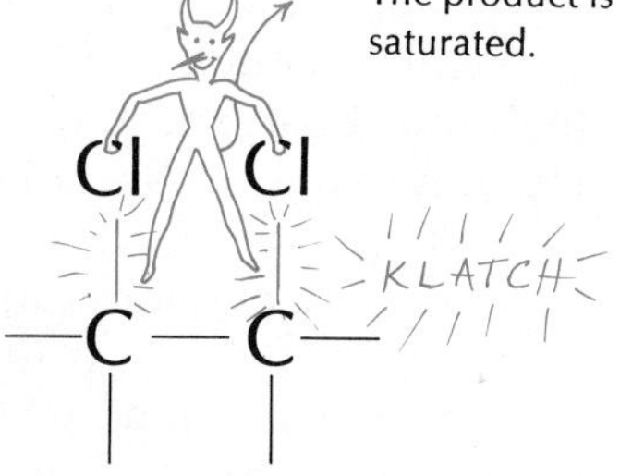
and here's the end of the reaction. The product is saturated.
Cl
Cl
KLATCH
—C—C—

have several double bonds and are liquids. To make solid shortening, polyunsaturated fats can be hydrogenated like this:

$$\underset{\substack{\text{Linoleic acid} \\ \text{(found in cottonseed oil)}}}{CH_3(CH_2)_4CH{=}CHCH_2CH{=}CH(CH_2)_7COOH} + 2\,H_2(g) \xrightarrow{\text{cat.}} \underset{\substack{\text{Stearic acid} \\ \text{(found in beef tallow)}}}{CH_3(CH_2)_{16}COOH}$$

Saturated fats have recently been linked to heart disease, such as hardening of the arteries. Emulsifying agents are now being used to turn unsaturated fats into solids while they still keep their double bonds.

Figure 18.6 illustrates addition to a double bond.

CHANGING CARBON'S OXIDATION NUMBER. If we look again at the pathway from ethylene to aspirin in Figure 18.1, we find the sequence ethyl alcohol to acetaldehyde to acetic acid. Here are the formulas for these, plus the oxidation number of the carbon atom that contains or is part of the functional group:

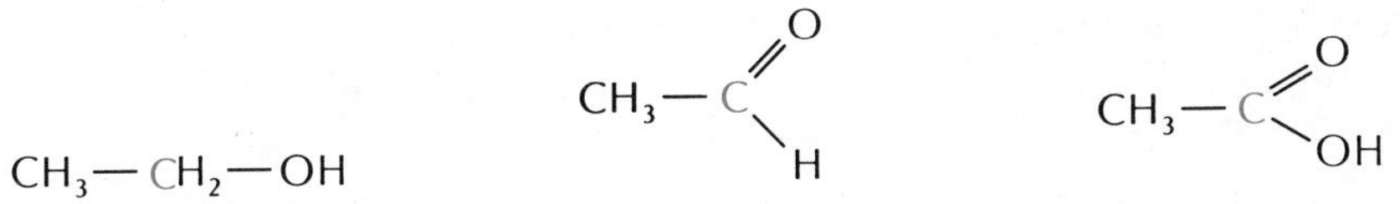

$$CH_3—CH_2—OH$$

Ethyl alcohol
oxidation number: 1−

$$CH_3—C(=O)H$$

Acetaldehyde
oxidation number: 1+

$$CH_3—C(=O)OH$$

Acetic acid
oxidation number: 3+

We know about ethyl alcohol and acetic acid, but acetaldehyde and the *aldehyde* functional group.

aldehyde

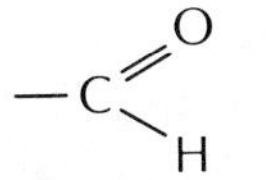

$$—C(=O)H$$

are unfamiliar. The carbon atom in the aldehyde group has an oxidation number in between that of ethyl alcohol's $—CH_2OH$ group and acetic acid's —COOH group. (In organic chemistry, "oxidation" usually means loss of hydrogen or gain of oxygen. Thus the $—CH_2OH$ group loses two hydrogens to become the —CHO group, which gains one oxygen to become the —COOH group.) Acetaldehyde is an important industrial chemical. The largest part of it is used to make acetic acid.

The sequence alcohol to aldehyde to acid is important in industry. The oxidizing agent that is the most practical, economical, and readily available is the oxygen in air itself. Different catalysts are used, depending on what reaction is desired. For instance, in the sequence of ethyl alcohol to acetic acid, these reactions are used:

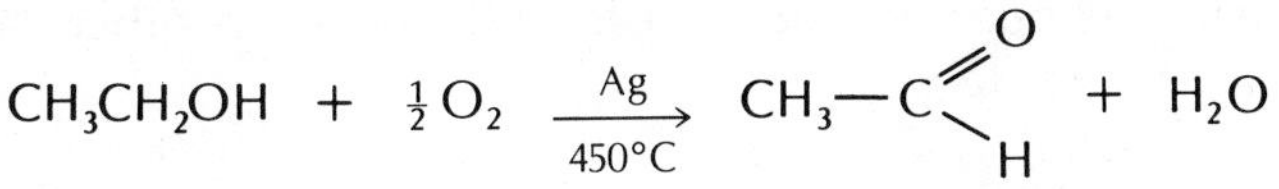

$$\underset{\text{Ethyl alcohol}}{CH_3CH_2OH} + \tfrac{1}{2}O_2 \xrightarrow[450°C]{Ag} \underset{\text{Acetaldehyde}}{CH_3—C(=O)H} + H_2O$$

$$\underset{\text{Acetaldehyde}}{CH_3—C(=O)H} + \tfrac{1}{2}O_2 \xrightarrow{Mn(C_2H_3O_2)_2} \underset{\text{Acetic acid}}{CH_3—C(=O)OH}$$

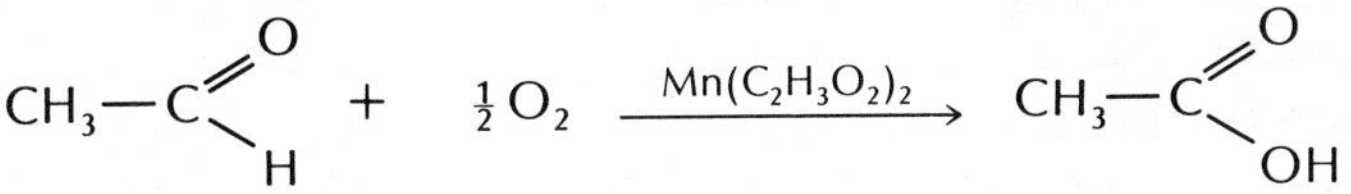

In Figure 18.1, the sequence of coal gas to methyl alcohol to formaldehyde leads to the formation of Bakelite resin. Again we use oxidation, but this time we stop at the aldehyde stage. We saw in Table 18.4 that methyl alcohol is made from coal gas. The methyl alcohol is then oxidized.

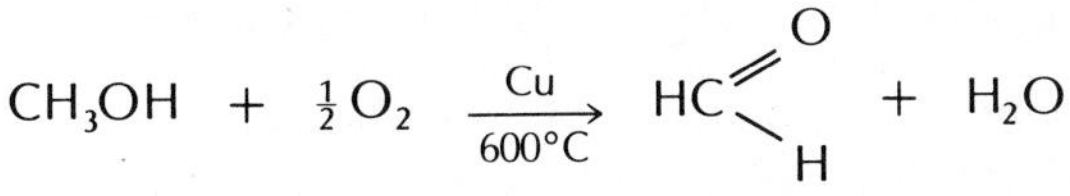

$$\underset{\text{Methyl alcohol}}{CH_3OH} + \tfrac{1}{2}O_2 \xrightarrow[600°C]{Cu} \underset{\text{Formaldehyde}}{HC(=O)H} + H_2O$$

About half the methyl alcohol produced is used to make formaldehyde, and, in turn, most of the formaldehyde is used to make the hard plastic Bakelite. Formaldehyde is a gas, so it's used mostly as a water solution called *formalin*. A minor use of formalin is to preserve biological specimens. formalin

The carbon atom in the ketone functional group has the same oxidation number as that in the aldehyde group. The difference between an aldehyde and a ketone can be seen by comparing the structures of acetaldehyde and acetone.

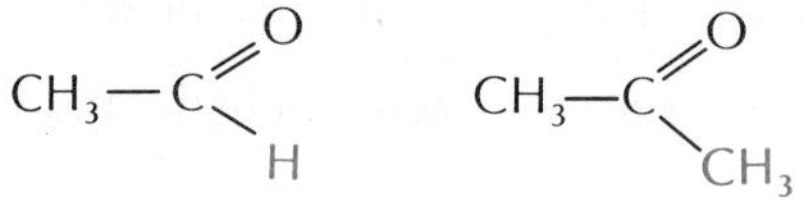

$$CH_3{-}C(=O){-}H \qquad\qquad CH_3{-}C(=O){-}CH_3$$

Acetaldehyde (an aldehyde) Acetone (a ketone)

We see that an alkyl group appears in a *ketone* where a hydrogen does in an aldehyde. Ketones can be obtained by oxidizing alcohols where the —OH group appears somewhere in the middle of the carbon chain rather than at the end. For instance, acetone is made commercially by oxidizing isopropyl alcohol. ketone

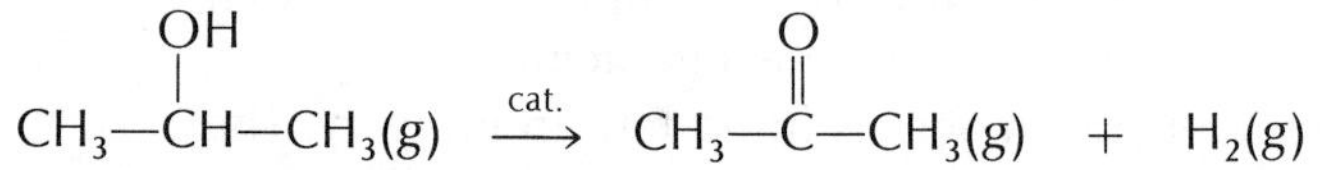

$$CH_3{-}CH(OH){-}CH_3(g) \xrightarrow{\text{cat.}} CH_3{-}C(=O){-}CH_3(g) + H_2(g)$$

If formaldehyde were oxidized still further, the product would be formic acid, HCOOH. Formic acid occurs naturally in the sting of ants and other insects. Although formic acid has some industrial uses, the demand for formaldehyde is far greater. Enough formic acid is obtained from other sources without oxidizing formaldehyde.

CONDENSATION REACTIONS AND THEIR PRODUCTS.

We've seen one example of a condensation reaction. The formations of detergents, aspirin, and many synthetic plastics and fibers also depend on condensation reactions.

In making synthetic detergents, alcohols with six or more carbon atoms are allowed to react with sulfuric acid. For example:

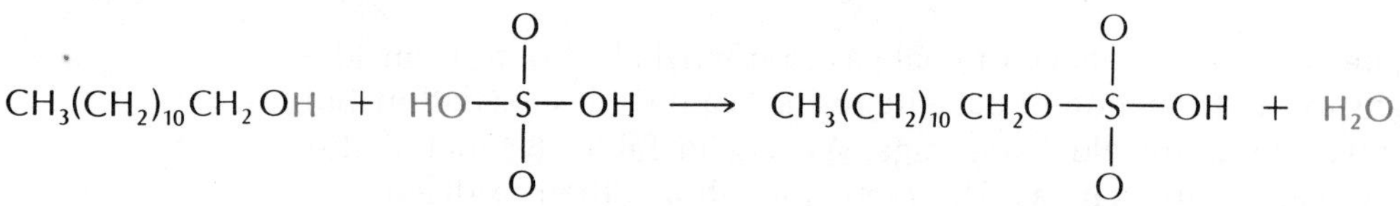

$$CH_3(CH_2)_{10}CH_2OH + HO{-}S(O)(O){-}OH \longrightarrow CH_3(CH_2)_{10}CH_2O{-}S(O)(O){-}OH + H_2O$$

Lauryl alcohol Sulfuric acid Lauryl hydrogen sulfate

The product of condensation between an alcohol and an acid is an *ester*. In this case, the product is a sulfate ester. Since sulfuric acid had two protons to begin with and one still remains, it could react with one more alcohol molecule in the same way, just as H_2SO_4 can react with one or two moles of NaOH. However, lauryl hydrogen sulfate is much more useful the way it is. To make a detergent, the remaining proton is neutralized with NaOH. ester

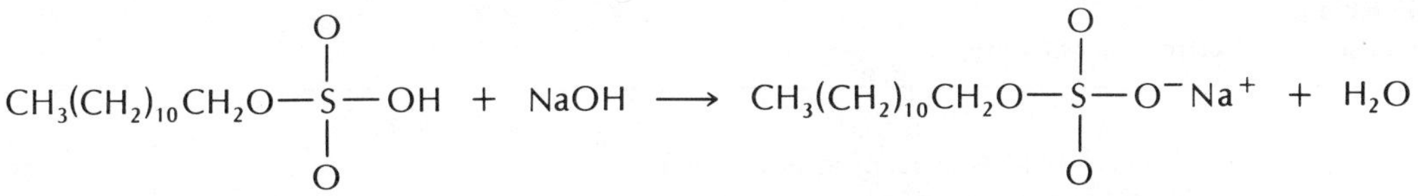

$$CH_3(CH_2)_{10}CH_2O-\overset{\overset{\large O}{|}}{\underset{\underset{\large O}{|}}{S}}-OH + NaOH \longrightarrow CH_3(CH_2)_{10}CH_2O-\overset{\overset{\large O}{|}}{\underset{\underset{\large O}{|}}{S}}-O^-Na^+ + H_2O$$

Sodium lauryl sulfate

Sodium lauryl sulfate is one of a class of alkyl sulfate detergents. These all have the long hydrophobic ends and the ionic hydrophilic ends that are necessary for detergent action, as we saw in Chapter 12. Sodium lauryl sulfate, one of the first synthetic detergents, is now commonly used as a detergent in toothpaste.

Lauryl alcohol is obtained from natural products, but many alcohols for making synthetic detergents are made from ethylene or propylene by complex processes. Another class of synthetic detergents is the benzenesulfonates.

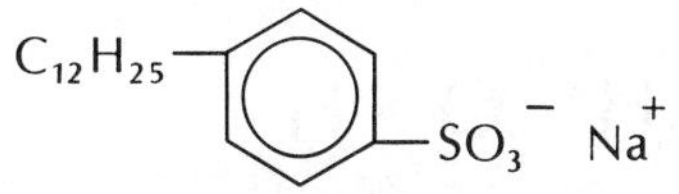

$$C_{12}H_{25}-C_6H_4-SO_3^-\ Na^+$$

Earlier detergents of this type had highly branched alkyl groups attached to the benzene ring. This made the detergent incapable of being broken down by microorganisms (nonbiodegradable), so that natural waters became full of soap suds. Now, processes have been developed for attaching straight chain instead of branched chain alkyl groups, and these detergents are biodegradable.

Like sulfuric acid and alcohols, carboxylic acids and alcohols form condensation products that are also esters. Aspirin (acetylsalicylic acid) is an ester of salicylic acid and acetic acid.

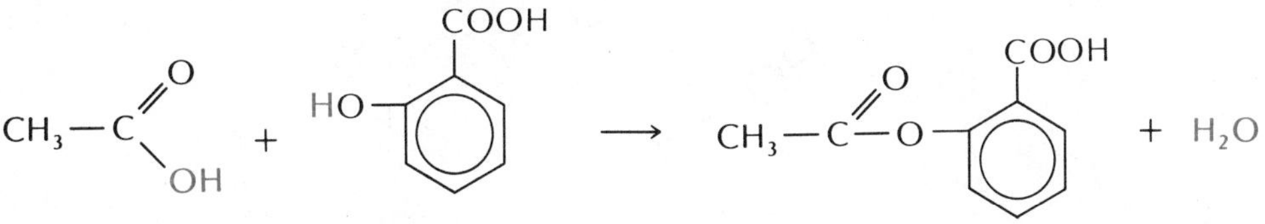

$$CH_3-C(=O)-OH + HO-C_6H_4-COOH \longrightarrow CH_3-C(=O)-O-C_6H_4-COOH + H_2O$$

Acetic acid — Salicylic acid — Aspirin

Salicylic acid has both a carboxylic acid and an alcohol functional group. Only the alcohol group reacts with acetic acid to form an ester group.

We find it convenient to show where an ester came from by condensing a water molecule from an acid and an alcohol, but esters usually aren't made this way. This is because the reaction is reversible, and for most esters the position of equilibrium lies far to the side of the acid and the alcohol. This principle is illustrated by the way aspirin works in the system. The real pain reliever is salicylic acid. When the aspirin passes through our intestines and blood stream, bases catalyze this reaction:

FIGURE 18.7
Formation and hydrolysis of esters

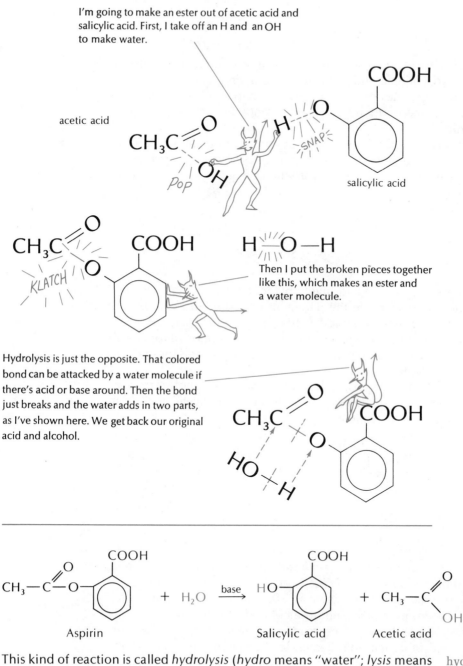

I'm going to make an ester out of acetic acid and salicylic acid. First, I take off an H and an OH to make water.

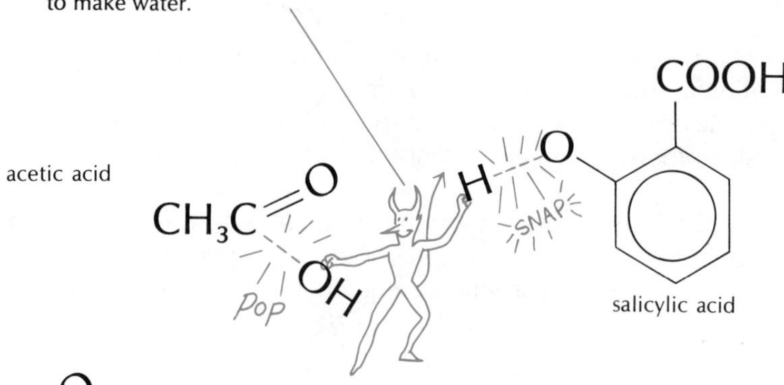

CH_3C O O COOH KLATCH $H—O—H$

Then I put the broken pieces together like this, which makes an ester and a water molecule.

Hydrolysis is just the opposite. That colored bond can be attacked by a water molecule if there's acid or base around. Then the bond just breaks and the water adds in two parts, as I've shown here. We get back our original acid and alcohol.

CH_3C O O COOH HO H

Aspirin ($CH_3—C(=O)—O—C_6H_4—COOH$) + H_2O $\xrightarrow{\text{base}}$ Salicylic acid ($HO—C_6H_4—COOH$) + Acetic acid ($CH_3—C(=O)OH$)

This kind of reaction is called *hydrolysis* (*hydro* means "water"; *lysis* means "breaking apart"). Here we have a good example of something we said earlier: interruptions in the carbon skeleton provide vulnerable spots where an organic molecule may be broken apart. The carbon skeleton of hydrolysis an ester is interrupted by oxygen. As a result, most esters can be hydrolyzed easily in the presence of water and an acid or a base. Figure 18.7 illustrates the formation and hydrolysis of esters.

If we can't make an ester from an acid and an alcohol, then how *do* we make one? Figure 18.1 shows a substance called acetic anhydride coming between acetic acid and aspirin in the chain of production. An *anhydride* is a condensation product of two carboxylic acid molecules. We arrive at the structure of acetic anhydride this way:

anhydride

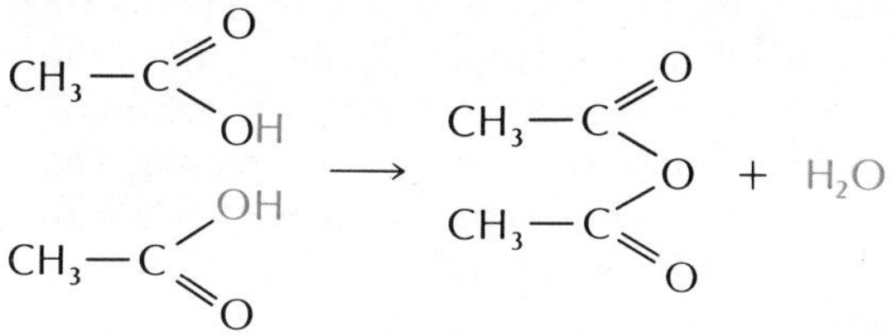

$$\mathrm{CH_3{-}C({=}O){-}OH} + \mathrm{HO{-}C({=}O){-}CH_3} \longrightarrow \mathrm{CH_3{-}C({=}O){-}O{-}C({=}O){-}CH_3} + H_2O$$

Anhydrides are very high-energy compounds that react with alcohols much more readily than their parent acids do. For this reason, anhydrides are often used to make esters. In the final step of making aspirin, acetic anhydride made from petroleum products is allowed to react with salicylic acid made from coal products. The acetic acid produced as a by-product is recycled and used over again.

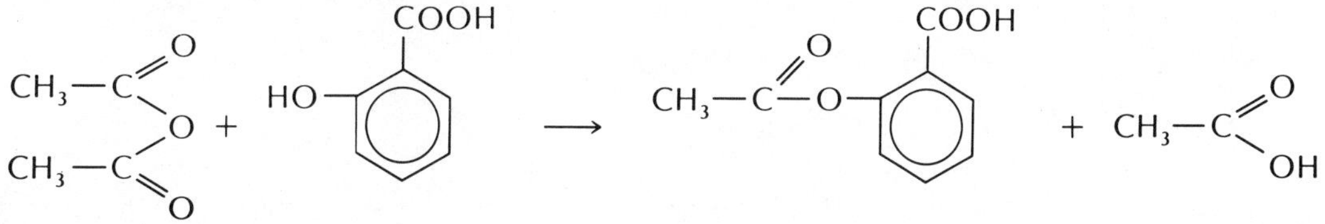

$$\underset{\text{Acetic anhydride}}{\mathrm{(CH_3{-}C({=}O))_2O}} + \underset{\text{Salicylic acid}}{\mathrm{HO{-}C_6H_4{-}COOH}} \longrightarrow \underset{\text{Aspirin}}{\mathrm{CH_3{-}C({=}O){-}O{-}C_6H_4{-}COOH}} + \underset{\text{Acetic acid}}{\mathrm{CH_3{-}C({=}O){-}OH}}$$

18.4 POLYMERS: STRUCTURE FROM CARBON ATOMS

Polymers are huge molecules that contain many repeating parts (*poly* means "many"; *mer* means "part"). Wool, hair, and rubber are natural polymers. Humans have imitated these natural polymers in synthetic polymers that have many different uses.

polymer

ADDITION POLYMERS. Painting a table with linseed oil or oil-based paint will eventually cause a hard surface to form on the table. These oils contain double bonds and form natural addition polymers when exposed to air. Synthetic addition polymers imitate them.

We've seen that substances can add to an alkene's double bond. Two or more alkenes can also add to one another's double bonds and form an *addition polymer*. The simplest synthetic addition polymer, polyethylene, comes from the reaction of many ethylene molecules with one another.

addition polymer

$$CH_2{=}CH_2 + CH_2{=}CH_2 + CH_2{=}CH_2 + \cdots \xrightarrow{\text{cat.}} \cdots CH_2CH_2CH_2CH_2CH_2CH_2 \cdots$$

Formation of polymers is called *polymerization*. The small molecules that react to produce the polymer are called *monomers*. Usually we find it convenient to write the polyethylene polymer like this: $-(CH_2CH_2)-_n$. The n stands for a large number (at least 1000) whose value isn't known exactly.

polymerization

monomer

If other groups are substituted for one or more of the hydrogen atoms on the ethylene monomer, different polymers result. All these—including polyethylene—are called *vinyl polymers*. Table 18.5 shows some common vinyl polymers. Polyethylene itself is the most important plastic made in the United States. Natural rubber and one of its synthetic relatives, neoprene rubber, can stretch because their long chains of molecules form coils. The coils straighten out when they're pulled and spring back when they're released. Notice in the table that some polymers, such as plastic wrap, are made from more than one monomer. These are called *copolymers*.

vinyl polymer

copolymer

POLYESTERS. We've already seen that an ester is a condensation product. A *polyester* is a condensation polymer. If we have a monomer that has an alcohol group on each end, and another monomer that has a carboxylic acid group on each end, then we can link them together in a continuous chain. The synthetic fiber Dacron is such a polymer.

polyester

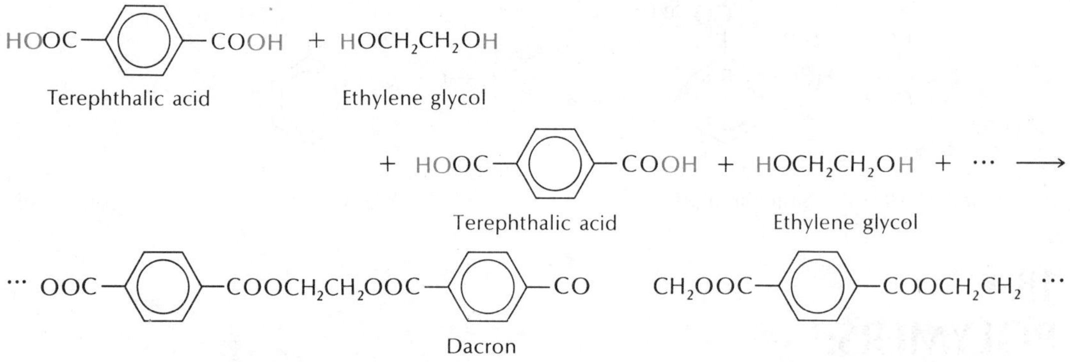

$$HOOC{-}C_6H_4{-}COOH + HOCH_2CH_2OH$$

Terephthalic acid Ethylene glycol

$$+ HOOC{-}C_6H_4{-}COOH + HOCH_2CH_2OH + \cdots \longrightarrow$$

Terephthalic acid Ethylene glycol

$$\cdots OOC{-}C_6H_4{-}COOCH_2CH_2OOC{-}C_6H_4{-}CO \quad CH_2OOC{-}C_6H_4{-}COOCH_2CH_2 \cdots$$

Dacron

To make a fiber, the polymer is extruded, or pushed through many small holes. The polymer can also be formed into sheets; then it's Mylar.

The ethylene glycol monomer is made from ethylene. The terephthalic acid monomer is made by oxidizing paraxylene.

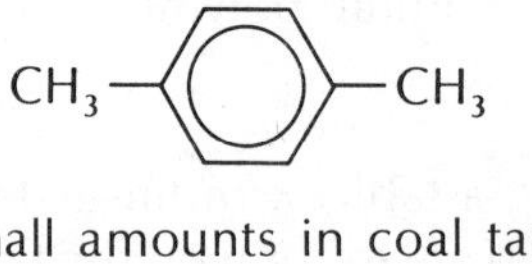

$$CH_3{-}C_6H_4{-}CH_3$$

Paraxylene is found in small amounts in coal tar, but it is in such great demand for making Dacron that ways have been found to produce it from petroleum.

NYLON. There are many kinds of nylons, but they're all condensation polymers. Nylon 66 (so named because both monomers have six carbon atoms) is made by the following reaction.

TABLE 18.5
Some vinyl polymers

Monomer	*Polymer*	*Name, Use*
Ethylene, $CH_2{=}CH_2$	$-(CH_2-CH_2)_n-$	Polyethylene. Used for squeeze bottles, toys, packaging.
Propylene, $CH_2{=}CH(CH_3)$	$-(CH_2-CH(CH_3))_n-$	Polypropylene. Used in bottles, pipes, valves, carpets.
Vinyl chloride, $CH_2{=}CH(Cl)$	$-(CH_2-CH(Cl))_n-$	Polyvinyl chloride, PVC. Used in vinyl tile, pipes, phonograph records.
Tetrafluoroethylene, $CF_2{=}CF_2$	$-(CF_2-CF_2)_n-$	Teflon. Used in cooking utensils, bearings, valves.
Acrylonitrile, $CH_2{=}CH_2(CN)$	$-(CH_2-CH(CN))_n-$	Polyacrylonitrile. Used to make Orlon and Acrilan textile fibers.
Styrene, $CH_2{=}CH(C_6H_5)$	$-(CH_2-CH(C_6H_5))_n-$	Polystyrene. Used for Styrofoam, toys, knobs.
Methyl methacrylate, $CH_2{=}CCH_3(COOCH_3)$	$-(CH_2-C(CH_3)(COOCH_3))_n-$	Polymethylmethacrylate. Lucite, Plexiglas. Used in transparent surfaces, furniture, jewelry.
Chloroprene, $CH_2{=}C(Cl)-CH{=}CH_2$	$-(CH_2-C(Cl){=}CH-CH_2)_n-$	Neoprene rubber.
Isoprene, $CH_2{=}C(CH_3)-CH{=}CH_2$	$-(CH_2-C(CH_3){=}CH-CH_2)_n-$	Natural rubber.
Vinyl chloride + vinylidene chloride, $CH_2{=}CCl_2$	$-(CH_2-CCl_2-CH_2-CH(Cl))_n-$	Plastic wrap (a copolymer).
Styrene + butadiene, $CH_2{=}CH-CH{=}CH_2$	$-(CH_2CH(C_6H_5)CH_2CH{=}CHCH_2)_n-$	Styrene-butadiene rubber. Used in automobile tires.

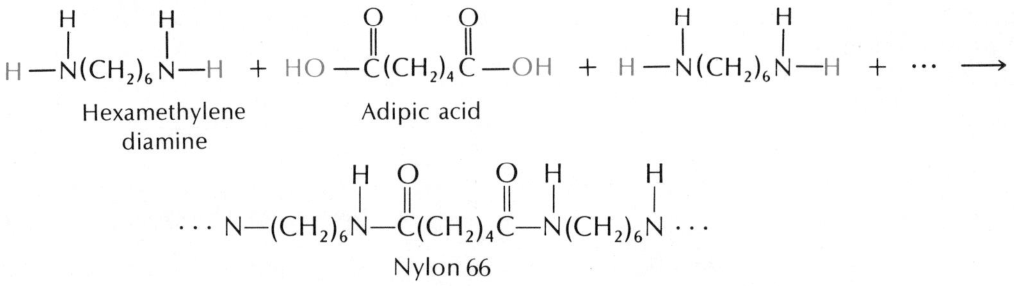

$$H{-}\overset{H}{\overset{|}{N}}(CH_2)_6\overset{H}{\overset{|}{N}}{-}H \;+\; HO{-}\overset{O}{\overset{\|}{C}}(CH_2)_4\overset{O}{\overset{\|}{C}}{-}OH \;+\; H{-}\overset{H}{\overset{|}{N}}(CH_2)_6\overset{H}{\overset{|}{N}}{-}H \;+\; \cdots \longrightarrow$$

Hexamethylene diamine   Adipic acid

$$\cdots\; N{-}(CH_2)_6\overset{H}{\overset{|}{N}}{-}\overset{O}{\overset{\|}{C}}(CH_2)_4\overset{O}{\overset{\|}{C}}{-}\overset{H}{\overset{|}{N}}(CH_2)_6\overset{H}{\overset{|}{N}}\;\cdots$$

Nylon 66

Hexamethylene diamine is an example of an amine that has two amino groups, as shown in Table 18.3. The condensation product between an amine and a carboxylic acid is called an *amide*. Nylons are all *polyamides*. As we'll see in the next chapter, proteins are polyamides, too. By using the amide linkage, nylon imitates the protein polymers wool and silk.

amide

polyamide

Both of Nylon 66's monomers are made from cyclohexane, which in turn is made by hydrogenating benzene from coal tar; nylon is a coal product.

BAKELITE. Bakelite is a condensation polymer between formaldehyde and phenol. We've already seen how formaldehyde is made from coal gas. Phenol is made from benzene; so both are coal products. In Bakelite, formaldehyde condenses with hydrogen atoms on the benzene ring of phenol.

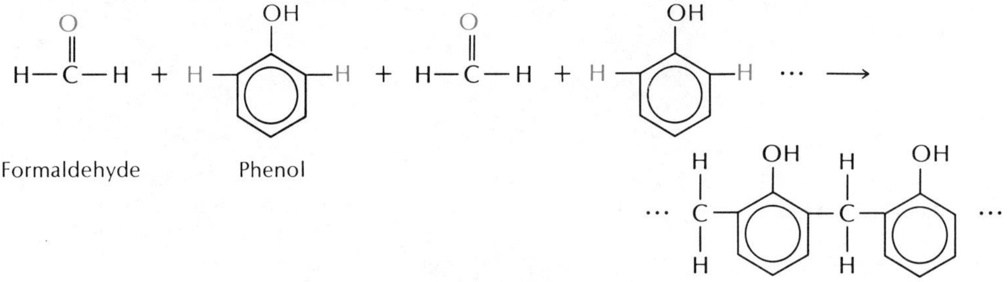

$$H{-}\overset{O}{\overset{\|}{C}}{-}H \;+\; H{-}C_6H_3(OH){-}H \;+\; H{-}\overset{O}{\overset{\|}{C}}{-}H \;+\; H{-}C_6H_3(OH){-}H \;\cdots \longrightarrow$$

Formaldehyde   Phenol

$$\cdots\;{-}\overset{H}{\underset{H}{C}}{-}C_6H_3(OH){-}\overset{H}{\underset{H}{C}}{-}C_6H_4(OH)\;\cdots$$

Condensation can also occur with hydrogen atoms that are across the ring from the OH group, as well as with those that are next to it. This lets the chains join, or *cross-link*. Bakelite is a very hard resin because it's highly cross-linked. A portion of its structure looks like this:

cross-link

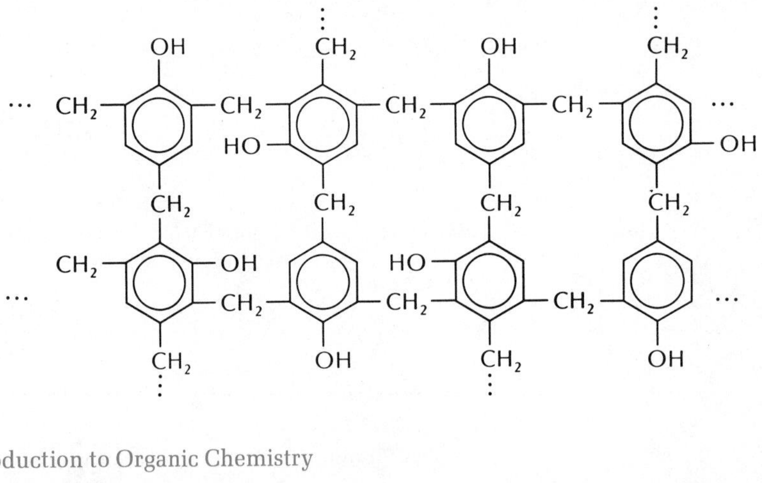

(Structure: two rows of benzene rings, bearing OH groups, joined by CH_2 bridges within each row and between the rows; chains continue ($\cdots$) at the edges.)

Bakelite is used for electrical insulation, knobs, switches, and plugs. Formica and Melmac are related polymers that are condensation products of formaldehyde with other substances. Cross-linking makes them hard and tough, too—suitable for kitchen counters and unbreakable dishes.

We've touched on the major types of polymers, but there are many others that are variations of the ones we've seen. The properties of polymers can be changed in a number of ways, including regulating the amount of cross-linking, the length of the carbon chain of monomers, and the nature of the groups attached to them.

REVIEW QUESTIONS

1. What are *organic compounds?* What are the two major sources of organic compounds, and where did they come from?

The Structures of Organic Compounds

2. What do we mean by the *structure* of a compound?
3. What is an organic compound's *carbon skeleton?* Give an example. What can happen where a substance's carbon skeleton is interrupted by another element?
4. What is the name for compounds that contain only carbon and hydrogen? What other elements are often present in organic compounds?
5. What is a *structural formula?* How is it like a Lewis structure, and how is it different? Give an example.
6. What is the difference between a *straight chain* and a *branched chain* hydrocarbon? Give examples of each.
7. What are *isomers?* Do isomers all have the same properties?
8. How do we write *condensed structural formulas?* Write one for any compound and tell how it is different from the structural formula.
9. Write condensed structural formulas for two ring compounds.
10. What do we mean by *saturated* and *unsaturated* compounds?
11. Why does carbon and no other element form the backbone of organic compounds?

Energy from Carbon Compounds

12. What is the difference between an *aromatic* and an *aliphatic* hydrocarbon? What is the main feature of an aromatic hydrocarbon?
13. What are *alkanes, alkenes,* and *alkynes?* How are they alike? How are they different?
14. What is an *alkyl group?* Give some examples.
15. What are the compounds in *coal gas?* Where does coal gas come from? Name one of its uses.
16. What is a substance's *heat of combustion?* How does structure affect a hydrocarbon's heat of combustion?
17. What is the purpose of petroleum distillation?
18. What are *reforming* and *cracking?* What are their purposes?

Function from Carbon Compounds

19. What are *functional groups?* Write condensed structural formulas for five, and give their names.
20. How are alcohols related to water? Give examples of some waterlike chemical and physical properties of alcohols.
21. What are *dehydration* and *condensation?* In equation form, give an example of each involving an alcohol.
22. How are ethers related to water? Why are they less like water than alcohols are?
23. Write formulas for a primary, a secondary, and a tertiary amine, and explain how they are related to ammonia.
24. What reactions of amines are like those of ammonia? Write an equation.
25. What is *substitution?* Show how substitution reactions are used to make one kind of Freon.
26. How are the raw materials ethylene and propylene obtained from petroleum? Why are ethane and propane not as useful as starting materials?
27. How is *addition* different from substitution? Explain how a reactant can add to a double bond, and give an example.
28. What is *hydrogenation?* How is it used? Give an example.
29. Arrange aldehyde, acid, and alcohol in order of increasing oxidation number of the functional group carbon atom, and write condensed structural formulas for their functional groups.

Write the equations for an industrial process that takes advantage of this sequence.

30. What is the difference between an aldehyde and a ketone? What type of compound must be oxidized to obtain a ketone?
31. What is the product of condensation between an acid and an alcohol? Write the formula of one. What is it used for?
32. What is *hydrolysis?* Write an equation for a hydrolytic reaction.
33. Write the structure of an *anhydride.* How are anhydrides useful?
34. Write the equation for the preparation of aspirin.

Polymers: Structure from Carbon Atoms

35. What are *polymers?* What are some natural polymers?
36. Give an example of an *addition polymer.* What is the *monomer?*
37. Name some *vinyl polymers* and their monomers.
38. What is a *copolymer?* Give an example.
39. What kind of a polymer is Dacron? Draw part of its structure.
40. Write part of the structure for Nylon 66 and identify the *amide* linkages.
41. What is a *cross-linked* polymer? Show part of the structure of one.

EXERCISES

Set A (Answers at back of book. Asterisks mark the more difficult exercises.)

1. Make hydrocarbons out of the following carbon skeletons by putting hydrogen atoms wherever they are needed.

```
a. C—C—C—C—C          c.  C=C—C—C

         C                    C
         |                 C     C
b. C—C—C—C            d.   |     |
         |                 C     C
         C                    C
```

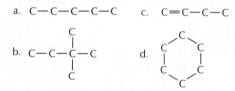

2. Write condensed structural formulas for the following.

```
      H  H       H  H  H
      |  |       |  |  |
a. H—C—C—————C—C=C—H
      |  |       |
      H  H       |
              H—C—H
                 |
                 H

         H
         |
         C        H  H  H
       //  \      |  |  |
b. H—C      C——————C—C=C—H
      |     ||     |
   H—C      C—H    H
       \\  /
         C
         |
         H

      H  H     H
      |  |     |
c. H—C=C—C=C—H
           |
        H—C—H
           |
           H
```

3. Write full structural formulas for the following condensed structural formulas.

```
    OH    OH    OH
    |     |     |
a.  CH2—CH2—CH2
```

Glycerin

```
           O
          //
b. CH3—C
          \
           O(CH2)4CH3
```

Pentyl acetate (banana flavoring)

```
       OH
       |
c. CH3CH—COOH
```

Lactic acid (acid in sour milk)

4. Identify the functional groups in each compound in Exercise 3.
5. Write as many isomers as you can for the formula C_6H_{14}.
6. Classify each of the following as an alkane, alkene, or alkyne.

a. $CH_2{=}CH{-}CH{=}CH_2$

Butadiene

b. [hexagon]

Cyclohexane

c. $HC{\equiv}CH$

Acetylene

7. Which compound in each of the following pairs would have the higher heat of combustion (kilocalories/gram)?

   a. $HC{\equiv}CCH_3$ or $CH_3{-}CH_2{-}CH_3$

   b. CH_3CH_3 or [cyclohexane]

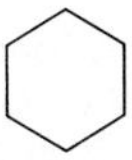

   c. $CH_2{=}C{=}CH_2$ or $CH_2{=}CH{-}CH_3$

8. Which compound in each of the following pairs would have the higher octane rating?

   a. $CH_3CH_2CH_2CH_2CH_3$ or [benzene]

   b. 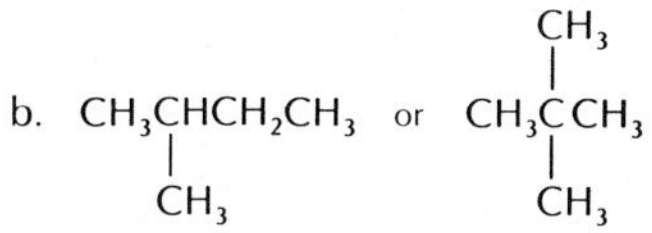 $CH_3CH(CH_3)CH_2CH_3$ or $CH_3C(CH_3)_2CH_3$

   c. CH_4 or $CH_3CH_2CH_2CH_3$

9. Which of the following would be expected to be more water soluble?
   a. CH_3OH or $CH_3(CH_2)_4OH$
   b. CH_3OH or CH_3OCH_3
10. State which compounds in the previous exercise would be expected to have the higher boiling point, and explain.
11. Write complete, balanced equations showing these.
    a. methyl and ethyl alcohols undergoing condensation with each other
    b. butyl alcohol undergoing dehydration
    c. propyl alcohol reacting with sodium metal
12. Write formulas for ethylamine, diethylamine, and triethylamine, and write an equation showing one of them acting as a base.
13. The first step in the manufacture of Freon 115, $CClF_2CF_3$, is the chlorination of ethane. Write the equation.
14. Substitute the correct compound for each question mark.

    a. $? + H_2O \not\rightarrow CH_3{-}CH(OH){-}CH_3$

       Isopropyl alcohol (rubbing alcohol)

    b. $CH_2{=}CH_2{-}CH_3 + Cl_2 \not\rightarrow ?$

       Propylene

    c. 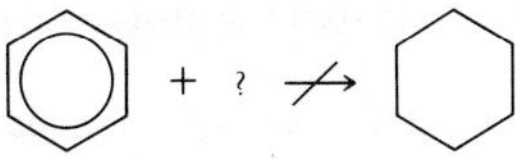 [benzene] $+ ? \not\rightarrow$ [cyclohexane]

    d. $CH_3{-}CH_2{-}CH(OH){-}CH_3 + \frac{1}{2}O_2 \not\rightarrow$

15. Fill in the blanks in the following table of related alcohols, aldehydes or ketones, and acids.

Alcohol	Aldehyde or Ketone	Acid
$CH_3{-}OH$	$H{-}CHO$	$H{-}COOH$
$CH_3(CH_2)_2{-}OH$	___	___
___	$CH_3(CH_2)_2C({=}O)CH_3$	None
___	___	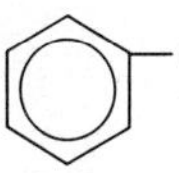 [benzene ring]$-COOH$

16. Write condensation products for the following pairs.

    a. $CH_3{-}CH_2{-}C({=}O){-}OH$ and $CH_3{-}CH_2{-}OH$

    b. $CH_3{-}OH$ and $H{-}C({=}O){-}OH$

17. Write formulas for the alcohol and the carboxylic acid that are parents of these esters.

    a. $CH_3{-}(CH_2)_{14}{-}C({=}O){-}O(CH_2)_9CH_3$

       Myricyl palmitate (beeswax)

    b. 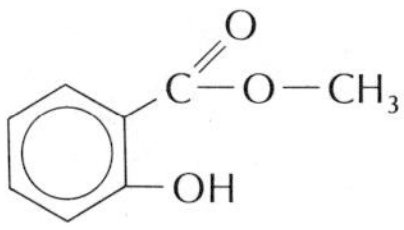 [benzene ring with $-C({=}O){-}O{-}CH_3$ and ortho $-OH$]

       Methyl salicylate (wintergreen)

18. Write equations for the hydrolysis of the esters in Exercise 17.
19. Oxalic acid, HOOC—COOH, is the simplest acid that has two carboxylic acid groups. Write an equation and show the structure of a polyester

that might be made from oxalic acid and ethylene glycol.

20. A natural material used to make nylon is sebacic acid, $HOOC(CH_2)_8COOH$, found in castor oil. Sebacic acid and hexamethylenediamine, $NH_2(CH_2)_6NH_2$, make nylon 610. Write an equation showing condensation and the product's partial structure.

*21. Melmac, a polymer similar to Bakelite, is made with formaldehyde and melamine. Melamine has this structure:

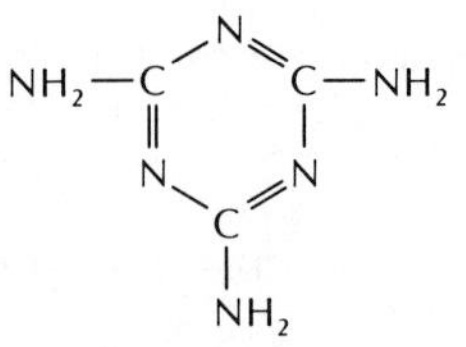

The $-NH_2$ groups each use one of their hydrogen atoms to condense with the oxygen of the formaldehyde. Write a partial structure of this polymer, showing cross-linking.

Set B (Answers not given.)

1. Make hydrocarbons out of these carbon skeletons by putting hydrogen atoms on wherever they are needed.
   a. 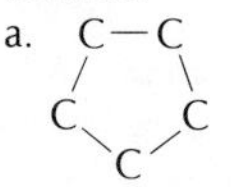 (five-membered ring of C atoms)
   b. 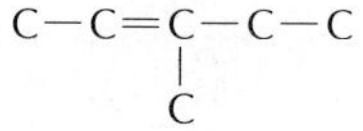 C—C=C—C—C with C on the third carbon
   c. C—C—C—C with C on the second carbon
   d. C=C—C=C

2. Write condensed structural formulas for these.
   a. 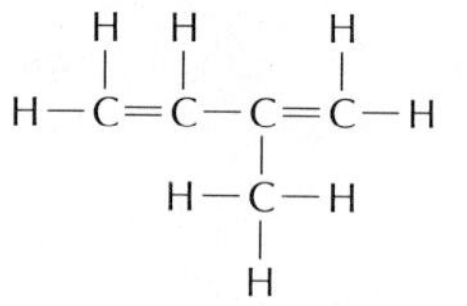 H—C(H)=C(H)—C(CH₃ as H—C—H)=C(H)—H
   b. 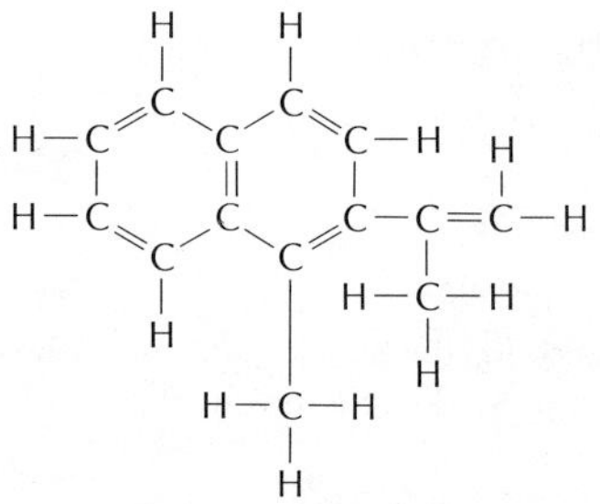 (naphthalene ring with H substituents, a —C(H)=C—H group bearing H—C—H, and an H—C—H group)

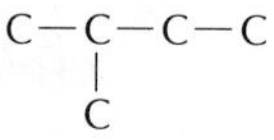

3. Write full structural formulas for these condensed structural formulas.
   a. 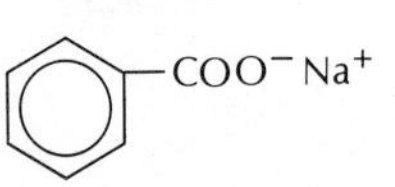 (benzene ring)—$COO^- Na^+$
   Benzoate of soda (preservative)
   b. 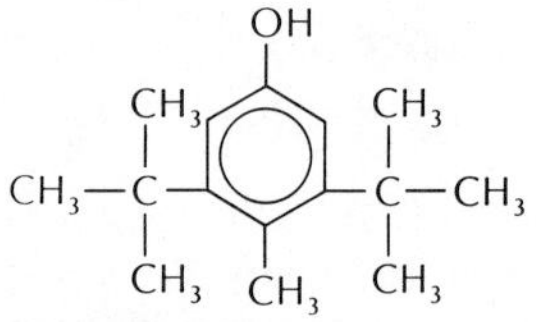 (benzene ring with OH, two $CH_3-C(CH_3)_2-$ groups, and CH_3)
   BHT (preservative)

c. CCl_2FCClF_2
Freon-113

4. Identify the functional groups in each compound in Exercise 3.

5. Which of the following are isomers?
   a. $CH_3-CH_2-CH_2-CH_2-CH_3$
   b. 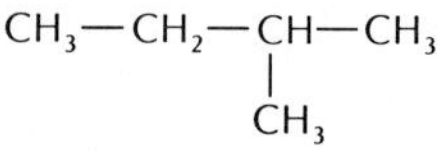 $CH_3-CH_2-CH(CH_3)-CH_3$
   c. 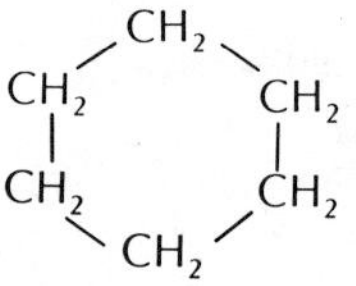 (ring of six CH_2 groups)
   d. 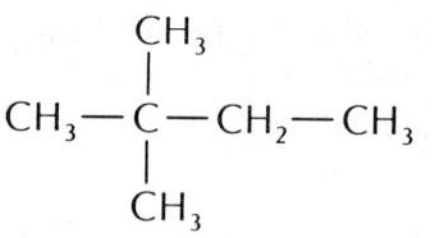 $CH_3-C(CH_3)_2-CH_2-CH_3$
   e. $CH_2=CH-CH_2-CH_2-CH_2-CH_3$
   f. 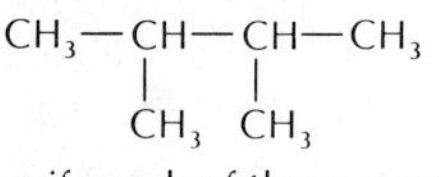 $CH_3-CH(CH_3)-CH(CH_3)-CH_3$

6. Classify each of these as an alkane, alkene, or alkyne.
   a. 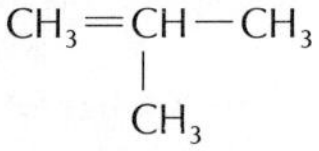 $CH_3=CH-CH_3$ with CH_3 on the second carbon
   Isobutene
   b.  (cyclohexene ring)
   Cyclohexene
   c. 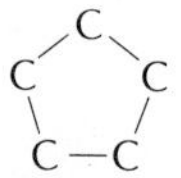 (five-membered ring of C atoms)
   Cyclopentane

7. Which compound in each of these pairs would have the higher heat of combustion (kcal/g)?
   a. $CH_2=CH-CH=CH_2$ or (benzene ring)

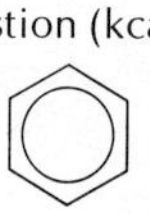

b. CH_3 (on a benzene ring) or CH_4

c. CH_3CH_3 or $CH_3{-}C(CH_3)(=CH_2){-}C{=}CH_2$

8. Which compound in each of these pairs would have the higher octane rating?

   a. $CH_3{-}CH_3$ or $CH_3(CH_2)_5CH_3$

   b. $CH_3(CH_2)_4CH_3$ or cyclohexane

   c. $CH_3(CH_2)_3CH(CH_3){-}CH_3$ or $CH_3{-}C(CH_3)(CH_3){-}CH(CH_3){-}CH_3$

9. Which of these would be expected to be more water soluble?
   a. $CH_3CH_2CH_2CH_2OH$ or CH_3CH_2OH
   b. $CH_3CH_2CH_2OH$ or CH_3OCH_3
10. State which compounds in the previous exercise would be expected to have the higher boiling point, and explain.
11. Write complete, balanced equations showing the following.
    a. methyl alcohol reacting with potassium metal
    b. propyl alcohol undergoing dehydration
    c. methyl alcohol undergoing condensation
12. Why can't tertiary amines undergo condensation reactions?
13. The plastic Teflon is made of tetrafluoroethylene, $CF_2{=}CF_2$, which is manufactured by this reaction:

$$\underset{\text{Chloro-difluoromethane}}{2\,CHClF_2} \xrightarrow{\Delta} CH_2{=}CF_2 + 2\,HCl$$

    Suggest a sequence of reactions that might be used to make chlorodifluoromethane from methane.
14. Substitute the correct compound for each question mark.

    a. $CH_3{-}CH_2{-}CH_2{-}OH + \frac{1}{2}O_2 \not\rightarrow ?$

    b. $? + \frac{1}{2}O_2 \not\rightarrow H{-}C({=}O){-}OH$

    c. $C_6H_5{-}CHO + \frac{1}{2}O_2 \not\rightarrow ?$

    d. $? + Cl_2 \not\rightarrow$ 1,2-dichlorocyclohexane (cyclohexane ring bearing two Cl)

15. Fill in the blanks in this table of related alcohols, aldehydes or ketones, and acids.

Alcohol	*Aldehyde or Ketone*	*Acid*
___	$C_6H_5{-}CH_2CHO$	___
$CH_3CH(CH_3){-}OH$	___	None
___	___	$CH_3CH_2CH_2COOH$

16. Write condensation products for these pairs.
    a. H_2SO_4 and $CH_3{-}(CH_2)_8{-}CH_2{-}OH$

    b. cyclohexyl${-}COOH$ and $CH_3{-}CH(CH_3){-}OH$

17. Write formulas for the alcohol and the carboxylic acid that are parents of these esters.

    a. $CH_3{-}C({=}O){-}O{-}CH_2{-}CH_3$

    Ethyl acetate (nail polish remover)

    b. $CH_3{-}CH_2{-}C({=}O){-}OCH_2{-}CH(CH_3){-}CH_3$

    Isobutyl propionate (rum flavor)

18. Write equations for the hydrolysis of the esters in Exercise 17.
19. Kodel is a polymer made from these substances. Write an equation and show the structure of this polymer.

    $HOOC{-}C_6H_4{-}COOH$ (para-benzene ring) $\quad HO{-}CH_2{-}C_6H_{10}{-}CH_2OH$ (para-cyclohexane ring)

20. Nylon 6 is made from just one monomer:

$$H_2N(CH_2)_5COOH$$

    Write an equation for the formation of Nylon 6 and write a part of its structure.

CHAPTER 19

Introduction to Biochemistry

LEARNING OBJECTIVES

After studying this chapter, you should be able to:

1. Supply a correct definition, example, or explanation for each of these:

biochemistry	denature
organelle	genetic defect
gene	substrate
nucleus	active site
nucleic acid	coenzyme
enzyme	carbohydrate
replication	monosaccharide
deoxy-ribonucleotide	disaccharide
	polysaccharide
transcription	lipid
nucleotide	triglyceride
protein	steroid
polypeptide	hormone
antibody	photosynthesis
amino acid	metabolism
genetic code	respiration
codon	mitochondrion
anticodon	vitamin
fibrous protein	digestion
disulfide bridge	excretion
globular protein	diuretic

2. Describe DNA structure and replication and write or recognize complementary DNA strands or pairs.
3. Describe RNA structure and transcription. List the three types of RNA and their functions. Write or recognize complementary DNA–RNA strands or pairs.
4. Recognize structures of nucleotides and their component parts (bases, sugar, phosphate) and describe how they are bonded in nucleic acids.
5. From its formula, recognize whether a given amino acid side chain is ionic, polar, nonpolar, or forms a disulfide bridge, and predict the effect of a given set of amino acids on protein structure and solubility.
6. Describe protein structure and synthesis.
7. Recognize or explain the difference in structure between starch and cellulose, and give the functional consequences of this difference.
8. Explain the biological cycle and the chemical importance of each component of it.
9. Recognize the structure of ATP and of hemoglobin and describe their roles in the respiratory chain. Explain the function of some vitamins in terms of their relationships to respiratory coenzymes.
10. Describe the digestion of starch, cellulose, protein, and fat.
11. Recognize the structures of adrenalin, vasopressin, and urea and explain their roles in metabolism and excretion.

A baby is born. Suddenly there is a human being where once there was not. And yet, as amazing as birth is, the events leading up to it are even more amazing. At the moment of conception, cells from each parent fuse to form a single fertile egg cell, destined to become a human being. From that single cell come arms, legs, eyes, nose, ears. From that single cell comes the ability to walk, talk, see, hear, think.

A tiny seed sprouts in the ground. Nourishing itself on simple inorganic materials in the soil and air and on the sun's energy, the seedling builds complex structures. After a time, there is a huge tree where once there was not. From a single cell contained in the seed come trunk, roots, stems, leaves, flowers. From a single cell comes the ability to produce, or synthesize, large organic molecules from inorganic compounds.

A tree and a human being are very different, but they and all other living organisms share the same ingredients. From a single cell of each comes the ability to grow, reproduce, and repair. Molecules within the single cell in a fertile egg or a seed determine that a tree will be a tree and a person a person. As the organisms grow, more molecules are formed. As more molecules are formed, the organisms grow. Life from molecules, and molecules from life—that's what *biochemistry* is about.

biochemistry

19.1 NUCLEIC ACIDS

Figure 19.1 is a diagram of a typical cell. The outside is a semipermeable membrane, which lets specific molecules pass into and out of the cell. Most of the inside is water, with many small molecules dissolved in it. Various solid bodies, called *organelles,* are dotted all around the cell like islands. Each organelle has a function.

organelle

FIGURE 19.1
A generalized animal cell

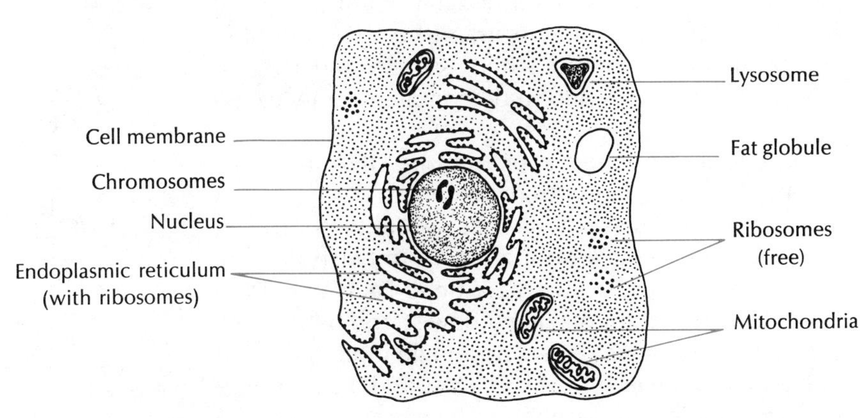

A newly formed cell comes with packages of information, called *chromosomes.* The information tells the cell how to make others just like it and also how to make cells with different functions—eye cells, for instance, or liver or leaf cells. The cell makes other cells by first making proteins. Some of the proteins will form the structural parts of the new cells, whereas others will control or regulate the production of other kinds of molecules that provide structure or function. Within the chromosomes are *genes*—directions for making individual protein molecules. A human being has forty-six chromosomes, which contain over a hundred thousand genes.

chromosome

gene

The chromosomes are located in the *nucleus,* the organelle that contains the cell's "brains." The molecules in the chromosomes that contain information, plus others found in the nucleus, are called *nucleic acids.*

nucleus

nucleic acid

DNA STRUCTURE AND REPLICATION. As soon as the cell is formed, long slender threads unravel from each chromosome. These threads are the nucleic acid *deoxyribonucleic acid* (DNA). The DNA in human chromosomes, originally contained in a nucleus only 8×10^{-4} centimeters in diameter, has a combined length of over 2 meters when it's all unraveled.

deoxyribonucleic acid (DNA)

DNA has two functions. The first, which it performs when the cell isn't dividing, is to direct the synthesis of proteins. Some of these proteins are catalysts called *enzymes.* Every reaction in a cell needs a special enzyme to make it go, and thus enzymes are second in command to DNA when it comes to getting things done in the cell.

enzyme

DNA's second function is *replication,* or copying itself. It stops directing protein synthesis and starts replicating as soon as the cell is ready to begin dividing. When the DNA has completely replicated itself, the two sets that result pack themselves back up into separate chromosomes and the cell divides. Each of the two new cells will have a copy of the directions.

replication

This is a double helix.

A closer look at a strand of DNA reveals that it's actually two strands twisted together in a double helix (or spiral). This structure of DNA was discovered in 1953 by James Watson and Francis Crick. DNA is a polymer whose monomers are *deoxyribonucleotides.* There are four kinds of these monomers, and we'll look at their structures a little later. Right now we'll abbreviate them as dA, dT, dC, and dG. The two strands in a DNA double helix are held together by hydrogen bonding between dA and dT and between dC and dG. Each DNA strand is said to be complementary to the other.

deoxyribonucleotide

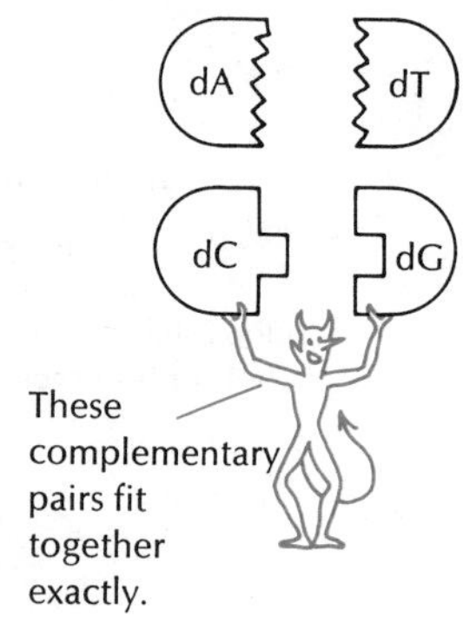

These complementary pairs fit together exactly.

The order in which these four monomers appear in the DNA molecule is the code that will later tell the cells how to make proteins. When DNA is replicated, this order must be copied exactly so that the right information gets into the new cells. To do this, the DNA double helix untwists and the two strands separate from each other, a little at a time. Each untwisted single DNA strand is used as a model for making a new strand, which will be complementary to its model. To make the strand, single monomers that are dissolved in the cell's water come up to the DNA strand. Each monomer hydrogen-bonds to its complement on the model strand. When two

FIGURE 19.2
DNA replication

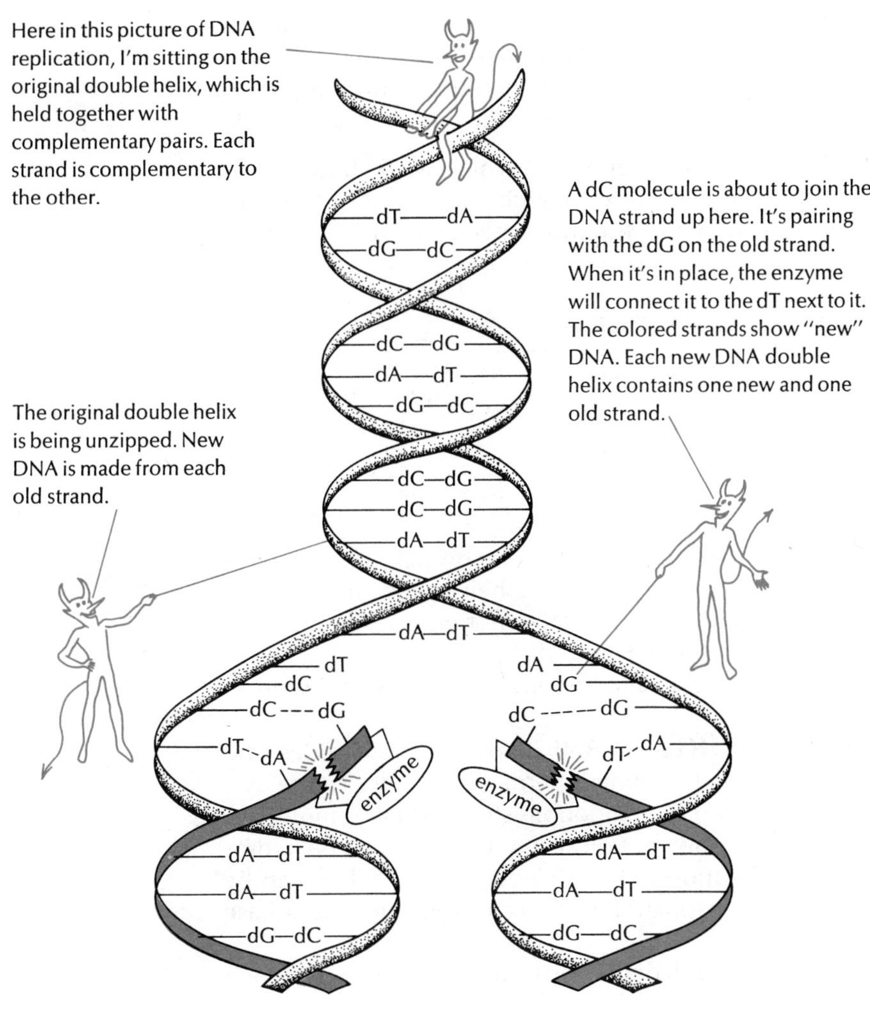
Here in this picture of DNA replication, I'm sitting on the original double helix, which is held together with complementary pairs. Each strand is complementary to the other.
A dC molecule is about to join the DNA strand up here. It's pairing with the dG on the old strand. When it's in place, the enzyme will connect it to the dT next to it. The colored strands show "new" DNA. Each new DNA double helix contains one new and one old strand.
The original double helix is being unzipped. New DNA is made from each old strand.
dT—dA
dG—dC
dC—dG
dA—dT
dG—dC
dC—dG
dC—dG
dA—dT
dA—dT
dT
dC
dC --- dG
dT-- dA
enzyme
dA
dG
dC ---- dG
dT-- dA
enzyme
dA—dT
dA—dT
dG—dC
dA—dT
dA—dT
dG—dC

new monomers are hydrogen-bonded next to each other on the model, then an enzyme comes along and fastens the monomers to each other, forming the new strand. Figure 19.2 illustrates DNA replication.

RNA STRUCTURE AND TRANSCRIPTION.

Ribonucleic acid (RNA) is the middleman between DNA and protein. It reads the orders on DNA and helps them to be carried out. Unlike DNA, which is carried along from one cell to another, RNA is made as it's needed. The process of making RNA is called *transcription,* because the directions on the DNA are being transcribed into a form that's used directly to make proteins.

ribonucleic acid (RNA)

transcription

RNA is a polymer made of monomers called *ribonucleotides*. There is only one kind of DNA, but there are three kinds of RNA. ribonucleotide

1. Ribosomal RNA (rRNA) goes into making ribosomes, one kind of organelle in Figure 19.1. *Ribosomes* are where protein synthesis occurs. ribosome
2. Messenger RNA (mRNA) carries the code for making a specific protein.
3. Transfer RNA (tRNA) fetches a protein monomer unit from somewhere in the cell and takes it to the ribosome where the protein is assembled.

As with DNA, there are four RNA monomers, which we'll abbreviate A, U, C, and G, with a different letter prefix for each kind of RNA. There are sets of complementary pairs between the monomer units of DNA and of RNA. These are:

DNA	*rRNA*	*mRNA*	*tRNA*
dA	rU	mU	tU
dT	rA	mA	tA
dC	rG	mG	tG
dG	rC	mC	tC

All RNA is single-stranded. It's made from DNA by pairing. The DNA double helix unzips and RNA monomers bind to their mates on one of the DNA strands. Then an enzyme bonds the RNA monomers together. The process of transcription is illustrated in Figure 19.3.

Figure 19.4 summarizes the relationships among DNA, RNA, and protein.

NUCLEOTIDE STRUCTURE. Both deoxyribonucleotides and ribonucleotides are included in the general term *nucleotide*. Figure 19.5 shows the chemical structure of part of a double-stranded DNA molecule, with the nucleotide monomers shown. Both DNA and RNA can be hydrolyzed to their individual nucleotides. The nucleotides themselves can be hydrolyzed to three parts: a phosphate ion, a sugar molecule, and a base molecule. nucleotide

The sugar is ribose in RNA and deoxyribose in DNA. Here are their structures:

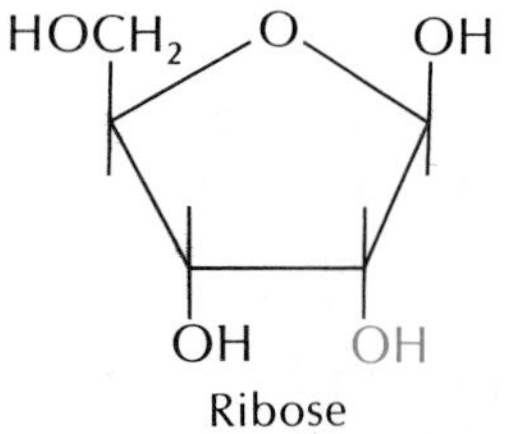

$HOCH_2$ O OH

OH OH

Ribose

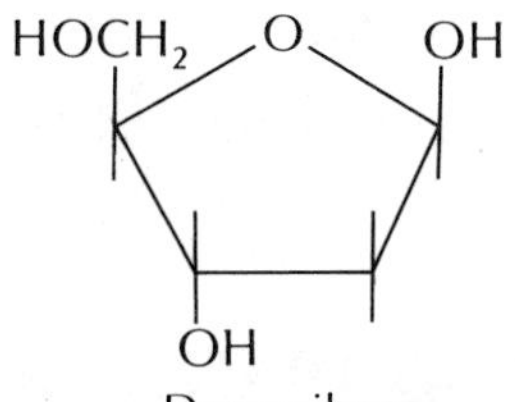

$HOCH_2$ O OH

OH

Deoxyribose

This shorthand notation is slightly different from that used in other ring compounds. If we imagine that we're looking at the ring sideways, then the groups at the ends of the straight lines can be seen as sticking up and down from the ring. If there is no group shown, we assume it's a hydrogen.

FIGURE 19.3
RNA transcription

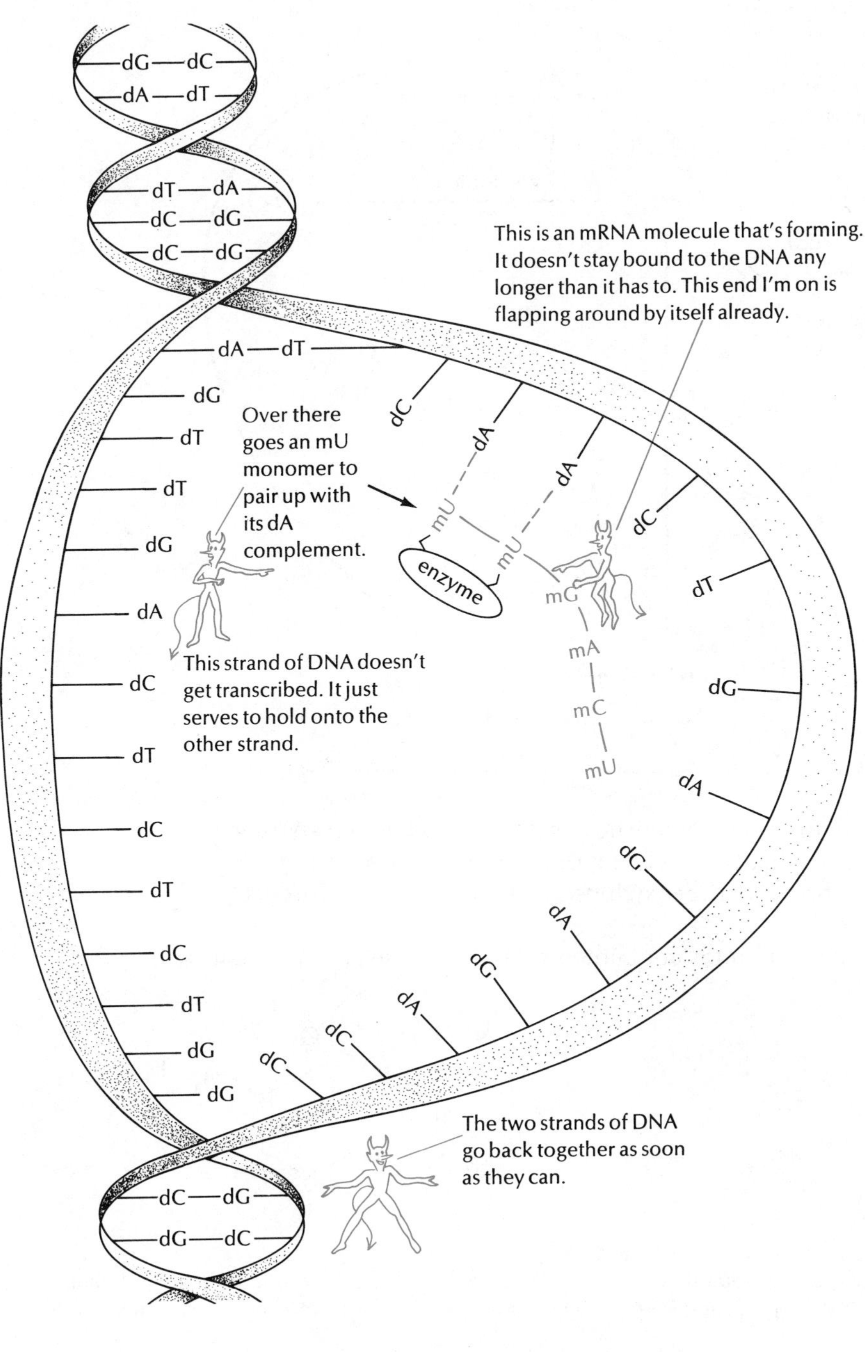

FIGURE 19.4
The relationships among DNA, RNA, and protein

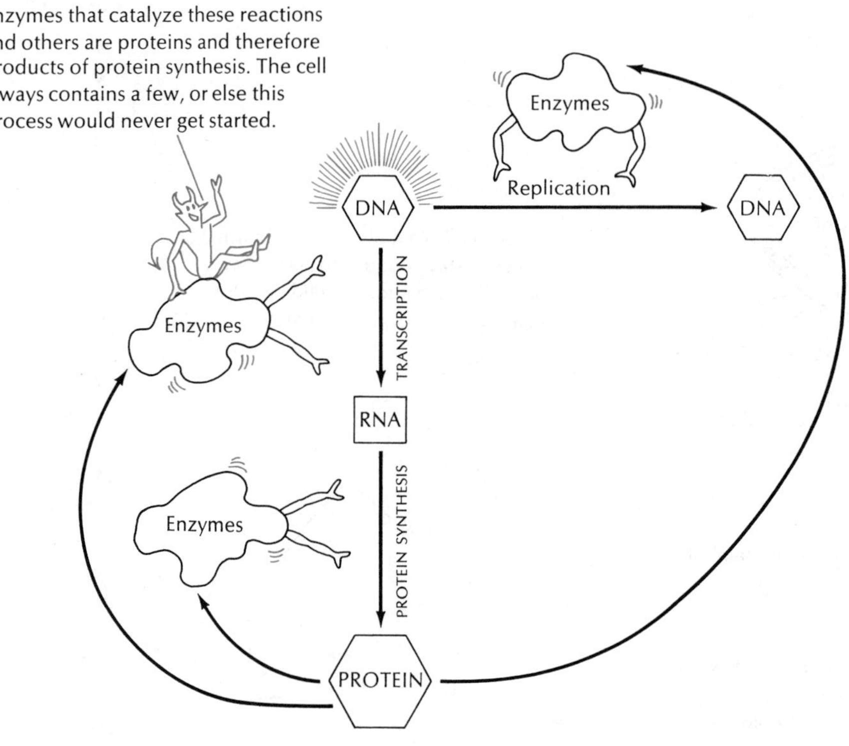

Enzymes that catalyze these reactions and others are proteins and therefore products of protein synthesis. The cell always contains a few, or else this process would never get started.
Enzymes
Replication
DNA
DNA
Enzymes
TRANSCRIPTION
RNA
PROTEIN SYNTHESIS
Enzymes
PROTEIN

This type of ring structure is characteristic of sugars. They contain an oxygen as part of the ring and usually an —OH group on nearly every carbon. The difference between ribose and deoxyribose is the presence or absence of the OH shown in color.

The bases are ring compounds containing nitrogen. These are the five in both DNA and RNA.

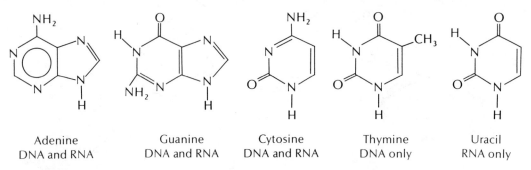

Adenine	Guanine	Cytosine	Thymine	Uracil
DNA and RNA	DNA and RNA	DNA and RNA	DNA only	RNA only

We can think of nucleotides as being formed from their parts by condensation. One example is:

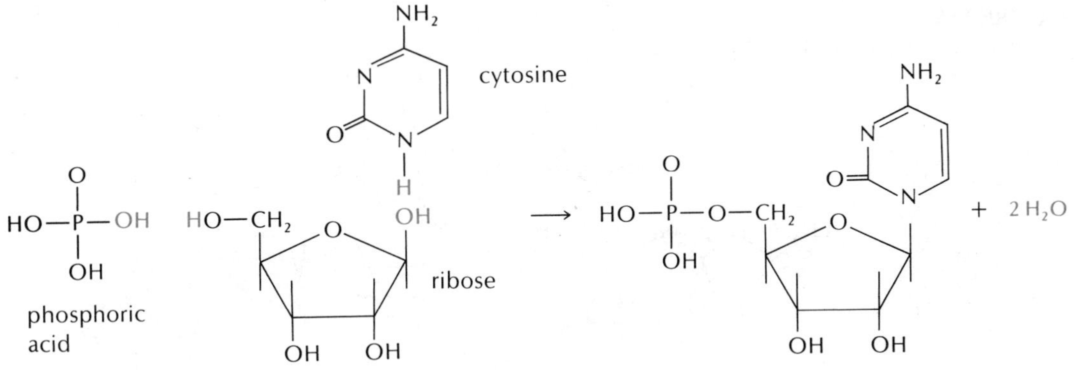

cytosine

phosphoric acid

ribose

$$HO-P(=O)(OH)-OH + HO-CH_2\text{-ribose-cytosine} \longrightarrow HO-P(=O)(OH)-O-CH_2\text{-ribose-cytosine} + 2\,H_2O$$

Cytosine monophosphate (CMP)

The correct names for all the nucleotides, and what they contain, are shown in Table 19.1.

The individual nucleotides join together by condensation, where colored atoms are shown, like this:

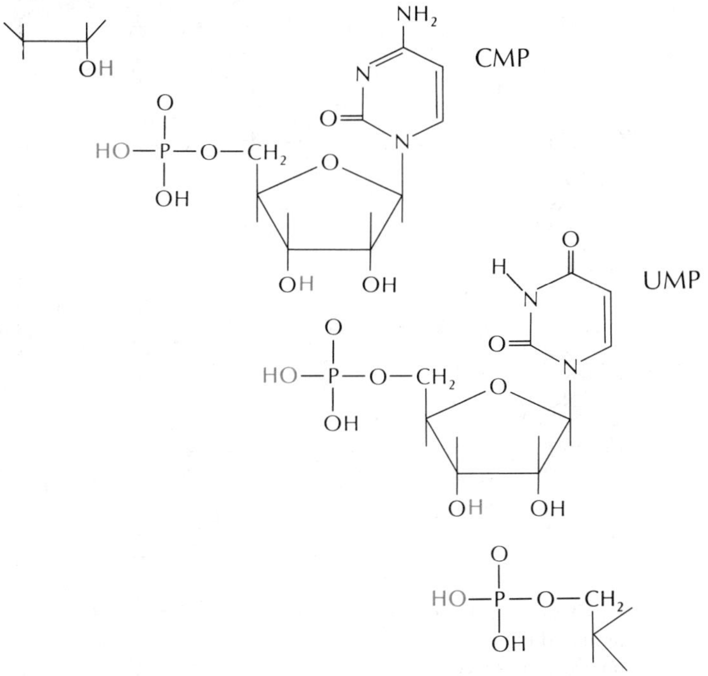

CMP

UMP

19.2 PROTEINS

The word "protein" comes from the Greek word "proteios" meaning "of first importance." *Proteins* are high-molecular-weight (from about six thousand to forty million atomic mass units) polymers whose monomers are connected by amide linkages. The same type of compounds having lower

protein

FIGURE 19.5
The chemical structure of DNA

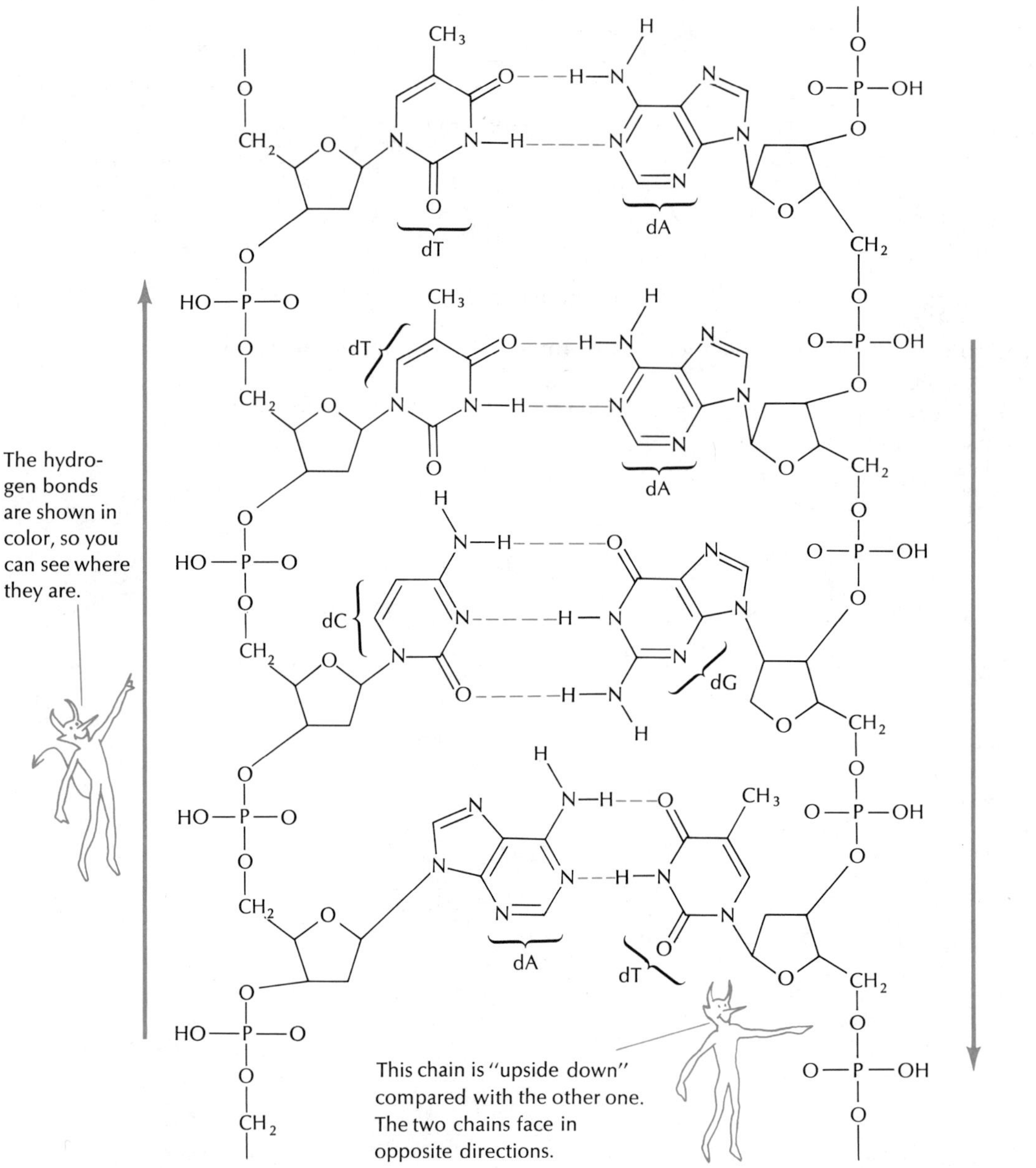

The hydrogen bonds are shown in color, so you can see where they are.

This chain is "upside down" compared with the other one. The two chains face in opposite directions.

molecular weight are called *polypeptides.* About 50 percent of the dry weight of most organisms is protein.

polypeptide

We've already been introduced to one important class of proteins: the enzymes. But proteins perform other functions besides catalyzing reactions.

TABLE 19.1
The composition of nucleotides

Name and Abbreviation[a]	*Where Found*	*Sugar*	*Base*
Adenosine monophosphate (AMP)	RNA	Ribose	Adenine
Cytidine monophosphate (CMP)	RNA	Ribose	Cytosine
Guanosine monophosphate (GMP)	RNA	Ribose	Guanine
Uridine monophosphate (UMP)	RNA	Ribose	Uracil
Deoxyadenosine monophosphate (dAMP)	DNA	Deoxyribose	Adenine
Deoxycytidine monophosphate (dCMP)	DNA	Deoxyribose	Cytosine
Deoxyguanosine monophosphate (dGMP)	DNA	Deoxyribose	Guanine
Thymidine monophosphate (TMP)	DNA	Deoxyribose	Thymine

[a]These abbreviations replace the A,C,G,U, and T, etc., used earlier in the chapter.

Proteins in the blood, called *antibodies,* protect us from disease by combining with foreign substances. Other proteins are organic buffers in the blood, supplementing the inorganic buffers we learned about in Chapter 16. The hemoglobin that carries our life-giving oxygen is mostly protein.

antibody

Besides function, proteins provide structure to the greater part of animal cells. All the organelles shown in Figure 19.1 contain varying amounts of protein. The outer packaging of the chromosomes is protein. Ribosomes contain protein as well as RNA. On a larger scale, skin, hair, and muscle tissue are mostly protein. Bone is a protein network filled in with calcium phosphate.

AMINO ACIDS AND PROTEIN SYNTHESIS. In Chapter 18, we saw that nylon is a polymer containing amide linkages. In fact, nylon is an imitation of wool, a protein fiber. Unlike nylon, which is made of just one or two monomers, proteins can be made from any of twenty-six monomers. The monomers in proteins are *amino acids* and have this general structure:

amino acid

```
NH2
|
CH—side chain
|
COOH
```

The amide linkages in proteins are formed by condensation between amino acid molecules, like this:

```
H       O          H       O          H       O
|       ‖          |       ‖          |       ‖
N—CH—C       +  N—CH—C       +  N—CH—C       + ··· ⟶
|   |    \OH     |   |    \OH     |   |    \OH
H   |            H   |            H   |
    side chain       side chain       side chain

          O               O               O               O
          ‖               ‖               ‖               ‖
··· NH—CH—C—NH—CH—C—NH—CH—C—NH—CH—C ···
       |               |               |               |
       side chain      side chain      side chain      side chain
```

As with any condensation product, the new linkages provide points where the substance can be hydrolyzed.

Each of the twenty-six amino acids has a different side chain: these can contain nonpolar (hydrocarbon) groups, polar groups, or ionic groups. The names and side chains of some amino acids are shown in Table 19.2. Side chains that have acidic (—COOH) or basic ($—NH_2$) groups react with each other or with water to form their ionic conjugate bases ($—COO^-$) or conjugate acids ($—NH_3^+$). In this way, side chains with ionic groups are formed.

protein synthesis

From these twenty-one monomer units, *protein synthesis* takes place in this sequence. (1) The messenger RNA (mRNA), freshly transcribed from DNA, fastens itself to a ribosome. (2) Elsewhere in the cell, each transfer RNA (tRNA) molecule binds an amino acid and carries it to the mRNA on the ribosome. (3) The tRNA molecules fasten themselves to the mRNA. (4) An enzyme forms the amide linkage between adjacent amino acids. (5) The mRNA moves along the ribosome. (6) After its amino acid is joined to the protein chain, each tRNA is released and goes to find an identical amino acid. (7) The process goes on until the whole protein chain is formed.

genetic code

There are at least twenty-six different tRNA molecules, each having a shape that lets it bind to one and only one kind of amino acid. The mRNA tells the tRNA where to put its amino acid by means of the *genetic code*. In the genetic code, a sequence of three nucleotides on mRNA—

TABLE 19.2
Amino acid side chains

Nonpolar

Name	*Abbr.*	*Side Chain*
Glycine	Gly	$-H$
Alanine	Ala	$-CH_3$
Valine	Val	$-CH-(CH_3)_2$
Leucine	Leu	$-CH_2-CH-(CH_3)_2$
Isoleucine	Ileu	$-CH(CH_3)-CH_2-CH_3$
Proline[a]	Pro	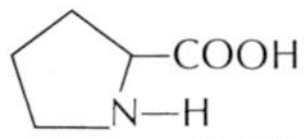 (ring) $-COOH$, $N-H$
Phenylalanine	Phe	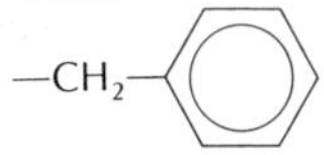 $-CH_2-$(benzene ring)
Methionine	Met	$-CH_2CH_2SCH_3$

Polar (Uncharged)

Name	*Abbr.*	*Side Chain*
Serine	Ser	$-CH_2-OH$
Threonine	Thr	$-CH(CH_3)-OH$
Cysteine	Cys	$-CH_2-SH$
Tyrosine	Tyr	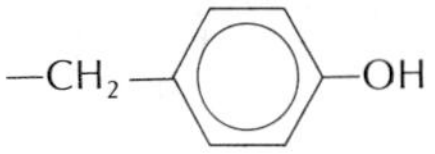 $-CH_2-$(benzene ring)$-OH$
Asparagine	AspN	$-CH_2-C(=O)-NH_2$
Glutamine	GluN	$-CH_2-CH_2-C(=O)-NH_2$

Ionic (Conjugate Acids Are +)

Name	*Abbr.*	*Side Chain*
Lysine	Lys	$-CH_2-CH_2-CH_2-CH_2-NH_2$
Arginine	Arg	$-CH_2-CH_2-CH_2-NH-C(=NH)-NH_2$
Histidine	His	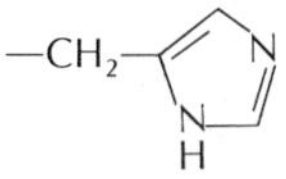 $-CH_2-$(imidazole ring: N, N—H)

Ionic (Conjugate Bases are −)

Name	*Abbr.*	*Side Chain*
Aspartic acid	Asp	$-CH_2-CO-OH$
Glutamic acid	Glu	$-CH_2-CH_2-CO-OH$

[a]Proline does not have the normal $-NH_2$ group, so its whole structure is shown.

called a *codon*—codes for a single amino acid. In addition to having a binding position for an amino acid, each tRNA molecule has a place where it binds to the mRNA. This binding position contains a sequence of three nucleotides—called an *anticodon*—that is complementary to the codon of that tRNA. In this way, each amino acid is added to the protein chain in its proper place. Figure 19.6 shows the process of protein synthesis.

codon

anticodon

PROTEIN STRUCTURE. Proteins in muscle and cartilage are examples of *fibrous proteins*. An overcooked pot roast shows just how fibrous muscle protein is. Fibrous proteins have straight chains and are water insoluble because they contain large numbers of nonpolar hydrophobic side chains. Single fibrous protein molecules twist themselves into a helix

fibrous protein

FIGURE 19.6
Protein synthesis

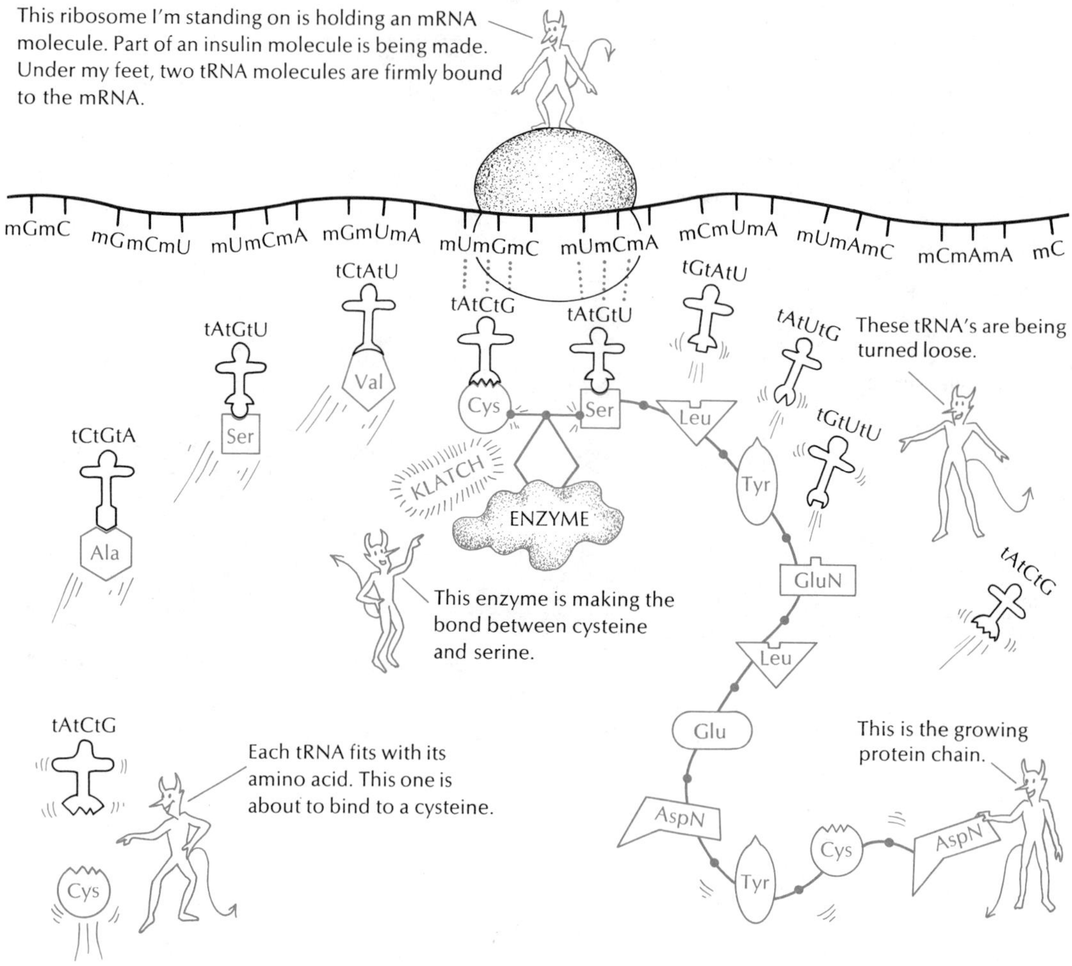

This ribosome I'm standing on is holding an mRNA molecule. Part of an insulin molecule is being made. Under my feet, two tRNA molecules are firmly bound to the mRNA.

mGmC mGmCmU mUmCmA mGmUmA mUmGmC mUmCmA mCmUmA mUmAmC mCmAmA mC

tCtGtA tAtGtU tCtAtU tAtCtG tAtGtU tGtAtU tAtUtG tGtUtU tAtCtG

Ala Ser Val Cys Ser Leu Tyr GluN Leu Glu AspN Tyr Cys AspN

KLATCH

ENZYME

These tRNA's are being turned loose.

This enzyme is making the bond between cysteine and serine.

Each tRNA fits with its amino acid. This one is about to bind to a cysteine.

This is the growing protein chain.

because of the hydrogen bonding between the —C=O and the —N—H groups in the chain. Several of these helixes twist together to form thicker strands, which in turn twist to form tough ropelike muscle fibers. The protein in hair, which is also fibrous, contains *disulfide bridges* between adjacent protein chains. disulfide bridge

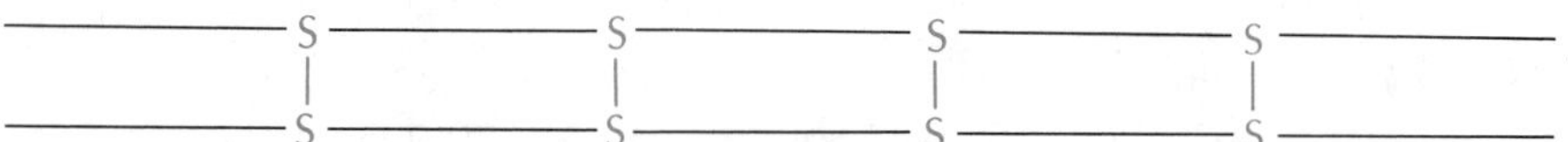

These disulfide bridges are created by the presence of cysteine, which has an —SH (*mercapto*) group. A very mild oxidizing agent can cause two adjacent —SH groups to form a disulfide bridge. mercapto

$$-SH + SH- \longrightarrow -S-S- + 2(H)$$

Disulfide bridges can be broken easily by reduction. Permanent waving reduces the disulfide bridges, then re-forms them again while the hair is molded into the desired shape.

Enzymes, hemoglobin, and antibodies are examples of *globular proteins*. These are roundish (like globes) and water soluble because of the large numbers of polar and ionic side chains they contain. Attractions among the polar and ionic side chains, plus disulfide bridges, give each globular protein the shape it needs to do its particular job. Figure 19.7 illustrates the difference between a fibrous and a globular protein. globular protein

Anything that destroys the shape of a globular protein also destroys its function. When this happens, the protein has been *denatured*. Table 19.3 shows some ways of denaturing proteins. denatured

Just as the meaning of a sentence depends on the words and their order, so the identity and function of a protein depends on the amino acids it contains and on their order. If a mistake is made, either in the kinds of amino acids or in their order, it may change the protein's shape or charge, or both. The disease sickle-cell anemia is caused by a mistake of one out of the three hundred amino acids contained in hemoglobin. An amino acid with a nonpolar side chain is wrongly substituted for an amino acid with an ionic one, changing the hemoglobin's charge and making it less soluble. The hemoglobin precipitates, and the red blood cells that contain it become lopsided. The cells clump together and block blood vessels, or sometimes break. These cells don't carry oxygen well, and the person becomes weak and less resistant to infections. This is an example of a *genetic defect*—a mistake in the cell's DNA, which is passed on by heredity and causes an incorrect protein structure. genetic defect

ENZYMES. The many biochemical reactions that happen in organisms have to take place fast. If we tried to carry them out in a test tube, they'd either go very slowly or not at all. Enzymes, like all catalysts, speed up the rates of reactions by lowering their activation energies.

Consider a factory with assembly lines. There are a lot of processes going on that involve just one thing, like putting on a nut, tightening a bolt, or taking out a screw. Usually these jobs are done by specialists. One kind of worker only puts on a certain nut. That worker has a special tool and a supply of nuts. The kind of worker who takes out a screw needs a special tool and a place to put the screws. The worker who tightens a bolt needs a special tool, but he or she doesn't need a supply or a place to put anything.

Enzymes are like the workers. The molecule that an enzyme is working on, which is like the piece of work on the assembly line, is called the *substrate*. The shape and charge of an enzyme cause it to hold onto its particular substrate like a vise. Somewhere near is the enzyme's "tool," called its *active site*. The active site is a functional group or groups on certain of the enzyme's amino acid side chains. substrate active site

Each reaction that happens in an organism needs at least one enzyme. Some enzymes can work on one reaction; others can work on several similar

FIGURE 19.7
Shapes of fibrous and globular proteins

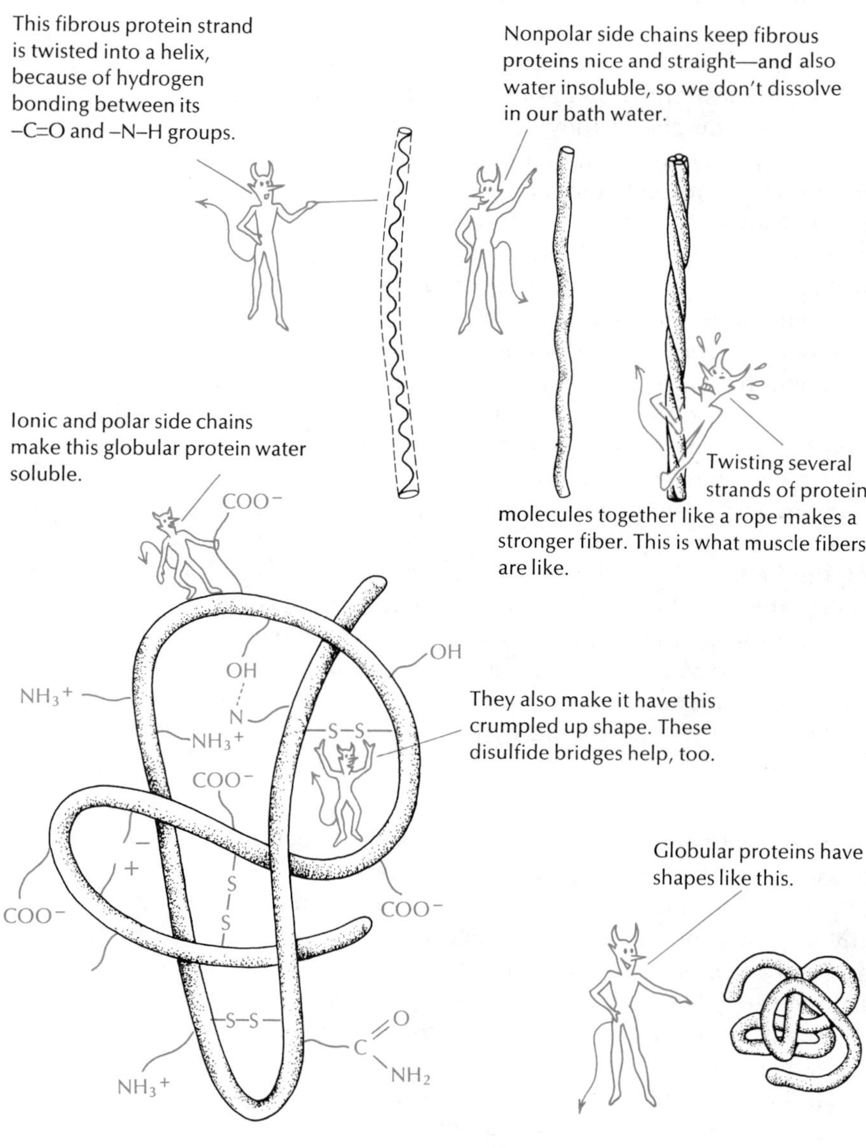

This fibrous protein strand is twisted into a helix, because of hydrogen bonding between its –C=O and –N–H groups.
Nonpolar side chains keep fibrous proteins nice and straight—and also water insoluble, so we don't dissolve in our bath water.
Ionic and polar side chains make this globular protein water soluble.
Twisting several strands of protein molecules together like a rope makes a stronger fiber. This is what muscle fibers are like.
COO⁻
OH
OH
NH₃⁺
N
NH₃⁺
S–S
COO⁻
S
S
COO⁻
COO⁻
S–S
O
C
NH₂
NH₃⁺
They also make it have this crumpled up shape. These disulfide bridges help, too.
Globular proteins have shapes like this.

reactions involving different substrates. The most common reactions involve either hydrolysis or condensation—the addition or removal of water molecules. Since cells are mostly water, there is always plenty of it around when and where an enzyme needs it.

However, many other reactions involve putting on or taking off hydrogen, electrons, methyl groups, sulfate, and amino or acetate groups. Since these

TABLE 19.3
Ways to denature proteins

Method	*Effect*	*Comments*
Add acid or base	Breaks up ionic attractions by destroying acidic or basic groups.	Milk curdles when lemon juice or vinegar is added. The casein (milk protein) is denatured.
Add concentrated salt or urea (NH_2CONH_2)	These interact with the protein's polar or ionic groups and prevent them from interacting with each other.	This kind of denaturing can sometimes be reversed just by washing the protein free of the salt or urea.
Heat	Breaks up most interactions.	Frying an egg denatures the albumin (egg protein).
Add alcohol	This disrupts the ionic and polar interactions through hydrogen bonding.	Alcohol is a good disinfectant because it denatures the proteins of bacteria and viruses.
Add heavy metal ions (Hg^{2+}, Pb^{2+})	React with the disulfide linkages and usually precipitate the protein.	Mercury and lead denature important enzymes and other proteins so that they no longer function.

aren't so readily available, the enzyme must have either a supply or a place to put them. This service is provided by compounds called *coenzymes,* which are small soluble nonprotein molecules that hold pieces of molecules for enzymes. Each coenzyme can hold only one of a certain kind of part or piece. However, a coenzyme can give its part to or take it from any enzyme that needs to use it or to get rid of it. If an enzyme's job is to take a piece off its substrate, it gives the piece to a coenzyme. That coenzyme can give the part to another enzyme in the cell that needs the part. Figure 19.8 illustrates the relationship among enzyme, substrate, and coenzyme.

coenzyme

19.3
CARBOHYDRATES AND FATS

Besides nucleic acids and proteins, living organisms contain carbohydrates and fats. A *carbohydrate,* meaning "hydrate of carbon," is usually a compound containing a carbon, hydrogen, and oxygen in the proportion of one C to one H_2O. (Since H_2O as such does not appear in these, the term "hydrate" is not strictly correct.) The simple sugar glucose is a carbohydrate made by plants to provide energy and stored by them as a polymer, starch. Another polymer of glucose is cellulose, used by plants as structural material. Many cell membranes and walls contain carbohydrates.

carbohydrate

Fats are made from carbohydrates in the bodies of both plants and animals. One type of fat is used for energy storage. Other types make up the large part of brain and nerve tissue in animals. Still other types make up parts of cell membranes and walls.

FIGURE 19.8
The relationship among enzyme, substrate, and coenzyme

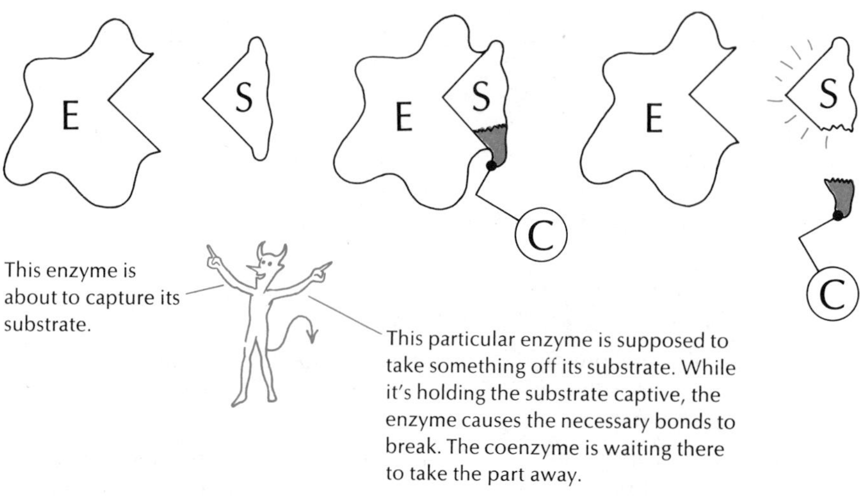

E
S
E
S
C
E
S
C
This enzyme is about to capture its substrate.
This particular enzyme is supposed to take something off its substrate. While it's holding the substrate captive, the enzyme causes the necessary bonds to break. The coenzyme is waiting there to take the part away.

MONOSACCHARIDES. Starch can be hydrolyzed to glucose, but glucose can't be hydrolyzed to any simpler carbohydrate. *Monosaccharides,* or simple sugars, are carbohydrates that can't be hydrolyzed any further. We've already seen two monosaccharides: ribose and deoxyribose. These have five carbon atoms. Glucose has six carbon atoms and exists in both these ring structures:

monosaccharide

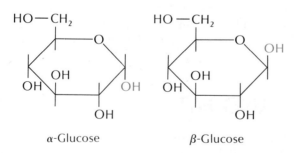

HO—CH_2 O OH OH OH OH

α-Glucose

HO—CH_2 O OH OH OH OH

β-Glucose

The difference between these two kinds of glucose lies in the position of the —OH group printed in color. All other —OH groups in glucose keep their positions—up or down—on the ring. Changing one or more of these will change the identity of the sugar.

Glucose is also called "dextrose," or "corn sugar." Its commercial use is almost exclusively in candy and sweeteners.

Two other important examples of monosaccharides, shown here, are galactose, which is found in milk sugar, and fructose, which is found in honey and fruits.

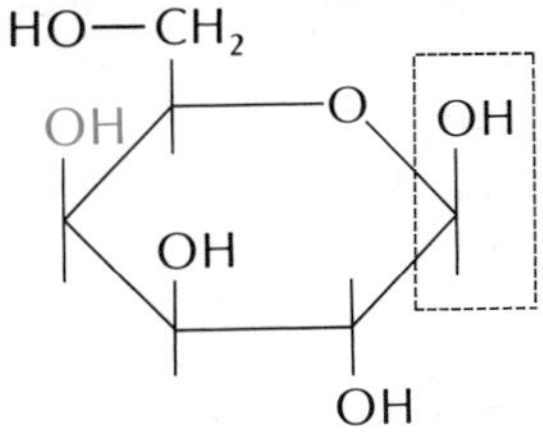

HO—CH_2
OH OH
OH
OH

Galactose

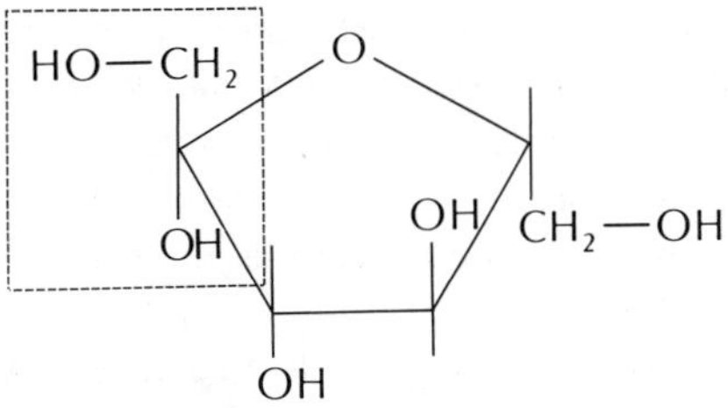

HO—CH_2 O
OH CH_2—OH
OH
OH

Fructose

These also exist in α and β forms by reversing the position of the groups in the boxes. Galactose differs from glucose in the position of the —OH group printed in color.

It is often convenient to abbreviate these ring structures by leaving out some or all of the —OH groups. When this is done, the sugar must be clearly specified.

DI- AND POLYSACCHARIDES. Plants make glucose and use it for energy. But they also store it for their own future needs—at night, when the sun's energy isn't available, or literally for a rainy day, when some of the sun's energy is blocked by clouds. Plants also store glucose in seeds and tubers so that their offspring will have energy to live on until they have leaves and can make their own glucose. We can store a lot of small, regularly shaped objects more efficiently by stacking them neatly than by throwing them in a heap. In the same way, monosaccharides can be stored more efficiently by condensing them, like this:

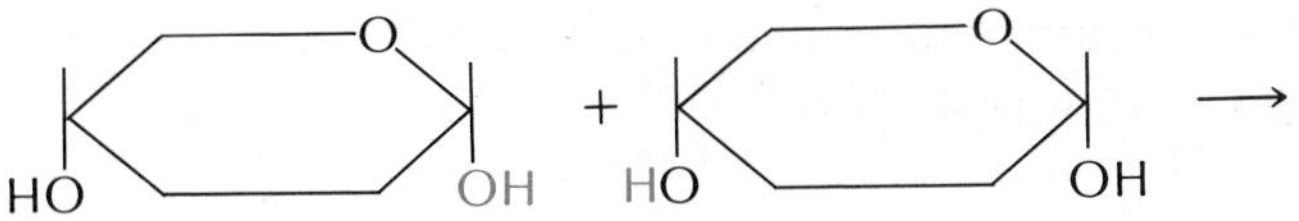

HO OH + HO OH →

α-Glucose α-Glucose

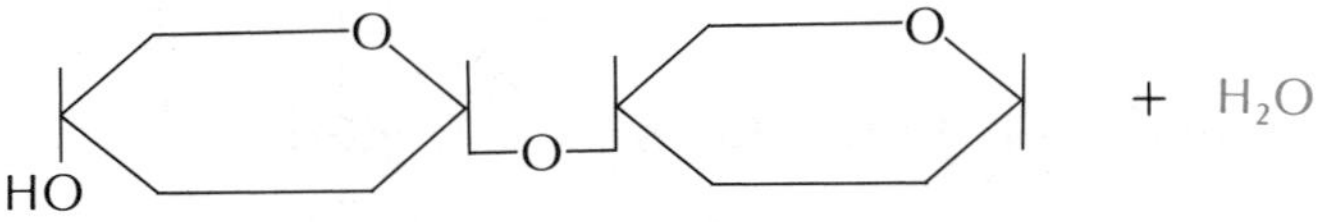

HO + H_2O

Maltose (malt sugar)

Two monosaccharides that are joined form a *disaccharide*. The point where the monosaccharides are joined is also where they can be hydrolyzed. Two other important disaccharides have these structures: disaccharide

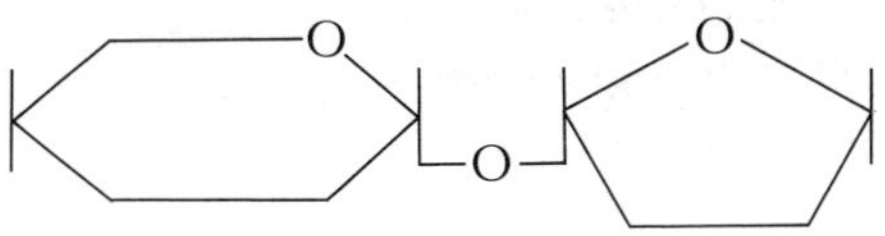

(glucose) (fructose)

Sucrose (table sugar, beet sugar, cane sugar)

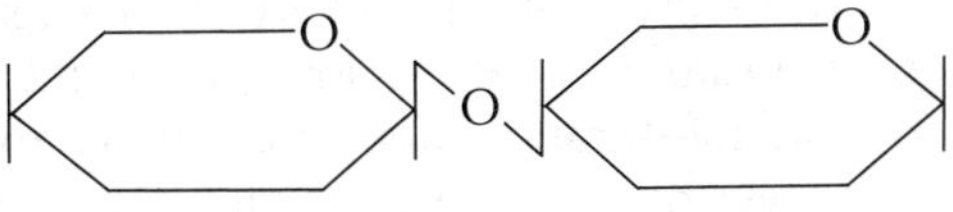

(galactose) (glucose)

Lactose (milk sugar)

Polysaccharides are composed of many monosaccharides and are an even more condensed way to store them. Plants store glucose in their seeds and tubers as starch, a polysaccharide made of α-glucose. Animals get glucose from plants and store it in their livers as glycogen, or animal starch. Plant starch and glycogen have very similar structures and are easily hydrolyzed by enzymes when the organism needs glucose. polysaccharide

Another way that plants put glucose together is in the polysaccharide cellulose, made of β-glucose. Unlike starch, cellulose is a building material and not an energy storehouse. Cellulose is the most abundant organic material of the plant world; it is to plants what fibrous proteins are to animals. Woody stems and tree trunks are mostly cellulose. The roughage we eat in celery, carrots, and leafy vegetables is cellulose. We use cellulose mostly in paper, made from wood, and in cotton, a natural fiber.

Cellulose is useful as a structural material because it isn't easily hydrolyzed, and most organisms don't have enzymes that will hydrolyze it at all. However, it can be hydrolyzed by heating with a strong acid, such as HCl. Depending on the cost of other sugar sources, this method is sometimes used as a commercial source of glucose.

Figure 19.9 shows the structures of starch and of cellulose.

LIPIDS. Biological substances that are soluble in ether—that is, fats and oils—are *lipids*. Like carbohydrates, they are both energy sources and building materials. Some of them also provide function. Some simple lipids are glyceryl esters, esters of glycerol with fatty acids (long straight chain organic acids). Glycerol has three —OH groups, so it can form esters with three fatty acids. The acids can be the same or different, but they're usually different. A typical glyceryl ester might have this structure: lipid

$$
\begin{array}{l}
CH_3-(CH_2)_{18}-C(=O)-O-CH_2 \\
\qquad\qquad\qquad\qquad\qquad\qquad\qquad\qquad\quad | \\
CH_3-(CH_2)_7-CH=CH-(CH_2)_7-C(=O)-O-CH \\
\qquad\qquad\qquad\qquad\qquad\qquad\qquad\qquad\quad | \\
CH_3-(CH_2)_7-CH=CH-(CH_2)_7-C(=O)-O-CH_2
\end{array}
$$

The ester linkages, indicated in color, provide points where the fat can be hydrolyzed. Such a lipid is a *triglyceride*, because each of the three —OH groups on glycerol is now part of an ester group. Most animal triglycerides contain saturated fatty acids and are fats (solids). Most plant triglycerides contain unsaturated fatty acids and are oils (liquids). Fats and oils are used as energy sources directly, or are stored in the form of fat globules. triglyceride

Other, more complex lipids may contain phosphate, monosaccharide, or other molecules. These are found in nerve and brain tissue and in cell membranes. One such lipid is the following (page 476), called a *lecithin* and found in nerve tissue: lecithin

FIGURE 19.9
Starch and cellulose structures

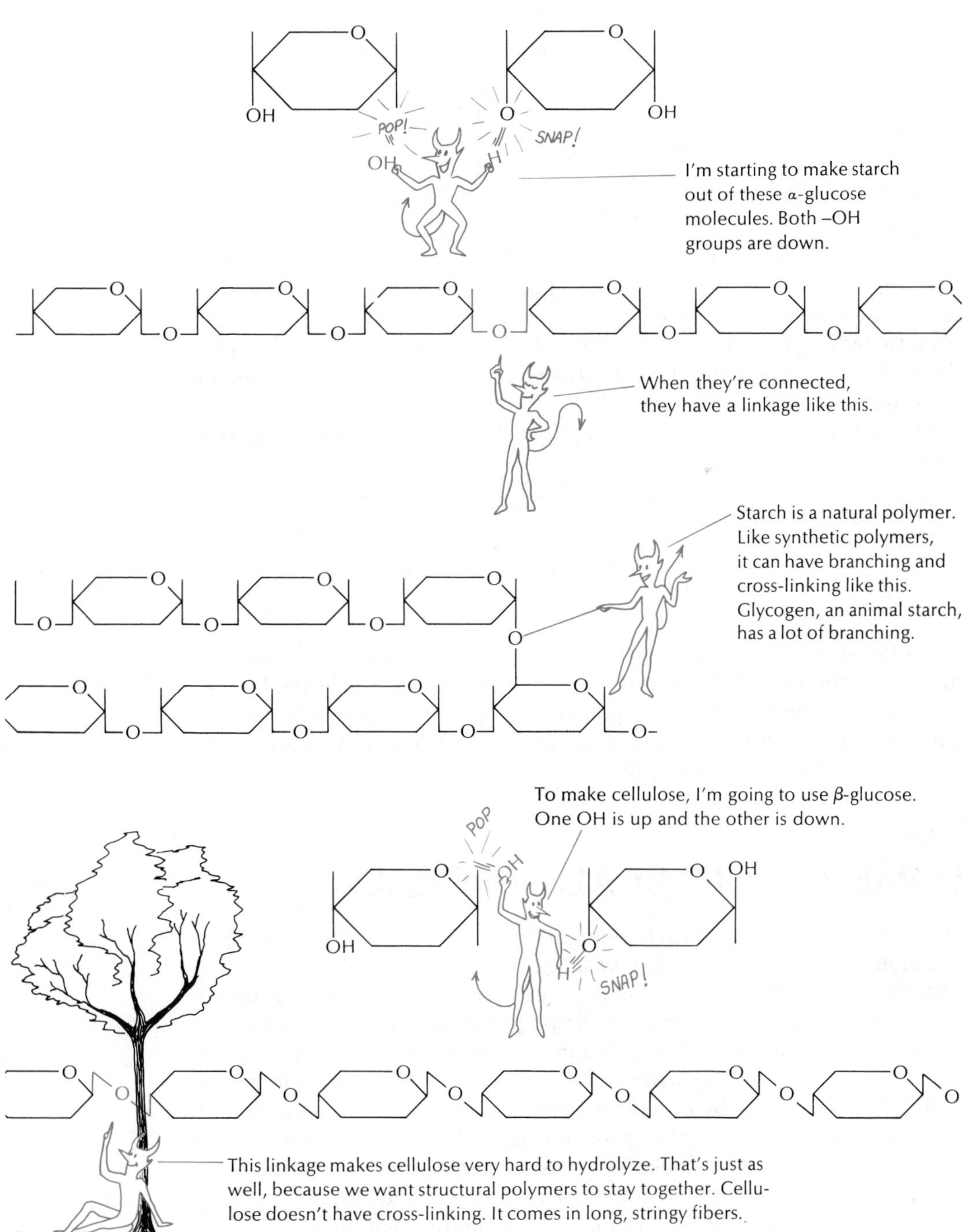

OH
O
O
OH
POP!
OH
H
SNAP!
I'm starting to make starch out of these α-glucose molecules. Both –OH groups are down.
When they're connected, they have a linkage like this.
Starch is a natural polymer. Like synthetic polymers, it can have branching and cross-linking like this. Glycogen, an animal starch, has a lot of branching.
To make cellulose, I'm going to use β-glucose. One OH is up and the other is down.
POP
OH
SNAP!
This linkage makes cellulose very hard to hydrolyze. That's just as well, because we want structural polymers to stay together. Cellulose doesn't have cross-linking. It comes in long, stringy fibers.

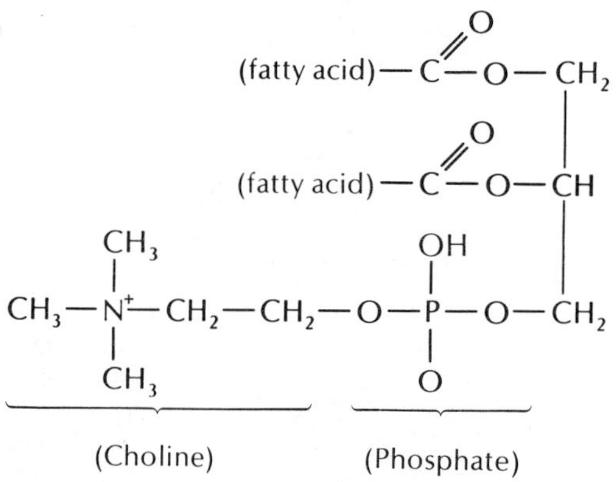

$$
\begin{array}{l}
\text{(fatty acid)}-\overset{\displaystyle O}{\overset{\|}{C}}-O-CH_2 \\
\qquad\qquad\qquad\qquad\qquad\quad | \\
\text{(fatty acid)}-\overset{\displaystyle O}{\overset{\|}{C}}-O-CH \\
\qquad\qquad\qquad\qquad\qquad\quad | \\
CH_3-\underset{\underset{\displaystyle CH_3}{|}}{\overset{\overset{\displaystyle CH_3}{|}}{N^+}}-CH_2-CH_2-O-\underset{\underset{\displaystyle O}{|}}{\overset{\overset{\displaystyle OH}{|}}{P}}-O-CH_2
\end{array}
$$

(Choline) (Phosphate)

The structure of a lecithin is similar to that of a triglyceride, but in place of one of the fatty acid groups there is a phosphate and a choline group. It is because these lipids are soluble in ether that ether is such an effective anesthetic. When we breathe ether, it dissolves in nerve cell lipids, changing them temporarily.

Another class of lipids includes the *steroids,* which have this basic structure: steroid

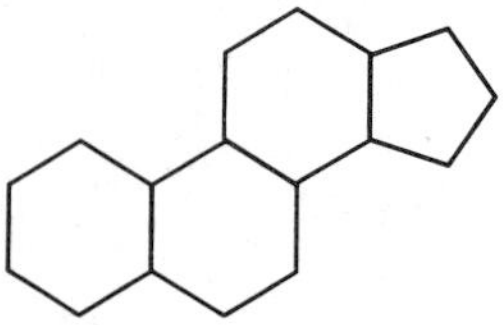

Table 19.4 shows the complete structures for some important steroids. One of them is cholesterol. The others are all *hormones,* relatively small molecules that are produced in one part of an organism and sent to another part. Their function is to trigger a reaction or reactions in the part of the organism to which they are sent. hormone

19.4 SOME BIOCHEMICAL PROCESSES

Only a certain number of each element's atoms exist on earth. For plants and animals to reproduce and grow, atoms must be recycled and used over and over again. The atoms that make up our bodies now are only temporarily ours. These atoms probably belonged at one time to prehistoric plants and animals, and they will belong to other organisms in the future. Figure 19.10 shows roughly how carbon and nitrogen are recycled in nature.

Plants make glucose by photosynthesis, and they use the energy from it to make amino acids out of inorganic nitrogen—ammonia and nitrates—in the soil. Then the plants make proteins for their own use. Herbivorous (plant-eating) animals eat the plants, taking amino acids from the plants' proteins and re-forming them into their own protein structures. Carnivorous (meat-eating) animals eat the herbivorous animals to obtain protein for their own structure and to store fat for their energy. All animals and plants

TABLE 19.4
The structures of some steroids

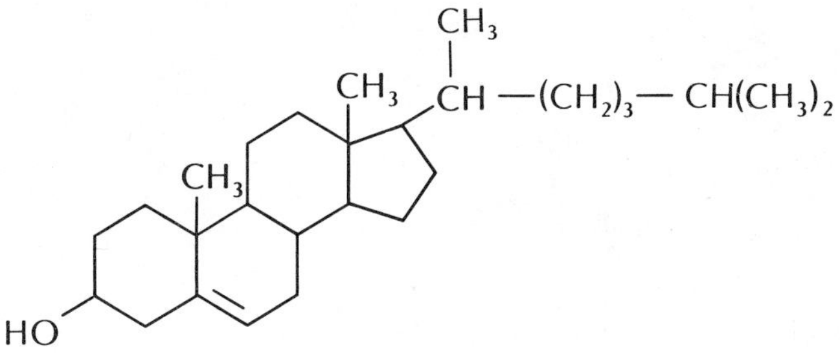 Cholesterol	A steroid alcohol, or sterol. Found in nearly all vertebrates, mostly in the brain and spinal cord, and in gallstones. Has been associated with hardening of the arteries. Function is probably to be made into steroid hormones and bile acids.
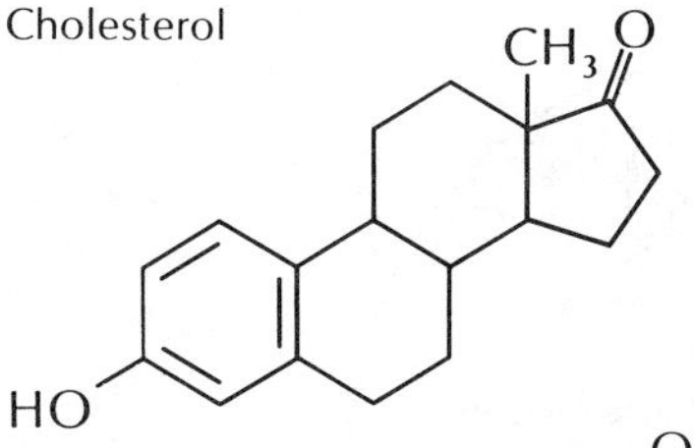 Estrone	A female sex hormone. Prevents release of eggs. Secreted in greater amounts during pregnancy.
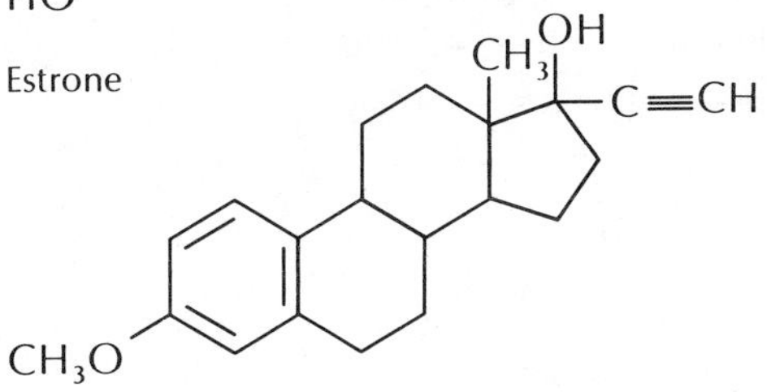 Mestranol	A synthetic hormonelike compound, used in birth-control pills. Imitates estrone and others by simulating pregnancy and preventing release of egg.
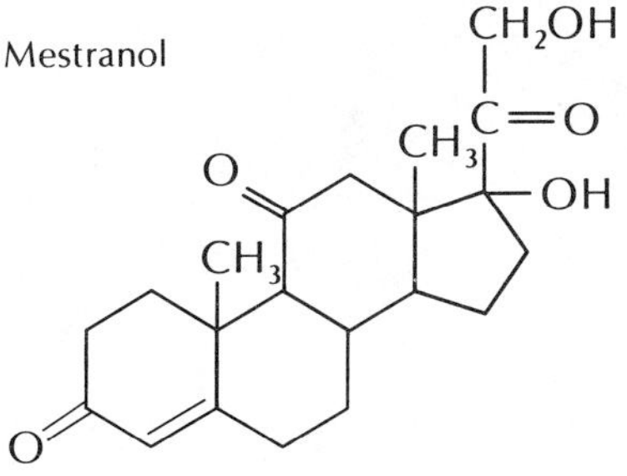 Cortisone	A hormone that regulates the use of carbohydrates. Also effective in relieving inflammation, such as that from arthritis.
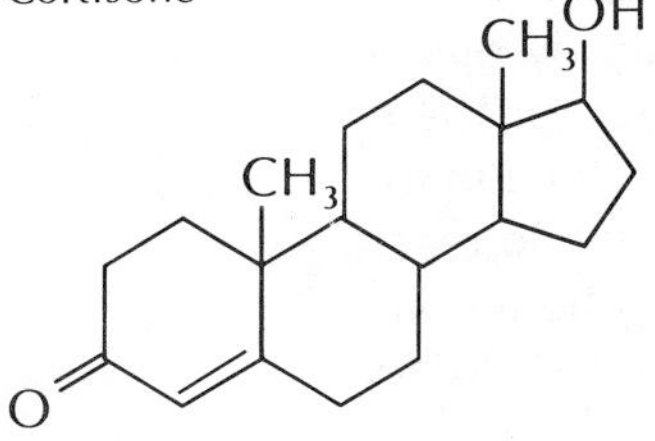 Testosterone	A male sex hormone. Regulates the development of reproductive organs.

FIGURE 19.10
The biological cycle of elements

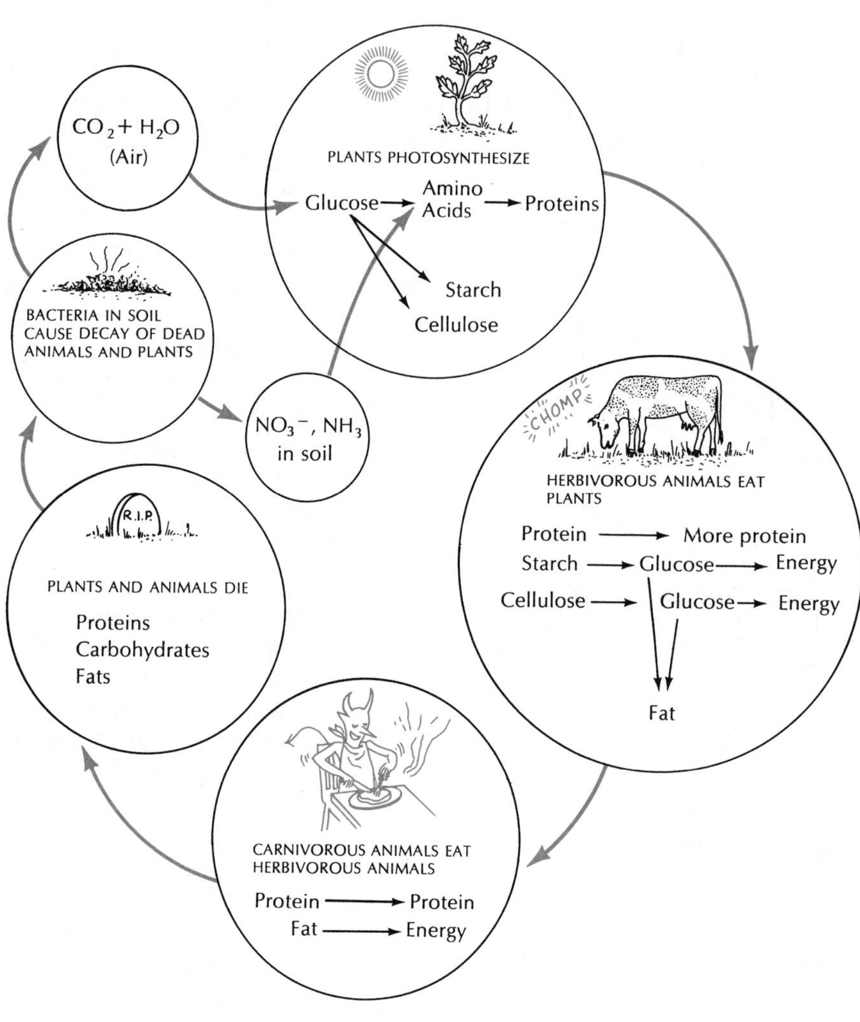

die, and microorganisms in the soil turn their complex organic structures back into simple inorganic compounds that can be used again by plants.

Animals can't make amino acids, so they depend on plants for them. Plants can't use organic compounds, so they depend on the bacteria in the soil to break them down. If decay didn't occur, the elements would reach a literal dead end as organic compounds, and the cycle would be incomplete. Our production of certain plastics and other substances that do not decay is beginning to interrupt this cycle.

Although the biological cycle of elements depends on each segment, it is logical to start at photosynthesis, the process that directly and indirectly furnishes all organisms with energy from the sun.

PHOTOSYNTHESIS. This equation for photosynthesis sums up a very complex series of biochemical reactions that happen in a green plant.

$$6\,CO_2(g) + 6\,H_2O(l) \longrightarrow C_6H_{12}O_6(aq) + 6\,O_2(g)$$

The important end result is that the product, glucose, contains seven C—H bonds, while the CO_2 from which it is made contains none. We saw in Chapter 18 that C—H bonds are the main energy source when organic compounds are burned, and it's the same with biochemical compounds. The energy that went into making glucose is stored mostly in its C—H bonds. Hydrogen to make these bonds is stripped off the water molecules, leaving their oxygen to be released to the atmosphere as the free element.

Only green plants can carry out photosynthesis, because the first step depends on absorbing the sun's energy by means of the green pigment chlorophyll.

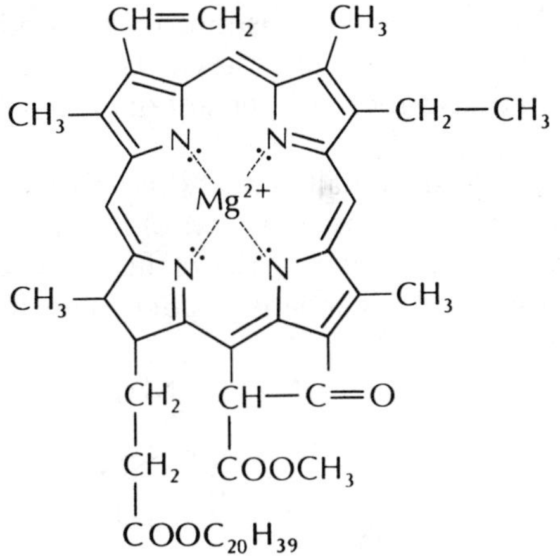

$CH{=}CH_2$ CH_3

CH_3 $CH_2{-}CH_3$

Mg^{2+}

CH_3 CH_3

CH_2 $CH{-}C{=}O$

CH_2 $COOCH_3$

$COOC_{20}H_{39}$

Chlorophyll *a*

This is one of several chlorophylls that have similar structure. In these compounds, Mg^{2+} acts as a Lewis acid—that is, as an electron-pair acceptor. The available electron pair on each nitrogen belonging to the ring system is donated to four coordinate covalent bonds. As we'll see a bit later, hemoglobin contains a similar system.

Scientists still don't understand all the processes that occur during photosynthesis. However, the initial reaction is the absorption of energy by chlorophyll.

$$\text{chlorophyll} + \text{energy} \longrightarrow \text{chlorophyll (higher energy)}$$

Then, the chlorophyll provides energy for removing hydrogen from water and giving it to a hydrogen-accepting coenzyme.

$$\text{chlorophyll} + H_2O + \text{coenzyme} \xrightarrow{\text{enzymes}} \text{chlorophyll} + O_2 + H_2\,\text{coenzyme}$$

After that, the hydrogen is passed around through many other coenzymes until it ends up bonded to carbon in glucose.

All of the other reactions we'll look at are involved in *metabolism*—an organism's use of raw materials from its environment to provide its energy and structural materials. metabolism

RESPIRATION. The controlled burning of fuel by cells to obtain energy is called *respiration,* and it is carried out by both plants and animals. Most respiration reactions occur in the *mitochondrion,* one of the cell's organelles shown in Figure 19.1. When the fuel is glucose, the result is almost the opposite of the photosynthesis reaction. respiration mitochondrion

$$C_6H_{12}O_6(aq) + 6\,O_2(g) \longrightarrow 6\,CO_2(g) + 6\,H_2O(g) + 277 \text{ kcal}$$

Hydrogen is stripped from the C—H bonds and made into water molecules, with a release of energy. Every day, the average adult generates about 2500 kilocalories from food, needs about 500 liters of oxygen, and generates about 400 liters of carbon dioxide. The heat of combustion of glucose is 690 kilocalories per mole. Since the respiration of glucose yields only 277 kilocalories, the human body manages to use about 40 percent of glucose's energy.

How do cells harvest this energy? Obviously not all as heat (though some is needed to maintain body temperature) or the organism would burn up. Most of the energy is captured in other chemical bonds. The main chemical substance that performs this function in all organisms is adenosine triphosphate (ATP), which has this structure:

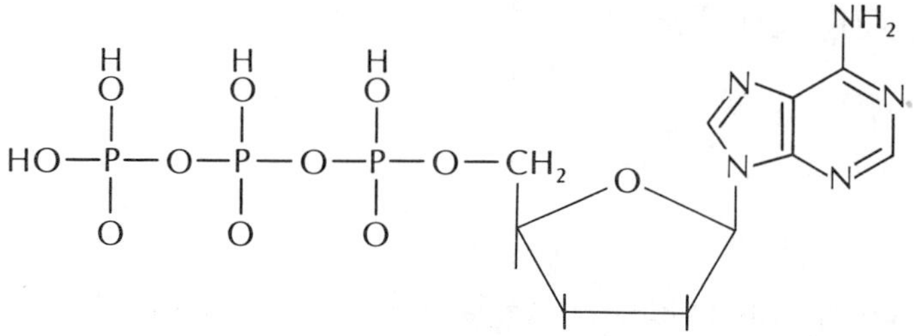

ATP is a coenzyme that stores and transfers energy and a phosphate group according to this reversible reaction:

$$ATP \rightleftharpoons ADP + PO_4^{3-} + 7.3 \text{ kcal}$$

Here, ADP (adenosine diphosphate) is simply ATP without one of its phosphate groups. Both ATP and ADP are abundant in cells. If a process releases energy (as respiration does), the energy is invested by driving the above reaction from right to left and forming ATP. We thus have an energy conversion factor: 7.3 kilocalories per mole of ATP. Expressing the energy from respiration in moles of ATP instead of kilocalories, we then write:

$$C_6H_{12}O_6(aq) + 6\,O_2(g) + 38\text{ ADP} + 38\,PO_4^{3-}(aq) \longrightarrow 6\,CO_2(g) + 6\,H_2O(g) + 38\text{ ATP}$$

Or, the respiration of 1 mole of glucose yields 38 moles of ATP.

Any kind of food that contains C—H bonds and can be broken down by an organism's enzymes can be used as an energy source in respiration,

FIGURE 19.11
Hemoglobin's structure and the respiratory chain

This is the heme, or nonprotein, part of hemoglobin. There are four of these in every hemoglobin molecule.

CH=CH$_2$ CH$_3$

CH$_3$ CH=CH

N N
Fe^{2+}
N N

CH$_3$ CH

CH$_2$ CH$_2$

CH$_2$ CH$_2$

COOH COOH

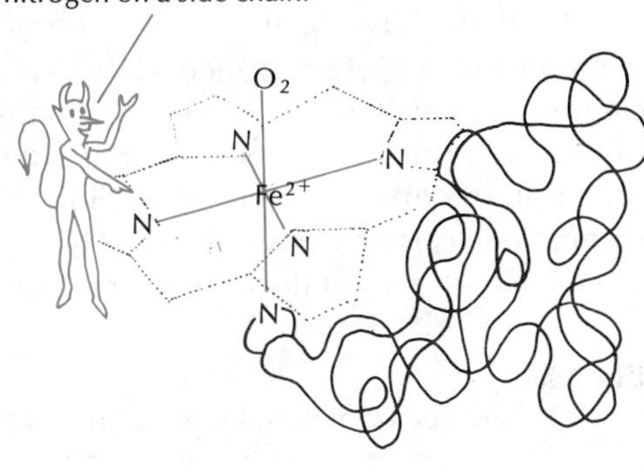

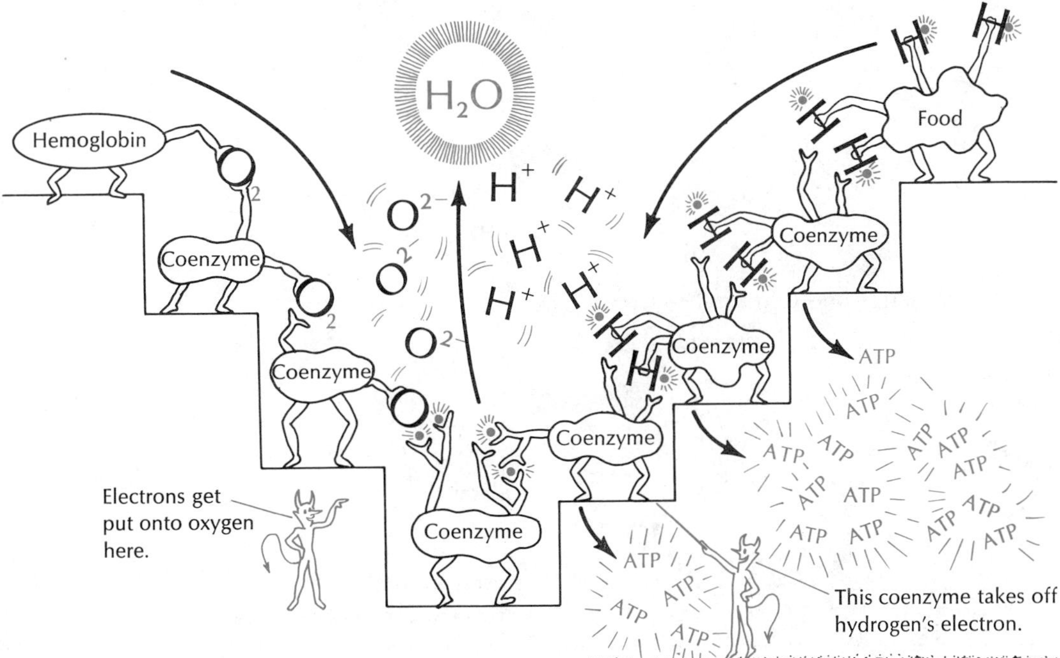

provided that it's soluble and can be carried to the cells by plant sap or blood. (Hemoglobin's role in respiration is shown in Figure 19.11.) Glucose, amino acids, and fatty acids are such substances. A 6-carbon-atom fatty acid yields 42 moles of ATP, so it's a slightly more efficient energy source than glucose. If animals take in more fuel than they need immediately for energy, they store the excess as fat. About 30 percent of the carbohydrates eaten by humans go to make stored fat.

We can think of respiration as happening from two extreme points: the C—H bonds in the food, and the O_2 we breathe. Getting these two together to form H_2O and release energy involves a long chain of reactions in which coenzymes pass hydrogen and electrons around. The reactions begin on one end with the binding of oxygen by hemoglobin. We've said that hemoglobin contains a protein (called globin) and a nonprotein (called heme). The heme part binds the oxygen to Fe^{2+} in the center of a ring system similar to that of chlorophyll. Fe^{2+} has six binding positions. Four are used by the nitrogens in the ring; one is used to bind the heme to the globin; and the remaining one is for binding oxygen. Hemoglobin's structure and a simplified version of the respiratory chain are shown in Figure 19.11.

TABLE 19.5
Some important coenzymes and their vitamins

Vitamin	*Coenzyme and Function*
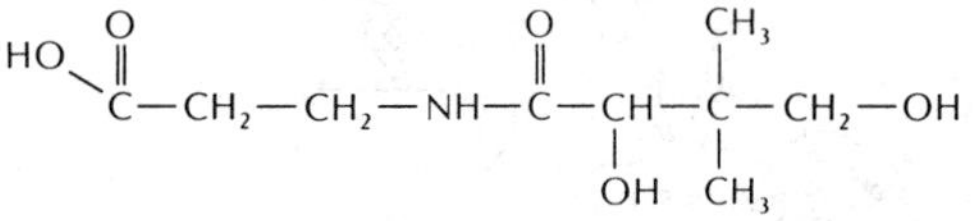 $HO-C(=O)-CH_2-CH_2-NH-C(=O)-CH(OH)-C(CH_3)_2-CH_2-OH$ Pantothenic acid	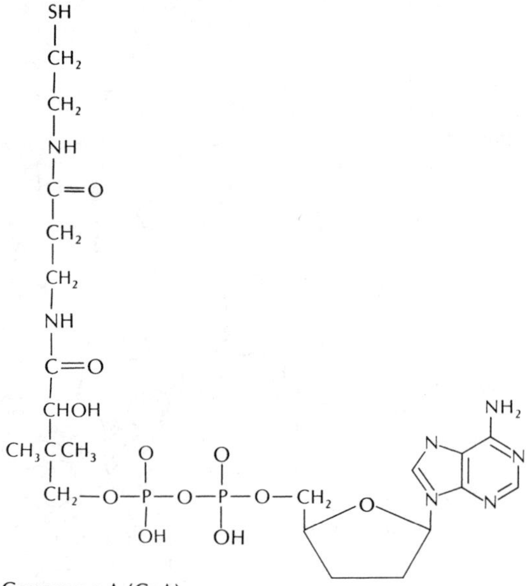 $HS-CH_2-CH_2-NH-C(=O)-CH_2-CH_2-NH-C(=O)-CHOH-C(CH_3)_2-CH_2-O-P(=O)(OH)-O-P(=O)(OH)-O-CH_2$-(ribose)-adenine ($NH_2$) Coenzyme A (CoA) Transfers acetyl group important in the metabolic processes.
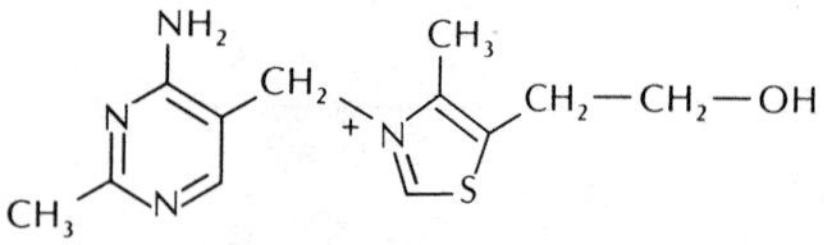 (pyrimidine: NH_2, CH_3)$-CH_2-$(thiazolium: ^+N, S, CH_3)$-CH_2-CH_2-OH$ Thiamine (vitamin B_1)	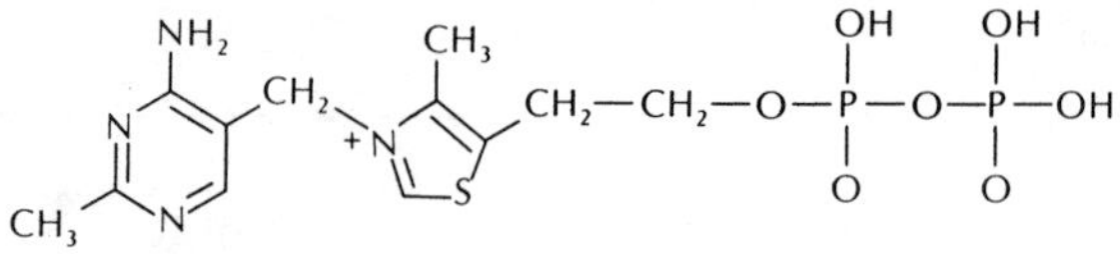 (pyrimidine: NH_2, CH_3)$-CH_2-$(thiazolium: ^+N, S, CH_3)$-CH_2-CH_2-O-P(OH)(=O)-O-P(OH)(=O)-OH$ Thiamine pyrophosphate (TPP) Also transfers acetyl group

Coenzymes are clearly very important in respiration. Many of the vitamins we need are used to make parts of these coenzymes. The word "vitamin" comes from "vital amine," because it was once thought that all vitamins were amines. Now *vitamin* means any substance that an organism needs but can't make itself and therefore must have in its diet. Table 19.5 shows a few important coenzymes and the vitamins that we need to make them. In addition to these, there are many other vitamins that aren't converted to coenzymes, and some whose exact functions aren't well known.

vitamin

DIGESTION. Cell respiration requires soluble food. The process of converting food from an insoluble form (such as starch or protein) to a soluble form (such as glucose or amino acids) is called *digestion*. All digestive reactions involve hydrolysis and are catalyzed by hydrolytic enzymes.

digestion

Digestion begins in the mouth, where the enzyme amylase in saliva

Vitamin	*Coenzyme and Function*
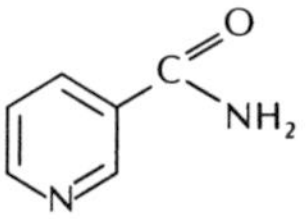 Niacin (nicotinamide)	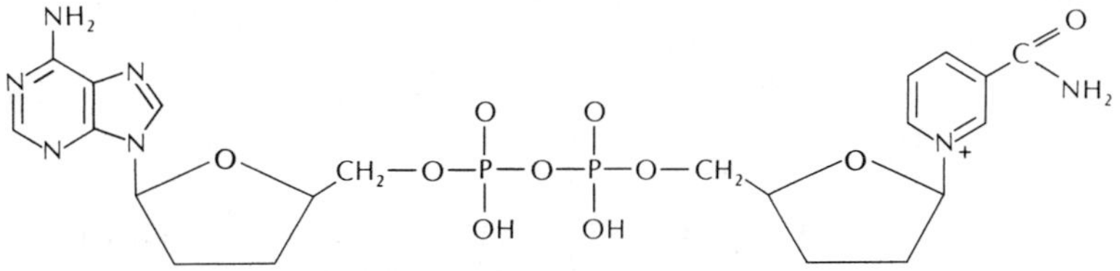 Nicotinamide adenine dinucleotide (NAD) A hydrogen-accepting coenzyme, important in the respiratory chain.
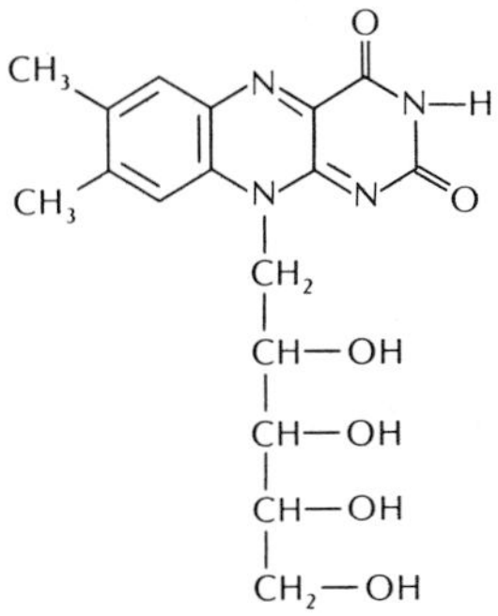 Riboflavin	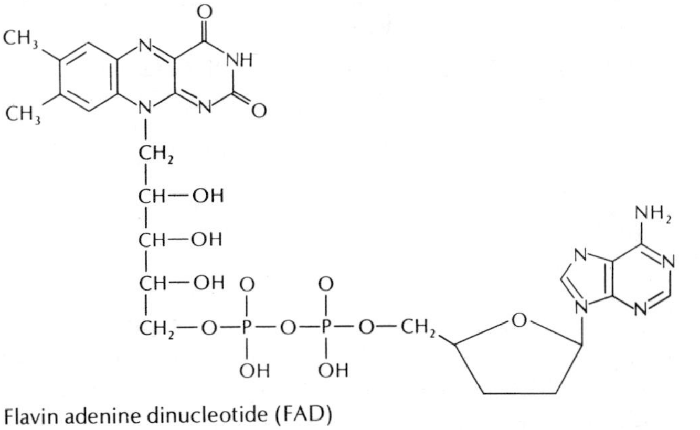 Flavin adenine dinucleotide (FAD) Like NAD, an important hydrogen-accepting coenzyme in the respiratory chain.

begins breaking starch molecules into smaller polysaccharides. In the stomach, the enzyme pepsin, along with the stomach's HCl, begins breaking protein chains down into smaller units. In the intestines, other enzymes complete the breakdown to glucose and amino acids, which pass through the walls of the intestines and into the blood. Bile salts from the liver act as emulsifying agents for fats and let them mix with fluid so that they, too, can be attacked by enzymes. Fatty acids and glycerol then also pass into the blood. Bacteria that are normal residents of the intestinal tract aid in the final digestion. Feces contain large quantities of dead bacteria, as well as indigestible substances such as cellulose.

The intestines of ruminating animals like cattle contain bacteria that can digest cellulose and make its glucose available to the animal. These bacteria, and others in the soil, are among the few organisms that have the enzyme *cellulase*, which catalyzes the hydrolysis of cellulose. Microorganisms in the digestive systems of some wood-eating insects, such as termites and carpenter ants, also have this enzyme.

cellulase

Besides digesting food they eat, organisms also digest food that's stored in their own bodies. Enzymes in a sprouting seed begin to digest its starch so the seedling can have glucose for its growth. Beer brewing takes advantage of this enzymatic process. To make malt, barley is moistened and allowed to sprout, then cooked and mashed. The enzymes have converted starch, an unacceptable food for yeast, into maltose, which yeast can convert into alcohol. Brewer's yeast is then added to the malt, along with other things. The resulting fermentation produces beer.

When our diets aren't supplying enough energy, our bodies call on their energy reserves by digesting their glycogen or stored fat. Fat is the most deeply stored of all the reserves, sort of like money tied up in stocks and bonds. Glycogen is a bit more available, like money in a savings account. Blood sugar (glucose) is still more available, like money in a checking account. And ATP is like having cash on hand. The amount of sugar in blood will last only a few minutes. The sugar from glycogen in the liver will last several hours. After that, if we're not taking in any food, we start tapping our reserve of stored fat. An average adult (not overweight) has enough reserve fat to last 30 to 40 days.

The enzymes that break down liver glycogen are controlled by the hormone *adrenalin*, which has this structure:

adrenalin

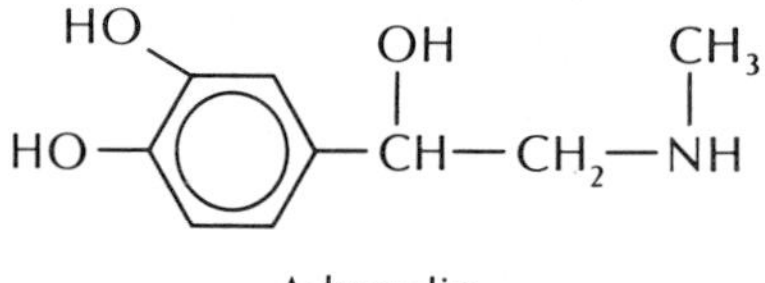

$$(HO)_2C_6H_3-\underset{\displaystyle |}{\overset{\displaystyle OH}{CH}}-CH_2-\overset{\displaystyle CH_3}{\overset{|}{NH}}$$

Adrenalin

In stressful situations, adrenalin is secreted and triggers the conversion of glycogen to glucose, probably by somehow activating a key enzyme. One molecule of adrenaline causes the release of about thirty thousand molecules of glucose. This gives the brain and muscles a lot of extra energy that might be needed in an emergency.

EXCRETION. A lot of the amino acids in cells are used to make new proteins, but the surplus is used for energy. The first step in converting an amino acid to energy is removal of the amino group by enzymes.

$$H_2\text{coenzyme} + NH_2\text{—}\underset{\substack{|\\ \text{side chain}}}{CH}\text{—COOH} \xrightarrow{\text{enzymes}} \underset{\substack{|\\ \text{side chain}}}{CH_2}\text{—COOH} + NH_3 + \text{coenzyme}$$

Now the amino acid has become a fatty acid and is available for respiration. (If enough fuel is already available, this fatty acid is combined with glycerol and stored as fat. Thus proteins, as well as fats and carbohydrates, can end up as fat if more is eaten than the body needs to use.)

The ammonia that results from the above reaction is toxic to cells and is immediately converted to urea.

$$2\,NH_3(aq) + CO_2(aq) \longrightarrow \underset{\text{Urea}}{NH_2\text{—}\overset{\overset{\displaystyle O}{\|}}{C}\text{—}NH_2(aq)} + H_2O(l)$$

The urea is returned to the blood. Urea isn't as toxic as ammonia, but it must still be removed or it will cause uremic poisoning. The process of removing the waste products of metabolism is called *excretion,* and in human beings is done by the kidneys and urine. Blood constantly filters through the kidneys, like water through a filter pump in an aquarium. The kidneys remove excess water and salt along with urea and various small ions. Over a 24-hour period, the same average adult we've been talking about will produce about 30 grams of urea, 15 grams of sodium chloride, and 15 grams of miscellaneous substances, contained in about 1.5 liters of urine. People on high-protein diets will produce more urea and should drink plenty of water so that the urea can be excreted frequently.

excretion

Besides urea, salt and other ions are removed to maintain a constant osmotic pressure of blood. In Chapter 12, we saw that larger amounts of dissolved solutes increase the osmotic pressure of a solution. The osmotic pressure of blood is maintained by a hormone called *vasopressin,* which consists of nine amino acids joined together.

vasopressin

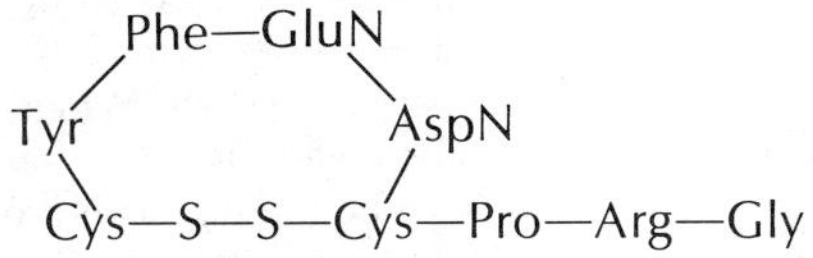

```
      Phe—GluN
     /        \
  Tyr          AspN
     \        /
     Cys—S—S—Cys—Pro—Arg—Gly
```

Vasopressin

When we eat salted popcorn or peanuts, we get thirsty because of the action of vasopressin. When a lot of salt is consumed, it goes into the blood, raising blood's osmotic pressure and triggering the secretion of vasopressin. Vasopressin represses the production of urine and puts the water back into the blood, lowering its osmotic pressure by diluting the salt. This is only a temporary method, though, and so the thirst response is also turned on.

With a fresh supply of water to further dilute the blood, the secretion of vasopressin is turned off and urine is produced normally.

Diuretics are substances that increase urine production. Frequent production of urine removes more solute from the blood and lowers its osmotic pressure. For this reason, diuretics are sometimes given to people with high blood pressure. diuretic

---

REVIEW QUESTIONS

Nucleic Acids

1. What are *nucleic acids?* In what part of the cell are they found?
2. What are *chromosomes?* What happens to the chromosomes after a cell divides?
3. What information is contained in DNA? What is a *gene?*
4. What is *replication?* Why must replication occur before a cell divides?
5. How do complementary pairs of monomers hold the two DNA strands together?
6. Describe, with the aid of a sketch, the process of DNA replication.
7. What is RNA? What are its monomers called?
8. Describe and sketch the process of *transcription.*
9. List the three types of RNA and describe the function of each.
10. Explain the relationship among DNA, RNA, and protein.
11. What are *nucleotides?* What are their three parts?
12. What are the differences between the nucleotides found in DNA and those found in RNA?

Proteins

13. What are *proteins?* What are some of their functions in organisms?
14. Give the general structure of an *amino acid.* Show how amino acids are linked in proteins.
15. What is meant by the *genetic code? Codon? Anticodon?*
16. Sketch and describe the process of protein synthesis.
17. Describe the structure of a *fibrous protein.*
18. What amino acid must be present to make a *disulfide bridge?*
19. What are *globular proteins?* What gives them shape and solubility? Why must they be water soluble?
20. What does it mean to *denature* a protein? What are some ways of denaturing proteins?
21. Why are enzymes needed in biological systems? What is an enzyme's *substrate?* It's *active site?*
22. What are *coenzymes?* Are they always needed? Explain.
23. Explain the relationship among enzyme, substrate, and coenzyme.

Carbohydrates and Fats

24. What is a *carbohydrate?* Name some carbohydrates.
25. What are *monosaccharides?* Give examples.
26. Name three *polysaccharides,* and describe each. Which is used for energy, and which for structure?
27. What are *lipids?* What kind of lipids are found in nerve tissue?
28. What are *steroids? Hormones?* Give some examples of steroid hormones.

Some Biochemical Processes

29. Sketch the biological cycle of elements. What is the key energy source for both plants and animals?
30. What is the first step in photosynthesis? Describe the role of chlorophyll and of coenzymes in photosynthesis.
31. What is meant by *metabolism?*
32. How do cells receive energy from *respiration?* How do they store and use energy?
33. What is the conversion factor between kilocalories and moles of ATP?
34. Do we get more energy from glucose or from fat? What happens to excess glucose or fat that isn't needed for energy?
35. Describe the two parts of hemoglobin. Show where the oxygen is bound.
36. Sketch and describe the respiratory chain.
37. What are *vitamins?* Show how some vitamins are related to coenzymes.
38. What is *digestion?* Why do some substances need to be digested?
39. Describe the digestive process in human beings. Why are bile salts needed?
40. Compare the energy availabilities of starch, glycogen, glucose, and ATP.
41. What happens to amino acids after they reach the cells?
42. What is *excretion?* Why do organisms need to excrete?
43. How does the hormone vasopressin maintain the blood's osmotic pressure?

EXERCISES

Set A (Answers at back of book.)

1. Suppose one strand of a small segment of DNA has this sequence:

   dA—dA—dT—dC—dT—dC—dG—dT—dA—dT—dC—dC—dA—dC—dG—dT—dC

   Write the sequence for the complementary strand.
2. Assume that the DNA segment in Exercise 1 is the strand that is transcribed into RNA. Write the sequence of the RNA that would be transcribed from it.
3. For this piece of mRNA:

   mU—mU—mG—mG—mG—mC—mA—mU—mA—mC—mG—mA—mU—mU

   write the sequence for the DNA strand that would code for its transcription.
4. Which of these two protein segments would be expected to be the more water soluble? Explain.
   a. —Gly—Ser—Asp—Ala—Arg—Cys—Glu—Thr—Val—
   b. —His—Phe—Cys—Ala—Gly—Gly—Ileu—Met—Thr—
5. Show where hydrogen bonds might form between amino acid side chains in this polypeptide:

   Ser—Gly—GluN—Gly—Met—Ala—Tyr—Thr
6. Using amino acids that can form disulfide bridges, ionic attractions, and hydrogen bonding, design polypeptide molecules that could have the following shapes.
   a. S-shaped
   b. H-shaped

   Use ten or fewer amino acids, and indicate what interactions are causing the structures.
7. The antidote for immediate poisoning with a heavy metal like lead or mercury is to take egg white and then induce vomiting. Why would this work?
8. Write the products of the hydrolysis of the lecithin shown on page 476.
9. Why is magnesium an important plant nutrient?
10. One of the processes by which nucleotides are synthesized in cells requires 9 moles of ATP. How many kilocalories is this?
11. In an aerobic sport, such as mountain climbing, a 100-kg person may use about 0.5 moles of oxygen in an hour. How many kcal of energy will this provide?
12. Muscle contains about 5×10^{-6} moles of ATP per gram. Assuming no other source of energy, calculate the amount of energy (kilocalories) available in a 500-gram muscle.
13. The earth's atmosphere contains about 2×10^{12} metric tons of CO_2. If this were all converted to glucose via photosynthesis, calculate the following.
    a. the metric tons of glucose that could be produced
    b. the moles of ATP this could yield in metabolism
    c. the energy equivalent in kilocalories
14. Structures of four vitamins are given below.

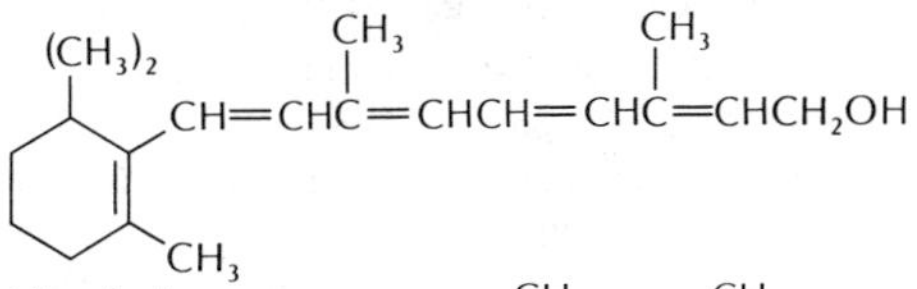

    Vitamin A

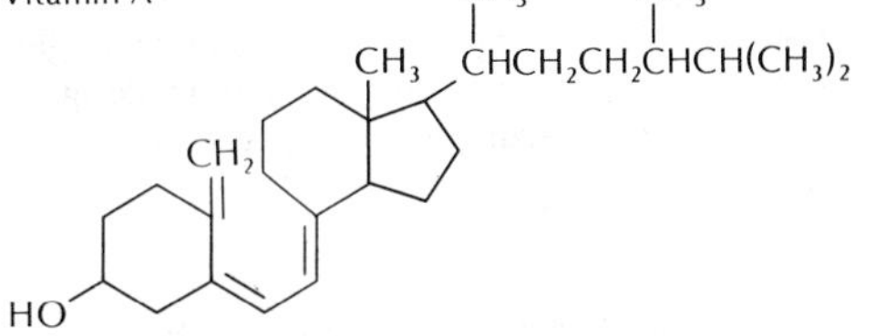

    Vitamin D

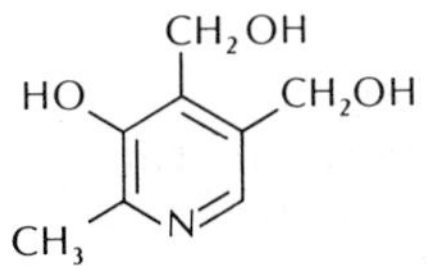

    Vitamin B_6

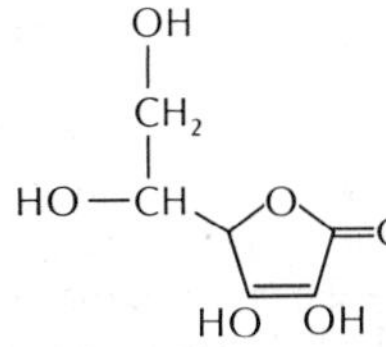

    Vitamin C

    Vitamins A and D tend to collect in fatty tissues, while vitamins B_6 and C tend to be excreted in urine. Suggest structural features of these vitamins that would explain this behavior.
15. Insulin, a protein hormone used for the treatment of diabetes, must be administered by injection. Explain why it is not effective if taken orally.
16. "Enzyme detergents" contained enzymes that hydrolyze proteins and thus destroy protein stains. However, these detergents had to be removed from the market because many people suffered respiratory irritation from breathing the dust. Suggest a reason for this undesirable side effect.
17. Alcohol represses the secretion of vasopressin. With this in mind, explain why people urinate frequently while drinking liquor and then wake up thirsty the "morning after."
18. How much urea would be produced by removing the amino acid group from 1 gram of glycine?

Set B (Answers not given.)

1. Write the sequence for the complementary strand of this piece of DNA:

   dG—dC—dA—dA—dT—dC—dG—dA—dT

2. Write the sequence of the RNA strand that would be transcribed from the DNA strand of Exercise 1.
3. Write the sequence for the DNA strand that would code for the transcription of this piece of mRNA:

   mA—mG—mA—mU—mC—mU—
   mU—mA—mC—mG—mA—mC

4. Consider these two protein fragments:

   Gly—Ala—Pro—Gly—Gly—Lys—Pro
   Asp—Glu—Lys—Val—His—Ser—Gly

   Which would most likely be part of a fibrous protein? A globular protein? Explain.
5. Show what would happen if this polypeptide were mildly oxidized:

   Gly—Cys—Ser—Glu—Gly—Ala—Cys—Met

6. Using amino acids that can form disulfide bridges, ionic attractions, and hydrogen bonding, design polypeptide molecules that could have these shapes.
   a. 8-shaped
   b. T-shaped
7. Arsenic and phosphorus are both in Group V of the periodic table. How might this fact account for arsenic's poisonous nature?
8. Write the products of the hydrolysis of the glyceryl ester shown on page 474.
9. House plants placed in dark corners tend to grow greener leaves than those placed in direct sunlight. Suggest a reason for this.
10. A good cross-country skier can use 15 Calories (kcal) per minute. How many moles of ATP would be used by the skier in an hour of skiing?
11. How many moles of oxygen would have to be breathed to produce the equivalent of 100 kilocalories of energy via the respiratory chain?
12. Calculate the number of kilocalories of energy that can be obtained from one gram of glucose.
13. Calculate the effect of 0.001 moles of adrenalin on a blood-sugar level that starts at 60 mg/ml. Assume a total blood volume of 5 liters. How much energy, in kilocalories, would be produced?
14. Trace the path of the following through the digestive process.
    a. the triglyceride

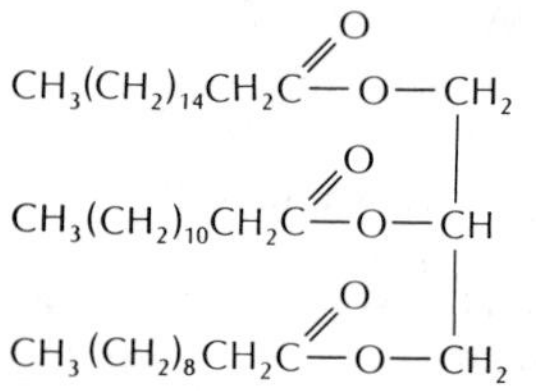

    $$\begin{array}{l} CH_3(CH_2)_{14}CH_2C(=O)—O—CH_2 \\ \qquad\qquad\qquad\qquad\qquad\quad | \\ CH_3(CH_2)_{10}CH_2C(=O)—O—CH \\ \qquad\qquad\qquad\qquad\qquad\quad | \\ CH_3(CH_2)_8CH_2C(=O)—O—CH_2 \end{array}$$

    b. table sugar (sucrose)
    c. muscle protein (meat)
    d. cellulose
15. Papain, an enzyme that hydrolyzes protein, is the active ingredient in meat tenderizer. Explain its function.
16. Does the body have a way of storing excess protein the way it does excess fat? What would happen to a person on a protein-free diet?
17. Vasopressin is sometimes used to prevent low blood pressure caused by shock following surgery. Explain how this might work.
18. Since fatty acids are slightly better energy sources than carbohydrates, why do we depend on carbohydrates and not on fats for energy? Explain what would happen to the biological cycle if suddenly no carbohydrates were available to animals.

APPENDIX A

Working with Numbers

A.1 ROUNDING OFF

Suppose you're buying a new electronic calculator to help you in this chemistry course. The price is $14.95. The local sales tax is 3.5%. Just for fun, you try out your new calculator to figure out how much the tax will be. When you push 14.95 × .035 =, the calculator gives an answer of 0.52325. What's this? The clerk looks at the tax table and says, "That's $14.95, plus 52 cents tax." Of course. Obviously you can't pay fractions of a cent, because cents are the smallest unit of money we have. So, the answer on the tax table has been rounded off to the nearest cent.

Many times in chemistry we'll need to round off numbers. *Rounding off* means eliminating all digits to the right of a specified place and then adjusting the numbers in the places to the left to reflect this change. The two rules for rounding off, along with some examples of those rules, are shown in Table A.1.

rounding off

TABLE A.1
Rules for rounding off, with examples

	Example	*Round to*	*Answer*
Rule 1. If the number to be eliminated is 5 or greater, add 1 to the number in front of it.	$3.489	Nearest cent	$3.49
	$3503.75	Nearest dollar	$3504
	93.99	One decimal place	94.0
	11595	Three digits	11600
Rule 2. If the number to be eliminated is less than five, drop it.	$4.5219	Nearest cent	$4.52
	$97.34	Nearest dollar	$97
	5.54	One decimal place	5.5
	0.05453	Three digits	0.0545

A.2 SIGNIFICANT FIGURES

WHAT THEY ARE. For any object that's being measured, the accuracy of the measurement is limited by the measuring device. You might get on the bathroom scale and say, "I weigh 125 pounds," but not, "I weigh 125.456 pounds." The lines on most bathroom scales only occur at every pound, and we can only guess at the nearest pound. The last figure—the nearest pound—is uncertain.

The last figure in any measurement is always uncertain. For instance, an automobile odometer usually gives mileage to the nearest tenth of a mile. If we want to know the mileage at any given time, we might look at the odometer and see something like this:

1	1	4	5	0	3/4

The last number is between 3 and 4. We can only guess at whether it's closer to the 3 or to the 4, and certainly we wouldn't know whether it's 11450.32 or 11450.36. The best we can do is guess at the first decimal place, but we do know that it's either a 3 or a 4. This figure is uncertain.

Significant figures are the figures in any measurement that we know accurately, plus one figure that is uncertain. If we decide that the last digit above is a 4, then 11450.4 is our number. This number contains six significant figures: the first five are known accurately and the last, 4, is uncertain but significant. This case is easy to understand, but some won't be. We need a set of rules for deciding how many significant figures a number has. Table A.2 lists and illustrates significant figure rules.

significant figure

MULTIPLYING AND DIVIDING WITH SIGNIFICANT FIGURES. Suppose that, in the course of trying out your new calculator, you do this operation: 7 divided by 3 times 3. Your calculator gives you the answer 6.9999999. This is puzzling, because you know that if you set it up as a fraction, the threes would cancel, like this:

$$\frac{7 \times \cancel{3}}{\cancel{3}} = 7$$

Clearly, the answer should be 7, but you got a six and all those nines instead. We can see why this happened if we take the operation in steps. Dividing 7 by 3 gives us 2.3333333 (the calculator cut it off at eight significant figures). If we then multiply 2.3333333 by 3, we of course get 6.9999999. The calculator didn't know that we were going to give it another 3 to cancel with the previous 3. We know, though, that we should round off this number to 7 and not keep it at 6.9999999.

The point is that when we're working with an electronic calculator, many times it'll give us an answer with far more places than we're justified in keeping. But how do we always know how many places we *are* justified in keep-

TABLE A.2
Significant figure rules, with examples

Rule	*Example*	*Number of Significant Figures*	*Reason*
1. Always begin counting significant figures with the first nonzero integer.	0.0000345	Three: 3, 4, and 5.	3 is the first nonzero integer.
2. All nonzero integers are significant.	1543	Four: 1, 5, 4, and 3.	All are integers.
3. Zeros.			
a. A zero is significant if it is not needed to fix the decimal point.	24.00	Four: 2, 4, 0, and 0.	Both zeros are significant. Writing 24 without them still places the decimal in the same spot.
	27,000	Two: 2 and 7.	These zeros are not significant because we do need them to fix the decimal point.
	0.004500	Four: 4, 5, 0, and 0.	The end zeros are significant. Writing 0.0045 without them still places the decimal in the same spot.
b. A decimal point placed after the end zeros in a number greater than one makes those zeros significant.	500.	Three: 5, 0, and 0.	Zeros before the point are significant. Without the point, they wouldn't be.
c. A zero between two significant figures is significant.	1.0087	Five: 1, 0, 0, 8, and 7.	The two zeros are between the 1 and the 8, both of which are significant.

ing? The example above is simple and obvious, but most are not. The rule for multiplying and dividing with significant figures is that the answer can't have more significant figures than the input number that has the fewest significant figures. When the calculator gives us more figures than this, we round off the calculator answer to the correct number of significant figures. Table A.3 illustrates how to use this rule.

ADDING AND SUBTRACTING WITH SIGNIFICANT FIGURES. When the president talks about the projected budget for the coming year, numbers like 52 billion dollars get thrown around right and left. These numbers are rounded numbers, because they're only an estimate.

TABLE A.3
Examples of multiplying and dividing with significant figures

Example	Calculator Answer	Limiting Input Number	No. Significant Figures	Corrected Answer
Rule: In multiplication or division, the answer may not have more significant figures than the input number that has the fewest significant figures.				
$2.36 \times 3.4 = ?$	8.024	3.4	2	8.0
$2.00 \times 0.003345 = ?$	0.00669	2.00	3	0.00669
$5.040 \times 3000 = ?$	15120	3000	1	20000
$6251 \times 93.0 = ?$	581343	93.0	3	581000
$\frac{11.95}{3.00} = ?$	3.9833333	3.00	3	3.98
$\frac{0.555}{9.070} = ?$	0.0611907	0.555	3	0.0612
$\frac{45.9}{15{,}000} = ?$	0.00306	15,000	2	0.0031
$\frac{5}{3.0} = ?$	1.6666666	5	1	2
$\frac{2100.0}{7.0000} = ?$	300	Either	5	300.00[a]

[a]Electronic calculators often do not give final zeroes that are significant. These must be added if they are needed, as they are in this case.

Suppose we then ask, "How will an additional expenditure of $3.47 for coffee cups affect the projected budget?" We might add it like this:

$$\begin{array}{r} 52{,}000{,}000{,}000 \\ 3.47 \\ \hline 52{,}000{,}000{,}003.47 \end{array}$$

Is this reasonable? No: 52 billion has only two significant figures, and we're trying to add 3.47 out there in the noise. 52,000,000,000 plus 3.47 is still 52,000,000,000.

In adding and subtracting, we round off the answer to the first column that contains an uncertain digit. In the operation above, we see that the second column (2) is the first one that has an uncertain digit, so our answer can't have any significant figures after that. Table A.4 illustrates this rule.

PURE NUMBERS. We've seen how to handle numbers used with measurements. Some things, though, can be counted exactly. These are *pure numbers.* If we counted four apples, the number would be exactly four, no more and no less. We can write the number as simply 4, or as 4.00, or with as many zeros after the decimal as we want or need in calculations. Numbers

pure number

TABLE A.4
Examples of adding and subtracting with significant figures

Example		*Column with Uncertain Digit*	*Answer*
Rule: In addition and subtraction, the answer must be rounded off to the first column that contains an uncertain digit.			
6.54 + 1.2766 = ?	6.54 1.2766 7.8166	Third column	7.82
9.5433 − 0.002 = ?	9.5433 − 0.002 9.5413	Fourth column	9.541
5400 + 0.122 = ?	5400 0.122 5400.122	Second column	5400[a]
155 + 100 + 206 = ?	155 100 206 461	First column	500
47,000 − 368 = ?	47000 − 368 46632	Second column	47,000
500. + 12 = ?	500. 12 512	Third column	512

[a]Note that the small addition is insignificant.

that are defined exactly are also pure numbers. An example is the atomic weight of carbon-12, which is defined as 12, or 12.00000. . . . We'll see later that some conversion factors are known as pure numbers because they're defined.

When we work with these pure numbers in problems, we may find that we want to alter the number of significant figures they seem to contain. This will be so when we're multiplying a pure number by a measured number that has limited significant figures. A pure number never limits the number of significant figures an answer can have.

A.3 NEGATIVE NUMBERS

One of the added bonuses of your new calculator is that you can use it to compute your checking account balance. Your computation might look something like this:

		Running Balance
Starting balance	134.56	
Electronic calculator	−15.47	119.09
Chemistry book	−17.95	102.14
Aspirin, No-Doz	−5.43	96.71
Room and board	−125.00	−29.29
New balance	−29.29	

Your calculator gives you an answer with a negative sign. What does that mean? It means that you're overdrawn. You'll have to go and quickly deposit at least 28.29 to cover yourself.

A calculator adds numbers algebraically. To *add algebraically* means to combine two or more numbers according to their signs (+ or −). If we're adding a positive number and a negative number, the result will be the difference between the two numbers. The sign of the answer depends on whether the positive number or the negative number is larger. algebraic addition

To illustrate this, imagine two towns: Town A, at elevation 2000 feet, and Town B, at 8000 feet. If we ask the question, "How much higher is Town B than Town A?" the answer is "6000 feet." But if we ask, "How much higher is Town A than Town B?" the answer is "−6000 feet." Our answer has a negative sign because Town A is *lower* than Town B. Positive and negative signs are an indication of direction. The difference in elevation between the two towns is always 6000 feet, but its sign can be + or − depending on which direction we take.

To *subtract* means to change the sign and add algebraically. Table A.5 illustrates various possibilities of addition and subtraction. subtraction

A.4 EXPONENTIAL NOTATION

WRITING NUMBERS AS EXPONENTIALS. Try multiplying Avogadro's number, 602,000,000,000,000,000,000,000, by anything. We can't even get that huge number onto a calculator. A calculator doesn't have enough windows. And why should it? We only need calculator windows for significant figures, and few measurement numbers even approach eight significant figures. All of those zeros are just excess baggage—important, of course, but something we can just count and set aside. We count them and keep track of them by writing them in *exponential notation.* We can write that huge number above as 6.02×10^{23}. The "10^{23}" is read, "ten to the twenty-third." The "23" is an *exponent* or *power* of ten. It tells us the number of times ten has to be multiplied by itself. To get the exponent, we count the number of places to the left that the decimal point must be moved (in this case, 23) in order to put it after the first significant figure (in this case, 6). exponential notation exponent (power)

Exponential notation can be used for very small numbers, too: for example, 0.00000000000000000000000167. In exponential notation, this is 1.67×10^{-24}. The "10^{-24}" is read, "ten to the minus twenty-fourth." To get

TABLE A.5
Examples of algebraic addition and subtraction

Operation	*Calculator Setup*	*Manual Setup*	*Explanation*
Add 3.4 to 4.3.	$3.4 + 4.3 = 7.7$ or $4.3 + 3.4 = 7.7$	3.4 $+\ 4.3$ 7.7	Signs of both numbers are (+).
Subtract 3.4 from 4.3.	$4.3 - 3.4 = 0.9$ or $-3.4 + 4.3 = 0.9$	4.3 $-\ 3.4$ 0.9	"Subtract 3.4" means that 3.4 becomes −3.4. Numbers may be given to calculator in any order.
Subtract 4.3 from 3.4.	$3.4 - 4.3 = -0.9$ or $-4.3 + 3.4 = -0.9$	-4.3 3.4 -0.9	Since the negative number is larger than the positive number, the answer is negative.
Add (−3.4) to 4.3.	$-3.4 + 4.3 = 0.9$ or $4.3 - 3.4 = 0.9$	4.3 -3.4 0.9	Numbers are combined according to their signs.
Subtract (−3.4) from 4.3.	$4.3 + 3.4 = 7.7$ or $3.4 + 4.3 = 7.7$	4.3 $-\ (-3.4)$ 7.7	"Subtract (−3.4)" means that − (−3.4) becomes (+3.4). We change the sign and add.
Subtract (−3.4) from (−4.3).	$-4.3 + 3.4 = -0.9$ or $3.4 - 4.3 = -0.9$	-4.3 $-\ (-3.4)$ -0.9	Again, a − (−3.4) becomes +3.4.
Add (−3.4) to (−4.3).	$-3.4 - 4.3 = -7.7$ or $-4.3 - 3.4 = -7.7$	-3.4 -4.3 -7.7	Both numbers are negative. Their sum is negative.
Add 3.4 to −4.3.	$3.4 - 4.3 = -0.9$ or $-4.3 + 3.4 = -0.9$	-4.3 $+\ 3.4$ -0.9	Again, answer is negative because negative number is larger than positive number.

this exponent, we count the number of places to the right that the decimal point must be moved (in this case, 24) in order to put it after the first significant figure (in this case, 1).

We see that here, too, positive and negative signs indicate direction. A negative exponent always means a number less than 1—that is, the original decimal point lies to the left of the first significant figure. A positive exponent always means a number greater than 1—that is, the original decimal point lies to the right of the first significant figure.

Numbers between 1 and 10 are usually written without exponents, but they could also be written as multiples of 10^0 ($10^0 = 1$). Whether the number is less than or greater than 1, we always put the decimal point after the first significant figures in the number. Thus we write "6.02×10^{23}" instead of "60.2×10^{22}" or "0.602×10^{24}," although they all mean the same thing.

SIGNIFICANT FIGURES AND EXPONENTIAL NOTATION. All digits in the coefficient of an exponential number are significant. The coefficient and other parts of an exponential number are shown below.

coefficient

exponent

exponential

$$\underset{\text{Coefficient}}{6.02} \times \underbrace{10^{\overset{\text{Exponent}}{23}}}_{\text{Exponential}}$$

Now we can show significant figures easily, by writing numbers in exponential notation. We also see that we can now specify significant figures for some numbers that we couldn't do any other way—for instance, expressing a number like 7,000 to two significant figures. We write this as 7.0×10^3, and there is no doubt as to how many significant figures this number has. We can't do it in its nonexponential form. Table A.6 shows some cases.

JUSTIFICATION OF EXPONENTIAL NUMBERS. An exponential number is *justified* when the decimal point of the coefficient appears directly after the first significant figure. 6.02×10^{23} is justified; 60.2×10^{22} is not. When we write numbers directly from the nonexponential form, they'll be justified. But often the answer that we get by adding, subtracting, multiplying, or dividing two or more exponential numbers will not be justified. The last step, therefore, is justifying the answer.

justification

To justify a number, first write the coefficient itself as an exponential. Then multiply the two exponentials together. Multiplying exponentials is easy: we just add the exponents. For example, to justify 669×10^{-2}, we write:

TABLE A.6
Exponential numbers, their nonexponential forms, and significant figures

Exponential Number	*Nonexponential Form*	*Number of Significant Figures*
3.47×10^3	3470	Three: 3, 4, and 7.
1.2×10^{-2}	0.012	Two: 1 and 2.
9.000×10^3	9000.	Four: 9, 0, 0, and 0.
6.720×10^5	67200	Four: 6, 7, 2, and 0. Note that this number cannot specify its significant figures in the nonexponential form.
8.14×10^{-4}	0.000814	Three: 8, 1, and 4.
1×10^2	100	One: 1.
1.000×10^2	100.0	Four: 1, 0, 0, and 0.

$$(6.69 \times 10^2)(10^{-2}) = 6.69 \times 10^0 = 6.69$$

We added the exponents like this: $2 + (-2) = 0$. We'll talk about multiplication and division of exponentials in the next section.

MULTIPLYING AND DIVIDING WITH EXPONENTIAL NUMBERS. To multiply exponential numbers, we multiply the coefficients and add the exponents algebraically. To divide exponential numbers, we divide the coefficient in the numerator by the coefficient in the denominator. We subtract the exponent in the denominator from the exponent in the numerator. Table A.7 shows multiplication and division of exponential numbers, with justification of answers.

Some calculators are equipped to handle exponentials. Instructions vary from one calculator to the other, so consult the instruction manual.

ADDING AND SUBTRACTING WITH EXPONENTIAL NUMBERS. To add and subtract exponential numbers, the exponents have to be the same. If they aren't, then we have to change one of the exponents so that it is the same as the other exponent. Then we add and subtract the coefficients, but we *don't* add and subtract the exponents. One rule of thumb for deciding which exponent to change is to change the exponent of the smaller number. This usually eliminates the need to justify the answer.

TABLE A.7
Multiplication and division of exponential numbers, with justified answers

	Coefficient Answer			
Example	*Calculator*	*Corrected*	*Exponent Answer*	*Full Answer*
$3.55 \times 10^4 \times 2.05 \times 10^3 = ?$	7.2775	7.28	$10^{(4+3)} = 10^7$	7.28×10^7
$4.32 \times 10^{-1} \times 1.19 \times 10^{-7} = ?$	5.1408	5.14	$10^{(-1-7)} = 10^{-8}$	5.14×10^{-8}
$6.4 \times 10^{-2} \times 9.88 \times 10^8 = ?$	63.232	63	$10^{(-2+8)} = 10^6$	63×10^6 $(6.3 \times 10^1)(10^6) = 6.3 \times 10^7$
$\frac{10.1 \times 10^7}{5.22 \times 10^2} = ?$	1.9348659	1.93	$10^{(7-2)} = 10^5$	1.93×10^5
$\frac{4.21 \times 10^{-4}}{8.22 \times 10^{-2}} = ?$	0.5121654	0.512	$10^{(-4-(-2))} = 10^{-2}$	0.512×10^{-2} $(5.12 \times 10^{-1})(10^{-2}) = 5.12 \times 10^{-3}$
$\frac{8.2 \times 10^3}{2.6 \times 10^5} = ?$	3.1538461	3.2	$10^{(3-5)} = 10^{-2}$	3.15×10^{-2}
$\frac{1.02 \times 10^{-2}}{9.75 \times 10^{-8}} = ?$	0.1046153	0.105	$10^{(-2-(-8))} = 10^6$	$0.105 \times 10^6 = (1.05 \times 10^{-1})(10^6) = 1.05 \times 10^5$

TABLE A.8
Addition and subtraction of exponential numbers

Example	Change	To	Method	Solution	Correct, Justified Answer
4.35×10^3 $+9.22 \times 10^3$?	No change. Both exponents are the same.			4.35×10^3 $+9.22 \times 10^3$ 13.57×10^3	1.357×10^4
2.79×10^3 -1.62×10^2 ?	1.62×10^2	$? \times 10^3$	$(1.62)(10^{-1})(10^3)$ $= 0.162 \times 10^3$	2.79×10^3 -0.162×10^3 2.628×10^3	2.63×10^3
6.44×10^{-4} -2.33×10^{-5} ?	2.33×10^{-5}	$? \times 10^{-4}$	$(2.33)(10^{-1})(10^{-4})$ $= 0.233 \times 10^{-4}$	6.44×10^{-4} -0.233×10^{-4} 6.207×10^{-4}	6.21×10^{-4}
8.65×10^5 $+2.28 \times 10^2$?	2.28×10^2	$? \times 10^5$	$(2.28)(10^{-3})(10^5)$ $= 0.00228 \times 10^5$	8.65×10^5 $+0.00228 \times 10^5$ 8.65228×10^5	8.65×10^5 [a]

[a] Note that 2.28×10^2 is insignificant compared with 8.65×10^5.

Changing an exponent is simply *un*justifying the number. We reverse the steps that we took to justify numbers. For instance, if we want to write 6.02×10^{23} as some number times 10^{20}, we do it this way:

$$6.02 \times 10^{23} = (6.02)(10^3)(10^{20}) = 6020 \times 10^{20}$$

Table A.8 shows problems in adding and subtracting, with and without changing exponents.

We notice that when exponents are very far apart relative to the numbers of significant figures in the coefficients, the smaller number becomes insignificant. What if you tried to do this: you get on the bathroom scale and weigh yourself. Then you get on the same scale with a letter, and you try to see how much more the letter weighs to know the postage. Ridiculous. The precision of a bathroom scale doesn't even come close to the weight of the letter, so you'd notice no difference. We don't usually add or subtract very large numbers with very small numbers, because the effect of the small number is lost.

A.5 APPROXIMATE ANSWERS

Another use you've discovered for your new electronic calculator is in the grocery store. You can add up the contents of your shopping cart as you go along, so that you don't spend too much money. But today, unfortunately, you accidentally left it at home. You have only \$7 to spend and you want to make sure you don't go over that amount. You estimate the cost of your groceries like this: One dozen eggs at 79¢—that's about 80¢; a head of lettuce, 49¢ (about 50¢)—\$1.30 so far. A box of crackers is 83¢ (about 80¢)—\$2.10 so far. One T-bone steak is \$3.46 (about \$3.50)—that's \$5.60 so far. You probably can get a pound of potatoes for 55¢ (about 60¢). That comes to approximately \$6.20. When you check out, the register slip says \$6.21, plus 22¢ tax: \$6.43. Whew, just made it.

In chemistry, we need to be able to do approximate calculations like that. It's handy if we don't happen to have a calculator, and it's useful even if we do. In doing a calculation that involves a long string of numbers, it's easy to push the wrong button in one or more of the operations, and get the wrong answer. We'd never know whether or not it happened unless we had some feeling for the sort of answer we were expecting. For instance, in the example above, you were expecting an answer of about \$6.20 more or less. If the clerk had said, "\$62.00, please," you'd have known that something was wrong with the cash register, the clerk, or your approximate answer.

To do approximate calculations, first simultaneously round off each number to one significant figure and express it as an exponential. Then compute the answer in your head, rounding off any intermediate steps to the nearest single significant figure. Cancel numbers or exponents above and below the fraction line wherever possible. If the calculator answer is way out of line with the approximate answer, you should check both the calculator answer and the approximate answer. Table A.9 shows how to calculate approximate answers and relate them to calculator answers.

A.6 LOGARITHMS

LOGARITHMS OF EXPONENTIALS. A *logarithm* is an exponent. When the number that has the exponent is 10, the exponent is a *base ten logarithm,* also called *common logarithm.*

logarithm

base ten logarithm

common logarithm

The number 100 can be written as the exponential 10^2, where 10 is the base, and 2 is the base ten logarithm of 10^2. The mathematical way of saying this is

$$\log_{10}10^2 = 2 \qquad (\log_{10}100 = 2)$$

We read this "log-ten of ten squared equals two" (or "log-ten of one hundred equals two"). We sometimes find it convenient to abbreviate logarithm as "log."

Another type of logarithm used in science is the "base e" or "natural" logarithm, abbreviated either $\log_e$, ln, or LN. We won't need to define or use natural logarithms. The only reason for mentioning them here is that many calculators come equipped with both common and natural logs, and we want to make sure that we use the common log function only. In this

TABLE A.9
Approximate answers

Example	*Approximate Answer*	*Calculator Answer*	*Conclusion*
$454 \times 22.4 = ?$	$5 \times 10^2 \times 2 \times 10^1 = 10 \times 10^3 = 1 \times 10^4$	$10169.6 = 1.02 \times 10^4$	OK
$\frac{1.986}{760} = ?$	$\frac{\overset{1}{\not{2}}}{\underset{4}{\not{8}} \times 10^2} = 0.25 \times 10^{-2} = 3 \times 10^{-3}$	$0.0026131 = 2.6 \times 10^{-3}$	OK
$22.4 \times \frac{749}{760} = ?$	$2 \times 10^1 \times \frac{\not{7} \times \not{10}^2}{\not{8} \times \not{10}^2} = 2 \times 10^1$	$22.075789 = 2.21 \times 10^1$	OK
$\frac{18.9}{111 \times 100.} = ?$	$\frac{2 \times 10^1}{1 \times 10^2 \times 1 \times 10^2} = 2 \times 10^{-3}$	$0.0017027 = 1.70 \times 10^{-3}$	OK
$\frac{1.45 \times 62.4 \times 298}{1.00 \times 756} = ?$	$\frac{1 \times \overset{3}{\not{6}} \times 10^1 \times 3 \times \not{10}^2}{1 \times \underset{4}{\not{8}} \times \not{10}^2} = \frac{9}{4} \times 10^1 = 2 \times 10^1$	$20384058 = 2.04 \times 10^7$ $35.665396 = 3.57 \times 10^1$	Something is wrong. Answers differ too much. Do calculator answer again. OK now.
$\frac{0.0821 \times 11.2 \times 40.0}{273 \times 2.98 \times 0.760} = ?$	$\frac{\not{8} \times 10^{-2} \times 1 \times 10^1 \times 4 \times 10^1}{3 \times 10^2 \times 3 \times \not{8} \times 10^{-1}} =$	$0.0594879 = 5.95 \times 10^{-2}$	OK

book we'll use only base ten (common) logs, and we'll adopt the shorthand notation of leaving out that designation, like this:

$$\log 10^2 = 2$$

We would say this "log ten squared equals two."

Just as we can have negative exponents, we can have negative logarithms. Like exponents, negative logarithms indicate a number that is less than 1:

$$\log 10^{-2} = -2$$

We can have fractional exponents, too. Thus,

$$\log 10^{1/2} = \tfrac{1}{2}, \text{ and } \log 10^{.235} = 0.235$$

LOGARITHMS OF EXPONENTIAL NUMBERS. In the section on exponential notation, we saw that an exponential number has a coefficient and an exponential, like this:

$$1 \times 10^2$$

When the coefficient is 1, the logarithm is the same as it was with the pure exponential.

$$\log(1 \times 10^2) = 2$$

It is the same because of a very useful characteristic of logarithms: *The logarithm of several numbers multiplied together is the sum of their individual logarithms.* That is,

$$\log(1 \times 10^2) = \log 1 + \log 10^2$$

Another important characteristic of logarithms that's worth remembering is that *the logarithm of one is zero.* We can say this mathematically:

$$\log 1 = 0$$

It follows that $\log(1 \times 10^2) = \log 1 + \log 10^2 = 0 + 2 = 2$. But why does $\log 1 = 0$?

Another way of writing "1," we saw in Section A.5, is "10^0"; therefore, $\log 1 = \log 10^0 = 0$.

When the coefficient of the exponential isn't 1, then we'll have a fractional logarithm. There are two ways to find the logarithm of such a number: We can use a logarithm table or a calculator.

FINDING LOGARITHMS WITH A TABLE.

Logarithms, like exponentials, have two parts, shown below:

$$\log 2.500 \times 10^3 = 3.\overbrace{3979}^{\text{mantissa}}$$

(characteristic: 3; mantissa: 3979)

The characteristic is a whole number and comes from the exponential. The mantissa is a fractional number and comes from the coefficient, like this:

$$\begin{aligned}\log 2.500 \times 10^3 &= \log 2.500 + \log 10^3 \\ &= \quad 0.3979 \quad + \quad 3 \quad = 3.3979\end{aligned}$$

We get the logarithm of 2.500 from Table A.10. In doing this, we must observe a rule of significant figures that applies to logarithms: *The mantissa of a logarithm must have the same number of significant figures as the number whose logarithm is being determined.* Thus the number 2.500 has four significant figures and so does the mantissa, 0.3979.

Here are the steps for finding the logarithm of a number using the log table.

Find the logarithm of 234.

Step 1. *Write the number as a justified exponential number.*

$$234 = 2.34 \times 10^2$$

Step 2. *Write its logarithm in two parts.*

$$\log(2.34 \times 10^2) = \log 2.34 + \log 10^2$$

TABLE A.10
Table of four-place logarithms

	0	1	2	3	4	5	6	7	8	9
1.0	.0000	.0043	.0086	.0128	.0170	.0212	.0253	.0294	.0334	.0374
1.1	.0414	.0453	.0492	.0531	.0569	.0607	.0645	.0682	.0719	.0755
1.2	.0792	.0828	.0864	.0899	.0934	.0969	.1004	.1038	.1072	.1106
1.3	.1139	.1173	.1206	.1239	.1271	.1303	.1335	.1367	.1399	.1430
1.4	.1461	.1492	.1523	.1553	.1584	.1614	.1644	.1673	.1703	.1732
1.5	.1761	.1790	.1818	.1847	.1875	.1903	.1931	.1959	.1987	.2014
1.6	.2041	.2068	.2095	.2122	.2148	.2175	.2201	.2227	.2253	.2279
1.7	.2304	.2330	.2355	.2380	.2405	.2430	.2455	.2480	.2504	.2529
1.8	.2553	.2577	.2601	.2625	.2648	.2672	.2695	.2718	.2742	.2765
1.9	.2788	.2810	.2833	.2856	.2878	.2900	.2923	.2945	.2967	.2989
2.0	.3010	.3032	.3054	.3075	.3096	.3118	.3139	.3160	.3181	.3201
2.1	.3222	.3243	.3263	.3284	.3304	.3324	.3345	.3365	.3385	.3404
2.2	.3424	.3444	.3464	.3483	.3502	.3522	.3541	.3560	.3579	.3598
2.3	.3617	.3636	.3655	.3674	.3692	.3711	.3729	.3747	.3766	.3784
2.4	.3802	.3820	.3838	.3856	.3874	.3892	.3909	.3927	.3945	.3962
2.5	.3979	.3997	.4014	.4031	.4048	.4065	.4082	.4099	.4116	.4133
2.6	.4150	.4166	.4183	.4200	.4216	.4232	.4249	.4265	.4281	.4298
2.7	.4314	.4330	.4346	.4362	.4378	.4393	.4409	.4425	.4440	.4456
2.8	.4472	.4487	.4502	.4518	.4533	.4548	.4564	.4579	.4594	.4609
2.9	.4624	.4639	.4654	.4669	.4683	.4698	.4713	.4728	.4742	.4757
3.0	.4771	.4786	.4800	.4814	.4829	.4843	.4857	.4871	.4886	.4900
3.1	.4914	.4928	.4942	.4955	.4969	.4983	.4997	.5011	.5024	.5038
3.2	.5051	.5065	.5079	.5092	.5105	.5119	.5132	.5145	.5159	.5172
3.3	.5185	.5198	.5211	.5224	.5237	.5250	.5263	.5276	.5289	.5302
3.4	.5315	.5328	.5340	.5353	.5366	.5378	.5391	.5403	.5416	.5428
3.5	.5441	.5453	.5465	.5478	.5490	.5502	.5514	.5527	.5539	.5551
3.6	.5563	.5575	.5587	.5599	.5611	.5623	.5635	.5647	.5658	.5670
3.7	.5682	.5694	.5705	.5717	.5729	.5740	.5752	.5763	.5775	.5786
3.8	.5798	.5809	.5821	.5832	.5843	.5855	.5866	.5877	.5888	.5899
3.9	.5911	.5922	.5933	.5944	.5955	.5966	.5977	.5988	.5999	.6010
4.0	.6021	.6031	.6042	.6053	.6064	.6075	.6085	.6096	.6107	.6117
4.1	.6128	.6138	.6149	.6160	.6170	.6180	.6191	.6201	.6212	.6222
4.2	.6232	.6243	.6253	.6263	.6274	.6284	.6294	.6304	.6314	.6325
4.3	.6335	.6345	.6355	.6365	.6375	.6385	.6395	.6405	.6415	.6425
4.4	.6435	.6444	.6454	.6464	.6474	.6484	.6493	.6503	.6513	.6522
4.5	.6532	.6542	.6551	.6561	.6571	.6580	.6590	.6599	.6609	.6618
4.6	.6628	.6637	.6646	.6656	.6665	.6675	.6684	.6693	.6702	.6712
4.7	.6721	.6730	.6739	.6749	.6758	.6767	.6776	.6785	.6794	.6803
4.8	.6812	.6821	.6830	.6839	.6848	.6857	.6866	.6875	.6884	.6893
4.9	.6902	.6911	.6920	.6928	.6937	.6946	.6955	.6964	.6972	.6981
5.0	.6990	.6998	.7007	.7016	.7024	.7033	.7042	.7050	.7059	.7067
5.1	.7076	.7084	.7093	.7101	.7110	.7118	.7126	.7135	.7143	.7152
5.2	.7160	.7168	.7177	.7185	.7193	.7202	.7210	.7218	.7226	.7235
5.3	.7243	.7251	.7259	.7267	.7275	.7284	.7292	.7300	.7308	.7316
5.4	.7324	.7332	.7340	.7348	.7356	.7364	.7372	.7380	.7388	.7396

	0	1	2	3	4	5	6	7	8	9
5.5	.7404	.7412	.7419	.7427	.7435	.7443	.7451	.7459	.7466	.7474
5.6	.7482	.7490	.7497	.7505	.7513	.7520	.7528	.7536	.7543	.7551
5.7	.7559	.7566	.7574	.7582	.7589	.7597	.7604	.7612	.7619	.7627
5.8	.7634	.7642	.7649	.7657	.7664	.7672	.7679	.7686	.7694	.7701
5.9	.7709	.7716	.7723	.7731	.7738	.7745	.7752	.7760	.7767	.7774
6.0	.7782	.7789	.7796	.7803	.7810	.7818	.7825	.7832	.7839	.7846
6.1	.7853	.7860	.7868	.7875	.7882	.7889	.7896	.7903	.7910	.7917
6.2	.7924	.7931	.7938	.7945	.7952	.7959	.7966	.7973	.7980	.7987
6.3	.7993	.8000	.8007	.8014	.8021	.8028	.8035	.8041	.8048	.8055
6.4	.8062	.8069	.8075	.8082	.8089	.8096	.8102	.8109	.8116	.8122
6.5	.8129	.8136	.8142	.8149	.8156	.8162	.8169	.8176	.8182	.8189
6.6	.8195	.8202	.8209	.8215	.8222	.8228	.8235	.8241	.8248	.8254
6.7	.8261	.8267	.8274	.8280	.8287	.8293	.8299	.8306	.8312	.8319
6.8	.8325	.8331	.8338	.8344	.8351	.8357	.8363	.8370	.8376	.8382
6.9	.8388	.8395	.8401	.8407	.8414	.8420	.8426	.8432	.8439	.8445
7.0	.8451	.8457	.8463	.8470	.8476	.8482	.8488	.8494	.8500	.8506
7.1	.8513	.8519	.8525	.8531	.8537	.8543	.8549	.8555	.8561	.8567
7.2	.8573	.8579	.8585	.8591	.8597	.8603	.8609	.8615	.8621	.8627
7.3	.8633	.8639	.8645	.8651	.8657	.8663	.8669	.8675	.8681	.8686
7.4	.8692	.8698	.8704	.8710	.8716	.8722	.8727	.8733	.8739	.8745
7.5	.8751	.8756	.8762	.8768	.8774	.8779	.8785	.8791	.8797	.8802
7.6	.8808	.8814	.8820	.8825	.8831	.8837	.8842	.8848	.8854	.8859
7.7	.8865	.8871	.8876	.8882	.8887	.8893	.8899	.8904	.8910	.8915
7.8	.8921	.8927	.8932	.8938	.8943	.8949	.8954	.8960	.8965	.8971
7.9	.8976	.8982	.8987	.8993	.8998	.9004	.9009	.9015	.9020	.9026
8.0	.9031	.9036	.9042	.9047	.9053	.9058	.9063	.9069	.9074	.9079
8.1	.9085	.9090	.9096	.9101	.9106	.9112	.9117	.9122	.9128	.9133
8.2	.9138	.9143	.9149	.9154	.9159	.9165	.9170	.9175	.9180	.9186
8.3	.9191	.9196	.9201	.9206	.9212	.9217	.9222	.9227	.9232	.9238
8.4	.9243	.9248	.9253	.9258	.9263	.9269	.9274	.9279	.9284	.9289
8.5	.9294	.9299	.9304	.9309	.9315	.9320	.9325	.9330	.9335	.9340
8.6	.9345	.9350	.9355	.9360	.9365	.9370	.9375	.9380	.9385	.9390
8.7	.9395	.9400	.9405	.9410	.9415	.9420	.9425	.9430	.9435	.9440
8.8	.9445	.9450	.9455	.9460	.9465	.9469	.9474	.9479	.9484	.9489
8.9	.9494	.9499	.9504	.9509	.9513	.9518	.9523	.9528	.9533	.9538
9.0	.9542	.9547	.9552	.9557	.9562	.9566	.9571	.9576	.9581	.9586
9.1	.9590	.9595	.9600	.9605	.9609	.9614	.9619	.9624	.9628	.9633
9.2	.9638	.9643	.9647	.9652	.9657	.9661	.9666	.9671	.9675	.9680
9.3	.9685	.9689	.9694	.9699	.9703	.9708	.9713	.9717	.9722	.9727
9.4	.9731	.9736	.9741	.9745	.9750	.9754	.9759	.9763	.9768	.9773
9.5	.9777	.9782	.9786	.9791	.9795	.9800	.9805	.9809	.9814	.9818
9.6	.9823	.9827	.9832	.9836	.9841	.9845	.9850	.9854	.9859	.9863
9.7	.9868	.9872	.9877	.9881	.9886	.9890	.9894	.9899	.9903	.9908
9.8	.9912	.9917	.9921	.9926	.9930	.9934	.9939	.9943	.9948	.9952
9.9	.9956	.9961	.9965	.9969	.9974	.9978	.9983	.9987	.9991	.9996

Step 3. *Evaluate the logarithm of the coefficient (the mantissa) from the log table.*

Find the number 2.3 in the lefthand column. Then read across to column 4. The number, 0.3692, is the mantissa, but we round it off to 0.369 because our number, 2.34, had only three significant figures.

$$\log 2.34 = 0.369$$

Step 4. *Evaluate the logarithm of the exponential (the characteristic).*
This is easy: $\log 10^2 = 2$.

Step 5. *Add the mantissa to the characteristic to get the log.*

$$0.369 + 2 = 2.369$$

Answer: 2.369.

To find the logarithm of a number less than 1, we follow exactly the same procedure, being careful not to lose track of the minus sign.

Find the logarithm of 0.00769.

Step 1. $0.00769 = 7.69 \times 10^{-3}$
Step 2. $\log(7.69 \times 10^{-3}) = \log 7.69 + \log 10^{-3}$
Step 3. From table, $\log 7.69 = 0.8859$, which we round to 0.886
Step 4. $\log 10^{-3} = -3$
Step 5. $\log(7.69 \times 10^{-3}) = \log 7.69 + \log 10^{-3} = 0.886 + (-3) = -2.114$
Answer: -2.114.

We see that, for a number less than one, the characteristic will always be one less in absolute value than the exponent. It is less because the log of the coefficient is positive and the log of the exponential is negative. Table A.11 gives some examples of finding logarithms using the table.

FINDING LOGARITHMS WITH A CALCULATOR. To do this, you must have a calculator with a log function. Consult the instructions that accompany your calculator for details about using the log function. If your calculator has a log function, finding the logarithm of a number is easy.

Find the logarithm of 234.

Step 1. *Punch in the number.*
Step 2. *Press log.*
Step 3. *Read the logarithm from the display.* Round to three places. The display reads 2.3692159, which we round to 2.369.
Answer: 2.369 (the same answer we got with the table).

Notice that, to find a log on a calculator, we don't have to write the number in exponential notation. At first, though, it may help to do so, to get an idea of what the characteristic should be. Above, if we got an answer of 7.5698321, we'd know it was wrong because the exponent was 2, and therefore the characteristic will be 2. Repeat all the examples in Table A.11 and check the answers obtained with the table.

TABLE A.11
Finding logarithms of numbers

Number	*Exponential Form*	*Separate Logarithms*	*Mantissa*	*Charac-teristic*	*Addition of Mantissa to Characteristic*		*Logarithm*
72.6	7.26×10^1	$\log 7.26 + \log 10^1$	.861	1	$0.861 + 1$	=	1.861
0.145	1.45×10^{-1}	$\log 1.45 + \log 10^{-1}$	.161	−1	$0.1614 - 1$	=	−0.839
2.49	—	log 2.49	.396	—	—		0.396
0.0000563	5.63×10^{-5}	$\log 5.63 + \log 10^{-5}$	.751	−5	$0.751 - 5$	=	−4.249
3460	3.46×10^3	$\log 3.46 + \log 10^3$	.539	3	$0.539 + 3$	=	3.539
—	6.02×10^{23}	$\log 6.02 + \log 10^{23}$	.780	23	$0.780 + 23$	=	23.780
125	1.25×10^2	$\log 1.25 + \log 10^2$	.0969	2	$0.0969 + 2$	=	2.0969
0.0081	8.1×10^{-3}	$\log 8.1 + \log 10^{-3}$	.91	−3	$0.91 - 3$	=	−2.09

FINDING ANTILOGARITHMS WITH A TABLE. Just as multiplication is the inverse of division, an *antilogarithm* is the inverse of a logarithm (an "un-logarithm"). That is,

antilogarithm

$$\text{antilog } 2.3692 = 2.34 \times 10^2$$

To find the antilogarithm of a number, we follow these steps.

Find the antilogarithm of 5.75.

Step 1. *Find the mantissa in the body of the log table.* If the exact number doesn't appear there, pick the closest value. Read the antilog from the horizontal line and vertical column.

We look for .75 in the body of the table. The closest we find is .7497. The antilogarithm is 5.62: we find 5.6 as the horizontal row, and 2 as the column, but 5.6 gives us the right number of significant figures.

Step 2. *Express the characteristic as an exponential.*

$$\text{antilog } 5 = 10^5$$

Step 3. *Put them together with a multiplication sign between.*

$$\text{antilog } 5.75 = 5.6 \times 10^5$$

Answer: 5.6×10^5.

Finding the antilogarithm of a negative log is a little trickier.

Find the antilogarithm of −3.593.

Step 1. *Subtract the logarithm from the next higher integer.* In this case, 4 is the next higher integer. $4 - 3.593 = 0.407$.

Step 2. *Find the antilogarithm of the result of Step 1 in the table.* This antilogarithm will be the coefficient of the result.

Here, we look for 0.407 on the table: we find .4065. The antilogarithm of this is 2.55. (It has 3 significant figures, which is correct here.)

TABLE A.12
Finding antilogarithms of logarithms

Positive Logarithms

Logarithm	*Mantissa*	*Antilog of Mantissa*	*Characteristic*	*Antilog of Characteristic*	*Antilogarithm*
1.597	0.597	3.96	1	10^1	3.96×10^1
10.56	0.56	3.6	10	10^{10}	3.6×10^{10}
0.875	0.875	7.50	—	—	7.50
3.116	0.116	1.31	3	10^3	1.31×10^3

Negative Logarithms

Logarithm	*Next Higher Integer*	*Difference*	*Antilogarithm of Difference*	*Exponential*	*Antilogarithm*
−2.395	3	0.605	4.03	10^{-3}	4.03×10^{-3}
−0.447	1	0.553	3.57	10^{-1}	3.57×10^{-3}
−9.87	10	0.13	1.35	10^{-10}	1.35×10^{-3}
−5.027	6	0.973	9.40	10^{-6}	9.40×10^{-6}

Step 3. *Use the integer from Step 1 as a negative exponent, and write the exponential.*

4 was our integer: 10^{-4}

Step 4. *Write the two parts together with a multiplication sign between.*

$$2.55 \times 10^{-4}$$

Answer: 2.55×10^{-4}.

Notice that we may always check our answer by taking the logarithm again and seeing whether it is the same as our original number. In the same way, we can check a logarithm by taking its antilog. Table A.12 gives some examples of logarithms and their antilogarithms.

FINDING ANTILOGARITHMS WITH A CALCULATOR.

To do this, we use the log and INV buttons.

Find the antilogarithm of 5.75.

Step 1. *Punch the number into the calculator.*
Step 2. *Press INV, then log.*
Step 3. *Read the answer from the display.*
We read 562341.32, which we round to three significant figures.
Answer: 5.62×10^5 (the same answer we got from the table).

Work all the examples in Table A.12 with a calculator and check the results.

EXERCISES

Note: Answers are given on p. 545.

1. Round off the following numbers as indicated.
   a. 3.489 to two decimal places
   b. 45.219 to one decimal place
   c. 67,432 to the nearest thousand
   d. \$3.42 to the nearest dollar
   e. 0.122 to two decimal places
2. State the number of significant figures in each of the following.
   a. 0.00123
   b. 6789
   c. 0.0070
   d. 9.87
   e. 8.040
   f. 98600
   g. 200.
   h. 101.0
3. Perform these operations, and correct the significant figures.
   a. $3.0 \times 4.00 =$
   b. $6.00 \times 10.0 =$
   c. $2.000 \times 0.0017 =$
   d. $400 \times 1.98 =$
   e. $12.00/2.0 =$
   f. $8.0/2.00 =$
   g. $9.54/2000 =$
4. Perform these operations, and correct the significant figures.
   a. $\begin{array}{r} 3.980 \\ +0.001 \\ \hline \end{array}$
   b. $\begin{array}{r} 54.09 \\ +0.230 \\ \hline \end{array}$
   c. $\begin{array}{r} 143.9 \\ -40. \\ \hline \end{array}$
   d. $\begin{array}{r} 777.12 \\ -98.1 \\ \hline \end{array}$
   e. $\begin{array}{r} 9.12 \\ +11.9760 \\ \hline \end{array}$
   f. $\begin{array}{r} 12.0 \\ -9.4711 \\ \hline \end{array}$
5. In arithmetic, we learned that $4 \times 4 = 16$. Is this always true when we're talking about scientific measurements? What circumstances would cause this equality to not be true?
6. Add algebraically.
   a. $8.2 - 4.3 =$
   b. $9.7 + 2.2 =$
   c. $6.2 - 10.2 =$
   d. $-3.3 + 2.1 =$
   e. $-4.6 + 9.0 =$
   f. $-6.6 - 4.7 =$
   g. $4.4 - (-1.9) =$
   h. $-(-3.2) - 2.2 =$
7. Subtract.
   a. 3.3 from 7.9
   b. 7.9 from 3.3
   c. 0.2 from (−7.7)
   d. −2 from −4
   e. −3 from 6
   f. 2.7 from (−1.3)
   g. (−9.8) from 3.3
   h. (−3) from (−5)
8. Write these numbers in exponential notation with the correct number of significant figures.
   a. 52800
   b. 0.00000897
   c. 21000
   d. 0.0000000650
   e. 980
   f. 6
9. State the number of significant figures in each of the following, and justify.
   a. 39.8×10^{7}
   b. 4.30×10^{4}
   c. 0.198×10^{-3}
   d. 101.0×10^{-3}
   e. 454×10^{2}
   f. 233×10^{-9}
10. Perform the following multiplications, and justify the answers (use the correct number of significant figures).
    a. $(1.03 \times 10^{-5}) \times (7.22 \times 10^{2}) =$
    b. $(4.87 \times 10^{9}) \times (9.22 \times 10^{10}) =$
    c. $(6.34 \times 10^{-2}) \times (2.37 \times 10^{-3}) =$
    d. $(5.61 \times 10^{22}) \times (6.02 \times 10^{23}) =$
11. Perform the following divisions, and justify the answers (use the correct number of significant figures).
    a. $\dfrac{4.18 \times 10^{8}}{2.63 \times 10^{2}} =$
    b. $\dfrac{5.99 \times 10^{2}}{9.81 \times 10^{6}} =$
    c. $\dfrac{7.23 \times 10^{-4}}{3.49 \times 10^{2}} =$
    d. $\dfrac{3.75 \times 10^{8}}{4.99 \times 10^{-2}} =$
12. Perform the following additions or subtractions, and justify the answers (use the correct number of significant figures).
    a. $\begin{array}{r} 1.22 \times 10^{-5} \\ -\ 3.45 \times 10^{-5} \\ \hline \end{array}$
    b. $\begin{array}{r} 4.68 \times 10^{4} \\ +\ 3.99 \times 10^{3} \\ \hline \end{array}$
    c. $\begin{array}{r} 2.97 \times 10^{-4} \\ -3.42 \times 10^{-3} \\ \hline \end{array}$
    d. $\begin{array}{r} 5.72 \times 10^{-9} \\ +4.61 \times 10^{-10} \\ \hline \end{array}$
    e. $\begin{array}{r} 8.53 \times 10^{-2} \\ -6.47 \times 10^{-3} \\ \hline \end{array}$
    f. $\begin{array}{r} 1.46 \times 10^{5} \\ +7.21 \times 10^{6} \\ \hline \end{array}$
13. Compute approximate answers.
    a. $0.0234 \times 34.6 =$
    b. $\dfrac{5670}{19.8} =$
    c. $22.4 \times 760 \times 273 =$
    d. $\dfrac{2.58 \times 88.7}{0.333} =$
    e. $\dfrac{5.95}{3.4 \times 75.2} =$
    f. $\dfrac{4.42 \times 189}{0.0012 \times 450} =$
    g. $\dfrac{79.8 \times 0.221 \times 899}{3.34 \times 454 \times 1.986} =$
14. By computing approximate answers, state which of the following calculator answers are incorrect.
    a. $98.6 \times 2.54 \times 454 = 113701.57$
    b. $\dfrac{22.4 \times 32 \times 23.0}{760 \times 2.2 \times 0.0821} = 69370302$

c. $\frac{44.0 \times 0.621 \times 8700}{77.2 \times 3 \times 88.0} = 11.663859$

d. $\frac{33.8 \times 0.11}{6.02 \times 10^{23}} = 0.6176079 \times 10^{-23}$

e. $\frac{1752 \times 1.67 \times 10^{-24} \times 888}{5.75 \times 98.0 \times 3.4 \times 10^{10}} = 15676.478 \times 10^{-34}$

15. For the correct answers in Exercise 14, express the calculator answer in justified exponential form with corrected significant figures.

16. Find the logarithms by inspection.

a. 1×10^4	e. 1×10^{-7}
b. 1	f. 100
c. 0.001	g. 1×10^5
d. 1×10^{-2}	h. 0.01

17. Find the logarithms, using a log table or a calculator or both as directed by your instructor.

a. 26.7	e. 876
b. 0.0821	f. 1.6×10^{-4}
c. 4.45×10^3	g. 6.78
d. 0.973	h. 11.7

18. Find the antilogarithms by inspection.

a. 4	c. 0
b. −2	d. −7

19. Find the antilogarithms, using a log table or a calculator or both as directed by your instructor.

a. −6.884	e. −2.31
b. 7.81	f. 1.9876
c. 0.566	g. 0.342
d. −0.628	h. −0.02334

APPENDIX B

Working with Units

In scientific calculations, each number usually has a label, which we call a unit. In mathematics, people work just with numbers, such as $2 \times 3 = 6$, but in science we usually want to know, "2 what? 3 what? 6 what?" The unit is the "what." A *unit* is a word or words that describe what is being measured or counted. It can be anything at all: 2 dollars, 2 atoms, 2 kilometers.

In Appendix A, working with numbers alone is discussed. In this appendix, we'll work with units that usually come with numbers. All the rules for working with numbers will be followed here. Appropriate sections of Appendix A should be consulted, if necessary, to review the rules about numbers.

B.1 METRIC UNITS

Scientists (including American and British scientists) use the metric system of units. The *metric system* is a decimal system of weights and measures based on the kilogram and on the meter. For us Americans, the metric system may at first seem more difficult than the English system, which we're used to. At the time of this writing, steps are being taken to convert the United States to the metric system, which really is much simpler and more convenient.

metric system

Our money system is a decimal system: 100 cents to the dollar. To change from one of these money units to the other, all we have to do is move a decimal point two places. Moving a decimal point two places to the right is the same as multiplying by 10^2, and moving a decimal point two places to the left is the same as multiplying by 10^{-2}. (For a review of exponential numbers, see Appendix A.4, p. 494.) The metric system is about as easy, once we've studied it awhile. Like our money system, we move a decimal point to change from one unit to another.

METRIC PREFIXES. In the metric system, prefixes tell us what to multiply a given unit by to get another unit, just as we know that we multiply dollars by 10^2 to get cents and cents by 10^{-2} to get dollars. Table B.1

TABLE B.1
Metric prefixes and their origins

Prefix	Symbol	Means Multiply by: Exponential	Rational	Origin	Original Meaning
exa-	E	10^{18}	1,000,000,000,000,000,000	Greek "hexa"	Six groups of three zeros
peta-	P	10^{15}	1,000,000,000,000,000	Greek "penta"	Five groups of three zeros
tera-	T	10^{12}	1,000,000,000,000	Greek	Monstrous
giga-	G	10^{9}	1,000,000,000	Greek	Gigantic
mega-	M	10^{6}	1,000,000	Greek	Great
kilo-	k	10^{3}	1,000	Greek	Thousand
hecto-	h	10^{2}	100	Greek	Hundred
deca-	da	10^{1}	10	Greek	Ten
deci-	d	10^{-1}	1/10	Latin	Tenth
centi-	c	10^{-2}	1/100	Latin	Hundredth
milli-	m	10^{-3}	1/1000	Latin	Thousandth
micro-	μ	10^{-6}	1/1,000,000	Greek	Small
nano-	n	10^{-9}	1/1,000,000,000	Greek	Very small
pico-	p	10^{-12}	1/1,000,000,000,000	Spanish	Extremely small
femto-	f	10^{-15}	1/1,000,000,000,000,000	Scandinavian	Fifteen
atto-	a	10^{-18}	1/1,000,000,000,000,000,000	Scandinavian	Eighteen

shows the metric prefixes expressed as exponentials as well as rational (fractional) numbers. For instance, the prefix *centi-* tells us to multiply by 10^{-2} (or to divide by 100). That's where the word "cent" comes from. The prefix *kilo-* means to multiply by 10^{3}, or 1000. The prefix *milli-* means to multiply by 10^{-3} (or to divide by 1000).

SI UNITS. The International Bureau of Weights and Measures in 1960 established a system of units to simplify communication among all scientists throughout the world. This is called the *International System of Units* (abbreviated SI, for the French Système International). It's constructed from seven basic units, given in Table B.2. The SI system recognizes other units derived from these, shown in Table B.3. Larger or smaller units from the basic or derived units should preferably be in multiples of three powers of ten. Thus the metric prefixes *deca-*, *deci-*, and *centi-* aren't technically approved SI prefixes. However, they were introduced before the SI system was established, and they are sometimes tolerated. Other units that were introduced before the SI system are still used by scientists, but some are recommended to be phased out. At the time of this writing, though, many of these units are still in use by chemists because of their greater convenience. Table B.4 shows commonly used units, both SI and non-SI. Values in terms of SI units are given, as well as useful conversion factors.

conversion factor

A *conversion factor* is a factor that tells how many of a given unit is contained in exactly one of another unit. If we say that there are 1000 grams in

TABLE B.2
Basic SI units

Unit	*Symbol*	*Physical Quantity*
Meter	m	Length
Kilogram	kg	Mass
Second	s (sec)	Time
Ampere[a]	A (amp)	Electric current
Kelvin	K	Temperature
Candela[a]	cd	Light intensity
Mole	mol	Amount of substance

[a]These units are not used in this book.

a kilogram, the conversion factor is 1000 grams per kilogram, written 1000 g/kg, or 10^3 g/kg. To find a conversion factor in kilograms per gram (kg/g), we just invert the previous conversion factor, like this:

$$\frac{10^3\ \text{g}}{\text{kg}} \text{ inverted is } \frac{1\ \text{kg}}{10^3\ \text{g}} = \frac{10^{-3}\ \text{kg}}{\text{g}}$$

We see that inverting a metric conversion factor changes the sign of the power of ten (Appendix A.4, p. 494). Conversion factors in both directions are given in Table B.4.

SIGNIFICANT FIGURES IN THE METRIC SYSTEM. Conversion factors within a system of measurement are defined numbers. One kilogram contains *exactly* 1×10^3 (or 1000) grams, because a gram is defined that way. These conversion factors are pure numbers. (See Appendix A.2, p. 490). Thus we can write 10^3 g/kg as 1.00×10^3 g/1.00 kg, or with as many zeros after the decimal as we want or need in calculations. Defined conversion factors never limit the number of significant figures that an answer may have.

TABLE B.3
Units derived from basic SI units

Unit	*Symbol*	*Physical Quantity*	*Definition*
Newton[a]	N	Force	$(\text{kg} \times \text{m})/\text{s}^2$
Pascal[a]	Pa	Pressure	N/m^2
Joule[a]	J	Energy	$(\text{kg} \times \text{m}^2)/\text{s}^2$
Coulomb[a]	C	Electric charge	$\text{A} \times \text{s}$
Cubic meter[a]	m^3	Volume	m^3

[a]These units are not used in this book.

TABLE B.4
Some commonly used metric units and their conversion factors

Basic SI Unit	*Conversion Factors*	*SI Conversion Factor*
Length: meter (m)		
1 kilometer (km) = 10^3 meters (m)	10^3 m/km 10^{-3} km/m	10^{-3} km/m
1 centimeter (cm)[a] = 10^{-2} meters (m)	10^2 cm/m 10^{-2} m/cm	10^2 cm/m
1 millimeter (mm) = 10^{-3} meters (m)	10^3 mm/m 10^{-3} m/mm	10^3 mm/m
1 micron (μ) = 10^{-6} meters (m)	10^6 μ/m 10^{-6} m/μ	10^6 μ/m
1 Angstrom (Å)[b] = 10^{-8} centimeters (cm)	10^8 Å/cm 10^{-8} cm/Å	10^{10} Å/m
1 nanometer (nm) = 10^{-7} centimeters (cm)	10^7 nm/cm 10^{-7} cm/nm	10^9 nm/m
Volume: cubic meter (m^3)		
1 liter (ℓ) = 10^{-3} cubic meters (m^3)	10^3 ℓ/m^3 10^{-3} m^3/ℓ	10^3 ℓ/m^3
1 milliliter (mℓ) = 10^{-3} liters (ℓ)	10^3 mℓ/ℓ 10^{-3} ℓ/mℓ	10^6 mℓ/m^3
1 milliliter = 1 cubic centimeter (cm^3)	1 mℓ/cm^3 1 cm^3/mℓ	10^6 cm^3/m^3
Mass: kilogram (kg)		
1 metric ton (t) = 10^3 kilograms (kg)	10^3 kg/t 10^{-3} t/kg	10^{-3} t/kg
1 gram (g) = 10^{-3} kilograms (kg)	10^3 g/kg 10^{-3} kg/g	10^3 g/kg
1 milligram (mg) = 10^{-3} grams (g)	10^3 mg/g 10^{-3} g/mg	10^6 mg/kg
Energy:		
1 kilocalorie (kcal)[c] = 10^3 calories (cal)	10^3 cal/kcal 10^{-3} kcal/cal	0.239 cal/J

[a]The centimeter is not an approved SI unit, but it is used in chemistry for convenience.
[b]The Angstrom is not an approved SI unit, and it is being replaced by the nanometer, which is ten times larger.
[c]The kilocalorie is not an approved SI unit, but it is still used by chemists.

B.2 NONMETRIC UNITS

Since we live in the United States, which uses nonmetric units, and we're studying chemistry, which uses metric units, we need to be able to convert between the two systems. In addition, there are many other conversions that we'll need to be able to do in chemistry.

These have all been exactly defined in terms of SI units. For convenience, we list these to only three significant figures. For example, the conversion factor between pounds and kilograms is exactly 0.45359237 kg/lb. We round this off to 0.454 kg/lb to make it more manageable. (For a discussion of rounding off, see Appendix A.1, p. 489). Thus these metric-nonmetric conversion factors aren't pure numbers, as the metric-metric conversion

TABLE B.5
Metric-nonmetric conversion factors

Metric Unit	*Nonmetric Unit*	*Conversion Factors*
Length		
centimeter (cm)	inch (in)	2.54 cm/in 0.394 in/cm
meter (m)	inch (in)	0.0254 m/in 39.4 in/m
meter (m)	foot (ft)	0.305 m/ft 3.28 ft/m
kilometer (km)	mile (mi)	1.61 km/mi 0.621 mi/km
Volume		
liter (ℓ)	pint (pt)	0.473 ℓ/pt 2.11 pt/ℓ
liter (ℓ)	quart (qt)	0.946 ℓ/qt 1.06 qt/ℓ
liter (ℓ)	gallon (gal)	3.78 ℓ/gal 0.264 gal/ℓ
Mass		
gram (g)	pound (lb)	454 g/lb 2.20×10^{-3} lb/g
gram (g)	ounce (oz)	28.4 g/oz 3.52×10^{-2} oz/g
kilogram (kg)	pound (lb)	0.454 kg/lb 2.20 lb/kg
Pressure		
pascal (Pa)	atmosphere (atm)	1.01×10^{5} Pa/atm 9.90×10^{-6} atm/Pa
pascal (Pa)	torr (torr)[a]	1.33×10^{2} Pa/torr 7.52×10^{-3} torr/Pa
pascal (Pa)	millimeter of mercury (mm Hg)[a]	1.33×10^{2} Pa/mm Hg 7.52×10^{-3} mm Hg/Pa
Energy		
joule (J)	calorie (cal)	4.18 J/cal 0.239 cal/J

[a]Torr and millimeter of mercury are equivalent.
Conversion factors useful to learn are printed in color.

factors are. We can't write 0.4540 kg/lb just because we need four significant figures. If more significant figures are needed, another table can be consulted. (We *can*, however, assume that the bottom unit has as many significant figures as the top one. Thus 0.454 kg/lb is the same as 0.454 kg/1.00 lb.) The conversion factors expressed in the table to three significant figures can limit the number of significant figures in an answer.

B.3 USING CONVERSION FACTORS

We've listed a lot of conversion factors, and now we'll see how to use them to solve conversion problems. First we need to look at some properties of conversion factors.

PROPERTIES OF CONVERSION FACTORS. There are three properties that we will consider here.

1. The *reciprocal* of a conversion factor means one over the conversion factor. A reciprocal can be calculated simply by inverting the conversion factor and dividing the numerator by the denominator. For instance, 2.54 cm/in inverted is 1.00 in/2.54 cm = 0.394 in/cm. The two columns of conversion factors in Table B.5 are reciprocals of each other. reciprocal
2. Conversion factors can be used right side up or upside down. Thus 0.454 kg/lb is the same as 1.00 lb/0.454 kg = 2.20 lb/kg.
3. Units in conversion factors must be multiplied, divided, and canceled just like numbers. Thus:

$$\frac{\text{cm}}{\text{in}} \times \text{cm} = \frac{\text{cm}^2}{\text{in}}$$

$$\frac{\cancel{\text{qt}}}{\text{l}} \times \frac{\text{pt}}{\cancel{\text{qt}}} = \frac{\text{pt}}{\text{l}}$$

WORKING PROBLEMS. Conversion factors are used to convert one unit to another. To solve a problem in conversion, first we decide what units are to be converted to what other units. Next, we choose appropriate conversion factors to accomplish the conversion. The rest is arithmetic.

Let's see how this works by following an example. In Europe, gasoline is sold by the liter. A typical tankful of gas might be 54.4 liters. We want to know how many gallons this is.

Step 1. *Decide what units are to be converted to what other units. Write the starting units on the left and the desired new units on the right.*
We want to convert 54.4 liters to gallons.

$$54.4\,\ell = \underline{\quad\quad}\ \text{gal}$$

Step 2. *Choose appropriate conversion factors.*
We need a conversion factor that contains liters and gallons. Since we want to get rid of liters and substitute gallons, our conversion factor should contain liters on the bottom and gallons on the top. The liters will cancel and we'll be left with gallons. From Table B.5, we see that 0.264 gal/ℓ is what we want.

Step 3. *Set up the problem with the conversion factors, check units, and solve the problem.*

$$54.4\,\cancel{\ell} \times \frac{0.264\ \text{gal}}{\cancel{\ell}} = 14.4\ \text{gal}$$

The unit check in Step 3 is a good way to check the correctness of your answer. (Another way is by approximate answers; see Appendix A.5, p. 498.) If the units aren't right, chances are the problem is set up wrong. For instance, suppose we chose the conversion factor 3.78 ℓ/gal and set it up like this:

$$54.4\,\ell \times \frac{3.78\,\ell}{\text{gal}} = \underline{\quad\quad}\ \frac{\ell^2}{\text{gal}}$$

Without doing the arithmetic, we know this is wrong because we wanted gallons, and instead we got ℓ^2/gal, which doesn't have any meaning in this problem, or anywhere else, for that matter.

However, we can use this conversion factor if we invert it, like this:

$$54.4\,\cancel{\ell} \times \frac{1.00\text{ gal}}{3.78\,\cancel{\ell}} = 14.4\text{ gal}$$

We get the same answer, with the units canceling as they should.

In the examples above and in the ones that follow, calculator answers are rounded off to the proper number of significant figures. Where convenient, exponential notation is used as well. (For a complete discussion of these topics, see Appendix A, pp. 489–498.)

EXAMPLE B.1: In Germany, there is no speed limit on the superhighway called the "autobahn." Speeds there often reach as high as 195 km/hr (and higher). How many miles per hour is this?

Solution:

Step 1: We want to convert 195 km/hr to miles per hour (mi/hr):

$$195\,\frac{\text{km}}{\text{hr}} = \underline{\quad\quad}\,\frac{\text{mi}}{\text{hr}}$$

Step 2: Only the kilometer part needs to change, so we can use the conversion factor 0.621 mi/km from Table B.5.

Step 3:

$$195\,\frac{\cancel{\text{km}}}{\text{hr}} \times 0.621\,\frac{\text{mi}}{\cancel{\text{km}}} = \underline{\quad\quad}\,\frac{\text{mi}}{\text{hr}}$$

The units are correct.

Answer: 121 mi/hr (frightening!). ■

EXAMPLE B.2: Convert 121 mi/hr in Example B.1 to centimeters per second (cm/sec).

Solution:

Step 1: We want to convert 121 mi/hr to centimeters per second (cm/sec).

$$121\,\frac{\text{mi}}{\text{hr}} = \underline{\quad\quad}\,\frac{\text{cm}}{\text{sec}}$$

Step 2: Here, both miles and hours change, so we need more than one conversion factor. We need to go from miles to kilometers to meters to centimeters, and from hours to minutes to seconds. To get rid of miles first, we choose the conversion factor 1.61 km/mi.

$$121\,\frac{\cancel{\text{mi}}}{\text{hr}} \times 1.61\,\frac{\text{km}}{\cancel{\text{mi}}}$$

This leaves us with kilometers, which we can change to meters and then to centimeters by using conversion factors from Table B.4: 10^3 m/km and 10^2 cm/m.

$$121\,\frac{\cancel{\text{mi}}}{\text{hr}} \times 1.61\,\frac{\cancel{\text{km}}}{\cancel{\text{mi}}} \times 10^3\,\frac{\cancel{\text{m}}}{\cancel{\text{km}}} \times 10^2\,\frac{\text{cm}}{\cancel{\text{m}}}$$

Now we've taken care of changing miles to centimeters. We still have to change hours to seconds. We use 1.00 hr/60.0 min and 1.00 min/60.0 sec.

$$121 \frac{\cancel{mi}}{\cancel{hr}} \times 1.61 \frac{\cancel{km}}{\cancel{mi}} \times 10^3 \frac{\cancel{m}}{\cancel{km}} \times 10^2 \frac{cm}{\cancel{m}} \times \frac{1.00\ \cancel{hr}}{60.0\ \cancel{min}} \times \frac{1.00\ \cancel{min}}{60.0\ sec} = \_\_\_ \frac{cm}{sec}$$

The units are correct, so we do the arithmetic and get the answer.

Answer: 5.41×10^3 cm/sec. ■

Most of the conversions we'll run into in this book won't need so many conversion factors.

Notice that we really need only one conversion factor for each type of physical quantity. Three convenient ones are shaded in Table B.5. These can be memorized, so that conversions can always be made even if a table isn't available. From them, we can even make conversion factors that we don't have.

EXAMPLE B.3: Calculate a conversion factor for converting feet to kilometers (ft/km).

Solution:

Step 1: We need to select our starting unit. Looking at Table B.5, we see a conversion between feet and meters, so that'll be what we'll use to convert to feet per kilometer.

$$3.28 \frac{ft}{m} = \_\_\_ \frac{ft}{km}$$

Step 2: We need a conversion factor to convert meters to kilometers. This is 10^3 m/km.

Step 3:

$$3.28 \frac{ft}{\cancel{m}} \times 10^3 \frac{\cancel{m}}{km} = \_\_\_ \frac{ft}{km}$$

The units are correct.

Answer: 3.28×10^3 ft/km. ■

Notice that no arithmetic was involved in this calculation. That was because we were converting one metric unit to another, and this kind of conversion involves only powers of ten. With a little practice and experience, we can do calculations like that in our heads. We can say, "Well, it takes a thousand meters to make a kilometer, so I multiply feet/meter by a thousand to get feet/kilometer." At first, though, it's best to write the problem out so we don't multiply where we should be dividing, or vice versa. We'll know whether it is set up right if the units cancel properly.

B.4 TEMPERATURE SCALES

Temperature conversions are a little different from the unit conversions of the last section. We've seen how one unit can be converted to another unit by multiplying by a conversion factor. In temperature conversions, addition and subtraction are involved as well.

THE KELVIN SCALE. The basic SI temperature scale is the *Kelvin scale,* and its units are called kelvins (K). The zero point on the Kelvin scale (0 K) is the point at which everything is frozen and all motion stops. On this scale, water freezes at 273.16 K and boils at 373.16 K. Room temperature is about 295 K. When we're working with gases, we use this temperature scale.

Kelvin scale

THE CELSIUS (CENTIGRADE) SCALE. The one that we often find more convenient to use for ordinary laboratory measurements is the *Celsius scale.* Its units are called degrees Celsius (°C). The zero point on the Celsius scale (0°C) is the freezing point of water, and 100°C is the boiling point of water. That's how the scale was defined. The difference between the freezing and boiling points of water was divided into one hundred equal parts, which are the degrees. That's why it's sometimes called the centigrade scale (*centi-* means "one hundredth").

Celsius scale

The size of a Celsius degree is the same as the size of a Kelvin. The distinction between the two scales is the difference in their zero points. 0°C is the same as 273.16 K and 0 K is the same as −273.16°C.

THE FAHRENHEIT SCALE. The scale we use in the United States to measure temperature, but isn't used in science at all, is the *Fahrenheit scale.* Its units are called degrees Fahrenheit (°F). On the Fahrenheit scale, water freezes at 32.0°F and boils at 212°F. The 0°F and 100°F points don't have any special significance to us. The size of a Fahrenheit degree isn't the same as the size of a Kelvin or a Celsius degree. A Fahrenheit degree is smaller than the other two.

Fahrenheit scale

Figure B.1 shows the three temperature scales side by side.

KELVIN-CELSIUS TEMPERATURE CONVERSIONS. We've said that the size of a Celsius degree is the same as the size of a Kelvin. Because their zero points are different, we convert between these two scales by adding and subtracting. To change from Kelvin to Celsius, we subtract 273.16 (the difference in zero points on these scales) from the Kelvin temperature. In the opposite direction, to convert from Celsius to Kelvin, we add 273.16 to the Celsius temperature. Strictly speaking, though, if we add or subtract two numbers, their units have to be the same. We can take care of this easily with the conversion factor 1°C/K, and formalize the conversions with these equations:

$$°\mathrm{C} = \mathrm{K} \times \frac{1°\mathrm{C}}{\mathrm{K}} - 273.16°\mathrm{C}$$

$$\mathrm{K} = °\mathrm{C} \times \frac{1\mathrm{K}}{°\mathrm{C}} + 273.16\ \mathrm{K}$$

In practice, we'd obviously just subtract or add the number. These are easy formulas to use, because they involve only addition or subtraction.

FIGURE B.1
Comparison of temperature scales

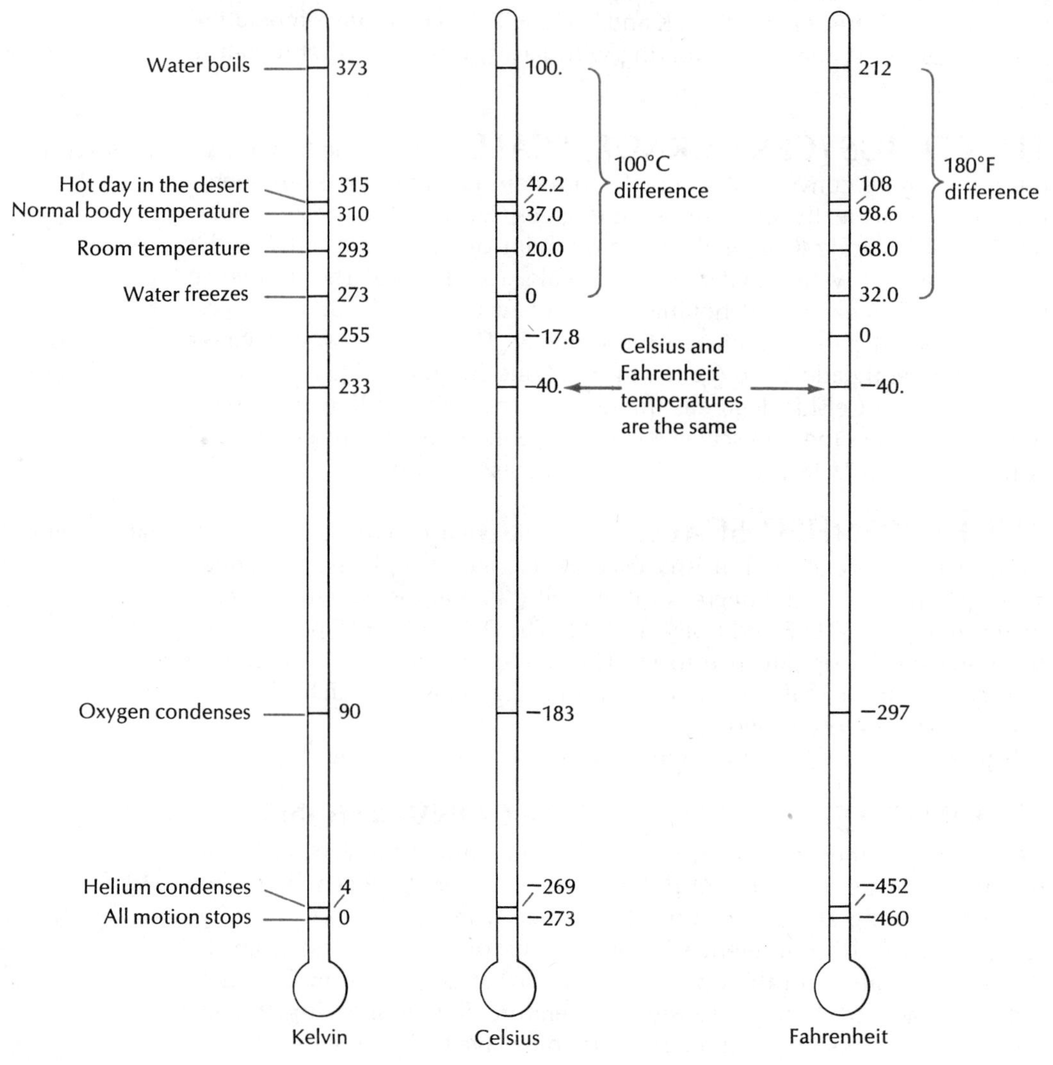

EXAMPLE B.4: Express the temperature 18.0°C in kelvins.

Solution: $$\mathrm{K} = 18.0°\mathrm{C} \times 1\frac{\mathrm{K}}{°\mathrm{C}} + 273.16\ \mathrm{K}$$

Answer: 291.2 K. ■

EXAMPLE B.5: Express the temperature 18.0 K in °C.

Solution: $$°\mathrm{C} = 18.0\ \mathrm{K} \times 1\frac{°\mathrm{C}}{\mathrm{K}} - 273.16°\mathrm{C}$$

Answer: −255.2°C. ■

CELSIUS-FAHRENHEIT TEMPERATURE CONVERSIONS. Looking at the Celsius and Fahrenheit temperature scales side by side (as in Figure B.1), we see that we need 180°F and only 100°C to measure the same difference between the boiling and freezing points of water. This distance is the same, no matter which scale we use to measure it. Using the distance as a standard of reference, we can get a conversion factor between Celsius and Fahrenheit degrees. There are 180 Fahrenheit degrees for every 100 Celsius degrees, or:

$$\frac{180°\text{F}}{100°\text{C}} = \frac{9°\text{F}}{5°\text{C}}$$

So, our conversion factor is 9°F/5°C, or 5°C/9°F (this conversion factor is a pure number). If we remember that Fahrenheit degrees are smaller, and that therefore it takes more of them to cover the same distance, we'll always get the conversion factor right. (See Figure B.2.)

FIGURE B.2
Derivation of conversion factor between Fahrenheit and Celsius degrees

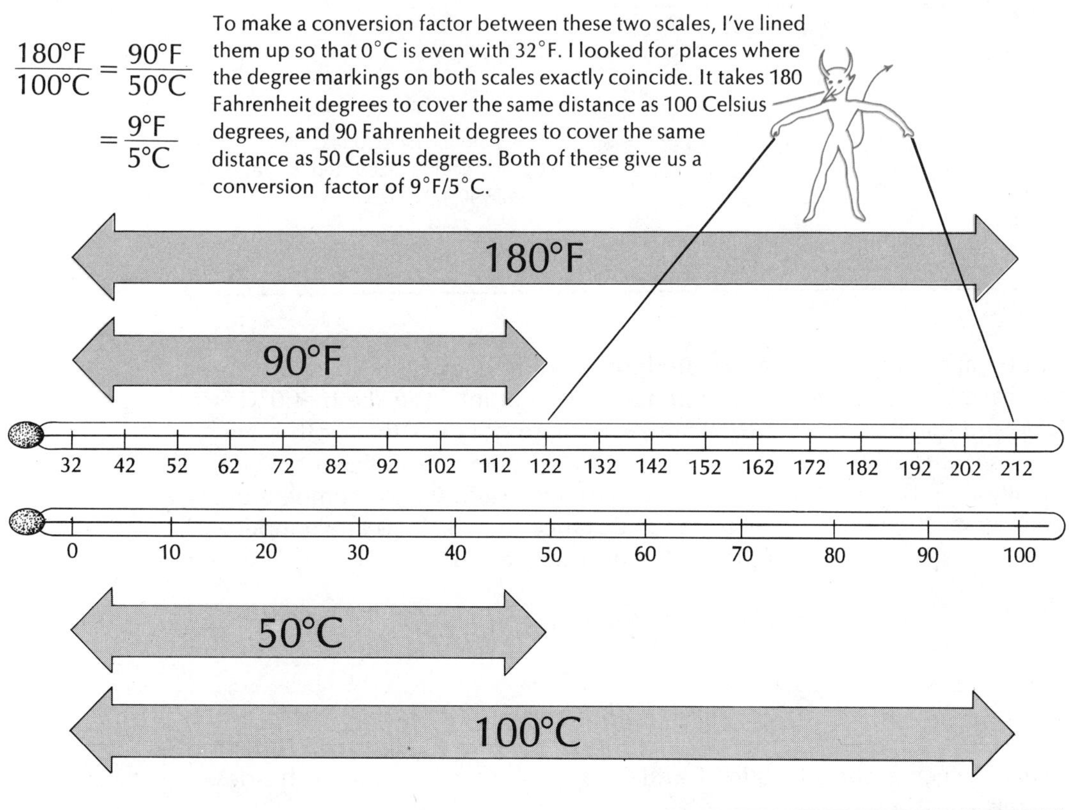

$\frac{180°\text{F}}{100°\text{C}} = \frac{90°\text{F}}{50°\text{C}} = \frac{9°\text{F}}{5°\text{C}}$
To make a conversion factor between these two scales, I've lined them up so that 0°C is even with 32°F. I looked for places where the degree markings on both scales exactly coincide. It takes 180 Fahrenheit degrees to cover the same distance as 100 Celsius degrees, and 90 Fahrenheit degrees to cover the same distance as 50 Celsius degrees. Both of these give us a conversion factor of 9°F/5°C.
180°F
90°F
32 42 52 62 72 82 92 102 112 122 132 142 152 162 172 182 192 202 212
0 10 20 30 40 50 60 70 80 90 100
50°C
100°C

FIGURE B.3
Add 40 degrees as the first step in Fahrenheit-Celsius conversion

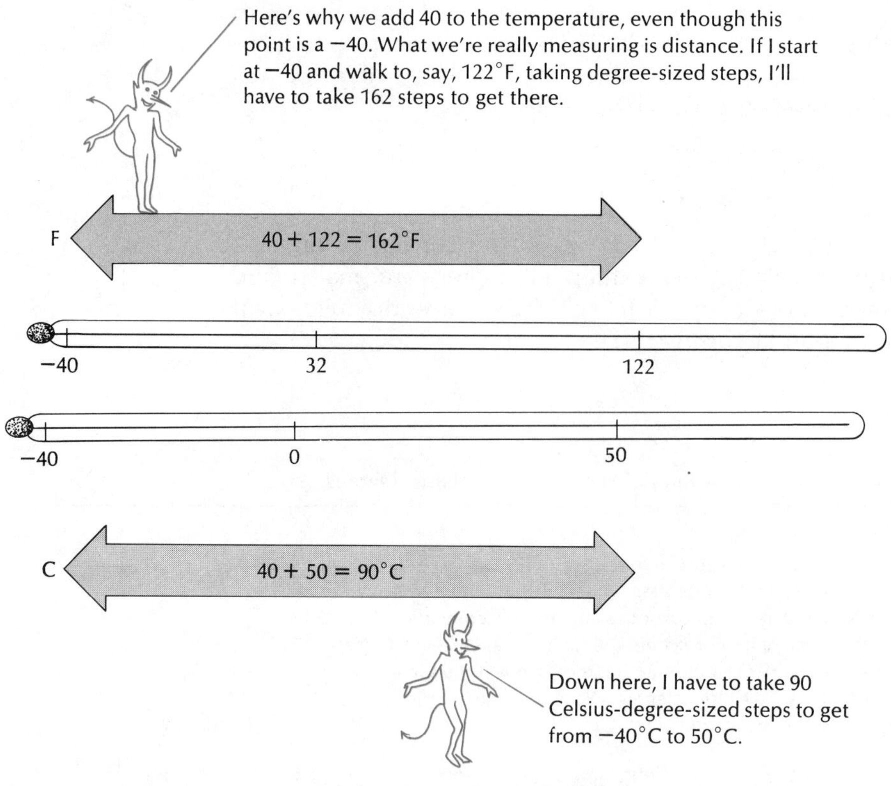

Here's why we add 40 to the temperature, even though this point is a −40. What we're really measuring is distance. If I start at −40 and walk to, say, 122°F, taking degree-sized steps, I'll have to take 162 steps to get there.
F
40 + 122 = 162°F
−40
32
122
−40
0
50
C
40 + 50 = 90°C
Down here, I have to take 90 Celsius-degree-sized steps to get from −40°C to 50°C.

But that's not all. We've accounted for the different degree sizes, but we also have to account for the difference in zero points. We see that 0°C is 32°F, and that 0°F is −17.8°C. Another thing to notice is that both scales have the same value at −40°. If we start at this common "zero" point, which is really a "−40" point, then we see that we have to add 40° to a temperature reading on either scale to measure the distance from −40 to any temperature. (See Figure B.3.) Then, these relationships are true.

$$°F + 40°F = \frac{9°F}{5°C}(°C + 40°C)$$

$$°C + 40°C = \frac{5°C}{9°F}(°F + 40°F)$$

Now, to get temperatures in °F and °C, we subtract 40° from both sides of each equation.

$$°F = \frac{9°F}{5°C}(°C + 40°C) - 40°F$$

$$°C = \frac{5°C}{9°F}(°F + 40°F) - 40°C$$

(Note: Both the 40° and the 5/9 are pure numbers, and may have as many significant figures as needed in a given instance.) This gives us these two working formulas for converting between Celsius and Fahrenheit temperatures. They're easy to remember, because they're the same except for the inversion of the conversion factor.

EXAMPLE B.6: Express the temperature 37.0°C in °F.

Solution: We want to convert °C to °F, so we use the first equation above.

$$°F = \frac{9°F}{5°C}(37.0°C + 40.0°C) - 40.0°F$$

$$= \frac{9°F}{5°C}(77.0°C) - 40.0°F$$

$$= 138.6°F - 40.0°F$$

Answer: 98.6°F (normal body temperature). ■

EXAMPLE B.7: Express the temperature 68°F in °C.

Solution: We want to convert degrees Fahrenheit to Celsius, so we use the second formula.

$$°C = \frac{5°C}{9°F}(68°F + 40°F) - 40°C$$

$$= \frac{5°C}{9°F}(108°F) - 40°C$$

$$= 60°C - 40°C$$

Answer: 20°C (room temperature). ■

The above formulas are easier to remember than the ones that are often given, which are:

$$°F = \frac{9°F}{5°C}(°C) + 32°F$$

$$°C = \frac{5°C}{9°F}(°F - 32°F)$$

We could rework Examples B.6 and B.7 with these formulas, and we'd get the same answers. The first pair of formulas show where the two temperature scales are equal. The second pair show that there is a difference of 32 Fahrenheit degrees between their zero points. Both show the 9°F/5°C conversion factor. Which pair we use is merely a matter of which we remember more easily.

EXERCISES

Note: Answers are given on p. 546.

1. Metric prefixes have permeated the financial world. Financiers use words like "kilobucks" and "megabucks." What do these units mean in terms of dollars?
2. Fill in the blanks.
   a. 1 deciliter ($d\ell$) = ____ liters (ℓ)
   b. 1 kilogram (kg) = ____ micrograms (μg)
   c. 5 cubic centimeters = ____ milliliters ($m\ell$)
   d. 500. kilograms (kg) = ____ metric tons
   e. ____ centimeters (cm) = 0.75 meters
   f. ____ liters (ℓ) = 500. cubic centimeters (cm^3)
   g. ____ grams (g) = 200. milligrams (mg)
   h. ____ meters (m) = 0.400 kilometers (km)
   i. 800. calories (cal) = ____ kilocalories (kcal)
3. Which of the following conversion factors may be treated as pure numbers?
   a. 1000 mg/g
   b. 0.621 mi/km
   c. 1 $cm^3/m\ell$
   d. 1.06 qt/ℓ
   e. 2.20×10^{-3} lb/g
   f. 60 sec/min
4. Write reciprocals of the conversion factors in Exercise 3.
5. Some of the following conversions are set up incorrectly. Select the incorrect ones and explain why they are incorrect.

   a. $15.7 \text{ cm} \times \dfrac{1.00 \text{ in}}{2.54 \text{ cm}} = 6.18 \text{ in}$

   b. $2.25\,\ell \times 0.946\,\dfrac{\ell}{\text{qt}} = 2.13 \text{ qt}$

   c. $121 \text{ lb} \times 0.454\,\dfrac{\text{kg}}{\text{lb}} = 54.9 \text{ kg}$

   d. $547 \text{ cm}^3 \times 10^3\,\dfrac{\text{cm}^3}{\ell} \times 2.11\,\dfrac{\text{pt}}{\ell} = 1.15 \times 10^6 \text{ pt}$

   e. $1.50 \text{ atm} \times 760\,\dfrac{\text{torr}}{\text{atm}} = 1.14 \times 10^3 \text{ torr}$

   f. $495 \text{ oz} \times 28.4\,\dfrac{\text{g}}{\text{oz}} \times \dfrac{1.00 \text{ g}}{10^{-3} \text{ kg}} = 14.1 \text{ kg}$

   g. $1.55 \text{ ft} \times 0.305\,\dfrac{\text{m}}{\text{ft}} \times 10^3\,\dfrac{\text{mm}}{\text{m}} = 4.27 \times 10^2 \text{ mm}$
6. You are having a physical examination in Europe, and the doctor asks you for your height in centimeters and your weight in kilograms. Perform the conversion.
7. You are driving through Europe, and Paris, your destination, is 115 kilometers away. How many miles do you have to go?
8. You put 40.0 liters of gas in the car you're driving. How many gallons is that?
9. The 40.0 liters of gas cost 45.0 francs. How many cents per gallon is that, if the exchange rate is 4.00 francs to the dollar?
10. Your car weighs 2.51 tons. How many kilograms is that?
11. The distance between the earth and the sun is 9.3×10^7 miles. How many kilometers is that?
12. A polio virus particle is about 34.5 nanometers in diameter. How many inches is that?
13. The speed of light is 3.0×10^{10} cm/sec. Convert this to miles/hour.
14. A whimsical unit among some scientists is "furlongs per fortnight." Convert this to miles per hour. (A furlong is 220 yards, and a fortnight is 14 days.)
15. Calculate a conversion factor that tells how many milliliters are in a cup ($m\ell$/c). (1 pt = 2 cups.)
16. Calculate a conversion factor for converting ounces to kilograms (kg/oz).
17. Calculate a conversion factor for converting centimeters to yards (yd/cm).
18. A good temperature for baking bread is 375°F. Express this temperature in Kelvins and in degrees Celsius.
19. Make the following temperature conversions.
   a. 34.0°C to kelvins
   b. 89.6°C to °F
   c. 125 K to °C
   d. 66°F to °C
   e. −90.0°C to kelvins
   f. −21.0°F to °C
   g. 22.0 K to °F
   h. 250 K to °C
20. A certain antifreeze claims to protect water from freezing down to −40°. Whether this is degrees Celsius or Fahrenheit is not specified. Which do you think it is? Explain.

APPENDIX C

Answers to Set A Exercises

Chapter 2

1. Yes. When added together, the masses and charges of the proton and electron give the mass (1) and charge (0) of the neutron.

2.

No. Protons	*No. Electrons*	*No. Neutrons*	*Atomic Mass Units*
2	____ 2	2	____ 4
____ 3	3	____ 3	6
3	____ 3	4	____ 7
4	____ 4	____ 3	7

3. a. 13 protons, 13 electrons, 14 neutrons; b. 18 protons, 18 electrons, 22 neutrons; c. 29 protons, 29 electrons, 35 neutrons; d. 79 protons, 79 electrons, 118 neutrons; e. 28 protons, 28 electrons, 31 neutrons; f. 27 protons, 27 electrons, 33 neutrons
4. a. silver, Ag; b. thallium, Tl; c. thorium, Th; d. uranium, U; e. zirconium, Zr
5. *a* is an isotope of hydrogen ($Z = 1$); *b*, *e*, and *g* are isotopes of lithium ($Z = 3$); *c*, *f*, and *h* are isotopes of beryllium ($Z = 4$); and *d* is an isotope of helium ($Z = 2$).
6. He-3, 1 neutron; He-4, 2 neutrons; He-5, 3 neutrons; 2 protons, 2 electrons for all three
7. Barium, zirconium, and aluminum are metals; oxygen and sulfur are nonmetals; arsenic, astatine, and antimony are metalloids.
8. a. Mg (all others are Group IA metals); b. Br (the only nonmetal); c. Ne (the only monatomic gas)
9. a. Na reacts violently with water, Mg does not; b. H is a gas, Li is a metal; c. Hg is shiny and metallic and Br is reddish-brown; d. Ca reacts violently with water; Mg does not; e. Si shatters when pounded with a hammer, Al is a malleable metal; f. Br is a reddish-brown liquid, Cl is a yellowish-green gas; g. N is a gas, P is a solid.
10. Na reacts easily with air, water, and skin.
11. Fe. The nonmetal Se does not conduct electricity; Ca is a metal but too reactive.
12. Cl, which is yellowish-green; N and Ar are both colorless.
13. a. Na; b. Mn; c. Hg; d. lithium; e. potassium; f. phosphorus; g. tin; h. copper; i. silver; j. Pb; k. Fe; l. bromine

Chapter 3

1. $1\ \cancel{\text{doz doughnuts}} \times \frac{12\ \cancel{\text{doughnuts}}}{\cancel{\text{doz doughnuts}}} \times \frac{225\text{ g}}{\cancel{\text{doughnut}}} = 2700\text{ g}$

2. $1\ \cancel{\text{mole He atoms}} \times \frac{6.02 \times 10^{23}\ \cancel{\text{He atoms}}}{\cancel{\text{mole He atoms}}} \times \frac{6.64 \times 10^{-24}\text{ g}}{\cancel{\text{He atom}}} = 4.00\text{ g}$

3. $2.5\ \cancel{\text{doz apples}} \times \frac{1.4\text{ kg}}{\cancel{\text{doz apples}}} = 3.5\text{ kg}$

   $2.5\ \cancel{\text{doz apples}} \times \frac{12\text{ apples}}{\cancel{\text{doz apples}}} = 30\text{ apples}$

4. $2.50\ \cancel{\text{moles H atoms}} \times \frac{1.01\text{ g}}{\cancel{\text{mole H atoms}}} = 2.53\text{ g}$

   $2.50\ \cancel{\text{moles}} \times \frac{6.02 \times 10^{23}\text{ H atoms}}{\cancel{\text{mole}}} = 1.51 \times 10^{24}\text{ H atoms}$

5. $1\ \cancel{\text{mole H}_2\text{ molecules}} \times \frac{2\ \cancel{\text{moles H atoms}}}{\cancel{\text{mole H}_2\text{ molecules}}} \times \frac{1.01\text{ g}}{\cancel{\text{mole H atoms}}} = 2.02\text{ g}$

6. a. $10.0\ \cancel{\text{moles Hg}} \times \frac{200.6\text{ g}}{\cancel{\text{mole Hg}}} = 2010\text{ g}$

   b. $4.00\ \cancel{\text{g Ca}} \times \frac{1\text{ mole}}{40.1\ \cancel{\text{g Ca}}} = 9.98 \times 10^{-2}\text{ moles}$

   c. $3.82 \times 10^{10}\ \cancel{\text{atoms}} \times \frac{1\text{ mole}}{6.02 \times 10^{23}\ \cancel{\text{atoms}}} = 6.35 \times 10^{-14}\text{ moles}$

   d. $2.50\ \cancel{\text{moles}} \times \frac{6.02 \times 10^{23}\text{ atoms}}{\cancel{\text{mole}}} = 1.51 \times 10^{24}\text{ atoms}$

7. $\frac{197.0\text{ g Au}}{\cancel{\text{mole}}} \times \frac{1\ \cancel{\text{mole}}}{6.02 \times 10^{23}\text{ atoms}}$

$= 3.27 \times 10^{-22}\text{ g Au/atom}$

8. $(0.196 \times 10.0\text{ g}) + (0.804 \times 11.0\text{ g}) = 10.8\text{ grams}$
9. $8 \times 10^{4}\text{ kg water} \times (25 - 10)°\text{C} = 1 \times 10^{6}\text{ kcal}$
10. $\frac{4.05\text{ kcal}}{0.566\text{ g}} = 7.16\text{ Cal (kcal)/g}$

$\cancel{1\text{ lb}} \times \frac{454\ \cancel{\text{g}}}{\cancel{\text{lb}}} \times \frac{7.16\text{ Cal}}{\cancel{\text{g}}} = 3250\text{ Cal}$

11. largest liquid range: uranium (30°C to 1980°C); smallest liquid range: neon (−248°C to −246°C)
12. Tungsten has the highest melting and boiling points.
13. Group IA metals tend to have lower melting points than Group IIA metals.
14. $2.5\ \cancel{\text{kg}} \times 0.113 \frac{\text{kcal}}{\cancel{\text{kg}} \times \cancel{°\text{C}}} (210 - 23)\cancel{°\text{C}} = 53\text{ kcal}$
15. $53\ \cancel{\text{kcal}} \times \frac{4.184\text{ J}}{10^{-3}\ \cancel{\text{kcal}}} = 2.2 \times 10^{5}\text{ J, or 220 kJ}$
16. highest specific heat: hydrogen (3.41); lowest specific heat: uranium (0.0276)
17. $\text{temperature change} = \frac{30\ \cancel{\text{cal}}}{0.031\ \cancel{\text{cal}}/(\cancel{\text{g}} \times °\text{C}) \times 5.2\ \cancel{\text{g}}}$

$= 190°\text{C}$

final temp. = 25°C + 190°C = 215°C. Because lead's melting point is 327°C, it is not hot enough to melt.

18. $\frac{5.49\text{ g}}{3.16\text{ cm}^3} = \frac{1.74\text{ g}}{\text{cm}^3}$ The metal is magnesium.
19. $10.5\ \cancel{\text{cm}^3 Br_2} \times 3.12 \frac{\text{g}}{\cancel{\text{cm}^3 Br_2}} = 32.8\text{ g}$
20. $4.00\ \cancel{\text{g He}} \times \frac{1\ \ell}{0.18\ \cancel{\text{g He}}} = 22\ \ell$
21. $\text{Temperature change} = \frac{952\ \cancel{\text{cal}}}{155\ \cancel{\text{g}} \times 1.00 \frac{\cancel{\text{cal}}}{\cancel{\text{g}} \times °\text{C}}}$

$= 6.14°\text{C}$

6.14°C + 20°C = 26°C

22. H_2, He
23. Seawater is more dense than pure water.
24. Oil has lower density and floats on water.
25. $2\ \cancel{\text{metric tons}} \times \frac{1000\ \cancel{\text{kg}}}{\cancel{\text{metric ton}}} \times \frac{1.00\ \ell}{0.79\ \cancel{\text{kg}}} = 3000\ \ell$
26. The heat would melt the substance. Later, when no heat was being supplied, the substance would solidify, giving off its heat of fusion.

Chapter 4

1. a. covalent; b. ionic; c. ionic; d. covalent; e. covalent; f. ionic
2. a. NCl_3; b. CF_4; c. SCl_2; d. B_2H_6
3. a. oxygen difluoride; b. chlorine iodide; c. dinitrogen tetrahydride; d. sulfur hexafluoride
4. a. hydrosulfuric acid, dihydrogen sulfide; b. hydriodic acid, hydrogen iodide
5. a. HBr; b. H_2Se
6. a. zinc chloride; b. lithium bromide; c. barium oxide; d. aluminum sulfide
7. a. CdF_2; b. $(NH_4)_2S$; c. K_2O; d. $SrCl_2$
8. a. covalent; dichlorine monoxide; b. ionic: zinc sulfide; c. ionic; Mg_3N_2; d. covalent: boron tribromide; e. covalent; P_4O_6; f. covalent; sulfur dioxide; g. ionic; calcium carbide; h. ionic; NH_4Cl; i. covalent; H_2Te; j. ionic; barium fluoride
9. a. mercury(II) iodide; b. lead(II) chloride; c. iron(III) oxide; d. tin(IV) oxide
10. a. CuCl; b. HgS; c. Co_2O_3; d. MnF_4
11. a. lithium hydride; b. beryllium oxide; c. iron (II) sulfide; d. nickel(II) chloride; e. potassium nitride; f. mercury(II) bromide
12. a. SnO_2; b. BaS; c. KBr; d. $CrCl_2$; e. CdF_2; f. PbI_2
13. a. sulfate ion; b. nitrite ion; c. perchlorate ion; d. permanganate ion
14. a. NO_3^-; b. CrO_4^{2-}; c. ClO^-; d. PO_3^{3-}
15. a. potassium perchlorate; b. iron(II) sulfate; c. ammonium dichromate; d. silver acetate; e. sodium phosphate; f. copper(II) carbonate
16. a. $Ca(ClO)_2$; b. NH_4NO_3; c. $BaCrO_4$; d. $Mg(OH)_2$; e. NaCN; f. $(NH_4)_2CO_3$
17. a. nitric acid; b. phosphoric acid; c. bromous acid; d. carbonic acid
18. a. H_2SO_4; b. HNO_2; c. HIO_3; d. H_3PO_3
19. a. lithium aluminum sulfate; b. potassium hydrogen carbonate; c. magnesium hydrogen sulfite; d. lithium hydrogen phosphate
20. a. NaH_2PO_4; b. NH_4HSO_4; c. $NaKSO_3$; d. $Al(HCO_3)_3$
21. a. binary ionic, Na_2O_2; b. binary ionic, Al_2S_3; c. ternary covalent (oxyacid), H_3PO_4; d. binary covalent, NO; e. binary covalent, NH_3; f. binary ionic, CdF_2; g. binary covalent, CO_2; h. ternary ionic, $CuSO_4$
22. a. hydrate, chromium(II) sulfate tetrahydrate (or 4-hydrate); b. ternary covalent (oxyacid), perchloric acid (or hydrogen perchlorate); c. mixed ionic, sodium potassium sulfite; d. binary covalent, phosphorus pentafluoride; e. binary covalent, silicon dioxide; f. ionic, ammonium cyanide; g. ternary ionic, barium chromate; h. binary covalent, sulfur trioxide
23. See table on p. 525.
24. See table on p. 525.

Chapter 5

1. a. correct; b. incorrect ($NaBr_2$ should be NaBr); c. incorrect (S should be Si); d. incorrect (O should be O_2); e. correct

Chapter 4, Problem 23

	OH^-	CN^-	SO_4^{2-}	PO_4^{3-}	HCO_3^-	NO_3^-	ClO_2^-	O_2^{2-}
NH_4^+	NH_4OH[a]	NH_4CN	$(NH_4)_2SO_4$	$(NH_4)_3PO_4$	NH_4HCO_3	NH_4NO_3	NH_4ClO_2	$(NH_4)_2O_2$[a,b]
Co^{2+}	$Co(OH)_2$	$Co(CN)_2$	$CoSO_4$	$Co_3(PO_4)_2$	$Co(HCO_3)_2$	$Co(NO_3)_2$	$Co(ClO_2)_2$	CoO_2
K^+	KOH	KCN	K_2SO_4	K_3PO_4	$KHCO_3$	KNO_3	$KClO_2$	K_2O_2[b]
Hg_2^{2+}	$Hg_2(OH)_2$[a,b]	$Hg_2(CN)_2$[b]	Hg_2SO_4	$(Hg_2)_3(PO_4)_2$	$Hg_2(HCO_3)_2$[b]	$Hg_2(NO_3)_2$[b]	$Hg_2(ClO_2)_2$[b]	Hg_2O_2[a,b]
Mn^{2+}	$Mn(OH)_2$	$Mn(CN)_2$	$MnSO_4$	$Mn_3(PO_4)_2$	$Mn(HCO_3)_2$	$Mn(NO_3)_2$	$Mn(ClO_2)_2$	MnO_2[c]
Pb^{4+}	$Pb(OH)_4$	$Pb(CN)_4$	$Pb(SO_4)_2$	$Pb_3(PO_4)_4$	$Pb(HCO_3)_4$	$Pb(NO_3)_4$	$Pb(ClO_2)_4$	$Pb(O_2)_2$
Sn^{2+}	$Sn(OH)_2$	$Sn(CN)_2$	$SnSO_4$	$Sn_3(PO_4)_2$	$Sn(HCO_3)_2$	$Sn(NO_3)_2$	$Sn(ClO_2)_2$	SnO_2[c]
Ca^{2+}	$Ca(OH)_2$	$Ca(CN)_2$	$CaSO_4$	$Ca_3(PO_4)_2$	$Ca(HCO_3)_2$	$Ca(NO_3)_2$	$Ca(ClO_2)_2$	CaO_2
Li^+	$LiOH$	$LiCN$	Li_2SO_4	Li_3PO_4	$LiHCO_3$	$LiNO_3$	$LiClO_2$	Li_2O_2[b]

[a]These compounds do not exist, but we may still write their formulas.
[b]Note that these formulas may not be simplified without destroying a polyatomic ion.
[c]These formulas are the same as for manganese(IV) oxide and tin(IV) oxide.

Chapter 4, Problem 24

	O^{2-}	F^-	N^{3-}	S^{2-}	Cl^-	Br^-	I^-
Zn^{2+}	ZnO	ZnF_2	Zn_3N_2	ZnS	$ZnCl_2$	$ZnBr_2$	ZnI_2
Li^+	Li_2O	LiF	Li_3N	Li_2S	$LiCl$	$LiBr$	LiI
Na^+	Na_2O	NaF	Na_3N	Na_2S	$NaCl$	$NaBr$	NaI
Al^{3+}	Al_2O_3	AlF_3	AlN	Al_2S_3	$AlCl_3$	$AlBr_3$	AlI_3
Ca^{2+}	CaO	CaF_2	Ca_3N_2	CaS	$CaCl_2$	$CaBr_2$	CaI_2
Fe^{3+}	Fe_2O_3	FeF_3	FeN	Fe_2S_3	$FeCl_3$	$FeBr_3$	FeI_3
Sr^{2+}	SrO	SrF_2	Sr_3N_2	SrS	$SrCl_2$	$SrBr_2$	SrI_2
Mg^{2+}	MgO	MgF_2	Mg_3N_2	MgS	$MgCl_2$	$MgBr_2$	MgI_2
K^+	K_2O	KF	K_3N	K_2S	KCl	KBr	KI
Cr^{2+}	CrO	CrF_2	Cr_3N_2	CrS	$CrCl_2$	$CrBr_2$	CrI_2
Hg^{2+}	HgO	HgF_2	Hg_3N_2	HgS	$HgCl_2$	$HgBr_2$	HgI_2

2. a. $Li + N_2 \longrightarrow Li_3N$
   b. $NH_3 + O_2 \longrightarrow NO + H_2O$
   c. $Zn + Cu(NO_3)_2 \longrightarrow Cu + Zn(NO_3)_2$
   d. $P + O_2 \longrightarrow P_4O_6$
   e. $(NH_4)_2S + HgBr_2 \longrightarrow NH_4Br + HgS$
3. a. balanced; b. unbalanced; c. balanced, d. unbalanced; e. unbalanced
4. b. $N_2 + O_2 \longrightarrow 2\,NO$
   d. $4\,Fe + 3\,O_2 \longrightarrow 2\,Fe_2O_3$
   e. $2\,H_2O_2 \longrightarrow 2\,H_2O + O_2$
5. The "balanced" equations given will be unbalanced when their formulas are corrected. a. H should be H_2, O should be O_2; b. FeBr should be $FeBr_2$, HS should be H_2S; c. H_2F_2 should be HF; d. Na_2Cl_2 should be NaCl; e. Cl_3 should be Cl_2; f. $Li_2(OH)_2$ should be LiOH; g. $AlCl_2$ should be $AlCl_3$; h. $PbNO_3$ should be $Pb(NO_3)_2$, PbCl should be $PbCl_2$; i. O_4 should be O_2
6. a. $2\,H_2 + O_2 \longrightarrow 2\,H_2O$
   b. $FeS + 2\,HBr \longrightarrow FeBr_2 + H_2S$
   c. $H_2 + F_2 \longrightarrow 2\,HF$
   d. $Na_2CO_3 + CaCl_2 \longrightarrow CaCO_3 + 2\,NaCl$
   e. $2\,BCl_3 \longrightarrow 2\,B + 3\,Cl_2$
   f. $Li_2O + H_2O \longrightarrow 2\,LiOH$
   g. $2\,Al + 6\,HCl \longrightarrow 2\,AlCl_3 + 3\,H_2$
   h. $Pb(NO_3)_2 + 2\,KCl \longrightarrow PbCl_2 + 2\,KNO_3$
   i. $KClO_4 \longrightarrow KCl + 2\,O_2$
7. a. $2\,P + 3\,Cl_2 \longrightarrow 2\,PCl_3$
   b. $Mg_3N_2 + 6\,H_2O \longrightarrow 3\,Mg(OH)_2 + 2\,NH_3$
   c. $4\,Fe + 3\,O_2 \longrightarrow 2\,Fe_2O_3$
   d. $Ni(OH)_2 + H_2SO_4 \longrightarrow NiSO_4 + 2\,H_2O$
   e. $Fe + 2\,AgNO_3 \longrightarrow Fe(NO_3)_2 + 2\,Ag$
   f. $BaCl_2 + (NH_4)_2CO_3 \longrightarrow BaCO_3 + 2\,NH_4Cl$
   g. $2\,NaNO_2 + H_2SO_4 \longrightarrow 2\,HNO_2 + Na_2SO_4$
   h. $2\,NaOH + CO_2 \longrightarrow Na_2CO_3 + H_2O$
   i. $2\,B + 3\,F_2 \longrightarrow 2\,BF_3$
   j. $2\,HNO_2 \longrightarrow N_2O_3 + H_2O$
8. a. $2\,NH_3 + H_2SO_4 \longrightarrow (NH_4)_2SO_4$
   b. $CH_4 + 2\,O_2 \longrightarrow CO_2 + 2\,H_2O$
   c. $N_2O_3 + H_2O \longrightarrow 2\,HNO_2$
   d. $BaCO_3 \xrightarrow{\Delta} BaO + CO_2$
   e. $Si + 2Cl_2 \longrightarrow SiCl_4$
9. a. "Solid calcium (or calcium metal) reacts with hydrobromic acid to yield aqueous calcium bromide and hydrogen gas (or gaseous hydrogen)."
   b. "Hydrogen iodide gas decomposes to form hydrogen gas and solid iodine."
   c. "Solid beryllium reacts with oxygen gas to form solid beryllium oxide."

d. "Aqueous manganese(II) nitrate reacts with aqueous sodium sulfide to yield a precipitate of manganese(II) sulfide and aqueous sodium nitrate."

e. "Liquid dichlorine heptoxide reacts with liquid water to form aqueous perchloric acid."

10. a. $4\ B(s) + 3\ O_2(g) \longrightarrow 2\ B_2O_3(s)$
 b. $Pb(s) + S(s) \longrightarrow PbS(s)$
 c. $N_2(g) + 3\ Cl_2(g) \longrightarrow 2\ NCl_3(g)$
 d. $H_2(g) + Br_2(l) \longrightarrow 2\ HBr(g)$
 e. $KOH(aq) + HNO_3(aq) \longrightarrow H_2O(l) + KNO_3(aq)$
 f. $ZnS(s) + H_2SO_4(aq) \longrightarrow H_2S(g) + ZnSO_4(aq)$
 g. $NH_4NO_2(s) \xrightarrow{\Delta} N_2(g) + 2\ H_2O(g)$
 h. $Cl_2(aq) + 2\ KBr(aq) \longrightarrow Br_2(aq) + 2\ KCl(aq)$
 i. $NaHCO_3(s) + HCl(aq) \longrightarrow NaCl(aq) + H_2O(l) + CO_2(g)$
 j. $BaCl_2(aq) + (NH_4)_2SO_4(aq) \longrightarrow 2\ NH_4Cl(aq) + BaSO_4(s)$

11. a. combination; b. double replacement; c. single replacement; d. decomposition; e. single replacement; f. combination; g. double replacement; h. double replacement; i. decomposition; j. combination

12. a. "Solid aluminum (or aluminum metal) reacts with oxygen gas to yield solid aluminum oxide."
 b. "Calcium iodide and mercury(II) nitrate react in aqueous solution to yield aqueous calcium nitrate and a precipitate of mercury(II) iodide."
 c. "Solid magnesium reacts with acetic acid to yield aqueous magnesium acetate and hydrogen gas."
 d. "Solid cobalt(II) sulfite decomposes thermally to yield solid cobalt(II) oxide and sulfur dioxide gas."
 e. "Solid iron(II) oxide reacts with solid carbon to yield solid iron and carbon monoxide gas."
 f. "Solid copper reacts with solid sulfur to yield solid copper(I) sulfide."
 g. "Silver nitrate reacts with potassium chromate in aqueous solution to yield a precipitate of silver chromate and aqueous potassium nitrate."
 h. "Strontium hydroxide and sulfuric acid react in aqueous solution to yield a precipitate of strontium sulfate and water."
 i. "Solid lead(II) nitrate decomposes thermally to yield solid lead(II) oxide and nitrogen dioxide gas."
 j. "Ammonia gas reacts with sulfuric acid to yield aqueous ammonium sulfate."

13. a. $4\ Li(s) + O_2(g) \longrightarrow 2\ Li_2O(s)$
 b. $2\ Be(s) + O_2(g) \longrightarrow 2\ BeO(s)$
 c. $N_2(g) + 2\ O_2(g) \longrightarrow 2\ NO_2(g)$
 d. $H_2O(l) + CaO(s) \longrightarrow Ca(OH)_2(s)$
 e. $H_2O(l) + SO_2(g) \longrightarrow H_2SO_3(aq)$
 f. $BaO(s) + SO_3(g) \longrightarrow BaSO_4(s)$
 g. $NH_3(g) + HC_2H_3O_2(aq) \longrightarrow NH_4C_2H_3O_2(aq)$
 h. $3\ Mg(s) + N_2(g) \longrightarrow Mg_3N_2(s)$

14. a. $FeCO_3(s) \xrightarrow{\Delta} FeO(s) + CO_2(g)$
 b. $2\ KHCO_3(s) \xrightarrow{\Delta} K_2CO_3(s) + H_2O(g) + CO_2(g)$
 c. $MgSO_4 \cdot 7\ H_2O(s) \longrightarrow MgSO_4(s) + 7\ H_2O(g)$

15. a. $2\ Li(s) + 2\ H_2O(l) \longrightarrow 2\ LiOH(aq) + H_2(g)$
 b. $Fe(s) + H_2SO_4(aq) \longrightarrow FeSO_4(aq) + H_2(g)$
 c. $SnCl_2(aq) + Fe(s) \longrightarrow FeCl_2(aq) + Sn(s)$
 d. $Br_2(l) + 2\ KI(aq) \longrightarrow I_2(s) + 2\ KBr(aq)$
 e. $Ba(s) + 2\ H_2O(l) \longrightarrow Ba(OH)_2(aq) + H_2(g)$
 f. $Cl_2(g) + CaBr_2(aq) \longrightarrow Br_2(l) + CaCl_2(aq)$
 g. $Mg(s) + 2\ HCl(aq) \longrightarrow MgCl_2(aq) + H_2(g)$
 h. $2\ AgNO_3(aq) + Cu(s) \longrightarrow 2\ Ag(s) + Cu(NO_3)_2(aq)$

16. a. $3\ H_2SO_4(aq) + 2\ Al(OH)_3(s) \longrightarrow 6\ H_2O(l) + Al_2(SO_4)_3$
 b. $H_2CO_3(aq) + 2\ NaOH(aq) \longrightarrow Na_2CO_3(aq) + 2\ H_2O$
 c. $CaO(s) + 2\ HCl(aq) \longrightarrow CaCl_2(aq) + H_2O(l)$
 d. $BaCl_2(aq) + H_2SO_4(aq) \longrightarrow BaSO_4(s) + 2\ HCl(aq)$
 e. $Fe(NO_3)_3(aq) + 3\ KOH(aq) \longrightarrow Fe(OH)_3(s) + 3\ KNO_3(aq)$
 f. $Pb(NO_3)_2(aq) + Na_2CrO_4(aq) \longrightarrow PbCrO_4(s) + 2\ NaNO_3(aq)$
 g. $CaCO_3(s) + 2\ HCl(aq) \longrightarrow CaCl_2(aq) + CO_2(g) + H_2O(l)$

17. a. double replacement:
 $BaCO_3(s) + 2\ HNO_3(aq) \longrightarrow Ba(NO_3)_2(aq) + CO_2(g) + H_2O(l)$
 b. double replacement:
 $Fe_2O_3(s) + 2\ H_3PO_4 \longrightarrow 2\ FePO_4(s) + 3\ H_2O(l)$
 c. single replacement:
 $Cu(s) + MgCl_2(aq) \longrightarrow$ N.R.
 d. double replacement:
 $PbCl_2(aq) + H_2S(aq) \longrightarrow PbS(s) + 2\ HCl(aq)$
 e. combination:
 $2\ Ni(s) + O_2(g) \longrightarrow 2\ NiO(s)$
 f. decomposition:
 $Li_2CO_3(s) \xrightarrow{\Delta} Li_2O(s) + CO_2(g)$
 g. single replacement:
 $Ag(s) + H_2SO_4(aq) \longrightarrow$ N.R.
 h. combination:
 $2\ Al(s) + 3\ Cl_2(g) \longrightarrow 2\ AlCl_3(s)$
 i. single replacement:
 $Mg(s) + 2\ HCl(aq) \longrightarrow MgCl_2(aq) + H_2(g)$

18. a. $Cu(OH)_2(s) \xrightarrow{\Delta} CuO(s) + H_2O(g)$
 b. $Ag_2SO_3(s) \xrightarrow{\Delta} Ag_2O(s) + SO_2(g)$
 c. $K_2S(aq) + I_2(s) \longrightarrow 2\ KI(aq) + S(s)$
 d. $(NH_4)_2S(aq) + 2\ HCl(aq) \longrightarrow H_2S(g) + 2\ NH_4Cl(aq)$
 e. $PCl_5(l) \longrightarrow PCl_3(l) + Cl_2(g)$
 f. $N_2O_3(l) + H_2O \longrightarrow 2\ HNO_2(aq)$
 g. $2\ C_2H_2(g) + 5\ O_2(g) \longrightarrow 4\ CO_2(g) + 2\ H_2O(g)$

19. $2\ ZnS(s) + 3\ O_2(g) \longrightarrow 2\ ZnO(s) + 2\ SO_2(g)$

$HgS(s) + O_2(g) \longrightarrow Hg(l) + SO_2(g)$
The first reaction involves burning a binary compound to yield oxides of both elements. The second is single replacement.

20. a. decomposition:
$CaCO_3(s) \longrightarrow CaO(s) + CO_2(g)$
b. combination:
$CaO(s) + H_2O(l) \longrightarrow Ca(OH)_2(aq)$
c. combination:
$NH_3(g) + CO_2(aq) + H_2O(l) \longrightarrow NH_4HCO_3(aq)$
d. double replacement:
$NaCl(aq) + NH_4HCO_3(aq) \longrightarrow NaHCO_3(aq) + NH_4Cl(aq)$
e. gas formation:
$NH_4Cl(aq) + Ca(OH)_2(aq) \longrightarrow NH_3(g) + CaCl_2(aq) + H_2O(l)$
21. $ZnO(s) + CO(g) \longrightarrow Zn(s) + CO_2(s)$
22. $2\ Al(s) + 3\ Ag_2S(s) \longrightarrow Al_2S_3 + 6\ Ag(s)$

Chapter 6

1. a. $\frac{1}{3}\ \cancel{\text{mole He}} \times \frac{6.02 \times 10^{23}\text{ He atoms}}{\cancel{\text{mole He}}} = 2.01 \times 10^{23}\text{ He atoms}$

b. $3.21\ \cancel{\text{g S}} \times \frac{1.00\ \cancel{\text{mole S}}}{32.1\ \cancel{\text{g S}}} \times \frac{6.02 \times 10^{23}\text{ S atoms}}{\cancel{\text{mole S}}} = 6.02 \times 10^{22}\text{ S atoms}$

c. $6\ \cancel{\text{moles Ar}} \times \frac{6.02 \times 10^{23}\text{ Ar atoms}}{\cancel{\text{mole Ar}}} = 4 \times 10^{24}\text{ Ar atoms}$

2. a. $\frac{1}{2}\ \cancel{\text{mole NO}} \times \frac{6.02 \times 10^{23}\text{ molecules NO}}{\cancel{\text{mole NO}}} = 3.01 \times 10^{23}\text{ molecules NO}$

b. $2\ \cancel{\text{moles SO}_2} \times \frac{6.02 \times 10^{23}\text{ molecules SO}_2}{\cancel{\text{mole SO}_2}} = 1.20 \times 10^{24}\text{ molecules SO}_2$

c. $\frac{1}{4}\ \cancel{\text{mole O}_3} \times \frac{6.02 \times 10^{23}\text{ molecules O}_3}{\cancel{\text{mole O}_3}} = 1.51 \times 10^{23}\text{ molecules O}_3$

3. a. $3.01 \times 10^{23}\ \cancel{\text{NO molecules}} \times \frac{1\text{ O atom}}{\cancel{\text{NO molecule}}} = 3.01 \times 10^{23}\text{ O atoms}$

b. $1.20 \times 10^{24}\ \cancel{\text{SO}_2\text{ molecules}} \times \frac{2\text{ O atoms}}{\cancel{\text{SO}_2\text{ molecule}}} = 2.40 \times 10^{24}\text{ O atoms}$

c. $1.51 \times 10^{23}\ \cancel{\text{O}_3\text{ molecules}} \times \frac{3\text{ O atoms}}{\cancel{\text{O}_3\text{ molecule}}} = 4.53\text{ O atoms}$

4. a. $1\text{ mole }\cancel{\text{tables}} \times \frac{4\text{ legs}}{\cancel{\text{table}}} = 4\text{ moles legs}$

b. $\frac{1}{2}\text{ mole }\cancel{\text{butterflies}} \times \frac{6\text{ legs}}{\cancel{\text{butterfly}}} = 3\text{ moles legs}$

$\frac{1}{2}\text{ mole }\cancel{\text{butterflies}} \times \frac{2\text{ wings}}{\cancel{\text{butterfly}}} = 1\text{ mole wings}$

$\frac{1}{2}\text{ mole }\cancel{\text{butterflies}} \times \frac{2\text{ antennae}}{\cancel{\text{butterfly}}} = 1\text{ mole antennae}$

c. $1\text{ mole }\cancel{\text{P}_4\text{ molecules}} \times \frac{4\text{ P atoms}}{\cancel{\text{P}_4\text{ molecule}}} = 4\text{ moles P atoms}$

d. $1\text{ mole }\cancel{\text{H}_3\text{PO}_4\text{ molecules}} \times \frac{3\text{ H atoms}}{\cancel{\text{H}_3\text{PO}_4\text{ molecules}}} = 3\text{ moles H atoms}$

$1\text{ mole }\cancel{\text{H}_3\text{PO}_4\text{ molecules}} \times \frac{1\text{ P atom}}{\cancel{\text{H}_3\text{PO}_4\text{ molecule}}} = 1\text{ mole P atoms}$

$1\text{ mole }\cancel{\text{H}_3\text{PO}_4\text{ molecules}} \times \frac{4\text{ O atoms}}{\cancel{\text{H}_3\text{PO}_4\text{ molecule}}} = 4\text{ moles O atoms}$

e. $\frac{1}{2}\text{ mole }\cancel{\text{Al}_2(\text{SO}_4)_3\text{ formulas}} \times \frac{5\text{ ions}}{\cancel{\text{Al}_2(\text{SO}_4)_3\text{ formula}}} = 2\frac{1}{2}\text{ moles ions}$

5. a. 2 N atoms, each 14.0 g/mole = 28.0 g/mole

b. 1 H atom, each 1.01 g/mole = 1.01 g/mole
1 N atom, each 14.0 g/mole = 14.0 g/mole
3 O atoms, each 16.0 g/mole = 48.0 g/mole
63.0 g/mole

c. 2 Na atoms, each 23.0 g/mole = 46.0 g/mole
1 S atom, each 32.1 g/mole = 32.1 g/mole
4 O atoms, each 16.0 g/mole = 64.0 g/mole
142.1 g/mole

6. a. $22.4\ \cancel{\text{g Ne}} \times \frac{1.00\text{ mole Ne}}{20.2\ \cancel{\text{g Ne}}} = 1.11\text{ moles Ne}$

b. Na: 23.0 g/mole
Br: 79.9 g/mole
NaBr: 102.9 g/mole

$107\ \cancel{\text{g NaBr}} \times \frac{1.00\text{ mole NaBr}}{102.9\ \cancel{\text{g NaBr}}} = 1.04\text{ moles NaBr}$

7. a. I_2: 253.8 g/mole

$2.00\ \cancel{\text{moles I}_2} \times \frac{253.8\text{ g I}_2}{\cancel{\text{mole I}_2}} = 508\text{ g I}_2$

b. H: 1.01 g/mole
Cl: 35.5 g/mole
O_3: 48.0 g/mole
$HClO_3$: 84.5 g/mole

$1.57\ \cancel{\text{moles HClO}_3} \times \frac{84.5\text{ g HClO}_3}{\cancel{\text{mole HClO}_3}} = 133\text{ g HClO}_3$

c. $0.12 \text{ mole Fe} \times \frac{55.8 \text{ g Fe}}{\text{mole Fe}} = 6.7 \text{ g Fe}$

8. a. $2 \times 1 = 2$ moles Fe
$2 \times 3 = 6$ moles O
$2 \times 3 = 6$ moles H
b. $3 \times 3 = 9$ moles Ca
$3 \times 2 = 6$ moles N

9. a.

$$\frac{2 \text{ moles } C_2H_2}{5 \text{ moles } O_2} \qquad \frac{1 \text{ mole } C_2H_2}{2 \text{ moles } CO_2}$$
$$\frac{1 \text{ mole } C_2H_5}{\text{mole } H_2O} \qquad \frac{5 \text{ moles } O_2}{4 \text{ moles } CO_2}$$
$$\frac{5 \text{ moles } O_2}{2 \text{ moles } H_2O} \qquad \frac{2 \text{ moles } CO_2}{\text{mole } H_2O}$$

b.

$$\frac{1 \text{ mole } Fe_2O_3}{3 \text{ moles CO}} \qquad \frac{1 \text{ mole } Fe_2O_3}{2 \text{ moles Fe}}$$
$$\frac{1 \text{ mole } Fe_2O_3}{3 \text{ moles } CO_2} \qquad \frac{3 \text{ moles CO}}{2 \text{ moles Fe}}$$
$$\frac{1 \text{ mole CO}}{\text{mole } CO_2} \qquad \frac{2 \text{ moles Fe}}{3 \text{ moles } CO_2}$$

10. a. $CuO + 2\,HCl \longrightarrow CuCl_2 + H_2O$

$$0.175 \text{ moles CuO} \times \frac{1 \text{ mole } CuCl_2}{\text{mole CuO}} = 0.175 \text{ moles } CuCl_2$$

b. $2\,Al + 3\,Cl_2 \longrightarrow 2\,AlCl_3$

$$0.175 \text{ moles Al} \times \frac{1 \text{ mole } AlCl_3}{\text{mole Al}} = 0.175 \text{ moles } AlCl_3$$

11. a. $$1.18 \times 10^{13} \text{ g } NH_3 \times \frac{1.00 \text{ mole } NH_3}{17.0 \text{ g } NH_3} = 6.94 \times 10^{11} \text{ moles } NH_3$$

b. $$6.94 \times 10^{11} \text{ moles } NH_3 \times \frac{1 \text{ mole } N_2}{2 \text{ moles } NH_3} = 3.47 \times 10^{11} \text{ moles } N_2$$

$$6.94 \times 10^{11} \text{ moles } NH_3 \times \frac{3 \text{ moles } H_2}{2 \text{ moles } NH_3} = 1.04 \times 10^{12} \text{ moles } H_2$$

12. a. $$785 \text{ g Pb} \times \frac{1 \text{ mole Pb}}{207.2 \text{ g Pb}} \times \frac{1 \text{ mole Zn}}{\text{mole Pb}} = 3.79 \text{ moles Zn}$$

b. $2\,AgNO_3 + H_2S \longrightarrow Ag_2S + 2\,HNO_3$

$$785 \text{ g } Ag_2S \times \frac{1 \text{ mole } Ag_2S}{247.9 \text{ g } Ag_2S} \times \frac{2 \text{ moles } AgNO_3}{\text{mole } Ag_2S} = 6.33 \text{ moles } AgNO_3$$

13. a. $$2.34 \text{ g } Cl_2 \times \frac{1 \text{ mole } Cl_2}{71.0 \text{ g } Cl_2} \times \frac{1 \text{ mole S}}{\text{mole } Cl_2} \times \frac{32.1 \text{ g S}}{\text{mole S}} = 1.06 \text{ g S}$$

b. $Ba(NO_3)_2 + (NH_4)_2CO_3 \longrightarrow BaCO_3 + 2\,NH_4NO_3$

$$2.34 \text{ g } (NH_4)_2CO_3 \times \frac{1 \text{ mole } (NH_4)_2CO_3}{96.1 \text{ g } (NH_4)_2CO_3} \times \frac{1 \text{ mole } Ba(NO_3)_2}{\text{mole } (NH_4)_2CO_3} \times \frac{261.3 \text{ g } Ba(NO_3)_2}{\text{mole } Ba(NO_3)_2} = 6.36 \text{ g } Ba(NO_3)_2$$

14. $$50.0 \text{ g } Cr_2O_3 \times \frac{1 \text{ mole } Cr_2O_3}{152 \text{ g } Cr_2O_3} \times \frac{1 \text{ mole } (NH_4)_2Cr_2O_7}{\text{mole } Cr_2O_3} \times \frac{252 \text{ g } (NH_4)_2Cr_2O_7}{\text{mole } (NH_4)_2Cr_2O_7} = 82.9 \text{ g } (NH_4)_2Cr_2O_7$$

15. $$5.00 \times 10^3 \text{ g } C_6H_{12}O_6 \times \frac{1 \text{ mole } C_6H_{12}O_6}{180.1 \text{ g } C_6H_{12}O_6} = 27.8 \text{ moles } C_6H_{12}O_6$$

$$27.8 \text{ moles } C_6H_{12}O_6 \times \frac{2 \text{ moles } CO_2}{\text{mole } C_6H_{12}O_6} \times \frac{44.0 \text{ g } CO_2}{\text{mole } CO_2} = 2.45 \times 10^3 \text{ g } CO_2$$

$$27.8 \text{ moles } C_6H_{12}O_6 \times \frac{2 \text{ moles } C_2H_5OH}{\text{mole } C_6H_{12}O_6} \times \frac{46.1 \text{ g } C_2H_5OH}{\text{mole } C_2H_5OH} = 2.56 \times 10^3 \text{ g } C_2H_5OH$$

16. a. $$5.00 \text{ t } H_2SO_4 \times \frac{1 \text{ t-mole } H_2SO_4}{98.1 \text{ t } H_2SO_4} \times \frac{1 \text{ t-mole } H_2S}{\text{t-mole } H_2SO_4} \times \frac{34.1 \text{ t } H_2S}{\text{t-mole } H_2S} = 1.74 \text{ t } H_2S$$

b. $$1.74 \text{ t } H_2S \times \frac{1 \text{ t-mole } H_2S}{34.1 \text{ t } H_2S} \times \frac{1 \text{ t-mole } H_2O}{\text{t-mole } H_2S} \times \frac{34.0 \text{ t } H_2O_2}{\text{t-mole } H_2O_2} = 1.73 \text{ t } H_2O_2$$

We get nearly the same answer in *a* and *b* because H_2S and H_2O_2 have very nearly the same molecular weights.

17. a. $$75.0 \text{ g Zn} \times \frac{1 \text{ mole Zn}}{65.4 \text{ g Zn}} \times \frac{148.5 \text{ kcal}}{\text{mole Zn}} = 170. \text{ kcal}$$

b. $$75.0 \text{ g } C_2H_6 \times \frac{1 \text{ mole } C_2H_6}{30.1 \text{ g } C_2H_6} \times \frac{736 \text{ kcal}}{2 \text{ moles } C_2H_6} = 917. \text{ kcal}$$

18. a. $$855 \text{ kcal} \times \frac{2 \text{ moles KCl}}{21.4 \text{ kcal}} \times \frac{74.6 \text{ g KCl}}{\text{mole KCl}} = 5960 \text{ g KCl}$$

b. $$855 \text{ kcal} \times \frac{1 \text{ mole CaO}}{42 \text{ kcal}} \times \frac{56.1 \text{ g CaO}}{\text{mole CaO}} = 1100 \text{ g CaO}$$

19. $$1.00 \text{ g } CH_4 \times \frac{1 \text{ mole } CH_4}{16.0 \text{ g } CH_4} \times \frac{192 \text{ kcal}}{\text{mole } CH_4} = 12.0 \text{ kcal (per g } CH_4)$$

$$1.00\,\cancel{g\,C_3H_8} \times \frac{1\,\cancel{mole\,C_3H_8}}{44.1\,\cancel{g\,C_3H_8}} \times \frac{489\text{ kcal}}{\cancel{mole\,C_3H_8}}$$
$$= 11.1\text{ kcal (per g } C_3H_8)$$

$$1.00\,\cancel{g\,C_4H_{10}} \times \frac{1\,\cancel{mole\,C_4H_{10}}}{58.1\,\cancel{g\,C_4H_{10}}} \times \frac{1270\text{ kcal}}{2\,\cancel{moles\,C_4H_{10}}}$$
$$= 10.9\text{ kcal (per g } C_4H_{10})$$

CH_4 is the best.

20. Hg: 200.59 $\quad Hg = \dfrac{200.59}{454.40} \times 10^2 = 44.14\%$

2 I: 253.81 $\quad I = \dfrac{253.81}{454.40} \times 10^2 = 55.86\%$

HgI_2: 454.40 $\qquad$ 100.00%

3 Ca: 120.24 $\quad Ca = \dfrac{120.24}{310.18} \times 10^2 = 38.76\%$

2 P: 61.94 $\quad P = \dfrac{61.94}{310.18} \times 10^2 = 19.97\%$

8 O: 128.00 $\quad O = \dfrac{128.00}{310.18} \times 10^2 = 41.27\%$

$Ca_3(PO_4)_2$: 310.18 $\qquad$ 100.00%

21. Sn: 118.69
2 Cl: 70.90
2 H_2O: 36.04

225.63

$$\% H_2O = \frac{36.04}{225.63} \times 10^2 = 15.97\%$$

Na: 22.99
2 C: 24.02
3 H: 3.03
2 O: 32.00
3 H_2O: 54.06

136.10

$$\% H_2O = \frac{54.06}{136.10} \times 10^2 = 39.72$$

$NaC_2H_3O_2 \cdot 3\,H_2O$ has the larger percentage of water.

22. a. Na: $\dfrac{41.82\,\cancel{g}}{22.99\,\cancel{g}/\text{mole}} = \dfrac{1.819\text{ moles Na}}{1.819\text{ moles Na}} = \dfrac{1.000\text{ mole Na}}{\text{mole Na}}$

O: $\dfrac{58.18\,\cancel{g}}{16.00\,\cancel{g}/\text{mole}} = \dfrac{3.636\text{ moles O}}{1.819\text{ moles Na}} = \dfrac{1.999\text{ moles O}}{\text{mole Na}}$

Answer: NaO_2

b. H: $\dfrac{1.48\,\cancel{g}}{1.01\,\cancel{g}/\text{mole}} = \dfrac{1.47\text{ moles H}}{1.461\text{ moles Cl}} = \dfrac{1\text{ mole H}}{1\text{ mole Cl}}$

Cl: $\dfrac{51.78\,\cancel{g}}{35.45\,\cancel{g}/\text{mole}} = \dfrac{1.461\text{ moles Cl}}{1.461\text{ moles Cl}} = \dfrac{1\text{ mole Cl}}{\text{mole Cl}}$

O: $\dfrac{46.74\,\cancel{g}}{16.00\,\cancel{g}/\text{mole}} = \dfrac{2.921\text{ moles O}}{1.461\text{ moles Cl}} = \dfrac{2\text{ moles O}}{\text{mole Cl}}$

Answer: $HClO_2$

23. $(NH_4)_3PO_4$ $\quad$ 3 N: 42.0
12 H: 12.1
PO_4: 95.0

149.1

$$\% PO_4 = \frac{95.0\text{ g } PO_4}{149.1\text{ g}(NH_4)_3PO_4} \times 10^2$$
$$= 63.7\%\text{ phosphate}$$

$Ca_3(PO_4)_2$ $\quad$ Ca: 120.
PO_4: 190.

310.

$$\% PO_4 = \frac{190.\text{ g } PO_4}{310.\text{ g } Ca_3(PO_4)_2} \times 10^2$$
$$= 61.3\%\text{ phosphate}$$

Ammonium phosphate is slightly better, but they are very close.

24. C: $\dfrac{85.59\,\cancel{g}}{12.01\,\cancel{g}/\text{mole}} = \dfrac{7.13\text{ moles C}}{7.13\text{ moles C}} = \dfrac{1\text{ mole C}}{\text{mole C}}$

H: $\dfrac{14.41\,\cancel{g}}{1.01\,\cancel{g}/\text{mole}} = \dfrac{14.3\text{ mole H}}{7.13\text{ moles C}} = \dfrac{2\text{ moles H}}{\text{mole C}}$

Empirical formula, CH_2 = 14.03 g/mole

$$\frac{56.08\,\cancel{g}/\text{mole (molecular weight)}}{14.03\,\cancel{g}/\text{mole (wt. of empirical formula)}} = 3.997$$

The molecular formula is 4 times the empirical formula: C_4H_8

25. $255\,\cancel{kg\,Fe} \times \dfrac{120.\,\cancel{kg\,FeS_2}}{55.8\,\cancel{kg\,Fe}} \times \dfrac{1.00\text{ kg ore}}{0.328\,\cancel{kg\,FeS_2}}$
$$= 1670\text{ kg ore}$$

26. a. 5.120 g − 4.970 g = 0.150 g O_2 lost from sample

$$.150\,\cancel{g\,O_2} \times \frac{1.00\,\cancel{mole\,O_2}}{32.0\,\cancel{g\,O_2}} \times \frac{2\,\cancel{moles\,HgO}}{\cancel{mole\,O_2}}$$
$$\times \frac{217\text{ g HgO}}{\cancel{mole\,HgO}} = 2.03\text{ g HgO in sample}$$

b. $\dfrac{2.03\text{ g HgO}}{5.12\text{ g ore}} \times 10^2 = 39.6\%\text{ HgO in ore}$

27. $1.00\,\cancel{kg\,SO_3} \times \dfrac{1\,\cancel{kg\text{-}mole\,SO_3}}{80.1\,\cancel{kg\,SO_3}} \times \dfrac{1\,\cancel{kg\text{-}mo\,H_2SO_4}}{\cancel{kg\text{-}mole\,SO_3}}$

$$\times \frac{1\,\cancel{kg\text{-}mole\,CaCO_3}}{\cancel{kg\text{-}mole\,H_2SO_4}} \times \frac{100.1\text{ kg } CaCO_3}{\cancel{kg\text{-}mole\,CaCO_3}}$$
$$= 1.25\text{ kg } CaCO_3 \text{ dissolved by the } SO_3$$

$$\frac{1.25\,\cancel{kg\,CaCO_3}}{545\,\cancel{kg\,CaCO_3}} \times 10^2$$
$$= 0.229\%\text{ of the statue dissolved}$$

28. a. $25.0\,\cancel{g\,H_2SO_4} \times \dfrac{1.00\,\cancel{mole\,H_2SO_4}}{98.0\,\cancel{g\,H_2SO_4}}$

$$\times \frac{1\text{ mole } BaSO_4}{\cancel{mole\,H_2SO_4}} = 0.255\text{ moles } BaSO_4$$

$$25.0 \text{ g } BaCl_2 \times \frac{1.00 \text{ mole } BaCl_2}{208 \text{ g } BaCl_2} \times \frac{1 \text{ mole } BaSO_4}{\text{mole } BaCl_2} = 0.120 \text{ moles } BaSO_4$$

$BaCl_2$ is the limiting reagent.

$$0.120 \text{ moles } BaSO_4 \times \frac{233 \text{ g } BaSO_4}{\text{mole } BaSO_4} = 28.0 \text{ g } BaSO_4$$

$$0.120 \text{ moles } BaSO_4 \times \frac{1 \text{ mole } H_2SO_4}{\text{mole } BaSO_4} \times \frac{98.0 \text{ g } H_2SO_4}{\text{mole } H_2SO_4} = 11.8 \text{ g } H_2SO_4 \text{ that have reacted.}$$

and $25.0 - 11.8 = 13.2$ g H_2SO_4 are left over

b. $$25.0 \text{ g } NH_3 \times \frac{1.00 \text{ mole } NH_3}{17.0 \text{ g } NH_3} \times \frac{2 \text{ moles } N_2}{4 \text{ moles } NH_3} = 0.735 \text{ moles } N_2$$

$$25.0 \text{ g } O_2 \times \frac{1.00 \text{ mole } O_2}{32.0 \text{ g } O_2} \times \frac{2 \text{ moles } N_2}{3 \text{ moles } O_2} = 0.521 \text{ moles } N_2$$

O_2 is the limiting reagent.

$$0.521 \text{ moles } N_2 \times \frac{28.0 \text{ g } N_2}{\text{mole } N_2} = 14.6 \text{ g } N_2 \text{ obtained}$$

$$0.521 \text{ moles } N_2 \times \frac{4 \text{ moles } NH_3}{2 \text{ moles } N_2} \times \frac{17.0 \text{ g } NH_3}{\text{mole } NH_3} = 17.7 \text{ g } NH_3 \text{ reacted}$$

so $25.0 - 17.7 = 7.3$ g NH_3 left over

29. $$21.3 \text{ g C} \times \frac{1.00 \text{ mole C}}{12.0 \text{ g C}} \times \frac{1 \text{ mole Zn}}{\text{mole C}} = 1.78 \text{ moles Zn}$$

$$123 \text{ g ZnO} \times \frac{1.00 \text{ mole ZnO}}{81.4 \text{ g ZnO}} \times \frac{1 \text{ mole Zn}}{\text{mole ZnO}} = 1.51 \text{ moles Zn}$$

ZnO is the limiting reactant.

$$1.51 \text{ moles Zn} \times \frac{65.4 \text{ g Zn}}{\text{mole Zn}} = 98.7 \text{ g Zn obtained}$$

$$1.51 \text{ moles Zn} \times \frac{1 \text{ mole C}}{\text{mole Zn}} \times \frac{12.0 \text{ g C}}{\text{mole C}} = 18.1 \text{ g C reacted}$$

so $21.3 - 18.1 = 3.2$ g C left over

30. theoretical yield:

$$500. \text{ kg } Ca(OH)_2 \times \frac{1.00 \text{ kg-mole } Ca(OH)_2}{74.1 \text{ kg } Ca(OH)_2} \times \frac{1 \text{ kg-mole } Mg(OH)_2}{\text{kg-mole } Ca(OH)_2} \times \frac{58.3 \text{ kg } Mg(OH)_2}{\text{kg-mole } Mg(OH)_2} = 393 \text{ kg } Mg(OH)_2$$

$$\% \text{ yield} = \frac{245 \text{ kg}}{393 \text{ kg}} \times 10^2 = 62.3\% \text{ yield}$$

31. theoretical yield:

$$5.2 \text{ t } Ca_3(PO_4)_2 \times \frac{1.0 \text{ t-mole } Ca_3(PO_4)_2}{310 \text{ t } Ca_3(PO_4)_2} \times \frac{1 \text{ t-mole } Ca(H_2PO_4)_2}{\text{t-mole } Ca_3(PO_4)_2} \times \frac{234 \text{ t } Ca(H_2PO_4)_2}{\text{t-mole } Ca(H_2PO_4)_2} = 3.9 \text{ t } Ca(H_2PO_4)_2$$

actual yield $= 0.523 \times 3.9 = 2.0$ tons $Ca(H_2PO_4)_2$

Chapter 7

1. a. a bowling ball on the fourth floor; b. an electron and a proton that are close together; c. two electrons that are far apart; d. an electron and a proton that are far apart; e. an electron in the first energy level (*K* shell); f. an electron in the *s* sublevel; g. two electrons close together that have opposite spin
2. a. 2; b. 2; c. 10; d. 6; e. 8; f. 2
3. a. 3rd energy level, *d* sublevel, 3 electrons in *d* sublevel; b. 2nd energy level, *s* sublevel, 2 electrons in *s* sublevel; c. 5th energy level, *p* sublevel, 4 electrons in *p* sublevel; d. 6th energy level, *f* sublevel, 10 electrons in *f* sublevel
4. a. Hg: (80) $1s^22s^22p^63s^23p^64s^23d^{10}4p^65s^24d^{10}5p^66s^25d^{10}4f^{14}$
   b. I: (53) $1s^22s^22p^63s^23p^64s^23d^{10}4p^65s^24d^{10}5p^5$
   c. Fe: (26) $1s^22s^22p^63s^23p^64s^23d^6$
5. a. $2p$; b. $4s$; c. $5s$
6. a. Zn: (30) $1s^22s^22p^63s^23p^64s^23d^{10}$
   b. Sn: (50) $1s^22s^22p^63s^23p^64s^23d^{10}4p^65s^24d^{10}5p^2$
   c. Xe: (54) $1s^22s^22p^63s^23p^64s^23d^{10}4p^65s^24d^{10}5p^6$
7. a. Al: (13) $1s^22s^22p^63s^23p^1$
   b. Ba: (56) $1s^22s^22p^63s^23p^64s^23d^{10}4p^65s^24d^{10}5p^66s^2$
   c. As: (33) $1s^22s^22p^63s^23p^64s^23d^{10}4p^3$
8. a. Wrong. The *p* sublevel can contain only 6 electrons.
   b. Correct.
   c. Wrong. There is no *p* sublevel in the first energy level.
   d. Wrong. The single orbital in the *s* sublevel can contain only 2 electrons.
   e. Wrong. After Ne-10, the third energy level, not the second, begins to fill.

Chapter 8

1. a. Group IIA, *s* block; b. Group IVA, *p* block; c. Group VIIA, *p* block; d. Group VB, *d* block
2. a. N, P, As, Sb, Bi; b. Eu, Am; c. Sc, Y, La, Ac; d. Cu, Ag, Au; e. Kr, Xe, Rn; f. H. Li, Na, K, Rb, Cs, Fr
3. a. *s*; b. *s*; c. *p*; d. *p*; e. *d*; f. *p*; g. *p*; h. *d*; i. *f*
4. a. H, He; b. B, C, N, O, F. Ne; c. Na, Mg; d. F; e. V; f. Ag
5. outer configuration s^1, 6th energy level: $6s^1$
6. Group VIIA

7. 14
8. $4s^2$
9. $5s^25p^2$, Group IVA
10. Group VIIA, Period 4
11. a. (Kr-36) $5s^2$; b. (Xe-54) $6s^1$; c. (Kr-36) $5s^24d^2$; d. (Ar-18) $4s^23d^{10}4p^4$; e. (Ar-18) $4s^23d^{10}4p^1$; f. (Kr-36) $5s^24d^{10}5p^5$; g. (Kr-36) $5s^24d^5$; h. (Ar-18) $4s^23d^6$
12. a. X· b. X c. ·X:
13. IA IIA IIIA IVA VA VIA VIIA VIIIA
E· E· ·E· ·E· ·E: ·E: :E: :E:
14. a. Rb; b. I; c. Sn
15. a. one *s* electron; b. two *s* electrons, one *d* electron; c. two *s*, two *d* electrons; d. one *s*, two *d* electrons
16. a. Na; b. Cs; c. F_2
17. a. Ca; b. K; c. Bi
18. a. Group IVA; b. Sn, Pb; c. $7s^27p^2$; d. ·X·; e. metal, more metallic than others in group; f. XO, XO_2, XCl_2, XCl_4; g. approximately 290; h. solid; i. more dense

Chapter 9

1. a. Li· + ·Ö: + Li· ⟶ Li^+ Li^+ $:Ö:^-$

b. Mg· + Mg· + Mg· + ·N· + ·N· ⟶ Mg^{2+} Mg^{2+} Mg^{2+} $:N:^{3-}$ $:N:^{3-}$

c. Al· + H· + H· + H· ⟶ Al^{3+} $H:^-$ $H:^-$ $H:^-$

d. Rb· + ·I: ⟶ Rb^+ $:I:^-$

2. a. Ca; b. O^{2-}; c. Br^-
3. a. the second ionization energy of K; b. the third ionization energy of Be; c. the second ionization energy of Li
4. a. Group VIIIA; b. Group IA
5. a. Na; b. Ar; c. Ca; d. N
6. :Cl:Cl: :Br:Br: :I:I:
7. :S::S:
8. a. H:P:H (H below P) b. :Cl:N:Cl: (:Cl: below N) c. :Br:O:Br:

d. :F:Se:F:

9. BF_3 can form a coordinate covalent bond with NH_3. We can substitute either nonbonding electron pair of water for that of NH_3, and write the formula:

H—Ö:B—F: with :F: above B, and H below O and :F: below B

10. 3 bonds/molecule × 93.4 kcal/mole/bond = 280 kcal/molecule/mole
11. a. H—H; b. C≡N
12. a. The H is bonded to the Cl and not to the O.
b. There are too many electrons (26 instead of 24).
c. The charge is omitted.
13. a. H—Ō—Cl—Ō| with |O| below Cl

b. |Ō—S=Ō| with |O| below S

c. $[\,|Ō—N=Ō|$ with |O| below N $]^-$

14. H_3BO_3: H—O—B—O—H (6 pairs), with O—H below B

B: 3 valence electrons
O_3: 18 valence electrons
H_3: 3 valence electrons
H_3BO_3: 24 electrons, 12 pairs

H—Ō—B—Ō—H (12 pairs), with |O| and H below B

15. H_2SO_4: H—O—S—O—H (6 pairs), with O above and O below S

S: 6 valence electrons
O_4: 24 valence electrons
H_2: 2 valence electrons
H_2SO_4: 32 electrons, 16 pairs

H—Ō—S—Ō—H (16 pairs), with |Ō| above and |O| below S

HSO_4^- (H_2SO_4 minus 1 H):

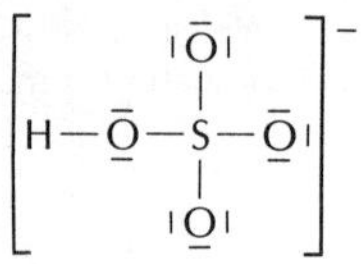

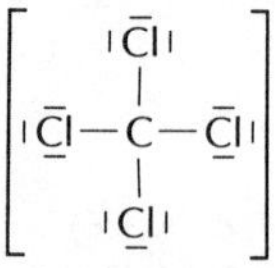

```
[      |Ō|      ]⁻
[       |       ]
[ H—Ō—S—Ō|      ]
[       |       ]
[      |Ō|      ]
```

SO_4^{2-} (HSO_4^- minus 1 H):

```
[     |Ō|     ]²⁻
[      |      ]
[ |Ō—S—Ō|     ]
[      |      ]
[     |Ō|     ]
```

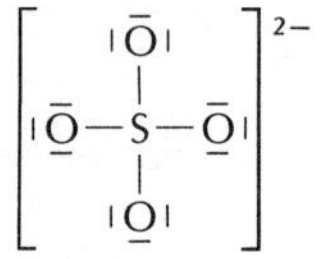

All have the same number of electron pairs.

16. SO_4^{2-}:

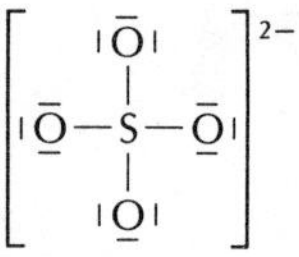

```
[     |Ō|     ]²⁻
[      |      ]
[ |Ō—S—Ō|     ]
[      |      ]
[     |Ō|     ]
```

(Exercise 15)

PO_4^{3-}:

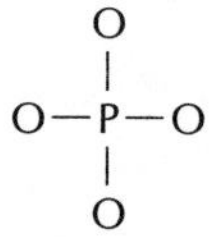

```
    O
    |
O—P—O   (4 pairs)
    |
    O
```

P: 5 valence electrons
O_4: 24 valence electrons
3−: <u>3</u> electrons
PO_4^{3-}: 32 electrons, 16 pairs

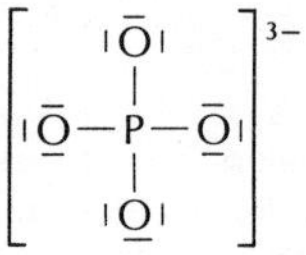

```
[     |Ō|     ]³⁻
[      |      ]
[ |Ō—P—Ō|     ]   (6 pairs)
[      |      ]
[     |Ō|     ]
```

ClO_4^-:

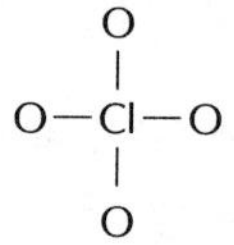

```
     O
     |
O—Cl—O   (4 pairs)
     |
     O
```

Cl: 7 valence electrons
O_4: 24 valence electrons
1−: <u>1</u> electron
ClO_4^-: 32 electrons, 16 pairs

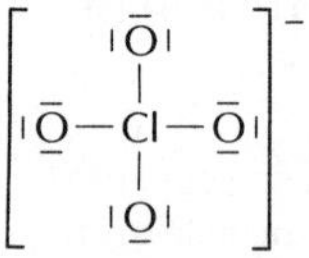

```
[     |Ō|      ]⁻
[      |       ]
[ |Ō—Cl—Ō|     ]   (16 pairs)
[      |       ]
[     |Ō|      ]
```

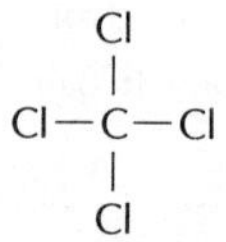

CCl_4:

```
      Cl
      |
Cl—C—Cl   (4 pairs)
      |
      Cl
```

Cl_4: 28 valence electrons
C: <u>4</u> valence electrons
CCl_4: 32 electrons, 16 pairs

```
[     |C̄l|      ]
[      |        ]
[ |Cl—C—Cl|     ]   (16 pairs)
[      |        ]
[     |Cl|      ]
```

SO_4^{2-}, PO_4^{3-}, ClO_4^-, and CCl_4 all have 16 electron pairs and 4 atoms attached to a central atom.

17. C_2F_4:

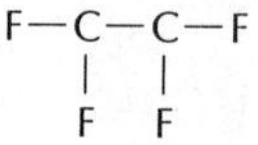

```
F—C—C—F
  |  |
  F  F
```

F_4: 28 valence electrons
C_2: <u>8</u> valence electrons
C_2F_4: 36 electrons, 18 pairs

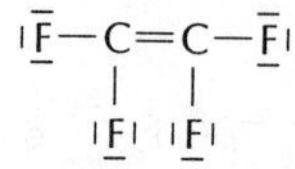

```
|F̄—C=C—F̄|   (18 pairs)
   |  |
  |F| |F|
```

18. a. OH^-: O—H

O: 6 valence electrons
H: 1 valence electrons
1: <u>1</u> electron
OH^-: 8 electrons, 4 pairs

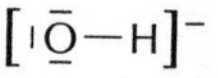

[|Ō—H]⁻ (4 pairs)

b. CN^-: C—N

C: 4 valence electrons
N: 5 valence electrons
1−: <u>1</u> electron
CN^-: 10 electrons, 5 pairs

[|C≡N|]⁻ (5 pairs)

c. S is in the same group as O, so SH^- has the same Lewis structure as OH^-:

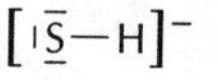

[|S̄—H]⁻ (4 pairs)

d. O_2^{2-}: O—O

O_2: 12 valence electrons
2−: <u>2</u> electrons
O_2^{2-}: 14 electrons, 7 pairs

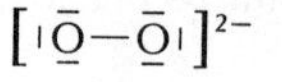

[|Ō—Ō|]²⁻ (7 pairs)

e. HCO_3^- is the same as CO_3^{2-} except that it has one less negative charge and a hydrogen:

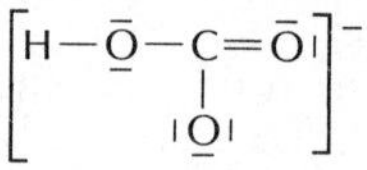

```
[ H—Ō—C=Ō| ]⁻
[      |    ]   (12 pairs)
[     |O|   ]
```

```
[ |Ō—C=Ō| ]²⁻
[     |    ]   (12 pairs)
[    |O|   ]
```

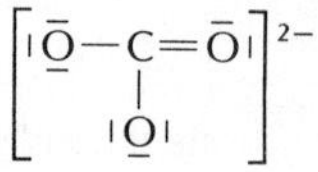

f. ClO^-: Cl—O

Cl: 7 valence electrons
O: 6 valence elecrtrons
1^-: 1 electron
ClO^-: 14 electrons, 7 pairs

$[\,|\bar{Cl}-\bar{O}|\,]^-$ (7 pairs)

19. a. Se, S, Cl, F; b. Rb, Sr, I, F
20. a. nonpolar; b. polar covalent; c. nonpolar; d. polar covalent
21. b. Ba $\leftrightarrow$ Br; d. Be $\leftrightarrow$ Se
22. a. From Exercise 21, Ba—Br and Be—Se fit this category. Others can be found.
b. Si—F (electronegativity difference 2.2)

Chapter 10

1. $20.\,\cancel{g} \times \frac{0.5\text{ cal}}{\cancel{g} \times \cancel{K}} \times 10.\,\cancel{K} = 100\text{ cal}$

$20.\,\cancel{g} \times 79.9\frac{\text{cal}}{\cancel{g}} = 1600\text{ cal}$

$20.0\,\cancel{g} \times 1.0\frac{\text{cal}}{\cancel{g}} \times 100.\text{ K} = 2000\text{ cal}$

$20.0\,\cancel{g} \times 540\frac{\text{cal}}{\cancel{g}} = \underline{10800\text{ cal}}$

Total: 14500 cal

2. 15.000 liters of water is equal to 15,000 kg of water

$7.5 \times 10^4\,\cancel{kg} \times 1.0\frac{\text{kcal}}{\cancel{kg} \times \cancel{K}} \times 5\,\cancel{K} = 7.6 \times 10^4\text{ kcal}$

$1.5 \times 10^4\,\cancel{kg} \times 79.9\frac{\text{kcal}}{\cancel{kg}} = 1.2 \times 10^6\text{ kcal}$

$1.5 \times 10^4\,\cancel{kg} \times 0.5\frac{\text{kcal}}{\cancel{kg}} \times 5\text{ K} = \underline{3.8 \times 10^4\text{ kcal}}$

Total: 1.3×10^6 kcal

3. $1\,\cancel{kg\,NH_3} \times \frac{327\text{ kcal}}{\cancel{kg\,NH_3}} = 327\text{ kcal}$

4. a. 380 torr; b. harder; c. longer; d. worse
5. A column 5.0 meters long and 1 cm^2 in area would have a weight of:

$500\,\cancel{cm} \times 1\,\cancel{cm^2} \times 1\frac{g}{\cancel{cm^3}} = 500\text{ g, or }0.5\text{ kg}$

Since a pressure of 1.04 kg/cm^2 is 1 atm,

$0.5\,\cancel{kg/cm^2} \times \frac{1\text{ atm}}{1.04\,\cancel{kg/cm^2}} = 0.48\text{ atm}$

1.0 atm (on water surface) + 0.48 atm = 1.48 atm. Answer: 1.5 atm

6. a. a gas; b. the first flat part on the left is the heat of fusion, the next is the heat of vaporization; c. solid; d. weaker (melting and boiling points lower than water's)
7. The alcohol, having a low vapor pressure, evaporates rapidly from the skin, removing heat from the skin and cooling it.
8. a. because increasing the temperature increases water's vapor pressure, and also the amount of water vapor that the air can hold; d. because changing the air around the clothes removes the water vapor from the immediate area; f. because, with it raining outside, the outside air already has as much water vapor as it can hold
9. Yes, by decreasing its pressure.
10. dry air
11. The ice would melt, until the two vapor pressures were the same.
12. a. It takes more energy to vaporize than to melt, because the attraction between particles in solid and liquid are about the same, whereas the attractions between particles in the liquid are far stronger than they are in the gas; b. More energy is needed to overcome the hydrogen bonding in H_2O.
13. a. H_2O is more highly hydrogen bonded than HF; b. SiO_2 is held together by stronger interactions (covalent bonding) than NaCl is (ionic bonding); c. It takes more energy to break an ionic bond (in NaCl) than to break a hydrogen bond (in H_2O); d. Hydrogen bonding can occur in NH_3, and not SO_2. It takes more energy to vaporize hydrogen-bonded substances.

Chapter 11

1. $100\,\cancel{miles} \times \frac{5\,\cancel{g\,NO}}{\cancel{mile}} \times \frac{1\,\cancel{mole\,NO}}{30\,\cancel{g\,NO}} \times \frac{22.4\,\ell\text{ NO}}{\cancel{mole\,NO}} = 400\,\ell\text{ NO}$

2. $2 \times 10^{10}\,\cancel{\ell\,NH_3} \times \frac{1\,\cancel{mole\,NH_3}}{22.4\,\cancel{\ell\,NH_3}} \times \frac{17\,\cancel{g\,NH_3}}{\cancel{mole\,NH_3}} \times \frac{1\text{ t }NH_3}{10^6\,\cancel{g\,NH_3}} = 2 \times 10^4\text{ t }NH_3$, or, more exactly, 1.52×10^4 t

3. $\frac{30.0\text{ g NO}}{\cancel{mole\,NO}} \times \frac{1\,\cancel{mole\,NO}}{22.4\,\ell\text{ NO}} = 1.34\text{ g}/\ell\text{ NO}$

$\frac{46.0\text{ g }NO_2}{\cancel{mole\,NO_2}} \times \frac{1\,\cancel{mole\,NO_2}}{22.4\,\ell\,NO_2} = 2.05\text{ g}/\ell\,NO_2$

NO_2 is more dense.

4. $1.35\frac{g}{\cancel{\ell}} \times \frac{22.4\,\cancel{\ell}}{\text{mole}} = 30.2\frac{g}{\text{mole}}$

The pollutant is probably NO (molecular weight 30.0 g/mole).

5. a. $\frac{0.469\text{ g}}{0.125\,\cancel{\ell}} \times \frac{22.4\,\cancel{\ell}}{\text{mole}} = 84.0\text{ g/mole}$

b. $\frac{84.0}{14.0} = 6$; C_6H_{12}

6. $500\,\cancel{\ell\,N_2} \times \frac{2\,\ell\text{ NO}}{\cancel{\ell\,N_2}} = 1000\,\ell\text{ NO}$

7. $1.2 \times 10^4\ \cancel{\ell\ \mathrm{NO}} \times \dfrac{4\ \ell\ \mathrm{NH_3}}{6\ \cancel{\ell\ \mathrm{NO}}} = 8.0 \times 10^3\ \ell\ \mathrm{NH_3}$

8. $250\ \cancel{\ell\ \mathrm{H_2S}} \times \dfrac{1\ \cancel{\mathrm{mole\ H_2S}}}{22.4\ \cancel{\ell\ \mathrm{H_2S}}} \times \dfrac{1\ \cancel{\mathrm{mole\ S}}}{\cancel{\mathrm{mole\ H_2S}}} \times \dfrac{32.1\ \mathrm{g\ S}}{\cancel{\mathrm{mole\ S}}}$

$= 360\ \mathrm{g\ S}$

9. $2.5\ \cancel{\mathrm{moles\ glucose}} \times \dfrac{6\ \cancel{\mathrm{moles\ O_2}}}{\cancel{\mathrm{mole\ glucose}}} \times \dfrac{22.4\ \ell\ \mathrm{O_2}}{\cancel{\mathrm{mole\ O_2}}}$

$= 340\ \ell\ \mathrm{O_2}$

10. $\text{theoretical yield} = \dfrac{\text{actual yield} \times 10^2}{\text{percent yield}}$

$= \dfrac{1.00\ \mathrm{kg} \times 10^2}{46.4\%} = 2.16\ \mathrm{kg}$

$2.16 \times 10^3\ \cancel{\mathrm{g\ NaHCO_3}} \times \dfrac{1\ \cancel{\mathrm{mole\ NaHCO_3}}}{84.0\ \cancel{\mathrm{g\ NaHCO_3}}}$

$\times \dfrac{1\ \cancel{\mathrm{mole\ NH_3}}}{\cancel{\mathrm{mole\ NaHCO_3}}} \times \dfrac{22.4\ \ell\ \mathrm{NH_3}}{\cancel{\mathrm{mole\ NH_3}}} = 576\ \ell\ \mathrm{NH_3}$

(CO_2 is the same, because it has the same mole ratio)

11. $V_2 = 2 \times 10^4\ \ell \times \dfrac{1.00\ \cancel{\mathrm{atm}}}{0.72\ \cancel{\mathrm{atm}}} = 3 \times 10^4\ \ell\ (30{,}000\ \ell)$

12. $P_2 = 1\ \mathrm{atm} \times \dfrac{1000\ \cancel{\ell}}{0.5\ \cancel{\ell}} = 2 \times 10^3\ \mathrm{atm}\ (2{,}000\ \mathrm{atm})$

13. $V_2 = 1\ \ell \times \dfrac{273\ \cancel{\mathrm{K}}}{923\ \cancel{\mathrm{K}}} = 0.3\ \ell$

14. Needles: $1.00\ \ell \times \dfrac{316\ \cancel{\mathrm{K}}}{273\ \cancel{\mathrm{K}}} = 1.16\ \ell$; density $= \dfrac{1.29\ \mathrm{g}}{1.16\ \ell}$

$= 1.11\ \mathrm{g}/\ell$

Bozeman: $1.00\ \ell \times \dfrac{244\ \cancel{\mathrm{K}}}{273\ \cancel{\mathrm{K}}} = 0.894\ \ell$; density

$= \dfrac{1.29\ \mathrm{g}}{0.894\ \ell} = 1.44\ \mathrm{g}/\ell$

Density of air in Needles is roughly 3/4 the density of air in Bozeman.

15. $T_2 = 298\ \mathrm{K} \times \dfrac{1.3 \times 10^4\ \cancel{\mathrm{atm}}}{713\ \cancel{\mathrm{atm}}} = 5.4 \times 10^3\ \mathrm{K}\ (5433\ \mathrm{K})$

$5433 - 273 = 5160°\mathrm{C}$, or $5200°\mathrm{C}$

16. $P_2 = 0.972\ \mathrm{atm} \times \dfrac{300.\ \cancel{\mathrm{K}}}{263\ \cancel{\mathrm{K}}} = 1.11\ \mathrm{atm}$

17. 23. $V_2 = 89.2\ \mathrm{cm^3} \times \dfrac{273\ \cancel{\mathrm{K}}}{295\ \cancel{\mathrm{K}}} \times \dfrac{0.978\ \cancel{\mathrm{atm}}}{1.00\ \cancel{\mathrm{atm}}} = 80.7\ \mathrm{cm^3}$

18. $V_2 = 533\ \mathrm{cm^3} \times \dfrac{760\ \cancel{\mathrm{torr}}}{745\ \cancel{\mathrm{torr}}} \times \dfrac{298\ \cancel{\mathrm{K}}}{273\ \cancel{\mathrm{K}}} = 594\ \mathrm{cm^3}$

19. $0.234\ \mathrm{atm} + 0.438\ \mathrm{atm} + 0.199\ \mathrm{atm} = 0.871\ \mathrm{atm}$

20. a. $0.957\ \mathrm{atm} - 0.0294\ \mathrm{atm} = 0.928\ \mathrm{atm}$

b. $V_2 = 44.0\ \mathrm{cm^3} \times \dfrac{0.928\ \cancel{\mathrm{atm}}}{1.00\ \cancel{\mathrm{atm}}} \times \dfrac{273\ \cancel{\mathrm{K}}}{297\ \cancel{\mathrm{K}}} = 37.5\ \mathrm{cm^3}$

21. $P_2 = 0.954\ \mathrm{atm} - 0.0261 = 0.928\ \mathrm{atm}$

$V_2 = 97.2\ \mathrm{cm^3} \times \dfrac{295\ \cancel{\mathrm{K}}}{273\ \cancel{\mathrm{K}}} \times \dfrac{1.00\ \cancel{\mathrm{atm}}}{0.928\ \cancel{\mathrm{atm}}} = 113\ \mathrm{cm^3}$

22. a. $5.5 \times 10^5\ \cancel{\mathrm{g\ CaC_2}} \times \dfrac{1\ \cancel{\mathrm{mole\ CaC_2}}}{64.1\ \cancel{\mathrm{g\ CaC_2}}} \times \dfrac{1\ \cancel{\mathrm{mole\ C_2H_2}}}{\cancel{\mathrm{mole\ CaC_2}}}$

$\times \dfrac{22.4\ \ell\ \mathrm{C_2H_2}}{\cancel{\mathrm{mole\ C_2H_2}}} = 1.9 \times 10^5\ \ell\ \mathrm{C_2H_2}$

b. $V_2 = 1.9 \times 10^5\ \text{liters} \times \dfrac{573\ \cancel{\mathrm{K}}}{273\ \cancel{\mathrm{K}}} \times \dfrac{1.00\ \cancel{\mathrm{atm}}}{2.00\ \cancel{\mathrm{atm}}}$

$= 2.0 \times 10^5\ \text{liters}$

23. $2 \times 10^4\ \cancel{\ell\ \mathrm{air}} \times \dfrac{0.05\ \ell\ \mathrm{NO}}{10^6\ \cancel{\ell\ \mathrm{air}}} = 1 \times 10^{-3}\ \ell\ \mathrm{NO}$

24. ideal gas equation, $n = \dfrac{PV}{RT}$:

$n = \dfrac{(0.976\ \cancel{\mathrm{atm}})(6 \times 10^5\ \cancel{\ell}\ \mathrm{SO_2})}{\left(0.0821\ \dfrac{(\cancel{\ell})(\cancel{\mathrm{atm}})}{(\mathrm{mole})(\cancel{\mathrm{K}})}\right)(473\ \cancel{\mathrm{K}})}$

$= 2 \times 10^4\ \text{moles}\ \mathrm{SO_2} = 2 \times 10^4\ \text{moles S}$

Alternative method:

$6 \times 10^5\ \ell\ \mathrm{SO_2} \times \dfrac{0.976\ \cancel{\mathrm{atm}}}{1.00\ \cancel{\mathrm{atm}}} \times \dfrac{273\ \cancel{\mathrm{K}}}{473\ \cancel{\mathrm{K}}}$

$= 3.38 \times 10^5\ \ell\ \mathrm{SO_2}$

$3.38 \times 10^5\ \cancel{\ell\ \mathrm{SO_2}} \times \dfrac{1\ \cancel{\mathrm{mole\ SO_2}}}{22.4\ \cancel{\ell\ \mathrm{SO_2}}} \times \dfrac{1\ \text{mole S}}{\cancel{\mathrm{mole\ SO_2}}}$

$= 2 \times 10^4\ \text{moles S}$

25. a. 2×10^4 moles H_2SO_4 (moles S = moles SO_2 = moles H_2SO_4)

b. $2 \times 10^4\ \cancel{\mathrm{moles\ H_2SO_4}} \times \dfrac{98.1\ \mathrm{g\ H_2SO_4}}{\cancel{\mathrm{mole\ H_2SO_4}}}$

$= 2 \times 10^6\ \mathrm{g\ H_2SO_4}\ (2 \times 10^3\ \mathrm{kg})$

26. $1.00\ \cancel{\mathrm{lbs\ glucose}} \times \dfrac{454\ \cancel{\mathrm{g}}}{\cancel{\mathrm{lb}}} \times \dfrac{1\ \cancel{\mathrm{mole\ glucose}}}{180\ \cancel{\mathrm{g\ glucose}}}$

$\times \dfrac{6\ \cancel{\mathrm{moles\ CO_2}}}{\cancel{\mathrm{mole\ glucose}}} \times \dfrac{22.4\ \ell\ \mathrm{CO_2}}{\cancel{\mathrm{mole\ CO_2}}} = 339\ \ell\ \mathrm{CO_2}$

$339\ \cancel{\ell\ \mathrm{CO_2}} \times \dfrac{100.\ \ell\ \text{air}}{0.031\ \cancel{\ell\ \mathrm{CO_2}}} = 1.1 \times 10^6\ \ell\ \text{air}$

Chapter 12

1. $\dfrac{0.352\ \mathrm{g}}{0.0500\ \ell} = 7.04\dfrac{\mathrm{g}}{\ell}$ Answer: $7.04\ \mathrm{g}/\ell$

2. $150\ \cancel{\mathrm{g\ sugar}} \times \dfrac{1.0\ \ell\ \text{corn syrup}}{1.1 \times 10^3\ \cancel{\mathrm{g\ sugar}}}$

$= 0.14\ \ell$ corn syrup, or $140\ \mathrm{m}\ell$

3. $250.\ \cancel{\mathrm{m}\ell\ \text{solution}} \times \dfrac{0.900\ \mathrm{g\ NaCl}}{100.\ \cancel{\mathrm{m}\ell\ \text{solution}}} = 2.25\ \mathrm{g\ NaCl}$

4. $22.5\ \mathrm{g}/\ell$

5. $\dfrac{1.44\ \text{g glucose}}{25.0\ \text{g solution}} \times 100 = 5.76\%$ by weight

6. $15.0\ \cancel{\mathrm{g\ HNO_3}} \times \dfrac{100\ \cancel{\text{g conc. acid}}}{72\ \cancel{\mathrm{g\ HNO_3}}} \times \dfrac{1\ \mathrm{m}\ell\ \text{conc. acid}}{1.42\ \cancel{\text{g conc. acid}}}$

$= 14.7\ \mathrm{m}\ell$ conc. acid

7. $\frac{5.0\text{ m}\ell\ H_2O_2}{500.\text{ m}\ell\text{ solution}} \times 100 = 1.0\%$ by volume

8. $1\ \cancel{\ell\text{ gin}} \times \frac{40\ \ell\text{ alcohol}}{100\ \cancel{\ell\text{ gin}}} = 0.4\ \ell$ alcohol

9. $\frac{45.0\ \cancel{\text{g KOH}}}{0.250\ \ell} \times \frac{1\text{ mole}}{56.1\ \cancel{\text{g KOH}}} = 3.21\ M$

10. $5.0\ \cancel{\text{g HCl}} \times \frac{1\ \cancel{\text{mole}}}{36.5\ \cancel{\text{g HCl}}} \times \frac{1000\text{ m}\ell}{0.30\ \cancel{\text{mole}}} = 460\text{ m}\ell$

11. $500\ \cancel{\text{m}\ell\text{ solution}} \times \frac{2 \times 10^{-17}\ \cancel{\text{moles } Ag_2S}}{1.00 \times 10^3\ \cancel{\text{m}\ell\text{ solution}}} \times \frac{247.9\text{ g } Ag_2S}{\cancel{\text{mole } Ag_2S}} = 2 \times 10^{-15}\text{ g } Ag_2S$

12. O_2: $\frac{4.5 \times 10^{-2}\ \cancel{\text{g}}/\ell}{32.0\ \cancel{\text{g}}/\text{mole}} = 1.4 \times 10^{-3}\ M$

CO_2: $\frac{0.145\ \cancel{\text{g}}/\ell}{44.0\ \cancel{\text{g}}/\text{mole}} = 3.30 \times 10^{-3}\ M$

NH_3: $\frac{320\ \cancel{\text{g}}/\ell}{17.0\ \cancel{\text{g}}/\text{mole}} = 19\ M$

13. Add water to the mixture and stir. NaCl is soluble and will dissolve, leaving the insoluble AgCl behind.

14. $Na^+ = \frac{10{,}760\ \cancel{\text{kg } Na^+}}{10^6\ \cancel{\text{kg seawater}}} \times \frac{1.03\ \cancel{\text{kg seawater}}}{\ell\text{ seawater}} \times \frac{1\text{ kg-mole}}{23.0\ \cancel{\text{kg } Na^+}} = 4.82 \times 10^{-4}\ \frac{\text{kg-mole}}{\ell}$

$= 4.82 \times 10^{-1}\ M\ (0.482\ M)$

$Mg^{2+} = \frac{1294\ \cancel{\text{kg } Mg^{2+}}}{10^6\ \cancel{\text{kg seawater}}} \times \frac{1.03\ \cancel{\text{kg seawater}}}{\ell\text{ seawater}} \times \frac{1\text{ kg-mole}}{24.3\ \cancel{\text{kg } Mg^{2+}}} = 5.48 \times 10^{-5}\ \frac{\text{kg-mole}}{\ell}$

$= 5.48 \times 10^{-2}\ M$

15. $15{,}000\ \cancel{\text{g swordfish}} \times \frac{0.5\text{ g Hg}}{10^6\ \cancel{\text{g swordfish}}} = 7.5 \times 10^{-3}\text{ g Hg}$

16. a. from Table 12.2, Cl_2 has a solubility of 6.3 g/ℓ.
Solubility of BrCl = 12 × 6.3 g/ℓ = 75.6 g/ℓ

$\frac{75.6\ \cancel{\text{g}}/\ell}{115\ \cancel{\text{g}}/\text{mole}} = 0.657\text{ moles}/\ell$ or $0.657\ M$

b. $BrCl + H_2O \longrightarrow HBrO + HCl$

17. a. HF (polar molecule); b. NO_2 (more polar than N_2); c. H_2SO_4 (O—H bond hydrogen bonds better than S—H bond); d. CH_3OH (can H bond, CH_4 can't)

18. Mayonnaise is an emulsion, and the excess ions in the air during an electrical storm could destroy the emulsion.

19. A colloid; it's not transparent.

20. The small salt ions go through the membrane and leave the larger protein molecules behind.

21. Ice is cloudy because of air bubbles trapped in the ice's crystal structure. This air was dissolved in the water before it froze. Boiling the water before freezing it drives a great deal of the dissolved air from the water.

22. a. $0.145\text{ g}/\ell \times \frac{350\ \cancel{\text{torr}}}{760\ \cancel{\text{torr}}} = 0.0668\text{ g}/\ell$

b. $6.3\text{ g}/\ell \times \frac{2.5\ \cancel{\text{atm}}}{1\ \cancel{\text{atm}}} = 16\text{ g}/\ell$

23. $P_2 = P_1\frac{S_2}{S_1} = 1\text{ atm}\frac{0.200\ \cancel{\text{g}/\ell}}{0.145\ \cancel{\text{g}/\ell}} = 1.38\text{ atm}$

24. When this has happened, the honey was originally a supersaturated solution of sugar in water. After the honey solution has stood for a time, the excess solute has crystallized out of it.

25. N_2, Ar, O_2 (in increasing order of boiling points)

26. $0.250\ \cancel{\text{kg water}} \times \frac{1.00\ \cancel{\text{moles alcohol}}}{\cancel{\text{kg water}}} \times \frac{46.1\text{ g alcohol}}{\cancel{\text{mole alcohol}}} = 11.5\text{ g alcohol}$

27. $\frac{5.54\ \cancel{\text{g glycerin}}}{0.125\text{ kg water}} \times \frac{1\text{ mole glycerin}}{92.1\ \cancel{\text{g glycerin}}} = 0.481$ molal

28. $\frac{27.0\ \cancel{\text{g methyl alcohol}}}{0.0730\text{ kg } H_2O} \times \frac{1\text{ mole methyl alcohol}}{32.0\ \cancel{\text{g methyl alcohol}}} = 11.6\frac{\text{moles methyl alcohol}}{\text{kg } H_2O} = 11.6\ m$

$T_f = 0.00°\text{C} - \left(1.84\frac{°\text{C}}{\cancel{m}}\right)(11.6\ \cancel{m}) = -21.3°\text{C}$

29. $m = \frac{-(-1.00\ \cancel{°\text{C}})}{1.84\frac{\cancel{°\text{C}}}{m}} = 0.543$ molal

$100.\ \cancel{\text{g } H_2O} \times \frac{0.543\ \cancel{\text{moles}}}{1000\ \cancel{\text{g } H_2O}} \times \frac{180.\text{ g glucose}}{\cancel{\text{mole}}} = 9.77\text{ g glucose}$

30. Yes; any soluble solute will lower the freezing point of the solution. However, care should be taken not to contaminate the ice cream with ethylene glycol.

31. a. Water leaves the snail's cells to dilute the salt, and the snail becomes dehydrated and dies.
b. Water passes from cells; trying to dilute the salt water.
c. Water passes into the cells of the celery.

32. To answer this question, compare each solution to a 0.9% (wt/vol) saline solution, which is isotonic with blood:

a. $0.154\ M\text{ NaCl} = \frac{0.154\ \cancel{\text{moles NaCl}}}{\text{liter solution}} \times \frac{58.5\text{ g NaCl}}{\cancel{\text{mole NaCl}}}$

9.01 g/ℓ, which is the same as a 0.9 wt/vol% NaCl solution. Nothing would happen to the red blood cells in this solution, isotonic with blood.

b. 0.154% wt/vol is much less than 0.9%, so the red blood cells would swell up and possibly burst.

c. 0.154 g/ℓ is the same as 0.0154 g/100. mℓ, much less than 0.9 g/100. mℓ. The red blood cells would be even more likely to burst than in part *b*.

Chapter 13

1. a. $HNO_3(l) + H_2O(l) \longrightarrow NO_3^-(aq) + H_3O^+(aq)$; completely
   b. $K_2O(s) + H_2O(l) \longrightarrow 2\,OH^-(aq) + 2\,K^+(aq)$; completely
   c. $HF(l) + H_2O(l) \longrightarrow F^-(aq) + H_3O^+(aq)$; partially
   d. $NH_3(g) + H_2O(l) \longrightarrow NH_4^+(aq) + OH^-(aq)$; partially
   e. $NaOH(s) \longrightarrow Na^+(aq) + OH^-(aq)$; completely
   f. $K_2CO_3(s) + H_2O(l) \longrightarrow HCO_3^-(aq) + 2\,K^+(aq) + OH^-(aq)$; partially
2. $S^{2-}(aq) + H_2O(l) \longrightarrow HS^-(aq) + OH^-(aq)$
3. a. $[H_3O^+] = 0.035\ M$; b. $[OH^-] = 0.15\ M$;
   c. $[H_3O^+] = 0.0021\ M$; d. $[OH^-] = 1.4 \times 10^{-13}\ M$
4. a. $[H_3O^+] = 1 \times 10^{-6}\ M$; $[OH^-] = 1 \times 10^{-8}\ M$
   b. $[H_3O^+] = 1 \times 10^{-5}\ M$; $[OH^-] = 1. \times 10^{-9}\ M$
   c. $[H_3O^+] = 1 \times 10^{-14}\ M$; $[OH^-] = 1 \times 10^{0}\ M$ $(= 1\ M)$
   d. $[H_3O^+] = 1 \times 10^{-8}\ M$; $[OH^-] = 1 \times 10^{-6}\ M$
   e. $[H_3O^+] = 1 \times 10^{-7}\ M$; $[OH^-] = 1 \times 10^{-7}\ M$
5. a. $[OH^-] = 1.00 \times 10^{-1}$; pH = 13, basic
   b. $[H_3O^+] = 1\ M$; pH = 0, acidic
   c. $[OH^-] = 0.010\ M$; pH = 12, basic
   d. $[H_3O^+] = 1.0 \times 10^{-4}\ M$; pH = 4, acidic
6. a. $[H_3O^+] = 3.6 \times 10^{-2}$, pH = 1.44, acid;
   b. $[OH^-] = 1.5 \times 10^{-1}$, pH = 13.18, base;
   c. pH = 4.12, acid; d. pH = 6.63, acid; e. pH = 8.06, base; f. pH = 13.30, base

7.

$[H_3O^+]$	$[OH^-]$	$[H_3O^+]$	$[OH^-]$
10^{-7}	10^{-7}	2.3×10^{-4}	4.3×10^{-11}
10^{-13}	10^{-1}	1×10^{-7}	9×10^{-8}
2.2×10^{-13}	0.045	0.12	8.3×10^{-14}

8. a. $[H_3O^+] = 5.5 \times 10^{-2}$; $[OH^-] = 1.82 \times 10^{-13}$
   b. $[H_3O^+] = 2.9 \times 10^{-7}$; $[OH^-] = 3.4 \times 10^{-8}$
   c. $[H_3O^+] = 2.9 \times 10^{-9}$; $[OH^-] = 3.4 \times 10^{-6}$
   d. $[H_3O^+] = 3.24 \times 10^{-11}$; $[OH^-] = 3.09 \times 10^{-4}$
9. a. pH = 7.30; $[OH^-] = 2.00 \times 10^{-7}$ b. yes
10. Trees and green plants in a forest produce large amounts of CO_2, which washes down with rain and makes the soil acidic.
11. a. red; b. blue; c. red; d. blue
12. Too high a pH means too basic; add an acid, a or e.

13.

Conjugate Acid	*Conjugate Base*	*Conjugate Acid*	*Conjugate Base*
$HC_2H_3O_2$	$C_2H_3O_2^-$	HSO_4^-	SO_4^{2-}
H_3O^+	H_2O	HCO_3^-	CO_3^{2-}
NH_4^+	NH_3	H_2O	OH^-
HF	F^-	HPO_4^{2-}	PO_4^{3-}
H_3O^+	H_2O	HNO_3	NO_3^-
HCO_3^-	CO_3^{2-}	NH_4^+	NH_3

14. a. $CO_3^{2-} + H_2O \longrightarrow HCO_3^- + OH^-$
   base, acid, conj. acid, conj. base
   b. $HSO_4^- + H_2O \longrightarrow SO_4^{2-} + H_3O^+$
   acid, base, conj. base, conj. acid
   c. $HF + H_2O \longrightarrow F^- + H_3O^+$
   acid, base, conj. base, conj. acid
   d. $S^{2-} + H_2O \longrightarrow HS^- + OH^-$
   base, acid, conj. acid, conj. base
15. H_2O (weakest), HClO, H_3PO_4, HSO_4^- (strongest)
16. a. acidic: $NH_4^+ + H_2O \longrightarrow NH_3 + H_3O^+$
   acid, base, conj. base, conj. acid
   b. basic: $PO_4^{3-} + H_2O \longrightarrow HPO_4^{2-} + OH^-$
   base, acid, conj. acid, conj. base
   c. no prediction
   d. basic: $C_2H_3O_2^- + H_2O \longrightarrow HC_2H_3O_2 + OH^-$
   base, acid, conj. acid, conj. base
   e. neutral
   f. basic: $O^{2-} + H_2O \longrightarrow OH^- + OH^-$
   base, acid, conj. acid, conj. base
17. a. as acid: $H_2PO_4^- + H_2O \longrightarrow HPO_4^{2-} + H_3O^+$
   acid, base, conj. base, conj. acid
   as base: $H_2PO_4^- + H_2O \longrightarrow H_3PO_4 + OH^-$
   base, acid, conj. acid, conj. base
   b. as acid: $HSO_3^- + H_2O \longrightarrow SO_3^{2-} + H_3O^+$
   acid, base, conj. base, conj. acid
   as base: $HSO_3^- + H_2O \longrightarrow H_2SO_3 + OH^-$
   base, acid, conj. acid, conj. base
18. $C_2^{2-} + 2\,H_2O \longrightarrow C_2H_2 + 2\,OH^-$
19. CN^- is a weak base, and reacts with water as follows:
$$CN^- + H_2O \longrightarrow HCN + OH^-$$
20. a. $H_2PO_4^-$, HSO_3^-; b. CaO, Na_2O; c. NaCl, KNO_3
21. a. base; b. base; c. base
22. a. Lewis acid; b. Lewis acid, Bronsted base; c. Lewis base, Bronsted base

23. Lewis acid: $AlCl_3$; Lewis base, Cl^-
24. Numbers of oxygen atoms not bonded to hydrogen are shown in parentheses. Stronger acid is underlined. a. $\underline{HClO_3}$(2), H_2SO_3(1); b. $\underline{HC_2H_3O_2}$(1), HClO(0); c. H_3PO_3(0), $\underline{H_2SO_3}$(1); d. HBrO(0), $\underline{HBrO_2}$(1); e. $\underline{HClO_4}$(3), H_2SO_4(2); f. $\underline{HNO_3}$(2), H_2SO_3(1)

Chapter 14

1. a. $ClO_4^- + H^+ + K^+ + OH^- \longrightarrow$ (strong acid: H^+; strong base: OH^-)
   $H_2O + K^+ + ClO_4^-$
   Net: $H^+ + OH^- \longrightarrow H_2O$
   b. $2\,HC_2H_3O_2 + Ca^{2+} + 2\,OH^- \longrightarrow$ (weak acid: $HC_2H_3O_2$; strong base: OH^-)
   $2\,C_2H_3O_2^- + Ca^{2+} + 2\,H_2O$
   Net: $2\,HC_2H_3O_2 + 2\,OH^- \longrightarrow$
   $2\,C_2H_3O_2^- + 2\,H_2O$
   c. $NH_3 + HC_2H_3O_2 \longrightarrow NH_4^+ + C_2H_3O_2^-$ (weak base: NH_3; weak acid: $HC_2H_3O_2$)
   Net: Same as above
   d. $Cl^- + H^+ + 3\,Na^+ + PO_4^{3-} \longrightarrow$ (strong acid: H^+; weak base: PO_4^{3-})
   $Cl^- + 3\,Na^+ + HPO_4^{2-}$
   Net: $H^+ + PO_4^{3-} \longrightarrow HPO_4^{2-}$
   e. $2\,H^+ + 2\,NO_3^- + Sr^{2+} + 2\,OH^- \longrightarrow$ (strong acid: H^+; strong base: OH^-)
   $2\,H_2O + 2\,NO_3^- + Sr^{2+}$
   Net: $2\,H^+ + 2\,OH^- \longrightarrow 2\,H_2O$
   f. $Na^+ + HCO_3^- + HF \longrightarrow$ (weak base: HCO_3^-; weak acid: HF)
   $Na^+ + F^- + H_2CO_3(CO_2 + H_2O)$
   Net: $HCO_3^- + HF \longrightarrow$
   $F^- + H_2CO_3(CO_2 + H_2O)$
2. $H^+ + HSO_4^- + Na^+ + HCO_3^- \longrightarrow$ (strong acid: H^+; weak base: HCO_3^-)
   $HSO_4^- + Na^+ + H_2CO_3(H_2O + CO_2)$
   Net: $H^+ + HCO_3^- \longrightarrow$
   $H_2CO_3(H_2O + CO_2)$
3. (These answers show only one of the possibilities in each case.)
   a. K_2SO_4:
   $2\,K^+ + SO_4^{2-} + Pb^{2+} \longrightarrow 2\,K^+ + PbSO_4(s)$
   Net ionic: $SO_4^{2-} + Pb^{2+} \longrightarrow PbSO_4(s)$
   b. $CaCl_2$:
   $Ca^{2+} + 2\,Cl^- + CO_3^{2-} \longrightarrow CaCO_3(s) + 2\,Cl^-$
   Net ionic: $Ca^{2+} + CO_3^{2-} \longrightarrow CaCO_3(s)$
   c. Na_2S
   $2\,Na^+ + S^{2-} + Cu^{2+} \longrightarrow 2\,Na^+ + CuS(s)$
   Net ionic: $S^{2-} + Cu^{2+} \longrightarrow CuS(s)$
   d. $Mg(NO_3)_2$:
   $Mg^{2+} + 2\,NO_3^- + 2\,F^- \longrightarrow MgF_2(s) + 2\,NO_3^-$
   Net ionic: $Mg^{2+} + 2\,F^- \longrightarrow MgF_2(s)$
   e. K_2CO_3:
   $6\,K^+ + 3\,CO_3^{2-} + 2\,Cr^{3+} \longrightarrow 6\,K^+ + Cr_2(CO_3)_3(s)$
   Net ionic: $3\,OH^- + 2\,Cr^{3+} \longrightarrow Cr_2(CO_3)_3$
   f. $ZnCl_2$:
   $Zn^{2+} + 2\,Cl^- + 2\,OH^- \longrightarrow Zn(OH)_2(s) + 2\,Cl^-$
   Net ionic: $Zn^{2+} + 2\,OH^- \longrightarrow Zn(OH)_2(s)$
   g. Na_2S:
   $2\,Na^+ + S^{2-} + Co^{2+} \longrightarrow 2\,Na^+ + CoS(s)$
   Net ionic: $S^{2-} + Co^{2+} \longrightarrow CoS(s)$
   h. $AgNO_3$:
   $Ag^+ + NO_3^- + Cl^- \longrightarrow AgCl(s) + NO_3^-$
   Net ionic: $Ag^+ + Cl^- \longrightarrow AgCl(s)$
   i. Na_3PO_4:
   $Al^{3+} + 3\,Na^+ + PO_4^{3-} \longrightarrow AlPO_4(s) + 3\,Na^+$
   Net ionic: $Al^{3+} + PO_4^{3-} \longrightarrow AlPO_4(s)$
4. a. $MgCO_3$:
   $2\,K^+ + CO_3^{2-} + Mg^{2+} + 2\,Cl^- \longrightarrow$
   $2\,K^+ + MgCO_3(s) + 2\,Cl^-$
   Net ionic: $CO_3^{2-} + Mg^{2+} \longrightarrow MgCO_3(s)$
   b. PbI_2:
   $Pb^{2+} + 2\,NO_3^- + Sr^{2+} + 2\,I^- \longrightarrow$
   $PbI_2(s) + 2\,NO_3^- + Sr^{2+}$
   Net ionic: $Pb^{2+} + 2\,I^- \longrightarrow PbI_2(s)$
   c. No precipitate; all possible compounds are soluble.
   d. $Co(OH)_2$ and CaF_2:
   $Co^{2+} + 2\,F^- + Ca^{2+} + 2\,OH^- \longrightarrow$
   $CaF_2(s) + Co(OH)_2(s)$
   Net ionic equation is the same.
   e. No precipitate; all possible compounds are soluble.
   f. HgS:
   $Hg^{2+} + 2\,Cl^- + 2\,NH_4^+ + S^{2-} \longrightarrow$
   $HgS(s) + 2\,Cl^- + 2\,NH_4^+$
   Net ionic: $Hg^{2+} + S^{2-} \longrightarrow HgS(s)$
5. (These answers show only one of the possibilities in each case.)
   a. $MgCl_2$ precipitates $Mg(OH)_2$ from NaOH; no reaction with Na_2SO_4
   $$Mg^{2+} + 2\,OH^- \longrightarrow Mg(OH)_2(s)$$
   b. NaF precipitates CaF_2 from $CaCl_2$; no reaction with $CoCl_2$
   $$Ca^{2+} + 2\,F^- \longrightarrow CaF_2(s)$$
   c. $Ba(NO_3)_2$ precipitates yellow $BaCrO_4$ from K_2CrO_4; no reaction with KCl
   $$Ba^{2+} + CrO_4^{2-} \longrightarrow BaCrO_4(s)$$

d. NaOH precipitates $Mg(OH)_2$ from $Mg(NO_3)_2$; no reaction with $Ca(NO_3)_2$

$$Mg^{2+} + 2\,OH^- \longrightarrow Mg(OH)_2(s)$$

6. $Ca_3(PO_4)_2(s) + 3\,H^+ + 3\,HSO_4^- \longrightarrow 3\,Ca^{2+} + 2\,H_3PO_4^- + 2\,SO_4^{2-}$

   The insoluble substance contains a negative ion that's a base (PO_4^{3-}).

7. a. $ZnS(s) + 2\,H^+ \longrightarrow Zn^{2+} + H_2S$

   b. Acid won't help, because the negative ion (NO_3^-) is not a base.

   c. $CaCO_3(s) + 2\,H^+ \longrightarrow Ca^{2+} + H_2CO_3$

   d. Acid won't help, because Cl^- is not a base.

   e. $CaO(s) + 2\,H^+ \longrightarrow Ca^{2+} + H_2O$

   f. $Al(OH)_3(s) + 3\,H^+ \longrightarrow Al^{3+} + 3\,H_2O$

8. a. CuS isn't an acid and can't be dissolved in strong base.

   b. $C_3H_7COOH + OH^- \longrightarrow C_3H_7COO^- + H_2O$

   c. $SO_3(g) + OH^- \longrightarrow HSO_4^-$

   d. NH_3 isn't an acid and can't be dissolved in strong base.

   e. $Ca_3(PO_4)_2$ isn't an acid and can't be dissolved in strong base.

   f. $HCl(g) + OH^- \longrightarrow Cl^- + H_2O$

9. CO_2, from air, is easily dissolved in NaOH:

$$CO_2(g) + 2\,NaOH(aq) \longrightarrow Na_2CO_3(aq) + H_2O(l)$$

10. a. $SO_4^{2-} + Ba^{2+} \longrightarrow BaSO_4(s)$

    b. $ZnS(s) + H^+ + HSO_4^- \longrightarrow Zn^{2+} + H_2S(g) + SO_4^{2-}$

    c. $2\,OH^- + SO_3 \longrightarrow SO_4^{2-} + H_2O$

    d. $NH_4^+ + OH^- \longrightarrow NH_3 + H_2O$

    e. $CaCO_3(s) + H^+ + HSO_4^- \longrightarrow Ca^{2+} + CO_2(g) + H_2O + SO_4^{2-}$

11. a. $HCO_3^- + H_3C_6H_5O_7 \longrightarrow H_2CO_3 + H_2C_6H_5O_7^-$
    $(CO_2 + H_2O)$

    b. Water allows the ions to come in contact with each other and react.

12. a. $1.00\,\ell \times \dfrac{0.500\text{ moles}}{\ell} \times \dfrac{142\text{ g}}{\text{mole}} = 71.0\text{ g } Na_2SO_4$

    b. $0.500\,\ell \times \dfrac{1.00\text{ moles}}{\ell} \times \dfrac{142\text{ g}}{\text{mole}} = 71.0\text{ g } Na_2SO_4$

    c. $0.100\,\ell \times \dfrac{0.100\text{ moles}}{\ell} \times \dfrac{170.\text{ g}}{\text{mole}} = 1.70\text{ g } AgNO_3$

    d. $0.250\,\ell \times \dfrac{0.0250\text{ moles}}{\ell} \times \dfrac{74.6\text{ g}}{\text{mole}} = 0.466\text{ g KCl}$

13. a. $V_1 = 500\text{ m}\ell \times \dfrac{0.25\,M}{12\,M} = 10.4\text{ m}\ell$ conc. HCl

    b. $V_1 = 1.00\,\ell \times \dfrac{0.10\,M}{17.5\,M}$
    $= 0.0057\,\ell\ (5.7\text{ m}\ell)$ conc. $HC_2H_3O_2$

    c. $V_1 = 250.\text{ m}\ell \times \dfrac{2.0\,M}{16\,M} = 31\text{ m}\ell$ conc. HNO_3

    d. $V_1 = 100.\text{ m}\ell \times \dfrac{6.0\,M}{12\,M} = 50.\text{ m}\ell$ conc. HCl

14. $V_1 = 1.00\,\ell \times \dfrac{0.016\,M}{12\,M} = 0.0013\,\ell\ (1.3\text{ m}\ell)$

15. a. $M_2 = 0.15\,M \times \dfrac{10.0\text{ m}\ell}{100.\text{ m}\ell} = 0.015\,M$

    b. $M_2 = 0.50\,M \times \dfrac{25.0\text{ m}\ell}{250.\text{ m}\ell} = 0.050\,M$

    c. $M_2 = 0.112\,M \times \dfrac{10.0\text{ m}\ell}{1000.\text{ m}\ell} = 0.00112\,M$

    d. $M_2 = 6\,M \times \dfrac{25.0\text{ m}\ell}{1000.\text{ m}\ell} = 0.15\,M$

16. a. $V_2 = 10.0\text{ m}\ell \times \dfrac{1.00\,M}{0.100\,M} = 100.\text{ m}\ell$

    b. $V_2 = 25.0\text{ m}\ell \times \dfrac{0.250\,M}{0.100\,M} = 62.5\text{ m}\ell$

    c. $V_2 = 5.0\text{ m}\ell \times \dfrac{6.0\,M}{1.0\,M} = 30.\text{ m}\ell$

    d. $V_2 = 50.0\text{ m}\ell \times \dfrac{3.0\,M}{0.5\,M} = 300.\text{ m}\ell$

17. a. $NaOH(aq) + HCl(aq) \longrightarrow NaCl(aq) + H_2O(l)$

$$25.0\text{ m}\ell\text{ HCl solution} \times \frac{0.233\text{ moles HCl}}{1000.\text{ m}\ell\text{ HCl solution}} \times \frac{1\text{ mole NaOH}}{\text{mole HCl}} \times \frac{1000.\text{ m}\ell\text{ NaOH solution}}{0.122\text{ moles NaOH}}$$
$$= 47.7\text{ m}\ell\text{ NaOH solution}$$

    b. $NaHCO_3(s) + NaOH(aq) \longrightarrow Na_2CO_3(aq) + H_2O(l)$

$$0.566\text{ g } NaHCO_3 \times \frac{1.00\text{ mole } NaHCO_3}{84.0\text{ g } NaHCO_3} \times \frac{1\text{ mole NaOH}}{\text{mole } NaHCO_3} \times \frac{1.00\,\ell\text{ NaOH solution}}{0.122\text{ moles NaOH}}$$
$$= 0.0552\,\ell\ (55.2\text{ m}\ell)\text{ NaOH solution}$$

18. $$50.0\text{ g palmitic acid} \times \frac{1\text{ mole palmitic acid}}{256\text{ g palmitic acid}} \times \frac{1\text{ mole NaOH}}{\text{mole palmitic acid}} \times \frac{1.00\,\ell\text{ NaOH}}{1.00\text{ moles NaOH}}$$
$$= 0.195\,\ell\text{ NaOH, or } 195\text{ m}\ell$$

19. $HC_2H_3O_2(aq) + NaOH(aq) \longrightarrow NaC_2H_3O_2(aq) + H_2O(l)$

$$28.1\text{ m}\ell \times \frac{0.122\text{ moles NaOH}}{1000.\text{ m}\ell}$$
$$= 3.42 \times 10^{-3}\text{ moles NaOH}$$
$$= 3.42 \times 10^{-3}\text{ moles } HC_2H_3O_2$$
$$\frac{3.42 \times 10^{-3}\text{ moles } HC_2H_3O_2}{0.0250\,\ell} = 0.137\,M$$

20. $H_2SO_4(aq) + Na_2CO_3(aq) \longrightarrow CO_2(g) + H_2O(l) + Na_2SO_4(aq)$

$$15.3\text{ m}\ell\ Na_2CO_3\text{ solution} \times \frac{0.246\text{ moles } Na_2CO_3}{1000.\text{ m}\ell\ Na_2CO_3\text{ solution}} \times \frac{1\text{ mole } H_2SO_4}{\text{mole } Na_2CO_3}$$
$$= 3.76 \times 10^{-3}\text{ moles } H_2SO_4$$

$$\frac{3.76 \times 10^{-3} \text{ moles } H_2SO_4}{0.0500\ \ell \text{ mine drainage}} = 0.0752\ M$$

21. $NH_3(aq) + HCl(aq) \longrightarrow NH_4Cl(aq)$

$$24.1\ m\ell \text{ HCl solution} \times \frac{0.0500 \text{ moles HCl}}{1000.\ m\ell \text{ HCl solution}}$$

$$\times \frac{1 \text{ mole } NH_3}{\text{mole HCl}} = 1.20 \times 10^{-3} \text{ moles } NH_3$$

$$\frac{1.21 \times 10^{-3} \text{ moles } NH_3}{0.0100\ \ell \text{ ammonia}} = 0.120\ M$$

22. $2\ NaOH(aq) + H_2SO_4(aq) \longrightarrow$

$$Na_2SO_4(aq) + 2\ H_2O(l)$$

$$45.7\ m\ell\ H_2SO_4 \text{ solution} \times \frac{0.206 \text{ moles } H_2SO_4}{1000.\ m\ell\ H_2SO_4 \text{ solution}}$$

$$\times \frac{2 \text{ moles NaOH}}{\text{mole } H_2SO_4} = 1.88 \times 10^{-2} \text{ moles NaOH}$$

$$\frac{1.88 \times 10^{-2} \text{ moles NaOH}}{0.0250\ \ell} = 0.753\ M \text{ NaOH}$$

23. a. $Ca^{2+} + 2\ F^- \longrightarrow CaF_2(s)$

b. $0.0282\ \ell\ CaCl_2 \text{ solution}$

$$\times \frac{1.00 \text{ moles } CaCl_2}{\ell\ CaCl_2 \text{ solution}} \times \frac{2 \text{ moles } F^-}{\text{mole } CaCl_2}$$

$$= 0.0564 \text{ moles } F^-$$

$$\frac{0.0564 \text{ moles } F^-}{0.0100\ \ell \text{ solution}} = 5.64\ M$$

c. $$6\ oz \times \frac{1\ pt}{16\ oz} \times \frac{0.473\ \ell}{pt} \times \frac{5.64 \text{ moles } F^-}{\ell}$$

$$\times \frac{19.0 \text{ g } F^-}{\text{mole } F^-} = 19.0 \text{ g } F^-$$

24. a. $0.5\ M \times 2 = 1\ N$; b. $1.00\ N$; c. $0.200 \times 3 = 0.600\ N$; d. $0.5 \times 2 = 1\ N$

25. a. eq. wt. = f.w./3 = 98.0/3 = 32.7 g/eq
b. eq. wt. = f.w./1 = 40.0 g/eq
c. eq. wt. = f.w./1 = 98.1 g/eq
d. eq. wt. = f.w./2 = 82.1/2 = 41.0 g/eq

26. a. $$98.0\ g \times \frac{1 \text{ eq}}{32.7\ g} = 3.00 \text{ eq}; \frac{3.00 \text{ eq}}{0.500\ \ell} = 6.00\ N$$

b. eq. wt. = f.w./2 = 74.1/2 = 37.0 g/eq

$$4.01\ g \times \frac{1 \text{ eq}}{37.0\ g} = 0.108 \text{ eq}; \frac{0.108 \text{ eq}}{\ell}$$

$$= 0.108\ N$$

27. a. $$N_2 = 0.315\ N \times \frac{37.8\ m\ell}{25.0\ m\ell} = 0.476\ N$$

b. $$N_2 = 0.175\ N \times \frac{15.2\ m\ell}{15.2\ m\ell} = 0.266\ N$$

28. a. $18\ M \times 2$ eq/mole = $36\ N$; b. $16\ N$

29. eq. wt. $H_2C_2O_4 \cdot 2\ H_2O$ = 126/2 = 63.0 g/eq

$$2.25\ g \times \frac{1 \text{ eq } H_2C_2O_4}{63.0\ g} = 3.57 \times 10^{-2} \text{ eq } H_2C_2O$$

$$= \text{eq NaOH}$$

$$\frac{3.57 \times 10^{-2} \text{ eq NaOH}}{0.0429\ \ell} = 0.832\ N$$

Check: $$N_2 = 9\ N \times \frac{90\ m\ell}{1000\ m\ell} = 0.81\ N.$$

The answer is reasonable.

30. $$28.7\ m\ell \text{ HCl solution} \times \frac{0.448 \text{ eq HCl}}{1000.\ m\ell \text{ HCl solution}}$$

$$= 1.28 \times 10^{-2} \text{ eq HCl} = \text{eq unknown}$$

$$\text{eq. wt.} = \text{g/eq} = \frac{5.34\ g}{1.28 \times 10^{-2} \text{ eq}} = 417 \text{ g/eq}$$

eq. wt. of strychnine = f.w./2 = 334/2 = 167 g/eq
The substance could not be strychnine.

Chapter 15

1.

	Oxidizing Agent	*Reducing Agent*
a.	S	Cu
b.	O_2	N_2
c.	H^+	Fe
d.	F_2	Cl^-
e.	H_2O	Zn
f.	$AgNO_3$	Mg

2.

	Oxidized Substance	*Reduced Substance*
a.	Cu	S
b.	N_2	O_2
c.	Fe	H^+
d.	Cl^-	F_2
e.	Zn	H_2O
f.	Mg	Ag^+

3. a. Oxidation happens at the anode, so Al is the anode. Reduction happens at the cathode, so Pb is the cathode.

b.

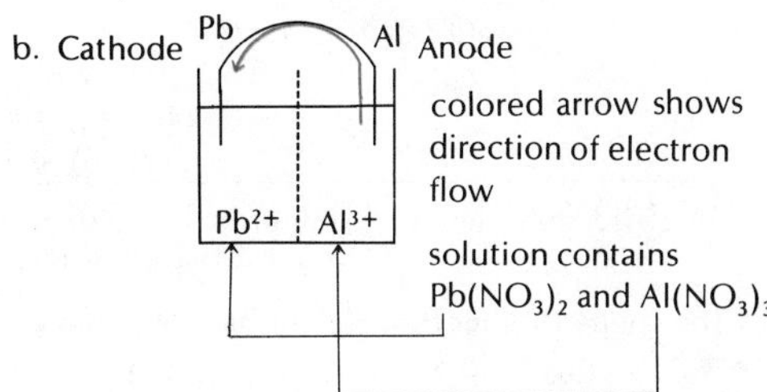

4. a. strong; b. weak; c. strong; d. strong; e. weak; f. weak; g. weak; h. weak; i. non; j. strong; k. weak

5. a.
$$4\ [Al^{3+}(l) + 3e^- \longrightarrow Al(l)]$$
$$3\ [2O^{2-}(l) \longrightarrow O_2(g) + 4e^-]$$
$$4\ Al^{3+}(l) + 6\ O^-(l) \longrightarrow 4\ Al(l) + 3\ O_2(g)$$

b.

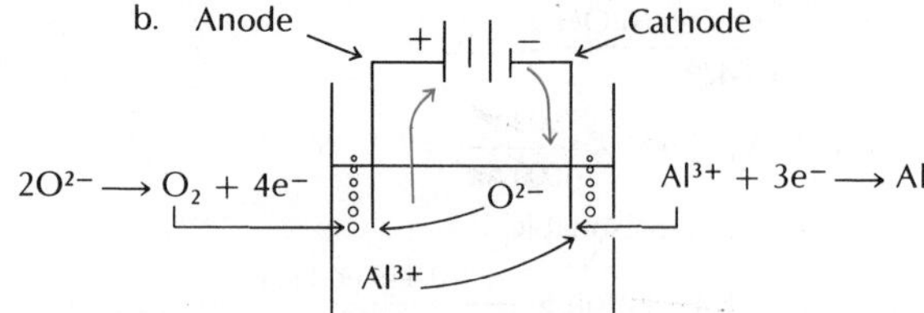

c. Oxygen gas bubbles off at anode. Aluminum metal plates out onto cathode.

6. a. no reaction

b.
$$2\,[MnO_4^-(aq) + 8\,H^+(aq) + \cancel{5e^-} \longrightarrow Mn^{2+}(aq) + 4\,H_2O(l)]$$
$$5\,[2I^-(aq) \longrightarrow I_2(aq) + \cancel{2e^-}]$$
$$\overline{2\,MnO_4^-(aq) + 16\,H^+(aq) + 10\,I^-(aq) \longrightarrow 2\,Mn^{2+}(aq) + 8\,H_2O + 5\,I_2(aq)}$$

c. no reaction

d.
$$3\,[Ag(s) \longrightarrow Ag^+(aq) + \cancel{e^-}]$$
$$NO_3^-(aq) + 4\,H^+(aq) + \cancel{3e^-} \longrightarrow NO(g) + 2\,H_2O(l)$$
$$\overline{3\,Ag(s) + NO_3^-(aq) + 4\,H^+(aq) \longrightarrow 3\,Ag^+(aq) + NO(g) + 2\,H_2O(l)}$$

e.
$$MnO_2(s) + 4\,H^+(aq) + \cancel{2e^-} \longrightarrow Mn^{2+}(aq) + 2\,H_2O(l)$$
$$2\,[Fe^{2+}(aq) \longrightarrow Fe^{3+}(aq) + \cancel{e^-}]$$
$$\overline{MnO_2(s) + 4\,H^+(aq) + 2\,Fe^{2+}(aq) \longrightarrow Mn^{2+}(aq) + 2\,H_2O(l) + 2\,Fe^{3+}(aq)}$$

f. no reaction

g.
$$Cr_2O_7^{2-}(aq) + 14\,H^+(aq) + \cancel{6e^-} \longrightarrow 2\,Cr^{3+}(aq) + 7\,H_2O(l)$$
$$3\,[H_2O_2(aq) \longrightarrow O_2(g) + 2\,H^+(aq) + \cancel{2e^-}]$$
$$\overline{Cr_2O_7^{2-}(aq) + 8\,H^+(aq) + 3\,H_2O_2(aq) \longrightarrow 2\,Cr^{3+}(aq) + 3\,O_2(g) + 7\,H_2O(l)}$$

h.
$$PbO_2(s) + SO_4^{2-}(aq) + 4\,H^+(aq) + \cancel{2e^-} \longrightarrow PbSO_4(s) + 2\,H_2O(l)$$
$$2\,Br^-(aq) \longrightarrow Br_2(aq) + \cancel{2e^-}$$
$$\overline{PbO_2(s) + SO_4^{2-}(aq) + 4\,H^+(aq) + 2\,Br^-(aq) \longrightarrow PbSO_4(s) + 2\,H_2O(l) + Br_2(aq)}$$

7. The copper(II) ions oxidized the aluminum metal:
$$3\,CuSO_4 + 2\,Al \longrightarrow Al_2(SO_4)_3 + 3\,Cu$$

8. Yes. Gold is above H^+.

9. $0.00 + 1.23\text{ V} = +1.23\text{ V}$

$\dfrac{4.07\,\cancel{V}}{1.23\,\cancel{V}} = 3.31,$

so 4 cells are needed for NaCl electrolysis

10.
$$NiO_2(s) + 2\,H_2O(l) + \cancel{2e^-} \longrightarrow Ni(OH)_2(s) + 2\,OH^-(aq) \quad +0.49\text{ V}$$
$$Cd(s) + 2\,OH^-(aq) \longrightarrow Cd(OH)_2(s) + \cancel{2e^-} \quad +0.81\text{ V}$$
$$\overline{NiO_2(s) + 2\,H_2O(l) + Cd(s) \longrightarrow Ni(OH)_2(s) + Cd(OH)_2(s) \quad +1.30\text{ V}}$$

11.
$$Ni(s) \longrightarrow Ni^{2+}(aq) + 2e^- \quad + 0.25\text{ V}$$
$$Ag^+(aq) + e^- \longrightarrow Ag(s) \quad + 0.80\text{ V}$$
$$0.25 + 0.80 = +1.05\text{ V}$$

12. $Na^+(l) + e^- \longrightarrow Na(l)$ -2.71 V
$Li^+(l) + e^- \longrightarrow Li(l)$ -3.01 V
Na would be obtained at the cathode, because its potential is less negative.
$2\,Cl^-(l) \longrightarrow Cl_2(g)$ -1.36 V
$2\,Br^-(l) \longrightarrow Br_2(l)$ -1.06 V
Br_2 would be obtained at the anode, because its potential is less negative.

13. The simplest way would be to add a few drops of phenolphthalein indicator to the water. According to the equations on page 348, hydroxide ions form at the cathode, so that the electrode around which a pink color began to develop would be identified as the cathode. Or the gases coming off the electrodes could be tested to see which is hydrogen (cathode) and which is oxygen (anode).

14. a. $3\,Fe^{2+}(aq) + 2\,Al(s) \longrightarrow 3\,Fe(s) + Al^{3+}(aq)$
$3\,Pb^{2+}(aq) + 2\,Al(s) \longrightarrow 3\,Pb(s) + 2\,Al^{3+}(aq)$

b. The second reaction would take place, because its positive potential ($-0.13 + 1.66 = +1.53$ V) is greater than that of the first reaction ($-0.44 + 1.66 = +1.22$ V).

15. $Zn^{2+}(l) + 2e^- \longrightarrow Zn(s)$ $V = -0.76$ V
$2\,Cl^-(l) \longrightarrow Cl_2(g) + 2e^-$ $V = -1.36$ V
$V = -2.12$ V

$$E = (-2.12\,\cancel{V})(2\,\cancel{\text{moles e}^-})\left(23.1\,\frac{(\text{kcal})}{\cancel{(\text{volts})}\cancel{(\text{moles e}^-)}}\right)$$
$$= -97.9\text{ kcal}$$

16. a. From Table 15.4, the potential is $+1.10$ V.
$$E = (1.10\,\cancel{V})(2\,\cancel{\text{moles e}^-})\left(23.1\,\frac{\text{kcal}}{\cancel{(\text{volts})}\cancel{(\text{mole e}^-)}}\right)$$
$$= 50.8\text{ kcal}$$

b. $\dfrac{4.07\,\cancel{V}}{1.10\,\cancel{V}} = 3.7$. We need 4 cells to provide enough voltage. Table 15.5 shows that 94 kcal of energy are needed for 1 mole NaCl. Four cells provide $4 \times 50.8 = 203.1$ kcal, more than enough energy

17. From Table 15.5, energy for NaCl is -188 kcal/2 moles e^-, and for $MgCl_2$, -172 kcal/2 moles e^-.
$$\frac{-188\text{ kcal}}{2\,\cancel{\text{moles e}^-}} \times \frac{1\,\cancel{\text{mole e}^-}}{\text{mole Na}} = -94\,\frac{\text{kcal}}{\text{mole Na}}$$
$$\frac{-172\text{ kcal}}{2\,\cancel{\text{moles e}^-}} \times \frac{2\,\cancel{\text{moles e}^-}}{\text{mole Mg}} = -172\,\frac{\text{kcal}}{\text{mole Mg}}$$

It takes more energy to produce a mole of Mg than a mole of Na.

18. a. H, 1+; O, 1−. b. K, 1+; Cr, 6+; O, 2−. c. Na, 1+; Cl, 5+; O, 2−. d. Hg, 1+; Cl, 1−. e. S, 4+; O, 2−. f. Mn, 2+; S, 4+; O, 2−. g. Ni, 4+; O, 2−

19.
$$2\,\cancel{OH^-} + Zn \longrightarrow Zn(OH)_2 + \cancel{2e^-}$$
$$H_2O + 2\,AgO + \cancel{2e^-} \longrightarrow Ag_2O + 2\,\cancel{OH^-}$$
$$\overline{Zn + H_2O + 2\,AgO \longrightarrow Zn(OH)_2 + Ag_2O}$$

20. a.
$$MnO_2 + 4\,H^+ + \cancel{2e^-} \longrightarrow Mn^{2+} + 2\,H_2O$$
$$2\,Cl^- \longrightarrow Cl_2 + \cancel{2e^-}$$
$$\overline{MnO_2 + 4\,H^+ + 2\,Cl^- \longrightarrow Mn^{2+} + Cl_2 + 2\,H_2O}$$

b.
$$5\,[Cl_2 + \cancel{2e^-} \longrightarrow 2\,Cl^-]$$
$$I_2 + 6\,H_2O \longrightarrow 2\,HIO_3 + 10\,H^+ + \cancel{10e^-}$$
$$\overline{I_2 + 6\,H_2O + 5\,Cl_2 \longrightarrow 2\,HIO_3 + 10\,H^+ + 10\,Cl^-}$$

c.
$$4\,H^+ + O_2^{2-} + \cancel{2e^-} \longrightarrow 2\,H_2O$$
$$2\,Cl^- \longrightarrow Cl_2 + \cancel{2e^-}$$
$$\overline{4\,H^+ + O_2^{2-} + 2\,Cl^- \longrightarrow Cl_2 + 2\,H_2O}$$

d.
$$5\,[H_2O_2(aq) \longrightarrow 2\,H^+(aq) + O_2(g) + \cancel{2e^-}]$$
$$2\,[MnO_4^-(aq) + 8\,H^+(aq) + \cancel{5e^-} \longrightarrow Mn^{2+}(aq) + 4\,H_2O(l)]$$
$$\overline{5\,H_2O_2(l) + 2\,MnO_4^-(aq) + 6\,H^+(aq) \longrightarrow 5\,O_2(g) + 2\,Mn^{2+}(aq) + 8\,H_2O(l)}$$

e.
$$2\,IO_3^-(aq) + 12\,H^+(aq) + \cancel{10e^-} \longrightarrow I_2(s) + 6\,H_2O$$
$$5\,[SO_2(g) + 2\,H_2O(l) \longrightarrow SO_4^{2-}(aq) + 4\,H^+(aq) + \cancel{2e^-}]$$
$$\overline{2\,IO_3^- + 5\,SO_2 + 4\,H_2O \longrightarrow I_2 + 5\,SO_4^{2-} + 8\,H^+}$$

Chapter 16

1. a. 1; b. 2; c. 1

2.

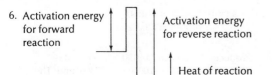

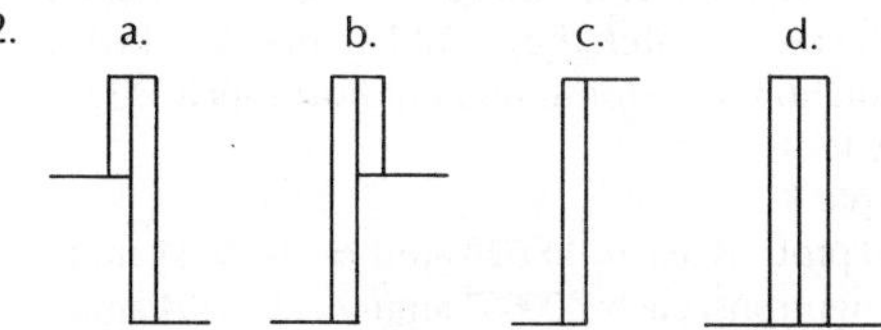

3. a. right; b. left; c. left; d. in the middle
4. a. reverse; b. forward; c. forward; d. neither
5. a. no effect; b. no effect; c. shift to the right; d. shift to the left; e. shift to the right; f. no effect

6. Activation energy for forward reaction

Activation energy for reverse reaction

Heat of reaction (17 kcal)

7. a. supply heat, reduce pressure, remove CO_2 as it forms; b. supply heat, reduce pressure, remove CO as it forms; c. remove heat, reduce pressure, remove C_2H_2 as it forms; d. increase pressure, remove heat as reaction proceeds; e. increase pressure, remove product as it forms

8. E_f = activation energy of forward reaction; E_r = activation energy of reverse reaction; H_r = heat of reaction

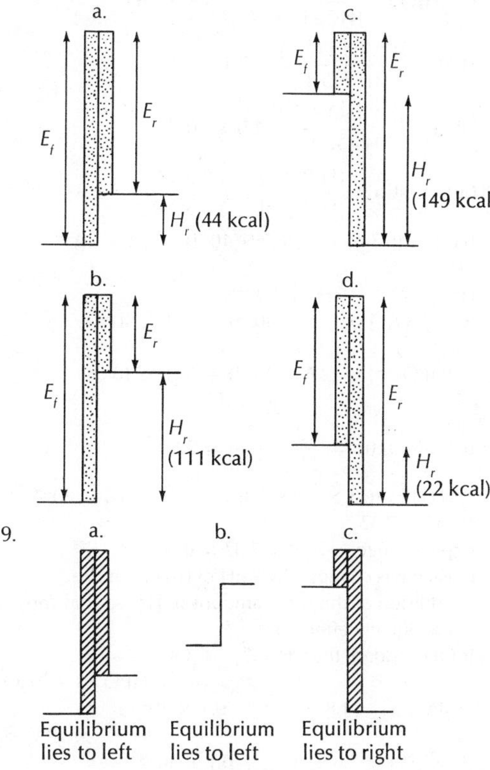

10. a. more rapid; b. more rapid; c. slower

11. a. $\dfrac{[NO_2][NO_2]}{[NO][NO][O_2]}$

b. $\dfrac{[NO][NO][NO][NO][H_2O][H_2O][H_2O][H_2O][H_2O][H_2O]}{[NH_3][NH_3][NH_3][NH_3][O_2][O_2][O_2][O_2][O_2]}$

c. $\dfrac{[H_2O][H_2O]}{[H_2][H_2][O_2]}$

12. a. products; b. reactants; c. products (only slightly)
13. a. exothermic; b. too little product would be obtained; c. begin the reaction at a high temperature, then remove heat as the reaction proceeds
14. a. $[Pb^{2+}][CrO_4^{2-}]$; b. $[Ba^2][SO_4^{2-}]$
15. A large excess of Cl^- (in a soluble chloride such as NaCl) will drive the equilibrium in the direction of Hg_2Cl_2.

16. $[Ag^+] = \frac{K_{sp}}{[Br^-]} = \frac{5.2 \times 10^{-13}}{1.0} = 5.2 \times 10^{-13}\ M\ Ag^+$

17. $[Pb^{2+}] = \frac{2.0 \times 10^{-16}}{1.0 \times 10^{-8}} = 2.0 \times 10^{-8}\ M\ Pb^{2+}$

18. In order of increasing strength:

HCO_3^-: $\frac{[H^+][CO_3^{2-}]}{[HCO_3^-]} = 4.7 \times 10^{-11}$

$HC_2H_3O_2$: $\frac{[H^+][C_2H_3O_2^-]}{[HC_2H_3O_2]} = 1.8 \times 10^{-5}$

H_3PO_4: $\frac{[H^+][H_2PO_4^-]}{[H_3PO_4]} = 7.1 \times 10^{-3}$

HSO_4^-: $\frac{[H^+][SO_4^{2-}]}{[HSO_4^-]} = 1.0 \times 10^{-2}$

19. buffer ratio $= \frac{[H_2PO_4^-]}{[HPO_4^{2-}]} = \frac{1.8 \times 10^{-4}}{2.6 \times 10^{-4}} = 6.9 \times 10^{-1}$

$[H^+] = (6.3 \times 10^{-8})(6.9 \times 10^{-1}) = 4.3 \times 10^{-8}$

pH = 7.37

20. $H^+ + HPO_4^{2-} \rightleftharpoons H_2PO_4^-$

a. $[H_2PO_4^-] = (1.8 \times 10^{-4}) + (0.1 \times 10^{-4})$
$= 1.9 \times 10^{-4}\ M$

$[HPO_4^{2-}] = (2.6 \times 10^{-4}) - (0.1 \times 10^{-4})$
$= 2.5 \times 10^{-4}\ M$

b. buffer ratio $= \frac{1.9 \times 10^{-4}}{2.5 \times 10^{-4}} = 7.6 \times 10^{-1}$

$[H^+] = (6.3 \times 10^{-8})(7.6 \times 10^{-1}) = 4.8 \times 10^{-8}$

pH = 7.32

c. pH change = 7.36 − 7.32 = 0.04

d. Each has changed by a pH of 0.04 with the addition of the same amount of H^+, so that they are equally effective.

21. If OH^- added, then $H_2PO_4^- + OH^- \longrightarrow HPO_4^{2-} + H_2O$

$[H_2PO_4^{2-}] = 1.8 \times 10^{-4} - 0.1 \times 10^{-4}$
$= 1.7 \times 10^{-4}\ M$

$[HPO_4^{2-}] = 2.6 \times 10^{-4} + 0.1 \times 10^{-4}$
$= 2.7 \times 10^{-4}\ M$

buffer ratio $= \frac{1.7 \times 10^{-4}}{2.7 \times 10^{-4}} = 6.3 \times 10^{-1}$

$[H^+] = (6.3 \times 10^{-8})(6.3 \times 10^{-1}) = 4.0 \times 10^{-8}$

pH = 7.40; pH change = 7.40 − 7.36 = 0.04

22. If $1.0 \times 10^{-4}\ M\ H^+$ added,

$[H_2PO_4^-] = (1.8 \times 10^{-4}) + (1.0 \times 10^{-4})$
$= 2.8 \times 10^{-4}$

$[HPO_4^{2-}] = (2.6 \times 10^{-4}) - (1.0 \times 10^{-4})$
$= 1.6 \times 10^{-4}$

The change is a large percentage of the total amounts of each ion initially present, so the buffer probably cannot withstand this change. To prove it, calculate the buffer ratio and the associated pH.

buffer ratio: $\frac{2.8 \times 10^{-4}}{1.6 \times 10^{-4}} = 1.75$

$[H^+] = (6.3 \times 10^{-8})(1.75) = 1.1 \times 10^{-7}$

The resulting pH is 6.96. The buffer cannot withstand the H^+ change.

23. A pH of 5.40 means $[H^+]$ of $4.0 \times 10^{-6}\ M$. Acetic acid (a) would be the best choice, since its K_a is closest to the value of the desired $[H^+]$.

Chapter 17

1. a. ${}^{235}_{92}U \longrightarrow {}^{4}_{2}He + {}^{231}_{90}Th$
b. ${}^{232}_{90}Th \longrightarrow {}^{4}_{2}He + {}^{228}_{88}Ra$
c. ${}^{239}_{94}Pu \longrightarrow {}^{4}_{2}He + {}^{235}_{92}U$
d. ${}^{223}_{88}Ra \longrightarrow {}^{4}_{2}He + {}^{219}_{86}Rn$
2. a. ${}^{32}_{15}P \longrightarrow {}^{0}_{-1}e + {}^{32}_{16}S$
b. ${}^{131}_{53}I \longrightarrow {}^{0}_{-1}e + {}^{131}_{54}Xe$
c. ${}^{60}_{27}Co \longrightarrow {}^{0}_{-1}e + {}^{60}_{28}Ni$
d. ${}^{24}_{11}Na \longrightarrow {}^{0}_{-1}e + {}^{24}_{12}Mg$
3. a. ${}^{30}_{15}P \longrightarrow {}^{0}_{1}e + {}^{30}_{14}Si$
b. ${}^{22}_{11}Na \longrightarrow {}^{0}_{1}e + {}^{22}_{10}Ne$
4. From radium, uranium, and other alpha emitters. An alpha particle picks up electrons from its surroundings and becomes a helium atom.
5. beta emission: ${}^{40}_{19}K \longrightarrow {}^{0}_{-1}e + {}^{40}_{20}Ca$
electron capture: ${}^{40}_{19}K + {}^{0}_{-1}e \longrightarrow {}^{40}_{18}Ar$
6. a. ${}^{130}_{52}Te + {}^{2}_{1}H \longrightarrow {}^{132}_{53}I$
b. ${}^{32}_{16}S + {}^{0}_{-1}e \longrightarrow {}^{32}_{15}P$
c. ${}^{239}_{94}Pu + {}^{4}_{2}He \longrightarrow {}^{242}_{96}Cm + {}^{1}_{0}n$
7. a. 1000 becquerels of an alpha emitter; b. 1000 becquerels of a gamma emitter
8. a. 1000, 1000; b. 500, 1000
9. Approximately 3 half-lives will have elapsed. After 1 half-life: 5 g left. After 2 half-lives: 2.5 g left. After 3 half-lives: 1.25 g left. About 1 g will be left after 15 years.
10. 16 counts/min × 0.501 = 8 counts/min. One half-life, or 5730 years.
11. A known quantity of phosphorus-32 could be fed to the plants. Later, they could be measured with a radiation detector to find out how much P-32 they took up.
12. sample B
13. 80 protons, each 1.00728 amu = 80.5824 amu
120 neutrons, each 1.00867 amu = 121.0404 amu
total 201.6228 amu
201.6228 − 199.9683 = 1.6545 amu
14. No. This means going uphill on the curve of Figure 17.4, expending energy instead of releasing it.
15. Fe
16. Reactants: Li-7 = 7.0160
proton = 1.00728
8.0233
Products: 2 He, each 4.00150 = 8.0030. Mass difference: 8.0233 − 8.0030 = 0.02303 amu. The reaction releases energy.

17. For a mole of reactants, there is a mass defect of 0.02303 g.

$0.02303 \text{ g/mole} \times 2.2 \times 10^{10} \text{ kcal/g} = 4.5 \times 10^{8} \text{ kcal}$

18. ${}^{235}_{92}U + {}^{1}_{0}n \longrightarrow {}^{97}_{40}Zr + 2\,{}^{1}_{0}n + {}^{137}_{52}Te$

19. No, because a single neutron could strike only one other nucleus.

20. ${}^{90}_{35}Br \longrightarrow {}^{90}_{38}Sr + 3\,{}^{0}_{-1}e$

Chapter 18

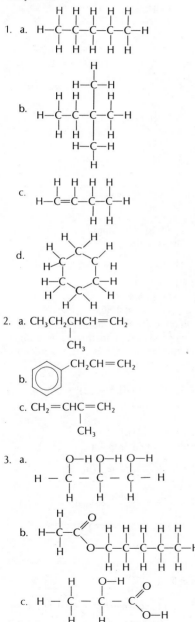

1. a.
$$\begin{array}{ccccccccccc} & & H & & H & & H & & H & & H \\ & & | & & | & & | & & | & & | \\ H & - & C & - & C & - & C & - & C & - & C - H \\ & & | & & | & & | & & | & & | \\ & & H & & H & & H & & H & & H \end{array}$$

b. (central carbon bearing two $-CH_3$ groups drawn in full)
$$H_3C{-}C(CH_3)_2{-}CH_2{-}CH_3$$

c.
$$\begin{array}{ccccccccc} & & H & & H & & H & & H \\ & & | & & | & & | & & | \\ H & - & C & = & C & - & C & - & C - H \\ & & & & & & | & & | \\ & & & & & & H & & H \end{array}$$

d. (cyclohexane, each carbon bearing two H): C_6H_{12} ring

2. a. $CH_3CH_2CH(CH_3)CH{=}CH_2$
$$\begin{array}{l} CH_3CH_2CHCH{=}CH_2 \\ \qquad\quad\;\; | \\ \qquad\quad CH_3 \end{array}$$

b. benzene$-CH_2CH{=}CH_2$

c.
$$\begin{array}{l} CH_2{=}CHC{=}CH_2 \\ \qquad\quad\; | \\ \qquad\;\; CH_3 \end{array}$$

3. a.
$$\begin{array}{ccccccc} & & O{-}H & & O{-}H & & O{-}H \\ & & | & & | & & | \\ H & - & C & - & C & - & C \;-\; H \\ & & | & & | & & | \\ & & H & & H & & H \end{array}$$

b.
$$H_3C{-}C(=O){-}O{-}CH_2{-}CH_2{-}CH_2{-}CH_2{-}CH_3$$
(all H atoms drawn in full)

c.
$$\begin{array}{ccccccc} & & H & & O{-}H & & O \\ & & | & & | & & /\!/ \\ H & - & C & - & C & - & C \\ & & | & & | & & \backslash \\ & & H & & H & & O{-}H \end{array}$$

4. a. 3 alcohol groups; b. an ester group; c. an alcohol group and an acid group

5. $CH_3CH_2CH_2CH_2CH_2CH_3$ $\quad$ $CH_3CH(CH_2)_2CH_3$ with CH_3 on the CH

$CH_3{-}C(CH_3)_2{-}CH_2CH_3$

$$\begin{array}{c} CH_3 \\ | \\ CH_3{-}C{-}CH_2CH_3 \\ | \\ CH_3 \end{array}$$

$CH_3CH{-}CH{-}CH_3$ with CH_3 on each CH $\quad$ $CH_3CH_2CH{-}CH_2CH_3$ with CH_3 on the CH

6. a. alkene; b. alkane; c. alkyne

7. a. $CH_3CH_2CH_3$ b. CH_3CH_3 c. $CH_2{=}CHCH_3$

8. a. benzene b. $CH_3C(CH_3)_2CH_3$
$$\begin{array}{c} CH_3 \\ | \\ CH_3CCH_3 \\ | \\ CH_3 \end{array}$$
c. CH_4

9. a. CH_3OH; b. CH_3OH

10. a. $CH_3(CH_2)_4OH$, because of higher molecular weight; b. CH_3OH, because of H bonding

11. a. $CH_3OH + CH_3CH_2OH \xrightarrow[\text{low temp.}]{H_2SO_4} CH_3OCH_2CH_3 + H_2O$

b. $CH_3CH_2CH_2CH_2OH \xrightarrow[180°C]{H_2SO_4} CH_3CH_2CH{=}CH_2 + H_2O$

c. $2\,CH_3CH_2CH_2OH + 2\,Na \longrightarrow 2\,CH_3CH_2CH_2ONa + H_2$

12. $CH_3CH_2NH_2$ Ethylamine $\quad$ $(CH_3CH_2)_2NH$ Diethylamine $\quad$ $(CH_3CH_2)_3N$ Triethylamine

$CH_3CH_2NH_2 + H^+ \longrightarrow CH_3CH_2NH_3^+$

13. $CH_3CH_3 + Cl_2 \longrightarrow CH_3CH_2Cl + HCl$

14. a. $CH_3CH{=}CH_2$ b. $CH_2ClCHClCH_3$
$$\begin{array}{l} Cl\quad\;\; Cl \\ |\qquad\; | \\ CH_2CHCH_3 \end{array}$$
c. 3 H_2 d. $CH_3CH_2C(=O)CH_3$

15. Blank spaces, left to right and top to bottom:

$CH_3{-}CH_2{-}CHO$; $CH_3{-}CH_2{-}COOH$

$$\begin{array}{c} \qquad\qquad\qquad OH \\ \qquad\qquad\qquad | \\ CH_3{-}(CH_2)_2{-}CH{-}CH_3 \end{array}$$

benzene$-CH_2OH$ $\quad$ benzene$-CHO$

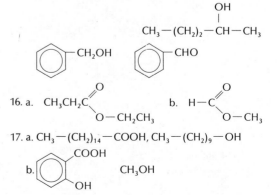

16. a. $CH_3CH_2C(=O){-}O{-}CH_2CH_3$ $\quad$ b. $H{-}C(=O){-}O{-}CH_3$

17. a. $CH_3{-}(CH_2)_{14}{-}COOH$, $CH_3{-}(CH_2)_9{-}OH$

b. benzene ring with $COOH$ and OH on adjacent carbons $\quad$ CH_3OH

18.

a. $CH_3(CH_2)_{14}C(=O)O(CH_2)_9CH_3 + H_2O \longrightarrow CH_3(CH_2)_{14}COOH + CH_3(CH_2)_9OH$

b. 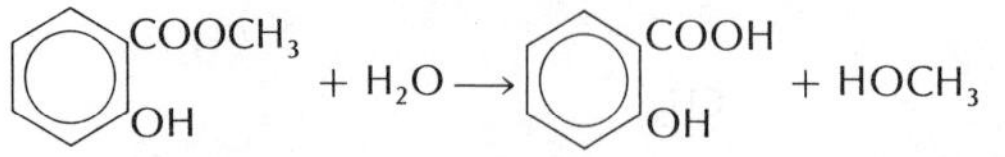 $C_6H_4(COOCH_3)(OH) + H_2O \longrightarrow C_6H_4(COOH)(OH) + HOCH_3$

19. $HO—CH_2CH_2—OH + HOOC—COOH + HO—CH_2CH_2—OH + HOOC—COOH + \ldots \longrightarrow$
$\ldots O—CH_2CH_2—OOC—COO—CH_2CH_2—OOC—COO \ldots$

20. $HOOC(CH_2)_8COOH + H_2N(CH_2)_6NH_2 + HOOC(CH_2)_8COOH + H_2N(CH_2)_6NH_2 + \ldots \longrightarrow$
$\ldots OC(CH_2)_8CONH(CH_2)_6NHOC(CH_2)_8CONH(CH_2)_6NH \ldots$

21. 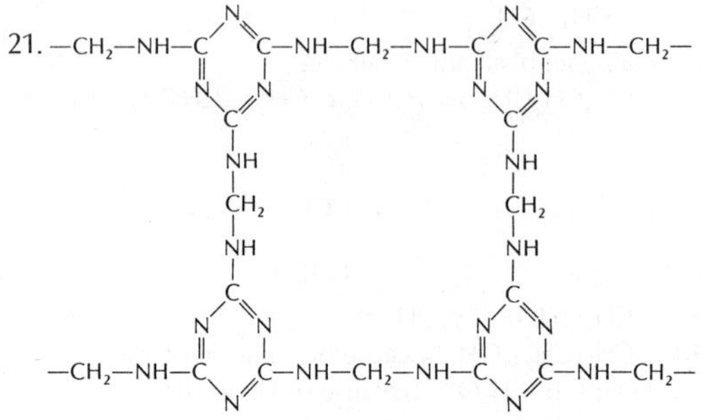 $—CH_2—NH—C_3N_3—NH—CH_2—NH—C_3N_3—NH—CH_2—$ (each triazine ring linked through $NH—CH_2—NH$ to a triazine ring of a second chain: $—CH_2—NH—C_3N_3—NH—CH_2—NH—C_3N_3—NH—CH_2—$)

Chapter 19

1. dT—dT—dA—dG—dA—dG—dC—dA—dT—
dA—dG—dG—dT—dG—dC—dA—dG
2. mU—mU—mA—mG—mA—mG—mC—mA—
mU—mA—mG—mG—mU—
mG—mC—mA—mG
3. dA—dA—dC—dC—dC—dG—dT—dA—dT—
dG—dC—dT—dA—dA
4. Protein segment *a* would be more water soluble, because it has more ionic and polar side chains than *b*.
5. Amino acids having polar sidechains are Ser, GluN, Tyr, and Thr. One possible system of hydrogen bonding:

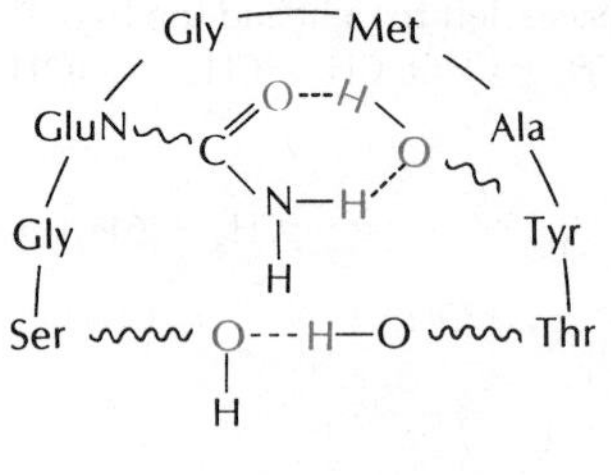

Gly — Met — Ala — Tyr — Thr; Gly — GluN — Gly — Ser; GluN side chain $C(=O)NH_2$ hydrogen bonded (O---H) to Tyr O—H; Ser O---H—O Thr

6. a. Gly, Lys, Thr, Asp, Val, Ser, Ala; H-bonding between Thr OH and Ser OH; Attraction between charged particles (Lys ⊕, Asp ⊖)

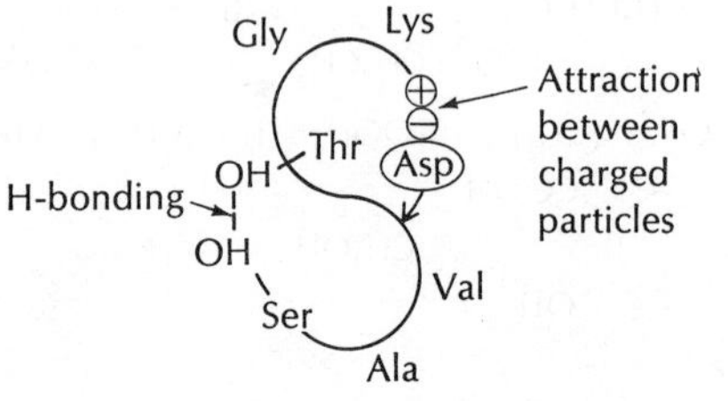

b. Ser — Thr (H-bonding); Cys —S-S— Cys (Disulfide bridge); Thr — Ser (H-bonding)

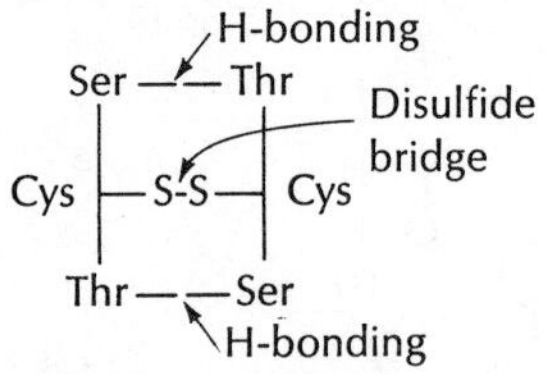

7. The protein in the egg white would react with the heavy metal ions and precipitate them, preventing them from getting into the bloodstream. Vomiting would then remove them from the stomach.
8. (fatty acid)$—COOH$; $(CH_3)_3\overset{+}{N}—CH_2—CH_2—OH$; H_3PO_4; $CH_2(OH)—CH(OH)—CH_2(OH)$
9. Mg is a constituent of chlorophyll.
10. $9\ \cancel{\text{moles ATP}} \times \frac{7.3\ \text{kcal}}{\cancel{\text{mole ATP}}} = 66\ \text{kcal}$
11. $0.5\ \cancel{\text{moles } O_2} \times \frac{227\ \text{kcal}}{6\ \cancel{\text{moles } O_2}} = 20\ \text{kcal}$
12. $500\ \cancel{\text{g muscle}} \times \frac{5 \times 10^{-6}\ \cancel{\text{moles ATP}}}{\cancel{\text{g muscle}}} \times \frac{7.3\ \text{kcal}}{\cancel{\text{mole ATP}}} = 2 \times 10^{-2}\ \text{kcal}$
13. a. $2 \times 10^{12}\ \cancel{\text{t } CO_2} \times \frac{1\ \cancel{\text{t-mole } CO_2}}{44.0\ \cancel{\text{t } CO_2}} \times \frac{1\ \cancel{\text{t-mole glucose}}}{6\ \cancel{\text{t-moles } CO_2}} \times \frac{180.\ \text{t glucose}}{\cancel{\text{t-mole glucose}}} = 1 \times 10^{12}\ \text{t glucose}$

b. 1×10^{18} g glucose $\times \dfrac{1 \text{ mole glucose}}{180. \text{ g glucose}} \times \dfrac{38 \text{ moles ATP}}{\text{mole glucose}} = 2 \times 10^{17}$ moles ATP

14. Vitamins A and D have very few or no polar groups and would be fat soluble rather than water soluble. Vitamins B_6 and C have many polar groups and would be water soluble rather than fat soluble.
15. Insulin is a polypeptide and thus would be digested by the body's enzymes. It will reach its destination intact only if it is injected directly into the blood.
16. The enzymes hydrolyzed the proteins in the respiratory passages.
17. Vasopressin represses the production of urine, so repressing vasopressin itself will *increase* urine production. As the effects of the alcohol wear off, vasopressin is again secreted and the thirst response is activated.
18. 1 g glyeine $\times \dfrac{1 \text{ mole glycine}}{75 \text{ g glyeine}} \times \dfrac{1 \text{ mole urea}}{2 \text{ moles glycine}} \times \dfrac{60 \text{ g urea}}{\text{mole urea}} = 0.4$ g urea

Appendix A

1. a. 3.49; b. 45.2; c. 67,000; d. \$3; e. 0.12
2. a. 3 (Rules 1 and 3a); b. 4 (Rule 2); c. 2 (Rules 1 and 3); d. 3 (Rule 2); e. 4 (Rules 2 and 3b); f. 3 (Rules 2 and 3a); g. 3 (Rules 2 and 3b); h. 4 (Rules 2, 3a, and 3c)

Calculator Answer	*Limiting No. of s.f.*	*Corrected Answer*
a. 12	2	12
b. 60	3	60.0
c. 0.0034	2	0.0034
d. 792	1	800
e. 6	2	6.0
f. 4	2	4.0
g. 0.00477	1	0.005

4. a. 3.981 (uncertain digit in 4th column); b. 54.32 (uncertain digit in 4th column); c. 104 (uncertain digit in 3rd column); d. 679.0 (uncertain digit in 4th column); e. 21.10 (uncertain digit in 4th column); f. 2.5 (uncertain digit in 3rd column)
5. This would be true if 4 and 4 are pure numbers. If they were measured numbers containing only one significant figure, the answer would be 20.
6. a. 3.9; b. 11.9; c. -4.0 (answer is negative because negative number was larger than positive number); d. -1.2 (see c); e. 4.4 (answer is positive because positive number was larger than negative number); f. -11.3 (both numbers were negative); g. 6.3 (two minuses equals a plus); h. 1.0 (3.2 becomes positive and larger than 2.2)
7. a. $7.9 - 3.3 = 4.6$
   b. $3.3 - 7.9 = -4.6$
   c. $-7.7 - 0.2 = -7.9$
   d. $-4 - (-2) = -4 + 2 = -2$
   e. $6 - (-3) = 6 + 3 = 9$
   f. $-1.3 - 2.7 = -4.0$
   g. $3.3 - (-9.8) = 3.3 + 9.8 = 13.1$
   h. $-5 - (-3) = -5 + 3 = -2$
8. a. 5.28×10^4; b. 8.97×10^{-6}; c. 2.1×10^4; d. 6.50×10^{-8}; e. 9.8×10^2; f. 6×10^0
9. a. Three. $(3.98 \times 10^1)(10^7) = 3.98 \times 10^8$
   b. Three. Number is justified.
   c. Three. $(1.98 \times 10^{-1})(10^{-3}) = 1.98 \times 10^{-4}$
   d. Four. $(1.010 \times 10^2)(10^{-3}) = 1.010 \times 10^{-1}$
   e. Three. $(4.54 \times 10^2)(10^2) = 4.54 \times 10^4$
   f. Three. $(2.33 \times 10^2)(10^{-9}) = 2.33 \times 10^{-7}$
10. a. $7.44 \times 10^{(-5+2)} = 7.44 \times 10^{-3}$
    b. $44.9 \times 10^{(9+10)} = 44.9 \times 10^{19} = 4.49 \times 10^{20}$
    c. $15.0 \times 10^{(-2-3)} = 15.0 \times 10^{-5} = (1.50 \times 10^1)(10^{-5}) = 1.50 \times 10^{-4}$
    d. $33.8 \times 10^{(22+23)} = 33.8 \times 10^{45} = (3.38 \times 10^1)(10^{45}) = 3.38 \times 10^{46}$
11. a. $1.59 \times 10^{(8-2)} = 1.59 \times 10^6$
    b. $0.611 \times 10^{(2-6)} = 0.611 \times 10^{-4} = (6.11 \times 10^{-1})(10^{-4}) = 6.11 \times 10^{-5}$
    c. $2.07 \times 10^{(-4-2)} = 2.07 \times 10^{-6}$
    d. $0.752 \times 10^{(8-(-2))} = 0.752 \times 10^{10} = (7.52 \times 10^{-1})(10^{10}) = 7.52 \times 10^9$
12. a. -2.23×10^{-5}
    b. $(3.99 \times 10^{-1})(10^4) = 0.399 \times 10^3$
    $$\begin{array}{r} 4.68 \times 10^4 \\ \underline{0.399 \times 10^4} \\ 5.08 \times 10^4 \end{array}$$
    c. $(2.97 \times 10^{-1})(10^{-3}) = 0.297 \times 10^{-3}$
    $$\begin{array}{r} 0.297 \times 10^{-3} \\ \underline{-3.42 \times 10^{-3}} \\ -3.12 \times 10^{-3} \end{array}$$
    d. $(4.61 \times 10^{-1})(10^{-9}) = 0.461 \times 10^{-9}$
    $$\begin{array}{r} 5.72 \times 10^{-9} \\ \underline{0.461 \times 10^{-9}} \\ 6.18 \times 10^{-9} \end{array}$$
    e. $(6.47 \times 10^{-1})(10^{-2}) = 0.647 \times 10^{-3}$
    $$\begin{array}{r} 8.53 \times 10^{-2} \\ \underline{-0.647 \times 10^{-2}} \\ 7.88 \times 10^{-2} \end{array}$$
    f. $(1.46 \times 10^{-1})(10^{-6}) = 0.146 \times 10^6$
    $$\begin{array}{r} 0.146 \times 10^6 \\ \underline{7.21 \times 10^6} \\ 7.36 \times 10^6 \end{array}$$
13. a. $(2 \times 10^{-2})(3 \times 10^1) = 6 \times 10^{-1}$
    b. $\dfrac{6 \times 10^3}{2 \times 10^1} = 3 \times 10^2$
    c. $(2 \times 10^1)(8 \times 10^2)(3 \times 10^2) = 50 \times 10^5$

d. $\dfrac{3 \times 9 \times 10^1}{3 \times 10^{-1}} = 9 \times 10^2$

e. $\dfrac{\overset{2}{\cancel{6}}}{\cancel{3} \times \underset{4}{\cancel{8}} \times 10^1} = .3 \times 10^{-1}$

f. $\dfrac{4 \times 2 \times 10^2}{1 \times 10^{-3} \times 5 \times 10^2} = \dfrac{8}{5} \times 10^3 = 2 \times 10^3$

g. $\dfrac{(8 \times 10^1)(2 \times 10^{-1})(\overset{3}{\cancel{9}} \times 10^2)}{(\cancel{3})(5 \times 10^2)(\cancel{2})} = \dfrac{24}{5} \times 10^0 = 5$

14. a. 20×10^4 (calculator answer correct); b. 1×10^2 (calculator answer incorrect); c. 1×10^1 (calculator answer correct); d. $.5 \times 10^{-23}$ (calculator answer correct); e. 2×10^{-31} (calculator answer incorrect)
15. a. 1.14×10^5; c. 1×10^1; d. 6.2×10^{-24}
16. a. 4; b. 0; c. −3; d. −2; e. −7; f. 2; g. 5; h. −2
17. a. 1.427; b. −1.0856; c. 3.648; d. −0.0119; e. 2.942; f. −3.80; g. 0.831; h. 1.0682
18. a. 1×10^4; b. 1×10^{-2}; c. 1×10^0; d. 1×10^{-7}
19. a. 1.31×10^{-7}; b. 6.4×10^7; c. 3.68; d. 2.36×10^{-1}; e. 4.9×10^{-3}; f. 9.718×10^1; g. 2.20; h. 0.9477

Appendix B

1. Kilobuck = 1000 dollars; megabuck = 1 million dollars.
2. a. 0.1 ℓ; b. 10^9 micrograms; c. 5 mℓ; d. 0.500 metric tons; e. 75 cm; f. 0.500 ℓ; g. 0.200 g; h. 400. meters; i. 0.800 kcal
3. a, c, and f
4. a. $\dfrac{10^3 \text{ mg}}{\text{g}} = \dfrac{1 \text{ g}}{10^3 \text{ mg}} = \dfrac{10^{-3} \text{ g}}{\text{mg}}$

   b. $\dfrac{0.621 \text{ mi}}{\text{km}} = \dfrac{1.00 \text{ km}}{0.621 \text{ mi}} = \dfrac{1.61 \text{ km}}{\text{mi}}$

   c. $\dfrac{1 \text{ cm}^3}{\text{m}\ell} = \dfrac{1 \text{ m}\ell}{\text{cm}^3}$

   d. $\dfrac{1.06 \text{ qt}}{\ell} = \dfrac{1.00\ \ell}{1.06 \text{ qt}} = \dfrac{0.943\ \ell}{\text{qt}}$

   e. $\dfrac{2.20 \times 10^{-3} \text{ lb}}{\text{g}} = \dfrac{1.00 \text{ g}}{2.20 \times 10^{-3} \text{ lb}} = \dfrac{454 \text{ g}}{\text{lb}}$

   f. $\dfrac{60 \text{ sec}}{\text{min}} = \dfrac{1 \text{ min}}{60 \text{ sec}} = \dfrac{0.0167 \text{ min}}{\text{sec}}$
5. b is incorrect. Units would be ℓ^2/qt, not qt. Conversion factor upside down. d is incorrect. Units would be $[(\text{cm}^3)^2\text{pt}]/\ell^2$, not pt. First conversion factor upside down. f is incorrect. Units would be g^2/kg. Last conversion factor upside down.
6. Everyone's height and weight will be different. Here are the setups:

$$? \cancel{\text{lb}} \times \frac{0.454 \text{ kg}}{\cancel{\text{lb}}} = ? \text{ kg}$$

$$? \cancel{\text{ft}} \times \frac{0.304 \cancel{\text{m}}}{\cancel{\text{ft}}} \times \frac{10^2 \text{ cm}}{\cancel{\text{m}}} = ? \text{ cm}$$

7. $115 \cancel{\text{km}} \dfrac{0.621 \text{ mi}}{\cancel{\text{km}}} = 71.4 \text{ mi}$
8. $40.0 \cancel{\ell} \times \dfrac{0.264 \text{ gal}}{\cancel{\ell}} = 10.6 \text{ gal}$
9. $\dfrac{45.0 \cancel{\text{francs}}}{10.6 \text{ gal}} \times \dfrac{100 \text{ cents}}{4 \cancel{\text{francs}}} = 106 \dfrac{\text{cents}}{\text{gal}}$, or $\dfrac{\$1.06}{\text{gal}}$
10. $2.51 \cancel{\text{tons}} \times \dfrac{2000 \cancel{\text{lb}}}{\cancel{\text{ton}}} \times \dfrac{1 \text{ kg}}{2.20 \cancel{\text{lb}}} = 2280 \text{ kg}$

    (2.28×10^3 kg)
11. $9.3 \times 10^7 \cancel{\text{mi}} \times \dfrac{1.61 \text{ km}}{\cancel{\text{mi}}} = 1.5 \times 10^8 \text{ km}$
12. $34.5 \cancel{\text{nm}} \times \dfrac{10^{-7} \cancel{\text{cm}}}{\cancel{\text{nm}}} \times \dfrac{1.00 \text{ in}}{2.54 \cancel{\text{cm}}} = 1.36 \times 10^{-6} \text{ in}$
13. $\dfrac{3.0 \times 10^{10} \cancel{\text{cm}}}{\cancel{\text{sec}}} \times \dfrac{1 \cancel{\text{m}}}{10^2 \cancel{\text{cm}}} \times \dfrac{1 \cancel{\text{km}}}{10^3 \cancel{\text{m}}} \times \dfrac{0.621 \text{ mi}}{\cancel{\text{km}}}$

    $\times \dfrac{3600 \cancel{\text{sec}}}{\text{hr}} = 6.7 \times 10^8 \dfrac{\text{mi}}{\text{hr}}$
14. $\dfrac{1 \cancel{\text{furlong}}}{\cancel{\text{fortnight}}} \times \dfrac{220 \cancel{\text{yds}}}{\cancel{\text{furlong}}} \times \dfrac{3 \cancel{\text{ft}}}{\cancel{\text{yd}}} \times \dfrac{1 \text{ mile}}{5280 \cancel{\text{ft}}} \times \dfrac{1 \cancel{\text{fortnight}}}{14 \cancel{\text{days}}}$

    $\times \dfrac{1 \cancel{\text{day}}}{24 \text{ hrs.}} = 3.72 \times 10^{-4} \dfrac{\text{mi}}{\text{hr}}$
15. $\dfrac{1 \cancel{\text{pt}}}{2 \text{ cups}} \times \dfrac{0.473 \cancel{\ell}}{\cancel{\text{pt}}} \times \dfrac{10^3 \text{ m}\ell}{\cancel{\ell}} = 237 \dfrac{\text{m}\ell}{\text{cup}}$
16. $\dfrac{0.454 \text{ kg}}{\cancel{\text{lb}}} \times \dfrac{1 \cancel{\text{lb}}}{16 \text{ oz}} = 2.84 \times 10^{-2} \dfrac{\text{kg}}{\text{oz}}$
17. $\dfrac{0.394 \cancel{\text{in}}}{\text{cm}} \times \dfrac{1 \text{ yd}}{36 \cancel{\text{in}}} = 1.09 \times 10^{-2} \dfrac{\text{yd}}{\text{cm}}$
18. $C = \dfrac{5°\text{C}}{9°\text{F}}(375°\text{F} + 40°\text{F}) - 40°\text{C}$

    $= \dfrac{5°\text{C}}{9°\text{F}}(415°\text{F}) - 40°\text{C}$

    $= 231°\text{C} - 40°\text{C} = 191°\text{C}$

    $K = 191 \cancel{°\text{C}} \times \dfrac{1 \text{ K}}{\cancel{°\text{C}}} + 273.16°\text{K} = 464°\text{K}$
19. a. $K = 34.0 \cancel{°\text{C}} \times \dfrac{1 \text{ K}}{\cancel{°\text{C}}} + 273.16 \text{ K} = 307.2 \text{ K}$
20. Both. At this temperature, both scales have the same reading.

APPENDIX D

Solubility Chart

	F^-	Cl^-	Br^-	I^-	S^{2-}	OH^-	NO_3^-	$C_2H_3O_2^-$	SO_4^{2-}	O^{2-}	CO_3^{2-}	PO_4^{3-}	CrO_4^{2-}
Ag^+	S	I	I	I	I	—	S	I	I	I	I	I	I
Pb^{2+}	I	I	I	I	I	I	S	S	I	I	I	I	I
Hg_2^{2+}	d	I	I	I	I	d	S	S	I	I	I	I	I
Hg^{2+}	d	S	s	I	I	d	S	S	I	I	I	I	d
Mg^{2+}	I	S	S	S	d	I	S	S	S	I	I	I	S
Cu^{2+}	S	S	S	S	I	I	S	S	S	I	I	I	I
Zn^{2+}	S	S	S	S	I	I	S	S	S	I	I	I	I
Fe^{2+}	s	S	S	S	d	I	S	S	S	I	I	I	I
Ca^{2+}	I	S	S	S	d	s	S	S	s	d	I	I	I
Ba^{2+}	s	S	S	S	d	S	S	S	I	S	I	I	I
Sr^{2+}	I	S	S	S	I	s	S	S	I		s	I	I
Co^{2+}	S	S	S	S	I	I	S	S	S	I	I	I	I
Ni^{2+}	S	S	S	S	I	I	S	S	S	I	I	I	I
Fe^{3+}	s	S	S	S	d	I	S	S	S	I	I	I	I
Al^{3+}	s	S	S	S	d	I	S	S	S	I	—	I	—
Cr^{3+}	I	S	S	S	d	I	S	S	S	I	I	I	I
Na^+	S	S	S	S	S	S	S	S	S	S	S	S	S
K^+	S	S	S	S	S	S	S	S	S	S	S	S	S
NH_4^+	S	S	S	S	S	—	S	S	S	—	S	S	S
Cd^{2+}	S	S	S	S	I	I	S	S	S	I	I	I	—

S = soluble in cold water (greater than 1 g/100 mℓ)
s = slightly soluble in cold water (between 0.1 and 1 g/100 mℓ)
I = insoluble in cold water (less than 0.1 g/100 mℓ)
d = decomposes in water

Index
With Key to Defined Terms

Page numbers of defined words are in boldface; page numbers of figures, tables, and boxes are in italics.

To the Owner of this Book

All of us who worked together to produce *Chemistry: An Introduction,* Second Edition, hope that you have enjoyed it as much as we did. If you did—or if you didn't—we'd like to know why, so we'll have an idea how to improve it in future editions.

School: ______________________

Course title: ______________________

Instructor's name: ______________________

Does the course have a lab? ______________________

1. Did you like the book? ______________________

2. **Content:** Was it too difficult? (If so, which parts?) ______________________

   Were all the chapters assigned? (If not, which ones weren't?) ______________________

   Which chapters did you like most? (Why?) ______________________

   Which chapters did you like least? (Why?) ______________________

3. Did you like Maxwell's Demon? (Why, or why not?) ______________________

4. **Illustrations:** Did you like them? (Why, or why not?) ______________________

   Were some illustrations particularly clear and helpful? (If so, which ones?) ______________________

   Were some illustrations confusing or unnecessary? (If so, which ones?) ______________________

5. Were the example problems worked out in the chapters useful to you? (Why, or why not?) ______________________

   Were particular problems confusing? (If so, which ones?) ______________________

6. Did you like the book's appearance (cover, layout, typeface)? ______________________

7. **Appendixes:** Did you use them? ______

8. **Review Questions:** Did you find them helpful? (Why, or why not?) ______

______

______

9. **Learning Objectives:** Did you find them useful? (Why, or why not?) ______

______

10. **Exercises:** Did your instructor assign them? ______

If not, did you use them on your own? ______

Did you find the Exercises helpful? (Why, or why not?) ______

______

Were there particular exercises you found confusing or too difficult? (If so, which ones?)

______

______

11. Do you feel your instructor should continue to assign this book? (Why, or why not?) ______

______

______

12. Will you keep this book? ______

13. Please add any comments or suggestions on how we might improve this book. ______

______

______

______

______

14. **Optional:** Your name: ______ Date: ______

Address: ______

15. May we quote you, either in promotion for this book or in future publishing ventures?

______ Yes ______ No

If you make it this far, we hope you'll follow through and mail the questionnaire to us at:

Chemistry: An Introduction, Second Edition
College Division
Little, Brown and Company
34 Beacon Street
Boston, MA 02106

Thank You!